Control Systems, Inc.
controlling your world since 1997

Dev DuRuz
General Manager

206.240.1344 Office
425.445.9845 Cell

16915 SE 272nd St., Ste. 154
Covington, WA 98042

www.controlsystems.ws
dev@controlsystems.ws

Web Based Energy Information and Control Systems: Case Studies and Applications

Web Based Energy Information and Control Systems: Case Studies and Applications

Compiled and Edited by
Barney L. Capehart, Ph.D., CEM, and
Lynne C. Capehart, JD

Associate Editors
Paul J. Allen
David C. Green

Library of Congress Cataloging-in-Publication Data

Capehart, B.L. (Barney L.)
Web based energy information and control systems : case studies and applications / compiled and edited by Barney L. Capehart and Lynne C. Capehart.
p. cm.
Includes bibliographical references and index.
ISBN 0-88173-501-9 (print) -- ISBN 0-88173-502-7 (electronic)
1. Power resources--Management. 2. Internet--Security measures. I. Capehart, Lynne C. II. Title.

TJ163.2.C366 2005
658.2'6--dc22

2005047318

Web based energy information and control systems : case studies and applications / compiled and edited by Barney L. Capehart and Lynne C. Capehart.

Published by The Fairmont Press, Inc.
700 Indian Trail
Lilburn, GA 30047
tel: 770-925-9388; fax: 770-381-9865
http://www.fairmontpress.com

Distributed by Taylor & Francis Ltd.
6000 Broken Sound Parkway NW, Suite 300
Boca Raton, FL 33487, USA
E-mail: orders@crcpress.com

Distributed by Taylor & Francis Ltd.
23-25 Blades Court
Deodar Road
London SW15 2NU, UK
E-mail: uk.tandf@thomsonpublishingservices.co.uk

Printed in the United States of America
10 9 8 7 6 5 4 3 2 1

0-88173-501-9 (The Fairmont Press, Inc.)
0-8493-3898-0 (Taylor & Francis Ltd.)

Table of Contents

Foreword

SURVIVAL OF THE FITTEST

There are clearly two classes of engineers today: those who "get" controls and those who do not. Those who get controls will have the keys to the future of the engineering field as the information-technology revolution continues its inexorable penetration into every nook and cranny of the buildings industry. Those who get controls will be able to respond to the needs of their clients, who, in turn, are responding to the needs of their clients—the occupants of buildings, purchasers of products.

Those who do not get controls will take an accelerated path to early obsolescence because they cannot cope with the web based energy information and control systems that are going into buildings today and tomorrow.

This book is written for those who "get" controls. It will not resolve the chasm between the individuals who are investing their time and resources to embrace technology developments and those who do not; instead, it will widen it. By making such a valuable compilation of experience and guidance from early adopters of new controls technology, this book has created a booster rocket for a larger body of engineers who want to apply it. Those who do not want to apply it are going to get left behind more quickly. This form of Darwinism is natural and inevitable.

Engineering controls has always been critical and complex, even when they were pneumatic. Criticality and complexity grew with time and technology, especially with the advent of direct-digital controls and then networked controls. With each increase in complexity, the field of engineers who would or could keep up has shrunk. Presently, it is common knowledge that the number of engineers, especially consulting engineers, who can design an integrated, interoperable building-automation system without depending on suppliers or their service-contracting arms to have a large role is disturbingly small.

It is easy to understand why. As the criticality and complexity of controls grew, engineering fees and billable hours available for controls on projects have shrunk. Control sequences, points lists, and other time-intensive, liability-packed tasks have fallen by the wayside. Engineering firms are so squeezed on profits, they are cutting training and travel budgets to the bone, making it difficult for their staffs to keep up to date.

Meanwhile, owners are seeking to maximize their facility investments while reducing staff and cutting other facility costs. Energy costs are rising for myriad reasons. Emissions from combustion equipment are being more tightly regulated. More facilities are putting in back-up power supplies and power-conditioning equipment. Pressures on municipal water supplies and the green-building movement are leading to automated controls on urinals, toilets, and faucets. There is much more metering and submetering going on today than yesterday—and data from meters, submeters, and controls are being integrated at the enterprise level locally, nationally, and even globally, through web based energy information and control systems. Data-driven decision making has come to the buildings industry.

The successful engineers will overcome all of these challenges. They know they have to. They will seek out and avail themselves to the resources that exist—such as this book—and get out of them what they need to understand and apply new web based energy information and controls systems.

Michael G. Ivanovich
Editor, HPAC Magazine

Foreword

The next few years of transition to web-based control of almost everything in our world will be interesting. Being in the large building automation industry for several decades has allowed me to witness previous radical changes. In the beginning automatic controls themselves were considered magic. Huff and puff (pneumatic controls) eventually evolved and finally gave way to electronic controls, which in the mid 70s and early 80s begat Direct Digital Control (DDC) or computerized control. Computerized control was a radical change that introduced many new concepts and players and quickly became the lowest cost method of doing business.

The industry changed rapidly with the new players making radical changes in the industry. The original players were forced to change, or they were simply overwhelmed and outpaced. The web-based energy information and control systems movement will be no different because it presents such an improved and simplified way to do business that it will be impossible for companies and other organizations to compete without building on the feature-rich internet and evolving web services. Web-based everything is changing the way we think and act and how facilities and electric generation will be controlled in the future.

I have been watching the evolution of web-based controls closely over the last five years, and I am pleased to see the rapid growth and acceptance of web-based technology in all areas of our large building automation industry. The growth and acceptance in the energy information and control arena has been truly amazing and hence this book has been written to describe these radical changes and to share evolving industry opinions about the development of this technology.

This book is the second in a planned three-book sequence, and concentrates on Case Studies and Applications of Web Based Energy Information and Control Systems. The first book presented the basics of Information Technology for Web Based EIS and ECS systems, and this volume now contains detailed case studies and applications which show how facilities and industries are using this web-based technology to improve energy efficiencies and reduce energy costs at their operations. In addition, authors describe how their systems improve the data collection process, and provide information to the facility operators that allows them to easily make decisions to improve the operation of their facilities. This assembly of evolving information in a one-of-a kind resource book links you to the dramatic changes that are occurring in facility energy use, and in the energy production and delivery industry itself.

This book presents many diverse industry views of the complex changes that are occurring in the energy information and control industry. It should help you understand the future of Web-Based Energy Information and Control Systems and the great changes occurring in our industry. I am a great fan of these two books because as much as I believe in the speed of the electronic media, the information must be organized and available in a logical form. We need links and strong connections to the traditional methods of learning to help kick start those that have lost connection with rapidly evolving web-based control. The format of this book allows a physical separation from the virtual side of web-based ways to what is reality. With this book and the Web you should have ready access to some of the most critical information on this new wave of web-based systems technology.

Ken Sinclair
Owner, AutomatedBuildings.com

Foreword

As we settle into the 21st century, we find ourselves at a crossroads that is shaping how we look at buildings and facilities in the decades to come.

Three major forces are converging; Information Technology is past the dizzying days of the nineties and is settling down to a pragmatic application of technology that is significantly contributing to the efficiencies of today's western organizations. Next, the internet is changing almost every aspect of our lives—entertainment, work, shopping, and information distribution. Our very social infrastructure is effected by the Internet! Finally, cheap broadband communications reach all corners of the known world via wired and wireless technologies.

Two significant common denominators result from the above forces for buildings: the first is that all buildings (with minor exceptions) are connectable together and to the enterprise organizations that own and have the responsibility to manage them. Only 5-10 years ago when the 20th century came to a close, the reliance on clunky and ineffective dial-up was the only means to access remote facilities. The information that flowed was limited by the media, and only critical data merited the "hassle" of dealing with such a cumbersome point-to-point communication. Organizations can now get real-time data from all of their facilities world-wide and command them in real-time. The impact of this cannot be underestimated.

The second common denominator is a de-facto common user interface for everything in our known universe. For years, building system integration vendors have tried to develop a common user interface to all things that occur in buildings, truly a difficult challenge. This has finally been satisfied by the "humble" Web Browser, a user interface that is common not only to all things that occur in buildings, but to all things—period.

These denominators provide buildings and facilities with an infrastructure for connectivity, a society that is comfortable with the use of a ubiquitous tool—the Web Browser, and a significant IT industry that has all of the skills necessary to make it all work.

The Web is the game-changer for the building and energy management industry. It is flexible, it is free (or at least has the perception of being so), it is a tool that everyone is comfortable using (since they buy books on it, order Pizza on it and conduct bank transactions with it). Lastly, the Web browser is infinitely flexible, just like a blank piece of paper.

To think that these trends and drivers are not going to change the very fabric of building and energy controls is foolhardy. The changes *are* happening now. We are currently in the midst of defining how the Web will transform the way buildings are managed by the professionals that are responsible for them, and how these changes will create change for the occupants of buildings.

Energy information and control systems are one area of building management that is adopting the web aggressively. By its very nature, this discipline needs to access information rapidly, remotely, from multiple sources and most importantly, in real time. Access to energy information also needs to be available to multiple people within organizations; from energy managers to corporate executives, people that are often based in disparate geographical locations. Clearly it is an ideal application for the web browser and the Internet.

This book explores the specifics of this trend. The implementation of the above vision by this industry is far from easy. The Web browser is only a starting place. The industry need to define how the Web is to be used for the tasks that are important to building owners, what types of servers need to be developed to serve up the information to the browsers, and more importantly what types of data will be presented, and how. Best practices and standards for creating web pages need to be developed that are effective for professionals and useful for the consumers of the information.

This book should be especially interesting to readers with the numerous case studies that explore many applications of the Internet and web based systems. These case studies should give the readers confidence that this is without doubt, a significant, cost effective, and developing area of energy management, and more particularly what is now termed enterprise energy management—the interaction between energy management at the building level to the enterprise systems that are running today's organizations.

Having a new "blank sheet of paper" is refreshing and powerful, and the difference between scribbles and a work or art on a canvas can sometimes be subtle. The challenge here is for the industry to leverage this powerful new tool for the ultimate benefit of more effectively managing buildings and energy usage.

Anto Budiardjo
Founder of BuilConn and BuilSpec

Preface

The capability and use of Information Technology and the Internet in the form of Web based energy information and control systems continues to grow at a very rapid rate. New equipment and new suppliers have appeared rapidly, and existing suppliers of older equipment are offering new Web based systems. Facility managers, maintenance managers and energy managers are all interested in knowing what problems and what successes are coming from the use of these web based systems, and need to be prepared for current and future installations of Internet based technologies in their facilities. Knowing what is being implemented at other facilities and knowing what is actually being accomplished is important information for the energy and facility managers if they are going to successfully purchase, install and operate complex, Web-based energy information and control systems..

The purpose of this book—*Case Studies and Applications of Web Based Energy Information and Control Systems*—is to document the operational experience with these web based systems in actual facilities and in varied applications. Web based systems have allowed the development of many new opportunities for energy and facility managers to quickly and effectively control and manage their operations. The case studies and applications described in this book should greatly assist all energy managers, facility managers, and maintenance managers, as well as consultants and control systems development engineers. These case studies and applications presented have shown conclusively that web based energy information and control systems are feasible, cost effective, and can create significant improvement in the energy related performance of a facility. Documented benefits include reduced energy costs, reduced maintenance costs, and reduced capital investments for these energy and maintenance savings. It is also clear that early adopters of the web based systems are seeing that it is giving them a competitive advantage over their non-adopting business and organizational peers.

Finally, I hope that all four of us on the Editorial Team have helped contribute to the successful application and implementation of new Web Based Energy Information and Control Systems in many of your facilities. It has been my pleasure to work with my Co-Editor and my Associate Editors on this important contribution to the IT education and training of working energy managers and facility managers. Ms. Lynne C Capehart ,Mr. Paul J Allen, Mr. David C Green have all played a major role in getting this book prepared and completed. My most sincere thanks go to each of these three people who have made my job much easier than it could have been. I also want to thank each of the 37 individual authors who have written material that appears in this book. Without their kind and generous help in writing these detailed chapters, this book would not have been possible. Each of these authors is identified in the alphabetic List of Authors following this preface.

Barney L. Capehart, Ph.D., CEM
University of Florida
Gainesville, Fl
November 2004

List of Authors

Paul Allen
Walt Disney World

Michael Bobker
Association for Energy Affordability

Michael R. Brambley
Pacific Northwest National Laboratory

David Brooks
Affiliated Engineers

Ron Brown
GridLogix Corporation

Anto Budiardjo
Clasma, Inc.

Barney L. Capehart
University of Florida

Lynne C. Capehart
Consultant

Bruce Colburn
EPS Capital Corp.

Gregory Cmar
Interval Data Systems

Rajesh Divekar
Al Systems

Khaled A. Elfarra
National Energy Corporation—Egypt

Keith E. Gipson
Impact Facility Solutions

David C. Green
Green Management Services

Daniel Harris
Association for Energy Affordability

Michael Ivanovich
HPAC Magazine

Safvat Kalaghchy
Florida Solar Energy Center

Sriniivas Katipamula
Pacific Northwest National Laboratory

Bill Kivler
Walt Disney World

Jim Lewis
Obvius Corporation

Fangxing Li
US ABB

Joe LoCurcio
Merck & Co.

Dirk E. Mahling
WebGen Systems, Inc.

John Marden
Honeywell Corporation

Jim McNally
Siemens Building Technologies, Inc.

Michael Kintner-Meyer
Pacific Northwest National Laboratory

Gerald R. Mimno
Advanced AMR Technologies, LLC

Naoya Motegi
Lawrence Berkeley National Laboratory

David E. Norvell
University of Central Florida

Mark A. Noyes
WebGen Systems, Inc.

Sarah E. O'Connell
ICF Consulting

William O'Connor
WebGen Systems, Inc.

Patrick O'Neill
NorthWrite Inc.

Richard Paradis
WebGen Systems, Inc.

Klaus D. Pawlik
Accenture

Mary Ann Piette
Lawrence Berkeley National Laboratory

Rich Remke
Carrier Corporation

Rich Rogan
Honeywell Corporation

Sandra Scanlon
Scanlon Consulting Services, Inc.

Osman Sezgen
Lawrence Berkeley National Laboratory

Blanche Sheinkopf
US DOE Energy Smart Schools

Travis Short
Performance Building Systems

Ken Sinclair
AutomatedBuildings.com

Greg Thompson
Power Measurement Corporation

Terrence Tobin
Power Measurement Corporation

Steve Tom
Automated Logic Corp.

Jason Toy
Advanced AMR Technologies

John Van Gorp
Power Measurement Corporation

Bill Von Neida
US Environmental Protection Agency

Rahul Walawalkar
Customized Energy Solutions

David S. Watson
Lawrence Berkeley National Laboratory

John Weber
SoftwareToolbox

Tom Webster
Lawrence Berkeley National Laboratory

Carla Fair-Wright
Cooper Compression

Gaymond Yee, Consultant
Lawrence Berkeley National Laboratory

Jeff Yeo
Power Measurement Corporation

A short biography for each author is provided in the section About the Authors that appears on page 523 at the end of the book.

Section One

Introduction

Chapter 1

Introduction to Case Studies and Applications of Web Based Energy Information and Control Systems

Barney L Capehart, Ph.D., University of Florida
Lynne C. Capehart, JD, Consultant
Michael Ivanovich, Editor, HPAC Magazine
David C. Green, Green Management Consultants

THE WAVE ROLLS ON

SINCE THE PUBLICATION of the first volume in this series on Information Technology for Energy Managers, our energy costs have risen dramatically and our energy supply chain has become quite tenuous. Only the relentless march of advancing technology has continued as a positive factor in our economy. But it is the utilization of new technologies that helps us to produce more efficient equipment, more efficient processes, and much more complex and effective control systems. It is this latter point that is addressed in this second book.

The technological wave of IT and web based energy information and control systems continues to roll on with increasing speed and intensity. In just one short year since the publication of the first volume in this series—*Information Technology for Energy Managers: Understanding Web Based Energy Information and Control Systems*—new web-based system supplier companies have come on the scene and many new, exciting applications and adoptions of web-based technology have taken place. What started as basic web-based energy information systems has expanded into web-based energy information and control systems, and finally to enterprise energy management systems.

Technological progress in Information Technology and use of the Internet and World Wide Web will continue to be made at a rapid rate. Applying these advancements to computerized facility and energy management systems requires the innovative skills of many people in both the IT and the Energy Management fields. If history in this area is a good indicator of what will happen in the future, we are all in for a fantastic ride on this new and powerful technological wave.

TECHONOMICS OF IT AND WEB-BASED ENERGY INFORMATION AND CONTROL SYSTEMS

The development and application of IT and web based systems for energy and facility management benefits from the three major laws that are contained in the framework of Techonomics, as discussed by Dr. H. Lee Martin in the Fall 2004 issue of *The Bent of Tau Beta Pi* [1]. Dr. Martin describes techonomics as a thought process for analyzing technology-driven trends in the economy, enabling people to spot and understand future business opportunities. He then defines techonomics in the following manner:

> "Techonomics is the study of trends in business and society resulting in observable economic change caused by the advancement of technology."

Traditional economic analysis relies on examining supply and demand to understand the prices of goods and services. Techonomics relies on examining the role of technology to supply goods and services at increasing quality, and decreasing prices; and to provide globally networked systems for new organizational models. Techonomics is exactly what is needed to understand the impact of this new technological wave of IT and web-based energy information and control systems.

The three basic laws of techonomics are as follows:

- First Law of Techonomics—Law of Computational Ubiquity
  The cost for equivalent computing performance halves every 18-24 months.

This means that for the same cost, the speed and capacity of computer chips doubles every 18-24 months.

- Second Law of Techonomics—Law of Global Information Networks
  The cost of locating information on the global network is diminishing exponentially as the number of users increases.

  This means that the cost to access a data point, or a page of information, is decreasing exponentially as the size of the network increases.

- Third Law of Techonomics—Law of the Innovation Economy
  As the cost of transactions diminish, optimum organization size reduces, thereby increasing the revenues generated per employee.

  This means that innovative technology companies and organizations using efficient outsourcing networks are finding ways to increase revenue and profits; while maintaining or reducing the number of employees. Outsourcing here means only moving the task to an organization outside the company; not necessarily out of the country.

The operation of these three laws can be considered as the explanatory drivers of the rapid development and application of IT and the use of web-based systems for energy and facility management. In addition, it explains the rapid movement to Enterprise Energy Management Systems. The first law—the law of increasing computer power and reduced cost—is responsible for the huge computational and processing capabilities of very low cost PCs and microcomputer controllers. This has provided the large scale computer power at reasonable cost that has allowed the hardware to be available for this use of IT.

Next, the second law—the law of exponentially decreasing costs of getting data or information from larger networks—is responsible for the large data bases both on-site and off-site that provide the inexpensive storage of data from meters and sensors in our energy and facility management systems. This has provided the huge base of data from which useful information to improve the operation of a facility can be produced and displayed to operational staff.

Finally, the third law—the law of organizations contracting in size and yet increasing their profits—is responsible for the decisions that companies and organizations are making to purchase and install these sophisticated IT and web-based systems. Companies today are reducing their workforce, and asking remaining employees to perform both their job and the job of another person that is no longer there. How can one person do the jobs of two people? By using systems such as we are discussing, where technology is providing the work result that would have come from the missing people. For example, through automatic alarming and automatic generation of work orders, the IT/web-based system can replace the person who checks a malfunctioning piece of equipment in a facility, and then comes back to generate a work order to fix the equipment.

The labor saving features of this new technology are what allows the reduction in the workforce while simultaneously maintaining revenue and profits for companies. As Keith Gipson said in his chapter, "Knowing how to effectively utilize their data network or enterprise management system is tantamount to gaining a competitive advantage over their counterparts in their businesses."

WIRELESS SYSTEMS IN BUILDINGS

One of the most fascinating and fastest growing sub-waves in this major technological wave of IT is the use of wireless sensors for buildings. Wireless technology has been available for well over a decade, but it is finally coming into cost effective use for reporting sensor values in buildings. Wireless sensors for room temperature are now cost competitive with hard-wired sensors; and projections are that wireless room temperature and room occupancy sensors will soon be available for under $10 each in larger buildings.

Battery powered sensors are small and accurate, and can be installed quickly and easily. Often called "peel and stick" sensors, this label is a true description of the time and effort needed to place many of these sensors in buildings. Future building control systems may well have wireless sensors as the majority of space property sensors. One of the chapters in this book is a technology report on the availability and use of these wireless sensors in buildings.

KEEPING UP WITH RAPIDLY CHANGING TECHNOLOGY

Michael Ivanovich, Chief Editor of *HPAC Magazine*, and author of one of the Forewords for this book, recently gave his view on the problem of controls engineers keeping up with rapidly changing technology. Michael says:

> As the chief editor of a leading engineering publication for the buildings industry, I am sensitive to the information and product needs of my readers, the hardware, software, and service products of suppliers, and the tidal surges of current events that tangentially touch upon the buildings industry or pave a road right through it. Sometimes, it is difficult to tell if a new development is significant or not—and much in the same way a spider has to determine if the vibrations traversing its web is a leaf or a fly, I have to make a call on what HPAC Engineering needs to cover.

Without question, the information-technology revolution washing over the HVAC controls and building-automation-systems sectors of the mechanical-engineering field is critically important. In fact, this revolution has been so overwhelmingly fast, broad, and deep, it has overtaken the educational resources of the buildings industry to cope with it. Several paradoxes exist that need to be resolved. Engineers coming out of college may have more computer expertise, but most do not have the practical HVAC experience needed to design controls systems. Veteran engineers may have the practical HVAC experience, but are generally resistant to learning the ins and outs of data communications technology required for engaging in networked-controls design at a level where they are independent of suppliers doing much of the engineering for them.

A revolution in controls education is something that HPAC Engineering has been promoting for some time, and which is the goal of this book, and its predecessor. The combined messages of these two books and HPAC Engineering, though consistent and rather blunt, bear repeating: Engineering controls is undoubtedly critical and complex. For example, in a large building, almost any significant project will involve hydronic control valves for heating and cooling, as well as dampers and actuators. These are very different types of controls, and although most controls are digital, pneumatic controls still are suitable for many applications. Now add boilers and burners, lighting, windows/blinds, security, and fire/smoke controls to the mix, and integrate everything with a building-automation system, providing for data archiving, analysis, and reporting at the local and enterprise levels. Whew! That's a lot of pieces and parts that are tied together internally and across a campus. Nowadays, that's not enough… web accessibility has to be included, sometimes for individual pieces of equipment, such as a critical motor or compressor. There's a lot more integration engineering for central control stations.

The networking isn't plug and play, and even networking media (physical layers) need to be determined (twisted pair, coaxial cable, wireless, etc.) and all sorts of means for accessibility: fixed or portable computer terminals, PDAs, tablets, and cell phones. Okay… now let's talk software for data aggregation, management, processing, and reporting. More of all this is on the horizon, not less. Tighter budgets means tougher decisions. Staffs are getting smaller, so there is greater reliance on automation and higher technologies.

Energy costs are rising because of uncertain supply. Emissions from combustion equipment are being more tightly regulated. More facilities are putting in back-up power supplies and power-conditioning equipment. Pressures on municipal water supplies and the green-building movement are leading to automated controls on urinals, toilets, and faucets and more metering and submetering.

All of this begs the question how can engineers expect to maintain their proficiency in heating, ventilation, air conditioning, and refrigeration and minimize the consumption of natural resources and the generation of pollution without regular training in controls. Without highly specific education and training, engineers cannot even responsibly outsource controls-related responsibilities, let alone perform them themselves.

Engineers have no control over how controls technologies evolve. Nor do they have control over occupants, weather, and events within the confines of a building. Nonetheless, they have a responsibility to ensure that the systems under their design, installation, operation, and maintenance purview meet performance requirements.

Performance is a word of action. And that's what controls are all about. Controls sense, process, and react. Engineers need to do the same or risk losing control over their destinies and those of their systems.

Therefore, control engineers need the information in this book and its predecessor, and to read these books carefully. They also need to go beyond these books and seek supplemental training. They need to take control on controls through continuing education and training.

What is an EIS like today and what will one be like in the future?

Dave Green provides the following views:

> Energy Information Systems (EIS) are starting to pop up everywhere now. Since more and more

meters and data collection devices connect to local area networks, the availability of data makes an EIS much easier to develop and use. Organizations might also opt to purchase a Commercial Off-the-Shelf (COTS) version of an EIS. Several companies are touting full-featured EISs that supplement the goals of any energy management team. Elutions Inc. has the *Active Energy Management Web* a web application to track and analyze utility data bundled with a service that collects and organizes the data. Interval Data Systems' *Energy Witness* software is a whole suite of modules designed to collect, organize and view utility data. Common features of the latest EISs are simplicity, drag and drop functionality to put data into a graph, virtual meters to track areas not distinctly covered by a single meter, propagation of reports by saving or emailing, drilling down through the data elements and interactive reporting. The components still missing from many of the EISs today are data integration *features* that collect data easier then the intensive programming components that are required now. COTS EISs now handle this by data integration *services* which no doubt add considerably to the cost of the software implementation. For the most part, data is only available in proprietary formats, ASCII, EDI, HTML or Extensible Markup Language (XML).

Looking forward, data integration with an organization's utility service provider(s) will continue to be a major challenge for EIS developers. Current options include utility data provided on tape disk, or via email, data provided on-line, and data recorded on parallel submeters. Before EISs can ever hope to contain data integration features that work like "plug and play" operating system features there needs to be more XML utility data exchange in the industry as a whole. Timely, accurate and efficient data collection is critical whether the data is coming from metering devices or utility companies. Data integration is a primary element in planning for an EIS whether the system is COTS software or developed in-house.

Data integration might also be the deciding factor in choosing between developing an EIS in-house or purchasing COTS software and services. The COTS option may serve the basic needs of many institutions and be more cost efficient if the data is already well organized and available. The COTS versions of EISs today contain many of the features required to meet those basic needs. On the other hand, custom EIS development is favorable to some since the requirements definition, EIS design, data acquisition and data organization is likely to be quite unique to any large complex requiring utility monitoring. Those development characteristics are also likely to change, perhaps frequently, with some large institutions. We can expect to see much more written about custom EIS design methodologies, data acquisition and data organization in the future. Custom development also provides the opportunity to add new features as the need arises.

The most needed features are likely data collection modules to draw utility data from various sources with little or no manipulation, configuration or manual intervention. The Environmental Protection Agency's National Energy Performance Rating System has made some progress in this direction by developing an XML schema to allow energy managers to upload data to them for scoring. This schema may be a start of standardized XML schemes for utility data transfer. For some complexes, the most convenient way to retrieve data is from the utility companies themselves. Hopefully, they will soon begin to publish data in XML format. Metering device manufacturers are also beginning to store and transfer data in XML format. XML will make it much easier for EISs to retrieve data and organize it into a useful database structure. The EIS interface itself may also improve by making reports more configurable, giving the user the ability to add flags, headings and columns in the order they desire. More comparison features would make it easy to compare one data set to another. Report subscriptions and alert subscriptions could be a big advantage in the future. It's not unreasonable to think that someday the reports themselves may be interchangeable from one EIS to another. GIS systems are already interested in linking to EIS reports to complement their existing data. Unfortunately, this may be difficult since EIS designs are taking many directions and few of them offer the opportunity to link directly to the reports.

The ultimate solution to the futuristic problem of interchanging EIS data may be an *Open-Source* EIS, an EIS that uses mostly open-source software. It might use XML data integration, MySQL for a database, PHP for a programming language and

Apache or IIS for a web server. Ideally the PHP code would allow developers to use different databases. This would allow the EIS development community to add features to the interface and adapt the data collection module to a wide variety of data sources. An EIS such as this may never come to be, but it is clear that it would likely be a huge success.

SPECIFYING DIRECT DIGITAL CONTROL SYSTEMS

ASHRAE took a major step forward when they created ASHRAE Guideline 13-2000, "Specifying Direct Digital Control Systems." This established a vendor neutral, professionally sanctioned framework for writing control system specifications, from sensor tolerances to graphical user interfaces. The guideline provides options and recommendations for the specifying engineer and by establishing an industry consensus it provides an excellent platform upon which to build a specification.

Automated Logic Corporation (ALC) has recently created a new on-line productivity tool called "CtrlSpecBuilder™" (Control Spec Builder) that is designed to help users write their control system specifications. Based upon the ASHRAE guidelines, it lets the user configure the general portion of the specification by answering questions about the scope of the project, communication protocol to be used, desired system-wide energy management features, and similar issues. CtrlSpecBuilder then allows the user to create sequences of control and points lists for thousands of typical HVAC systems through a simple menu-driven interface..

What makes CtrlSpecBuilder different from the guide specs published by other control manufacturers? There are four major differences:

1. It's based upon the ASHRAE guidelines.
2. It's open and non-proprietary.
3. It provides options (not mandates) for BACnet, Web-based systems, and other new technologies
4. It's an on-line tool that lets you configure a sequence of control for every piece of equipment in your project.

This web site is available to all users, and can be accessed at *www.CtrlSpecBuilder.com*.

CONTENT OF THIS VOLUME

The content of this book—the second volume in a three volume series—is devoted to case studies and applications of web-based energy information and control systems. The first book—*Information Technology for Energy Managers: Understanding Web Based Energy Information and Control Systems*—had a goal of introducing energy and facility managers to the new area of Information Technology, and its use in web-based energy and facility management systems. The purpose was to explain many of the terms of information technology as they applied to building and facility automation systems, as well as explaining the types of systems and types of tasks that could be performed by these web-based systems. A number of chapters then described the kinds of web-based systems that were being used at various facilities.

This present book assumes that readers have already become familiar with the content of the first book, and are now interested in the practical, real world capabilities and cost effectiveness of these web-based systems. Addressing these issues with detailed case studies and application descriptions is the goal of this book. In addition, this book opens up the area of enterprise energy management, and provides a beginning look at what this broader and more far reaching technology can do for both individual facilities as well as collections of facilities.

Part of the goal of this book is to serve as a general resource of information on web based energy and facility information and control systems. Since this is such a dynamic area, and since things are changing so rapidly, a unique chapter has been added at the front of the book. This is a chapter outlining the web resources that are available for reading about and learning about web-based energy information and control systems. The author of this chapter is Ken Sinclair, who is the owner of the website AutomatedBuildings.com. Ken has made a commitment to keep this resource chapter on his website, and keep it updated for the foreseeable future. Thus, it appears in static form in this book; but also in a dynamic form that can be accessed at any time from the Internet.

Next is a series of case studies of web-based energy information systems. The first is an example of a recently developed energy bill reporting system for schools in Orange County, FL. This EIS helps the schools there keep track of their use and cost, and they can see a comparison with other schools in the county based on their Energy Use Index. This benchmark helps schools determine where their use falls when compared to similar schools in the county. It is a great screening tool to use to help identify problem schools and to get some benefits from the simple feedback of energy cost information to the particular school. Special thanks go to Paul Allen at Disney World in Orlando, FL for initiating the use of this EIS, and to the Florida Solar Energy Center for implementing the data collection process and hosting the web server for this EIS.

The next case study centers on an open energy information system developed at the University of Central Florida through the efforts of David Norvell, Energy Manager at UCF. This is a great example of a completely open system using open source software and open protocols. It can easily be duplicated at other campuses or facilities. Other case studies are given in chapters dealing with multi-family buildings, and retail stores.

Following the web-based EIS case studies is the section with eight chapters on EIS Applications. The applications are similar to short case studies, but in general do not have as much detail on the construction or operation of the web-based EIS. Three of the Applications involve EIS systems for collecting energy data and producing performance metrics or benchmarks to assess the operating efficiency of facilities. A fourth Application adds to this area, as well as offers a broad range of other benefits and uses for EIS systems.

A unique Application is given in this section dealing with Computerized Maintenance Management Systems, and discussing the relationship between CMMS systems and Energy Management Systems or EIS systems. Carla Fair-Wright incorporates a wealth of experience in this chapter, and makes a clear case for integrating the features of both CMMS and EIS systems. Another Application focuses on the use of virtual metering to assist in the cost allocation and use in facilities, particularly schools. The next Application shows how EPA's Energy Performance Rating, used in the Energy Star Program, is accomplished through XML data transfer using third party hosts. The final application involves using a web based energy management system with a GIS system.

The Web-Based ECS Case Studies section begins with a comprehensive and detailed account of using an EIS/ECS at a Wellness Center at the University of Miami. This case study should help many potential users get a much better idea of what these systems will do, and how they accomplish these cost saving actions. Another case study details the Machine-to-Machine system to monitor and control the demand response to several commercial buildings which are provided real time utility price signals; while another provides details of facility management savings obtained from a commercial web service.

One of the ECS case studies is directed toward mission critical buildings such as data centers, hospitals, financial centers, and telecommunications facilities. This chapter gives the reader both problems and solutions for dealing with web based energy management systems in mission critical buildings. The final case study is for a web based energy monitoring and control system at a large pharmaceutical manufacturing complex with over 100 buildings. Much of this work was done on site by the controls staff, and shows how much of the development work on these web based systems can be done inhouse.

In the web based ECS applications section, Gaymond Yee and Tom Webster from Lawrence Berkeley Lab provide two chapters which provide excellent reviews of the technology and progress in using ECS/EIS systems. These non-vendor studies provide detailed, unbiased results on the benefits and costs of these web based systems. Load forecasting is an important part of all EIS and ECS systems, and Jim McNally presents his development of a new method for load forecasting and shows its features which include the ability to do short term load forecasting.

The next ECS application shows how existing energy management systems at Walt Disney World in Orlando were upgraded to interface and integrate with other web based EIS/ECS systems at that large facility. This Application should help a lot of facilities that have a number of different systems from different vendors. The last two Applications in this section discuss the planning and requirements for DDC and web based systems in universities and K-12 schools.

In the next section of the book, a collection of chapters address the hardware and software tools and systems for data input, data processing and display for EIS and ECS systems. The first is a comprehensive review of wireless sensor applications in building operation and management by two Staff members and a consultant to the Pacific Northwest National Laboratory This chapter describes wireless sensing technologies, and their applications in buildings to provide the data required to cost effectively operate and manage facilities at peak efficiency. Next, a chapter is devoted to examining the interoperability of industrial automation using web based technology for manufacturing control systems. It shows how opportunities exist for sharing of technologies, ideas and techniques from manufacturing control with facilities management applications to benefit the manufacturing business as a complete entity.

Two more chapters provide introductory coverage of net centric architectures for web based systems, and the ABCs of XML. XML is a key element of most of the integrated web based systems and particularly the enterprise energy management systems. The final two chapters provide overall guidance for obtaining the requirements for the design of an EIS and lay out the principles of EIS web page design for mining the data from electric and gas utilities.

The next section of the book gives the reader an exposure to the newly emerging area of enterprise energy management systems. At the most simplified level, an enterprise energy management system consolidates all energy related data (sources, costs, control, and monitoring points) from a facility, or group of facilities, into a data warehouse and provides tools to access and interact with the data to better manage the operation of the facility. The first chapter in this section provides a comprehensive and detailed description of what a state of the art enterprise energy management system is, what it should do, and how it accomplishes these functions. The chapter also contains a complete, detailed case study of such a system in a hospital.

The next chapter concentrates on the problem of providing high quality data for enterprise energy management systems, and how to perform quality control checks on the large amounts of data coming in to these systems. The next chapter provides a detailed description of the use of SOAP in a building monitoring and control system, and how this is related to the use of XML. The last chapter in this section expands our horizon by looking at an overall facility control and management program in Egypt, especially for the hotel industry in Egypt.

The last section of the book presents some views on future opportunities and directions for web based EIS and ECS systems. This section starts with a chapter from Keith Gipson, who holds one of the original patents on enterprise systems, and gives us his view of the history of enterprise systems. He concludes his chapter with some projections about where we are headed in using enterprise energy and resource management systems. Following this chapter is Toby Considine's contribution on what is needed to make the present day web based energy information and control systems ready for the enterprise task. He makes the case that we have a long way to go, and tells us that not only are the present Building Automation and Control Systems not ready for the enterprise; they are barely even ready for the operator.

The last regular chapter introduces readers to an analogy between building new cars and building new buildings. Comparing the degree of use of computer controls and computer information systems in new cars, versus what is typically found in new buildings leads to a conclusion that we have a long way to go to make our new buildings operate nearly as efficiently and effectively as our new cars.

The last chapter in the book is the conclusion, which tries to compare where we are today in the use of web based energy information and control systems, to where we need to go, and where we will most likely go in the future. We are well on our way to enterprise energy management systems, and the next logical step is to go to enterprise resource management systems. These two topics will form the major focus for a third volume in this book series, and will complete the series with a detailed examination of the use of enterprise systems, and cover additional application areas of commissioning, measurement and verification, and integration with other building systems.

The Appendix at the end of the book contains the Glossary, which has been compiled from Glossaries supplied by a number of the chapter authors. This Glossary has a great many of the acronyms and terms used in numerous chapters in the book. This Glossary should be one of the first sources checked to get more information about the meaning of a technical term or an acronym.

The Appendix also contains a short biographical description for each author.

ACKNOWLEDGMENTS

First and foremost, I would like to acknowledge the organizational, technical and editorial help provided by my Co-Editor Lynne C. Capehart, and my Associate Editors Paul J. Allen and David C. Green. The success of our first book in this series was greatly due to their efforts, and I believe the same success will follow from this book, and their tremendous efforts to make this volume as relevant and informative as the last one. My wife Lynne has worked tirelessly again to read and edit each of the 39 chapters contained in this book. Because of her there is a much more consistent format and style—as well as better readability—of the various chapters written by different authors. Without her, I would probably not have been able to complete this book in any reasonable amount of time.

Thanks go to Ken Sinclair, Michael Ivanovich and Anto Budiardjo, who were each kind enough to take their time and write a Foreword for this book. All three are active leaders in the educational movement to promote the understanding and use of web based energy information and control systems. Last, but not least, I would like to thank each of the 55 other authors who have contributed chapters to this book. The field of Information Technology itself is far broader than any one person can hope to fully understand. When the application of energy information and control systems is added to this, there is simply no one person who can blend these two areas together and present detailed information about each of the relevant pieces and topics. Thus it is only possible to produce a book on these diverse subjects with the help of many knowledgeable and skilled authors writing chapters in their individual areas of expertise. All of the authors have full time jobs that require the majority of their attention. Writing a chapter for this book means taking time each night and each weekend to create something that is related to their day-time work, but which still requires a separate effort to put together their ideas and explanations for the benefit of others. This time consuming task has been generously undertaken with the only reward being the knowledge that they have helped other people find their way in this complex field of IT and Web-based energy information and control systems. My most sincere compliments and thanks to each of you.

Finally, it should be noted that most of the book chapters are individual, stand-alone contributions to the book. As such, there is subject matter that is repeated in several of the chapters to make them stand-alone pieces. I have not attempted to remove this duplication, since I believe that it is useful in this context, and that also it is beneficial to have different authors explain some of the same concepts in different words. I find it valuable myself to read things on the same topic that are written by different authors so that I can get a good feeling for whether I understand the overall message being presented by each author. Also, the stand-alone style allows readers to go into the book at almost any point and read a chapter without having to read all of the preceding chapters.

The editors are pleased to have been able to compile this excellent selection of chapters—most of which have been specially written for this book—to further the application of these web-based energy information and control systems. Readers who have already implemented some of these EIS/ECS systems at the enterprise energy management level, or who have used them in Building Commissioning, Building Systems Integration or in Measurement and Verification projects, are encouraged to submit case studies for the third volume in this series to be published in 2006.

Chapter 2

Web Resources for Web Based Energy Information and Control Systems

Ken Sinclair
Owner, AutomatedBuildings.com Website

INTRODUCTION

THIS CHAPTER is intended to provide more perspective on the topics covered in this book. Here we provide a short abstract of the related articles published on our web site, which have direct linkage to the actual articles where you can gain more information. The web-based media is the message and we hope that providing web access to this information will amplify the power of the book with access to related information that has only been published electronically to date. I have been amazed and impressed with the widespread recognition of the need for a comprehensive, yet basic and readable book that introduces the topic of Information Technology for energy managers in a way that is understandable to the average person working in the energy and facility management area but is not an IT trained professional. The first book, *Information Technology for Energy Managers: Understanding Web Based Energy Information and Control Systems*, achieved this goal. Now, this sequel, ***Case Studies and Applications of Web Based Energy Information and Control Systems***, provides a discussion of the experience that has been obtained by some of the early adopters of this new IT based technology, as well as the experience from developers and researchers in this area.

This chapter also lives as a dynamic document on our AutomatedBuildings.com web site. The on-line chapter is dynamic because we will add new articles to these sections as they become available. Thus the on-line version will retain currency, a feature inherently unavailable in the printed version. To get access to the website with the active links to these articles, go to *www.automatedbuildings.com/education/*

This dynamic chapter is organized into four sections:

1. On-line information on Web Based Energy Information and Control Systems
2. Evolving Communication Standards and Protocols
3. Wireless's Rapid Evolution
4. Learning more about Web Based Control

All sections provide a brief abstract of the full article posted on the AutomatedBuildings.com web site. Once you are on-line interacting with this chapter of the book you will be able to activate the links to the actual articles, as well as other web resources.

1. ON-LINE INFORMATION ON WEB BASED ENERGY INFORMATION AND CONTROL SYSTEMS

Transforming the U.S. Electricity System, Rob Pratt, Pacific Northwest National Laboratory
Presented at "Bringing the Electricity System into the Information Age" symposia, Feb/04

Over the last decade, leading-edge industries have been using real-time information, e-business systems and market efficiencies to minimize the need for inventory and infrastructure while maximizing productivity and efficiency. However, the energy system has yet to make those advancements, or reap their benefits. The electric power grid is full of massive and expensive infrastructure that is generally underutilized. To meet growing demand, utilities continue to put up more iron and steel, and pass the cost on to their customers, who have little say in the matter. To meet the load growth projected by the U.S. Energy Information Agency, $450 billion of new electric infrastructure must be added by 2020 if we continue this "business-as-usual" approach.

Request for Participation—Summer 2004—Automated Demand Response Test for Large Facilities

Lawrence Berkeley National Laboratory (LBNL) is recruiting energy and facility managers of large facilities to participate in the 2004 Automated Demand Response (Auto-DR) research project. This project builds on the methods used in LBNL's 2003 Auto-DR tests.

The California Energy Commission and LBNL are studying the ability of facilities to reduce electricity demand temporarily through implementation and testing of Auto-DR. Auto-DR is being evaluated in terms of its potential to flatten out the grid load shape on peak days, help avoid blackouts like those that occurred in California (2001) and the Northeast (2003), and lower costs to ratepayers. Demand response has been identified as an important element of the State of California's Energy Action Plan, which was developed by the California Energy Commission (CEC), California Public Utilities Commission (CPUC), and Consumer Power and Conservation Financing Authority (CPA).

Merging Information Technology and Energy

In recent years EUN has postulated many variations on the energy industry future. This discussion defines energy future as "Gridwise," a topic that will recur as EUN chronicles the new development. Gridwise refers to the national electricity system and architecture addresses information technology, networking and the Internet. Gridwise is an initiative to stimulate development and adoption of an intelligent energy system that enables more effective use of the U.S. Electric System. This will result in significant opportunities for energy efficiency, but of equal importance it can result in a more reliable Grid. Gridwise is being sponsored by the new Department of Energy Office of Electricity Transmission and Distribution (DOE O-ETD). It is the first new office created within the DOE in years, which demonstrates the importance of this issue. Pacific Northwest National Labs has acted on behalf of DOE to select a team of national experts for the Gridwise Architecture Board. EUN readers will be able to stay informed on Gridwise through this author, who has been invited to sit on the Board.

Viking Energy—A Platform for Demand Response

Along with the challenges created by deregulation, The Department of Energy (DOE) has estimated that the demand for electricity will grow by approximately 2% annually through the year 2020. Other countries will be experiencing similar or higher growth rates as their economies become more reliant on energy. Unfortunately, as demand for energy goes up, generation units are reaching their plant maturity and being retired. In fact, almost half of the current generating capacity in the U.S. originates from units approaching their maturity. Although new generation continues to go on-line each year, hundreds of thousands of megawatts of capacity (368,000 MW in the U.S.) will soon be taken off line. Constructing new generation is difficult and, in many cases, units approved for construction have been deserted due to either unfavorable market conditions or the "Not In My Back Yard" (NIMBY) syndrome. As a result, generation growth cannot keep pace with demand.

"Do-it-yourself" energy information systems in a Web-based world

Manufacturers of metering products have introduced new metering and data collection products that allow any electrical contractor or building owner to provide Web-based energy information to customers and tenants cost-effectively.

Owners and managers of commercial and institutional properties are increasingly challenged to maintain profit margins in the face of high vacancy rates and more competition for tenants. This competitive pressure, combined with volatility in energy rates and the specter of deregulation in the electric industry has resulted in a rapidly growing demand for more timely and accurate energy information. Historically, getting this type of information has been expensive and the installation of energy information systems has been left to specialists such as systems integrators or building automation contractors. In response to the growing demand in this market, manufacturers of metering products have introduced new metering and data collection products that allow any electrical contractor or building owner to provide Web-based energy information to customers and tenants cost-effectively.

Why Connections to Our Clients' Web-Based Enterprise Are Important

Our clients' business models are evolving to enterprise based solutions as the easiest and lowest cost way of doing business.

I am often asked "why is web-based enterprise control so important?" I answer that all our clients' business models are evolving to enterprise based solutions as the easiest and lowest cost way of doing business. We have several great articles written by industry experts for our AutomatedBuildings.com web site on this exact topic and rather than inflicting you with my opinions and words, I am going to provide you with three direct extracts that I feel provide great insight to this matter

---

Facilities Are Poised To Become New Sources of Business Information

Automation vendors can now play by the same rules as IT and use modern integration technologies to transform the closed languages of their automation systems into the open and pervasive integration languages of IT.

Building automation and controls manufacturers recognize the unique needs of property managers and are beginning to acknowledge the power of XML Web Services to provide seamless, open interoperability between automation systems and corporate business applications. XML Web Services, the same powerful integration framework that has been embraced by software providers and IT departments everywhere, is also perfectly suited to transform facility sub-systems into valuable sources of new business information.

Networked Building Control Enhances Demand Responsiveness

The convergence of IT and building systems technology has produced secure, cost-competitive products that are more effective for demand response than non-networked control.

The proliferation of the Internet and information technology (IT) hasn't stopped at the outside of buildings—it's actually changing the way that buildings are operated. Facility managers and energy service providers (ESPs) are beginning to reap the benefits of networked building control—the practice of integrating building management systems (BMSs) with corporate intranets or the Internet. One of the main advantages of networked systems is that facility managers can control the operation of buildings scattered across a campus or across the U.S. It's no longer necessary to physically travel to numerous buildings to control equipment. And because these networked building systems are easier to use, facility managers are more likely to detect equipment problems before a total failure can occur.

Energy 2004—Are we really on line?

What has been heralded by the Buildings and Energy Industry as Convergence, System Integration and Internet Digital Control™ is in direct alignment with trends in E-business as a whole

The title of this column may seem to ask a silly question. Particularly for those who attended the recent AHR Expo or attended the XML Symposium in January. XML (Extensible Markup Language) may be a new term to many readers, and there is good reason for that, it is a fairly high-level software tool that the average user would not see. What is more important than the tool itself is the impact that its use will have on the energy business. For those who did not hear about the XML event, it was a world-class gathering, sponsored by the Continental Automated Building Association (CABA) *www.caba.org* and Clasma *www.clasma.com,* the sponsor of BuilConn. The XML symposium presented a unique opportunity to meet with industry leaders and talk about Internet based applications that will define the future of the energy and buildings industry. Again, XML is a software tool being used to develop Internet services that shape building automation and energy management over the next decade. But before getting too far down the road, it is worthwhile to stop and reevaluate the energy industry in the context of Information Technology at large. This would entail revisiting the basic premise for Energy On-line, and the accompanying chart provides an excellent context for that discussion. This chart was developed by Gartner, Inc. a business and market analysis think tank, with operations worldwide.

A Wake-Up Call and a No-Brainer

Ask any economist worth his salt what the total impact on American industry would be if our gross use of energy were to decrease by 15% over the next 10 years.

Anyone involved with development, design, construction or operations of commercial or industrial facilities will tell you that automation is an absolutely essential component of a successful energy conservation/management plan. But ask any member of that group how many facilities under 50,000 GSF have any level of automation and the answer is probably very few. Below 5,000

GSF, almost non-existent. You would have to look long and hard to find a million plus GSF facility without a building automation system. I am not talking about time clocks here. I am talking about automation systems capable of performing fairly sophisticated functions that would include interaction with external systems using the latest technology. The simple reasons for this are cost and complexity. Your basic Mom & Pop deli, the local dentist, the branch bank or the retail proprietor will only be able to afford automation if it comes to them in the form of an embedded appendix of those products they rely on to conduct their basic business. Their HVAC, refrigeration and lighting systems etc. Oh, and by the way, the man machine interface and networking of these devices to an IP connection will need to be as simple as unpacking the box, connecting the 110 and Internet. Local connectivity to the individual appliances and systems will almost certainly be a wireless solution.

Ending the Blackout Blues

We need (REALLY need) to improve the way our industry delivers more advanced technologies to our building construction projects.

The experts are only half right. What they are missing is the second and equally important cause of this tragic event. The truth is, the same type of outdated industry structure and practices plague the energy conversion industry that operates on the building side of the electric meter—our HVAC industry. And our industry's failure to come to grips with it has resulted in HVAC electric energy use to be about double what is easily achievable, making our industry every bit as responsible for this tragedy as the utilities. That we have not yet had to bear the blame is a gift, and should be seen as a wake-up call of our own—that we need (REALLY need) to improve the way our industry delivers more advanced technologies to our building construction projects. And perhaps we can do so if we can avoid the glare of publicity while we work to more efficiently utilize the increasingly precious energy resources available. It won't be a simple process, but I'd like to cite here two simple things all of us can do right now that will most certainly move us in the direction of real improvement.

Protection during Electrical Outages;
Power Quality Everyday

Electrical Power Quality changes hourly, these changes affect nearly every system's reliability in your facility and the bottom line of your business.

While most are concerned about "the other" electrical service issue, Power Reliability (e.g. power outage), the occurrence of such an event is infrequent as compared to Power Quality changes. An outage does have an extended negative effect on your bottom line however, one that is beyond the loss of productivity during the period of the outage. This extended negative effect has to do with the damage to computer, network, building automation, process control and many other systems. This type of damage can be immediate but often is not; it appears in the hours and days after a power outage event but can be prevented with the proper Power Quality device.

If Buildings Were Built Like Cars—The Potential for Information and Control Systems Technology in New Buildings Barney L. Capehart, University of Florida Harry Indig, KDS Energy, Lynne C. Capehart, Consultant

The purpose of this paper is to compare the technology used in new cars with the technology used in new buildings, and to identify the potential for applying additional technology in new buildings. The authors draw on their knowledge of both new cars and new buildings to present a list of sensors, computers, controls and displays used in new cars that can provide similar and significant opportunities for our new buildings. Some thoughts on how this new technology could be integrated into new buildings are also discussed. The authors hope that calling attention to using new car technology as a model for new building technology will stimulate recognition of the potential for new buildings, and ultimately lead to the implementation of similar technological improvements in new buildings.

2. EVOLVING COMMUNICATION STANDARDS AND PROTOCOLS

LonWorks and BACnet System Solution on a Chip

Both protocols have become well established and manufacturers are working to offer building systems supporting both platforms.

The building control market is at the edge of major

growth. It is being steered by the recent upturns in the economy and the desire for more control of buildings for the purpose of better optimization of energy, better service for the customer and added security. To respond to these new market opportunities companies are working on standard building control products based on LonWorks ANSI/EIA-709.1 and BACnet ISO 16484-5 protocols. Both protocols offer interoperability of products between manufacturers. The industry can now focus on two major platforms instead of 15 or so proprietary platforms. At one time there were major battles to see which protocol would win out in the industry. It is becoming clear that both protocols have become well established and manufacturers are working to offer building systems supporting both platforms.

---

oBIX Building Blocks

Simple modules. Complex modules. Each self contained. Each with a well-defined interface for interoperability.

Best practices in software systems today is to develop smaller modules, each provably able to perform its limited internal functions and operations correctly. These systems interact with other modules through well defined, highly abstracted and loosely coupled interfaces, not intimate programming interactions. There are three very big effects of this approach.

1. Each module can be swapped out or upgraded for enhanced performance of its single function without re-developing the entire system.
2. Functional modules can be distributed not only across computer systems, but across corporations, as modern ERP systems span logistical chains across companies, countries, and continents; this is done without committing to a single system everywhere.
3. Downtime in any one system does not imply downtime for all.

---

oBIXTM Evolves at AHR Expo

Short for Open Building Information Xchange, oBIX is an initiative to define XML and Web Services-based standards for exchanging building systems information with each other and enterprise systems.

That was the case with oBIXTM at the AHR Epxo recently concluded in Anaheim. Amidst the sunny Orange County setting of theme parks, hotels and freeways, oBIX was the subject of much buzz, discussion, and debate. Short for Open Building Information Xchange, oBIX is an initiative to define XML and Web Services-based standards for exchanging building systems information with each other and enterprise systems.

At a press conference that coincided with the AHR Expo kickoff, oBIX chair Paul Ehrlich, Business Development Leader, Trane, and others gave key industry editors background on the oBIX guideline and status on oBIX developments since its germination at BuilConn in April 2003.

Reporters and observers heard updates on the work of four task groups; Data and Services, Security, Network Management, and Marketing. In an effort that has become obligatory in the early stages of any business initiative, the oBIX Marketing task group has fashioned a vision statement to furnish clarity of purpose; **building systems working together for the enterprise**.

XML Spells Connection to the Future

An open letter to the Building Controls Industry turned into a reality at the consortium's first meeting at BuilConn in Dallas. That open letter, which we ran on the AutomatedBuildings.com website, proposed that we establish a consortium that would work on the creation of a guideline for use of XML and Web Services in building automation and control applications. I was able to attend this groundbreaking meeting in Dallas and the sense of purpose and cooperation in the room was amazing.

Let's all give special thanks to the BuilConn folk for hosting this meeting, and to the CABA organization for giving the newly formed consortium a home.

The first paragraph of the open letter reads:

"The Building Controls industry has made great strides over the last 10 years in the creation of communications standards. Both BACnet and LonTalk are now viable, commercially accepted solutions that provide owners with open communications. Yet while we have made great progress in these areas as an industry, there has been an emergence of a larger, more globally accepted standard created by the world of Information Technology. In particular the broad acceptance and ever lowering cost of Ethernet/TCP/IP/XML communications is finding its way into our industry."

Two Separate But Very Important Industry Acronyms XML & oBIX

XML is widely used by the IT community and considered by many as the most important enabling technology for the future of integrated and intelligent buildings, XML brings to fruition the convergence of building systems and the IT infrastructure. These IT-based technologies are drastically changing the buildings industry as they enable cooperation between disparate approaches to open systems, a vital issue within the industry.

"As possibly the ultimate integration mechanism for buildings, XML will revolutionize the buildings industry," Ron Zimmer, CABA President and CEO, said. "This symposium will provide the HVAC professional with ample XML and Web Services education and a practical expectation of their adoption rate within the industry."

Ehrlich: oBIX stands for Open Building Information Xchange, and it is an industry-wide initiative to define XML—and Web Services-based mechanisms to present building systems-related information on TCP/IP networks such as the Internet.

Ethernet: The Common Thread to Total Building Systems Integration

Ease of setup/configuration/use as well as connectivity options are key factors when considering a device server for your building automation application.

Ethernet Made Easy

Ethernet is a low cost, high speed, widely deployed, universally accepted medium for local area and wide area networks. Layer on top the TCP/IP protocol, the most common office networking language (as well as the language of the Internet) and you have the initial ingredients of an open, more easily integrated system. Of course, no solution is perfect and Ethernet has its drawbacks. However, most of Ethernet's weaknesses have been aptly addressed making it increasingly accepted in scenarios where it may have been rejected in the past. For example, critics have pointed to its lack of rugged components, non-determinism, and vendor acceptance. But, industrial grade switches, cables and connectors compensate for these fallbacks, and vendors are quickly adopting Ethernet as the new communication standard.

Open Systems Standards

This is the fifth of 9 articles where we are introducing the cost benefit for interoperability and that gained from the procurement technique achievable with open systems and choice. This article sets out some of my opinions, a collection of web-based technical data and then a current market summary opinion.

- object-oriented programming has gained widespread acceptance/preference as an alternative to flat data structures

The contributing editor acknowledges the following extract from a recent Automated Logic white paper as a useful contribution to the object-orientated programming point:

Since BACnet and EIB objects and LonMark functional profiles are information models and XML is a modeling language, we could express these high level information models in XML and in so doing make them compatible with the emerging Web services architecture. Because of the flexibility of XML and the web services architecture, these high level models could be expanded to include other types of facility-related (but not necessarily building automation-related) information. If each building automation protocol developed its own XML model, however, we would have similar but incompatible system models. Today's problems of translating from one protocol to another at the building controller level would become tomorrow's translation problems at the Web services level. What's needed is a unified system model, in XML, that can be used by any building automation protocol.

The contributing editor suggests the oBIX initiative at OASIS is a good thing to support but the editor also acknowledges the BACnet work in their extension of their objects to XML (and to KNX). What is needed is a unified model, oBIX can achieve this task the quickest and especially with help from BACnet such that unification at XML schema level is the best option for the end-user and supply chain.

Building Controls and BACnet, IT and XML.

A report from the BIG-NA conference; BACnet, IT and XML are hot items.

The convergence of Building Controls and IT, BACnet and XML are coming, and they're coming quick! The

BACnet Interest Group—North America (BIG-NA) and the BACnet Manufacturer's Association (BMA) jointly hosted the BACnet Conference and Expo on October 5th, 6th & 7th, 2003 at the Kingsgate Marriott Hotel and Conference Center, on the campus of the University of Cincinnati, Cincinnati, Ohio. The Conference featured in-depth education, demonstrations, applications and hand-out materials on using the BACnet building automation and control network standard within educational institutions focusing on campus/district/and global facilities management systems and was attended by construction and facility managers and engineering consultants representing many colleges, universities, and global corporations.

The Niagara Framework: Measuring up to Open

What exactly is an "open system"? What are the elements... ingredients... issues?

From the customer perspective an open system should provide freedom of choice—the freedom to choose the best products, the best manufacturers, the best contractor agencies and the best service providers. The freedom to adapt as his business needs change, and freedom to push forward on his timetable and to adopt different technologies as needed over time. Open technologies should help the industry better meet the demands of its' customers. In today's markets, customers need to be agile. Some of the enterprise's largest assets—their facilities, are not. Agile businesses will survive the onrush of global competition, Internet marketing, shrinking boundaries... whatever the future brings. Agile building control solutions are necessary to allow these enterprises to accomplish their goals.

Integrating OPC into Building Automation—
The Latest Trend

OPC is a viable solution for building automation and is available today.

OPC is an established standard that enables integrators to connect disparate systems together, creating robust solutions and providing true interoperability; while at the same time reducing implementation time and costs. In addition, OPC enables a fully scalable solution for future changes and expansion. No longer are integrators tied, or locked in, to a single vendor. The data has now been freed and the ability to choose from an abundance of options is sitting well with integrators. Integrators are now able to deploy control systems and applications, regardless of vendor, and build best-of-breed solutions. Building automation is no longer dominated by a few large companies. The playing field is now open for many developers to offer far more advanced and superior solutions, while reducing the total cost of ownership.

Protocol war yields to productive peace

Ironically by narrowing the choice of control protocols the industry can now invest more in well understood, and supported systems that are easy to specify, install, integrate and modify.

Some business sectors have managed to get further than the automobile industry. The IT world has the de-facto Wintel (Windows plus Intel) standard in which (assuming you don't have an Apple, Unix or Linux computer) you can swap files between different systems without a problem. This standard came about by pure commercial pressure. Whether or not you agree that a few companies should impose their technology on an industry, there are plenty of examples, which show it to be vital to industry and market expansion. VCRs, electricity supply, telephones and railways are examples of innovations, which did not really begin to expand until one technical standard rose to dominance.

What could UPnP possibly mean to Building Systems? Report from the UPnP Summit, Cannes, France October 28/29 2003

The UPnP forum is a voluntary, international, open organization for companies and individuals formed in 1999. The first fact to surprise me was that here are no membership fees. The size of the membership and who they are surprised me. Currently the 625 members are drawn from North America, Europe and Asia. The Steering Committee is composed of Axis Communications, Broadcom, Canon, Inc., Echelon, Hewlett-Packard, IBM, Intel, Lantronix, LG Electronics, Metro Link, Microsoft, Mitsubishi, Philips, Pioneer, Ricoh, Samsung, Siemens, Sony, Thomson and most recently, Pelco.

3. WAKING UP TO WIRELESS, WIRELESS'S RAPID EVOLUTION

Waking Up to Wireless

Wireless... WAKE UP it is here and it is now!

I am writing this column just after returning from BuilConn in Dallas. The haze of mega information is starting to clear and I want to tell you about one of the strong trends I saw at the show. Wireless... WAKE UP it is here and it is now!... AND it will again change the shape of our industry. Approximately 10 to 15% of the folks at the conference were in the wireless industry. Wireless is now cheaper, runs forever on batteries, can even be self-generating, and can now organize itself into self healing and repeating networks. Let me provide you connection to some of the new thinking that will appear in our building automation products in the near future.

Wireless Mesh Sensor Networks

Enable Building Owners, Managers, and Contractors to Easily Monitor HVAC Performance Issues

As the name implies, wireless mesh sensor networks are:

1. Comprised of wireless nodes. A node in this type of network consists of a sensor or an actuator that is connected to a bi-directional radio transceiver. Data and control signals are communicated wirelessly in this network and nodes can easily be battery operated.

2. Arranged in a networking topology called "mesh." A typical mesh network topology is shown in Figure 1. Mesh networking is a type of network where each node in the network can communicate with more than one other node thus enabling better overall connectivity than in traditional hub-and-spoke or star topologies. State-of-the-art mesh networks often have some of the following characteristics:

    a. They are self-forming. As nodes are powered on, they automatically enter the network.

    b. They are self-healing. As a node leaves the network, the remaining nodes automatically re-route their signals around the out-of-network node to ensure a more reliable communication path.

    c. They support multi-hop routing. This means that data from a node can jump through multiple nodes before delivering its information to a host gateway or controller that may be monitoring the network.

The self-forming, self healing, and battery operable attributes of a mesh sensor network make it ideal for temporary environmental monitoring applications in a wide range of facilities.

Autonomic Wireless Building Networks

The key to reliable ad hoc wireless networking for building automation applications and protocols is an Autonomic mesh network: an Autonomic network is one that self-configures, self-heals, self-regulates, and understands when to apply these parameters.

Wireless technology for building automation applications has developed rapidly and commodity radio technology has helped accelerate adoption. Wireless technology provides the benefits of wire replacement for reduced deployment and maintenance costs; increased scalability; and has enabled application development. As pointed out by Helmut Macht, chief technology officer for Siemens Building Technologies Group, "The innovations in wireless communication allow more and more wired communication to be replaced." In addition, wireless is an enabler for many hard-to-reach or hard-to-wire applications typically found in large old buildings, museums, factory floors, and remote job sites; and also in rapidly developing countries like China and India where the cost of wiring anything (labor, routes through old neighborhoods, etc.) is just about cost-prohibitive.

The ZigBee Alliance review of the ZigBee Alliance which is leading the new wave in wireless networking.

Mission: The ZigBee Alliance is an association of companies working together to enable reliable, cost-effective, low-power, wirelessly networked, monitoring and control products based on an open global standard.

Objective: The goal of the ZigBee Alliance is to provide the consumer with ultimate flexibility, mobility, and ease of use by building wireless intelligence and capabilities into every day devices. ZigBee technology will be embedded in a wide range of products and applications across consumer, commercial, industrial and government markets worldwide. For the first time, companies will have a standards-based wireless platform optimized for the unique needs of remote monitoring and control applications, including simplicity, reliability, low-cost and low-power.

What is GridWise?

GridWise is an entirely new way to think about how we generate, distribute and use energy. Using advanced communications and up-to-date information technology, GridWise will improve coordination between supply and demand, and enable a smarter, more efficient, secure and reliable electric power system.

GridWise seeks to modernize the nation's electric system—from central generation to customer appliances and equipment—and create a collaborative network filled with information and abundant market-based opportunities. Through GridWise, we can weave together the most productive elements of our traditional infrastructure with new, seamless plug-and-play technologies. Using advanced telecommunications, information and control methods, we can create a "society" of devices that functions as an integrated, transactive system. GridWise optimizes resources by integrating all elements of the electric system and allowing them to work together in smarter, faster ways. Taking advantage of new and existing technologies, GridWise integrates the energy infrastructure, processes, devices, information and markets into a collaborative arrangement that allows energy to be generated, distributed and consumed more efficiently.

4. LEARNING MORE ABOUT WEB BASED CONTROL

Connecting Convergence August Supplement Engineered Systems

In our fourth supplement "**Connecting Convergence**" prepared for Engineered Systems Anto and I identify the major trends that are fueling the connection of real time data to our client's enterprise. For the most part convergence has happened and what is left to do is sort out who will be the players and which standards will be used in providing these essential connections. Anto's views are well expressed in his portion of the supplement **Convergence or Divergence? Which way to Enterprise Building management?** He outlines the drivers and trends and then analyzes what is going on to move us forward in delivering value to building owners and operators.

Convergence or Divergence

While the industry buzz in the past couple of years has been the idea of convergence, now is a good time to review what is really going on in the industry delivering building systems. To do this, let us look at the drivers and trends and then analyze what is going on to move us forward in delivering value to building owners and operators. Marketing Convergence—Engineered Systems November Supplement

Technology developments are clearly the most visible driver in our industry. In the past decades we have seen the advent of DDC (Direct Digital Control), the proliferation of networked controls and even the creation of open system standards. These changes significantly affected the internal workings of the industry. They provided new features, flexibility, cost savings and even the freedom of choice between systems to control and manage a building. They did not however, change the relationship between building systems and the owners in any significant manner; building systems remained entities unto themselves.

Convergence Will Happen

In the past we did not realize that our industry's direction and our personal lives would be so effected by information convergence. We cannot wait until convergence occurs and then get involved with how it gets marketed, because it will be too late; the marketing plan will not include our industry. Our industry's presence in creating a marketing convergence plan changes everything. We as an industry bring new concepts and tools to the convergence table in the form of "real time information." Our industry's business is collecting, acting on, and distributing real time data such as temperature, pressures, energy usage, client comfort, humidity, IAQ, video, security card ID's.

Why Is There A Need For Marketing?

But some kind of mass communication must occur in order for convergence in the building systems industry to gain a stronger foothold and industry-wide acceptance. Building owners are aware that this convergence is happening, and they're out there looking for answers. And convergence is certainly talked about and written about, but for the most part the message of the remarkable benefits of truly integrated systems is still missing its target. So what does the building systems industry need to do to be heard?

Selling Integration and Convergence

The Buildings Industry is undergoing massive change, and its impact on automation and engineered systems will be profound. My conservative estimate is that 75% of today's "control contractors" will cease to exist, or dramatically change their business model over the next decade.

Controlling Convergence—Engineered Systems April Supplement

Knowledge gained from the digital office explosion has helped building owners understand the potential savings of cost effective web-based upgrades to insure and communicate client comfort at the lowest energy and manpower costs. These solutions are being applied to new buildings but attractive paybacks are also available with the upgrading of existing DDC automation to web-based solutions. Moving to the next level of advanced digital buildings for owners is achievable now because of our automation industry's present position, which includes a myriad of web-based solutions. This position allows us as an industry to lead the world's dynamic/interactive data integration revolution. Working examples now available are Dynamic Energy Control/Accounting, Indoor Air Quality Reporting, Client Comfort Communication & Control interfaces. The transforming of our conservative large building automation industry is necessary to increase our visibility to building owners, designers and decision makers. This new visibility is forcing us to reinvent, restructure and repackage ourselves to create an exciting identity that will make web-based information and real time interactions come alive with seamless connections to the clients' enterprise.

Tom responded with a powerful article, which we have included in this supplement called *"Convergence: What Is It, What Will It Mean, And When Will It Happen?"*

Over the last several years, a remarkable transition has begun in the building controls industry. We see it in what are called Web based control systems that permit operation over standard Web Browsers. In itself, a Web based system is not really much to be excited about. User-friendly human interfaces for building controls have been around a long time. The only real direct advantage of a Web based control interface is that because it uses standard Web browser software, no special software package is required for a computer to operate as a human interface terminal for the control system. Very few control systems really use more than a couple of human interface terminals anyway, so many in the industry are asking, "What's all the fuss about?"

Facility Operation Evolution or Revolution *Guide to Web-Based Facilities Operations*—Doing more with less by using Web-based anywhere information to amplify your existing building operational resources.

The creation of the new model of data and information and support anywhere provides an excellent opportunity to simplify and increase the ease of access to critical building operations. As the "DDC Revolution" exposed the complex relationship between controls and equipment the "Net Revolution" will expose our lack of understanding of the complex cost/comfort/safety issues of today's buildings. This will cause us to economically invest in the need for increased knowledge to operate today's complex buildings cost effectively.

The Web-based presentation of static and dynamic building information of what is actually required to achieve the best of client comfort, while maintaining excellent Indoor Air Quality for the least cost per square foot will be an eye opener for all. As these dynamic models evolve, reflecting the realities of day-to-day building operation, management will be required to re-evaluate the real cost of having less than the best possible operation people; they will arrive at the conclusion that the correct operation team can greatly add to the company's bottom line by enhancing the art of providing desirable cost effective buildings.

The new Web-based media access and presentation has the power to simplify and teach while greatly aiding the comprehension of building operations. The concept of creating a website for each complex allows critical building information to be located in one organized place accessible to all. The inter-weaving of the actual dynamic building information allows contractors and consultants to quickly understand complex building problems.

Procurement of the Latest and Greatest

The request for proposal approach allows active solicitation of the innovative approach.

The request for proposal approach allows active solicitation of the innovative approach. We solved this problem over 20 years ago when the Direct Digital System started to replace pneumatic controls. The problem then was that old ways were hard to change and to reap the benefit of these new technologies the traditional design needed to be retrofitted as soon as the system was installed. To really capture the power of DDC a new approach was needed. The control part of a building represented a small fraction of the total cost and assembly was left to a fragmented group with no concept or care to how the building owner may wish to use the system. The solution was to follow the IT industry procurement model and to buy the building controls much the same as an owner would buy his IT enterprise system. In purchasing IT systems the fact that it all fit together and worked was more important than the lowest cost. Feature, functionality and fit ruled the procurement process.

One of our sixteen is missing...
or can Integration be specified?

Integration presents challenges to the construction procurement process on numerous levels. The industry is demanding integration throughout the enterprise from building automation, fire alarm, access and video surveillance for security to Enterprise Energy Management, Metering and Maintenance Management. Integration goes beyond independent building systems such as those listed above, because it requires commonality with building application infrastructure including: hardwired and wireless LANs', central databases and even network-client software among others. In the last few months, I have reviewed more than a dozen Requests for Proposals and Specifications asking vendors to provide systems that integrate many or all of the above systems. Some of these have been solicitations independent of construction and some have been part of construction specifications. Set aside the flaws that might exist in any of these specifications because of a lack of understanding of the independent systems themselves or the complex interaction that must be achieved between all of these systems and the host of generations thereof. The fundamental issue raised here is that those solicitations that are done as part of a new construction process creates divisions between the various trades, tasks and scopes of work to be carried out.

Web-Based Automation

Access to real-time information anywhere, anytime through an Internet-enabled automation system is the real value of this technology.

Automatedbuildngs.com is a focal point for discussion of control technologies in general, and by extension related Internet technology due to the e-zine's unique vantage point on the industry. Having recently returned from the AHR show in Chicago it seems appropriate to address this juxtaposition of Automation and the Internet, as it seemed to be a significant factor every booth. "Native TCP/IP, Internet-Ready, IP enabled and Internet Control are among the related buzzwords that were used, along with a host of references to Web Browsers and other Internet based features. At the same time, many attendees were asking; what does it mean and how do you cut through the hype to see if it brings any value? The logical expansion of building automation has been to move from Heating Ventilation and Air Conditioning (HVAC) control to Direct Digital Control (DDC) and integration of fire and security, and potentially more. The next step for DDC was Internet access, but Web-based Automation or "Internet Control" goes beyond simple access. Internet control automates facilities, HVAC and processes, while expanding the scope of control to the enterprise level, thus using the Internet to convert a control system into a management information system.

Energy and Wireless Internet... what's the connection?

Wireless Internet is yet one more tool to use in the continual evolution toward Real-time Energy Management.

One of most exciting new technologies today is Wireless Internet Service. It became evident to me how quickly this technology is growing when my family recently visited four Universities, all of which touted wireless Internet access. This Internet technology is called Wireless Fidelity (EI-FI) and a recent *New York Times* article noted that one can walk down any street in lower Manhattan continually surfing the Web. WIi-FI is the reason and it is so pervasive that computers with Intel's centrino" mobile technology or an "802.11b card" can literally hop from one WI-FI network to another.

Real-time Energy Dashboard™ the next new thing in Building Automation

As both author and a System Integrator this entire topic has been an ongoing focus. The result of those efforts is a product called the Real-time Energy Dashboard™. Please note that this is not intended to be an advertisement, but this concept is so new that it is easiest to describe it by showing a real example, and of course it also answers the question above. The concept of dashboards has been growing in popularity for several years in the Information Technology business. Much as Ken discussed in his Engineered Systems treatise, the dashboard can be, in effect, a home page for a building. However it has the potential to be much more! The simple fact is that there is data trapped, and sometimes hoarded, in microcosms throughout any organization. Facility managers have data that only they usually get to see, and so does accounting, finance, purchasing, etc. In most cases these and other groups are more than happy to share the information that they have, but in order to get at it a user must be proficient with special hardware and software, as well as go to someone's office. The dashboard concept is based upon the definition of Web Services, which simply stated is a process that makes it possible for information to be shared between many different systems, that was previously not shared and not available, via the Internet.

Economy Drives Convergence

The first advantage of convergence is installed cost. It's less expensive for the BAS to use multi-function cabling and technology already installed for other enterprise networking applications: email, sales data, collaboration tools, and others. What's more, enterprise networks are correctly seen as mission critical. As a result, these networks usually are high-performance, capable of carrying a vast amount of application data at high speed with good security and reliability.

Access and reach are two more advantages that enterprise networks offer BAS systems. By their nature, enterprise networks reach out to points throughout a company or institution where a traditional BAS, until very recently, had limited or no means to communicate. When a BAS uses an enterprise network, the BAS and its users benefit from this expanded reach. Just as Simone in Fresno can send email with a spreadsheet attached back to the corporate office in Toledo, the BAS in Fresno can automatically share energy usage reports and other operating data with the Toledo office. The corporate facilities manager now has an unprecedented ability to keep tabs on increasingly dispersed holdings in real time, a great benefit to controlling costs in an energy-sensitive economy.

CONCLUSION

This chapter contains a short list that describes references covering four major areas of importance to understanding the use of web-based energy information and control systems. The chapter can also be found as a dynamic chapter at the following website: *www.automatedbuildings.com/education.htm*. The website will be continually updated in these areas to help those wishing to have the latest information on those subjects.

Section Two

Web Based EIS Case Studies

Chapter 3

The Utility Report Cards: An Energy Information System for Orange County Public Schools

Paul Allen
David Green
Safvat Kalaghchy
Bill Kivler
Blanche Sheinkopf

ABSTRACT

THE *UTILITY REPORT CARDS* (URC) program is a web-based Energy Information System that reports and graphs monthly utility data for schools. The program was developed and prototyped using Orange County Public Schools (OCPS) utility data. Each month, a web-based report is automatically generated and emailed to school principals and staff to examine their school's electricity usage (energy efficiency) and to identify schools with high-energy consumption for further investigation. The easy-to-use web-style report includes hyperlinks to (1) drill-down into further meter details, (2) show graphs for a 12-month comparison to prior-year data, (3) filter the data to show selected schools, and (4) re-sort the data to rank schools based on the data selected. The URC is also intended for teachers and students to use as an instructional tool to learn about school energy use as a complement to the energy education materials available through the U.S. Department of Energy's EnergySmart Schools program (ESS). To run the URC, go to *www.utilityreportcards.com* and click on "URC Live."

WHY UTILITY REPORT CARDS?

The URC was created to help OCPS staff understand and therefore manage their utility consumption and associated costs. The URC program was designed as a web-based system to take full advantage of email and web-browser technologies. The URC allows each school principal to become aware of how his/her school is performing relative to a projected benchmark and to other schools of similar design and capacity. Giving awards to schools that improve performance from prior-year levels could create a spirit of competition with the opportunity to recognize success. Those schools identified as high-energy users become the focus of attention to determine the reasons for their consumption level and ultimately to decrease the energy used. All of this is done by using the monthly utility data that is provided electronically at minimal or no–cost by the utilities

The URC public/private partnership includes:

Orange County Public Schools, FL (OCPS)

The Orange County public school system is the 14th largest district out of more than 16,000 in the nation and is the 5th largest in Florida. The OCPS operates 152 schools with over 160,000 students. [1]

Florida Solar Energy Center (FSEC), Cocoa, FL

A nationally recognized energy research institute of the University of Central Florida, FSEC developed the URC program and provides monthly utility data translation and maintenance. This approach provides a standard report format and consistency between the utilities and the school district. [2]

Orlando Utilities Commission and Progress Energy, Orlando, FL

These two local utility companies provide electric utility service to OCPS and electronically send the utility billing data to FSEC on a monthly basis. The utilities benefit from the URC by the positive public relations and outreach associated with helping schools manage their energy usage. [3]

State of Florida Energy Office, Tallahassee, FL

This agency provided the initial funding to FSEC for development and implementation of the URC. [4]

U.S. Department of Energy's EnergySmart Schools (ESS), Washington, DC

The ESS is a program that works with schools to reduce utility consumption/costs throughout the United States. The URC provides ESS with an Energy Information System tool for school districts that will complement energy education programs already offered through the EnergySmart Schools. Additionally, through the existing national network of ESS partnership districts, the URC can be easily replicated and offered to schools nationwide. [5]

Walt Disney World Company, Lake Buena Vista, FL

Walt Disney World assisted OCPS in streamlining their facilities management processes and reducing utility costs. Walt Disney World has developed web-based energy information systems and provided technical support for the development of the URC. [6]

URC DATA COLLECTION

Data Sources

The utility data used in the URC program is based on the monthly electric billing data from Orlando Utilities Commission and Progress Energy. The OCPS Superintendent authorized the utilities to release the OCPS utility data to FSEC, and each Utility Company was asked to provide OCPS utility data from the last two years. Both utilities agreed to provide the OCPS data electronically to FSEC at no-cost based on the positive public relations generated.

The URC also uses information from the Common Core of Data (CCD), a comprehensive national statistical database of information about all public schools and school districts. CCD is compiled annually by the U.S. Department of Education's National Center for Education Statistics. One component of data that was missing was the school square footage. A supplemental table was made for the square footage of each school. This was needed to calculate the consumption per square foot utility efficiency benchmarks for each school. Additionally, since the school's square footage could change as a result of school construction projects, the square footage table is updated monthly to account for school expansions.

Data Transfer

FSEC established an account on its server where the utility data files were transferred automatically using File Transfer Protocol (FTP). Although each utility had the OCPS utility data available electronically in its utility billing system in an ASCII delimited file format, there was no consistency in the data output formats provided to FSEC. Therefore, FSEC created a custom program called URC_DPP (URC Data Processing Program) that processed each utility's data file separately and loaded the data into a common database. A future enhancement for the URC would be to develop (specific?) standards for the format and method of delivery for electronic monthly utility data.

On a monthly basis, the utilities electronically transmit to FSEC the OCPS utility data, which then adds it to the URC relational database. FSEC sends an email to a designated email address at OCPS with a copy of the current month URC embedded in the email message (the URC format for February 2004 is shown in Figure 3-2). OCPS then forwards this email to their internal email distribution list for principals and staff. This makes it easy for users, since all they need to do is click on the hyperlinks in the URC email to produce graphs and detailed reports.

At this point, only electric utility data is being tracked. In future enhancements, the URC could include additional utilities, such as water and natural gas.

Relational Database

The Oracle relational database management system was used for the URC, primarily due to FSEC extensive use and experience with Oracle databases. The database tables used to store the URC information include:

- School Information: constructed from the CCD database
- District Information: constructed from the CCD database
- School Building Information: square footage and other building related data
- School Contact Information: contact information for school officials
- Utility Contact Information: contact information for utility personnel
- Service Type: types of services offered by the utilities
- Meter Information: school where located, meter ID, meter description
- Meter Readings: monthly reading from the meters
- Weather Information: Cooling Degree Day and Heating Degree Day

URC WEB-PROGRAM FUNCTIONALITY

The URC program is designed to be informative, intuitive and flexible for all users. It takes advantage of extensive use of hyperlinks that create graphs and detailed reports from an overall summary report listing all schools. Users are able to view graphs and see the electric consumption patterns by simply clicking on hyperlinks in the URC web page. They can also modify the reports by using hyperlinks to sort and filter the data. Hyperlinks provide the users with instant results, making the application exciting to work with and easy to understand.

The URC home page (*http://www.utilityreportcards.com*) serves two functions: (1) contains hyperlinks that provide information about the URC and its development partners and (2) it provides a launching point to run the URC program. Figure 3-1 shows the URC home page.

When the User selects the "URC Live" hyperlink, a summary report (see example in Figure 3-2) shows totals for each school type in the entire school district—primary (elementary) schools, middle schools, high schools, etc.—as the rows in the report. The data presented in the columns includes the electric consumption (kWh), the cost (dollars) and the efficiency (Btu/sqft). To provide meaningful comparisons to the same time in the prior year, the URC program divides the data by the number of days in the billing period to produce per day values. The URC also allows the User to change the values shown to per month figures which is just the per day values multiplied by 30 days. Figure 3-2 shows the overall school district summary report for February, 2004.

The URC program interface makes the program easy to use considering the enormous amount of data available. The top down approach lets Users view different levels of data from the overall district-wide summary to individual meters in a single school. For example, the User can click on a school type to display the details for each school. Clicking on the "On" link in the legend will toggle the color flags on or off. Figure 3-3 shows the result of clicking on the hyperlink "Regular High School": a report on all high schools in the database for February 2004. The report is sorted based on the percent change in kWh usage from the prior year levels. The schools that changed the most are at the top of the list. Re-sorting the schools is accomplished by simply clicking on the column title. To produce the report shown in Figure 3-3, the original report sorted by Efficiency Percent Change was re-sorted by clicking on the Consumption Percent Change column.

Graphing is accomplished by clicking on any current period value in the report. To display a 12-month graph for Apopka Senior High School, clicking on the number 22,399 would produce the graph shown in Figure 3-4. Note that the consumption levels are significantly higher than prior year levels for this school.

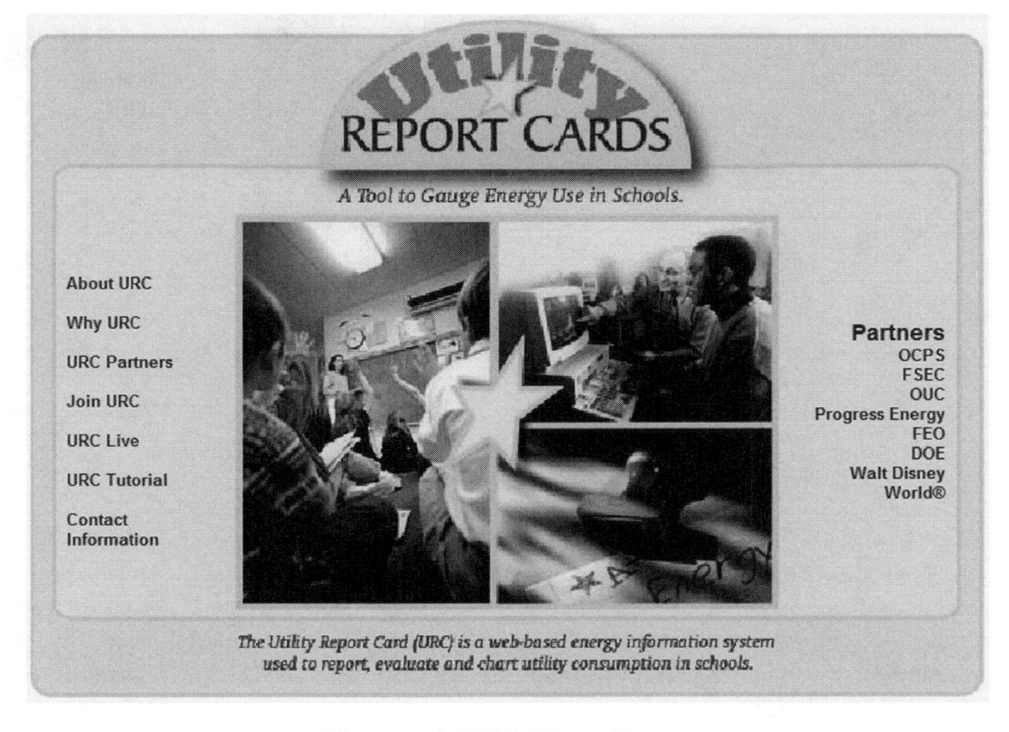

Figure 3-1. URC Home Page

UTILITY REPORT CARDS									
Electric Per Day	ORANGE COUNTY PUBLIC SCHOOLS					<< February 2004 >>			
	Consumption (kWh/day)			Cost ($/day)			Efficiency (Btu/sq ft/day)		
School Type	Current Period	Previous Period	Percent Change	Current Period	Previous Period	Percent Change	Current Period	Previous Period	Percent Change
Regular Primary School	178,073	181,162	-2 %	$ 13,923	$ 12,012	16 %	146	149	-2 %
Regular Middle School	121,735	124,550	-2 %	$ 9,119	$ 7,984	14 %	139	142	-2 %
Regular High School	168,196	167,136	1 %	$ 12,135	$ 10,121	20 %	161	160	1 %
Special Other	6,351	6,092	4 %	$ 479	$ 404	19 %	143	137	4 %
Vocational High School	5,910	6,572	-10 %	$ 469	$ 448	5 %	106	118	-10 %
Grand Total ORANGE COUNTY PUBLIC SCHOOLS	480,266	485,511	-1 %	$ 36,125	$ 30,969	17 %	148	150	-1 %

On — Denotes Increase From Previous Year
On — Denotes Decrease From Previous Year by ◂ 0 % ▸ or more

Figure 3-2. Overall School District Summary Page

UTILITY REPORT CARDS									
Electric Per Day	ORANGE COUNTY PUBLIC SCHOOLS					<< February 2004 >>			
Regular High School	Consumption (kWh/day)			Cost ($/day)			Efficiency (Btu/sq ft/day)		
SCHOOL	Current Period	Previous Period	Percent Change	Current Period	Previous Period	Percent Change	Current Period	Previous Period	Percent Change
APOPKA SENIOR HIGH SCHOOL	22,399	20,360	10 %	$ 1,777	$ 1,375	29 %	211	192	10 %
CYPRESS CREEK SENIOR HIGH SCHOOL	16,932	19,245	-12 %	$ 1,176	$ 1,108	6 %	153	174	-12 %
EVANS HIGH SCHOOL	17,108	16,526	4 %	$ 1,229	$ 1,022	20 %	156	151	4 %
OAK RIDGE HIGH SCHOOL	12,940	14,669	-12 %	$ 908	$ 855	6 %	148	168	-12 %
OLYMPIA HIGH SCHOOL (FORMERLY DR. PHILL	13,441	13,119	2 %	$ 980	$ 805	22 %	118	116	2 %
ROBERT HUNGERFORD PREPARATORY HIGH SCHOOL (FORMERL	5,525	5,727	-4 %	$ 424	$ 393	8 %	192	199	-4 %
TIMBER CREEK HIGH SCHOOL	15,133	15,379	-2 %	$ 1,081	$ 924	17 %	134	136	-2 %
UNIVERSITY HIGH SCHOOL	20,713	19,056	9 %	$ 1,482	$ 1,131	31 %	163	150	9 %
WEST ORANGE HIGH SCHOOL	21,375	21,898	-2 %	$ 1,473	$ 1,261	17 %	190	194	-2 %
WINTER PARK HIGH SCHOOL	22,631	21,157	7 %	$ 1,604	$ 1,247	29 %	168	157	7 %
Regular High School	168,196	167,136	1 %	$ 12,135	$ 10,121	20 %	161	160	1 %

Figure 3-3. High School Summary Report sorted by percent change in consumption from prior year.

Focusing in on the reasons for this increase should be the next step for the OCPS facility personnel. Once it is determined the increase is not due to other factors such as increased enrollment or extreme weather, personnel can consider making adjustments to the energy management system controls to turn the trend around. Clicking on the "Off" link in the legend of the graph will toggle on and off a display (above the graph) of the actual data for each month.

Users can also look at schools that are performing better than prior year kWh levels. In Figure 3-3, Cypress Creek Senior High School showed the largest percent change from prior year levels in February 2004. Clicking on the number 16,934 next to Cypress Creek Senior High

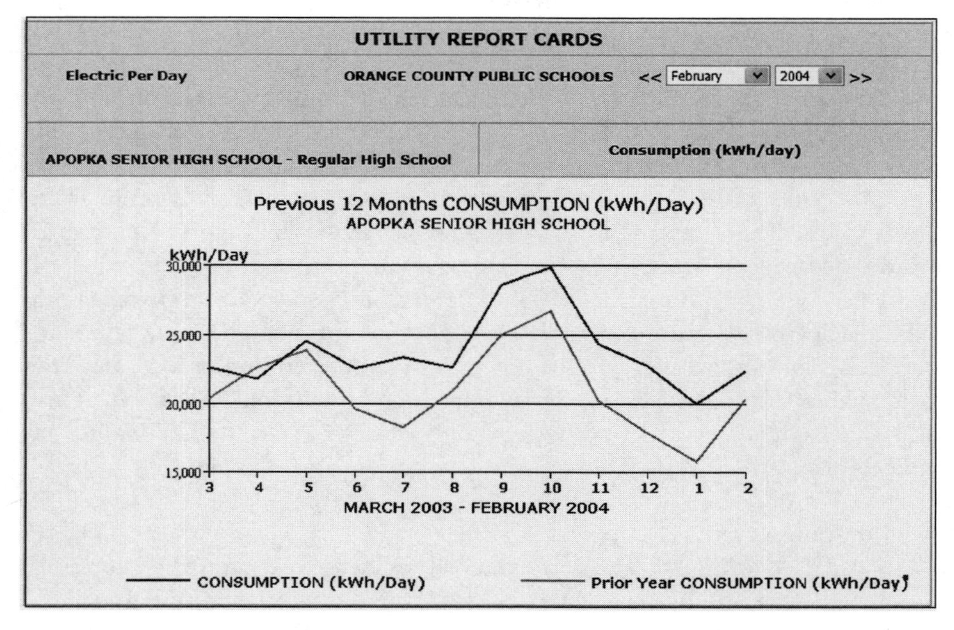

Figure 3-4. Apopka Senior High School 12-Month kWh graph.

The URC program has several other functions. Clicking on the school name produces a report that shows data for each electric meter at the school. Clicking on the efficiency column of the report in Figure 3-3 re-sorts the report and shows which schools are more efficient on a per square foot basis. In the case of OCPS, the newer schools have proven to be more efficient than the older schools. Readers are encouraged to try out the URC functionality themselves by visiting the URC website at *http://www.utilityreportcards.com/*. The driving forces behind all of this functionality are dynamic web programs and relational database tables.

School produced the graph shown in Figure 3-5 below. This graph shows that Cypress Creek has significantly lowered its kWh consumption from the prior year. Again, focusing on the changes that produced these reductions will help OCPS facilities management replicate these best practices in other schools. The user can click the "Back" button on their browser to return to the previous page.

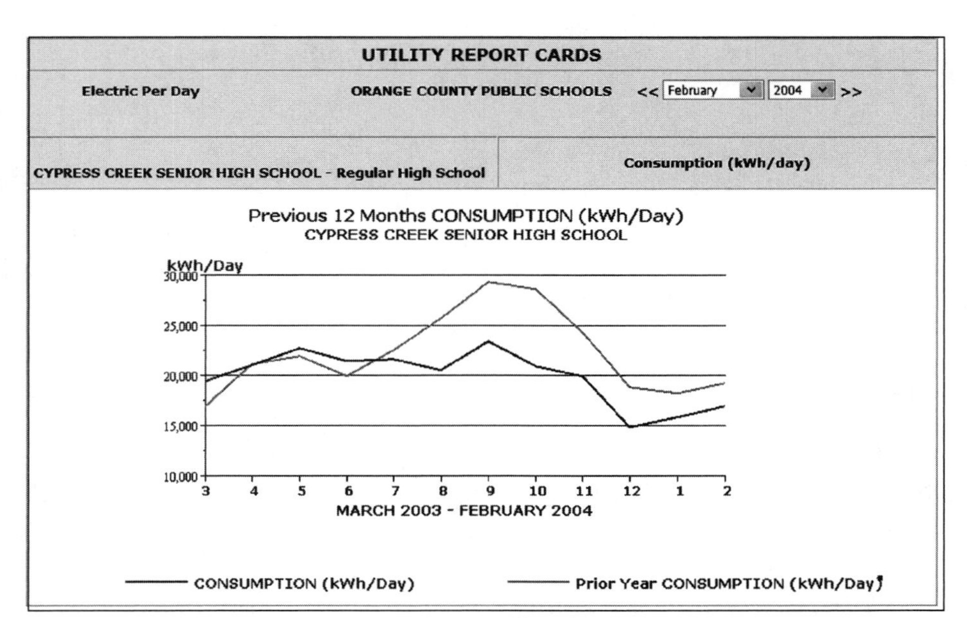

Figure 3-5. Cypress Creek Senior High School 12-Month kWh graph.

URC WEB-PROGRAM DETAILS

Two web programs make up the application interface. One is for reporting and one is for graphing. The programs are written in PHP (Hypertext Preprocessor). "PHP is a widely used Open Source general-purpose scripting language that is especially suited for Web development and can be embedded into HTML" according to PHPBuilder.com. This means that the PHP application itself is not for sale. It is developed and supported solely by volunteers. What makes it an attractive choice is that the developers' code is part of the HTML page itself. PHP is ideal for connecting to databases and running SQL queries, to return dynamic content to a web page. [8]

Using PHP, the reports are designed to accommodate both the novice and the expert user. Hyperlinks in the reports pass values back into the same two programs in a recursive manner using the CGI query string. These values are processed by a set of predetermined rules built into the program by the developer to incrementally change the appearance of the reports to suit the user. The hyperlink construction includes messages using the *onmouseover* event to explain the action of the hyperlink. Total and sub-total lines provide summary information. Data that shows an increase or decrease from the previous year is flagged with a different background cell color. The user can define *percent criteria* for marking decreases from the prior year because this is helpful in tracking progress toward a particular goal such as 5% decrease from previous year. Graphs are created using KavaCharts, a collection of Java applets available from Visual Engineering at *http://www.ve.com*. [9] The following section describes some of the technical programming details in the order they are executed. We hope these details will be useful for readers with some experience in program development.

Reporting Program

1. Retrieving Values

The program algorithm begins by setting up variables to define the background colors for the heading, sorted heading column, data cells and flagged data cells. The HTML table width is set to 670 pixels for portrait printing (see below).

Next, the program retrieves the values passed in from the various hyperlinks using the CGI query string or sets them to default values. Values for the time period of the report are first. The default month is set to last month since that is likely the most recent full month of data. The program retrieves values associated with the design of the reports such as the column to sort by and the direction of sort (ascending or descending). There are also values associated with turning the flags on and off and setting the percent criteria. Another value controls whether or not the report is normalized to an *average per day* basis or simply reports the totals per month. Data reported on an *average per day* basis is more meaningful since each meter and month has a unique number of billing days. Other values are associated with what data to show such as consumption, cost, efficiency or all three. There are values to define filtering of the data. The user can *drill down* in the report to look at data for only one school type or one school. Retrieving CGI values in PHP is done as at the bottom of this page.

2. Using the Values

The retrieved values help to design the report. If the flag1 value is "off," then the program sets the background color for flagged cells to the normal background color. It does the same for the flag2 value. If a school *type and level* hyperlink was selected, a name value pair such as "typelevel=11" is sent back into the program. So a *$filterby* variable is set to "typelevel" and a *$filter* variable is set to "11." This makes it easy to add filters to the SQL query. The same process is used if a school hyperlink is selected. The *$filterby* variable is set to "school" and the *$filter* variable is set to the school ID of the selected school. Similarly, sorting by a particular column is passed in by one of the values and assigned to a *$sortby* variable. Then the program changes the heading color for that column to some color other then the normal color so that users will know which column is

```
$cellcolorheading = "#CCCC99";        //Color of heading background.
$cellcolorsorted = "#CCCCCC";         //Color of sorted column heading background.
$cellbgcolor = "#FFFFFF";             //Color of data background.
$cellcolorflag1 = "#FF9933";          //Color of data background flagged for increase from previous year.
$cellcolorflag2 = "#99FF00";          //Color of data background flagged for decrease from previous year.
$tablewidth = 670;                    //Set the table width to 670 pixels for portrait printing.
```

```
//Get the date variables or set them to default values.
if(!$month) {                         //Check for a $month variable, if there is none,
      $month = $_POST['month'];       //then get the value from the POST query string.
}
if(!$month) {                         //Check for a $month variable, if there is none,
      $month = $_GET['month'];        //then get the value from the GET query string.
}
if(!$month) {                         //Check for a $month variable, if there is none,
      $month = date("m") - 1;         //set the current month to last month.
      if ($month == 12) {             //If last month is December,
                                      $year = date("Y") - 1; }//then set the $year variable to last year.
```

the sorted column. A *$dir* variable controls the direction of the sort, either ascending or descending (see first program below).

3. *Querying the Database*

Now the program can connect to the database using the appropriate connection settings required by the particular server being used and query the database for data. The program must construct a query string to do this.

First, the program creates a *select list* of the fields to return from running the query. It is important to *sum* the data for each field since there may be more than one meter for each school. If the *average per day* value is set to "yes" then the sum is divided by the number of billing days in the month. Next, the query joins the tables together using an SQL join or where clauses. Tables are typically together using key fields such as *school ID*, *meter ID*, etc. Then conditions are used to filter the data for the time period, school type and level, or school selected using SQL *where* clauses. The *group by* clause is used to sum the correct values. If a *total row* was selected then *group by* [school type and level]. If a school was selected then *group by* [meter], otherwise, *group by* [school]. You can use the *order by* clause to sort by school type and level, then apply the selected sort column and sort direction passed in as described above. The program runs the query and stores those results in a data array. The second program below gives a simple version of an SQL query to show consumption for one school.

4. *Displaying the Report*

The program then displays the main heading. On the report title, a hyperlink is created to reset the sort and filter and display the default report. Also, a hyperlink is available to toggle back and forth between *average per day* values and *total per month* values. The hyperlink to switch from *average per day* to *total per month* is created as shown below. Other hyperlinks are created in a similar manner. Notice that all values required to produce the desired report design are passed back into the program in each and every hyperlink (at bottom).

The program creates variables to hold the previous month and next month values. If the current month is January, then it sets the previous month to December of the previous year. If the current month is December then it sets the next month to January of the next year. Then it creates a hyperlink to change the month to the previous month as well as a hyperlink to change the month to the next month. The current month and year are displayed between the two hyperlinks (<< and >>).

```
if ($flag1 <> "yes") {$cellcolorflag1 = $cellbgcolor;}     //If flag1 is turned off, change cell color to normal background color.

if ($flag2 <> "yes") {$cellcolorflag2 = $cellbgcolor;}     //If flag2 is turned off, change cell color to normal background color.

if ($sortby == "consumption") {           //If sorting by consumption,
$cellcolorkwh = $cellcolorsorted;         //set the consumption heading background color different then the others.
} else {                                  //Otherwise,
$cellcolorkwh = $cellcolorheading;        //set the consumption heading background color to the normal heading background color
```

```
Select   sum(c.consumption) as consumption,
         a.School_name, b.meter_id, a.school_id, a.type_code, a.level_code
From school_info a, meter_info b, reading_info c
Where c.school_id = $filter
Group by b.meter_id, a.school_name, a.school_id, a.type_code, a.level_code
Order by $sortby $dir
```

```
//Create to a hyperlink to switch between average per day and total per month
echo "<a href=query.php?school=$school&sortby=$sortby&dir=$dir&$filterby=";
echo urlencode($filter)&content=$content&district=urlencode($district);
echo "&month=$prevmonth&year=$prevyear&flag1=$flag1&flag2=$flag2";
echo "&dpercent=$dpercent&perday=no ";
echo "onmouseover=\"window.status='Select This Link to Show Total Values for the Month'; return
true\" >";
     //Add a message to explain the hyperlink.
echo "Day"; }
```

5. ***Looping through the Records***

Before looping through the data records returned from the query, the program initializes a variable to hold the value of the current school *type and level*. Variables for totals and the grand totals are initialized as well. While looping through the data records the program checks for conditions to display sub-headings and total rows at the appropriate place.

It creates string descriptions of the school type code and school level code. If this record is a new school *type and level* the program adds a total line (if not the first record) then it displays the school *type and level* as well as the sub-headings.

The total line has a hyperlink to show only the total lines when selected. The data values for total consumption, cost and efficiency for each school type and level have hyperlinks which link to a program that generates a graph of the values for the last 12 months. Other values for prior year and percent difference are displayed without hyperlinks.

Next, the sub-heading row is added with hyperlinks to generate a sort by column; school (or meter), consumption, previous year consumption, percent difference, etc. The hyperlinks act as toggle switches changing the sorting back and forth between ascending and descending.

Now the program displays the detailed data lines. The first cell is the school name as well as a hyperlink to display the data for just the school. The next cell shows the consumption and is a hyperlink to graph consumption for the last 12 months. The prior year consumption and percent difference are displayed without hyperlinks. The same is done for cost and efficiency.

After displaying the data line, the current school *type and level* is assigned to the variable created for that purpose. Then the totals and grand totals are updated. The program then goes to the next record.

After looping through all of the data, a total line is added in the same manner as before for the last school *type and level*. Then the grand total line with hyperlinks to show only the grand total and to graph the grand total values for the last 12 months is added. Below the report are toggle hyperlinks for turning flags on or off and adjusting the percent decrease criteria by 5% in either direction. The final report is shown in Figure 3-8.

Graphing Program

1. ***Querying the data***

The graphing program algorithm is nearly the same as the reporting program algorithm. The first exception is that data is queried for the last 12 months including the current month rather than the current month alone.

```
//Filter for all records less than or equal to the current month
and greater than the same month last year.
$query = $query. " and ((a.bill_month <= $month and
a.bill_year = $year) or (a.bill_month > $month and a.bill_year
= $pyear)) ";
```

Secondly, all data are sorted by year and month to produce a data display and trend graph of the last 12 months. Figure 3-9 shows the graphic that is produced.

```
$query = $query. " order by a.bill_year, a.bill_month ";
//Sort by year and month
```

2. ***Displaying the data***

While looping through the data, aside from displaying the values as in reporting, text strings are created to assign to the Java applet parameters that produce the graph. The KavaChart applet parameter requires a comma-separated format of values.

```
<param name=dataset0yValues value='2602, 2966.66, 3276.36,
2562, 2715.625, 3271.03, 3651.33, 3145.625, 2992.41, 2726,
2541.76, 3626.20'>
```

Figure 3-6. URC Title Rows

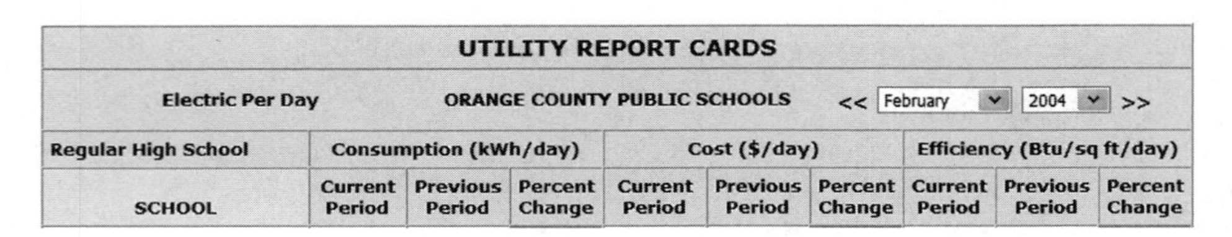

Figure 3-7. URC Column Row

UTILITY REPORT CARDS									
Electric Per Day	ORANGE COUNTY PUBLIC SCHOOLS				<< February 2004 >>				
Regular High School	Consumption (kWh/day)			Cost ($/day)			Efficiency (Btu/sq ft/day)		
SCHOOL	Current Period	Previous Period	Percent Change	Current Period	Previous Period	Percent Change	Current Period	Previous Period	Percent Change
APOPKA SENIOR HIGH SCHOOL	22,399	20,360	10 %	$ 1,777	$ 1,375	29 %	211	192	10 %
CYPRESS CREEK SENIOR HIGH SCHOOL	16,932	19,245	-12 %	$ 1,176	$ 1,108	6 %	153	174	-12 %
EVANS HIGH SCHOOL	17,108	16,526	4 %	$ 1,229	$ 1,022	20 %	156	151	4 %
OAK RIDGE HIGH SCHOOL	12,940	14,669	-12 %	$ 908	$ 855	6 %	148	168	-12 %
OLYMPIA HIGH SCHOOL (FORMERLY DR. PHILL	13,441	13,119	2 %	$ 980	$ 805	22 %	118	116	2 %
ROBERT HUNGERFORD PREPARATORY HIGH SCHOOL (FORMERL	5,525	5,727	-4 %	$ 424	$ 393	8 %	192	199	-4 %
TIMBER CREEK HIGH SCHOOL	15,133	15,379	-2 %	$ 1,081	$ 924	17 %	134	136	-2 %
UNIVERSITY HIGH SCHOOL	20,713	19,056	9 %	$ 1,482	$ 1,131	31 %	163	150	9 %
WEST ORANGE HIGH SCHOOL	21,375	21,898	-2 %	$ 1,473	$ 1,261	17 %	190	194	-2 %
WINTER PARK HIGH SCHOOL	22,631	21,157	7 %	$ 1,604	$ 1,247	29 %	168	157	7 %
Regular High School	**168,196**	**167,136**	1 %	**$ 12,135**	**$ 10,121**	20 %	**161**	**160**	1 %

On — Denotes Increase From Previous Year

On — Denotes Decrease From Previous Year by ◂ 0 % ▸ or more

Figure 3-8. Completed URC Report

A toggle hyperlink is added to turn the data display on or off.

URC MEDIA EVENT

On April 5, 2004, the URC was unveiled to OCPS and the nation in a media event held at Citrus Elementary School in Ocoee, Florida. U.S. Secretary of Energy, Spencer Abraham along with dignitaries from the URC partners and a group of fifth grade students attended the event. The media covered the event through television, radio and newspaper reports to inform the public about the URC.

Secretary Abraham said the report cards would allow schools to save money while also teaching children to be responsible energy consumers. The goal, Abraham said, is education. "If we have more money to spend on students, that means more teachers and more equipment," he said. "If we can save on the energy side and spend it on the student side, in our opinion, that's great." [7]

Schools can use this as just one tool in their arsenal to have better visibility of their energy costs," said Bill Kivler, director of engineering services for the Walt Disney World Co. To run the URC, go to *www.utilityreportcards.com* and click on "URC Live."

CONCLUSION

The U.S. Department of Energy and others have identified the need to help our nations' schools lower their energy consumption costs. The URC provides a way for schools to save money on energy consumption. Some day students may use the URC to learn to become efficient consumers themselves. The URC is a challenging and cooperative effort to collect utility data from various sources and put it all together in meaningful reports and graphs. In addition, the URC is designed as a comparison and analysis tool providing many different "views" into the data quickly and easily using its interactive web reporting features. The URC was developed using some of the most current database and open-source programming tools available. Hopefully, this approach will allow the URC to continue to help reduce energy costs far into the future.

At the time of this writing, the URC was just unveiled so it is too early to measure the impact the URC will have on reducing utility consumption for OCPS. Based on the initial positive comments, we think the URC will help OCPS focus attention on school energy consumption. One of the main benefits of the URC is that the school principals will know their own school's energy usage pattern. This knowledge allows the principals to take appropriate actions to focus on the reasons for the increases and ultimately return their schools to their normal consumption levels.

The URC reporting format is also not limited to schools. A facility manager could use the URC to list any similar group of facilities that are separately metered. Buildings on a university campus, resorts hotels in a city, supermarkets in the same geographic area are a few examples that come to mind.

The future for the URC looks bright. Future URC enhancements include (1) reporting other school districts (2) adding other utilities in addition to electricity (3) adding weather information (4) developing a standard for utility data transfer and (5) integrating the EPA's

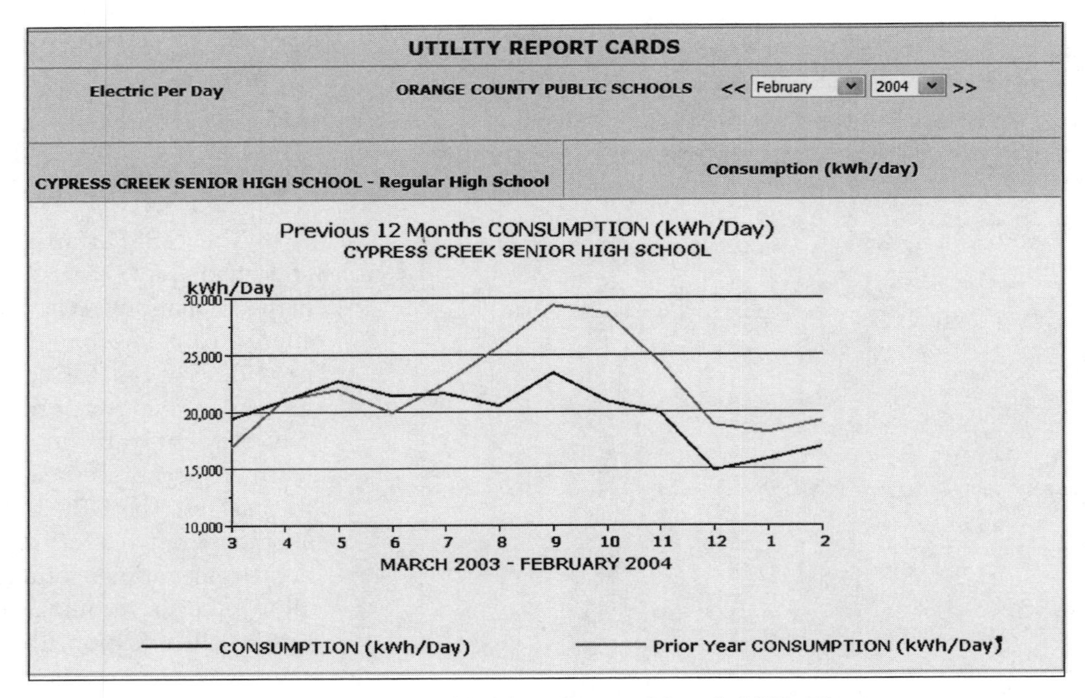

Figure 3-9. Graphic showing 12-Month kWh Usage

Figure 3-10. Left to right: U.S. Secretary of Energy Spencer Abraham, and Kym Murphy, Walt Disney Company Senior Vice President of Corporate Environmental Policy, at OCPS media event April 5, 2004.

Portfolio Manager Energy benchmarks for each school. School Districts that are interested in establishing their own URC are encouraged to find out more by visiting the URC website at *http://www.utilityreportcards.com.*

References

[1] Orange County Public Schools, Orlando, Florida; *http://www.ocps.k12.fl.us/*, Internet page accessed April, 2004.

[2] Florida Solar Energy Center; "A Research Institute of the University of Central Florida"; *http://www.fsec.ucf.edu/*, Internet page accessed April, 2004.

[3] Orlando Utilities Commission, "OUC The Reliable One"; *http://www.ouc.com*, Progress Energy, People, Performance, Excellence"; *http://www.progress-energy.com*, Internet page accessed April, 2004.

[4] State of Florida Energy Office; *http://dlis.dos.state.fl.us/fgils/agencies/energy*.html, Internet page accessed April, 2004.

[5] U.S. Department of Energy's EnergySmart Schools; "Rebuild America, Helping Schools Make Smart Choices About Energy"; *http://www.rebuild.org/sectors/ess/index*.asp, Internet page accessed April, 2004.

[6] Walt Disney World Company; "Disney's Environmentality"; *http://www.disneysenvironmentality.com*, Internet page accessed October, 2004.

[7] "Report Cards Track Schools' Utilities"; Orlando Sentinel; 4/6/2004.

[8] The PHP Group; "PHP Manual, Preface"; *http://www.phpbuilder.com/manual/preface.php;* Internet page accessed April, 2004; last updated, 10/26/2002.

[9] Visual Engineering; "KavaChart, The Complete Solution for Java-Based Charting"; *http://www.ve.com/*; Internet page accessed April, 2004.

ACKNOWLEDGMENTS

I would like to thank Robert J. Lewis for assuring me that it is ok to use recursive hyperlinks in web applications. The technique is not applicable to all projects, but it works well for the URC. Robert is a talented and accomplished programmer whose innovations inspire me greatly. —*David Green*

Chapter 4

Open Energy Information System

David E. Norvell PE, CEM
Energy Manager
University of Central Florida

ABSTRACT

THE POTENTIAL for savings with the installation of a comprehensive Energy Information System (EIS) has been well documented in the industry. However, only a small fraction of organizations has implemented a comprehensive EIS system. Many have hesitated due to the high up-front costs and long-term commitments, which are associated with the commercially available systems. These groups have simply not been able to justify the expense of a commercial EIS system.

The University of Central Florida (UCF) was one such group. Although cognizant of the tremendous energy savings that could be realized with a campus-wide real-time EIS system, the university was restricted by a limited budget. The solution was for the UCF energy department to develop a low cost, fully featured alternative—the Open Energy Information System (OEIS).

The OEIS was designed to match or surpass the features of commercially available EIS tools at a fraction of the cost. The OEIS is inexpensive, robust and web enabled. The system is easily accessible and understandable to both the energy manager and the individual building managers.

The methodology as well as the hardware and software components of the OEIS is presented here to enable the reader to reproduce the system in their particular campus environment. The end product is easily accessible and understandable to both the energy manager and the individual building managers within the organization. Although the options that were initially available had seemed quite limited (particularly within current budget constraints), the implemented OEIS at UCF has proven to be a tremendous EIS success. Its outcome may, in fact, have a long-lasting impact on the automated meter industry.

BACKGROUND

Why OEIS?

The University of Central Florida is a major metropolitan research university growing quickly to provide for the needs of an increasing student population. Like many college campuses, UCF is faced with tightening utility budgets and increasing utility costs. In the past the various colleges did not see their utility costs, as the Physical Plant department has always been responsible to fund these costs for the university Therefore, the department heads of the various colleges had not been engaged with the energy consumption of their buildings. Many studies[1] have shown that when people become aware of their energy use, a reduction in energy consumption of 10% or more can be realized through simple voluntary changes in behavior. This 10% reduction at UCF amounts to over $700,000 annually.

In early 2003, the university management decided that greater awareness of energy use and costs would bring about increased interest in conservation. The UCF energy staff was tasked to provide real-time interval meter data for a 140-building campus occupying 6 million square foot for a very modest cost. The one most frustrating aspect of real-time sub-metering has always been the high cost per point, where a point consists of one utility meter in one building. Most buildings have more than one utility meter resulting in several points per building. The UCF campus has approximately 500 meters (points), which we needed to monitor.

We received estimates from contractors to install a comprehensive system of real-time monitors on our existing electro-mechanical meters in excess of $600,000. In addition to the up-front costs, monthly maintenance

[1]http://www.rebuild.org/sectors/SectorPages/PartnershipView.asp?MktID=2&OrganizationID=1169

costs and a long-term contract would be required to warehouse and retrieve the data. These high costs would result in a lengthy payback period for the monitoring system itself, which would delay any future energy projects.

Nearly all of our buildings have a set of mechanical utility sub-meters, which record the following: electric power, natural gas, chilled water usage, and potable water usage. These meters are owned and maintained by the university; they are separate from the master meters provided by the utility company. Our staff must read each of these meters each month. The UCF Physical Plant uses this information to track the energy usage throughout the campus.

At UCF we have both educational and non-educational buildings. Physical Plant invoices the non-educational buildings each month according to their utility usage. This process of manual meter readings and invoicing is very labor intensive. Furthermore when the meters in the field fail, they may go undetected for a month or more.

The UCF Energy staff felt they could develop their own energy information system in-house. Open Energy Information (OEI) system was created to provide timely, accurate data to the UCF energy professionals and building managers.

OEIS DESIGN GOALS

The UCF Energy Staff felt the OEIS needed to be accurate, dependable, and easy to use. We wanted the faculty and staff to grow to depend on the system for up-to-date, reliable energy information. We wanted them to have the ability to budget and forecast their utility costs. In addition we wanted the OEIS to provide immediate feedback to the building managers as they attempt to change the behaviors of the building occupants. Behavior changes may include turning off lights and computers when they are not needed. The system had to be easily accessible at anytime, from anywhere.

The OEIS had to be very low cost. At the time of inception, there was only very limited funding available. We knew that if we could build and successfully operate the initial system with only a handful of meters attached, we could easily scale the system to cover the entire campus.

After the OEIS was proven successful on a small scale, we completed a few quick energy projects, using the OEIS for verification. These energy projects quickly produced savings in the utility budget. These documented savings were then transferred from the utility budget to be used to fund the continued implementation of the OEIS throughout the entire campus.

Our last design goal was to collect short (5 minute) interval data. There are many reasons the energy staff felt this was an important design feature:

- When implementing energy optimization sequences on a building automation system, feedback can be obtained instantaneously from real-time interval data. This was invaluable as we worked to optimize the condenser water system at the chiller plant.

- During commissioning of a new building or re-commissioning of an existing building, the results of tweaking and tuning the HVAC equipment can be quickly realized and quantified.

- Changes in schedules can be quickly reflected and accurately verified by building managers and administrators in real dollar savings.

- As the sub-meters in the buildings age and begin to fail, the system can alert the appropriate individuals for corrective action.

These desired features were divided and prioritized. Keeping these basic goals in mind, the OEIS was created.

HARDWARE INFRASTRUCTURE

The OEIS is not unlike many of the commercially available EIS services. The system is composed of both software and hardware components. The hardware can be divided into the personal computer (PC), the remote network engine devices, the network connection, and the energy meter. Figure 4-1 shows a schematic diagram of the basic layout of the OEIS network.

Personal Computer

At the heart of the OEIS is a basic desktop PC. All of the system features are executed in software within this central unit.

The OEIS does not require a high performance computer. Since all of the applications on the PC are Open source packages, there is much less software overhead than with other commercial proprietary products. This allows the PC to maintain good performance while

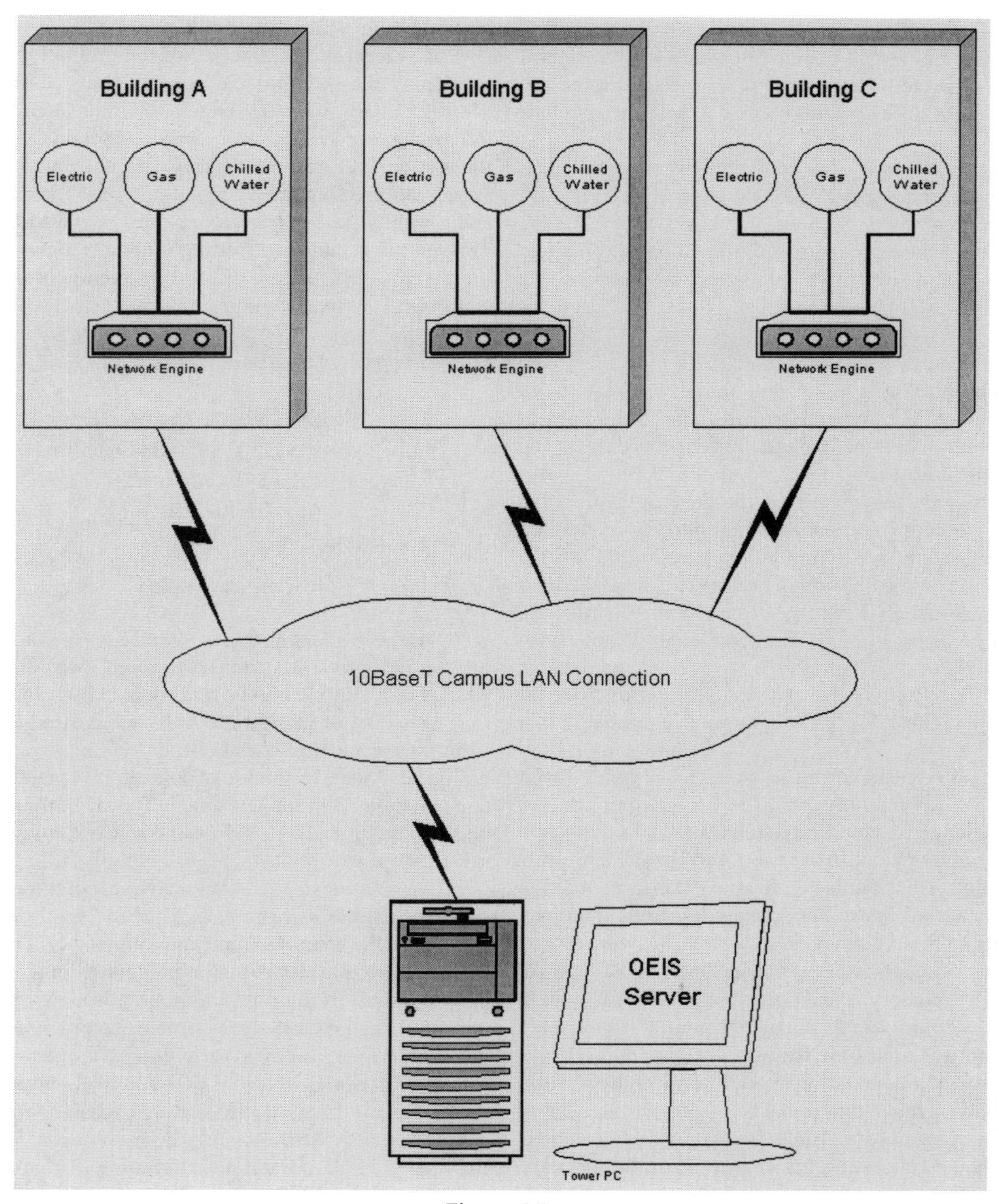

Figure 4-1

using only minimum resources. For example, the PC we are using was obtained at no cost through the campus surplus program. It is several generations old, runs at 1000 MHZ, and previously used an old version of the Windows operating system. With these limited resources, we are servicing approximately 60,000 hits per month to the OEIS.

The central PC works with the network engine devices in the field to obtain the energy data. Through the campus network, the OEIS PC must contact each building every 5 minutes, 24 hours a day. Although the OEIS can log data at less than 1-minute intervals, we believed that 5 minute intervals would provide sufficient resolution for our needs. A small change in this interval window generates a large impact on the size of the database, required data storage devices and the network traffic.

The PC also provides the web server functionality. This is the primary gateway by which system users and administrators obtain the requested energy data. The PC also provides the email servicing functionality. This feature is the other way users may receive timely energy data.

Remote Network Engines

Each building is fitted with an individual network engine device. The network engines take the pulse energy data from the meters and present the data to the campus network for retrieval by the OEIS PC.

Each of these network engine devices is, in fact, a small single-board embedded computer. An embedded computer system is a computational engine built for a single task or a tightly knit set of functions. Embedded systems are used in nearly all consumer electronics, home appliances, industrial process control, and automobiles.

The functions of the network engine devices are very simple. These functions consist of logging and displaying the data.

The devices have between 2 and 4 pulse input channels. When the input wires of a particular channel detect a change of state in the dry contacts in a meter, a counter register within the Network Engine is incremented. This counter with a date stamp is stored in battery protected RAM. Periodically the RAM data may be written to Flash memory for more permanent storage. This memory retains its contents in the event of a power outage. The memory is arranged in a circular fashion. In a circular memory layout the oldest reading is deleted to make room for each new reading.

Many of our existing electric and chilled water meters have a set of dry contacts, which can be used to provide a pulse output. The pulse output is a cost effective method of acquiring the energy consumption data. The OEIS integrates the consumption data to provide demand data or the rate of power consumption. Most of the natural gas and potable water meters on our campus are mechanical dial type. These meters must be retrofitted with some type of optical transducer. The transducer converts the mechanical action into an electrical pulse.

The pulses from the building meters are read and stored in the network engines. Each network engine can read multiple pulse inputs and typically accommodate all the meters within the building.

A second major function of the network engine is to display the accumulated counter data. Just like the central PC, the network engines contain a fully functional web server. The web server functionality allows the network engine to display the current data in a simple static web page when requested by the OEIS. The static web page can be viewed with any web browser. This feature is handy for setting up and trouble-shooting the network. See Figure 4-2 for a screen shot of one of our typical network engines.

```
<?xml version="1.0 ?>
- <VALUES>
    <count0>4.34101e+06</count0>
    <count>2.30369e+06</count1>
    <count2>395</count2>
    <count3>?350</count3>
  </VALUES>
```

Figure 4-2

The web page in the network engine is formatted into an Extensible Markup Language (XML) document. XML is a markup language much like HTML and is used to simply help describe data. XML was created as a platform for exchanging data between digital devices. Data within an XML document is defined by tags. The data tags are defined in the Document Type Definition (DTD) or XML Schema. The DTD or Schema is designed to be self descriptive.

For the most part, the software executed by the network engine devices is a "canned" application installed by the manufacturer and cannot be field modified. Some of the network engine devices however may be configured in the field to obtain additional features.

During the OEIS development we purchased these network engines from a variety of manufacturers. We purchased the Echelon ILON-100 Web Server. The ILON-100 has many useful features when compared to other similar devices. Among the many capabilities are data logging, alarming, and scheduling. It also contains a LonTalk network interface. We used this product in locations where there were many meters to be read or where there were existing LonTalk devices which we wanted to read.

There is also a new network engine, the PTW100 from activecircuit.com, which sells for less than $200. Although the PTW100 has very limited features, it contains all the functions needed for the OEIS and is equipped with 4 pulse inputs. We have begun to use these units exclusively and have had good success.

Another available network engine that is being used successfully at other facilities is the AcquiLite from Obvius. The AquiLite product contains 4 pulse inputs and can access data via the LAN or modem.

Energy Meters

Once the network engine devices are installed, it is necessary to connect the pulse output from the energy meters to the inputs of the network engines. The energy meter must provide a set of dry contacts. The dry contacts change states on intervals related to the energy consumption.

We researched many commercially available energy meters. We found several competing products which we purchased and installed during the early stages of the OEIS development.

The first energy meter we connected was the Wattnode device. This product is available in a pulse output model and a LonTalk model. We purchased the LonTalk model for our two central energy plants. The LonTalk model allowed us the flexibility to network all of the Wattnodes in the plant into one network. This LON network was then connected to a LonTalk web server. Having a local LON network reduced our installation cost at the plants because we only needed one network engine.

The second energy meter we purchased was the Veris model H8050 pulse output energy meter. This device looks similar to a split core current transducer. The Veris unit also connects to the electric lines. With these connections in place the unit computes the actual energy being consumed by the equipment or building. The Veris unit contains a single set of dry contacts. These dry contacts provide a pulse, which is equated to the energy usage, to the network engine device. The Veris power meter also provides a set of dip switches which allow the installer to adjust the amount of energy represented by each pulse on the output contacts.

The final energy meter which we connected to was the Osaki 9000K1 series mechanical power meter. The advantage of the Osaki meter was it was already installed at the service entrance of nearly all of our buildings. The disadvantage to the Osaki is the low frequency of the pulse. Although the very slow pulse rate does not affect the accuracy of the OEIS data, it does reduce the resolution. The lowered resolution may mean it takes longer to see system changes when they are implemented. Also the graphs do not show short frequent starts and stops of major equipment.

As the integration of the OEIS continues throughout the UCF campus, we are using both the Veris and the Osaki products.

Network Connection

The OEIS at UCF primarily uses the existing campus LAN network to provide the communication backbone between the central PC and the network engine devices. There are a number of reasons that we believe the existing LAN is the best choice for the OEIS integration. The network is already in place, operational and readily available. The campus has a team of professionals who maintain the network. These advantages result in much lower network connection costs for the OEIS project.

The existing network is readily accessible. As at most campus-type facilities (industrial and commercial as well as educational), the LAN at UCF is accessible in nearly every room in each building. There is also wireless coverage throughout most of the outdoor campus common areas. The campus LAN is therefore the ideal communications backbone for the OEIS project.

To make sure our project was successful, we needed to inform the IT group at our facility and get their agreement before implementing this system. This was not a problem because the network engines utilize very little bandwidth since each network engine is only contacted at 5 minutes intervals.

During our first phase of the OEIS, we found it advantageous to use another network in addition to the LAN. We planned to monitor the two central energy plants. To finely tune the equipment and control scheme, we needed to sub-meter each of the major pieces of equipment individually. To meet this goal we took a different approach than we applied to the rest of the buildings. We set up a separate local LonTalk network in each of the plants. This approach allowed the data from all the individual equipment in the plant to be collected into one network engine. Since the equipment in each energy plant is in that plant's building, this solution is much more economical than using multiple network engines.

The only network connection challenge we have faced involved two of our four utility service entrances. These utility service entrances are located about 50 feet apart, in a remote corner of the University property. To provide a network connection to these, a wireless Ethernet bridge was installed. This bridge is composed of two sets of radio transceivers and antennae. One set was installed on the Baseball building (about a mile away) and physically connected to the campus network. The other set was installed in a weatherproof box at the utility service entrance. This Ethernet Bridge has proved to provide a reliable network connection for this site.

SOFTWARE

Operating System

All modern computer systems utilize an operating system to organize and coordinate the software and hardware components into a single, integrated unit. The OEIS uses the Linux open-source operating system in place of a more popular proprietary commercial system.

The primary advantage to open-source software is that there exist extremely powerful packages readily available, completely free-of-charge. There is also never a charge for updates or licensing agreements. Linux is licensed under the General Public License (GPL) agreement. The GPL license allows anyone to use, view, understand and modify the source code without restrictions. There are no site licenses or legal agreements to sign as with the proprietary operating systems. Availability of the actual source code associated with open-source software also empowers the user to get to the root of technical problems quickly and efficiently.

Linux is also a very stable operating system. It is known to perform well on a wide range of hardware and processors. Linux is distributed by several organizations. We chose to use Redhat Linux. The Redhat installation is quite simple and for a fee, support is readily available.

Support for the Linux operating system is also available through commercial distributors, consultants, and by a very active community of users and developers. Support can be accessed through the web and through local Linux users groups.

The Linux operating system is exceptionally stable. When properly configured Linux systems will generally run until the hardware fails or the system is shut down. It is not uncommon for Linux applications to run continuously for hundreds of days without failure. Linux does not need to be rebooted periodically to maintain performance levels. It doesn't freeze or slow down over time due to "memory leaks." Linux was developed over the Internet by a group of programmers and therefore has strong support for network functionality.

Data Acquisition Application

A common method to complete large software projects is to divide the tasks into smaller more manageable components. After compiling the project design goals, we divided the project into small manageable tasks, which could be stand-alone mini-applications. These mini-applications were then prioritized. Our first task was to generate a small program to automatically read the energy meters. The small program or script needed to go to each of the network engines in the various buildings, retrieve the data, and store the data in the database.

The script starts by querying the database of buildings. It finds out which meters are to be read and where the meters are located. The program then looks at the IP address of the network engine at that location and reads the energy data. Once the data has been collected it is permanently stored into the database residing in the OEIS PC.

The Linux operating system provides a number of languages, which can be used to write custom programs to automate various tasks. We chose the Practical Extraction and Report Language (PERL) to write the small "script" programs to accomplish many of the tasks in the OEIS. PERL is freely distributed under the GPL agreement. PERL has powerful text manipulation functions and an interface to seamlessly connect to the OEIS database. PERL can be very useful for "gluing" together different applications. The popularity of PERL means that numerous code examples are available on the web. Quite useful PERL programs can be short. In fact most of the OEISPERL programs are less than a dozen lines in length.

When the data acquisition application requests the energy data from the network engines in the buildings, they present their most recent data as a simple web page according to a predefined XML schema.

Database

Once the energy data is obtained from the remote network engine devices, it is stored into a relational database. The OEIS uses the MySQL open-source relational database, which is freely distributed under the GPL agreement..

A relational database stores data in tables, which are related to each other. A database is a collection of tables. A table consists of a list of fields. Each record in a table has the same structure or same fields.

A structured query language (SQL) database like MySQL allows the user to form and pose complex questions of a database. MySQL has low connection overhead which results in fast, efficient operation. Thus, the MySQL application was the ideal choice for the OEIS database.

The various tables in the OEIS database include the following: buildings, meters, readings, and users. The advantage of a relational database is that you do not have to have duplicate information in the tables. One powerful function of a relational database is the ability to link a field in a table to a field in another table.

The OEIS is designed to primarily look at energy at the building level. The building table contains the building number, building name, building manager's name, and contact information. Sometimes there are multiple meters of the same utility type in the same building. An example would be sub-meters for the lighting, HVAC loads and plug loads within a building. This can be accommodated by placing multiple entries in the meter table for that building.

The meter table describes the meters attributes. The building number relates the meter table to the building table. The meter table contains the meter number, IP address, type, location, multiplier, unit of measurement, and square footage. One of our typical buildings would have 3 different entries in this table. The building would contain an electric meter, a gas meter, and a Btu meter to measure chilled water consumption.

The readings table is the largest table. Readings from each meter are stored as separate records in the reading table every 5 minutes. The readings table contains the meter number (which relates back to the meter table), the reading time and the reading value.

The users table contains identification and contact information for all building managers and anyone else who would like to receive email reports and updates from the OEIS. The users table stores the contact name, title, email address and phone number of the registered user.

Administration

To complete a major software project the administrative functions must be considered.

Once each day the OEIS prepares a compressed backup (or archive) file. The automated backup feature uses a script to determine which files are to be saved. The script can easily be modified to include any files the system administrator desires. In our case, the archive file contains the complete database, reports and the web documents. The archive file can be stored on recordable CDs, DVDs or transferred (over the LAN) to a sister computer in the system. In our case, the file is transferred, over the LAN, to a main frame tape backup system. Finally, the OEIS sends the administrator an email confirming the successful backup.

Another system feature automatically monitors the OEIS and alarms the administrator of abnormal conditions. The system monitor notifies the administrator, via cell phone, of system failures including database problems, meter failures, and LAN problems.

A necessary administration feature for any EIS system is database administration. The MySQL database can be administered through PHPmyAdmin. PHPmyAdmin is a web based front-end for MySQL. The application allows users to execute many administrative functions including creating and dropping databases, altering tables, maintaining tables, editing fields, and executing SQL statements....

In addition to the OEIS system administrator functions, there is a system configuration feature for the lay user. THE OEIS has a user interface where individuals can create their own personal account. Once the account is created an individual can chose to receive automated email with daily or monthly reports on various building on campus. The user also has the ability to customize the look of their OEIS front page.

Costs

A noteworthy advantage to the OEIS is the cost. The commercially available energy information systems are known for their inherent high up-front costs as well as lengthy monitoring agreements. These costs not only lengthen the payback period for the monitoring system but they also use up some of the funding for the energy efficiency projects which are revealed once the system is implemented.

In contrast, the OEIS can be implemented at a very modest cost. Powerful personal computers can now be obtained for under $500. The network engine devices range from $200 to $600 each. The vast majority of the OEIS software was obtained at no cost through the General Public License Agreement. The remainder of the software used to "glue" the applications together was written in–house.

The largest cost component associated with the OEIS is the initial purchase and installation of the network engines. Each building must have a LAN connection with a static IP address at the location where the network engine is installed. There is a cost associated with installing this connection. Although the large UCF campus installation is not yet complete, the first few buildings to be monitored by the OEIS have produced significant savings in energy. These savings are being used to fund the additional network engines and network connections required to complete the campus-wide OEIS.

Before developing the OEIS the University looked at comparable energy monitoring systems. The prices of commercial EIS installations are related to the number of points to be monitored. Each utility meter in each building is considered one point. Our campus has approximately 500 meters (points), which need to be monitored. We received estimates in excess of $600,000 from contractors to install a comprehensive system of real-time monitors on a portion our existing electro-mechanical meters. In addition to excessive up-front costs there are recurring monthly maintenance costs associated with the commercially available systems.

The OEIS has been implemented on the UCF campus for an average cost of $300 per point. When the system is fully implemented throughout the campus, the total cost should be approximately $150,000. This is a substantial savings over the commercial EIS systems.

OEIS FEATURES

Figure 4-3 shows a schematic diagram of the various OEIS features.

Web Interface

An early requirement of the OEIS was a completely web-enabled system. All user interfaces as well as system maintenance and administration are accessible through a standard web browser. To accomplish this requirement, the OEIS PC had to contain a web server application.

The OEIS utilizes the Apache open-source web server software solution. Apache is one of the most popular open-source applications. Apache is currently the number one web server software solution in the world. Apache is very stable and has been known to operate for years without downtime.

To assist in managing the web based front-end, the OEIS utilizes an open-source content manager program called Postnuke. Postnuke helps to organize the web site and takes care of many administrator functions. Postnuke allows the site administrator to make changes, additions and deletions to the web site by simply logging into the site through their web browser. Postnuke also provides an attractive organized wrapper for the web site. The content manager application affords convenience and flexibility to the administrator, but it is not a necessary component for the OEIS. The front page for the OEIS web site could simply contain a list of buildings or meters which currently accessible through the OEIS.

An important component of any energy information system is the graphing capability of the system. There are many open source graphing and charting applications available. The OEIS uses the open source JPGraph. JPGraph is an object oriented graphics library for PHP.

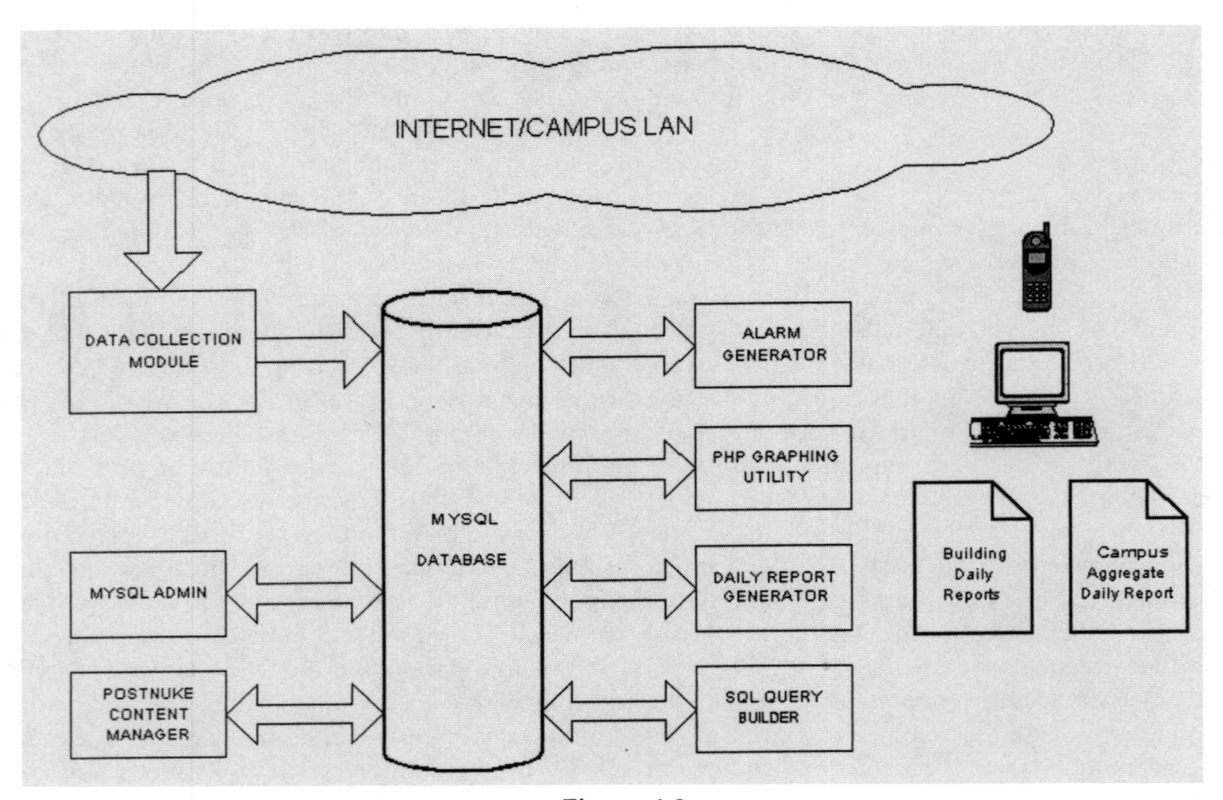

Figure 4-3

Once the OEIS main menu page is loaded, the user is prompted to choose a period of time as well as the type of energy data they wish to view. Once the user makes these selections the OEIS presents an interactive table of meters which meet the selection criteria. The table also presents various energy metrics associated with these meters. The interactive table can be sorted by any of the columns in the table with one mouse click.

The various table columns present the building number, building name, energy consumption, energy demand, and other energy metrics. These metrics allow the user to compare the energy performance of various buildings. By sorting these metrics the user can quickly find the least efficient and most efficient buildings for a particular period of time.

Once the user chooses a building meter in the table, the OEIS builds a graphical chart of the energy data within the requested time period. All graphs are customized by the user's choices.

In addition to the graphical representation, the user can choose to download the data in a comma-delimited file. This allows the user to archive the data and also import into a spreadsheet to perform additional analysis.

Daily Reports

One of the more popular features of the OEIS is its ability to automatically generate periodic reports. Using the Postnuke content manager front-end, users are able to create an account and then choose which buildings they wish to receive automated energy reports on. Each day, shortly after midnight, the database tables are automatically queried to analyze each building's performance for the prior day. A report is generated for each building (see Figure 4-4), showing the energy performance data for the previous day. As the reports are generated they are automatically emailed to the appropriate building managers as well as other users who have requested them for their review.

The report not only contains the consumption and peak usage for the period but also relates these numbers to a past period. With this short table the user can quickly see whether the buildings energy performance is improving or declining. Some of the additional energy metrics available on the report include consumption, demand, the Energy Utilization Index (EUI) and the Energy Cost Index (ECI). The buildings are also ranked, according to their efficiency, against other campus buildings in the system.

In addition to the individual building reports, the campus administrators receive a summary of the performance of all campus buildings in the system as well as the central energy plant reports. At the end of the month each of the users also receives a summary of that month's performance.

System Alarms

The Open Energy Information System has the capability to broadcast alarms to chosen recipients via email. Users can chose various criteria which produce a warning or alarm email when conditions meet certain criteria.

The OEIS monitors the database to assure proper operation. In the past our billing department would be unaware of a meter failure for up to 2 months. An inter-

Good morning Tom Jones,
This is your daily energy report for the Library for Thursday, 8/22/2004

Library	UNITS	Daily Use	Monthly Use	Daily Average	% Variance	Daily Cost	Monthly Cost	Projected Cost
Electric	KWH	9250	155360	7438	+9.4	$ 555	$ 9322	$ 12711
Chilled Water	TONH	7560	147534	6544	+9.4	$ 680	$ 13278	$ 18106
Natural Gas	THERM	169	3605	188	-10	$ 152	$ 3245	$ 4425
Potable Water	GAL	167	3402	154	-7.8			
TOTAL						$ 1189	$ 25845	$ 35242

Energy Metric	UNITS	Daily Average	% Variance
Energy Cost Index (ECI)	Dollars/SQF/YR	$ 5.66	+8.4
Energy Utilization (EUI)	KBTU/SQF/YR	126	+8.4
Chilled Water Utilization	TONH/CDD	1	
Peak Demand	KW	640	+5.4
Peak Time	Time	14:25	

See Real-Time Data for your Building.

Figure 4-4

mittent meter failure would take much longer to be discovered. Now, when the monitor detects that a meter is not responding to a request, the OEIS can send an email message to the meter maintenance personnel within minutes of a meter failure. Repairs can begin nearly immediately.

Results

Substations

The initial phase of the OEIS included the four electric service entrances and the two central chiller plants. The four electric service entrances provide all the electric power to the university, each are equipped with a pulse metering system provided by the utility. These four service entrances provide up to 20 megawatts of power to the University.

We wanted to monitor the main electrical service to the university for a number of reasons. The accounting section of Physical Plant wanted the ability to forecast the electric power bills during the current month. The University's electric utility bill can fluctuate more than $100,000 each month. The accountants would have more flexibility allocating their expense dollars throughout the month, if we could predict the cost during that month.

We also wanted to predict peak consumption periods. If we could see a peak period coming, we might be able to shed some of the less critical loads for a short period of time. By lowering the peak consumption, we could realize significant savings.

Finally we wanted to see any campus wide net changes in electrical consumption resulting from the optimization of the two chiller plants. The following example illustrates this need. An engineer trying to optimize the chiller plant might slow the secondary chilled water pumps, which would result in increased chiller plant efficiency. As a result of this change, tertiary pumps in the individual buildings may need to increase in speed, which results in increased consumption in the buildings. By monitoring the master meters, the net increase or decrease in campus efficiency can be quantified.

Chiller Plants

The largest single component of energy consumed on the UCF campus is by the chilled water production facilities. These facilities include two central chiller plants located opposite each other in our circular shaped campus. The two plants produce more than 12,000 tons of chilled water during peak conditions. The chilled water produced supplies a 3-mile district cooling loop, which cools nearly all of the buildings on the main campus. Upon completion of the monitoring systems at the electrical substations, the installation of monitoring equipment at the two central chiller plants was of primary importance. We hoped to optimize the two plants by implementing the latest chiller plant control strategies.

Unlike the other buildings, we felt that much of the individual equipment in the chiller facilities needed to be monitored separately. The plan involved monitoring the electrical consumption of each chiller individually, the secondary chilled water pumps, the primary chilled water pumps, and the cooling tower fans and pumps. Each of these pieces of equipment was individually metered and monitored. A LonTalk network was used bring all these meters to one central network engine located at each of the two plants.

Once the metering was in place, we began to implement various control strategies. Some of these strategies included tuning the PID loops on the cooling tower fans, optimal starting and stopping of chillers, and optimizing parallel pumping lead/lag routines on the secondary pumps. By utilizing the real-time data from the OEIS, we were also able to make program sequence changes and see the energy results immediately.

As we tuned the plants, we watched the plant performance through the OEIS. If it seemed that the new sequence was less efficient than the previous, we would analyze each component energy consumer, adjust the sequence, and return to monitoring. This was repeated until we felt we had achieved our goal.

In addition to the electrical sub-metering in the chiller plants, we also monitored the chilled water leaving the two plants. With the combination of the electrical consumption and the chilled water production, we created virtual meters within the OEIS. These virtual meters log the ratio of the energy being consumed by the plant to the energy going out of the plant in the form of chilled water. This ratio represents the overall efficiency of the chiller plant.

This feature has become the most valuable feedback tool we now have to tune the chiller plants. The chiller report, produced by OEIS, reports the kW per ton and the EER each day. This allowed us to see the efficiency and compare them to the daily environmental conditions. The environmental conditions include such things as the daily high temperature, daily low temperature, and cooling degree days.

If the chiller plant is not operating efficiently as a whole, we can now drill down to determine which component is operating inefficiently, using the OEIS. The system has repeatedly reduced the chiller operating cost by flagging equipment that has failed or is overridden.

Many challenges arose as we began to optimize the chiller plants. The dynamics of the two plants include opposing pumping energy. Our dual pipe chilled water piping layout has one plant pumping chilled water against the other plant. Using the OEIS, we identified energy savings by stopping one plant during lower load conditions. This small change resulted in $42,800 savings each year.

Another application of the OEIS system was found when we identified a number of air handlers on campus, which had 3-way valves. These 3-way valves allowed supply chilled water to mix with return chilled water. This mixing resulted in a lower chilled water return temperature back to the plants. We contracted to have these 3-way valves replaced with 2-way valves, which resulted in an increased return chilled water temperature at the plant. By increasing the temperature difference, between the supply and return, we reduced secondary water flow and ran chillers at more efficient operating conditions. This reduction was documented by the OEIS and amounted in a savings of more than $9,000 per year.

Monitoring the chiller plants has allowed us to test optimal control strategies and get real-time feedback of changes. We have shaved 15% from the yearly operational costs with sequence changes to the existing chiller plant controls system.

Library

The UCF library is the oldest building on the main campus. Like many of the larger buildings on campus, the Library already had utility meters in place. These meters were installed to do electrical and chilled water Btu submetering for departmental billing using manual entry of data into a spreadsheet.

Connecting the existing UCF owned electrical sub meters and chilled water Btu meters installed in the campus library into the OEIS was easy. With this data, we were able to calculate the total Btus per square foot and determine room for improvement.

There was also much insight gained after plotting only a one-week load profile (see Figure 4-5). It was immediately apparent that the actual energy use of the building did not correlate with the actual pattern of building use. This indicated a likely savings potential during the unoccupied and limited occupancy periods. We investigated the building equipment and found that there were no equipment schedules, no set-backs, and no control over the outside air dampers. Additionally the building was serviced by 4 very large air handler units. These units were constructed with inlet guide vanes to control the flow rates through the system. The inlet guide vanes were rusted in position causing them to be inoperable.

After purchasing and installing four VFDs for the major air handlers, we modified the building sequence of operation and modulated the outside air dampers based on CO_2 levels in the return air. The result of these changes is a documented savings of more than $7000 per month in utility costs. All of the costs associated with the building changes were repaid within 3 months.

Teaching Academy

Our experience with the new teaching academy

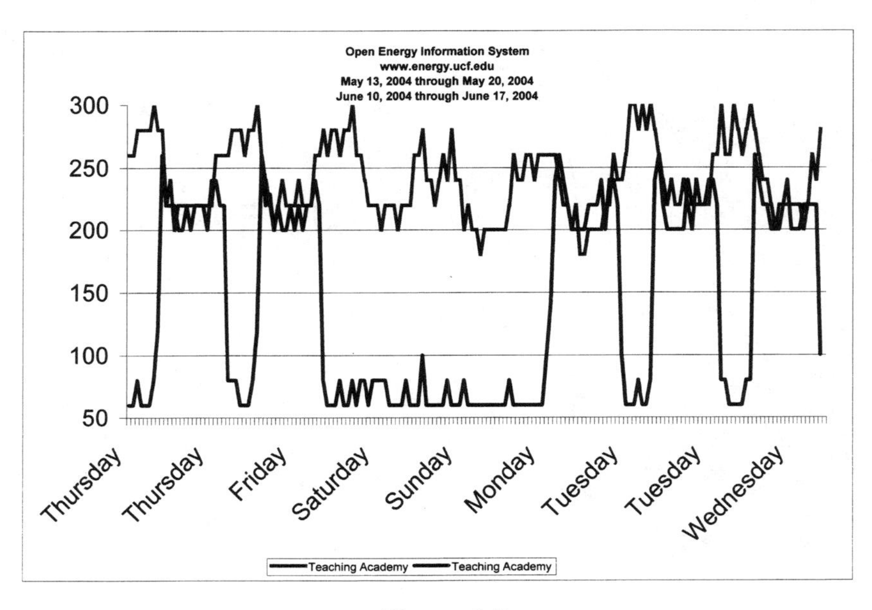

Figure 4-5

shows the importance of building commissioning and the value of using an EIS in this process. The teaching academy, a 67,735 square foot building designed to hold 1800 future teachers, was recently completed. The building's half moon exterior is nearly all glass with a large glass atrium in the center.

We were interested in how energy efficient the building was with this large solar load. The building was equipped with the latest control systems and mechanical equipment. The control system included CO_2 monitoring, humidity monitoring, VFDs and air flow stations on the air handler equipment. We were sure that the building would be reasonably efficient.

The building is connected to the campus district cooling loop. The mechanical system consists of seven air handlers. Each zone is equipped with variable air volume terminal units with electrical reheat. To adequately monitor the building we needed to measure the electrical consumption and the chilled water consumption.

We expected the daily load profile to fit the typical office space profile, which follows the daily solar load curve. The initial data (see Figure 4-6) coming from the OEIS lead us to momentarily believe the monitoring equipment was reading backwards!

Although the daily load curve was evident through the OEIS, it appeared inverted. After only one day of monitoring the electrical and chilled water graphs revealed a significant increase in energy consumption at night during unoccupied time. During the heat of the summer day and full occupancy, the building was consuming less energy than at night. As with other commercial buildings, we expected the electrical energy consumption to increase during the day due to solar load.

Armed with the inverted load curve, we began to look at the building more closely. We setup a host of trends, using the building automation system. After some investigation we determined the building control system was overcooling the space at all times. Because the space was overcooled, the building control system was also warming the space with the primary air handler heat and the secondary VAV heat as needed to maintain the comfort of the occupants. We began by resetting all of the temperature set-points in the building. We also increased the dead-band (difference between the heating and cooling set-points). We were careful to retain the de-humidification functions, which were designed into the building. These changes maintained space comfort while dramatically reducing the re-heat consumption.

More red flags appeared as we began to receive the daily building reports from the OEIS. The reports showed that the MMBtu per square foot was more than double what we had expected. At this point we continued to investigate the building's automation system. We found several discrepancies between the original sequence of operations specified by the design engineer and the actual sequence being followed by the BAS.

Instead of simply implementing the design sequence, we modified it to improve the building's performance. We found that there was an insufficient dead-band in many of the temperature set-points. We also found that the building occupancy schedule did not correlate with the actual operational times.

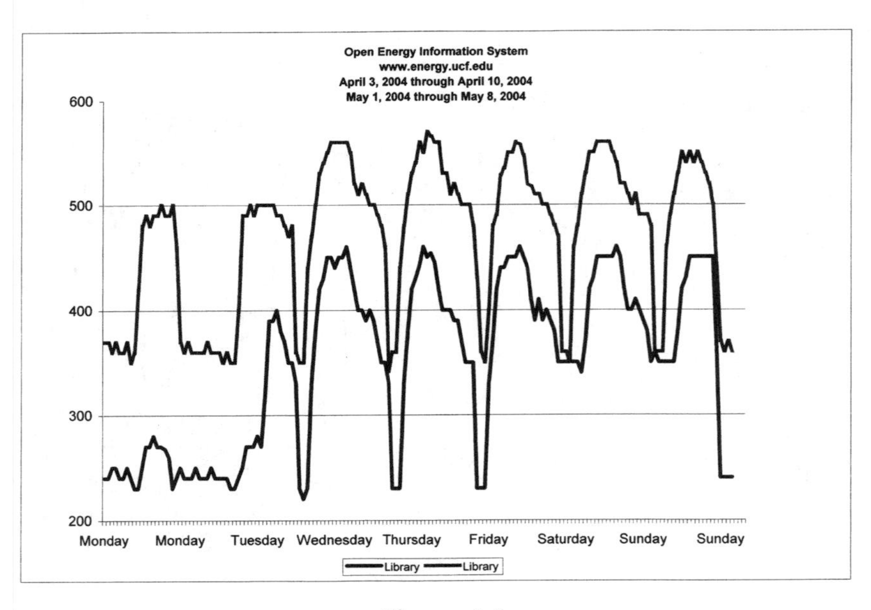

Figure 4-6

As a result of these changes, we have documented savings in utility costs exceeding $4000 per month. All of the costs associated with the building changes were repaid within 2 months. Since the building is nearly new, these changes will have a tremendous impact on the life cycle cost.

PLANNED SYSTEM ENHANCEMENTS

The OEIS is less than 2 years old. The system is operational, stable and provides invaluable data to dozens of users on a daily basis. System upgrades and additional features are nearly limitless. The open-source platform on which the OEIS stands allows features to be added and modified very quickly.

The OEIS developers recently asked users for feedback based on their individual experiences using the system. Among the most common user comment is that the current user interface is difficult to understand, complex, and not intuitive.

Based on this feedback, a simplified front-end for building managers, and other non-energy professionals is being created. The goal of the simplified interface is to minimize the steps the user must execute, while at the same time, providing the most understandable and comprehensive picture of the buildings performance. The new front-end will also default to the most common type of request. Most users will simply choose their building of interest and then tell the system to draw the graph.

Along with this new simplified front-end, tutorials and help screens are planned to help new users understand the various energy metrics in the reports. There is also a tutorial planned for the charting functionality of the system.

Additional energy metrics will provide greater insight into the behavior of particular buildings. The additional metrics may include the building occupied and unoccupied schedule. Also more weather-related metrics will be used to normalize the data. The system currently logs outside air temperature and relative humidity at 5 minute intervals.

Perhaps the most promising feature planned for the OEIS is a web application, which aids users in formulating and executing complex SQL queries to the database. The real power of the relational SQL database will not be realized until this application is fully implemented. For example, a query might ask the database to provide a list of all buildings showing the electrical consumption, normalized to the number of students enrolled in the classes in each building, only looking at Fridays in August, and sort this list in ascending order according to energy consumption. This example is illustrative of the nearly limitless possibilities as to the variation and complexity of the queries which could be made to the OEIS database.

The OEIS developers envision energy students and researchers from anywhere in the world, viewing and querying the various buildings. The students would be able to access real world energy data on actual buildings and develop class projects around this site. If they find nuances in the way the buildings behave which the on-site energy managers have not seen, the students (or other off-site users) can communicate their findings back to the site through email to the web master or energy manager.

Currently the OEIS allows only the system administrators to execute the SQL queries. The administrator must be fluent in the various SQL key words. The open-source web-based application phpMyAdmin is used for this function currently. PhpMyAdmin lacks the security features, which must be integral to the new application, to prevent users from causing damage to the database. Therefore, before users are allowed to initiate SQL queries, we will have to improve our security system.

An automated demand response feature is also on the drawing boards. About 80% of our large campus buildings are controlled by a building automation system using the BacNet protocol. Our goal is to allow the OEIS to automatically control major equipment on campus through a BacNet interface over the campus LAN.

There is currently an open-source project to develop a Linux based open-source Bacnet interface. Our development team has joined this project to help complete the development and begin testing the interface. Once the interface is complete, we would like to integrate it into the OEIS system. This automated load curtailment feature would primarily be used to lower our peak demand. In addition, building managers would be able to access the major equipment in their buildings and adjust schedules to match occupancy. One final byproduct of the feature would allow faculty and staff the ability to log onto a secure web site and adjust the temperature set points in their offices and classrooms.

There are also additional features planned for the charting module of the OEIS. We would like the user to be able to quickly choose their building and be presented with a graph of the previous 30 days. The image would load into their browser as a client-side image. The user would be able to use the mouse to select a specific portion of the graph and "drill-down" or zoom in on the area of interest. These interactive graphs would allow the user to quickly move through a lot of data through the intuitive use of their mouse. The current OEIS graphs are server-side images, which cannot be manipulated once created.

As long as there is interest in a open source EIS, the OEIS will continue to be updated and improved. The features grow and the processes get more streamlined each month. The OEIS web site will contain information on upgrades and enhancements.

CONCLUSION

With an energy information system, the building manager has a real-time sense of how well his particular building is performing. When the system is web-enhanced, the building manager can easily compare his building to other buildings on his own campus and to other facilities across the country.

A low-cost energy information system for organizations with multiple buildings is not only feasible, but also practical. One can easily be constructed with reasonable costs for both hardware and software. The direct benefits of such a system are substantial and the payback is rapid. The UCF system is expected to monitor more than 500 meters and cost about $150,000 when fully implemented. To date, we have achieved an annual savings attributable to the OEIS of at least $184,000.

References

University of Central Florida
Open Energy Information System
www.energy.ucf.edu

Apache Web Server
Open Source Server
www.apache.com

MYSQL
Open Source
www.mysql.com

JPGraph
Open Source Graph Library
www.aditus.nu/jpgraph

Redhat
Linux Distributer
http://www.redhat.com/

Postnuke
Open Source Content Manager
www.postnuke.com

Chapter 5

Information, Behavior and the Control of Heat in Multifamily Buildings: A Case Study of Energy Information System Use in Building Performance Improvement

Daniel Harris, Manager of Building Controls Development
Michael Bobker, Director of Strategic Planning
Energy Management Training Center
Association for Energy Affordability,
New York, New York

INTRODUCTION

ALTHOUGH OFTEN OVERLOOKED, behavioral change is a part of virtually all energy management projects. We may think that it is just a building system or component that we expect to perform differently. But building performance improvement is inextricably entwined with human behavior change and must be recognized as part of energy management. In doing so, we must additionally become more acutely aware of the relations between behavior and information, the latter as essential feedback to both control and change loops.

Energy management is never a single step process. Informational feedback guides corrective actions and next steps. Therefore the integration of IT-based energy information systems into our project work is a most encouraging trend.

Energy management itself can be considered a form of behavior based on information. Actions are taken based on analysis of energy use. Information is required to evaluate the realized effectiveness of the actions after they are in place. Even when we think that a capital investment and new equipment is driving the change in energy performance, the changes generally push maintenance staff to add new awareness and practice to their routines, especially if new systems are to perform optimally. Their perceptions and actions can determine whether improved performance is achieved for any length of time. Persistence of savings is an essential aspect of projects originally justified based on savings projections that stretch out for years.

The response of building occupants is another aspect of behavioral change that energy management projects must take into account, especially when key comfort factors such as temperature, air flows or lighting are altered. Occupant acceptance is a major prerequisite for project success. Conversely, occupant complaint is a major reason for the undoing of system adjustments or the manual by-pass of control systems.

IT itself can be a powerful driver of behavioral change. Certain IT applications, so-called "killer apps," such as cell phones, PDAs, web-shopping sites, have immediate and widespread appeal. People's behavior changes readily to adapt these new products. More specialized applications will have narrower appeal, without the weight of mass-market forces to encourage their adoption. Some may even be viewed negatively. One wonders how the maintenance worker feels about the beeper that is now connected to automated system alarms. These applications are more difficult to introduce successfully. In these cases, careful user orientation and education needs to be part of our planning and project support.

Adding web functionality to building Energy Management and Building Automation Systems (EMS and BAS) has advanced rapidly into the commercial office buildings sector. Web-based communication and programming is replacing proprietary protocols, so that building controls and HVAC data no longer stand-alone as functional islands but are brought into the realm of the enterprise IT department.[1] Behavior and communication between corporate Engineering, Facilities, and IT departments will surely soon be addressed by a much-needed primer. Systems are evolving for integration beyond the individual building level out through the utility network.[2] The functionality of Internet communications for real-time, system-wide utility demand-side management (DSM) by real-time building response is being explored and demonstrated by researchers in California and New York State.[3] New behaviors are evolving between utility coordinators, facility managers, and intermediary service providers, with web-based IT as the infrastructure for such new behavioral opportunities.

THE PROJECT CONTEXT

This chapter examines the use of an IT layer included as part of an EMS installed in an electrically heated multifamily apartment complex. The work builds upon earlier work in similar apartment complexes by Lopes and Hirschfeld (2002). AEA has pursued the application of advanced controls with IT monitoring within the multifamily housing sector, starting with room unit air-conditioning and electric heating, with goals of improving temperature control and enabling peak demand limitation and participation in utility demand-response programs. The technology choice for this development work is a radio frequency (RF) wireless communications system that has been successfully deployed as a sub-metering solution in a substantial base of NYC apartment buildings by Intech21 of Glen Cove, NY.

Electric heating has been a focus for energy management systems within the multifamily housing sector because of its expense. The promise of "electricity too cheap to meter" that informed this infrastructure choice in the 1960's through early 1970's was never realized. Especially when combined with master-metering, where the resident has no direct financial responsibility for energy used, room zone controls become largely useless—they will commonly be found at the highest settings, ironically along with window-opening. Energy audits in the NYC area have found apartment temperatures typically in the 80-85°F range. Heating operating costs have become a significant aspect of financial distress for managers.

In this situation, various upgrading possibilities are explored. Converting the domestic hot water from electricity to fossil fuel (usually natural gas) is relatively easy and cost-effective since the original water heating equipment is usually centrally located and distribution piping can remain unchanged. The building is able to maintain its favorable electric-heating rate with this conversion. Space heating is more difficult to address.

Converting to fossil fuel for space heating requires installing a completely new distribution system, such as a hydronic piping and baseboard or running gas piping throughout the building to supply decentralized through-the-wall units. Expense and disruption are major barriers even if the economics are marginally acceptable (10-15 year payback) with capital costs on the order of $3-4,000 per apartment. Economic calculation must take into account that the remaining electric loads will shift to a more expensive rate classification.

This led to consideration and eventual adoption of a controls-based approach to optimize performance of the existing electric space heating. Installed costs of $1,500 per apartment were estimated with a simple payback of 6 years. Actual costs were slightly higher, approaching $1,800 per apartment for a limited size pilot phase. Projected savings accrued from (a) reduction of apartment temperatures (b) night set-back and (c) coordinated duty-cycling to reduce peak load and demand charges.

The complex of 931 apartments in 14 medium-rise (5- to 7-story) buildings serves a low-income and working population. The first phase of the project, which is reported on here, encompassed 169 apartments in three buildings. The rest of the complex is scheduled in a second project phase, authorization of which was contingent upon performance of the first phase. While the purpose of this chapter is not to review the performance of the project[4], it should be

[1]This trend, which has been called "convergence" is amply documented in the monthly columns by Ken Sinclair in *Engineered Systems*, 2002-2004, and on his website *www.AutomatedBuildings.com*

[2]Trends in SCADA functionality include moving down towards end-use load monitoring and control. Utility and ISO demand response programs are early examples of such "smart grid" evolution. For a recent evaluation of opportunity and options at the residential level see Arthur Rosenfeld's chapter 3 "Customer Response to Dynamic Pricing" in Borenstein, et al. 2002. For a seminal popular treatment of smart-grid visioning developed through EPRI, see Silberman 2001.

[3]See work by Mary Ann Piette's lab at Lawrence Berkeley National Laboratory, such as Motegi et al. 2003. In New York State, Peter Douglas of NYSERDA has managed a portfolio of projects under the heading "Enabling Technology for Customer Demand Response." See for example, Lopes and Hirschfeld 2002

[4]For a detailed review of project performance at the end of the first year of operation see Lempereur, Bobker and Harris *Proceedings of the International Conference on Enhanced Building Operation* (ICEBO) 2004, organized by the Texas A&M University Energy Systems Laboratory.

noted that actual cost and performance after the first heating season have been sufficiently close to projections for the second phase to be authorized.

The project team consisted of AEA, acting as developer and system designer, technology vendor Intech21 which provided the wireless networking devices, assumed responsibility for communications and established the website for data acquisition and storage and an electrical contractor who fabricated and installed the control panels during the summer of 2003. The team worked together under AEA leadership to commission the communications and control functions, starting in October 2003. P-L Housing management and maintenance staff were involved in tenant communications, monitoring, and complaint response with an eye towards eventual full operational turnover of the system.

The project was developed and partially funded through the NYS Weatherization Assistance Program, which uses funds from the US DOE, and the NYSERDA Assisted Multifamily Program, both of which require matching funding from the owner.

THE COMMUNICATIONS AND CONTROL SYSTEM

The system integrates an RF wireless communication local network within the buildings with heating control elements and Web-based data for monitoring and management. Power (sub)meters in each apartment communicate in a two-way mode with a building access point installed on the roof of each building. This system operates at the 902-928MHz-frequency band dedicated for non-licensed Industrial Scientific Medical (ISM) applications in the USA. Because the housing complex consists of multiple buildings, another set of signal interfaces and RF transceivers using 802-11b wireless protocols is installed for building-to-central office communication. A server, installed in the central Management Office receives and stores data and transmits control instructions using this network. The server is connected to a Wide Area Network using broadband Internet (DSL connection). A Graphic User Interface (GUI) uses a web browser to allow remote access to data and control-setting parameters. Individuals in other locations, accessing website with secure ID and password privileges, may use the GUI to monitor data and control operations with the same convenience as on-site personnel. The off-site server provides third-party managed data storage and management via an SQL database.

The power meter in each apartment is incorporated into a control panel box installed near the circuit breaker panel. This panel includes a solid state relay for each electric baseboard circuit, actuated by a dry-contact closure from the power meter. The meter board includes a clock, power measurement and kWh-conversion circuitry, and receives input from up to two remote temperature sensors.

The first phase apartments were selected on the basis of having a single, shared utility master meter that includes registration of monthly peak demand. This meter was also upgraded to a fully digital interval meter and equipped with pulse-outputs to interface in real time with the new monitoring and control system at the system server.

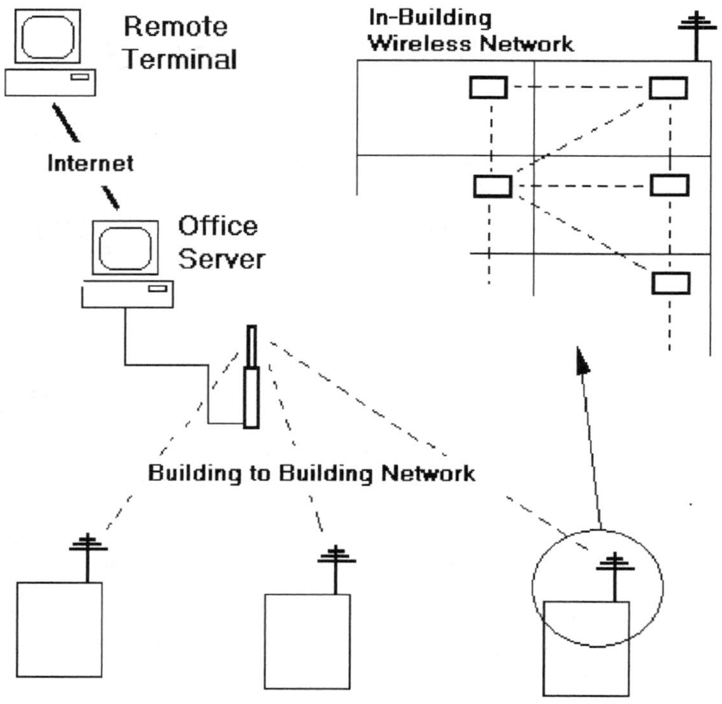

Figure 5-1. Wireless network infrastructure

PROJECT GOALS AND BEHAVIORAL CHANGE

The project's goal was to reduce heating cost by a projected 20% by:

(a) reducing overheating by imposition of remotely set and monitored setpoints;
(b) providing a night setback, and
(c) limiting peak demand, especially in less-than-peak shoulder months.

It was understood at the outset that the targeted reduction in apartment temperatures could cause complaint from residents. Management's response to complaints would, therefore, significantly impact the savings outcome. We worked with an implicit model that went from environmental (controlled variable) change through intervening behavioral change variables to project outcomes:

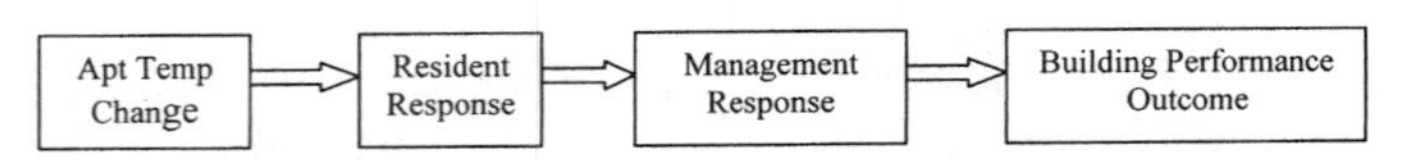

Figure 5-2. Project Change Model

We recommended a *gradual introduction* of temperature reduction as the system was made operational and the heating season began. Based on previous experience in the introduction of apartment temperature regime change, one of the great advantages of the system seemed to be its ability to allow repeated, incremental change of setpoints without requiring apartment access.

Management, however, preferred to "draw a line in the sand" and tell residents that now that temperatures could be monitored remotely they would receive legally required heat and no more. The change of some 10 degrees in most apartments was dramatic and, predictably, led to a barrage of "no-heat" calls.[5] Since maintenance staff were not used to this volume of calls nor to the troubleshooting procedure now required, the new system was held responsible for aggravation and added maintenance work. In fact, roughly a third of the complaint calls were traceable to operational problems with the new system, while a large majority of complaints were objections to the new temperature regime.[6]

Management and, to an even greater extent, maintenance staff did not take well initially to use of the website. They relied on the project developer, AEA, working in an on-going commissioning role to look at temperatures and, when deemed necessary, adjust setpoints.[7] This was probably for the best as it resulted in a comprehensive record of resident and management responses that could, eventually be correlated against energy performance. Nevertheless, *it is almost certainly the case that the control function would have been widely disabled in response to resident complaint if it had not been possible, via the IT functionality, to show a reliable, continuous and verifiably accurate record of apartment temperature data.*

---

[5]We will never know if our strategy might have avoided most complaints. But we did confirm our prediction that the hard-line approach would result in a variety of behavioral responses, of which the no-heat complaint was only one. From the combined temperature and power profiles in individual apartments we were also able to see ice applied to sensors, the use of electric space heaters, and the use of gas ovens as supplementary heating sources. Eventually several boxes were pried open when it was discovered that the panels would fail in the "heating on" mode. And in one case the lead maintenance mechanic wired bypasses around the controlling relays. All of these behaviors might have been reduced if tenants had the possibility of overriding the system for short periods of time. Receiving a tenant complaint usually resulted in having an operator increasing the temperature setpoint permanently in a zone. This response answered the complaint but led to savings deficiency. End-user interface should be explored and designed to match building performance with residents' needs and the use of local "over-rides" can be debated and considered in the context of the control strategy and goals.

[6]The new system complicated the logic in responding to a no-heat complaint. Previously maintenance staff had simply to check the breaker and the baseboard thermostat and if these were properly operative, replace the baseboard element. Now staff had to know if the apartment temperature actually required heat in relation to the setpoint; a temperature measurement was now necessary as well as looking at the control status on the system screen in the management office. We provided instructions, a written procedure with AEA phone support, and staff training but this set of behavioral modifications has also proved slow and somewhat difficult.

[7]Monitoring apartment temperatures on a daily basis and dealing with end-users expectations have built our experience in understanding the mechanism of an electric building as a whole. Analysis of historical data and trends provides solid information on building response and tenants behaviors. The system has learned to predict accurately the electric demand based on outdoor and ambient temperature conditions. Furthermore, tenants have also understood that the heat in their apartment is not unlimited. Window openings do not occur at 75°F as often as at 85°F therefore limiting the maximum temperature was definitely the driving force behind usage savings.

"SIGNATURES": MAKING BEHAVIORS OBSERVABLE VIA IT

In addition to making convincing data available to support the behavioral acceptance of the control function, the IT-based Energy Information system can also provide a valuable window into resident behaviors. This window is created by the ability to automatically generate time-series and multivariate graphs down to the apartment level. To the seasoned eye, the patterns become "signatures" of various behaviors. This concept and interpretive skill is a key element of working effectively with energy information systems. Examples of this interpretive process are provided below.

As we have suggested, the dramatic change in apartment temperature regime caused a certain amount of resident response to become visible as complaint calls. A variety of further resident "resistance" behaviors that would otherwise have been invisible became observable by interpretation of IT graphical plots. *The graphical representation of data, rather than tabular data, is key to this interpretive process, as it depends on viewing* ***both*** *temperature and power along a time axis.* In fact, the ability to show these behaviors graphically to management became a key factor in confirming the project's performance and obtaining approval to proceed with roll-out through continuing phases of installation.

Being able to see this otherwise invisible behavior within apartments provides a most valuable tool for management to respond to resident behaviors and, in so doing, optimize control acceptance and energy performance. In some cases, the in-apartment behaviors constitute health or safety hazards and suggest counseling and educational interventions. Finding these "resistance" behaviors suggests specific apartments where a setpoint approach for a more gradual acclimatization process can ultimately result in greater acceptance.

SENSOR COOLING

The temperature sensors report to the thermostat and are easily noticeable for their purpose. A wall bracket, the size of a single gang light switch bracket, is mounted on the wall in the common areas. The bracket is louvered and responds quickly to changes in temperature. Most if not all of the tenants either figure out what the bracket is or are told by neighbors. The concept of cooling the sensor is a quite simple and natural solution to the problem of not enough heat.

Figure 5-3 shows a typical apartment where the tenant is using ice, or other cold material, to fool the system into providing more heat. Note that the method is quite effective and typically provides demand for several hours and an apartment temperature above the normal setpoint. This behavior can be seen by the sudden fall of the temperature plot, immediately followed by full electric load and, probably after the cooling material has been removed, a return of the sensor back to room temperature.

If this behavior is persistent and widespread, it suggests a programming response that turns heat off in response to a sudden, extreme temperature drop.

GAS STOVES

Many tenants at Lambert Houses told us that they used the gas stove in previous years to supplement the baseboard heaters. Some apartments have drafty windows and AC sleeves and the hallways are unheated.[8] These conditions may contribute to inadequate heating when the outdoor temperature is sufficiently low. Most tenants who indicated this behavior open the oven and place a pot full of water inside. They leave the oven open and as it heats the air the water evaporates, providing warm, humid air.

This is a very dangerous method for heating a closed space and deaths have occurred as a result of people using gas stoves for space heat. The levels of carbon monoxide can build up rapidly and the combusted air makes for an uncomfortable and unhealthy environment in the apartment. Being able to identify this practice in specific apartments is a most significant safety finding that suggests maintenance response for acceptable heat along with education and counseling about heating safety.

Figure 5-4 shows an apartment where the tenant is using the gas stove for heat. Note that the temperature increases while the electric load drops off. The heat of

[8]The initial audit recommended window replacements and insulation of air conditioner sleeves. These measures were not implemented because of budget restriction. We believe that their implementation as well as any measure that could improve building envelope tightness would have reduced the need to increase temperature setpoints in a significant number of apartments. Leaky windows interfere with the convective flow of apartment air, which is the only carrier for electric baseboard heating. An additional improvement might have been the replacement of all outdated baseboards with the communication and control modules built-in.

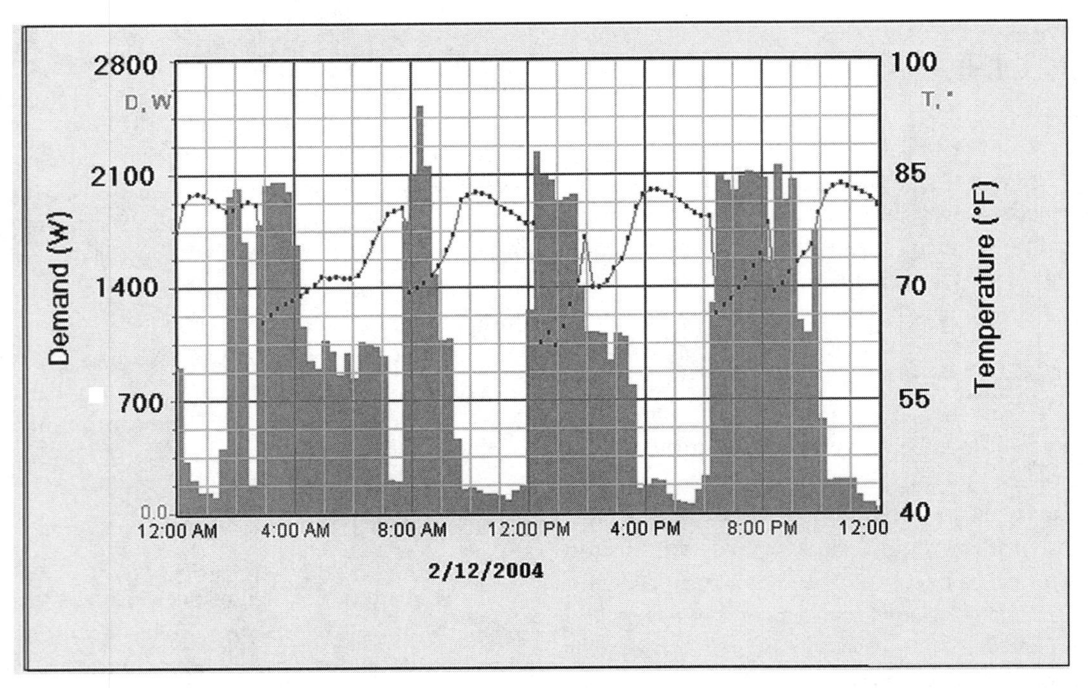

Figure 5-3. The pattern above is indicative of sensor cooling. The bar graph is apartment electrical demand and the line graph is apartment temperature. Courtesy of Intech21.

the gas stove brings the zone temperature above the set point and the electric heat turns off.

SUPPLEMENTAL ELECTRIC SPACE HEATING

A logical problem with the system is that a tenant can plug in a free-standing space heater that would replace the baseboard function and eliminate the savings from turning off baseboard elements. Note in the Figure 5-5 plot how the controlled baseboards are off with the apartment at its setpoint(s) but the new load of space heaters being plugged in becomes apparent along with the apartment temperature rising above the previously stable setpoint. The use of space heaters increases the apartment temperature past the setpoint thus recreating the condition that the system was installed to eliminate.

The use of spot heaters might, to a certain extent, be viewed as a useful tenant adaptation to the system. Most tenants are unlikely to purchase enough space heaters to overheat the apartments; for even a small apartment it would require purchasing at least two or three space heaters of 1500W and perhaps as many as six to eight heaters for the larger duplex apartments. Rather, a tenant is more likely to have one heater used to address a cold spot that is not adequately sampled by the sensor, such as a bedroom with its door closed at night. In this case a space heater may provide necessary spot heating. If not viewed by the resident as too much of a bother, the limited use of a stand-alone spot heater might actually improve overall acceptance of the new control regime.

TAMPERING WITH THE SENSOR WIRING OR CABINET DOOR SWITCH

Another response was to try to gain access to the installed equipment in an attempt to tamper with the system and restore constant electricity to the baseboards. There were two options: Damage the sensor or modify the electrical wiring.

In this system if a temperature measurement is not available due to damage to the sensor or sensor wiring then the temperature on the GUI will read "FAIL" and the system will automatically switch on all the relays until the temperature reading is restored. Some tenants discovered this and it was necessary for building man-

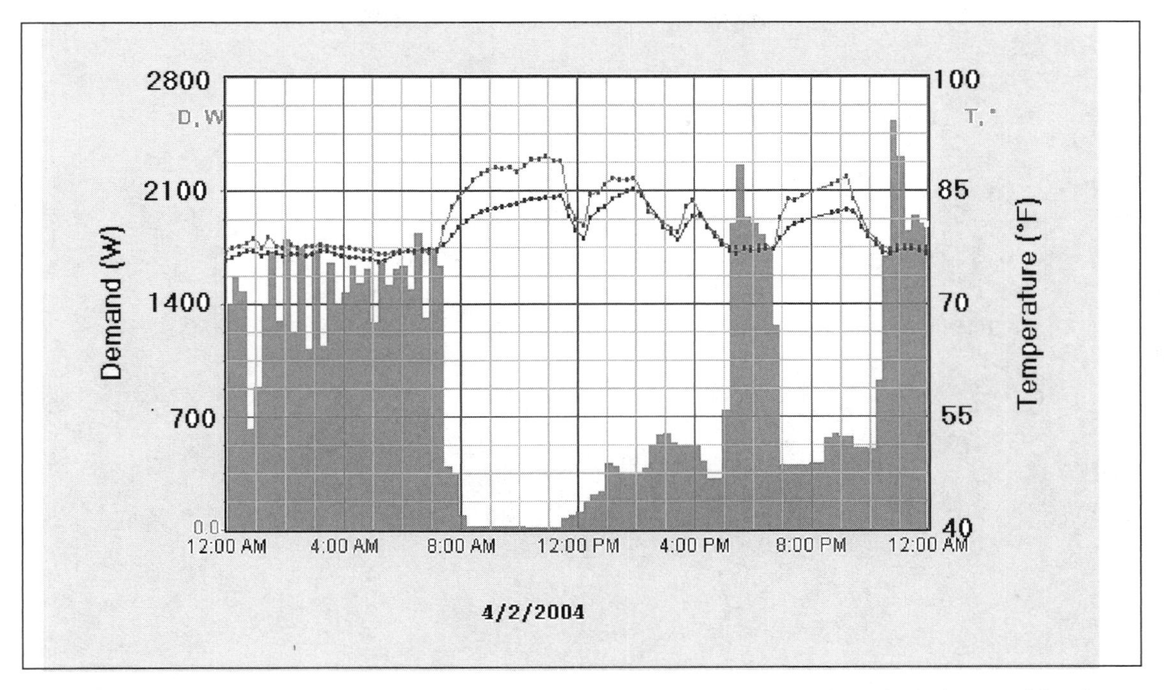

Figure 5-4. The data showing a temperature increase without an electrical demand increase. Courtesy of Intech21.

agement to make repairs.

Changing programming in a tampered unit to "FAIL = HEAT OFF" would be an attractive response to this uncooperative behavior.

SEEING THE BUILDING'S BEHAVIOR AS A WHOLE

While seeing resident behavior "in the small" is held forth in this chapter as an important technique for performance optimization, building-wide data visualization is also most significant in understanding the potential for control of the entire population of terminal elements.

The level of detail in the data makes it possible to observe cycling patterns both at the apartment level and in aggregate at the master-meter, in terms of usage and demand responses to temperature and other factors. Duty-cycling is evident in response to outdoor temperatures, as heating elements turn on and off to maintain the setpoint. At less than peak cold, heaters are able to meet their load more quickly and thus cycle more rapidly. This pattern contributes to incidental and random fluctuations at the master-meter level of monitoring. As peaks in apartments happen to coincide large "Rogue Waves" occur that can unintentionally set the monthly peak demand. This is important because a 15-minute maximum peak demand sets the electric demand charge for the month, which can be as much as 30-40% of the total electric cost. At Lambert heater-cycling driven fluctuations and peaks of 25-50 kW were common. The Con Edison demand charge for fully bundled service, monthly October through April, is $15 per kW (charges are substantially higher for summer months). Figure 5-7 shows a 24-hour graph of the building demand at the master meter where the smallest time interval is five minutes.

Note that the demand is constantly fluctuating due to heating switching and elevators. A large duration peak occurs at 4:30, 6:30, and somewhat at 11. There was no outdoor temperature change during this time and as heaters are run automatically there should be only a minimal tenant behavior phenomenon. The early morning peaks correspond to the lowest temperatures for the day and can thus set the monthly maximum demand.

Examining typical apartment profiles, such as that in Figure 5-8, confirms that in moderate weather heater cycling is short duration, as would be expected since these are on/off rather than modulating devices. As a

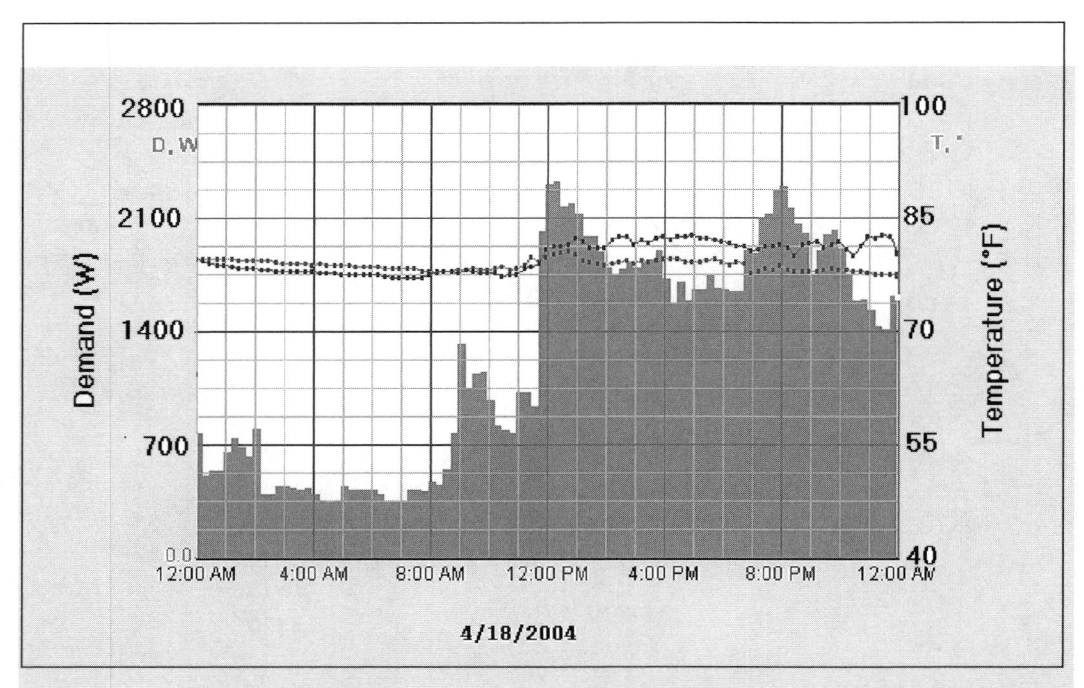

Figure 5-5. Use of an electric space heater: space temperature increases above setpoint with corresponding increase of consumption and demand. Courtesy of Intech21.

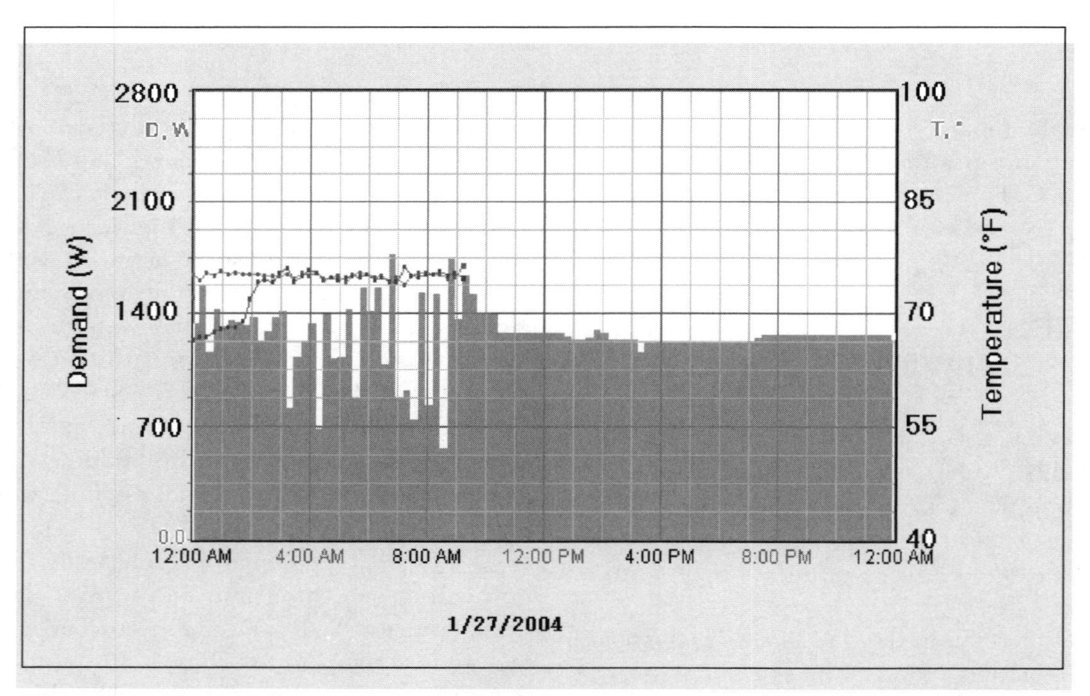

Figure 5-6. End of the temperature line shows the sensor failure, with subsequent "continuous-on" power. Courtesy of Intech21.

result, it is apparent that short-duration spikes can be avoided by short delays in enough individual heaters to maintain a pre-set peak. It can also be seen that the step-up from a night heating set-back will, if not properly managed, also set an unnecessary spike.

Figure 5-8 shows the demand and temperature in a typical apartment for a 24-hour period with the smallest interval being 15 minutes. Note the cycling of the heaters and the change to a night setback setpoint temperature of 74°F versus the daytime setpoint of 76°F. The average

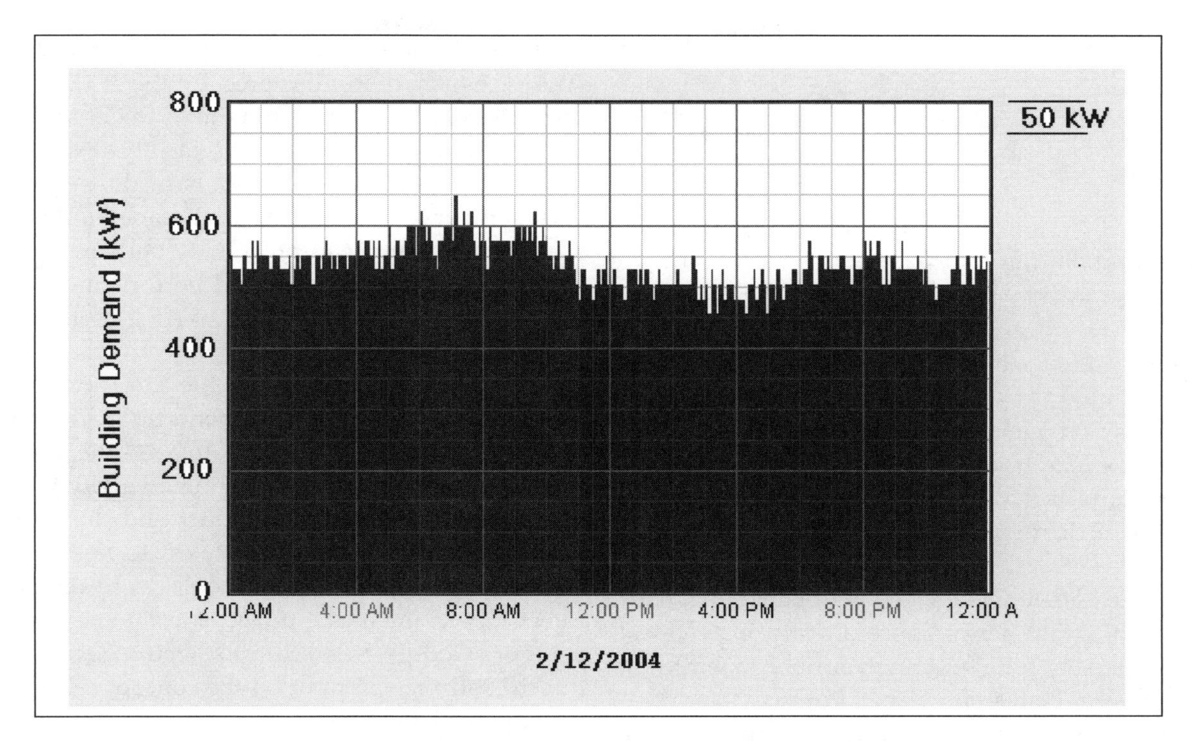

Figure 5-7. Graph of demand at the master meter for a 24-hour period. Courtesy of Intech 21.

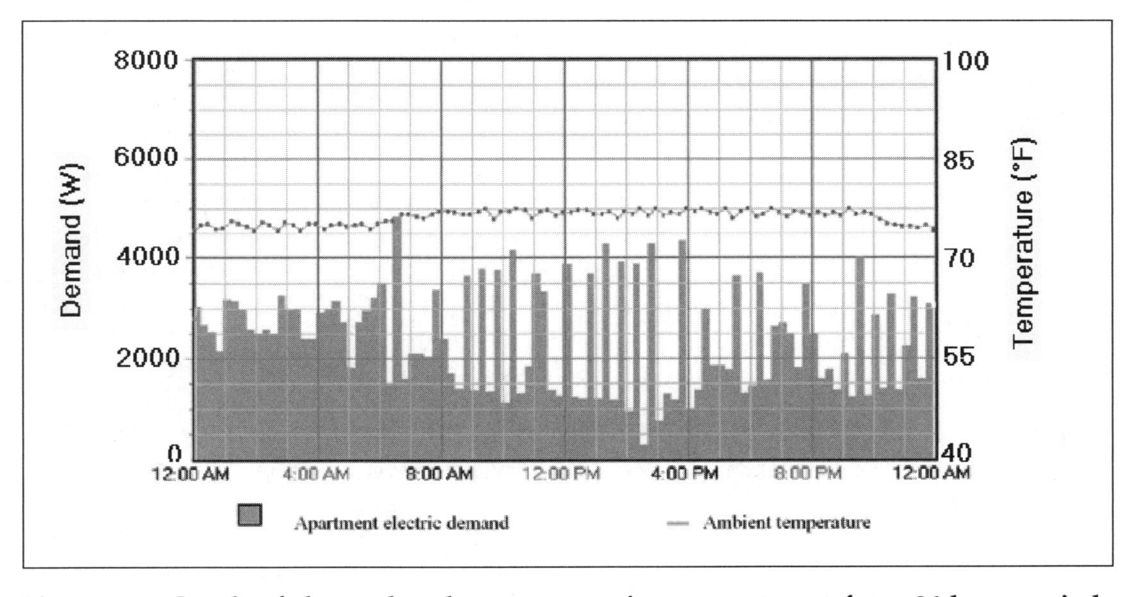

Figure 5-8. Graph of demand and temperature in an apartment for a 24-hour period. Courtesy of Intech21

outdoor temperature was 28°F. Note also how the energy consumption decreases during the course of the day as the outdoor temperature increases. The heaters come on for shorter durations as the set point is more easily achieved with a smaller temperature difference between room and outdoors. The dead band for this apartment was 1.0°F

It should be noted that the principle of peak-demand limiting exemplified here can be applied to any large population of energy-using terminal devices. This is the basis of AEA's approach to demand-response through room unit air-conditioning control. ***But this operation requires IT networking and coordination of the device population*** that is provided by an energy information system that is integrated with the controls.

CONCLUSION

We will surely see an expansion of digital information and controls in the residential sector. The network electric grid concept suggests a world of communicating, interactive devices down to the end-use level. But we will not get there in one leap but rather through an extended, continuing series of integrations in which human behaviors will play a significant part. In fact, one of IT's important strengths is that it provides superior feedback, in the form of information, to monitor the change process. It is this last aspect that we have emphasized.

In so far as IT systems implement improved building control, we sometimes mistakenly believe that they will improve building performance directly. In the ideal case, these changes will be imperceptible to building occupants or, even better, effortlessly improve their comfort and happiness. As we have argued in this case study, however, behavior often provides intervening variables between the technology and the desired project outcomes. It is important for to pay attention to this behavioral dimension, informed by our new information capabilities and tools, at the intersection of energy management and IT project design.

ACKNOWLEDGMENTS

The authors would like to recognize various parties for their vital support of the project: David Hepinstall, Executive Director of AEA, the New York State Division of Housing and Community Renewal (DHCR) Weatherization Assistance Program, in particular the New York City program office, the New York State Energy Research and Development Authority (NYSERDA) and the NYSERDA Assisted Multifamily Program (AMP) team. The project was made possible by funding from both the New York State Weatherization Assistance Program (WAP), which is administered by the NYS DHCR, and by NYSERDA AMP. Matching investment and staff support has been provided by the facility owner, who, for reasons of confidentiality is not named. An energy audit was performed by Dominique Lempereur, formerly of AEA, following the protocols of both programs, that provided the basis of cost and performance to develop the project. Francis Rodriguez of AEA performed admirably both in developing the project through the WAP contract he administers and as the construction manager interface to building management. Brian Tivnan and his firm Riverdale Electric Services played the essential role of installation contractor, bringing a variety of skills and efficiencies into both the product and the project. Last but by no means least, the wireless technology, website and web-based monitoring tools were developed by Intech21, serving as the project's technology vendor and partner; George Bilenko and Victor Zelmanovich of Intech21 strongly supported the concept and application development from the outset.

References

Borenstein, Severin, Michael Jaske and Arthur Rosenfeld *Dynamic Pricing, Advanced Metering and Demand Response in Electricity Markets* CSEM Working Paper 105, University of California Energy Institute, Center for the Study of Energy Markets October 2002

Lopes, Joseph and Herbert Hirschfeld *Multifamily Dual System Energy Management and Submetering* New York State Energy and Research Development Authority (NYSERDA) Project 5036, Final Report 02-08 July 2002

Motegi, Naoya, Mary Ann Piette, Satkartar Kinney and Karen Herter *Web-based Energy Information Systems for Energy Management and Demand Response in Commercial Buildings* California Energy Commission, Public Interest Energy Research Program LBNL-52510 April 2003

Silberman, Steve "The Energy Web" *Wired* July 2001

Chapter 6

Smart and Final Food Stores: A Case Study in Web Based Energy Information and Collection

Richard Rogan and John Marden
Honeywell Integrated Energy Services

ABSTRACT

SMART AND FINAL warehouse food stores teamed with Honeywell Building Solutions to develop and implement an energy strategy that includes a sophisticated energy information and control system. The system incorporates satellite-based data collection of detailed operational measures, centralized alarming and control, standardized local overrides, centralized energy use analysis, maintenance dispatch and a systematic approach to benchmarking, analysis and investments for efficiency improvements.

After the initial pilot phase, this system is being rolled out to all stores and has delivered annual energy savings of $2.3 million, utility bill error savings of $60K/year and operational and Accounts Payable savings of $750K/year.

BACKGROUND

Founded in 1871 in downtown Los Angeles, Smart and Final Inc. operates 231 non-membership warehouse stores for food and food-service supplies in California, Oregon, Washington, Arizona, Nevada, Idaho, and northern Mexico. Each facility has a mix of refrigeration, freezer, warehouse, retail and office spaces requiring its own mix of energy end-uses.

In 2001, Smart and Final Food Stores (S&F) experienced significant energy cost increases based on rising energy prices. S&F was also expanding, adding to a significant increase in operational costs. These cost increases were adding an unplanned $3M million/year to the cost of operating S&F facilities with no end in sight. Grocery stores typically operate on 3-4% profit margin, so changes in operating costs can have a significant impact on profitability.

To deal with these price increases, S&F executives had to gain a better understanding of energy use across their facilities—beyond the information available in their monthly utility bills or the static solution of walk-through energy audits of all facilities. Initial walk-through energy audits of the stores pointed out that lights were being left on overnight, air handlers were blocked with product, and the HVAC and refrigeration setpoints were being manually overridden. S&F wanted better centralized monitoring, reporting and control to first intimately understand their energy costs and then actively manage the major facility systems that drive energy costs. In addition, S&F saw the opportunity to:

- Reduce the costs of refrigerator and freezer downtime
- Reduce the costs of emergency maintenance calls
- Develop an increased, centralized insight and analysis into individual store operations to look for other opportunities for cost saving or operational improvements
- Use a systematic approach to cost-effectively invest in the most effective efficiency improvements
- Document and store food storage temperature data for FDA reporting requirements

Grocery stores spend the majority of their energy budget on refrigeration, lighting and HVAC systems. Improving the efficiency of these systems can lower energy costs, but changes to any one of these systems can affect the performance of others. Any improvements must take into account the interaction of these systems and their operational settings in relation to facility use and ambient conditions.

New technologies allowed S&F to gain greater insight into store operations through web-enabled energy monitoring and control. Through centralized control and monitoring of all facilities, S&F could operate each facility at peak efficiency and also determine which facilities were the best candidates to explore for capital investments to improve the efficiency of energy-using systems. Such a system would require monitoring of energy use, environmental conditions, lighting levels, refrigerated case temperature, freezer temperature, ambient temperature and ambient humidity. The system would also require the ability to control the major energy-using systems without negatively affecting store operations or the customer buying experience. An integrated solution had to seamlessly interact with lighting systems, HVAC controls, refrigeration controls and security systems. While the system did not require true real-time monitoring and control, an effective solution did require constant communications to respond to changes in conditions and maximize efficiency.

S&F operates a satellite-based wide area network connecting all their stores with Headquarters. This existing network would be used to enable the system.

THE SYSTEM SELECTION PROCESS

Before committing all sites to any specific solution, S&F required a process to design, test, improve and implement a system that could meet their energy-saving needs cost-effectively. Honeywell Energy Solutions proposed a process that would pilot the solution at a limited number of stores, test the system for cost and effectiveness, make any necessary improvements and eventually roll out to all U.S. sites. Honeywell's process was selected because Honeywell could deliver on all aspects of the process, from the pilot program through data collection, analysis, alarming, control and energy-saving retrofits.

During the pilot phase, the team chose to focus on twelve stores with locations having different weather patterns and utility rates. Honeywell used utility tracker software to scan in historical billing data for all stores. These billing data were then used to ID and implement an energy consumption-monitoring solution at 12 stores. These stores were tracked for 6 months to show results based on consumption rather than energy costs because energy prices were increasing. As a result of this pilot program, the system was improved, standardized and rolled out to stores across the U.S. Figure 6-1 depicts the layout of S&F's web-based energy information control system.

SYSTEM DESCRIPTION

Data are collected through a LON-based communicating meter allowing each device to poll and store 20+ variables. Refrigeration data and refrigeration controls use a CPC refrigeration protocol. The system incorporates other LON devices depending on store and location, including the Honeywell HVAC XL10 constant volume ASU controller and Honeywell XL10 input/output modules to monitor lighting levels and control lighting contacts. A Honeywell Web's Controller connected to S&F's satellite-based network is used to collect all data.

A Web Supervisor in S&F's Headquarters currently brings together data from 150 stores. The data are organized graphically, noting trends and alarms and allowing for setpoint monitoring and control.

The data are transferred once daily to the Honeywell Atrium service. Honeywell Atrium is a service bureau that collects and interprets data into meaningful recommendations. Atrium provides reporting, data mining, and benchmarking to determine "normal" operating conditions. The Atrium system and team of analysts determine sites that are out of specifications and supply recommendations for improvements. At Atrium, energy data are continually analyzed. Standard reports are delivered monthly, and ad hoc reports on the data are available at any time.

From Atrium, reports are provided to Honeywell Energy Engineers who can focus resources on high energy consumption facilities. Honeywell works with the S&F Energy Manager to determine causes for high consumption and to benchmark a store against its previous performance. The analysis adjusts for temperature, humidity, etc. to determine if a store is creeping out of control.

Standardized control criteria place each store into one of three modes at any one time:

1. Employee-occupied—lower lighting levels, slightly lower comfort levels.
2. Business Hours—Merchandise-highlighting lighting levels, maximum comfort.
3. Unoccupied—lights out and lowest comfort level.

Going forward, S&F has used the data collected from the Atrium system to invest in cost-effective equipment efficiency retrofits when other related construction

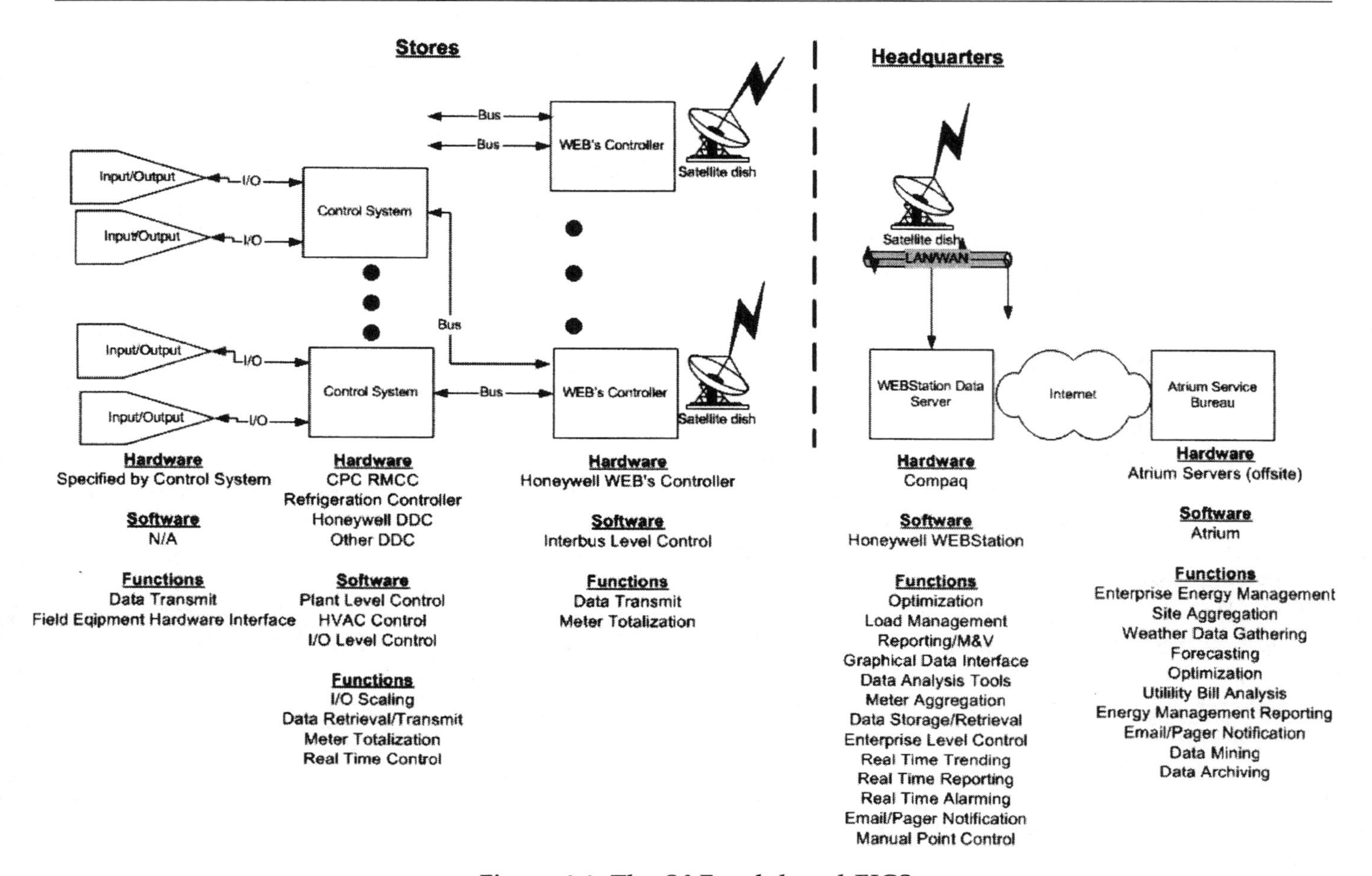

Figure 6-1. The S&F web-based EICS

projects are undertaken. Lessons learned from the project have been incorporated into new store design. New store designs are also better integrated. For example, new stores have a more advanced XL10 thermostat controller allowing for better zone control, control of individual devices and economizer control. Also, while original lighting designs allowed for a 50% setting to save energy while restocking, the lighting distribution pattern was not set up optimally for 50% lighting. Many of the lights turned off at this setting were needed to safely maneuver in the warehouse setting, so the 50% setting was rarely used. New store designs take the energy-saving 50% mode into account and allow for optimal coverage at 50% to allow for employee usage mode. New standards were also established for installing skylights.

IMPLEMENTATION ISSUES AND LESSONS LEARNED

Installation Approach

After the first pilot sites, the team made changes to the installation approach to bring the installed cost down. The pilot sites had originally used a Honeywell contractor to install the web supervisor, the web controller, and the LON-based devices; to program the system; and to create graphics. They used S&F's refrigeration contractors to work on the CPC controller. Using two contractors required an additional level of coordination. To eliminate problems with coordination, the refrigeration contractors now do the complete installation. The refrigeration contractors commission the system with support from a Honeywell partner who does the graphics and logic programming. Honeywell and the refrigeration contractor then jointly commission each system, using cell phones to commission the system point by point. As each point is tested, the refrigeration contractor confirms the physical results of each control point. This approach does not require all parties to be at the site; it also provides better testing of the web-based system and reduces installation costs by 10-15%

Data Formatting

All data flow through Smart & Final's satellite-based Wide Area Network. The satellite network has inherent delays that can create problems with the "look" of the data because the satellite network is designed to confirm information packets individually. This creates a data lag since the network cannot confirm an entire data stream. This type of problem is less apparent in data-collection only.

A hard-wired network solution like ISDN, cable or DSL would have been the best way to handle data, but S&F liked the satellite system since a single vendor can support it. To compensate, the data capture and reporting were adjusted to compensate for the problems encountered in the satellite-based system.

Lighting and Security

When initially implemented, all lights were turned off during the unoccupied mode. The lights stayed off until the building entered a programmed employee-occupied period. Therefore, security cameras did not have enough light during unoccupied periods to capture an individual on tape during a security breach. Now the lighting system is tied into the security systems and all lights are turned on in the event of a security breach. There are also manual lighting overrides on-site which allow for unprogrammed occupancy needs. These manual "momentary light switches" also change the environmental conditions back to the employee-occupied mode of operations for two hours.

ECONOMIC, TECHNICAL AND OPERATIONAL RESULTS

Energy-Saving Improvements that were implemented include:

- Space-temperature control by cycling the air handlers based on occupancy and allowing the temperature to creep up when the store is not occupied.

- Occupancy sensors for non-retail areas of the store to control lighting when areas are not in use.

- Anti-sweat on glass doors with a system heater to keep fog off the glass. The original heaters were designed for "worst case" conditions. The majority of the year, glass doors do not need the full door heat load. The system is now monitoring ambient conditions and cycling the resistance heat as needed to keep fog and sweat off the glass doors for better marketing.

- Utility tracker service: SF previously had two employees that processed utility bills. As part of this solution, utility bills are re-directed to Honeywell Atrium where they are received, scanned, web-posted and entered into an energy tracking system to validate the bills against historical bills, validate against monitored energy use, confirm utility calculations and follow up on estimated meter reads. Atrium reports on actual costs per store and feeds electronic files to accounts payable for approval and payment. This system makes these individual stores more productive and allows for some savings from reduced billing errors.

- Forklift battery charging stations: invested incentive dollars into additional chargers. Do not run any chargers on peak.

- Variable speed drives. Rather than cycling fans off and on to reduce peak usage, variable speed drives were used. All the fans were left on and fan speeds were controlled. This caused less stress on the fan, lowered peak energy use, and also lowered overall energy use.

- Condensing system that also senses outdoor conditions: This allowed operating at a lower condensing pressure when ambient conditions are right instead of always maintaining a fixed condensing system pressure which assumes a "worst case" for ambient temperature.

The EICS has resulted in many operational improvements beyond it significant energy savings:

- It has fulfilled FDA reporting requirements on food storage temperatures. Data are easily saved for one year as required.

- It monitors refrigeration equipment. The system continually monitors compressors, condensing systems, the cooled environment, refrigerated case temperature, energy use and ambient conditions. These points are monitored by Honeywell's ISO 9002-certified monitoring center in Atlanta 24/7. When any point enters alarm mode, the Honeywell team looks at the data to determine if an immediate response is required. They then contact the store manager and the service contractor for that location. Refrigeration contractors also have access to this refrigeration monitoring system and use the data for troubleshooting. This has reduced service costs and improved uptime for the equipment.

Equipment Service costs have been reduced for several reasons:

1. Dispatching a maintenance person to determine problems is not always necessary.
2. Implementation of energy efficiency strategies means equipment can run less often—which lowers energy costs and extends equipment life
3. Centralized ability to override the automatic control of systems. Lighting and HVAC systems were originally on pre-set schedules, so S&F would have to send a person to the site to override the timer settings for relamping, special stocking, etc. This can now be done centrally.
4. Benchmarking between stores allows continual improvement. Managers can apply lessons learned

COST

Total cost to date: At Honeywell's suggestion, Smart and Final added a Corporate Energy Manager to its staff. Including the cost of the energy manager, S&F has invested roughly $8 million to date for equipment and services. This was offset by nearly $1 million in rebates from the big three California utilities.

RESULTS

The energy strategy and system has resulted in an annual energy savings of $2.3 million, a reduction in utility billing errors of $60K/year and operational and accounts payable overhead savings of $750,000 per year. Some of the data on savings for S&F stores is shown below. Figure 6-2 shows the monthly savings from a lighting retrofit in store number 601. Figure 6-3

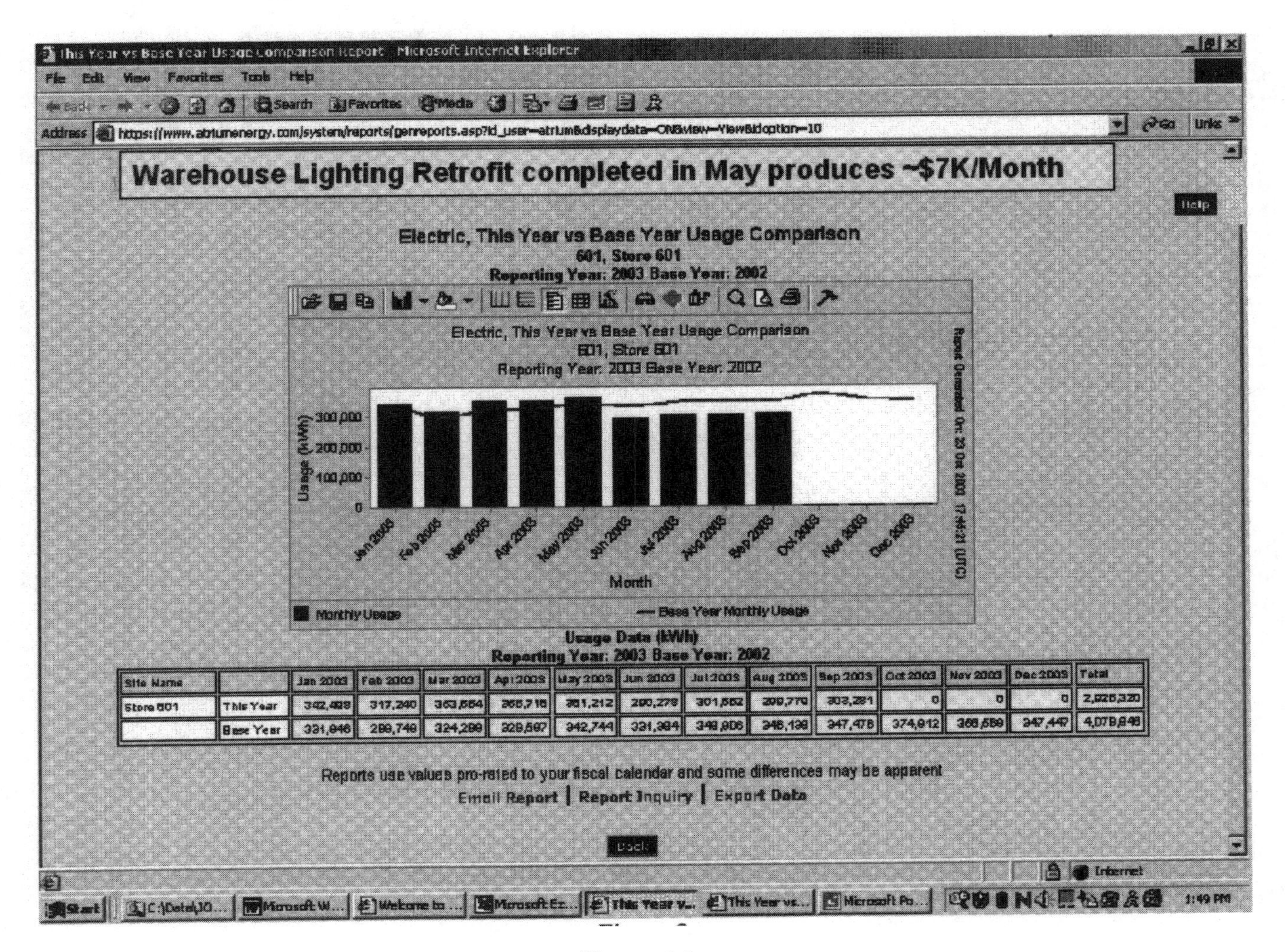

Figure 6-2.

shows the electric bill savings in 2002-2003 compared to the baseline year of 2000-2001. Figure 6-4 shows the projected and actual costs for 2002 and 2003; and the revised projected cost for 2004 with the expected savings. When fully implemented, the strategy should deliver more than $5 million in annual energy savings—a 20% reduction in S&F's $25 million annual energy expenditures.

Roll-up of EIS Savings

Smart and Final All Stores
Electric Bill Data

kWh/Day Total	*Baseline Year Sep 00 - Aug 01*	*Current Year Sep 02 - Aug 03*	*Baseline Adjustment*	*Change$*	*Change%*
District 1	227615	203611	-14796	4140,844	-16%
District 2	232973	188423	963	-$165,212	-19%
District 3	143108	125899	15948	-$120.360	-21%
District 5	211785	197921	21825	4129.551	-16%
District 6	71453	61237	666	-$39,502	-15%
District 8	138904	129517	12839	-$80.680	-15%
District 10	0	0	0	$0	0%
District 12	136042	118778	4972	-$80,717	-16%
District 13	0	0	0	$0	0%
District 16	135863	107655	5700	4123,086	-24%
District 18	179621	149084	6767	4135,414	-20%
All Stores	**14773641**	**1282125**	**844761**	**-$1,015,366**	**-18%**

Notes:
Includes 12 months of performance data for 52 stores
Includes 6 months of performance data for 26 stores
Projected annualized savings associated with these stores is ~$1.4M

Figure 6-3.

CONCLUSION

S&F stores are now more comfortable and safer. Equipment lasts longer and is easier to service. Each site has a better understanding of its energy costs and now has the right intelligence to implement additional improvements as new technologies and higher energy prices come their way.

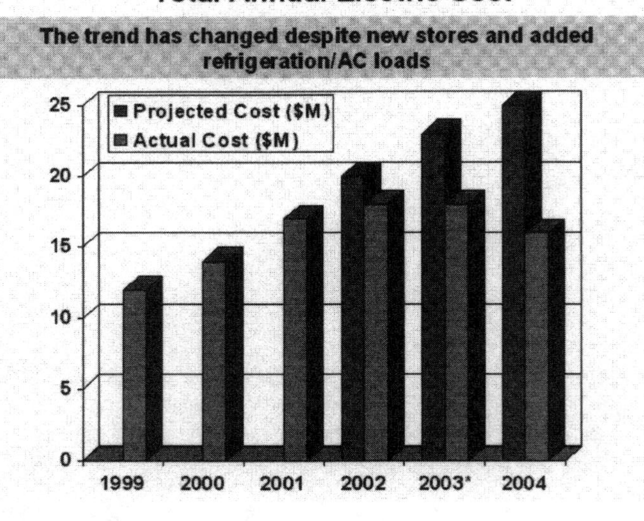

Figure 6-4.

Section Three

Web Based EIS Applications

Chapter 7

Measured Success: Constructing Performance Metrics for Energy Management

John Van Gorp
Power Measurement Corp.

INTRODUCTION

What gets measured, gets managed—Peter Drucker

OVER THE LAST several decades there has been increasing interest and activity in the field of energy management. A Lawrence Berkeley National Labs (LBNL) study of energy efficiency projects completed by US energy service companies over a ten-year period shows that total project spending has increased from roughly $500 million in 1990 to more than $2 billion in 2000 [1].

Energy management practice has traditionally focused exclusively on technologies that increase the energy efficiency of key energy-consuming processes and equipment. Although there is little doubt that upgrading equipment and processes is a key ingredient to increased energy efficiency, there have always been concerns that traditional deployment practices have not resulted in consistent (and long-term) energy savings. While the LBNL study mentioned above notes a steady increase in energy efficiency project spending over time, it also acknowledges the wide variation in typical energy savings.

Many energy managers are familiar with the challenges that are associated with these wide variations in energy savings. Projects designed to reduce energy consumption can involve significant capital investment and changes to operating procedures, which means such projects will inevitably be elevated for executive review and approval. Many different projects will be competing for funding at this level, and executives will expect and demand at least an assessment of the risk involved in realizing the projected savings. Such an assessment can be challenging for energy efficiency projects because consumption is often strongly linked with variable factors like outdoor temperature and production volume, making it difficult to attribute the savings realized by these projects.

Executive teams familiar with modern management practices such as those found in ISO 9000 and Six Sigma programs will often insist that energy efficiency projects follow the management philosophy outlined in those programs. These quality management programs highlight the importance of measuring baseline performance, setting goals and tracking performance against those goals. This approach can be adopted and applied to energy management practice, and in fact there are standards and best practices designed to increase the performance of energy efficiency projects and make the savings realized more predictable and repeatable. The International Performance Measurement and Verification Protocol (IPMVP), for example, provides best-practice methods for measuring and verifying the results of energy efficiency projects in commercial and industrial facilities [2]. MSE 2000, an energy management standard developed by the Georgia Institute of Technology and accredited by ANSI, specifies a management infrastructure for increasing energy efficiency and reducing costs [3]. Both highlight a more *strategic* approach to energy management.

STRATEGIC ENERGY MANAGEMENT

As the field of energy management matures, the knowledge gained from thousands of energy efficiency projects is driving a transition from traditional tactical practices (one-time "build-and-forget" projects) to more comprehensive best practices (involving active management throughout the lifetime of the project). This strategic approach to energy management is endorsed by a number of international organizations, including Energy Star (US), Natural Resources Canada (Canada) and Action Energy (UK). Although there are subtle differences between the energy management strategies proposed by these organizations, Table 7-1 highlights the main elements found in all of them.

Like many modern management practices, the strategic energy management approaches described by these organizations highlight the need for an information system to set goals, track performance and communicate results. The MSE 2000 energy management recommends the use of an information system in several areas, and provides several sample reports in its appendix [3]. Several documents from Action Energy, including *Introducing Information Systems for Energy Management* [4] and *Monitoring and Targeting in Large Companies* [5] provide rich detail about the role that information systems play in strategic energy management.

A strategic energy management approach that includes an energy information system has the power to increase energy savings above and beyond the savings realized by traditional tactical practices alone. One US DOE paper studied energy efficiency projects at more than 900 buildings and found that projects which implemented best practices in measurement and verification realized higher savings (both initially and over time) than comparable projects, yielding an additional return on investment of nearly 10% [6].

One key aspect of these management approaches is their focus on setting goals and measuring performance against those goals. It is not uncommon to see organizations build a comprehensive information system to support their quality management program, and similar information systems can be built to support energy management programs. These energy information systems can be designed to collect relevant data, compensate for external factors such as weather and production volume and provide the supporting information required to monitor the performance of energy management projects and keep them on track.

Figure 7-1 shows the typical components that form a modern energy information system. Intelligent, microprocessor-based devices measure energy use at key points within one or more facilities and communicate this data back to head-end software via a communications network. The software archives this data, processes it as required and presents it to the user in a variety of ways (e.g. using a web browser or by sending messages to wireless devices).

DATA OVERLOAD

Although it would seem clear that an energy information system can play an important role in reducing

Corporate Commitment	An effective strategic management plan requires a strong commitment to continuous improvement throughout the organization.
Evaluate Current Performance	Conduct an inventory and energy audit, and then create a profile and baseline of energy use at all key points.
Set Performance Goals	Energy performance goals provide direction for decision-making and serve as a foundation for tracking and measuring success.
Action Plan	The action plan drives and guides everyone in the organization to focus and prioritize their energy efficiency efforts.
Educate and Motivate Participants	The ultimate success of a plan will depend on the motivation and capability of the managers and employees implementing its components
Evaluate Ongoing Performance	Sustaining improvements in energy performance and guaranteeing long-term success of a plan requires a strong commitment to continually evaluate performance.
Communications Strategy	A communications strategy provides the framework for promoting energy management efforts throughout an organization.
Recognition Strategy	Identifying and communicating the contributions of all participants provides a solid foundation on which to build a successful energy management strategy.

Table 7-1. Common elements of several strategic energy management approaches

energy consumption and costs (and maintain those savings over time), it is also true that such systems can overwhelm their users with the volume of data they generate. The cost-per-monitored-point within energy information systems is steadily decreasing and it is becoming cost effective in a number of applications to build systems with hundreds of monitored points. Such systems can, however, become unusable without careful consideration of what data to collect, how often to collect it and how to present the data collected. All too often an energy information system is simply configured to capture as much data as possible, as quickly as it can, "just in case it is needed." If only a handful of monitored points are involved, this "catch everything" approach will simply make finding useful information in the data inconvenient; if several hundred monitored points are involved, it becomes impossible to find anything of value at all!

A well-designed energy information system starts by considering what "nuggets" of information are required to support the key goals of a strategic energy management plan. Modern business management practice refers to such nuggets of information as *performance metrics*, and these metrics are normally defined well in advance of any data collection in order to determine the scope of data collection activities. The sections below describe how this same approach can be applied to the selection of performance metrics that support energy management practice.

PERFORMANCE METRICS

Although it is often tempting to start planning an energy information system by considering what data to collect, it is more important (and usually more difficult!) to start by considering how the information system will support key goals in the energy management plan. If these goals are the best expression of what an organization hopes to achieve in managing their energy, then the first step is to convert those goals into performance metrics that can be measured and tracked.

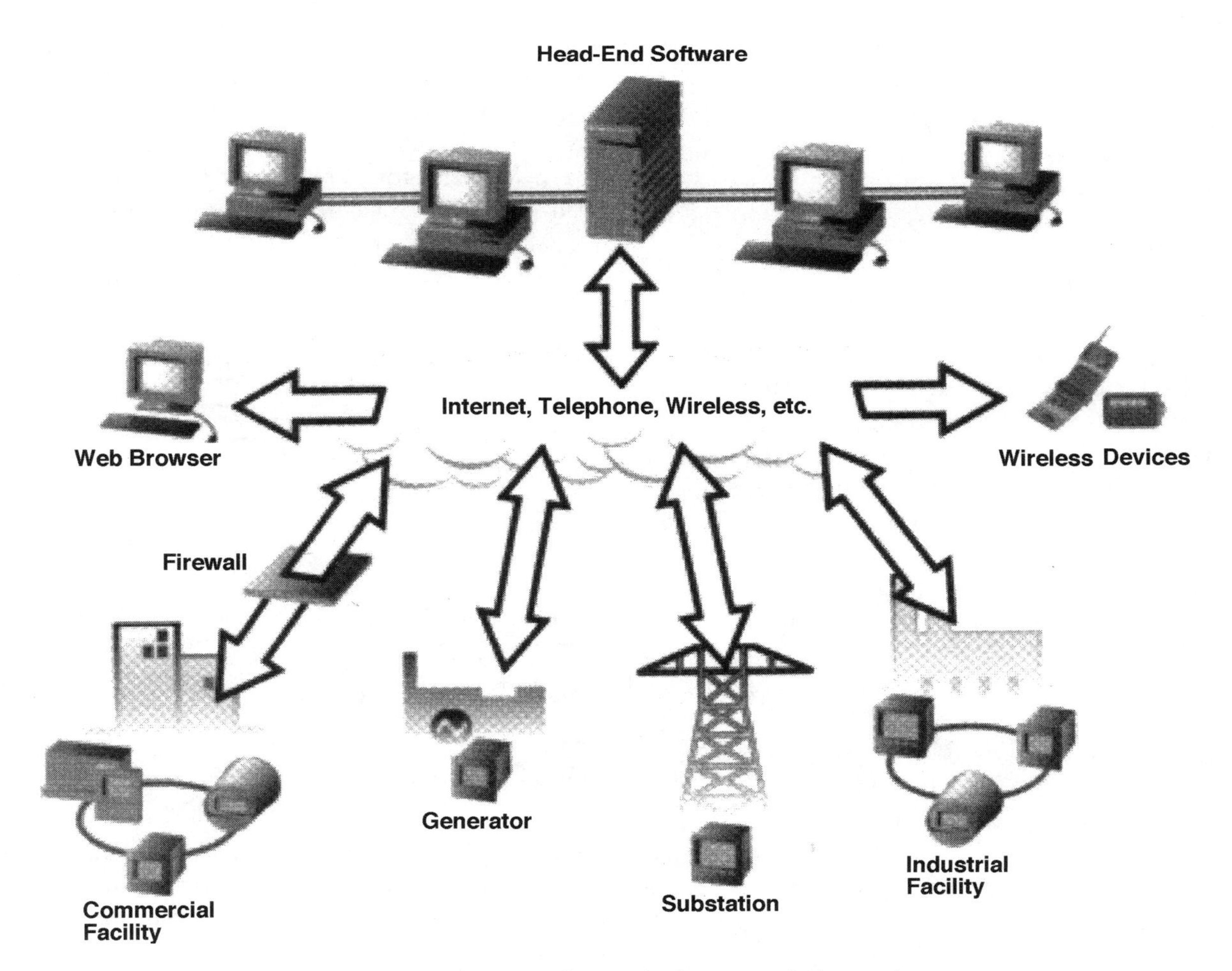

Figure 7-1. Components of a typical energy information system.

As an example, consider this goal statement from Executive Order 13123, *Greening the Government Through Efficient Energy Management,* targeted at federal industrial and research facilities:

> *Through life-cycle cost-effective measures, each agency shall reduce energy consumption per square foot, per unit of production, or per other unit as applicable by 20 percent by 2005 and 25 percent by 2010 relative to 1990.*

This goal states the key measurement of interest (energy consumption) and provides target levels and timeframes (20% by 2005, 25% by 2010) as well as a baseline year (1990) as a reference. If data regarding total energy consumption and production volume were available for a particular facility, the following sample performance metric definition could be used:

1. The baseline energy consumption for 1990 will be determined using consumption data from electricity and natural gas utility bills for the entire facility (in units of MBtu).
2. The baseline production volume for 1990 will be determined using production data from the facility Manufacturing Resource Planning (MRP) information system (in units of tons).
3. The baseline measurement for 1990 will be energy consumption per unit of production, expressed as MBtu/ton. This value will be reduced by 20% to set the 2005 target (expressed in MBtu/ton) and reduced by 25% to set the 2010 target (expressed in MBtu/ton).
4. Metering shall be installed on electric and natural gas service points to measure total facility energy consumption. The metering data collected will be converted to MBtu and summed on a monthly basis. These monthly values shall be aggregated into an annual sum for reporting against the target levels.
5. Production volume will be reported by the MRP information system on a monthly basis. These monthly values shall be aggregated into an annual sum for reporting against the target levels.
6. At the end of each year, the energy consumption and production data described in items 4 and 5 will be combined to generate the energy consumption performance metric. This metric will be combined with others into the annual energy management performance report and presented to the executive team.

Note: this chapter defines 1 MBtu as 1000 Btu.

A performance metric definition like the one above provides the foundation required to determine what data to collect, how often to collect it and how to present it. This sample definition is also careful to state the assumptions made so that everyone involved understands exactly what is being measured. Note, however, that the sample above is designed to provide only a summary performance metric that can be reported as part of an annual energy management performance report. This example definition could be further expanded to selectively provide the richer details that an energy manager would need to investigate deviations from the target goal and help keep the energy management plan on track.

Continuing with the example above, consider the expanded performance metric breakdown for energy consumption shown in Figure 7-2. This breakdown exposes different levels of additional detail that an energy manager can "drill down" into in order to understand what is driving the behavior of the defined performance metric. In this example, energy consumption is broken down into electricity and natural gas consumption (measured on a monthly basis). These measurements are further broken down into major consuming categories (such as motors, heaters and an "other" category for electricity). In addition to breaking down consumption by type and category, greater detail may be offered in the form of shorter measurement intervals, with monthly totals broken down into daily or even hourly intervals. The criteria for selecting which measurement details to

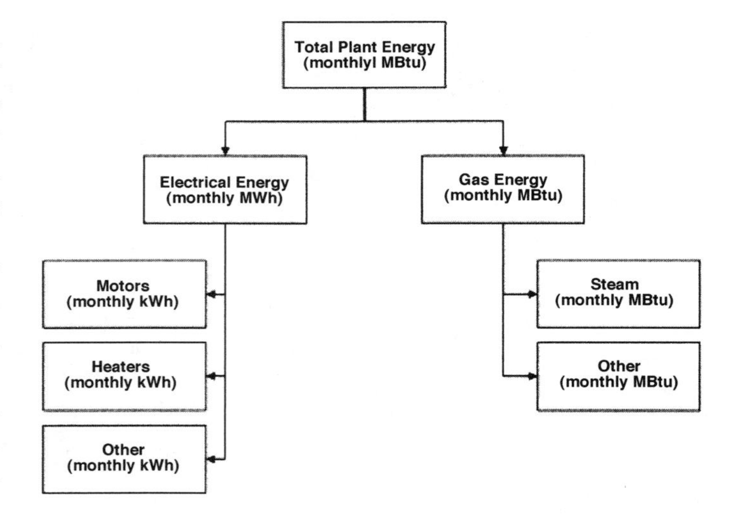

Figure 7-2. Expanded performance metric breakdown for energy consumption, showing breakdown of energy consumption into (a) electric and gas, on a monthly basis, then (b) breakdown electric (by motors, heaters and others) and gas (steam boiler and other).

highlight is determined by understanding the underlying drivers of the performance metric and knowing which details will give energy managers the information they need to correct deviations from target goals.

Once performance metrics have been defined and any supporting detailed measurements selected, the next step is to determine how the required data will be collected.

DATA COLLECTION

Compared to the potential volume of data that many energy information systems can generate, the volume required to support defined performance metrics can easily be an order of magnitude less. This is not to say that energy information systems should never collect detailed data at all; it is more accurate to say that such an information system should be designed to capture just the right amount of detailed data required to accomplish the primary goals of the system.

The data that a performance metric design process (like the one in the previous section) might specify tends to fall into one of two main categories:

- *Static data such as facility floor space and equipment ratings.* This type of data is often collected as part of an initial energy audit of a facility and is typically used to normalize measurements for benchmark comparisons.

- *Dynamic data such as energy consumption, external temperature and production volume.* This type of data needs to be collected at regular intervals and processed to generate the desired performance metrics.

Although both categories of data need to be collected, parameters in the dynamic data category tend to be more expensive to manage because there is some continuous effort involved in acquiring and processing this data. This category of data will also take up the vast majority of the total storage space in an energy information system. The cost and effort associated with dynamic data would suggest that selecting which data to collect should be done with care. Modern building or industrial automation systems may make it tempting to add a large number of measurements "just in case they are needed," but unless it serves a purpose in supporting the performance metrics described above, collecting this data will only consume unnecessary cost and effort.

Once the measurement parameters required have been selected, there are a variety of potential data sources to consider:

- *Energy consumption from utility bills.* Energy consumption totals can be keyed in from utility bills, but many utilities will also offer billing data in electronic form to their larger accounts. In addition to energy consumption totals for a billing period, some utilities will also offer load profile and other interval data.

- *Energy consumption from "shadow" metering.* If detailed energy consumption data is required but not available from the utility, a facility can install a separate meter at the utility service point to "shadow" the utility meter.

- *Energy consumption from sub-meters.* Data for an expanded performance metric breakdown of energy consumption can be obtained by installing meters on major loads or points within an energy distribution system.

- *Energy consumption from existing automation systems.* Some building and industrial automation systems have the ability to integrate with basic energy meters and transducers, acquire data from these devices and communicate it to an energy information system.

- *Temperature data from publications.* Public sources such as the National Oceanic and Atmospheric Administration (*http://www.noaa.gov*) and the Meteorological Service of Canada (*http://www.weatheroffice.ec.gc.ca/*) publish a variety of historical temperature data.

- *Temperature data from live, on-line sources.* Weather services such as Weather.com (*http://www.weather.com*) offer live access to current and forecasted temperatures.

- *Temperature data from local measurements.* A variety of products exist that allow you to take your own temperature measurements, ranging from inexpensive thermometers to sophisticated weather stations with interval data logging.

- *Production data from existing automation systems.* Nearly all manufacturing organizations record production volume using some form of information system, ranging from "process historians" in process control systems to shipment data in material resource planning (MRP) systems.

To better illustrate the process of selecting parameters for data collection, consider the Executive Order 13123 example described above. In that example we specified the key performance metric (energy consumption per unit of production) as well as a recommended breakdown of additional details that would support energy managers in tracking this performance metric. With this information in mind, a sample specification of the parameters and associated data sources required could include the following:

- **Shadow meters** shall be installed at both the electricity and natural gas service entrance points. The electric meter shall be configured to capture electrical consumption (in kWh) at hourly intervals. The gas meter shall be configured to capture gas consumption (in cubic feet) at hourly intervals. This data will be retrieved from the meters to the central energy information system at least once per day.

- **Production volume data** shall be imported from the material resource planning (MRP) system by the energy information system, at least once per day. Production volume (in tons) will be transformed into hourly intervals before being imported into the energy information system.

- **Electrical power meters** shall be installed on the circuits feeding production lines 1 and 2 to monitor electrical energy consumption. A gas meter will be installed on the main boiler to monitor gas consumption. These meters will be configured to record at hourly intervals. This data will be retrieved to the central energy information system at least once per day.

- **Time synchronization** between all energy consumption data and production data shall be within 1 minute or less. This synchronization is required in order to track changes in energy consumption with changes in production volume.

- **Data will be backed up** on a daily basis, and hourly data older than 12 months will be archived and trimmed from the database on a yearly basis. All hourly data will be aggregated to daily totals and the database will maintain a 3-year "buffer" of these daily totals.

This specification matches the metrics defined in the Performance Metrics section above and goes further to provide details about what kind of data is being collected, how often it will be collected and how it will be managed. This specification also attempts to balance the benefits of collecting some detailed data (i.e. some parameters at hourly intervals and some sub-metered points) against the additional cost and effort this involves.

Once the required data parameters have been defined and sources for these parameters selected, the next step is to build basic models that highlight the relationship between energy consumption and the primary driver of that consumption.

BASIC MODELING

Modeling energy consumption is a critical step on the path to constructing performance metrics that accurately reflect the impact of actions taken to manage energy. Modeling building or process energy usage normally involves determining the relationship between energy consumption data and some variable (such as temperature or production activity) that represents the *primary driver* of that energy consumption. For buildings, there is normally a direct relationship between the energy consumed by a building and the outdoor temperature. For production processes where energy use is largely determined by the physics of the process (such as heat-based and chemical processes), there is normally a direct relationship between the energy consumed and production volume.

The process of building basic models involves the following steps:

1. Select a *baseline set* of energy and primary driver data from the historical data collected;
2. Create and test a *baseline model* of energy vs. the primary driver; and
3. Create one or more *target models* to track the performance of an energy management plan.

Figure 7-3 illustrates the process and data flow involved in selecting a baseline data set and building both baseline and target models. Each of these steps is described in more detail below.

It is important to note that the modeling process described here is quite basic and will not generate robust energy consumption models in all circumstances. More sophisticated techniques for modeling industrial and commercial energy consumption are available, one example being the change-point models described in ASHRAE RP-1050, *Inverse Modeling Toolkit: Numerical Algorithms*.

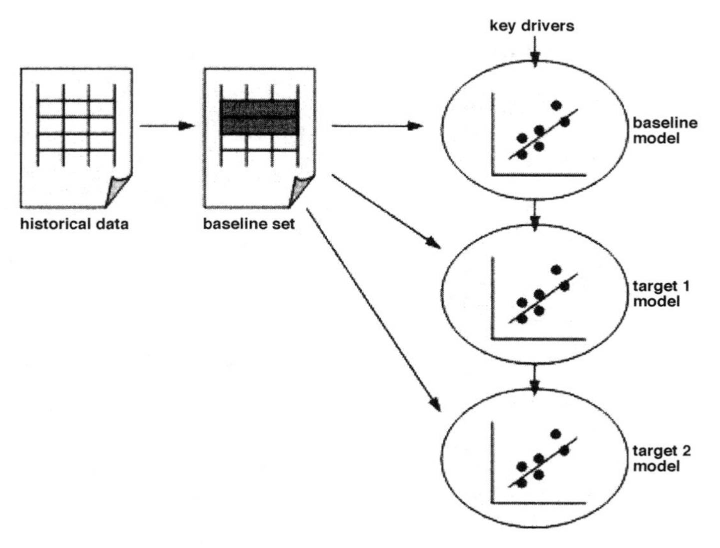

Figure 7-3. Diagram that illustrates the process and data flow of building baseline and target models.

BASELINE DATA SET

The data set selected over a defined length of time to represent the energy consuming behavior of some load (which may be a building or manufacturing process) before an energy management plan is implemented forms the *baseline data set*. This data set normally consists of two main variables: the energy consumption of a load and the primary driver associated with that energy consumption (which can include parameters such as temperature and production volume). The data collected for both of these variables will be represented in a common interval; for example, if energy consumption is totaled on a daily basis, for example, then production volume or temperature data needs to be aggregated into a daily value as well.

To obtain the most accurate model possible, the length of this baseline period should encompass the time period required for the load being studied to cycle through its entire operating range. In the case of a building, the baseline period will normally be at least one year in length to capture the energy consuming behavior of the building across all seasons. In the case of a production line, the baseline needs to be long enough to capture normal variations in production volume.

BASELINE MODEL

Once the baseline period for energy consumption has been established and the key driver data have been selected, the next step is the creation of a *baseline model* that will highlight the relationship between these two variables. A "visual" method of building this model involves the following steps: (a) create a scatter plot of baseline energy consumption and key driver data, and (b) plot a line that is the "best fit" for the points on the scatter plot. In many cases, there is a strong linear relationship between energy consumption and the key driver, and the equation for this "best fit" line can be easily determined through *linear regression* analysis.

To see how this modeling process works, consider the scatter plot of electrical energy consumption (in MWh/day) vs. production volume (in tons/day) shown in Figure 7-4. In this example, the energy consumption and production volume data set for the baseline have been plotted on a chart, with production volume on the x axis and the associated energy consumption on the y axis. Modern spreadsheet software makes the creation of such charts quite straightforward—the sample chart shown in Figure 7-4 was created using Microsoft Excel.

The chart in Figure 7-4 also includes a straight line that best fits the points on the scatter plot, along with the equation (in the form of $y = mx + b$) that describes this line. In this equation, the constant m represents the slope of the line and the constant b represents the intercept of the line. The *correlation coefficient* (R-squared) indicates the strength of the relationship between energy consumption and production volume, where a value of 1.0 indicates a perfect correlation between variables. Modern spreadsheet software can be used for this linear regression analysis and generate the straight line equation and correlation coefficient.

Although linear regression analysis of data within

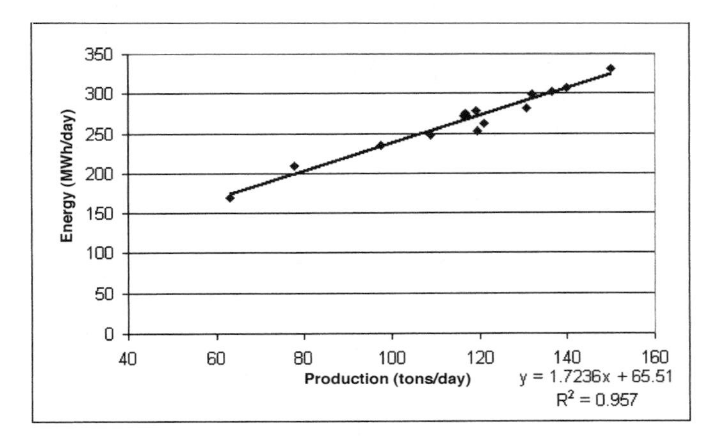

Figure 7-4. Scatter plot of energy consumption vs. production volume for some industrial process. Plot should show individual points and straight line of best fit (with associated equation).

a baseline data set will generally result in a straight line model with a high R-squared value, there are cases where the variables involved may not have a strong, linear relationship. The following Action Energy (*http://www.actionenergy.co.uk*) guides offer detailed information about the creation of baseline models and examples that demonstrate how to interpret non-linear results:

- Degree Days for Energy Management [GPG 310]
- Monitoring and Targeting in Large Companies [GPG 112]

TARGET MODELS

The role of a baseline model in performance metrics is to provide a reference model that describes the energy consumption before energy management activities are implemented. The role of *target models* in performance metrics is to provide the "yardsticks" by which the success of energy management activities will be measured. These models are constructed by applying the key goals embedded in the performance metrics identified above against the baseline model to generate the reference model that ongoing measurements will be compared against.

The following examples demonstrate how two typical goals can be converted into target models:

- *Reduce energy consumption to 20% of 1990 levels by 2005*. Given energy consumption and primary driver data for 1990 we can select a baseline period and create a baseline model of energy consumption, as described above. The straight line equation for this baseline model is in the form of $y = mx + b$, and by reducing the slope and intercept constants by 20% we can create the straight line equation for the 20% energy reduction model.

- *Reduce energy consumption to best practice levels within one year*. Best practice metrics are often expressed in normalized units relevant to a particular industry or application (e.g. MWh/ton for a particular manufacturing process). Assuming these normalized metrics hold true across a range of primary driver values (e.g. varying tons of production) then a simple straight line equation can be crafted to create a target model.

- *Figure 7-5* demonstrates the energy reduction target model concept by combining a sample baseline model with the Executive Order 13123 target reductions. The 20% and 25% reduction models can be expressed as straight line equations, and these models can be used to generate a variety of charts and displays that energy managers can use to actively track the performance of an energy management plan.

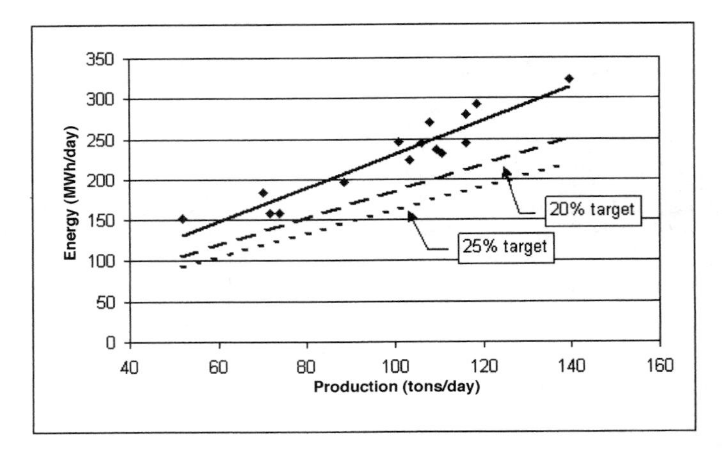

Figure 7-5. Scatter plot of energy consumption vs. production volume showing individual straight lines for the baseline model and the 20% and 25% reduction targets from Executive Order 13123.

Once baseline and target models have been constructed they can be used, along with current measured data, to track the performance of an energy management plan.

TRACKING PERFORMANCE

Our final step in constructing performance metrics that support an energy management plan is to build information displays using the data we have collected and energy consumption models we have created. The displays we will create typically fall into one of two main categories:

- *High-level overviews of a performance metric*. These concise views are designed to help an energy manager "see the forest for the trees" and are meant to provide a general indication of energy management performance.

- *Detailed drill-down view of the data behind the performance metric*. These views work in concert with high-level overviews of performance metrics, providing additional details about the behavior of the data behind the performance metrics. These details can help an energy manager understand why an energy management plan is starting to go off track.

There are a variety of ways to display performance metric information and detailed data, and the choice of which display method to use depends on what information is being conveyed and how this information will be used. Some examples include:

- *Performance metrics in a table.* A table is often the best way to organize and display the high-level target numbers that support a particular performance metric. A metric specifying a target goal of 10 kWh/square foot/year within 4 years, for example, might be shown as a table of declining target values for each of the four years.

- *Performance metrics in a bar chart.* Current and past performance can be visually compared against target goals using a bar chart. Such a chart may show month-by-month actual values vs. target values for the performance metric.

- *Drill-down data in a time-series chart.* To gain an understanding of what is driving a particular performance metric value, a time-series chart can provide a detailed view of the data behind the performance metric. Both actual and target values can be plotted over time to help an energy manager see where any deviations from the plan are taking place.

To complete our Executive Order 13123 example above, consider the information displays shown in Table 2 and Figures 6 through 8. Table 7-2 shows possible performance metric target reductions over five years that will bring energy consumption to the 20% reduction goal in the sixth year. Figure 7-6 is a bar chart comparing these target reductions against measured values on a year-by-year basis. The measured value for 2003 exceeded the target reduction for that year, and Figure 7-7 is a bar chart that breaks out this year into individual months. Finally, it is apparent from the bar chart in Figure 7-7 that the measured energy consumption exceeded the target reduction in February, and Figure 7-8 provides a detailed time-series chart showing measured vs. target consumption on a day-by-day basis. It is also clear from this time-series chart that much of the deviation from the target goal occurred on the third and fourth days of that month. This chart allows an energy manager to focus attention on events that occurred on those days and identify what contributed to this excessive energy use.

It is important to note that the information displays described above, moving from Table 7-2 to Figures 7-6 through 7-8, progress from a high-level overview of energy management performance on to increasing levels of detail. By reviewing high-level performance metrics first and drilling down into details only when there are deviations from target goals, an energy manager can avoid searching though thousands of data points to find the few that are of interest. This is not to say, however, that the data captured while the performance metrics are on track are without value; this data can be used for a variety of other tasks, including the development of operating "profiles" for monitored equipment.

It should also be noted that there can be different audiences within an organization for the different information displays described above. All stakeholders participating in the energy management plan will be interested in the high-level performance metrics, and the core energy management team will be the primary audience for detailed drill-down views that help them understand deviations from planned target levels. It is also likely that the energy manager will make use of both the high-level performance metric displays and select de-

Table 7-2. Possible performance metric target reductions over five years.

Year	*Target (MBtu/ton)*
2000	27
2001	25
2002	24
2003	23
2004	22
2005	21

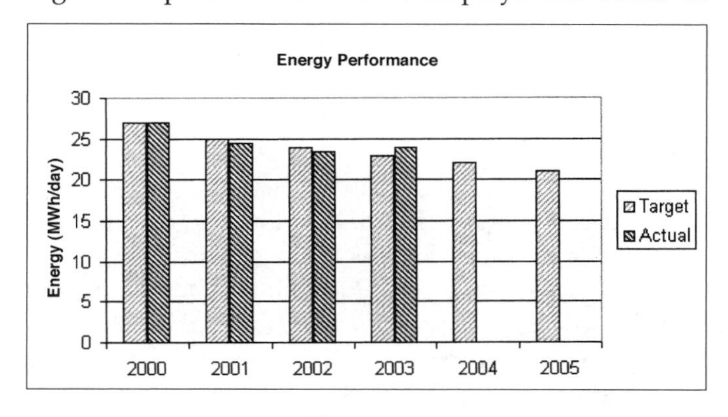

Figure 7-6.

Figures 7-6 though 7-8: Information displays illustrating the use of high-level performance metric overviews and detailed drill-down views.

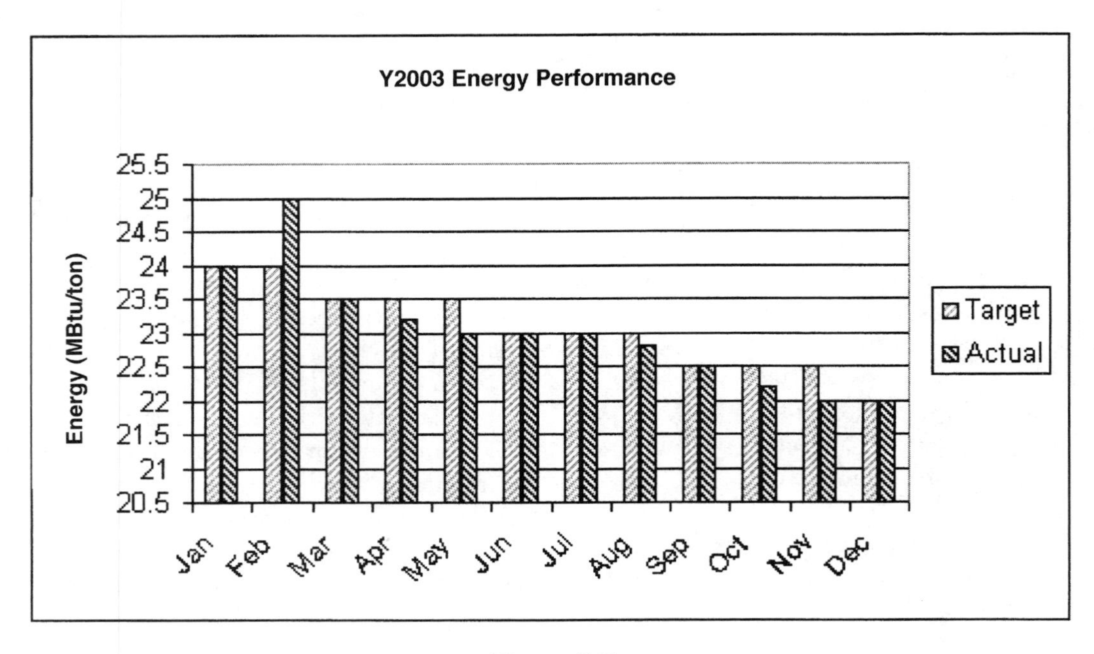

Figure 7-7.

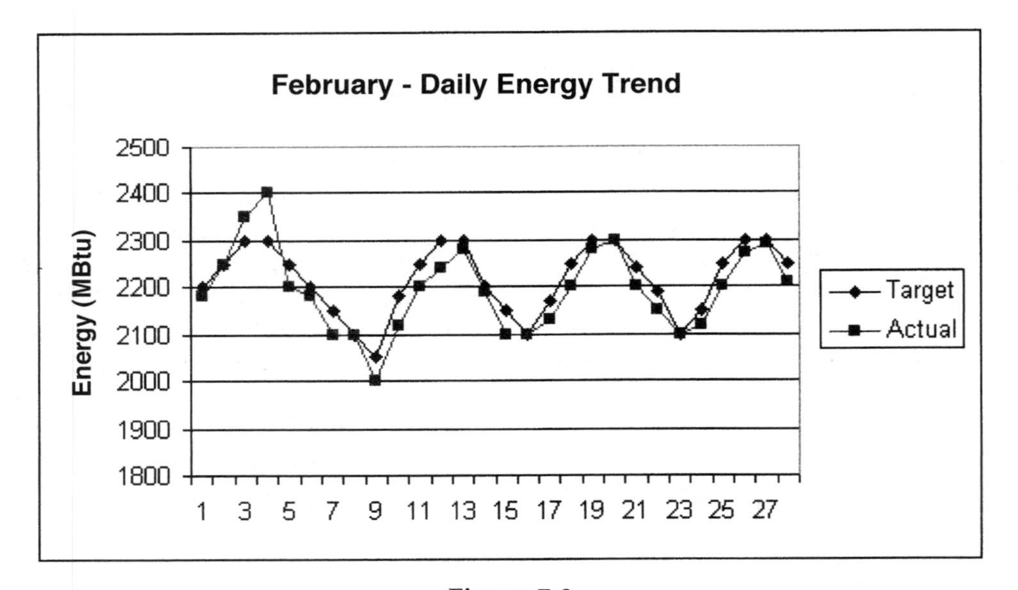

Figure 7-8.

tailed displays when presenting updates to executives.

One item not yet discussed is the breakdown of high-level goals into target values that can be used in the more detailed drill-down information displays to drive energy management initiatives. This process is much like the budgeting process that most organizations are familiar with. Few companies would set an annual budget and only check actual vs. planned spending at the end of the year; most instead break down the budget into monthly expenditures for different categories and groups within the organization. The process for breaking down a high-level energy consumption goal is quite similar, with the total "energy budget" being allocated to different processes or equipment over the course of a year.

CONCLUSIONS

Energy management practice has traditionally put greater emphasis on the technology involved in energy efficiency efforts than it has on the management of those efforts. There is no question that new technology plays an important role in helping organizations increase their energy efficiency, but it is also true that projects can see increased (and more consistent) savings by adopting the performance management approach integral to modern quality and energy management programs.

Information systems are becoming a key part of modern energy management practice, especially as the hardware and software components that make up these systems become more widely available. In the past such energy information systems were often prohibitively expensive, but advances in recent years have been steadily decreasing the cost involved to monitor an increasing number of measurements. As the costs involved in automating data collection continue to drop, the "total cost of ownership" for these systems will increase on the data management and information processing side of the equation. The value of future energy information systems will not be in the quantity of data they can collect, but rather in the quality of insight they can deliver.

The discussion and examples in this chapter focused primarily on targeting reductions in energy consumption, but the performance metric approach can also be applied to several other aspects of managing energy systems and equipment. Examples illustrating how this approach might be extended include the following:

- *Energy cost.* Utility rates can be applied to the performance metric process described above to track and manage energy costs. Models can be used to compensate for changes in primary drivers (such as temperature and production volume) to more accurately calculate the savings realized by energy efficiency projects.

- *Energy reliability.* Measurements related to the "health" of equipment can be added to an energy information system and used to generate performance metrics focused on energy system reliability. A basic implementation with real-time monitoring would quickly identify failures and assist with root cause analysis, but more advanced implementations could leverage sophisticated equipment models and measurements to help predict failures before they occur.

- *Forecasting.* Once energy consumption has been modeled, future consumption can be estimated based on projected values for the primary driver of consumption (such as temperature or production volume). Risk assessments for energy consumption can be generated by applying a range of expected primary driver values.

Energy management practice has seldom held much mind share with executives even though energy is often critical to the operation of their organization. Greater mind share can be won when energy management activities are cast more in business terms than in technical terms, and their contribution to the bottom line highlighted.

References

[1] C. Goldman et al., Market Trends in the U.S. ESCO Industry: Results from the NAESCO Database Project, LBNL-49601, Lawrence Berkeley National Laboratory, May 2002, *http://eetd.lbl.gov/ea/EMS/EMS_pubs.html*

[2] IPMVP 2001 [Volume I], International Performance Measurement & Verification Protocol, *http://www.ipmvp.org*

[3] MSE 2000: A Management System for Energy, Georgia Tech, *http://www.industry.gatech.edu/energy/*

[4] Introducing Information Systems for Energy Management, GPG 231, Action Energy, *http://www.actionenergy.co.uk*

[5] Monitoring and Targeting in Large Companies, GPG 112, Action Energy, *http://www.actionenergy.co.uk*

[6] G. Kats et al., Energy Efficiency as a Commodity: The Emergence of an Efficiency Secondary Market for Savings in Commercial Buildings, *Proceedings of the 1996 ACEE Summer Study*, 1996, Vol. 5, pp. 111-122

Chapter 8

Using Standard Benchmarks in an Energy Information System

Gerald R. Mimno
General Manager
Advanced AMR Technologies, LLC

Jason Toy
Northeastern University '05

ABSTRACT

ONCE INTERVAL DATA from an electric meter are captured, the question soon arises "What does my load profile data tell me?" There are some common benchmarks that can answer this question. These are: 1) The Energy Star building rating system. 2) Utility industry class load profiles and 3) Weather normalization based on regression analysis which reveals the baseline energy consumption of a facility Benchmarks give an owner or manager a frame of reference on how efficient a facility is; where, when, and how energy is being wasted; and what action items might reduce the monthly energy bill.

INTRODUCTION

For the last several years we have been engineering, manufacturing, and installing Internet based Energy Information Systems (EISs) for very typical high schools, factories, hospitals, retirement homes, quarries, municipal offices, recreation facilities, and many others. The EIS provides real time data every 15 minutes on a web browser. The owner or manager can select a wide variety of reports showing daily, weekly, or monthly load profiles. The cost of equipment to get an on-line signal is about $1,000, roughly half for metering and half for the wireless equipment to get the signal on the Internet. The equipment is often installed and managed by Energy Service Companies (ESCOs) who charge $75 to $90 month to provide a live interval data service.

Alternatively, many utilities offer a monthly load profile service. They pick data up from smart commercial electric meters such as the GE kV2 which has digital output and can store and download 15 minute data during the monthly read into a hand-held device. Data are available a few days later on the Internet for $30 per month. If you want to connect your phone to the meter, the utility will supply a meter modem for a few hundred dollars. The utility will call your meter at midnight and provide the previous day's interval data on the web for about $50 a month. In many systems, you can initiate a call and get a reading in about half an hour. We recommend making the effort to get "near real time" data on the web continuously. If your meter is less than ten years old, your utility can likely install a plug-in card for Internet service costing about $500 or your utility might provide a new meter with an Ethernet jack for about $1200.

Figure 8-1 shows the amount of energy used in a small office every 15 minutes over four days. Note that the area under the curve represents Kilowatt Hours (kWh) or the amount of energy used during a day or month. This is represented by the kWh charge on the monthly bill. To reduce the kilowatt hour charge, you need to reduce the area under the curve.

The peak of the curve represents the kilowatt demand charge (kW) or the highest amount of power used in a fifteen minute period. A mark of efficiency is a low night load. A mark of inefficiency is a high peak load. Another consideration is "coincident peak demand." Does your facility draw its peak power at the same time as the electric grid system peaks? There will be advantages and financial incentives to moving your peak off the system peak.

Residential customers do not pay demand charges, but virtually all commercial customers do. Demand represents the peak capacity of the electric system. The utility says, "You need to pay a demand charge because I

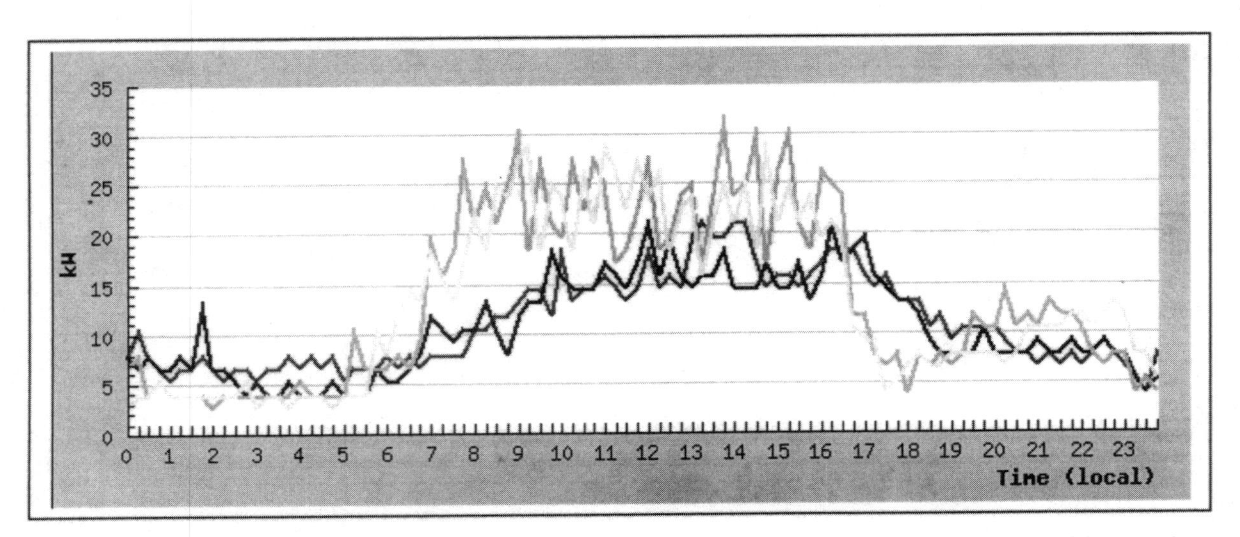

Figure 8-1. Load profiles from an Energy Information System or Utility Load Profile Service

had to build the system large enough to meet your peak need." Some utilities use a ratchet system. Even if you only reach a peak for 15 minutes, you pay the demand charge every month for a year. Other utilities charge only for the demand you reached in the highest 15 minutes of the month. In New England for example, a typical kilowatt hour costs $0.12. A typical kilowatt demand costs $12.00 per month. While it is not listed on the bill, a suburban house will have a peak of about 3 kW. In many utilities, small commercial customers have a demand less than 200 kW; large customers are over 200 kW. Very large users may have a demand of 1,000 kW or one megawatt. Generally about a third of the commercial electric bill is in the demand charge.

Many factories want to see their load profile so they can tell how to reduce energy cost. One method to accomplish this is to create a single facility system including meter pulses, Ethernet connection, computer, and software for about $3,500. The Independent System Operator (ISO) in New England offers such a package to participants in its summer load management program. Another approach is to collect data over cellular or traditional phones and send the data through the phone network to an off-site energy management network and then to the facility's web browser. The public telephone company charge to convey interval data can be quite costly depending on the type of service contract. Monitoring one meter at 15-minute intervals represents 2880 calls per month. We recommend you connect your meters to the Internet to eliminate recurring interval charges. The Automatic Meter Reading (AMR) industry offers many suitable wireless and power line carrier technologies you can use to connect to the Internet.

Your monthly utility bill includes your meter reading and energy use but this information has limited value in managing your operations. ESCOs can find ways to save five, ten, fifteen percent or more of the energy used in any facility, but facility managers seldom do so on their own. To date, the market penetration for commercial interval data is less than one percent, but the Internet has recently provided the means to make obtaining interval data practical and we expect the use of interval data will expand

Why don't more facility managers find value in interval data? What they need is a frame of reference. It is hard to look at a load profile and understand what it means. It is like a map without a scale or north arrow. The reaction of many of our customers is, "Well that looks interesting, but what does it mean?" They also question whether they should pay for a load profile service since they can't attribute a value to the data. Utilities offering load profile services at $90, $50, or even $30 per month have had very few customers sign up.

In our experience, there are three stages required to manage facilities better, reduce waste, and conserve energy. These stages and technologies needed to manage energy and reduce waste are shown in Table 8-1.

This chapter discusses how to meet the second requirement. If you have successfully completed the arduous journey placing real time interval data on the web, the next step is to find a frame of reference or **benchmark** for that data. Benchmarks will show how your facility compares to similar properties. They can also help identify where your waste and inefficiency is located, which will help you find opportunities to save money. In the past, this analysis was available on a cus-

Table 8-1. Three stages and technologies needed to manage energy and reduce waste.

Stage 1	
Show me my load.	Automated Meter Reading System
Stage 2	
Benchmark me.	Energy Information System
Stage 3	
Save me money and do it for me.	Energy Control System

tom basis or through expert outside analysis. Our interest is to find ways to automate this analysis and make it available at little or no cost beyond the basic monthly cost of an interval data service.

BENCHMARKS

There are several useful benchmarks you should consider. The Environmental Protection Administration (EPA) has expanded the Energy Star rating program from residential products to commercial buildings and has developed a web based program for ranking the efficiency of facilities. [3] Another benchmark is provided by power companies. For decades, utilities have studied and published "Class Load Profiles." Class load profiles show the typical consumption of residences, small and large commercial buildings, industrial facilities, and even street lighting. Utilities recognize that within a class, many customers operate in a similar way. Therefore the class load profile can be used for forecasting demand, setting tariffs, and settling payment of wholesale power contracts. Comparing your own load profile to the class load can offer a valuable benchmark.

Another reference point is the extensive studies made on the relation between weather and the consumption of electricity. An industry rule of thumb is that weather explains 30 % of the changing demand for power. Without taking account of the weather, it is difficult to compare this month's energy use to last month or last year. Removing the transient effects of weather reveals the baseline energy consumption of a facility.

As one vendor, we have incorporated these benchmarks in our Energy Information software. We use an XML link from our server to the Energy Star database. We have mined the web both for class loads in the public domain and for weather analysis. With a few key strokes, users can bring up a load profile of their own facility, use weather data to normalize the load profile, and then benchmark this data through Energy Star or their own utility's class load profile. We describe the details below. Our purpose is to make an initial pass at analyzing a load profile and to set the stage for a more detailed design of energy management and control systems to mitigate waste, reduce monthly bills, and conserve energy.

Benchmarking Architecture In An EIS

An EIS and benchmarking system has the following parts:

1) Automatic Metering System. If your utility does not provide interval data, have an electrician install a shadow meter downstream from the revenue meter. The shadow meter will use Current Transformers (CTs) and a transducer which outputs Watt Hour pulses or Kilowatt hours. We have had very good experience with shadow metering using the Square D Veris Hawkeye [1] and Ohio Semitronics Inc. WL55 [2]. The Veris is quick to install and costs about $600. The Ohio Semitronics unit is exceptionally accurate and also economical. The CT's and transducer for a 3 phase system total about $275. A benchmarking system needs some additional data inputs. Vendors offer software which contains links to a variety of databases including EPA's Energy Star database [3], libraries of utility class load profiles, and weather data. Live hourly weather data are available from the bigger airports. [4] The Energy Information Administration (EIA) [5] offers tables with 30 years of weather normalization data for every region in the US. You can also get useful temperature data from a building management system (BMS).

2) *Networking*. You should use the Internet to collect data for your benchmarking. A good benchmarking process needs access to websites which have historical data, libraries, and live weather information. Just as benchmarking is a continual process, so is the need to continually update the benchmarks. There are always additional good sites to find and link to your benchmarking process.

   You should also use the Internet to distribute the results of your benchmarking to all the relevant parties in your energy management organization.

These include facilities managers, budget managers, operations people, power suppliers, and interested users such as teachers and administrators. We have found that the first and heaviest users of interval data are operations people who manage the performance of retirement homes, hospitals, factories, commercial rental property, and quarries. They use live interval data to monitor their operations and like the convenience of checking on the facility from wherever they may be.

3) *Server and Database*. Interval data from hundreds of points in an operation comes in over the Internet as packets with a header, contents, and checksum. The server sorts the incoming data by such categories as packet type, business unit, ID, date & time stamp, and inserts data in the proper form into the EIS database.

The database is the heart of the benchmarking system. The databases frequently chosen for Energy Information Systems are Oracle, Microsoft SQL Server, and the open source MySQL. The advantages of open source (including the Linux operating system) are that a user group with talented software people can extend the features of the software on their own and make this available to other members of the group. This is particularly helpful for users, such as universities, electric coops, or schools, who share many common problems.
Alarm automation software monitoring the database can trigger alarms and make telephone calls using audio files, send emails, send Short Messaging Service (SMS) messages over cell phones, and chase recipients to their after hours and weekend addresses. Alarm incidents are logged in the database with a record of who responded to or canceled an alarm. The alarm automation program also monitors conditions in the field and can issue an alert, for example when a meter or device exceeds specified ranges such as the run time or start-stop cycles on a pump.

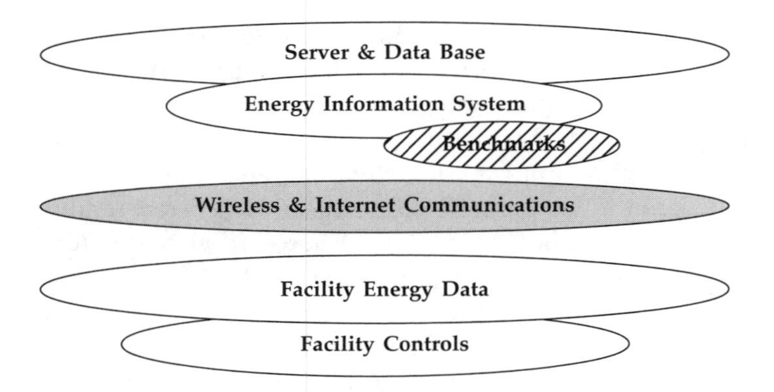

Figure 8-2. Data architecture in an Energy Information System

4) *The Browser*. Data from the data base is presented to users in reports they see on their web browsers. The browser presents a dashboard with a few key functions displayed and access to much more information available by drilling deeper into the database. There are many advantages to browser-based software. Upgrades to the browser software can be made in the server and are then immediately available to all users. Administrative and support functions can also be accessed by any authorized user anywhere. New meters can be enrolled and addresses and telephone numbers entered from the field without returning to the office. Tech support personnel can diagnose and fix many problems remotely without field calls. There are many EIS products on the market. Most present some form of live data on a running basis. This data can then be printed as reports or downloaded into Excel for further processing by the user.

In a benchmarking system, the customer brings up their own facility on a charting window, and then adds additional information from the benchmark features. For example, the benchmarks for School A will typically show School A's load profile compared to a population of other similar schools, the class load profile for an average school, and School A's load profile with variations due to temperature stripped out.

5) *The Control System*. Some vendors incorporate control systems into the EIS while others maintain a control system in the facility and have a separate EIS to reflect the performance of the control system remotely. Historically the facility control industry is a Tower of Babel but progress is being made on common protocols and standardized interfaces. Some of the simplest technologies are already a standard such as pulse outputs or the 4-20 MA industrial control protocol. Anyone desiring to tie an EIS to the control system is going to have to contend with different products that have incompatible outputs. The usual way to cope with this is to attach a relatively smart and flexible digital communications device to a relatively dumb and inflex-

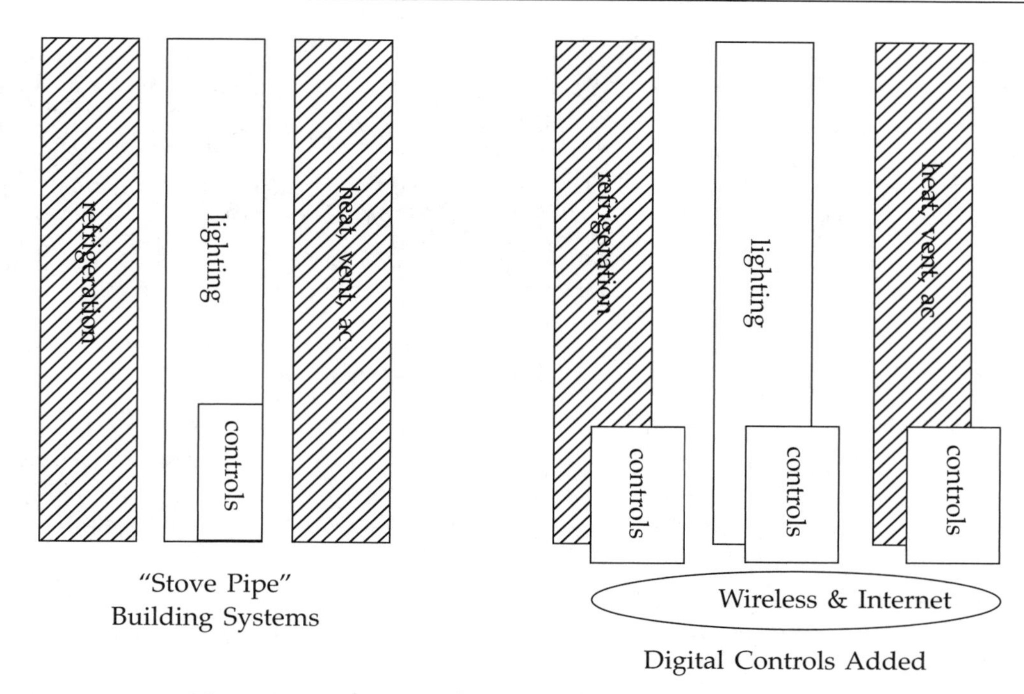

Figure 8-3. Adding digital controls to a facility

ible control or instrument. The digital device mimics the signal the instrument expects and packages the results in a communications packet in TCP/IP which can be sent or received over the Internet.

Benchmarking is not going to tell you how to set or reset your controls. The benchmark information is too general. But a benchmark will tell you where to look and where to make more detailed analyses. Is your inefficiency problem in night load? Is it in peak load? Are you way off from a standard or just a little off? These are the answers you can expect to find in your benchmarks.

ENERGY STAR BENCHMARK

The first benchmark we want to discuss is a *score* from the Energy Star National Energy Performance Rating System. Energy Star is an EPA program long associated with the sticker rating on refrigerators and boilers. Five years ago, Energy Star rolled out a program for benchmarking whole buildings including offices, schools, supermarkets, hospitals, and hotels. Energy Star has benchmarked 15,000 buildings representing 2.5 billion square feet. This is 14% of the eligible market. [6] The system is gradually being extended to more building types and will include higher education, residence halls, warehouses, banks, retail, and additional commercial building types.

You can get a score for your building by filling out a web form at *www.energystar.gov/benchmark*. Energy Star has made detailed analyses of thousands of buildings and the program uses this data to evaluate some relatively simple data you supply about your own facility. You provide figures from the building's utility bills, the building size and type of construction, the year it was built, and some operational data such as hours of operation and number of employees and computers. After comparing your building against the national stock of similar buildings, the Energy Star program returns an efficiency score between 1 and 100. The program takes account of weather, hours of use, occupant density, and plug load. The EPA has found the distribution of buildings along the efficiency scale is pear shaped. The majority of buildings is clustered at low efficiency while about ten percent reach out to the high efficiency "stem" of the pear. The best buildings are 4 times more efficient than the worst.

The Energy Star score is an excellent example of a benchmark. The single number could be described as an "efficiency motivation index." EPA makes recommendations for utilizing the score (see Table 8-2).

The National Energy Performance Rating System is only the first step in an ongoing Energy Star support network. Once a building has been rated, the building manager can create an on-line portfolio to track improvements and performance over the years. EPA also has a Financial Value Calculator to assess alternative energy improvement strategies on financial indicators including payback, market value, and earnings per share.

CLASS LOAD PROFILES

The second benchmark for comparison is Class Load. A class load is a 24 hour load profile for a typical utility customer in a specific category. Utilities conduct load research to determine patterns of utilization and then construct class loads. Class loads represent the pattern of use for each different tariff in the utility's rate structure. The residential class load for large apartment houses will be different from the commercial class load for small commercial customers and the commercial class load for offices. The class load shows how your facility compares to the average customer in your class. The class load does not show the absolute amount of kWh but rather the pattern of use.

Often the class load is described in a table of percentages. The percentage of power used between midnight and 12:15, the percentage between 12:15 and 12:30, etc. Multiplying your total use by the percentages will result in a load profile for your facility. The EIS will then chart your actual use and compare it with the class load rendered to a common scale. We recommend a 15 minute profile giving 96 intervals per day. An hourly profile does not show enough detail and five minutes shows unnecessary detail for most people. While we are talking about electric loads, class loads also apply to gas, steam, fuel oil, and water.

Who has class loads and where can we find them? They are part of the public record at Public Utility Commissions where they are used in rate proceedings. Many utilities publish their load research. They are also produced by Energy Service Companies, power marketers, large users, and in wholesale power settlement. Another good source is the US Government Energy Information Administration. [5] A search of the web is the most practical source. Excel is the standard software used for organizing, downloading, and charting a class load.

Loads change for a number of reasons. Daily activity varies. Weather varies. Processes and uses vary in a facility. However, there are also common patterns among users. These are represented in class loads and are typically divided into residential, small commercial, large commercial, and industrial categories. In addition, class loads are provided for weekdays, weekends, and sometimes holidays. Class loads are accurate enough that a lot of wholesale power is purchased and paid for on the basis of class load. The class load shows the timing of the delivery of power and charges for capacity or peak power.

Built in 1969, this 500,000 sq.ft. high school has an Energy Star rating of 8.

Built in 2001, this 190,000 sq.ft. middle school has an Energy Star rating of 32.

Figure 8-4. Energy Star scores

Table 8-2. Benchmark scores and strategies.

SCORES BELOW 50: INVESTMENT STRATEGIES	— Buildings in this range need new equipment. — Replacement equipment can be amortized by substantial savings in monthly utility costs. — New operational practices will also have a substantial impact on the bottom line.
SCORES BETWEEN 50 AND 75: ADJUSTMENT STRATEGIES	— Concentrate on simple, low cost measures such as improved operations and maintenance practices. — Upgrade equipment for additional savings.
SCORES BETWEEN 75 AND 100: MAINTAIN BEST PRACTICES	— Buildings with these scores represent best practices of design, operations and maintenance. — Slacking off will lower the building's score.

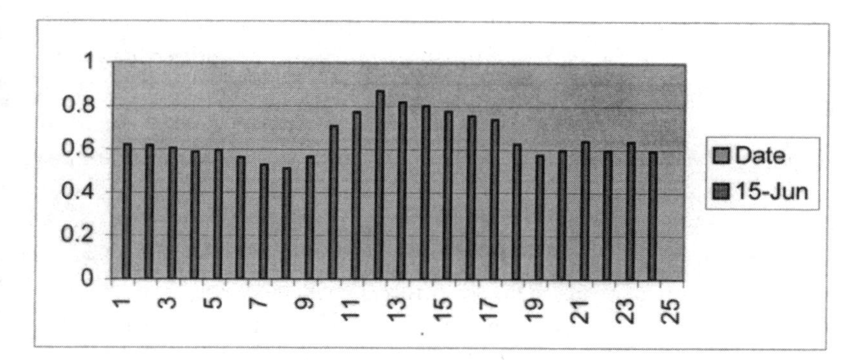

Figure 8-5. Sample Class Load Profile

In examining a facility load and comparing it to a class load, we want to look at a number of items. These are night load, day load, seasonal load, peak summer load, and the spring and summer shoulder months. In comparing your load profile to the class load, look for three things:

Base load percent (night load/day load)
Peak to base ratio (day load/night load)
Coincident peak

The base load is the amount of power always on. It is the area of the rectangle at the base of the load profile. Typically in a single facility your base load will be about 33% of your day load. The area of the base load is also defined as your "load factor." It is the percent of power you use 24/7. You should compare your facility base load with the class base load.

The peak-to-base ratio is the number of times the height of your day load exceeds the height of your base load. A small facility will use three times more power at the height of the day than the minimum used at night. A large facility will use 30% more at the daily peak than at night.

The third factor to note is the coincident peak. The entire state of California peaks at 3:30 p.m. Your facility will show a peak at some time during the 24 hours of the day. Does your facility peak at the same time as the class load? If so, you have a coincident peak.

Comparing these numbers to the class load will begin to show you if you should look for inefficiencies (i.e., if you are wasting kWh), and if you should try to reduce your peak demand (i.e., if you are wasting kW).

If your facility has a relatively high base load, you will typically find that people in your facility are unnecessarily leaving a lot of lights, fans, computers, and other equipment on. Note that the class load is an aggregation of many facilities and will look relatively high and flat. The smaller your facility, the smaller you would expect your base load to be when compared to the class load. Schools are often found to have a relatively high base load. School utility bills may be paid by the superintendent's office, custodians have little informa-

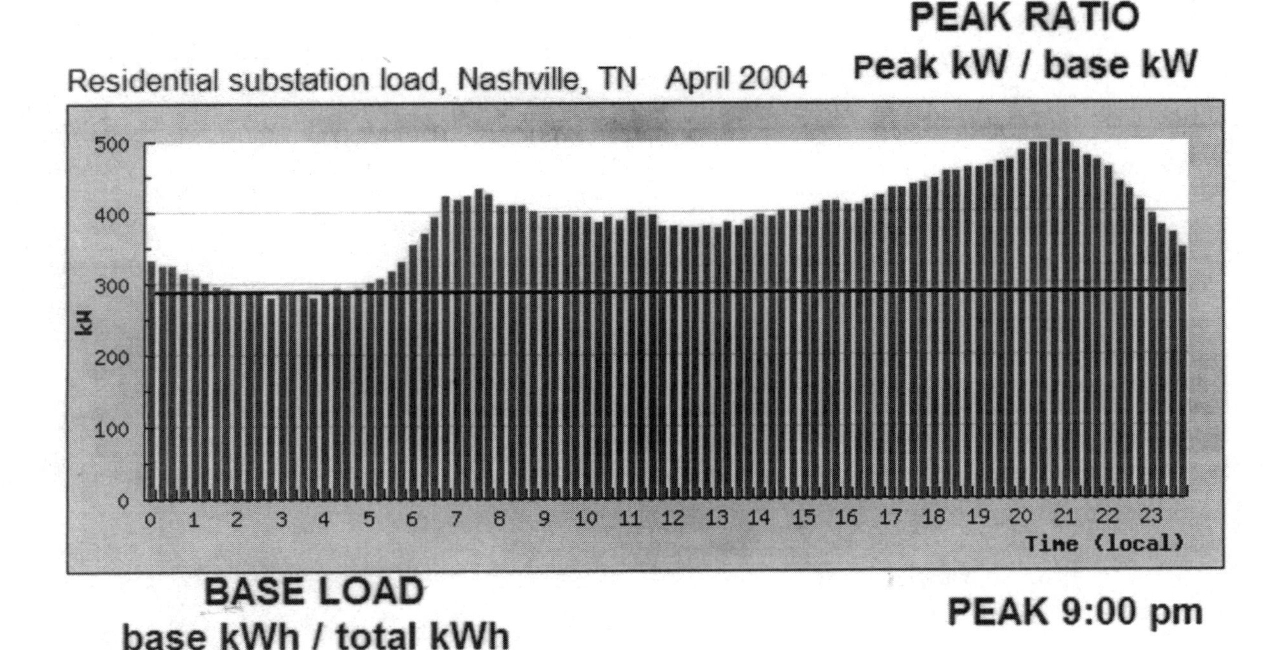

Figure 8-6. Ratios important in analyzing a load profile

tion about cost, teachers have their own priorities, the whole building is left on to accommodate a few night meetings, and the operating cost of the building drifts up. Even in buildings with an energy management system, controls and time clocks can be overridden or defeated, negating the advantages of the system. A comparison with class load may alert you to this.

The peak to base ratio is a measure of what is turned on during the day. The smaller the facility, the steeper the rise is expected. A comparison with class load will suggest some things to look for. The expected range of variation may be between 130 percent and three times, (day to night load). The class load (an aggregation) should show less variation than your facility. The steeper the facility peak, the more the small facility will be penalized in peak demand charges. A very sharp peak indicates more equipment and HVAC may be turned on at the peak than necessary and that HVAC might be coasted through the peak to reduce peak demand charges.

The coincident demand measure is the degree to which your facility peaks at the same time as the class load. A coincident peak is a sign of expensive power. Increasingly, commercial customers are going to be exposed to the real time price of power. The more power you use on-peak, the higher the price you will pay. On the flip side, if your peak is not coincident, you have the opportunity to use a Time of Use (TOU) tariff to lower your cost. There is a conflict between consumers and generators on peak power. Consumers want cheap power. Generators want a flat load. The mediator between these two is in many parts of the US the Independent System Operator, (ISO). The ISO takes bids from generators usually in a day ahead and hour ahead mar-

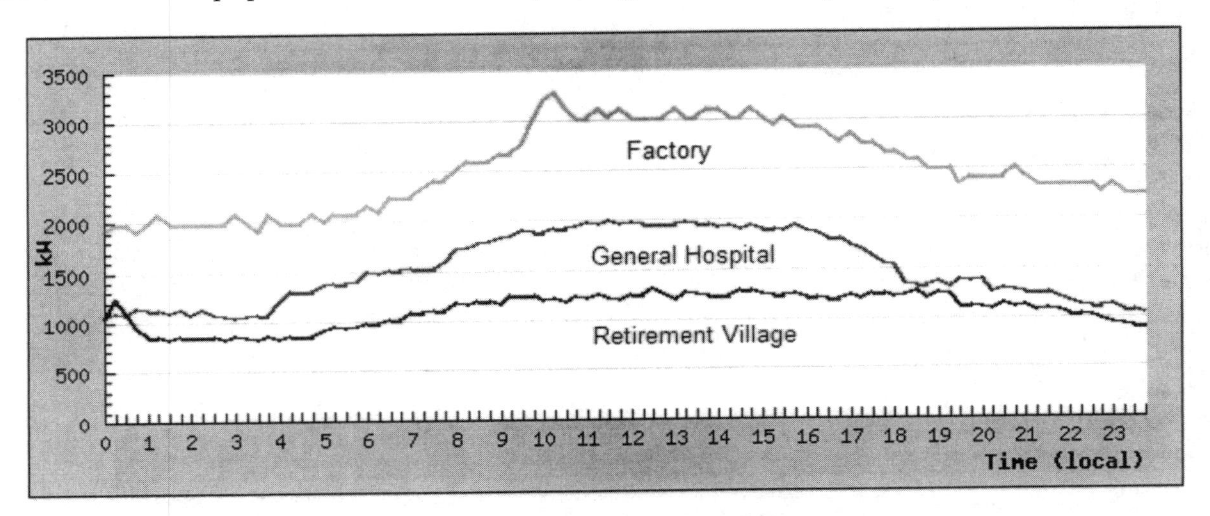

Figure 8-7. Base load analysis

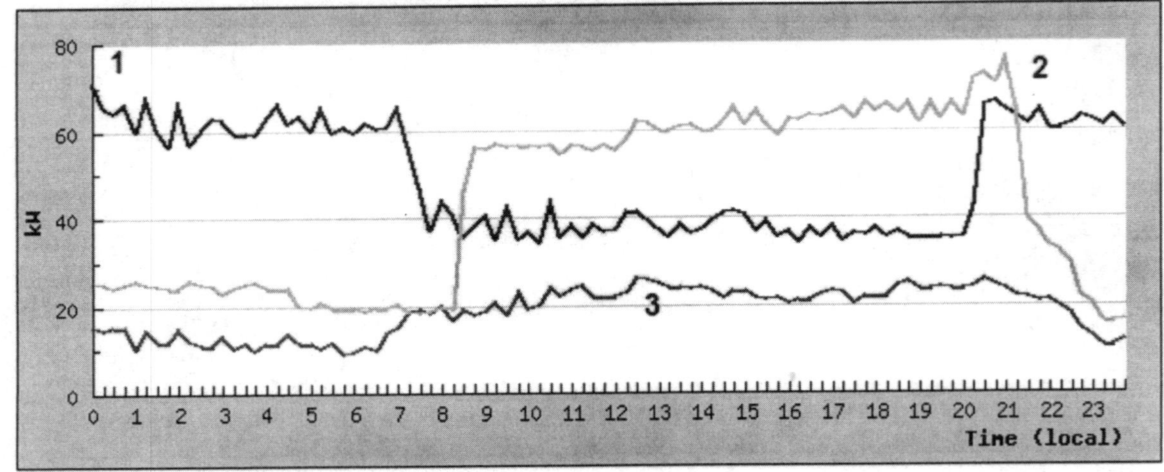

Figure 8-8. Peak load analysis

ket and builds a load stack taking the cheapest generator first and then adding higher bids as the demand increases. A few times a year we can expect a Critical Peak Pricing Incident in which demand outstrips supply and the price of wholesale power may shoot from $50/MWh to $500/MWh. If your peak demand can be moved, you may want to participate in one of the many new load management programs which pay the market price for reducing load. New York and California both offer about a dozen new load management programs. They also support "demand response aggregators" who can take your small response and combine it with the efforts of many others to reduce megawatts of demand. Peak reduction is the equivalent of building more generation and demand response aggregations will be treated by system operators as if they were generators bidding a price into the load stack. Your ability to participate in demand response programs will be increasingly rewarded. Comparing your facility to your class load should give you an opinion of how well you might be able to participate in load management. If your peak is very high, it is likely you have uncontrolled operations and you can add controls to mitigate this.

WEATHER NORMALIZATION

Many electric bills now show consumption for the past 13 months. A few even provide the average monthly temperature. This gives consumers a crude measure of the annual pattern of consumption. But is this month's bill comparable to the same month last year? The hours of daylight will be the same, but the economy can change, the utilization of a facility can change, and the weather can change. Energy managers need a means to factor out the variables and determine the baseline energy use. The baseline then becomes the standard against which efficiency improvements can be measured. Why is this needed? We can't do much about sunrise or the economy, but we can try to remove the effects of weather from the monthly bill and develop a measurement scale that shows us if we can make changes that will save energy and if so, how much can we save. Every business is used to paying the monthly utility bill but many are reluctant to spend "additional" money on conservation. Knowing the baseline will offer a true measure of energy savings and help calculate what efficiency measures are worth buying and what savings are possible on the monthly bill. Without baseline information, businesses are not likely to implement any conservation measures.

A single bill is useless in determining an underlying pattern of energy use. Since the bill is virtually the only information most people have, the majority of energy users are in the dark. With a little work, they can chart quarterly and annual patterns and get a sense of their annual pattern of consumption. Typically this is represented by load profiles showing a band of consumption representing the four seasons. The wider the band, the greater the weather effect. The Energy Information Administration (EIA) goes a step further by publishing a 30 year data set of temperature information [5].

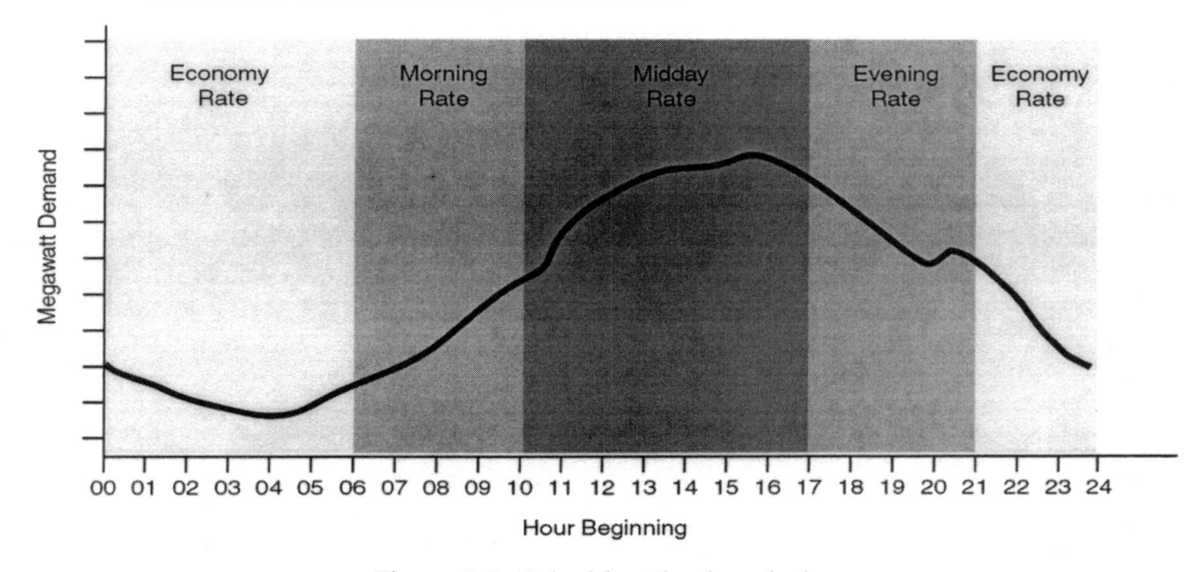

Figure 8-9. Coincident load analysis

The data are presented as Heating Degree Days (HDD), and Cooling Degree Days (CDD). By convention the data sets sum the deviation in Fahrenheit and duration in hours from 68 degrees (HDD) and 72 degrees (CDD)

Weather normalization is the statistical process of removing the variability of weather from energy consumption. Regression analysis is used to relate energy use to temperature; multiple regression analysis is used to relate energy use to multiple variables such as changes in day/night occupancy and activities. Regression analysis commonly requires highly paid consultants or expensive and data intensive software. Our interest was in producing a simple and automated process which would take the "first cut" at removing the effects of weather. This is the process we have implemented in our live Energy Information System.

The data we work with are 15-minute interval data collected by wireless from a commercial electric meter and temperature data provided hourly on the Internet from the nearest airport. Both sets of data are channeled through the Internet into a web server running an SQL data base. The raw material is normally plotted over the 24 hour period.

The next step is to plot the two variables against each other and to use a simple Microsoft Excel™ regression program to determine a linear relation between the two variables.

The third step is to use the equation developed in the regression to replot the load profile both in its original form and as it has been modified by the regression equation. Analysis of weekend data would require an additional data set because the utilization of the building is different. This may seem like a lot of computation, but it is accomplished quickly, automatically, and out of sight in the data base. To normalize a load profile, the user clicks on "Show Normal Plot." This compares the raw data plot with the processed data on the screen or in

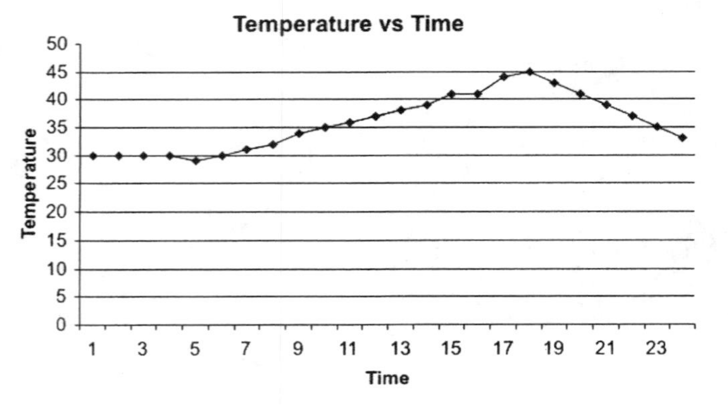

Figure 8-10. Charting hourly temperature from the nearest airport

a report. We provide the normalized data at no additional charge in our EIS platform. The automated process does not cost us any more and we want to encourage our users to plan and implement conservation strategies.

In using this technique we have made several simplifying assumptions. The first is to view the load profile as 96 independent readings (one for each 15 minutes in the 24 hours). This means that our process uses 96 regression equations, one for each 15 minute interval in the day. In this way we regard each interval of the day as a fixed point which varies only according to the temperature, and we assume this variation has a linear relationship. For example, if the actual summer temperature is higher than the normal summer temperature at a specific time of day, then we assume that more cooling energy will be used at that time. At some point a user may want to hire a consultant to do a more complex analysis. This type of analysis would represent the load profile as a quadratic equation with additional variables for each inflection point and would require a year or more of data and powerful analytic tools to give confidence to the results.

We use our simple system because the data loading is automatic and it can give useful results within days. The longer it runs, the more the scatter plots will yield a discernible pattern. Effectively we are measuring the number of lights turned on, the number of computers running, and the processes and machinery running in the building at any given time in the 24 hour cycle. What we remove is the run time of the heating or air conditioning equipment. We can then look at both plots—the "normal load" and the variable load—and devise different strategies to address each. To improve the normal load we need to consider how much lighting or process equipment such as heaters or air compressors may be left idling when not needed because no one takes the time to turn them off. To improve the variable load we need to consider when this load is on, how efficient it is, and whether it could be re-timed to operate off peak. When there is a lack of insulation, consumption will vary more in response to weather. It is also possible to see how a change in operations will affect cost. Holding to a narrow comfort range despite noticeable heat or cold will require more energy. Allowing more variation in response to changes in weather will be more economical. The normalized data will also let us compare what happens on average days and what happens on days of extreme heat or cold. One strategy may be appropriate for the 1900 regular hours of the work year and another strategy for the 100 extreme hours. The comparison be-

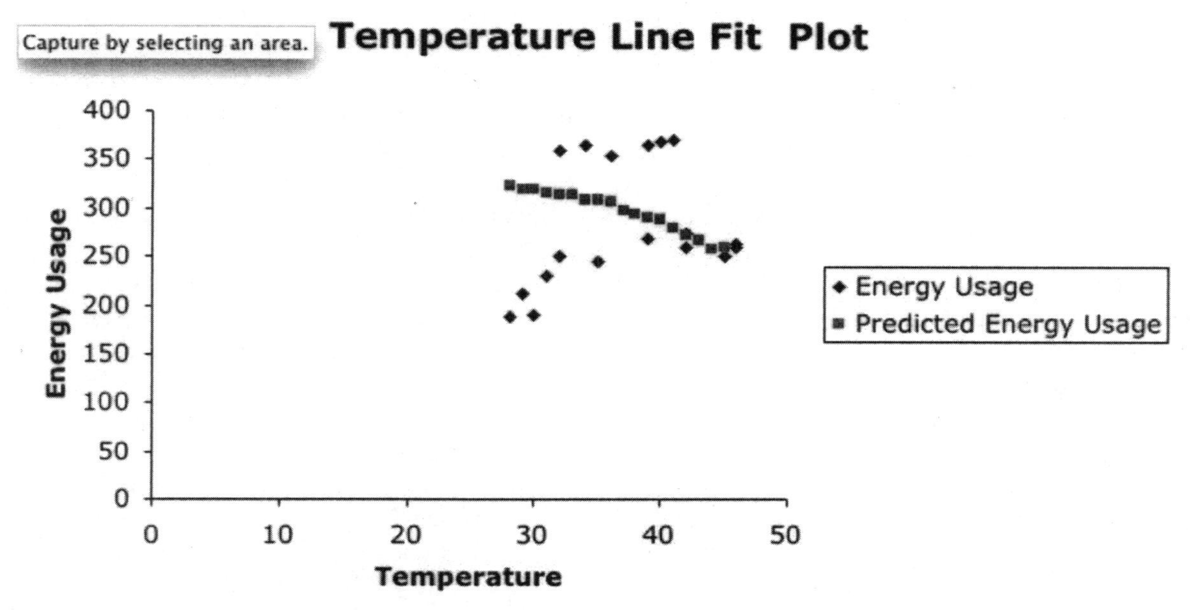

Figure 8-11. Finding the regression equation

tween your regular use and normalized use will begin to point out the places to find savings.

Using Microsoft Excel™ for Regression Analysis

Microsoft Excel™ has some easy to use tools for plotting linear regressions. First you need to define your independent variable and your dependent variable. Put the independent variable (temperature) in the left column of an Excel spreadsheet. Put the dependent variable (kilowatt hours used) in the right column. Select the charting function, select the two data columns with their headings, and use the chart wizard to plot the data. Next select the chart, click on a data point, and then choose Chart > Add Trend line. Excel will produce the trend line and give you the equation for the linear regression. You can experiment with several different trend lines according to whether your data are best represented by a linear, a logarithmic, exponential, or moving average trend line.

In normalizing the utility data based on weather, look for a linear relationship in the data. If your historical data set runs between 30°F and 50°F, you cannot use this data to normalize energy consumption at 60°F. You

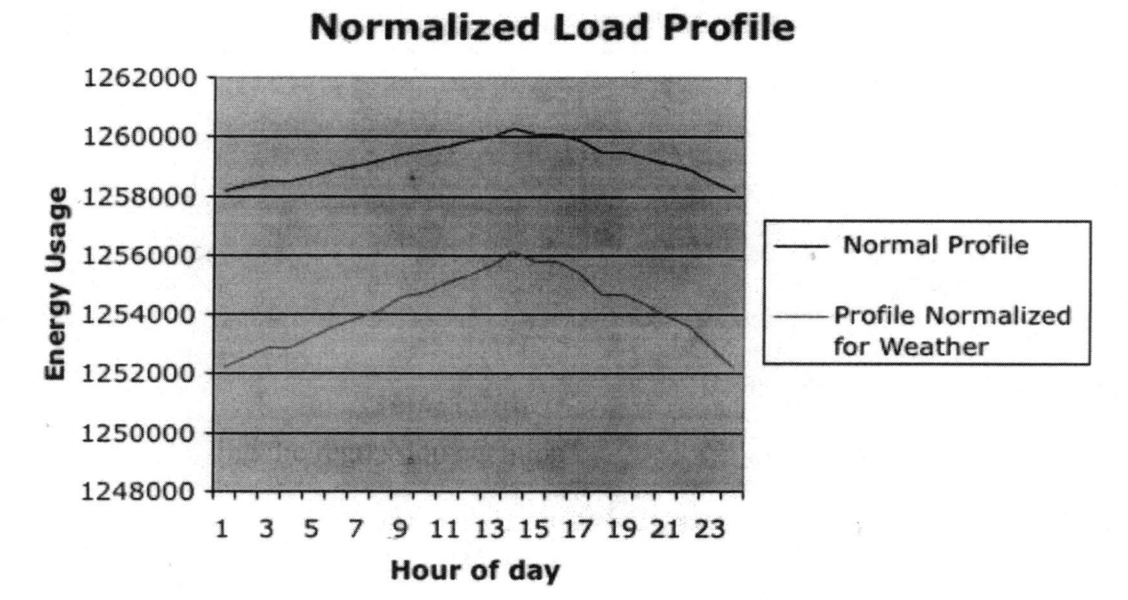

Figure 8-12. The normalized load profile

must stay within the bounds of the data you have collected. If the temperature rises to 60°F, collect data for a few days and make a new plot and use linear regression for that part of the season. We do not try to make one equation fit all circumstances. Rather we apply many different equations over small ranges of our normalization: one equation for weekdays, one for weekends, others for heating season, cooling season, and shoulder months. The software knows what day of the week and season you are normalizing and picks an appropriate linear relation for that period.

Excel also calculates R^2 values which measure the fit between the plotted points and the linear equation. The closer all the points lie to the line, the larger the R^2 value. An R^2 value of 0.0680 means that 68% of the variation in the data is explained by the linear relationship. 32% is due to some other factor plus random error. The R^2 value also shows the "Standard Error of Regression." An R^2 of 0.068 indicates that 68% of values plotted lie within one standard error of regression. Looking at the plots outside this range—the outliers—may also prove valuable as something in the building may be causing unusually high use when the temperature changes. For example, in the shoulder months, the heating and cooling systems may both be running at the same time.

To normalize an interval reading, the software inserts the kW and the temperature into the appropriate equation and returns a value for the "normalized" kW. This is then plotted along with the raw meter data. The narrower the range between the meter data and the normalized data, the less weather effects the operation of the building, and conversely, the wider the range, the greater the effect of the weather. The normalized load profile gives you the energy use baseline for the building. You will know you are saving energy when you make a change and you see a reduced baseline.

CONCLUSION

The amount of waste in electric consumption is large, and those who pay for it can find a better use for their money. Monthly bills are inadequate in pointing out where this waste occurs and how it can be prevented. Inexpensive digital systems are now available, and they can be used with the Internet to provide better information about electric consumption especially in the form of the facility load profile. However the mass market for EISs has never developed. We think that one reason is that bill payers and facility managers cannot interpret the information in a load profile and ascribe little value to it.

We have made a first step in answering this misperception by automating benchmarks and including them as an added feature in an EIS. The user can display the facility load profile and then click down to add additional frames of reference from which to view the data. The Energy Star score shows how the user's facility compares to other facilities of the same type. If your facility receives a percentile score of 30, you know that 70 percent of other facilities of the same type are more efficient then yours. You also know that a lot of energy efficiency capital improvements could be funded from savings in your monthly bill.

Class load profiles offer another perspective on your consumption with regards to the size of your base load, peak load, and coincident demand. A careful comparison of the facility load profile with the class load profile should give you some insight into operational changes that can reduce your facility demand.

Finally, normalization reduces the ever changing variations caused by weather. These changes prevent any meaningful comparison of the May bill to the June bill, or to the May bill for the previous year. Using regression analysis, weather factors can be removed revealing a normalized load. A wide spread between normalized load and the actual load suggests looking for more efficient heating and cooling. A narrow spread means the focus should be on how occupants use the building and whether wasteful practices, policies, or habits can be changed.

The idea of automated benchmarking in the EIS is in its early stages. As facility managers begin to recognize the value of the information they get from automated benchmarking, they will be able to reduce their energy cost and energy use. The technology exists to increase energy conservation by double digits. Providing automated benchmarks is a first step in starting that process.

References

[1] Square D Veris Hawkeye *http://www.veris.com/products/pwr/8000.html*
[2] Ohio Semitronics Inc. WL55 *http://www.ohiosemitronics.com/pdf/wl55.pdf*
[3] *www.energystar.gov*
[4] *http://weather.noaa.gov/index.html*
[5] *www.EIA.gov*
[6] AEE Globalcon Proceedings, Chapter 14, Energy Star, Boston MA, March 2004

Chapter 9

Using WAGES Information for Benchmarking and Operational Analysis

Jim Lewis
Obvius Corporation

ABSTRACT

THIS CHAPTER EXAMINES the use of WAGES (Water, Gas, Electricity and Steam) information as a valuable tool in analyzing the efficiency of operations in one or more buildings and setting priorities for facility investments. Web-based systems make gathering and analyzing this information a simple task for most facility managers.

INTRODUCTION

Most owners and managers of commercial and industrial facilities are familiar with and recognize the value of submetering within their buildings, particularly as it relates to the use of energy consumption and costs for traditional purposes such as:

- Cost allocation to departments or projects;
- Tenant submetering to assign costs to new or existing third party tenants;
- "Shadow" metering of utility meters to verify the accuracy of bills and the quality of the power being delivered;

Many owners are not, however, aware of the potential for using energy information to monitor the performance of both automated and non-automated energy-consuming systems and equipment within the building. Energy consumption information gathered from a variety of sensors and meters within one or more buildings provides not only verification of the efficiency of the equipment, but also the management of the systems by the users.

This chapter examines the practical use of energy information and the hardware and software needed to apply the information to everyday use. The prevalence of Internet access makes this information easily available to most facility managers.

SOME BASIC CONCEPTS

The term "WAGES information" as it is used in this chapter refers to data gathered from a variety of sources within the building that relates to:

1. Quantity of energy and/or water consumed during a particular interval
2. The time period during which the energy was consumed by one or more systems
3. The relationship between the energy consumed, the time it was used and the operation of the building's systems

Measurement of WAGES information consumed during a particular interval, whether from primary meters or from secondary metering and sensing devices is the first line of defense for the building owner wishing to monitor operations. Proper selection and installation of sensors and meters in critical areas provides the most valuable and timely source of operations verification.

THE VALUE OF TIMELY AND ACCURATE INFORMATION

There is an old saying in the business world: **"If you don't measure it, you can't manage it."** To better understand this time-tested adage, imagine yourself in the following scenario: after many years of struggling to make sound investment choices with your IRA or 401(K) money, you have finally decided to invest the money with a stockbroker that your brother-in-law highly recommends. You show up with your check in hand and ask the broker how he is going to invest your funds and how you can follow the progress your money is making.

He informs you that he believes in a diversified portfolio, with a mix of stocks, bonds, CD's, etc. to spread the risk of your investment. This sounds very encouraging of course, and a very practical way to invest, but you feel compelled to ask what tools are available for determining the performance of the different investments to make adjustments that may be appropriate as time goes on and circumstances change. In other words, how are you and the broker going to measure the performance?

The broker informs you that the way he handles these things is to look at the total amount in the account at the end of the month (give or take a week or two), subtract the total amount at the start of the month and then calculate the return on investment for that month.

> "Sounds simple enough, but how will we know what portion of the gain (or loss) came from bonds or stocks or whatever?" you ask.
>
> "You won't," is the broker's simple reply.
>
> "Okay, then how will we measure the performance of our stock portfolio versus, say, the Dow or NASDAQ or some mutual funds?"
>
> "You won't," he replies again.
>
> "So what you're saying is that we lump everything in the portfolio together and take a once-a-month snapshot that only tells us how much we have gained or lost, but not how we got there?"
>
> "That's about it…"

Of course, in this day of on-line investing, real-time quotes and Web access to financial information, no investor in his or her right mind would even remotely consider investing under this scenario. We take for granted the ability to monitor any or all of our investments and to make changes as necessary to insure the best bang for the buck. In other words, measuring the performance in a timely and useful manner allows us to manage the assets.

Or consider the following: how many of us would be willing to simply send our children off to school the first week of September and be content to get no information on their progress till May? To make matters worse, imagine that the only report card you got in May showed simply the average grade for all your child's classes combined, with no indication of what the grade was in math or science.

These examples seem ridiculous to us in today's world, but many building owners and managers find themselves in a similar situation in managing the operations of the buildings. The only way to know if something isn't working properly is when the failure results in uncomfortable conditions for the building occupants or when the monthly utility bills arrive and are significantly higher than expected.

Remember, **if you don't measure it, you can't manage it**. Using utility bills that arrive weeks after the energy was used and lump all of the building's systems into one account is hardly a timely and accurate measure of building performance and efficiency.

When your August electric bill is 20% higher than your budget, how do you determine the cause? Was it lights left on, HVAC systems operating inefficiently, new equipment installed in a tenant space, hotter weather than normal or some other cause?

Demand charges often make up half of your electric bill. When does the demand occur? Is it a short term spike or is the demand profile relatively flat? Is it something you can manage to limit the cost?

When your November gas bill is 30% higher than last year, was it because the weather was colder or did the hot water heater fail to shut off night? Is there a leak in the system, or are you simultaneously heating and cooling occupied spaces?

How much of the steam your central plant produces is lost to leaky steam traps? How much of the water your building uses goes to irrigation and landscaping and how much is used for makeup water to the boiler?

Most municipalities bill for sewer charges based on the water bill. As sewer charges become a greater part of the WAGES cost for many buildings, how much of the water you bring in actually reaches the sewer? How much goes to landscaping or cooling towers or evaporation from water features?

HOW DOES WAGES INFORMATION HELP IN OPERATIONS?

The previous section of this chapter touched on several operations-related issues that many building owners and managers are concerned about, but how does gathering information about energy and other utilities contribute to more efficient building operation? We will look at two examples from real buildings being monitored by Obvius through its web based monitoring system.

Example One—Retail Store

The first example is a retail store located in the Northeastern United States. This example provides a very clear case for the value of submetering and timely monitoring of electricity usage. The customer became concerned about electric bills that were higher than historical usage and called in a consultant to review the building's operations and make recommendations.

After looking at the operations, one of the primary opportunities identified for saving electricity was in the lighting systems for the warehouse/operations area of the store. The lighting consisted of a mix of incandescent and fluorescent tubes and it was determined that based on the operating hours of the store, a conversion to more efficient lighting would generate significant savings and an attractive payback.

The owner of the store and the energy consultant determined that since this retrofit would likely be used as a pilot project for all of the company's stores, it would be a good idea to provide some measurement and verification of the savings realized. The energy consultant provided two alternatives:

1. A snapshot view of the consumption before and after using simple hand-held tools like a multimeter and an amp clamp; or
2. Installation of a monitoring system to measure the actual power consumed by each of the circuits on 15-minute intervals. This option, while more expensive than the first option, would obviously provide much more accurate feedback on the success of the installation and would also have the added benefit of providing near real-time access to the data using a Web browser.

After reviewing the two options, the customer decided to install electrical submeters on the lighting circuits before the retrofit to establish a baseline and then to leave the submeters in place for a period of time after the installation to verify the exact savings realized from the changes. The meters would be monitored and interval data would be recorded with date/time stamps using an AcquiSuite Data Acquisition Server (DAS) from Obvius (see Figure 9-1).

The data gathered by the DAS would be sent each night to the Building manager's on-line website (*http://www.buildingmanageronline.com*) hosted by Obvius so that the data would be available the next morning via the Internet.

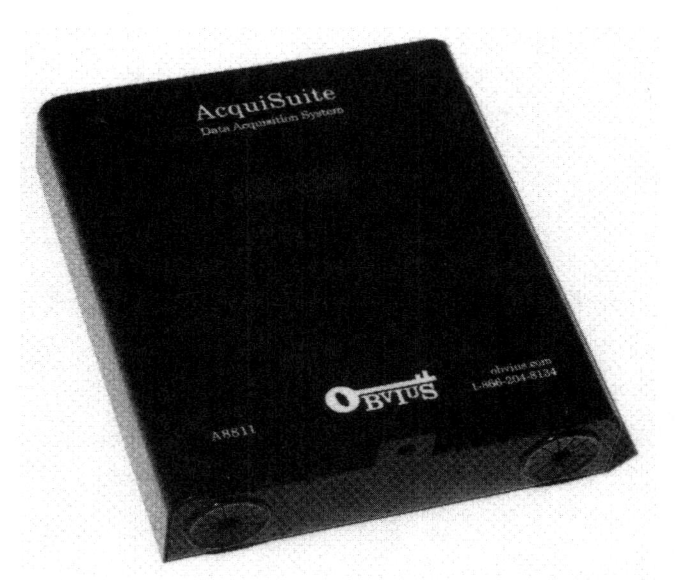

Figure 9-1. AcquiSuite Data Acquisition Server

The installation of the monitoring required the following:

- AcquiSuite DAS (A8811-1) from Obvius to monitor and record the data from all the submeters;
- Enercept submeters (H8035) from Veris Industries (see Figure 9-2) to connect to each of the 12 lighting circuits to be monitored;
- Ethernet connection to the existing store LAN to provide a path for sending the data to the host server;
- Labor to install the electrical devices and the wiring

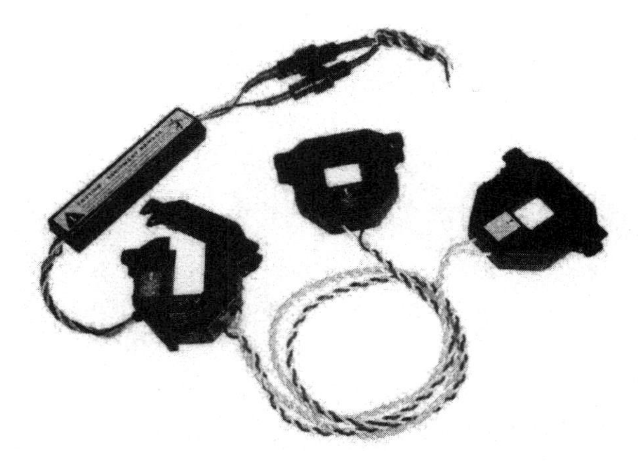

Figure 9-2. Retrofit electrical submeters

The meters were installed and connected via an RS485 serial cable to the DAS. The DAS gathered data from each meter and stored the kW information on 15 minute intervals in non-volatile memory. Every night, the data was uploaded to the Building Manager On-line (BMO) server site where it was automatically stored in a MySQL database. Once the time-stamped data was stored in the database, it could be accessed by any authorized user with a Web browser such as Internet Explorer.

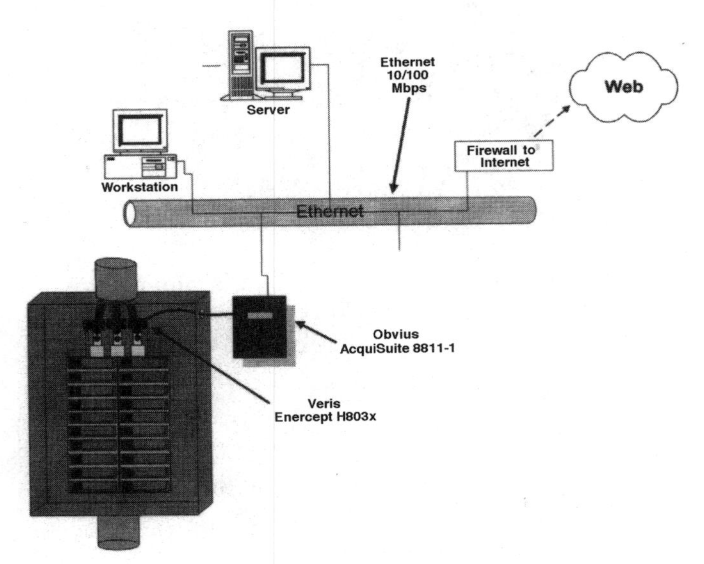

Figure 9-3. Meter installation and network connection

The monitoring system ran for several weeks to establish a baseline for energy usage before the lighting retrofit was done, and then ran for a period of time after the installation to verify the savings from the installation. To begin the verification process, the energy consultant and the store management reviewed the kW data for the month that included the benchmark period, the installation period and a few days after the installation. The data they saw are shown graphically in Figure 9-4.

This graph represents approximately 15 days (Feb. 1 through Feb. 15) and thus has just over 11 days of pre-retrofit (baseline) data and another 4 to 5 days of post-retrofit data. As the graph clearly indicates the reduction in energy usage by this lighting circuit (one of 12 modified in this retrofit) was almost 70%. Needless to say, the results were very well received and both the owner and the energy consultant were quite pleased to see just how much energy had been saved.

After the excitement over the results died down, however, the owner noticed an interesting anomaly. This was a retail store with typical retail hours (approximately 10 AM to 9 PM), but the graph would appear to indicate that the lights were operating 24 hours per day every day. Reviews of the graphs for the other 11 lighting circuits showed a similar pattern, with each showing

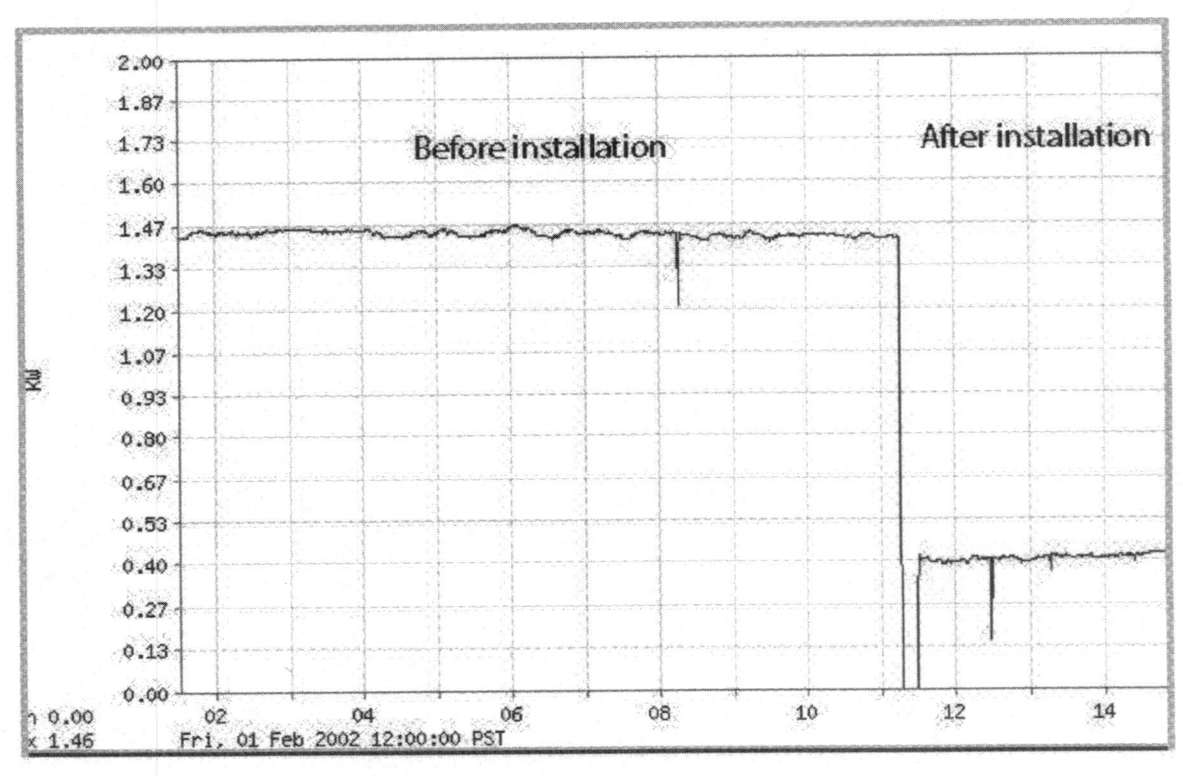

Figure 9-4. Lighting retrofit energy graph

significant reductions in loads, but also showing that the lights in the non-retail areas of the store were all running continuously.

The electrical contractor was sent back to the store and found that the control panel that operated the lighting circuits was programmed correctly to match the occupancy hours, but the relay in the panel designed to turn the circuits on and off had failed and was continuously on. The lighting relay was replaced and the next read of the graph showed the pattern in Figure9-5.

As this second graph clearly indicates, once the relay was replaced and the control panel functioned properly, the pattern of energy usage alternates between normal consumption (on) and no consumption (off), matching the operating hours of the store itself. Since this is only a 1.5 kW load, it might at first glance appear that the cost of leaving the lights on is not significant, but let's calculate the cost of leaving the lights on for 10 hours per day when the store is unoccupied:

$$(1.5 \text{ kW}) \times (10 \text{ hrs/day}) = 15 \text{ kWh per day}$$

$$(15 \text{ kWh/day}) \times (\$0.10/\text{kWh}) = \$1.50/\text{day}$$

$$(\$1.50/\text{day}) \times (365 \text{ days/yr}) = \$547.50/\text{yr}$$

$$(\$547.50/\text{yr/circuit}) \times (12 \text{ circuits}) = \mathbf{\$6{,}570/yr}$$

The wasted energy in this example is only part of the total waste since it does not reflect the reduced life of the bulbs and ballasts or the added cost of cooling the building due to the heat from the lights.

This example provides a very clear indication not only of the value of monitoring energy consumption to monitor operations, but also just how easy it is to spot and correct malfunctions. Anyone reading this graph can immediately spot the problem (although not necessarily the root cause) without the need for sophisticated analytical tools or experience in energy analysis or engineering. The owner knew that his store did not operate 24 hours a day, but a quick glance at the data in this chart led him to question why the lights were on all the time. It's this sort of rudimentary analysis that provides a significant portion of the savings on energy. The cost for installing the hardware for metering this circuit was approximately $1,800.

This example highlights the value of energy monitoring on several levels, any or all of which may be important to the building owner:

1. **Highlighting of problems**—as we see in this case, the owner was able to identify incorrect operation of the lighting systems relative to the operation of the store

2. **Verification of energy savings**—the other key point to this graph is that it shows the actual savings this retrofit generated. If the owner had relied solely on the utility bills to verify savings, the savings from the retrofit would have been overstated by more than $6,000 because it would have included the savings attributable to correcting the existing problem with the control panel. In this case, the owner was planning to use the data from this pilot project to determine the payback of rolling out a similar retrofit to hundreds of stores and the overstated savings would have been impossible to duplicate in other locations.

3. **Supervisory monitoring of control systems**—in this case, the lighting control panel appeared to be

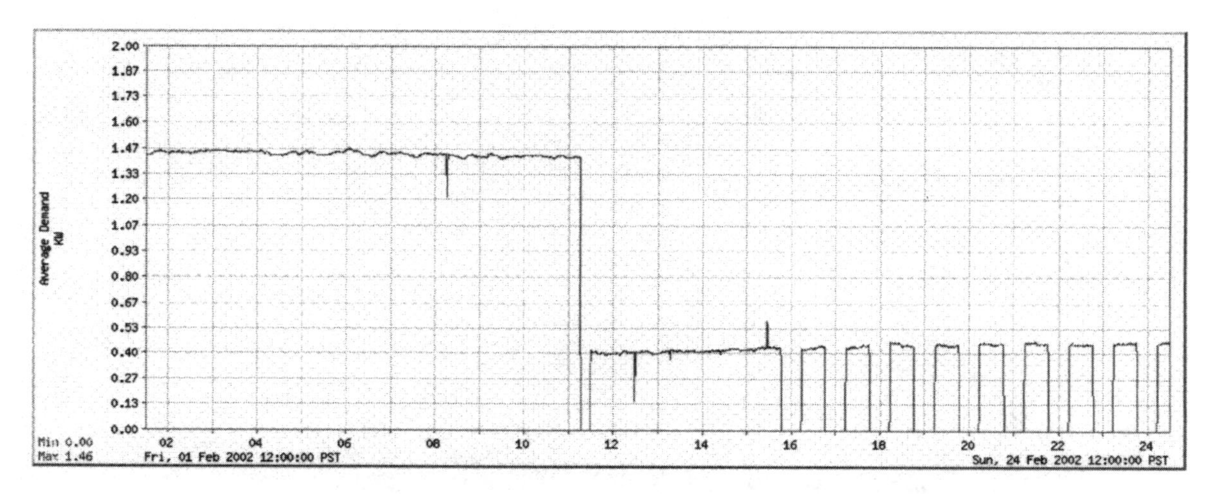

Figure 9-5. Lighting energy after relay replacement

functioning as designed, and the cursory review by the electrical contractor prior to the retrofit gave no indication of any problem. It was only after the issue was made clear by the energy data that the problem was identified and corrected

In the example of this store, everyone assumed that the system was operating properly because the lighting control panel in the store was properly programmed, and no one was concerned (or accountable) for the lights remaining on when they left. The only way this problem was identified and corrected was through the use of proper monitoring equipment.

Example Two—Fast Food Chain

Another example of the use of graphical energy data to monitor operational activities comes from a fast food restaurant in the Northeast. In this case, the data are relatively new and there is little information about cause and effect, but the data from this restaurant provide insight into how the energy manager can use energy consumption as an indicator of potential operational problems.

Figure 9-6 below shows the electrical demand for one panel in a fast food burger chain. This is just one of several meters located in the facility and does not include any HVAC or hot water heating loads, all of which are monitored on other meters.

This graph provides an excellent illustration in the starting point for the energy or facility manager using energy data as an indicator of potential operations problems. What we see is one week's electrical demand profile for one panel in the restaurant. The first thing that this graph makes apparent is that there is a significant (i.e., more than double) increase in the kW on Wed afternoon (right side of graph). The normal load of about 8 kW that is fairly consistent throughout the rest of the week suddenly jumps to nearly 20 kW for at least several demand intervals on Wednesday.

The first thought would be that this might be weather-related, but this is unlikely since we already know that this particular circuit does not serve the HVAC systems and thus is unlikely to be the result of an A/C compressor kicking on or resistance heat in a package unit turning on. At this point, a little detective work would be in order, starting with the usual suspects:

- Freezers/coolers—if the freezers are on this circuit, is it possible that the doors were left open for some period? Perhaps the deliveries of frozen goods are always on Wednesday afternoons and the delivery personnel leave the door propped open. If so, it might be worthwhile to make adjustments to the way deliveries are handled or to install additional

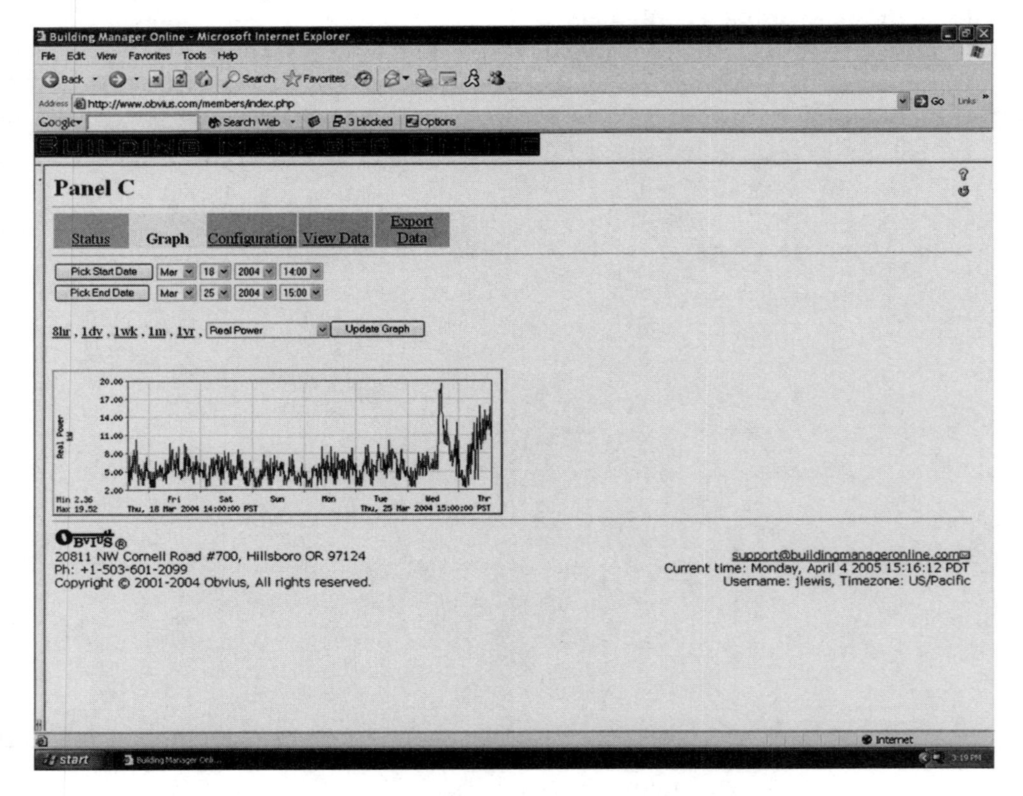

Figure 9-6. Electrical demand for fast food chain

hardware to permit easy entry to the freezers, but still save energy

- Fryers—is it possible that the fryers were being cleaned and the oil changed? If so, it might be that all the units were starting cold and came on at full demand at the same time. If this is the case, a new procedure to stagger the starts might solve much of the problem
- Maintenance/renovation—is there construction or remodeling work going on that might have caused the spike?
- Failed controls—are the controls for some high electrical demand equipment failing to limit the demand? Has the staging failed in some kitchen equipment, causing the equipment to come on at full load?

In the previous example, the concern was over consumption (kWh) charges that develop from having a low demand load that is left on for extended periods. In this case, however, the problem relates to demand (kW) charges that the utility levies for the maximum electrical demand during an interval (typically 15 minutes). Unlike kWh charges which are relatively low and build up to significant amounts over time, the demand charges are incurred in just a few minutes and are typically several dollars per kW. In this case, the customer could be facing a penalty of $40 to $50 per month (or more) for the next 12 months as a result of the demand from this one panel.

It is also entirely possible that the higher demand is in fact typical for this panel and that the lower kW seen earlier in the week was the result of equipment that was down for repair or some other explanation. The key point here is that the energy data does not explain what happened, but provides clues that can then be related in a timely manner to the operation of the facility and potentially save significant dollars.

HOW DOES IT WORK?

In order to get the information from the building and make it available in a user-friendly Web format, there are two major pieces that must be in place:

1. On-site hardware and software—each facility being monitored needs to have the meters and other sensors to actually measure the desired parameters and a data acquisition device to gather and store the data and to communicate it to a remote Web server
2. Application service provider (ASP)—the ASP provides two primary functions: first, an Internet accessible connection to each of the locations being monitored for uploading the interval data; and second, a Web based user interface for viewing information from all the locations using a Web browser such as Internet Explorer or Netscape.

ON-SITE HARDWARE AND SOFTWARE

The equipment required for specific monitoring projects will obviously vary depending on the application. Each job will require some type of data acquisition server or logger to log data from the meters and sensors and to communicate with a local or remote host server to upload the interval data gathered. Some representative applications and the hardware required:

1. **Monitoring existing utility meters with pulse outputs**
   - AcquiSuite or AcquiLite data acquisition server[1]
   - Ethernet or phone line

2. **Monitoring submeters**
   - AcquiSuite data acquisition server;
   - Ethernet or phone line
   - Electrical submeters with Modbus output for each circuit or system to be submetered
   - Gas meters with pulse or analog (4 to 20 mA) output
   - Flow meters with pulse or analog output

3. **Monitoring chiller plant efficiency**
   - AcquiSuite data acquisition server;
   - Ethernet or phone line
   - Electrical submeters with Modbus output for each chiller to be monitored
   - Electrical submeters with Modbus output for each cooling tower to be monitored
   - Electrical submeters with Modbus output for each chilled water supply pump system be monitored
   - Liquid flow meter
   - Chilled water supply temperature sensor
   - Chilled water return temperature sensor

[1]Acquisite and Acquilite are trade name products of the Obvius Corporation.

4. Monitoring indoor air quality

- AcquiSuite data acquisition server;
- Ethernet or phone line
- Analog (4 to 20 mA) output sensors to monitor:
  - — Temperature
  - — Humidity
  - — NO_x
  - — CO
  - — CO_2

Since the DAS is capable of monitoring any analog output device, the installation can be readily customized to meet the needs of a particular application including:

- Submetering of loads within the building for cost allocation, verification of operations or tenant billing;
- Runtime monitoring to verify that loads are shutting down when scheduled
- Monitoring flows such as chilled water to allocate costs from central heating and cooling plants to individual buildings
- Monitoring flows to sewer services to verify sewer charges based on water usage
- Measurement and verification of energy saving retrofits
- Supervisory monitoring of facilities to benchmark energy usage and provide feedback on operations

The on-site software required is typically contained in the DAS and should require little if any customization. Since the DAS is a Linux-based Web server, interface requires only a standard PC on the network with the DAS. Any changes or modifications are made using only a web browser (see Figure 9-7).

Figure 9-7 shows the typical setup page for a DAS. Using any web browser, any user with supervisory access can change alarm settings or make other adjustments (logging intervals, equipment names, multipliers, etc.) either locally or remotely. These changes can be made either directly to the DAS on the network or remotely through the ASP site hosting the data.

One of the major changes in installation of submeters today is the ability of the DAS to automatically recognize most meters (the metering equivalent of plug and play in the computer world). This capability (written into the software in the DAS) means that the

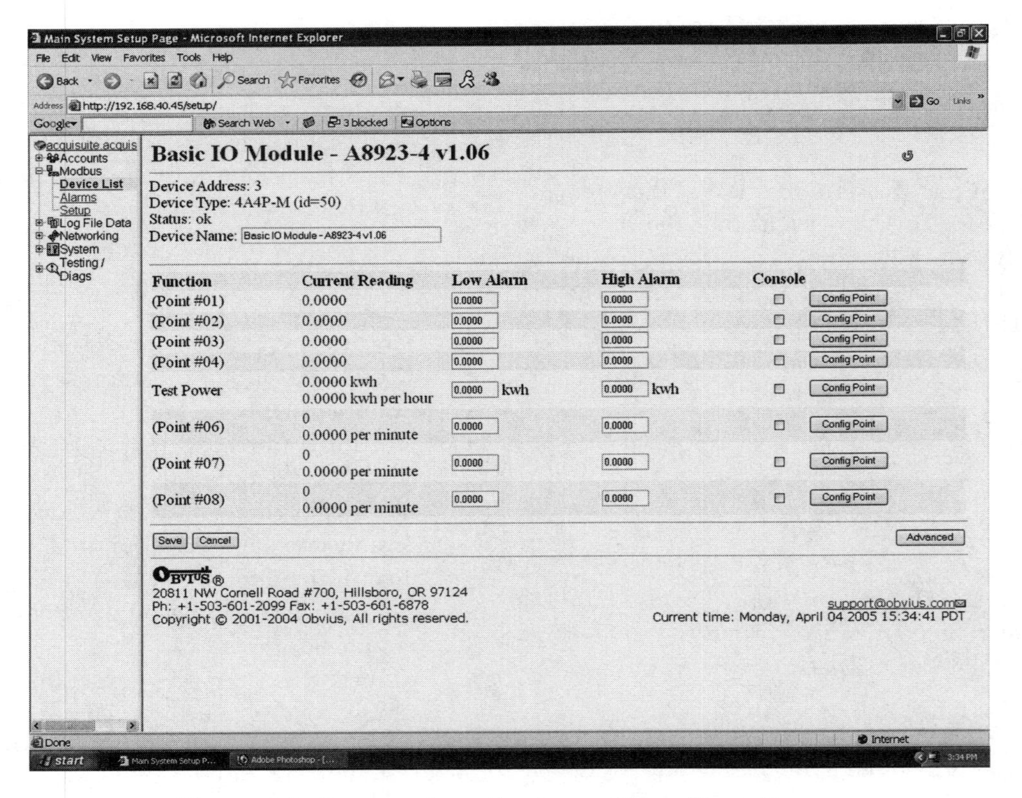

Figure 9-7. Browser based setup and configuration

installer does not have to map the points from each of the different meters, providing a significant savings in labor since many meters today have 60 or more points. The plug and play connectivity also means that meters from multiple manufacturers can be combined on a single serial port to meet different needs.

As an example, in a campus environment, there are needs for several different types of electrical meters within the campus. Primary service will typically include requirements for measuring not only energy consumption, but also power quality information such as power factor, phase imbalance and harmonic distortion. The DAS has the flexibility to automatically recognize a variety of meters, from high-end power quality meters to simple submeters for measuring kW and kWh.

For example, the primary service at a campus might have a meter with a variety of inputs such as the example below:

In contrast, the same campus would likely deploy a number of less expensive energy-only meters where the purpose is primarily to monitor energy consumption and provide cost allocation to departments and third party occupants:

Once the setup and configuration of the DAS is complete, the unit logs energy or power information on user-selected intervals, and that information is stored by the DAS until it is uploaded to a remote or local server for viewing and interpretation. Upload options include automatic upload via modem or LAN or manual upload using a Web browser on the network with the DAS. Locally uploaded data can be stored and viewed with a variety of standard programs including any ODBC database or spreadsheet programs such as Excel. In the example shown below, interval data for one month is displayed in an Excel chart.

The information contained in the chart above can be used in a variety of ways:

- Cost allocation of consumption and/or demand charges if this is a submeter for a tenant or department
- Comparison to historical demand profiles to indicate changes over time that might indicate that operational inefficiencies are developing
- Prioritization of capital expenditures to reduce demand profiles and costs

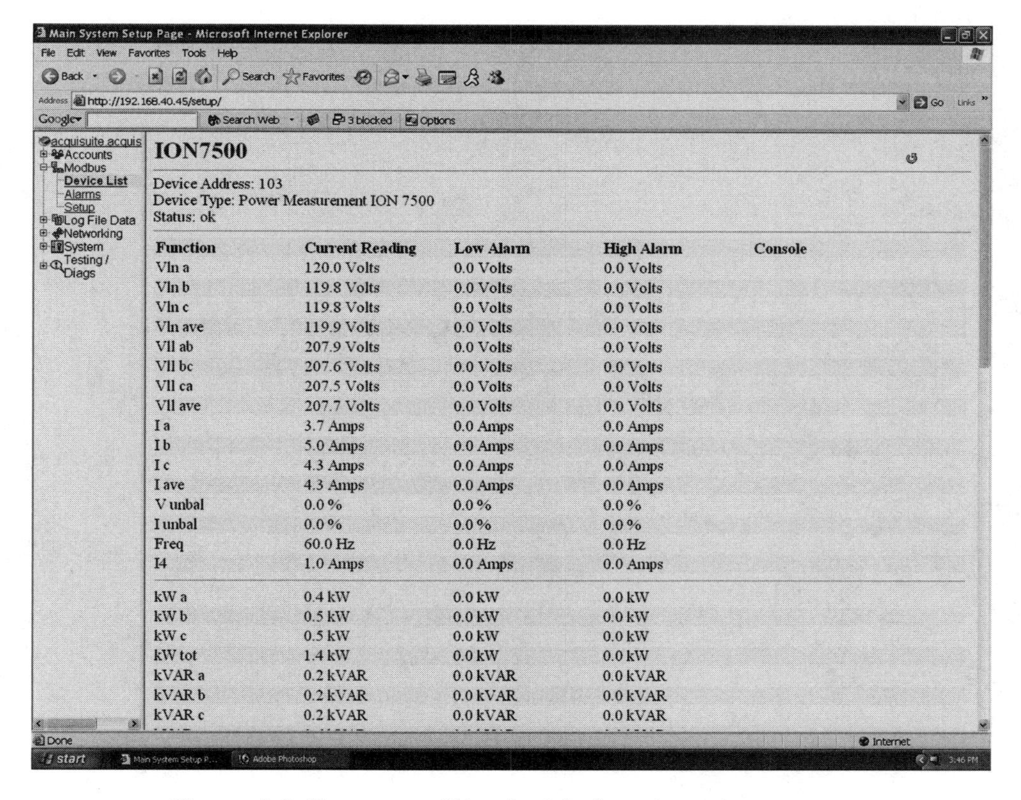

ION7500

Device Address: 103
Device Type: Power Measurement ION 7500
Status: ok

Function	Current Reading	Low Alarm	High Alarm	Console
Vln a	120.0 Volts	0.0 Volts	0.0 Volts	
Vln b	119.8 Volts	0.0 Volts	0.0 Volts	
Vln c	119.8 Volts	0.0 Volts	0.0 Volts	
Vln ave	119.9 Volts	0.0 Volts	0.0 Volts	
Vll ab	207.9 Volts	0.0 Volts	0.0 Volts	
Vll bc	207.6 Volts	0.0 Volts	0.0 Volts	
Vll ca	207.5 Volts	0.0 Volts	0.0 Volts	
Vll ave	207.7 Volts	0.0 Volts	0.0 Volts	
I a	3.7 Amps	0.0 Amps	0.0 Amps	
I b	5.0 Amps	0.0 Amps	0.0 Amps	
I c	4.3 Amps	0.0 Amps	0.0 Amps	
I ave	4.3 Amps	0.0 Amps	0.0 Amps	
V unbal	0.0 %	0.0 %	0.0 %	
I unbal	0.0 %	0.0 %	0.0 %	
Freq	60.0 Hz	0.0 Hz	0.0 Hz	
I4	1.0 Amps	0.0 Amps	0.0 Amps	
kW a	0.4 kW	0.0 kW	0.0 kW	
kW b	0.5 kW	0.0 kW	0.0 kW	
kW c	0.5 kW	0.0 kW	0.0 kW	
kW tot	1.4 kW	0.0 kW	0.0 kW	
kVAR a	0.2 kVAR	0.0 kVAR	0.0 kVAR	
kVAR b	0.2 kVAR	0.0 kVAR	0.0 kVAR	
kVAR c	0.2 kVAR	0.0 kVAR	0.0 kVAR	

Figure 9-8. Power quality electrical meter setup screen

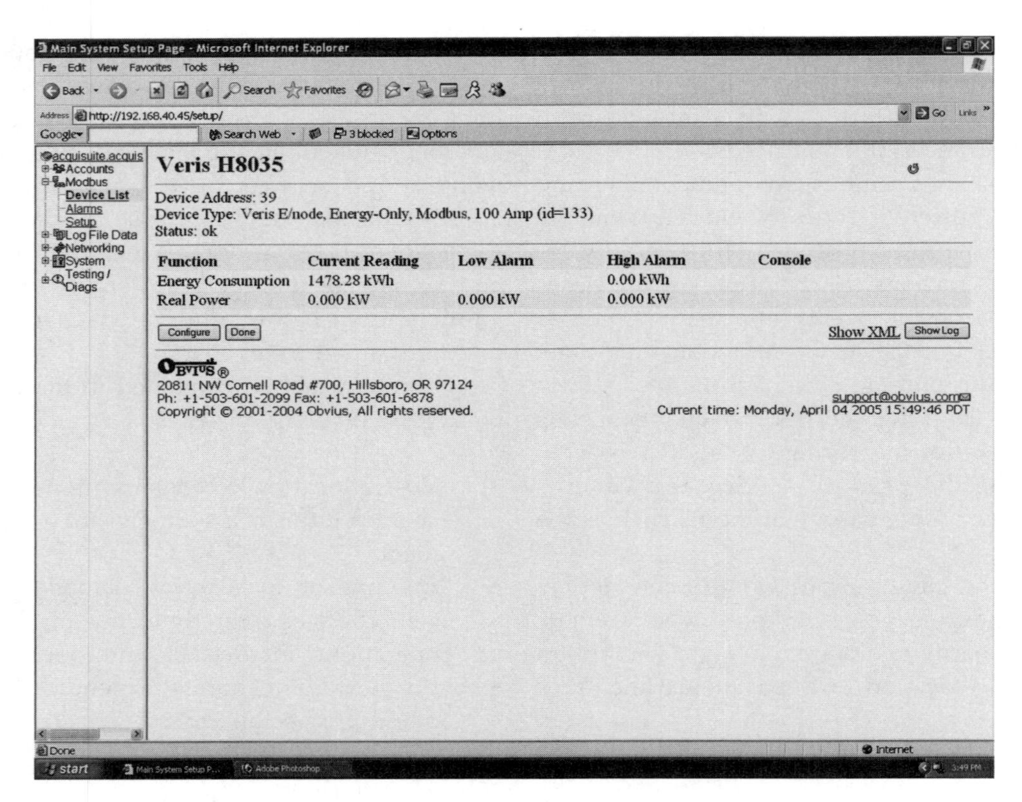

Figure 9-9. Energy only submeter (demand and consumption)

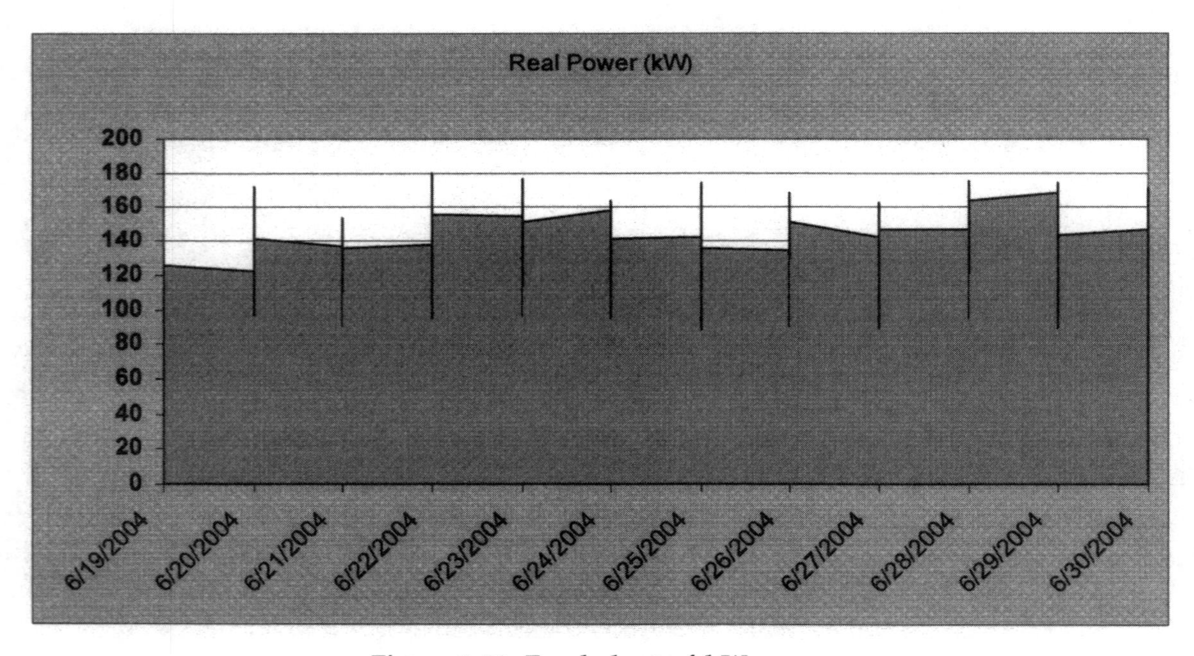

Figure 9-10. Excel chart of kW usage

PUTTING IT ALL TO WORK

Now that we have seen how data can be gathered and displayed, the key question is: what do we do to turn this information into actions that will save energy?

For most facility managers, the answer is best illustrated in reviewing four strategic approaches to the use of energy information. It is important to note that each of these strategies builds on the prior strategies and thus can be implemented in turn as part of an overall energy strategy:

1. **Installation of submetering equipment**—the key here is that not only are the meters installed, but the information gained from these meters is reviewed and communicated to building users responsible for turning off lights, HVAC, computers, etc. It has been conclusively shown that if occupants know that the usage is being tracked, energy reductions of up to 2.5% will be realized (this is generally known as the Hawthorne effect).

2. **Allocation of costs to tenants and other users**—The next logical step is to add a reporting component to the monitoring and begin to hold tenants and other occupants accountable for the energy use in their space. Whether this cost allocation is to in-house departments or third party tenants, the presentation of a "bill" for energy usage will prompt further reductions in energy usage as occupants can associate a cost to their activities. The expected savings from this approach are typically 2.5 to 5.0%.

3. **Operational analysis and performance reviews (the Building Tune-up Process)**—The next level of implementation involves a regular, comprehensive review of the performance of the equipment in the building, with particular emphasis on scheduling and occupancy. Questions to be asked and studied include whether the control systems and HVAC equipment are functioning as designed, whether the schedules for occupancy match the actual usage, and whether there are unusual loads or demands on the system. Using this information, the operation of the facility can be fine-tuned to provide occupant comfort with a minimum of waste. This data can also be used to identify areas for potential investment in energy retrofits and to pinpoint opportunities for maintenance and repair in a timely manner. Savings from the BTU process are generally in the 5.0 to 15.0% range with a very small investment in time and money.

4. **Continuous Commissioning**—Originally developed by the Energy Studies Lab at Texas A&M, the concept of Continuous Commissioning carries the use of energy information to the next level. This strategy involves the use of energy information as a tool for continuously refining the operations of building systems for maximum efficiency. Many (if not most) systems are installed and commissioned in a less than optimal manner, and then are generally not fine-tuned to meet the needs of the occupants with the best efficiency. Continuous Commissioning uses energy information in conjunction with specialized software to identify and correct deficiencies in operations on a regular ongoing basis with the involvement of building personnel and occupants a key element. Studies have shown that Continuous Commissioning provides savings of up to 45% of a building's energy usage as compared to buildings where no monitoring is performed.

WHERE DO I START?

Hopefully this chapter has provided at least a prima facie case for the value of gathering and using energy information for operational analysis. For the building owner or manager contemplating an energy information program, but not sure where to begin, there's good news on the technology front. Historically, the investment in time and money to implement a program like this was substantial and provided a significant hurdle to getting started as design, installation and integration costs were prohibitive. Changes in technology today allow the building owner to take a "do-it-yourself" approach to energy information and to use existing building resources for implementation. Highly scaleable hardware and Internet-based data hosting make installation of submetering products a project that can be accomplished by in-house personnel or any local electrical contractor.

For most building owners, the scalability of these systems means that they can start with the most valuable metering projects and expand the system as needed and the minimal investment required allows them to do a test program with minimal risk. In general, the best approach is to begin submetering at the highest levels and then add additional metering equipment as savings opportunities are identified. For example, the facility manager of a campus would likely start by submetering each building, identifying those facilities where energy use is highest and adding additional submeters to those buildings to isolate and correct problems.

Energy information is a valuable tool and the most important first step in any energy strategy. As stated earlier, **"If you don't measure it, you can't manage it."**

Chapter 10

The Power of Energy Information: Web-enabled Monitoring, Control and Benchmarking

Rahul Walawalkar, CEM, CDSM, Carnegie Mellon University
Bruce Colburn, Ph.D., P.E., CEM, EPS Capital Corp.
Rajesh Divekar, M.Tech., AI Systems

ABSTRACT

WITH THE RAPID expansion of the use and capabilities of the internet, it is only natural that the allure of remote data monitoring and control for multiple locations using vast quantities of data has evolved. This can be implemented as part of a comprehensive energy management strategy by facility owners with multiple sites. New software techniques and display systems are discussed for EMCS and industrial SCADA units and advanced meter reading equipment for utility billing. In the past energy managers had to be content with monthly energy billing data as the sole basis for determining energy optimization strategies. Now with the advent of web-enabled real-time solutions, facility energy managers will be able to modify the energy optimization strategies to respond to supply side cost reduction opportunities as well as the facility's normal demand side management features. Impacts of advanced meter reading for utility charges and the technology available today are presented, along with case studies. From this work, the potential value for monitoring for greenhouse gas emissions and predictions of future technology trends are presented.

INTRODUCTION

Due to the explosion of technology solutions in remote data transmission, storage and analysis, there has been a quantum leap in available technology solutions for Energy Managers to assist in operating cost control. Whereas some years ago there was a trend towards automating the office buildings and factories by tying electronic systems together, today due to security concerns there is a push towards independent systems for protection. Until recently energy managers had to use reactive energy optimization strategies, because the only energy data available to them was the monthly billing data made available 2-6 weeks after the actual energy was consumed. This is changing rapidly with the surge in web-enabled IT solutions that provide real-time energy information on the facility demand side as well as from the utility supply side. The Energy Manager today can have an energy management control system (EMCS) or Supervisory Control and Data Acquisition (SCADA) unit for each of their facilities. Such systems seem to constitute the bulk of the building and industrial controls today, due to the distributed nature and reduction of installation costs; these generally form the basis of local facility management capabilities for the majority of operations on the demand side. Over the past few decades the control and reporting capabilities of all EMCS have steadily improved. Now, with the deregulation of the energy industry in some states, the energy managers also have access to the supply side energy information (wholesale real-time energy prices, automated meter data and price responsive load management signals) that can open up new avenues for optimizing the energy costs for a facility. This energy information exchange is facilitated through new advanced meter reading (AMR) infrastructure. Still, at the core of the ability to optimize the energy costs is the requirement to understand the fundamental energy information and utilize it for fine tuning the facility operation. It has been said that "it is difficult to manage what you do not measure." [1] Now, due to the drop in electronic metering point costs, it may be possible to gather too much data for direct evaluation, so new software preprocessing tools may be required. Until now the trend has been that more points are better, since the total cost to implement an EMCS or SCADA has continued to drop. To assist operations personnel, new software diagnostic tools and information display systems have been developed in order to "interpret" the mountains of data available.

The information needs of energy customers are increasingly complex. They include load usage profiles in real-time, automated reporting, energy benchmarking tools, alarm functions, guidance on market developments, assistance in bidding and negotiating energy supply offers. Providing customers with this information presents great opportunity for energy companies and challenges energy managers to become informed consumers. With better information and the ability to communicate with their customers and influence energy usage patterns, utilities can optimize their energy supply/delivery planning, operations and provide new choices for their customers. One of these choices is price responsive load management (PRLM), under which the end user is both informed of costs and provided opportunities for controlling load through two-way communication (something not widely available on a cost-effective basis in past years). Empowering the management of these Demand Response (DR) resources is the Internet and on-line trading, both of which have brought "market transparency" to the forefront in the electric power industry. On-line energy trading for both electricity and gas is attractive to end-users due to its speed and efficiency. The increased use of these on-line trading systems allows for more liquidity and price transparency and provides energy companies the ability for superior real-time risk management. The Internet gives energy managers the ability to view real-time price signals and the facility's current energy usage. This makes energy managers more sensitive to perceived differences in the mix of utility cost variables and allows them to make more informed operational decisions for managing their energy costs. It also intensifies competition between energy suppliers, making deregulation work as envisioned.

DATA UTILIZATION METHODS FOR ENERGY MANAGERS

Web-enabled IT solutions have already found application in a wide spectrum of tools that enhance performance in Planning, Operations, Maintenance and Training. The need for such information systems is highlighted in the context of the growing consciousness about energy conservation and sustainable development. These systems are developed using a host of enabling core technologies such as those related to Data Communications, Centralized and Distributed Databases, Decision Support, Real-time Computing, Power System Analysis, Geographical Information Systems, Graphics and Multimedia, Distributed Process Control, Simulation, Forecasting and Enterprise Resource Planning (ERP).

Following the installation of a SCADA system in an industrial plant or EMCS at multiple commercial sites, the effective and simultaneous monitoring and control of an entire campus as well as individual facility presents additional energy cost control opportunities [2]. In the past, this wide array of real-time monitoring and control capability was not generally cost-effective. As simple an idea as this is, it has often not been accomplished due to various techno-commercial obstacles. Reaping these cost savings benefits requires planning and a vision from energy managers. Web-based controls and monitoring capabilities can provide the larger users a forum for knowledge sharing across the locations from which all affiliated sites can benefit from better management practices and benchmarking [3]. Experienced energy managers know that sharing energy consumption information with end users helps to achieve better control over energy costs through the natural competition as well as knowledge sharing that results from such an exercise. A basic benchmarking approach is shown in Figure 10-1 which follows the practical steps in gathering, analyzing and ultimately acting upon the results. Whether a site actually uses the web for closed loop control or not, the simple act of information sharing in real-time (or near real-time) as well as the ability to interact with users can easily lead to better energy and cost control. A good argument could be made that direct, closed loop control with the Internet as the communication medium can be advantageous in terms of hierarchical control, but may or may not be a good idea in isolation.

In using the Internet or an intranet as part of an information processing system, there are a number of practical issues to be considered with long distance energy systems control and monitoring:

1. Problems with internet connectivity can cut off information interchange for short durations. Any control system must therefore be robust in operating with missing data for some periods

2. Unless properly designed to take care of security issues, the web is not a secure access medium and data streams could be hacked or attacked

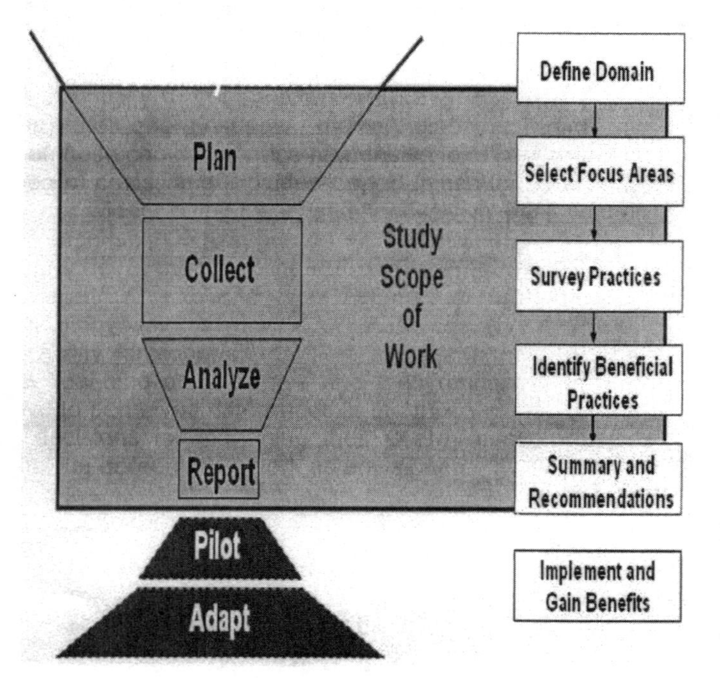

Figure 10-1. Benchmarking methodology

3. While developing and deploying web-based or web-enabled EMS systems, the ability of an outsider to deliberately alter data in a manner which could cause harm must be considered [4] and appropriate data security steps should be implemented. This generally means not to tie all the computer controls and data systems together in one single system, which is the antithesis of trends 10 years ago

For these reasons, most of the new control systems use distributed control technology, with periodic control parameter updates from a central unit, but with the ability to continue on locally without any input from a remote host. This configuration provides system integrity should some loss of connectivity occur, since local controls and monitoring are still functional. The growth of the Internet and the increasing availability of bandwidth have made web-based information sharing a common feature for building control systems. These web-enabled systems make it possible for any authorized end user to access energy information and control functions from any location and not merely from some central control station. At the same time, there are alternative means to receiving data, other than via the Internet, thus the use of web-based data exchange techniques is mainly for cost control and convenience.

There are many means to utilize the wealth of data available through monitoring and control of energy usage. These include:

(a) Energy usage analysis,
(b) Energy information monitoring of average usage, and peak demands,
(c) Actual closed loop controls,
(d) Benchmarking using energy data to determine short-term and long-term efficiencies,
(e) Automatic diagnostic evaluation via software tools.

These activities are discussed further below.

Energy Usage Analysis

Energy managers can use various IT tools for performing analysis of the energy billing history, long-term trend analysis of usage, regression analysis by kWh and kW for weather normalization and for uncovering other dependencies for the whole facility. These tools can include simple spreadsheet based tools or sophisticated energy information systems. An example of this is shown in Figure 10-2 which displays time, kW and kWh for a facility in a single graph. Similar graphs can be created using Microsoft Excel® based 3D load-profiling tool developed by Energywize with funding from NYSERDA. [5,6] Such analysis allows identification of patterns of the facility energy behavior and forecasting of future monthly costs. This tracking function can aid in understanding operating characteristics or anomalies requiring further investigation. Various energy metering information service providers help energy managers track and analyze their electrical, gas and water consumption from a single circuit to an entire facility—using the convenience of the Internet. The information is presented both in tabular as well as graphical format using load profiles and energy usage data to help users identify energy cost savings opportunities. Multi-site and aggregated load profiles enhance the energy manager's ability to proactively negotiate the best possible rates in the deregulated energy markets. Various on-line tools such as ABB's On-line Energy Profiler allow the interval meter data to be presented in customized reports on a variety of bases, including 15-minute intervals, hourly, daily, month-to-date and year-to-date, to assist in a more detailed, comprehensive analysis. Energy managers can utilize the load profiles to anticipate seasonal/daily usage patterns (day of the week/same day different year etc.) and select appropriate control algorithms in the EMS/SCADA systems to optimize the energy usage.

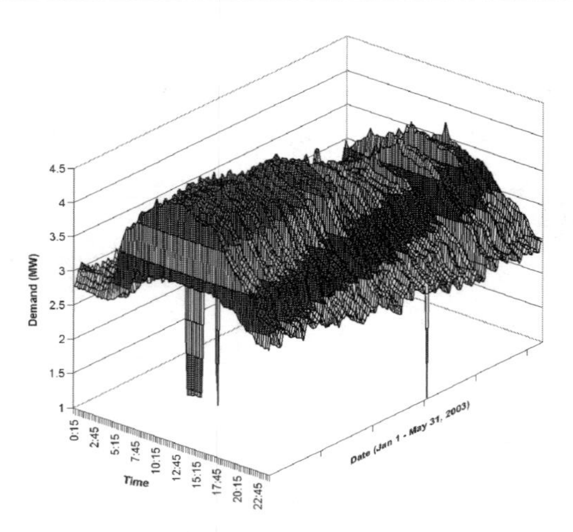

Figure 10-2. Sample 3D load profile using 15-min demand interval data.

Monitoring

Easy access to real-time usage information as well as design parameters opens up various avenues for energy managers for monitoring energy costs. Various utilities and independent service providers are developing simple web-enabled tools that could assist residential as well as commercial and industrial (C&I) customers to take control of their energy usage. Figure 10-3 shows a simple on-line energy calculator that allows end users to select appliances from a list and view the expected consumption as well as energy cost[7,8,9]. In Europe, some utilities are deploying meters that display the real-time energy costs in addition to the total energy consumption. With advanced IT systems energy managers can monitor daily comparison of current load profiles with expected load shapes based on historical data. The usage changes can be expressed in daily/weekly/monthly energy units or cost. This could include the option to compare data from multiple meters, via graphical and statistical on-peak/off-peak usage information, including minimum, maximum and average load shapes in comparison with day-by-day performance. From this the energy manager could compare the targeted savings from retrofits, thus enabling strong monitoring and verification (M&V) procedures and "what-if" scenarios while adjusting equipment control and operation. Remote energy monitoring and tracking via the Internet has become one potential low-cost method for gathering data rapidly. In the past, energy conscious organizations had their facilities personnel take daily readings of electric meters, fuel oil use (with dip sticks), natural gas readings, water meter readings and related chilled water or steam readings manually and then phone them in to a location where a technician tabulated the information and knowledgeable managers analyzed the conditions. Although certainly useful, it was time consuming and prone to errors. With electronic systems, and internet data transfer, monitoring can be further automated and be more accurate. One example of this is Project SAVE in the European Union (EU), where the City of Kolding, Denmark, led the way in using an internet monitoring system for all its municipal buildings. EMCS and SCADA systems can be used to breakdown the costs, which can then be internally charged back, thus aiding in a very accurate allocation of operating costs to the various cost centers, instead of merely treating utility

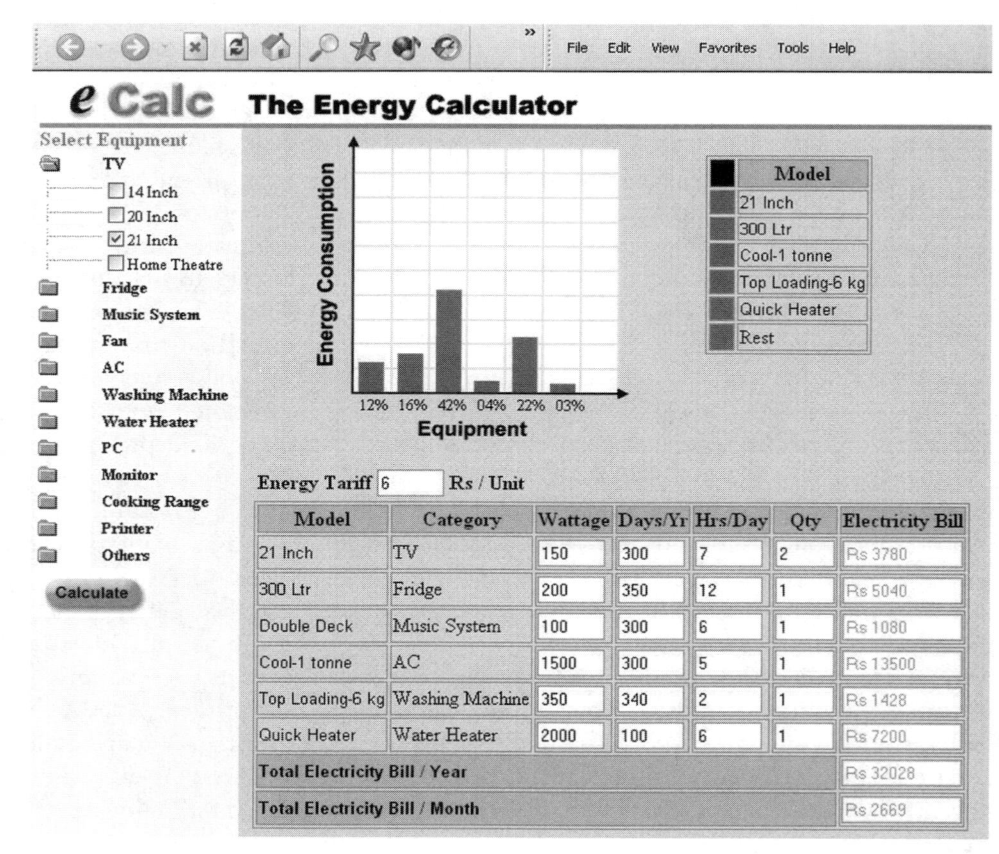

Figure 10-3. eCalc on-line energy calculation tool to help end users monitor energy usage of different equipment.

costs as "overhead." Many innovative businesses are now using the power of data analysis to begin directly charging back energy costs to specific work areas, thus reducing general overhead charges and allowing better management decisions.

Controls

With deregulation of energy markets, it is possible for real-time control of the cumulative total electric peak of instantaneous peaks for all facilities owned by a client, regardless of their location within a given electrical transmission system. This real-time ability for load control could be incorporated with limited two way communications to reduce coincident demand charges either from the power grid or the local utility. Energy management controls not only includes the actual modulation of systems to maintain the set points desired, but also includes the trending of multiple control points shown in convenient "month-at-a-glance" format. This allows the end user to determine if the units are operating properly. Trend-log analysis with respect to a system's design and sequence of operation, comparisons across different units, end use equipment schedule monitoring with respect to building occupancy schedules and similar activities are all possible. Such trending of data can also be in pseudo real-time, so that obvious directions of trends can be captured early. As part of the expansion of control technology, higher level "control" algorithms are being developed that build upon learning control theory, which has been in various stages of development since the 1960's. Application of this theoretical approach to practical HVAC systems is newer and is evidenced by some of the work by university researchers with an industrial applications background [10,11]. The premise here is that, in addition to having fixed control loops (even if some of the elements are adaptive loop tuning coefficients), that learning control allows improvement above the well defined, fixed loops available in most SCADA and EMCS packages installed by equipment vendors. The understanding of these control loops, in principle, is no different than has been developed in the past except for the ability to directly account for coupling of loops and interaction. Hence their practical use is not necessarily a problem for experienced HVAC technical personnel. The sole value of utilizing more complicated algorithms is to obtain more energy efficient solutions, which are robust with respect to changes in the systems themselves. When coupled with the diagnostic tools that are being developed, the control loops can react better to changes in parameters and yet extreme outliers can be flagged for remedial action by skilled technicians. The authors have seen numerous instances where, due to control valve or damper leakage, a given control loop could not effectively work well within the constraints of the turn down ratios of the actuator and sensor systems and so would eventually exhibit extremes of operation, almost acting like a "snap action" bang-bang controller. The usual solution, if even identified, is for the HVAC technician to disable the loop without documenting that he has done so.

One successful case study on web-based energy controls is the retail corporation Staples Inc. After the year 2000 California Energy Crisis, Staples implemented an energy management plan to protect over 100 California stores from steep demand charges and rolling blackouts. Staples utilized the installation of wireless control technology which allowed energy managers to automatically reduce lighting and HVAC loads at selected stores communicating through the Internet. They even have web-enabled utility meters. This not only led to significant savings in demand charges during peak periods, but also improved the regional power reliability. The ability to curtail and verify load reductions using interval meter data also enabled Staples to participate in demand response programs offered by the California ISO. Data are now archived so that Energy Managers can analyze use patterns, and by reviewing demand levels, Staples is able to verify that the paging signals are successfully reaching the facilities and controlling the targeted equipment [12].

Evaluating daily or monthly energy use data can identify deficiencies in control consistency, even if a sophisticated EMS is part of the program. Weather regression is one means to evaluate such patterns, as shown here in Figure 10-4. This graph compares seven sites of

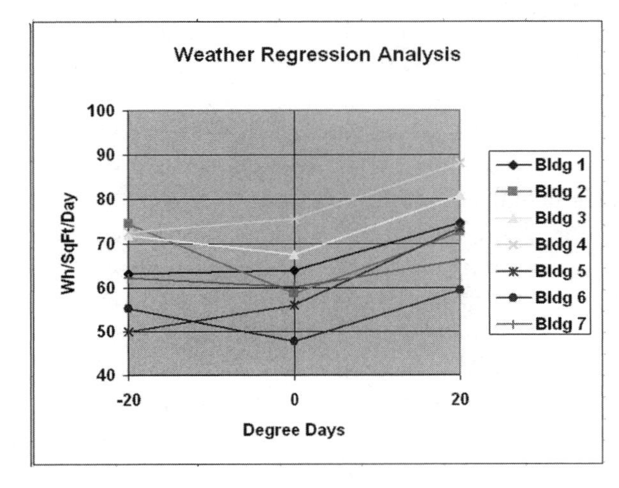

Figure 10-4. Weather regression analysis of electrical consumption on a square footage normalized basis.

a company on a normalized electrical use basis, to determine how the systems respond to weather conditions, based on heating and cooling degree days (HDD/CDD). The data are then used to determine whether the equipment is operating efficiently as a system. Part of the value of a control diagnostic tool is that deficiencies can be more rapidly identified and logged, so that remedial action can be taken and the appropriate systems returned to full, quality operation. This then yields a more optimum energy efficient solution in the long run. The data transfer requirements and analysis tools for such, can lie either within distributed modules at a site, at a central station at a site, or as part of a central data evaluation system at a remote site addressed through internet communications. The "hands off" approach promised by these new technologies can further assist Facilities Management and Production Management operations which seem to get leaner each year in manpower, simply due to continuing competitive cost pressures.

Benchmarking

Benchmarking, although usually utilized as an after-the-fact data analysis tool, is very powerful if the overall goals are to improve all operating areas and identification of the existing situation is the first step of this process. Energy managers can use benchmarking to target performance improvements or to make intelligent power purchase decisions in the new de-regulated energy marketplace. It can be used to develop relative measures of energy performance, track changes over time and identify best energy management practices. Benchmarking can involve comparing the Energy Utilization Index (EUI) of a facility to similar facilities and geographic locations, based on various parameters—kWh, costs, MMBtu fuel consumptions (as well as consumption profiles, akin to electrical demand charts). Normalization of the energy data can be performed by billing period, sq. footage, weather relationships (such as CDD and HDD) and/or customer-specific product units. Ranking against industry standards and/or standards developed from a customer's own building set can then be performed on a daily, weekly, monthly or annual basis, and comparisons made from year to year with weather and square footage adjustments. Grouping hourly load profiles can help in identifying typical operating practices.

Benchmarking ties in well with the concept of (M&V) procedures championed by the ESCO community to provide validity to retrofit actions taken on a guaranteed performance basis. The ESCO industry was aided by the NAEMVP developed in the 1990's and culminating today with the International Performance Measurement and Verification Protocol (IPMVP) standards, pushed by energy industry leaders, including Dr. Shirley Hanson [13]. Similarly, the concept of benchmarking allows not only a comparison of all sites to each other, but allows comparison to the past and also to "world class" standards which then push the envelope forward as to what reductions in operating costs and energy usage can occur, as in Figure 10-5. The US EPA Energy Star program offers Portfolio Manager as one such method of comparing one facility to a group of similar facilities with similar operating characteristics. Energy managers can rate the performance of buildings on a scale of 1-100 relative to similar buildings nationwide using Energy Star's national energy performance rating system. The rating system accounts for the impacts of year-to-year weather variations, as well as building size, location and

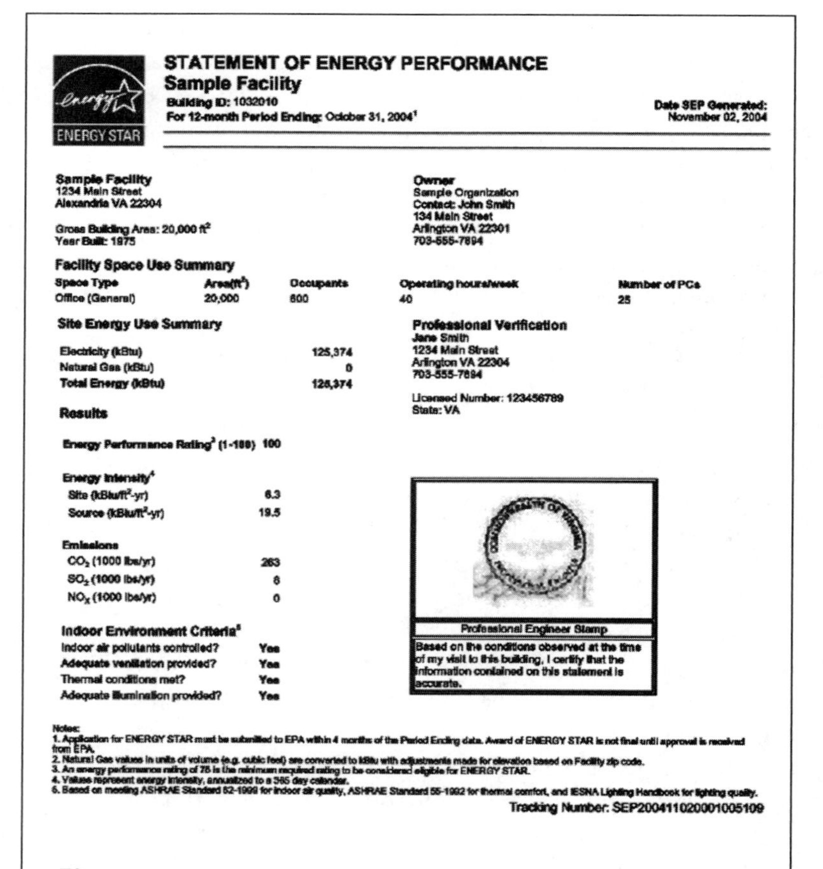

ENERGY STAR

STATEMENT OF ENERGY PERFORMANCE
Sample Facility
Building ID: 1032010
For 12-month Period Ending: October 31, 2004[1]

Date SEP Generated: November 02, 2004

Sample Facility
1234 Main Street
Alexandria VA 22304

Gross Building Area: 20,000 ft²
Year Built: 1975

Owner
Sample Organization
Contact: John Smith
134 Main Street
Arlington VA 22301
703-555-7894

Facility Space Use Summary

Space Type	Area(ft²)	Occupants	Operating hours/week	Number of PCs
Office (General)	20,000	600	40	25

Site Energy Use Summary

Electricity (kBtu)	125,374
Natural Gas (kBtu)	0
Total Energy (kBtu)	125,374

Professional Verification
Jane Smith
1234 Main Street
Arlington VA 22304
703-555-7894

Licensed Number: 123456789
State: VA

Results

Energy Performance Rating[3] (1-100) 100

Energy Intensity[4]

Site (kBtu/ft²-yr)	6.3
Source (kBtu/ft²-yr)	19.5

Emissions

CO_2 (1000 lbs/yr)	263
SO_2 (1000 lbs/yr)	6
NO_X (1000 lbs/yr)	0

Indoor Environment Criteria[5]

Indoor air pollutants controlled?	Yes
Adequate ventilation provided?	Yes
Thermal conditions met?	Yes
Adequate illumination provided?	Yes

Professional Engineer Stamp

Based on the conditions observed at the time of my visit to this building, I certify that the information contained on this statement is accurate.

Notes:
1. Application for ENERGY STAR must be submitted to EPA within 4 months of the Period Ending date. Award of ENERGY STAR is not final until approval is received from EPA.
2. Natural Gas values in units of volume (e.g. cubic feet) are converted to kBtu with adjustments made for elevation based on Facility zip code.
3. An energy performance rating of 75 is the minimum required rating to be considered eligible for ENERGY STAR.
4. Values represent energy intensity, annualized to a 365 day calendar.
5. Based on meeting ASHRAE Standard 62-1999 for indoor air quality, ASHRAE Standard 55-1992 for thermal comfort, and IESNA Lighting Handbook for lighting quality.

Tracking Number: SEP200411020001005109

Figure 10-5. Energy Star statement of performance generated using the Portfolio Manager

several operating characteristics. Buildings with Energy Star rating of 75 or greater may qualify for the Energy Star certificate [14]. Portfolio Manager can generate a Statement of Energy Performance that communicates information about a building's energy performance in a format that is both understandable and easy-to-use in business transactions. The Statement of Energy Performance can help energy managers formalize performance expectations to support leasing, building sales, appraisals, insurance, staff management and energy/O&M service contracts [15].

Careful evaluation of typical operating load shapes enable the energy managers to make accurate, predictable and measurable adjustments to operational practices. This technique utilizes normalized metrics to allow useful comparisons of such parameters as Btu/sq.ft./year, Btu/pound product manufactured, watt-hours per square foot per operating day, normalized baseline and weather dependency factors to allow comparison of commercial sites throughout the world on a per unit basis, allowing direct comparison. Using web-based monitoring and benchmarking systems, energy managers in large corporations or stores can not only compare their facility's performance with historical data, but can compare the performance in real-time with other facilities all over the world, as shown in Figure 10-6.

Diagnostics

Typically, the EMCS has controlled air handlers, chillers, boilers, exhaust fans, space temperatures, outside air condition and related HVAC comfort matters. It is altogether fitting that this be the root of the EMCS, which has been an extension of the old, stand-alone pneumatic controls going back to about 1880. However, with some exceptions, this same system, with increased computational power, can be used for much more. Diagnostic capabilities allow categorizing operating data into classes of energy usage (or wastage), which would allow transmission of interpretive data for many sites quickly and easily for comparison, or for simultaneous local site evaluation by technically trained end users. Some new software diagnostic tools have been developed to address this, which use learning control and pattern recognition techniques [16]. This also allows for projection of problems and notification via cell phones and WAP techniques to perform human evaluation rapidly and easily. This is a logical extension of voice activated telephone call up techniques for equipment failure detections that have been available over the past 20 years. Some of this identification may come about in the future with expert systems and data mining tools that can automatically identify trends and detect problems from the noisy data.

The authors are personally familiar with the benefits of simply being able to analyze data remotely in a logical format. This can be achieved using basic computerized pattern recognition procedures as well as statistical data calculations, to quantify, analyze and track the outliers, trend effects and efficiency factors can then be quantified, analyzed, and tracked. Such information can be used in analyzing problems, solutions, new cost savings opportunities and related factors, all from a desk, whether that desk be at home, across the world, or in a building or a corporate headquarters Facility Engineering Department [17]. Simply examining the daily electric profiles of a facility can be very telling, as shown in Figure 10-7.

Part of the problem with remote internet data tracking is that there could be the need for vast amounts of data in near-real time, if all information were automatically transferred. Just as with current SCADA and EMCS data transfer protocols from local control and monitoring modules that are distributed throughout a

Figure 10-6. Benchmarking Comparison of Energy Efficiency and Water Usage

Building	*Conditioned Sq. ft.*	*Electric Btu/Sq.Ft.Yr*	*Nat Gas Btu/Sq.Ft.Yr*	*Total Btu/Sq.Ft.Yr*	*Wtr/Sewer Gal/Sq.Ft.Yr*	*Electric $/kWh*	*Nat Gas $/MCF*	*Wtr/Sewer $/K Gallon*
1	4,411,494	83,866	13,389	97,255	10.7	$0.0446	$7.042	$4.75
2	602,181	91,948	4,404	96,353	4.1	$0.0486	$7.597	$5.90
3	214,156	83,060	649	83,709	40.6	$0.0492	$8.593	$3.70
4	291,092	70,044	13,952	83,996	91.8	$0.0876	$8.542	$0.83
5	427,000	75,409	25,638	101,047	17.2	$0.0547	$4.392	$2.14
6	523,558	66,647	8,433	75,080	24.0	$0.0835	$11.180	$4.66
7	366,183	95,875	14,799	110,674	41.2	$0.0505	$9.395	$6.41

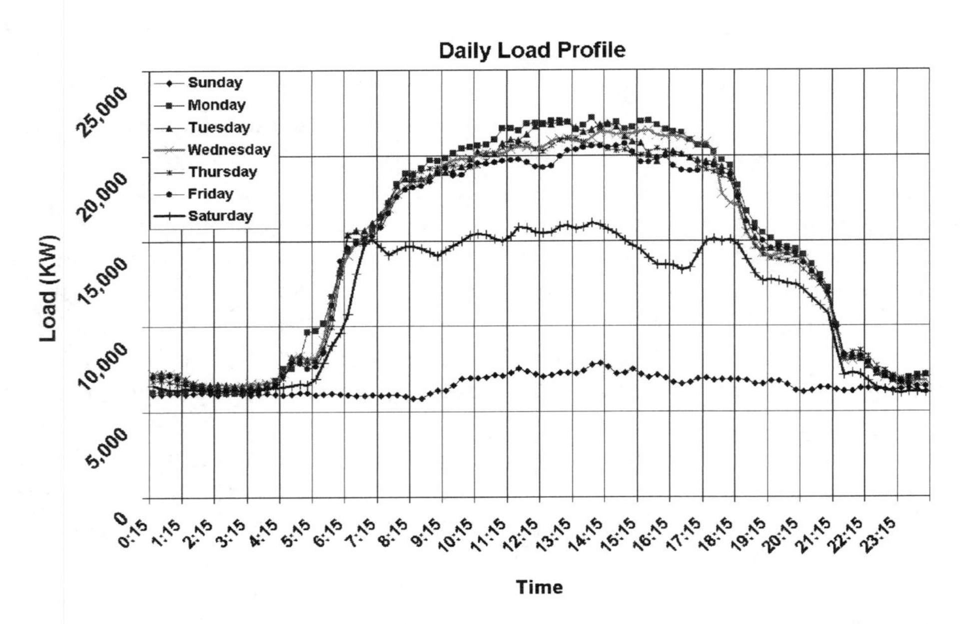

Figure 10-7. Daily load profile for analysis of building operating characteristics

facility and only transmit limited update data to a central location, a further refinement of that step has begun with respect to central data tracking of multiple facility sites for energy efficiency. Many of these tools are based on the whole building diagnostics (WBD) method, wherein the total utility bill such as the electric meter data is used on line to identify and detect particular actions, which can then be evaluated in light of energy efficiency goals. Tools include ENFORMA, EEM Suite, Whole Building Diagnostician, PACRAT, Analysis Tools for Built up Fan Systems and Universal Translator. However, only the WBD outside air economizer module and PACRAT have automatic diagnostic capabilities. All, however, are a step in the direction of newer, pre analysis data tools, which offer the potential to transmit only particular result data in many cases in real-time. Other information could be compiled, summarized, compressed and forwarded off line, along with more critical real-time data for analysis, tracking and cost control [18]. Such software tools and new diagnostic tools currently being developed, may become part of a series of "toolkits" which can be used in performing an additional level of evaluation and analysis without human interface. However, ultimately it still requires trained, experienced technical personnel to ascertain how to use the operational data to develop new energy cost savings techniques. There have been sufficient success stories, however, to demonstrate that these diagnostic tools can be of value [19,20].

INFORMATION TECHNOLOGY AND ELECTRIC UTILITY DEREGULATION

The internet and electricity deregulation were made for each other. Without the Internet to collect and disseminate information in near real-time and to make it easier to compare rates and procure energy, the idea of electrical deregulation would be far less persuasive and effective. At the same time, without deregulation to open power electricity, the Internet's ability to collect information and enable transactions would be of less useful to the energy supply side markets. The deregulated energy marketplace is providing the energy industry as well as energy managers with new opportunities, challenges and risks. As the market continues to develop in various US states, the key to maintaining a competitive edge is the development and implementation of technologies with real-time capabilities. This applies to electricity or natural gas purchases, although the majority of the work has been focused in the electrical field. Now more than ever, advanced technologies related to metering, energy management and billing are allowing energy managers to leverage future opportunities and minimize financial risks [21]. In past 3-4 years, various ISOs and RTOs have spent about $350 million to put together information technology infrastructure, with IT operating expenses comprising 15 to 26 percent of ISO revenue. The local distribution companies also must make their own IT investments to meet the data-communication requirements of these ISOs [22]. Under FERC's Standard Market Design (SMD), ISOs and RTOs are setting up Day-Ahead and Real-Time Energy Markets that consist of voluntary, bid-based, security-constrained markets in which generators and energy suppliers are free to engage in bilateral transactions. Market participants can follow market fluctuations as they happen and make informed decisions rapidly. Participants can respond to high prices and bring resources to the region in times of high demand.

As an example, consider the PJM Interconnection—a regional transmission organization (RTO) that coordinates the movement of wholesale electricity in all or parts of Delaware, Illinois, Indiana, Kentucky, Maryland, Michigan, New Jersey, Ohio, Pennsylvania, Tennessee, Virginia, West Virginia and the District of Columbia, serving more than 25 million people. The company dispatches more than 143,000 megawatts of generation capacity through 350 market participants and has administered more than $28 billion in energy and energy-service trades since the regional markets opened in 1997. PJM's wholesale electricity market is similar to a stock exchange, establishing a market price for electricity by matching supply with demand. On-line eTools makes trading easy for members/customers by enabling them to submit bids and offers and providing them with continuous real-time data. A number of the eTools—OASIS (Open Access Same-Time Information System), EES (Enhanced Energy Scheduler), eSchedules, eCapacity, eMTR, eFTR (Financial Transmission Rights), and eData—are grouped together to create eSuite, which allows authorized users to access these eTools with a single user ID and password. PJM keeps markets fair by making prices transparent through eTools [37]. The integration of this pricing information automatically into the EMCS arena is still an area being developed.

Energy procurement is one of the most rapidly developing e-commerce areas in deregulated energy markets and can have equal impact on energy cost control as the traditional energy efficiency and energy management systems. On-line energy trading is attractive to end-users because of its speed and efficiency. The Internet gives energy managers the ability to view real-time price signals and their respective energy usage and to compare energy resources, services and markets. This allows energy managers to make more informed operational decisions and manage their energy costs. It also intensifies competition between energy supplies, making deregulation work as envisioned. The ultimate beneficiaries of price transparency are, in effect, the end use customers.

AMR TECHNOLOGY FOR ENERGY COST CONTROL IN THE 21ST CENTURY

To the old adage about "managing and measuring" is the corollary that you cannot manage well what is not made available in a timely manner. Until recently, the metering and billing methods employed by most of the utilities in North America and Western Europe had not changed appreciably in about 80 years. With the regulatory changes happening in energy markets, it is inevitable that new metering and control strategies would emerge. Traditionally, electric usage has not been able to react to rapid price changes in wholesale energy markets as the communication and market systems did not exist on a cost-effective, wide scale. Due to the inherent limitations in conventional meter data collection technologies and geographic constraints, traditionally utilities could not take a "one-size-fits-all" approach required for widespread AMR implementation. These limitations have led utilities to deploy traditional electromagnetic electric meter systems for the mass market and advanced AMR for the large C&I segment.

In recent years utilities looking to capitalize on AMR technology are committed to replacing, upgrading or retrofitting existing meters along with the billing system upgrades. With new AMR and electronic systems providing capabilities at the individual meter level, there now exists the ability to collect and present metering information to the end user in a timely manner and influence the energy use patterns. Utilities can use such technology at large scale to effect changes by end use customers, in a manner that can both benefit the users and the local utility. The downside to these opportunities is that the end user could end up with higher utility costs than before if the user is not agile, trained and cognizant of the circumstances. So, along with cost savings opportunities is the risk of price increases.

The 7th edition of *The Scott Report:* AMR Deployments in North America [23] indicates that:

- Total AMR shipments grew by 18.3% from 2001 to 2002
- 49,311,372 AMR units were installed in North America through January 1, 2003.
- More new units were shipped to electric utilities than to water and gas utilities combined.
- 75.1% of all AMR units shipped in 2002 were to projects that were initiated in 2001 or earlier.

As upgrades to existing metering infrastructure are being considered, the opportunity exists to look 'beyond the billing' and address other meter data requirements. Measuring billing data needs information on real energy, kWh; reactive energy, kVAR and maximum demand data. The next step is to obtain data that can tell both the supplier and the consumer about the quality and reliability of the supply. With the technology available today, metering infrastructure can be easily transformed into a virtual energy information network used by both the energy consumer and the utility. To make a customer communication network work, utilities are deploying metering solutions with two-way communications capabilities. Key features of the modern metering systems include active and reactive energy measurement, AMR functions, TOU contract management functions, remote connect/disconnect for load control, fraud detection/anti-tampering functions, prepayment, demand management and potential development of value added services. This data allow companies to analyze, diagnose and mitigate a variety of power quality issues. Power quality problems are costly to both energy supplier and consumer and direct equipment damage is only a small portion of the overall disturbance-related costs. Unplanned outages and abnormal line losses can add significantly to the costs. Historically, power quality problems have been identified, analyzed and corrected with off line means, but now the technology exists to measure harmonics in real-time and to adjust for it with active harmonic filters for example.

Enel Distribuzione, the Italian power electricity utility with more than 30 million customers, launched a project in 2001 to replace all 30 million electromechanical meters with a new system. The Telegestore project is expected to be completed by 2005 at a cost of 2 billion Euros and will cover approximately 30 million customers. The Telegestore includes a remote meter reading system, a customer management system and a potential 'value added services' delivery system. The project started in June 2000 with the development of a remote metering management system which used the low-voltage distribution grid as a data carrier, in combination with the public telecommunication network. It includes a completely electronic and integrated meter, instead of a traditional electromechanical meter integrated with an external electronic communication device [24]. More of this type of activity can be expected in the future.

UTILITY BILLING AND ENERGY COST CONTROL USING THE INTERNET

Until very recently, the use of computerized control in energy management was limited to optimizing energy usage. Issues of utility rates had generally been factored into the algorithms and procedures previously selected for use within the computer by engineering consultants familiar with the opportunities in a given utility service area. With the advent of real-time pricing (RTP), such past procedures may be inadequate. Development of energy metering systems, which gather data in ways that accurately reflect the changing cost structures of the utilities, is a first step in this process [25]. In competitive energy markets, some negotiated energy supply contracts, do not penalize or reward demand side management measures directed at peak electric demand control. So the strategy for saving electrical energy costs could change as the contract structure changes. Hence the future missions of energy management may be more focused on a data management structure, which can be easily adapted to large variations in cost control. Also, in the future, contract windows of a day or a week may become commonplace, so the actual utility price structure may be a "constant-variable" to quote an old joke in engineering and therefore more robust control designs may be required.

This also points up the absolute necessity for Purchasing Department Personnel to communicate and plan closely with the O&M Facilities personnel before signing a supply side energy contract, since the two functions of energy cost procurement and energy management must be coordinated if investment funds are to be optimized. The authors have seen recent cases where such planning and coordination was not performed and the so-called "benefits" of a good supply side contract virtually negated millions of dollars annually in benefits from carefully planned energy investments. Historically, the connection between these two activities was much more decoupled, in that small supply side contract cost reductions, with the same rate structure had only minimal impact on the payback periods of energy cost reduction measures (ECRMs). However, with the near total deregulation in place in much of North America and Europe today and with much of the world coming close behind, these issues can no longer be performed in a near vacuum. Although one may think this is obvious, it has occurred recently (2004) in a number of large corporations, so it is becoming more common. In some ways this trend for highly variable utility rate structures is only a natural consequence of the electronic and internet revolution, where information can now be disseminated in near real-time, with a high degree of accuracy and great computing power is available throughout the marketplace inexpensively. Coupled with the continuing cost reductions in hardware, it was only a matter of time before these changes became practical and prevalent.

Over the years, the EMCS or SCADA systems were designed to not only gather data, but control key systems for energy efficiency. This has worked well and has been a hallmark of the controls industry. However, with the advent of electrical deregulation and to a lesser extent gas deregulation, the possibility looms that all of the energy efficiency features could be overshadowed if variable rate tariffs are not accurately accounted for. In past decades, a rate tariff was prepared by a utility and submitted for approval to a regulatory agency and after approval scheduled for a future date to go into effect, providing ample time for end users to be aware of these changes. Such rate tariffs might not go into effect for a year and then last for a subsequent 10 years. In the future, this may no longer be the case, due to the speed of communications possible and the pressures of competitiveness. However, with the advent of real-time metering and more information technology, it may loom in the future that a rate tariff may only become a structure with hourly "coefficients" to formulas being adjusted. Also, these numbers might not be specifically set in advance, so if an end user was not "on the ball," they could find large penalties in the ensuing monthly utility bills and of a magnitude greater than the overall energy efficiency otherwise garnered through control optimization from the EMCS or SCADA unit.

Just as the IT systems helped energy efficiency optimization in real-time, it can also assist in adapting to innovative tariffs. Therefore, it is simply a case of "moving to a higher plane," much as the computer control upgraded what had been small, local pneumatic or 24 v electric on/off controls. Instead of revolutionary, it is a continuing of the evolution of electronics and control strategies, in ways and means that help a situation. In this case, it is to avoid cost increases, or to optimize the possible energy cost reductions. Undoubtedly, however, this has a complicating effect on calculations of operating cost reductions due to energy conservation measures implemented. It also places a higher burden on the reliability of equipment, since a short period "goof" on the part of the electronics could result in a large penalty to the Owner for electric demand control for instance.

In California, a joint initiative by the California Energy Commission (CEC), the California Public Utilities Commission (CPUC) and the California Power Authority (CPA) is trying to establish a technology and pricing policy foundation that links customer rates with the market price for energy [26]. Preliminary results from a $10 million statistically designed state-wide pricing pilot implemented in 2003 by utilities demonstrated customer demand and energy impacts that substantially exceeded most expectations in terms of demand reduction. The project involved notifying customers using telephone/fax systems, as well as allowing customers to monitor their real-time energy consumption data online. Such an approach could have helped California avoid power blackouts such as the Year 2000 problems. Interestingly, preliminary results show that residential users will respond to price controls if given the opportunity and real-time data; this contradicts the prevailing attitude that residential customers are not good DSM candidates.

In Chartwell's 'Guide to E-Business in the Energy Industry, 2001', researchers found that utilities are offering more services via their Web sites [27]. Some of the ways utilities now leverage the Internet to improve customer service include on-line billing, requests for service changes, providing on-line account history, 24/7 energy usage data, on-line management of account information and allowing customer to enter meter readings. To deliver real value to top energy customers, suppliers are utilizing advanced billing systems that go beyond processing load data into billing determinants to include key capabilities necessary for serving metered energy customers in competitive markets [7]. Such systems offer summary/consolidated billing, true conjunctive billing, bills for innovative rates, billing for ancillary services and new products and services, bar coding, management reporting, direct import of interval load data, and direct export of accounting information.

INFORMATION DISPLAYS

Indirect beneficiaries of high technology electronics and the exponential expansion of the Internet are technologies which offer alternative means of data display. For the first few decades of digital computer use, data was displayed almost exclusively in paper printout form. This was followed by various CRT displays (almost always in black and white) and then over time by various color displays which were ergonomically developed to maximize information interpretation by the human operator interface with a minimum of effort. Just as the computer screen was a transition from paper printout, so are newer technology data displays. One of these is called "Oreb, by Ambient Devices, Inc. of Cambridge, MA, which is a plastic device which changes color based on data input to it—no computer screen, no actual numerical data displays. Although it is in the early stages of use, it likely represents one more new "information display" system which is likely to alter the way in which we analyze and interpret energy use data and energy operating characteristics.

Over the last twenty years, major improvements in data display technology have been developed, allowing at a glance, the interpretation of physical data being gathered by an EMCS or SCADA system. When coupled with learning control and adaptive control techniques, the information display, instead of being units of use such as kW/ton of chiller, kW of lighting, run hours of air handlers, pounds per hour of boiler output steam and similar physical energy parameters, the displays can show comparisons of usage to norms, which change with time. Looking rapidly at various color displays providing "digital" data feedback (good/bad/indifferent) type responses, allows much more information to be rapidly digested by the human operator. Such methods have been developed into new software approaches based on earlier pioneering work [28].

A first step at providing more information to the end user has come from the electric utilities, who historically provided a one or two page summary electric bill within about 10 days of the end of a billing period, representing about 40 days from the start of the period just reported. In recent years, many utilities, either internally or with outside assistance, have developed procedures for end users to secure either real-time information, or summarized electronic historical data along with basic computer diagnostic tools for analyzing the energy usage and patterns [29]. This data empowerment then allows the customers to adapt their use of energy through virtual on-line feedback. Figure 10-8 shows an on-line Energy Speedometer™, currently being developed by eMeter. Using real-time data, the Energy Speedometer™ allows the end users and energy managers to see how fast they are consuming electricity in real-time.

One example of the tremendous power of simple techniques is at the Boston Medical Center in downtown Boston, where the Director of Facilities, Mr. Thomas Tribble, P.E. requested as part of an energy efficiency retrofit project that a single accounting of all the electric meters associated with chilled water production in the central plant be gathered together and displayed as a single metric "kW/ton" for the entire plant. This was set up throughout the 20,000 point EMS unit and a simple, large LED panel prominently displayed where all the relevant utility personnel could see it, including the Director of Facilities. As a result, the operators learned quickly that if they did not respond rapidly to high-level indications, they would almost immediately receive a phone call from the Director to find out what caused the loss of efficiency and what steps were underway to rectify the problem. This simply demonstrates the great power of some basic metrics that all parties in the area of interest can easily understand. Sometimes it is these simple overall metrics that can be best utilized, if associated with trained personnel who have a vested interest in optimizing energy efficiency.

DATA INTERCHANGE ON THE INTERNET

A part of pre-transmission data processing is determining the necessary amount of data to be transmitted. As data bandwidths get cheaper, the situation may change, but in general it is still useful to assume that the more pre-processing of data the better. This assumes that if additional, specific data must be transmitted in real-time, it can be effected on a focused basis, as if additional channels of information were available, but not nearly enough to cover all the data points available. In this way, the techniques of data pre-processing can improve over time and yet as data bandwidth gets cheaper, the combination may allow use of more data points. This has happened already, as can be seen by comparing systems installed in 1975, 1990 and now. The number of points in EMS systems has dramatically increased and the cost per point greatly decreased (even in absolute dollar terms) and that technology has allowed more processing power

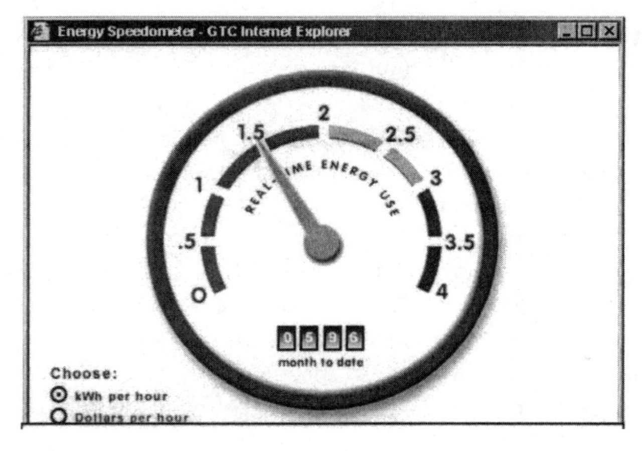

Figure 10-8. eMeter's Energy Speedometere™

to be pushed out to what used to be merely data gathering panels (DGPs). Even if all the data collected at a site could be cheaply transmitted over the Internet to a central "processing station," one is still faced with the mounting problem of data overload. This simply means that as technology prices have dropped, one is motivated to add more data channels since it is inexpensive; but at some point having a human being do a consistent physical analysis of all this raw data becomes problematic and somewhat self-defeating [30]. Just as the military aircraft field has developed HUD displays and other forms of summary data display without distracting the pilot, the energy industry is also following, albeit with much more severe cost constraints.

Interestingly, a new problem that has become more important lately is that large amounts of energy use data gathered by an EMCS or SCADA have sufficiently different time tags that it can negatively affect learning control algorithms or simultaneous peak demand control of a series of widely scattered sites. Therefore, an additional level of software pre-processing may have to be instituted in some cases to account for consistent data times, even if they are rationally interpolated. This problem has essentially come about because in the past, much data was analyzed off-line and now we are much more geared towards real-time evaluation and control strategy changes.

Today energy managers use a variety of web-based services that facilitate procuring electricity on-line. The only way to realistically coordinate this complex process is to implement a centralized computer system to manage all of these transactions. All the companies that offer services in this market have their computer systems to manage data and complete transactions, which must be coordinated to make the market work. To address this problem and simplify the requirements for customers to purchase power through any number of different energy service providers, many states are mandating the use of internet technology for Electronic Data Interchange (EDI). General Electric (GE) found out how beneficial it was to purchase commodities on the Internet, and in 1997 started to use the internat for soliciting supplies, including energy. As a result, today the majority of GE standardized purchases are all handled on line, including utility bulk purchases. GE alone handles billions of dollars a year in transactions through the Internet, allowing a small handful of highly trained experts to interpret data instead of having an army of clerical people handle volumes of paperwork. Subsequently the Internet data exchange process has greatly reduced clerical employment, but multiplied the value of "a learned individual" in analyzing, diagnosing and tracking critical data.

Electric utilities use power system models for a number of different purposes. For example, simulations of power systems are necessary for planning and security analysis. Power system models are also used in actual operations, e.g., by the EMCS used in energy control centers. An operational power system model can consist of thousands of classes of information. In addition to using these models in-house, utilities need to exchange system modeling information, for both planning and operational purposes, e.g., for coordinating transmission and ensuring reliable operations. However, individual utilities use different software for these purposes, and as a result the system models are stored in different formats, making the exchange of these models difficult. In order to support the exchange of power system models, utilities needed to agree on common definitions of power system entities and relationships.

To support this, EPRI has developed a Common Information Model (CIM). The CIM specifies common semantics for power system resources, their attributes and relationships. The CIM can represent all of the major objects of an electric utility as object classes and attributes, as well as their relationships. CIM uses these object classes and attributes to support the integration of independently developed applications between vendor specific EMC systems, or between an EMCS and other systems that are concerned with different aspects of power system operations, such as generation or distribution management. The CIM is specified as a set of class diagrams using the Unified Modeling Language (UML). The base class of the CIM is the Power System Resource class, with other more specialized classes such as Substation, Switch and Breaker being defined as subclasses [22]. The North American Electric Reliability Council (NERC, an industry-supported organization formed to promote the reliability of electricity delivery in North America) has adopted CIM/XML as the standard for exchanging models between power transmission system operators. The CIM/XML format is also going through an IEC international standardization process.

PRICE RESPONSIVE LOAD MANAGEMENT TECHNIQUES AND THE INTERNET

Prior to electric deregulation, timely information on detailed pricing breakdowns and consumption patterns was generally not readily accessible to end users. Customers were typically not able to determine accurately how much they were paying for energy until they received the bill the following month after the usage had

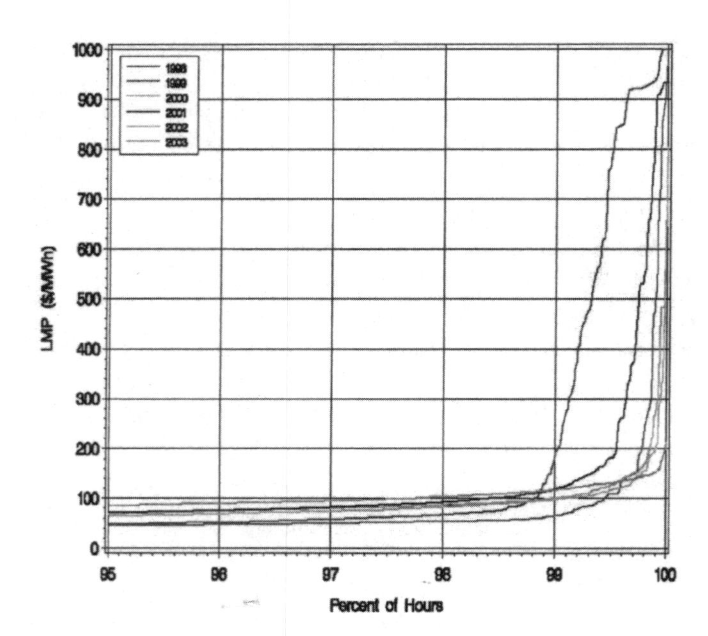

Figure 10-9. PJM hourly load duration vs. LMP curve[37]

been set, unless they had separately installed an electrical sub metering system. With electric market transparency, the demand side of the market can take part on more of a real-time basis. In Figure 10-9, the hourly load duration curve for the PJM Interconnections shows that from 1998 to 2003, the highest 15% of load on that system is used only 2% of the time, or less than 200 hours per year. The PJM price curve shows that this 2% of time also corresponds to time when Locational Marginal Prices (LMPs) have maximum volatility with price fluctuating between $100/MW to $1000/MW [37]. The customer with demand response capabilities now has the opportunity to get paid as much as an energy generator would get paid for selling electrical generation into the market, helping make it a more level playing field.

Demand response (DR) programs provide incentives to end users to reduce the electric load in response to system reliability or market conditions, mostly designed for customers with large loads who can make significant reductions at the utility's request (large commercial complexes or industrial plants). DR is favored by many utilities because it addresses their primary problem of bridging the gap between wholesale prices and retail prices and ensuring reliability of supply during these periods. Utilities would rather give a customer an incentive to reduce its load during such events rather than pay the customer to permanently reduce its overall load, which can result in lost revenue for the utility. By using direct load control or pre-programmed optimization strategies in an EMCS system, Curtailment Service Providers (CSPs) can respond to load reduction request by the ISOs and then distribute the incentives to program participants. This data can be best made available through the Internet. These programs offer incentives that are directly linked to the real-time LMPs. In 2002 it was estimated that the market potential for price responsive demand response programs will be $10 billion by 2005, opening up a market for DR infrastructure of $1.1 billion [31]. The potential benefit from demand response was estimated at $15 billion in benefits in 2001 according to McKinsey Company. A finding by the Edison Electric Institute suggests that a 5 % demand reduction can reduce market prices by 50 % and thus provide a significant check against the exercise of market power during those times.

The recent growth in AMR installations for cost-saving reasons unrelated to retail pricing has reduced some of the existing barriers to more efficient pricing and offers the potential for utilities to offer new services to their mass-market customers. For example, Gulf Power Company in Florida is expanding its critical peak TOU pricing program, known as GoodCents Select, finding that customers like both the opportunity to save money and the flexibility provided by the program's communication and pre-programmed control technology. During 85% of the hours of the year, Gulf Power customers see prices that are lower than under their standard tariff. Only in peak periods and the infrequent (1% of hours) critical periods do they face higher prices. Gulf Power has seen substantial load reductions during peak and critical peak periods, producing bill savings to customers and cost savings to the utility.

Florida Power & Light Co. (FPL) operates a simple and voluntary direct load control program that has proven to be highly successful. FPL uses a Two-Way Automatic Control System (TWACS) power-line carrier system. Using a portfolio of DSM programs, including interruptible rates for large power customers and a predominantly residential load-control program, FPL and its customers have successfully reduced demand by over 3000 MW. This reduction has allowed FPL to avoid building approximately 10 new 400-MW power plants. Of that total, 1000 MW of peak demand savings can be directly attributed to FPL's Load Management System (LMS) [36]. To achieve this, FPL generally avoids using load management to curtail air-conditioning loads and instead controls water heaters and pool pumps, unless capacity needs are critical.

The value of the Internet is that it allows data to be collected and transmitted economically over a long distance between various sites of a single customer group.

While it has been possible in the past to do that with dedicated phone lines or dedicated communications cables, the benefits were often not justified by the costs. Now that long-distance phone lines are no longer a required expense as the major data transmission media, and electric deregulation provides a further economic incentive, the benefits of data collection can more than outweigh the costs for supply side cost control, let alone the traditional demand side control.

IT FOR SECURING GREEN EMISSIONS CREDITS

With the growing focus on the environmental impact of development, and efforts to build sustainable facilities, future energy managers will need to implement green design techniques that help reduce direct/indirect Green House gas (GHG) emissions from the facility. Various proposed mechanisms such as the (Global) Kyoto Protocol or the (US) Climate Stewardship Act are trying to develop incentives for early action by allowing emission trading. Implementing GHG emissions trading is complicated. In an effectively designed market, emitters with low cost options for reducing emissions receive revenue from emitters with higher cost emission reduction options. The effectiveness of emissions trading in meeting environmental objectives, however, depends very much on the regulatory environment in which they function. For example, caps on emissions at a national and international level, such as the "assigned amounts" mandated by the Kyoto Protocol, would ensure that trading does not lead to a net increase in emissions over time. In such a cap and trade system, therefore, the seller offers a "surplus" of emission reduction units that have been achieved in excess required by the seller's cap.

A first step is establishment of a registration program to more accurately and reliably measure, report and track GHG emissions using on-line GHG registry programs. In an improved registry program, a company would establish a baseline consisting of current aggregate emissions from all major GHG sources under its control in the United States. Gross emissions on an annual basis could be compared to this established baseline. A reliable registry would make it possible to provide "baseline protection" for companies taking action now to reduce their emissions. These entities could be assured that—in the event of future controls involving the allocation of emissions allowances or requiring emissions reductions—they would not be penalized for reductions already achieved voluntarily. The improved registry program could also provide a mechanism to recognize the emissions reductions resulting from companies manufacturing more efficient or carbon-saving products. Finally, it could ensure that GHG reductions and sequestration offsets are of sufficient integrity that they can be traded and sustain their value in future years. This registry would include reductions and offsets achieved outside of the United States, in both developed and developing countries. In this manner, both gross and net (reductions and offsets) emissions would be recorded[32].

On the other side there are industry based initiatives such as 'Chicago Climate Exchange' (CCX) to help establish mechanisms for facilitating emission trading. The CCX is a greenhouse gas (GHG) emission reduction and trading pilot program for emission sources and offset projects in the United States, Canada, Mexico and Brazil. CCX is a self-regulatory, rules based exchange designed and governed by CCX Members. These members have made a voluntary but legally binding commitment to reduce their emissions of greenhouse gases by four percent below the average of their 1998-2001 baseline by the year 2006. The CCX Trading Platform is an internet-accessible marketplace based on SUN Java technology that is used to execute trades between Account Holders.

ON-LINE TRAINING FOR ENERGY MANAGERS

An important element that has been neglected in the commercial sector and has been well executed in the industrial sector, is the training of the operators and personnel who directly relate to energy cost control (facilities personnel in the commercial sector and production process operators in the industrial sector). Time and again the authors have seen large EMCS in the commercial sector being used by persons with almost no understanding of the data gathered, or how the end use pieces of equipment operate as a "system." Generally, this commercial sector deficiency is not due to lack of intelligence on the part of the operators, it is simply lack of management focus. It is strongly recommended that training, training and training be emphasized, so that the power of all the capital investment can be harnessed for economic good, as well as achieving environmental conditions. Simply achieving proper environmental conditions has become very important in the US as more lawsuits have been filed dealing with indoor air quality, from CO_2 levels, mold and mildew and even strict interpretation of contractual obligations of temperature and humidity levels within a space.

The Internet provides significant new functionality in transmitting information and providing forums for knowledge exchange. On-line training utilizes web-enabled IT solutions to communicate and collaborate. Various vendors such as Siemens as well as ISOs such as PJM are providing learning resources to enable energy managers to become familiar with their EMS systems and its operations. The on-line training is not only limited to the training material available on a website, but with the advances in telecommunication technologies, vendors and providers can offer real-time interactive training. Various universities are also offering specialized courses targeted at energy and energy managers, through on-line courses. The New York Institute of Technology is one of such universities that offers energy managers on-line courses. These e-learning programs can be deployed quickly and economically, without complex installation costs and a high upfront investment in hardware and technical staff. On-line Training enhances training resources, eases budget constraints and eliminates logistical barriers, allowing vendors to scale training programs to reach more people more often and far more easily.

PROJECTIONS OF TECHNOLOGY DIRECTIONS

Based on the current pace of technology change, the development and deployment of new hierarchical software and hardware systems can be expected. The Internet itself is speeding up with the second generation web and most commercial and a large number of residential users are switching to DSL or cable for broadband, so the backbone of high speed communication is being put in place. In many cases there is already two-way communication. From the topics covered previously, plus the trends that are underway, the direction of the Internet can only help energy managers. Some of the expected near future effects of technology include:

- Software designed to integrate existing EMCS and SCADA functions with supply side EIS.
- More expert system diagnostics tools, akin to what has evolved in the high performance aircraft and space industry. More diagnostic and pattern recognition software additions to the energy control packages, allowing for troubleshooting and prediction of problems and recommended courses of action. Such tools would not replace experienced users, but would augment and enhance them, given the increasingly complex EMCS/SCADA systems being implemented today.
- Although available in a crude form today in most EMCS software, additional program features allowing changing of control during specific event conditions such as peak demand charges changing the level of peak shaving, based on RTP data provided. This may likely involve some human machine interaction, but for larger systems this is not a concern.
- More two way communications for metering and load control with the electric companies not only for the larger users but also for the residential users.
- New electric and natural gas metering technologies, since natural gas is also going through demand control issues (independent of peak winter conditions as has historically been the case).
- New trend towards monitoring and control of individual pieces of equipment through its own Wi-Fi or similar hardware/software packages, allowing standardized interface through RS232, IEE488 or standard computer interface protocols. This could change the character of the existing EMCS industry [33].
- Interaction capabilities with state level/national database for GHG emission reporting as well as benchmarking and exchange of emission reduction credits.
- More push for ISO 14001, LEED and other certifications dealing with energy efficiency and energy procurement as additional environmental standards and greenhouse gas controls come into play [34].
- Continuing opportunities for ESCOs, CSPs and energy managers to work together to control and solve energy efficiency problems in facilities [35], since these new technologies require even more focus than the past to not only reap the rewards of cost savings, but to assist in avoidance of utility cost increase on the supply side.

References

[1] *Information Technology for Energy Managers*, edited by B.L. Capehart, AEE Fairmont Press, 2003

[2] "Energy Efficient and Affordable Small Commercial and Residential Buildings Research Program, D. Chassin, et al., *Batelle Pacific Northwest Division, Report PNWD-3317*, August 2003.

[3] "EMCS and Time-Series Energy Data Analysis in a Large Government Office Building, M. Piette, et al., *Proceedings of the 9th National Conference on Building Commissioning*, May 9-11, 2001, Cherry Hill, NJ

[4] "The Web Within the Web," *IEEE Spectrum*, February 2004.

[5] " Where Smart Meters Make Sense," *A Primer on Smart Metering*, NYSERDA, Fall 2003.

[6] "New Excel-Based Techniques to Visualize Energy Profiles with Interval Meter Data" Lindsey Audin, *Proceedings of World Energy Engineering Congress*, October 2003, Atlanta, GA USA.

[7] "Use of Information technology in the field of Energy Management"; R. Walawalkar, *DEED Research Report*, American Public Power Association, 2003.

[8] "Role of IT in Spreading the Energy Awareness and Popularizing the Efficient Designing Practices," R. Walawalkar, *Proceedings of National Seminar on Energy Efficiency and IT Industry*, Council of Energy Efficiency Companies in India; 1999.

[9] "IT Enabled Energy Efficiency," R. Walawalkar and M. Mittal, Proceedings of International Congress on Sustainable Development—Energy Conservation and Pollution Control, Mumbai, India, 2001.

[10] "MIMO Robust Control for Heating, Ventilating and Air Conditioning (HVAC) Systems" M. Anderson, P. Young, D. Hittle, C. Anderson, J. Tu and D. Hodgson, in *Proceedings of the 41st IEEE Conference on Decision and Control*, Las Vegas, Dec. 10-13, pp. 167-172, 2002.

[11] "An Experimental System for Advanced Heating, Ventilating and Air Conditioning (HVAC) Control," M. Anderson, P. Young, D. Hittle, et al., *IEEE Transactions on Control Systems Technology*.

[12] "Enhanced Automation Case Study 5, HVAC/Lighting Controls/Retail Chain: Staples Inc.," *A Success Story From the California Energy Commission*, California Energy Commission, 2002.

[13] "International Performance Measurement and Verification Protocol (IPMVP)," vols 1-3, *www.ipmvp.org*.

[14] "Portfolio Manager: Assess Building Performance," Energy Star Website, *https://www.energystar.gov/portfoliomanagertour*, EnergyStar, *2003.*

[15] "Introduction to the Statement of Energy Performance," *www.energystar.gov*, EnergyStar, 2003.

[16] "Pattern –Recognition Based Fault Detection and Diagnostics," M. Brambley, et al., *Report for Architectural Energy Corporation*, Feb 2000 for the California Energy Commission, PIER Program

[17] "Web-based Energy Information Systems for Large Commercial Buildings," N. Motegi, Mary Anne Piette, *Lawrence Berkeley National Labs Report 49977*, May 2002.

[18] "Comparative Guide to Emerging Diagnostic Tools for Large Commercial HVAC Systems," H. Friedman, M. Piette, *Lawrence Berkeley National Labs report No. 48629*, May 2001.

[19] "Web-Based Energy Information Systems for Energy Management and Demand Response in Commercial Buildings," S. Selkowitz, *Lawrence Berkeley Lab Report No. 500-03-097-A13*, Oct. 2003.

[20] "Web-Based Energy Monitoring System Provides Real-time Mobile Notification and Fault Alerts," *Enetics Inc. Brochure*,2003.

[21] "Advanced Metering, Energy Management and Billing Systems—The Link to Customer Service Innovation," Kellogg L. Warner, *Metering International*, 01/2003.

[22] "Real-Time IT in Electric Markets," Jill Feblowitz, *The Utilities Project Volume 3*, AMR Research, 03/2002.

[23] " International AMR Deployments" 3rd Edition, *The Scott Report*, 2002.

[24] "Enel Telegestore project is on track"; Vincenzo Cannatelli, *Metering International*, 01/2004.

[25] "Remote Monitoring," D. Smith, *Power Engineering*, March 2004, pp. 24-30.

[26] "California Initiative May Mandate Advanced Metering," Roger Levy, *Metering International*, 01/2004.

[27] "Guide to E-Business in the Energy Industry," *Chartwell Energy Industry Report*, Chartwell, 2001.

[28] "Artificial Neural Networks Demonstrated for Automated Generation of Energy Use Predictors for Commercial Buildings," J. Kreider, X. Wang, *ASHRAE Transactions*, vol. 97, part 2, 1997, pp. 775-779.

[29] "Utility Data Web Page Design-An Introduction," D. Green, P. Allen, *Strategic Planning for Energy and the Environment*, vol. 23, No. 2, Fall 2003, pp. 10-25.

[30] "More to Data Aggregation than Addition," W. Causey, *Energy Business & Technology Magazine*, vol. 5, no. 5, June 2003, pp. 55-57.

[31] "Economic Demand Response: Increasing Margins on Commercial and Industrial Customers," Jill Feblowitz, *The Utilities Project Volume 2*, AMR Research, o1/2002

[32] "Key Elements of a Prospective Program: Tracking and Reporting Greenhouse Gas Emissions," *The U.S. Domestic Response to Climate Change*, Pew Center on Global Climate Change.

[33] "Web-Enabled Communication Applications for Remote Access and Equipment Monitoring," *Sensors*, vol. 21, No. 4, April 2004, pp. 14-19

[34] "ISO 14001 Standard Environmental Management system—A Voluntary Mandate and a basis for regulatory Relief," W. Wang, *Business Standards at e-Magazine*, 2001.

[35] "ESCOS and In-House Energy Managers: A Winning Team," by B. Colburn and R. Walawalkar, *Strategic Planning for Energy and The Environment*, to be published 2004.

[36] "FPL On Call—1,000 MW and Fifteen Years Later," Ed Malemezian, *Proceedings of Metering America*, 03/2004.

[37] *State of the Market: 2003,* A report by PJM Interconnection Market Monitoring Unit, 2004

Chapter 11

Computerized Maintenance Management Systems (CMMS): The Evolution of a Maintenance Management Program

Carla Fair-Wright, Service Maintenance Planner
Cooper Compression

ABSTRACT

THERE ARE TWO approaches to acquiring a Computerized Maintenance Management System (CMMS) system: build or buy. Cost and time are key considerations for any project but are crucial factors to assess when selecting a CMMS solution. Managers faced with evaluating any CMMS system must make the build vs. buy decision early in the process. While many businesses choose commercial off-the-shelf solutions, there are distinct advantages to building an application in-house.

This chapter will discuss the software development process and address some of the primary issues managers face when considering an in-house CMMS solution. It will illustrate the key principles for successfully upgrading a CMMS application from a monolithic application model to n-tier architecture capable of supporting http protocol. Both the technical and business aspects of the project will be addressed.

The lessons learned in this case study include: (1) The need to understand that to build smart, flexible, and integrated software there must be reliable and robust tools available, (2) the need to get the cooperation of all interested parties in a large complex project, (3) the importance of project management, and (4) the obstacles faced when incongruent systems are required to communicate.

INTRODUCTION

Inadequate maintenance of energy-using systems is a major cause of energy waste in both the Federal government and the private sector. Energy losses from steam, water and air leaks, uninsulated lines, maladjusted or inoperable controls, and other losses from poor maintenance are often considerable. Good maintenance practices can generate substantial energy savings and should be considered a resource. Moreover, improvements to facility maintenance programs can often be accomplished immediately and at a relatively low cost.[1]

Good maintenance practices are wrapped in a blanket woven from a few simple concepts. First, the equipment in operation must be suitable for the job. Second, there should be an adequate supply of spare parts and a skilled labor force to install them. Finally and most importantly there should be a preventive maintenance program in place which utilizes a good CMMS.

Commercially available software packages do not inherently integrate with existing energy management and control systems (EMCS) as well as property management systems. For an organization running heterogeneous software systems, finding a compatible product can be very costly and time consuming. Customization of some software packages can almost equal the price of the program itself. One way to handle the problem is to develop the software in-house.

In the next few pages the particulars of CMMS software planning and design will be discussed. The goal in writing this chapter is to demystify the process of software construction and provide the reader with a better understanding of CMMS systems and how they work. The goal is not to teach programming, but there are various software topics that are specific to the development process which will be discussed. Also, there are many acronyms in this chapter. They are defined before being used and will also be listed alphabetically in the glossary at the end of this book.

This chapter is divided into four sections: a historical view of equipment maintenance at Cooper Cameron Corporation, a general description of CMMS applications, a technical examination of the application itself—internal structures and data specifications, and testing and deployment strategies.

EQUIPMENT MAINTENANCE AT COOPER CAMERON—HISTORICAL VIEW

Facility Background

Cooper Cameron Corporation is a leading international manufacturer of oil and gas pressure control equipment, including valves, wellheads, controls, chokes, blowout preventers and assembled systems for oil and gas drilling production and transmission used in onshore, offshore and subsea applications, and provides oil and gas separation equipment. Cooper Cameron is also a leading manufacturer of centrifugal air compressors, integral and separable gas compressors and turbochargers. With annual sales of approximately $1.2 billion, the company is divided into three divisions: Cameron, Cooper Cameron Valves, and Cooper Compression. Cooper Cameron operates in over 115 countries around the world with headquarters in Houston, Texas.

The Maintenance Technology Services (MTS) department is part of the Services organization of Cooper Compression. MTS provides customers with value-added, integrated equipment operating and maintenance service solutions, generally in the form of fixed price, risk assumptive long term operating and/or service agreements with varying performance incentive structures.

Service agreements, typically in the form of operating and/or maintenance contracts, can cover total (i.e., scheduled and unscheduled) maintenance, full or partial scheduled maintenance, operations, or maintenance management assistance to customer organizations. These agreements typically cover power and compression equipment with coverage options for related support systems as well as the facilities within which such equipment is installed.

Service agreements offer multiple benefits to Cooper Compression's customers:

- Maintenance-related equipment downtime is minimized so that equipment availability and productivity is maximized
- Maintenance expenditures are defined and controlled throughout the life of the agreement
- Customer fleet and asset management is optimized
- Risk, as well as direct and indirect operating and maintenance costs, is transferred to a qualified, knowledgeable service provider
- The service provider absorbs the peaks and valleys inherent in maintenance resource deployment

The goal of the MTS department is to help the client improve his competitiveness and productivity with effective maintenance management technology. A properly designed service agreement provides quality parts, experienced service technicians, machine shop services and the accumulation of years of experience in predictive maintenance, planning and scheduling.

MTS develops client-specific maintenance plans for both Cooper and non-Cooper manufactured equipment. Every Service Maintenance Planner knows from experience that optimized maintenance planning and scheduling are the foundation of a cost-effective maintenance program. With a good predictive maintenance program in place, there are major improvements in equipment availability, reliability, safety and performance.

Maintenance Management System

Rusty Creekmore, Director, Maintenance Technology Services, Cooper Compression has said:

> For Cooper Compression to be successful in providing operating and maintenance services for our customers' power and compression equipment, we must be able to maximize equipment availability, reliability, productivity and safety at the lowest possible cost. We rely on predictive maintenance technology to accomplish these goals, and our proprietary Computerized Maintenance Management System was developed and is being enhanced specifically to provide Cooper Compression with that competitive advantage.

In the mid 1980s the predecessors to today's Maintenance Technology Services group were unable to find a suitable CMMS software package that met the unique requirements of power and compression equipment. Therefore, they developed an in-house system called simply *Maintenance Management System* or "MMS." Over the years the software proved to be a valuable asset supporting Cooper's field personnel on numerous service agreements. With subsequent enhancements, MMS became the primary tool to process, analyze, and interpret equipment operating data supplied to it from handheld data collectors and from equipment control systems based on operating systems such as Invensys' Wonderware. In the process of several reorganizations in the early 1990s, Cooper's IT department assumed control of the MMS software, but a shortage of computer personnel skilled in Relational Database Management Systems (RDBMS) programming led to limited development and maintenance support. Ultimately, the software languished and by 2001 was in critical need of a major upgrade. Some of the underlying drivers for the upgrade program were:

- The software ran on Microsoft's Windows 98 operating system, which is considered to be obsolete in today's computer technology environment.
- Much of the interface was carried over from the original DOS version with only minor Windows-based enhancements.
- The system architecture was top-down structured and not object-oriented programming (OOP), which is the current industry standard.
- The software could not communicate with the newly installed Plan Maintenance (PM) module of the company's SAP R/3 enterprise management software.
- The software contained irregularities and inconsistencies that were a constant source of discontent.

Unfortunately, not all of Cooper Compression's senior managers understood the software application's critical role in Cooper's operating and maintenance business segment and this presented a challenge in obtaining the resources required to update or replace MMS. Ultimately, however, with the support of Cooper's executive-level Services management, a small project team was tasked to find or develop an appropriate system that would meet the operating and maintenance information requirements for the future.

The MTS Service Maintenance Planners assigned to this team found the CMMS systems that were commercially available at that time were limited in functionality. Their conclusion was supported by several authorities. According to Labib, the new generations of CMMSs are complicated and lack user friendliness. These systems are difficult for either production operators or maintenance engineers to handle. The author describes these systems as more accounting and/or IT oriented than engineering based.[2]

Based on their investigation, the Service Maintenance Planners decided to upgrade the current software which had been written in Microsoft FoxPro version 2.6a. The program had a number of analysis and reporting capabilities as well as equipment operating data collection, filtering, analysis, trending and reporting functions far superior to anything the Planners had found in the current marketplace. Few, if any, of its competitors had attained anything approaching this level of sophistication. The main reason for using FoxPro was its ability to handle the enormous volumes of operating and maintenance data used in the predictive analytical modules. The upgrade was written using the object-oriented relational database management system. Microsoft Visual FoxPro version 8.0.

OVERVIEW OF CMMS

Before we discuss the architecture and functions of a CMMS, we should have an understanding of what we are attempting to achieve in the software design process. What is exactly is a CMMS? What are the benefits of using it?

Prior to 1970, facilities seldom focused on identifying and managing the maintenance function. The primary focus was on direct manufacturing, operations and materials; generally, maintenance was looked upon as a cost of doing business. During the period from 1970 to 1980, there was more focus on documenting maintenance as it relates to preventive maintenance and equipment uptime. Machine maintenance is viewed as a business process that can be augmented by computer software. The role of the maintenance planner and scheduler expanded as it is became necessary to organize maintenance functions into quantitative tasks.. As we moved into the mid 1980's there was an increasing focus on preventive maintenance as well as predictive maintenance. The concept and practice of Total Productive Maintenance (TPM) began to receive attention and become implemented which involved the self directed work force [3]. Traditional maintenance regimes are normally reactive and produce high levels of unplanned expenditures. Technology-driven regimes are more proactive and will typically increase planned maintenance cost but will reduce unplanned and overall maintenance cost while providing increased equipment productivity and profitability.

Significant savings can be achieved by identifying and preventing expensive equipment breakdowns before they occur; an added benefit is that the efficiency of routine maintenance tasks improves. In fact, many insurance companies have recognized that proper use of a

CMMS reduces the chance of costly insurance claims and will reflect this through reduction of premiums. Depending upon the client, unplanned equipment downtime can cost a plant $1 million per day or more.

Studies have shown that optimized reliability and maintenance programs can increase operational availability up to 10%, decrease maintenance costs by 10-20%, and reduce both the number and severity of unplanned process interruptions. [4]

A technology-driven predictive maintenance program can not only decrease maintenance costs but, more importantly, improve operational availability and efficiency, thereby potentially enabling a client to achieve significant financial improvements to the bottom line. An added bonus is that the plant or facility becomes safer because required maintenance is performed on schedule. But the greatest potential utility of CMMS is its data storage capability related to all equipment operating and maintenance functions. For example, optimal use of a CMMS can alert an organization to under-performing or high-operating-cost equipment. It can highlight labor inefficiencies and can provide essential data for Root Cause Failure Analysis to eliminate chronic equipment failures. With this type of information at hand, plant and facility management can make better, more accurate data-driven decisions.

What Does a Computerized Maintenance Management System (CMMS) Do?

CMMSs are often perceived to be no more than a means of scheduling maintenance work. While preventive maintenance scheduling is normally part of a computerized system, most of them are capable of much more. Virtually all aspects of a maintenance department's work can be managed by the modern, integrated software packages. [5]

The following is a list of basic functions all CMMS applications are designed to perform:

1. Managing Assets
2. Scheduling Planned Maintenance
3. Recording Unplanned Maintenance
4. Creating and Validating Work Orders
5. Allocating Labor Resources
6. Monitoring Conditions
7. Reporting Statistics

Managing Assets

This is the entry point for starting a CMMS program. Gathering and storing asset information is an important component in building the maintenance plan. Most CMMS applications will require data on functional category, location, serial number, model number, make, and description of equipment as a minimum to set up the plan. Optional information can range from identifying information (such as an assigned asset barcode number), engineering drawings, or digitized pictures or diagrams of internal systems and components.

Scheduling Planned Maintenance

Tactics win battles, but strategy wins the war. Tactics in the maintenance planning business are the day-to-day operations. The strategy is the maintenance plan, a list of short and long-term objectives. The maintenance plan is simply a listing of all the preventive maintenance (PM) activities for each piece of monitored equipment. This list tells the plant manager or reliability engineer when and how often a specific set of predefined tasks should be performed.

The sample report from MMS shown in Figure 11-1 tells the maintenance planner what maintenance type or level is due for this particular type of piece of equipment and the specific hour meter reading at which the service should be performed. The report's header information consist of a description of the equipment, the period of the scheduled maintenance (1680 hours or 70 days), and the amount of time we will allow for overlap. Just like the odometer of a car, the number of hours a unit has been in operation provides an interval to base maintenance on. The term overlap refers to the time between scheduled maintenance events.

Activities are set by intervals that are either calendar time-based, operating time-based or operating cycle-based. Just as an automobile manufacturer might suggest that engine sparkplugs be changed every 100,000 miles, a machinery maintenance plan may specify that industrial engine sparkplugs be changed every 4,000 operating hours. The Maintenance Planner may import these maintenance task descriptions from an external application or enter them manually. The maintenance plan is the heart of a maintenance program. But as you will see later, monitoring equipment condition is its soul.

Recording Unplanned Maintenance

The history of a machine is used to analyze and improve performance and streamline expense. Therefore, a CMMS must be able to record unplanned work (repairs and/or breakdowns). What, when, and why a failure has occurred is important information that can be used to re-evaluate present maintenance practices or identify possible operating, process or engineering problems. Analysts can review the repair information and

REPORT DATE: 10/11/04

Cooper Compression - Customer Maintenance Services
MAINTENANCE MANAGEMENT SYSTEM
MAINTENANCE FORECAST

EQUIPMENT CODE: CEMFSACTCPAAAAAAAAAA
EQUIPMENT DESC: SUPERIOR 16SGTB/W74 COMPRESSION UNIT
MAINTENANCE NOW DUE OR DUE WITHIN: 1680 HOURS, 70 DAYS FROM TODAY'S DATE
OVERLAPPERCENTAGE: 80%HOURS, 80%DAYS

LEVELS INCLUDED: ABCD

	|<------- HOURS ------>|			|<------ DAYS ------>|		
LEVEL	**LAST**	**NEXT**	**DUE IN**	**LAST**	**NEXT**	**DUE IN**
A	N/A	N/A	N/A	05/10/04	10/11/04	0
A	117056.90	117392.90	0.00	N/A	N/A	N/A
A	N/A	N/A	N/A	05/10/04	10/25/04	14
A	117056.90	117728.90	336.00	N/A	N/A	N/A
A	N/A	N/A	N/A	05/10/04	11/08/04	28
A	117056.90	118064.90	672.00	N/A	N/A	N/A
A	N/A	N/A	N/A	05/10/04	11/22/04	42
B	N/A	N/A	N/A	04/28/04	11/22/04	42
A	117056.90	118400.90	1008.00	N/A	N/A	N/A
A	N/A	N/A	N/A	05/10/04	12/06/04	56
A	117056.90	118736.90	1344.00	N/A	N/A	N/A
B	116783.90	118783.90	1391.00	N/A	N/A	N/A
A	N/A	N/A	N/A	05/10/04	12/20/04	70
A	117056.90	119072.90	1680.00	N/A	N/A	N/A

Oil is checked every 14 days or 336 hours of operation

Calibrations and alignments are check every 42 days or 1000 hours of operation

Figure 11-1. Scheduled Maintenance Report from MMS

fashion questions to further refine the maintenance plan. For example, a planner may find by reviewing unscheduled work orders that the cost of changing a component during the preventive maintenance (PM) activity to prevent a certain type of failure is less costly than making unscheduled repairs and incurring unscheduled process downtime and loss of equipment production.

Creating and Validating Work Orders

The work order will detail estimated and capture actual labor hours for service or repairs, the labor skills required, type of work (repair or PM), instructions for performing the work, required parts, tools, expendable materials, and associated accounting costs. Some CMMS applications can generate formal work orders and some can not. But all CMMS programs must be able to validate the work orders for any asset within the database. Validation is necessary because parts and labor are charged to the work order. This information is a critical part of preventive maintenance; it determines the allocation of man, material, and time resources.

The information in a work order can be essential in diagnosing the health and life expectancy of the asset. Even if the application is not used to create the work order, it should save the critical actual "as performed" elements of the work order into a maintenance history file. This can be done via an interface to the company enterprise system or by importing data pulled from it.

Allocating Labor Resources

Keeping track of maintenance and repair labor costs is another benefit of a CMMS. Labor time and cost data can be captured via a variety of technologies including: badge-based, biometric, touch-screen kiosk, mobile/wireless, telephone/Interactive Voice Response (IVR), and PC/Web-based devices. Once collected, this information can be ported to the CMMS or keyed in manually.

Monitoring Conditions

Is it possible to predict potential failure and prevent costly repairs? Yes, it is possible—although only 31% of companies operating in the United States are doing so.[6] Studies done as recently as late 2000 still show reactive maintenance as the predominant mode of maintenance in the United States. Reactive maintenance, or the "run it till it breaks" maintenance mode, is the easiest maintenance model to implement. But, it usually results in the greatest overall equipment costs over time.

Condition monitoring is the beloved companion of Predictive or Condition-Based Maintenance (CBM) techniques. It refers to the use of advanced technologies to determine the condition of a mechanized area of a plant or equipment. It is based on the idea that identifying certain changes in the condition of a machine will indicate that some potential failure may be developing.

Once the physical characteristics are pinpointed that identify the normal condition of a machine, they can then be measured, analyzed, and recorded to reveal certain trends. Sensors may be used in sampling operational parameters. These sensors can detect flow rate, motion, weight, quantity, phases of electrical power and much more. The sensors usually come with switch contacts in an opened or closed state. A change in state alerts the operator to possible problems in the plant or facility.

Another method used to gather status information includes the use of embedded equipment control software and equipment monitoring software. At set intervals a reading is taken from the monitoring points and transmitted to a database with the origin, date, and time. Of course, not all points can be monitored by software. Figure 11-2 shows a snapshot of the upload screen from MMS. The software is flexible enough to retrieve information from a variety of sources.

If the signal exceeds or falls below a pre-determined range, or the rate of change of the measured parameter exceeds a pre-determined rate, then an alarm status is returned. But, this simple alarm system is not the only technology applied, as seen in Table 11-1.

Statistical Reporting

A CMMS is only valuable if you can access the information in it. It should be able to provide immediate answers in a repair situation or intricate details for a long-term equipment utilization, availability and reliability, cost or condition analysis. In fact, one of the most significant functions of a CMMS is the report generator.

Reports enable the analyst to make efficient use of the extensive information contained within a CMMS for fault analysis, costing and work statistics. Most systems supply a number of pre-defined report queries with customization capabilities. A good system will provide powerful ad-hoc querying with multiple report output formats, including Word, Excel, HTML and PDF. A diffi-

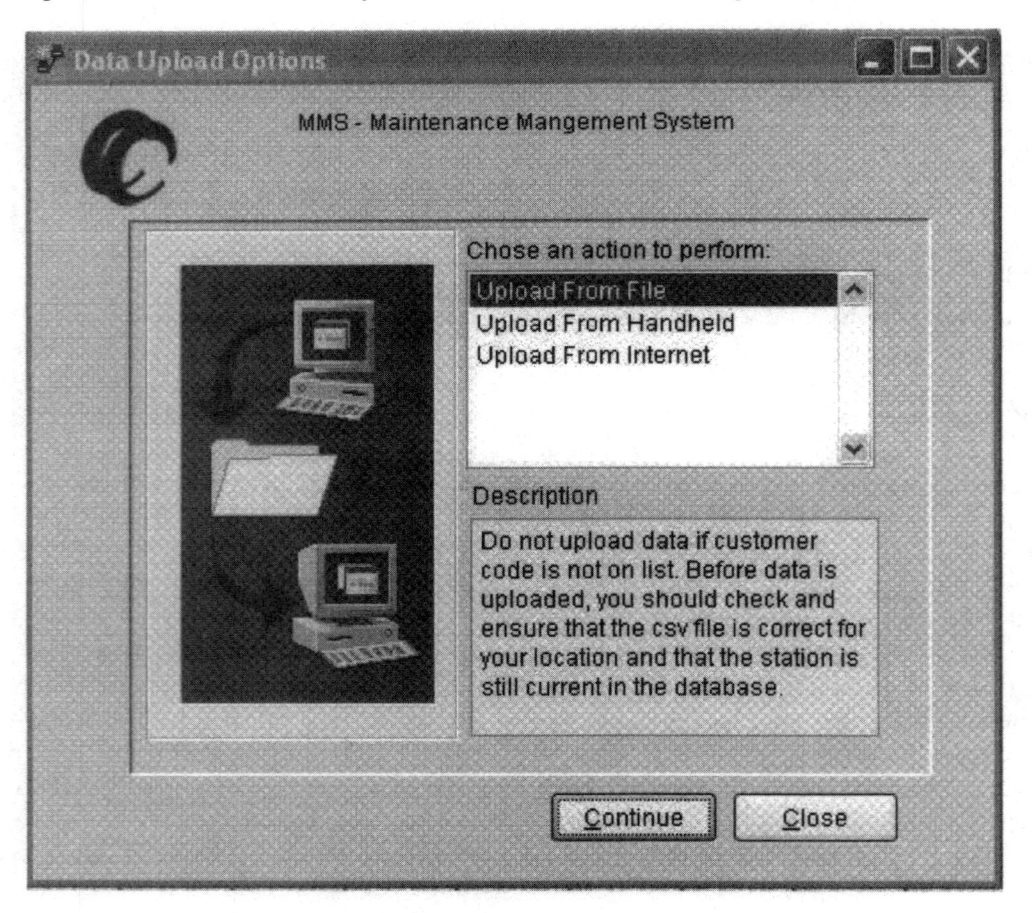

Figure 11-2. Upload Screen from MMS

Table 11-1. Condition Monitoring—adapted from Rockwell Automation[7]

Condition Monitoring Tools	
Vibration Analysis	The Vibration Collection and Analysis process monitors the response of the equipment to internal and external forces being applied. The response is measured by a general-purpose transducer at the pump and motor bearings and is passed to an analyzer for interpretation. The Analysis can provide early indications of problems such as machinery imbalance, misalignment, bearing wear, worn gears, etc. The results can be downloaded to the CMMS or asset management software for analysis and recommendations.
Temperature Analysis	There are a variety of devices used to measure temperature. These devices will be designed to measure either surface or gaseous temperature. Normally, thermocouples or templugs are used to determine surface temperature from inside an engine
	Fine wire resistance thermometer, Infrared, Spectroscopy, Hot Wire Anemometry (HWA), or 3-Colour pyrometry are used for gaseous temperatures. The concept of a thermometer and infrared are familiar to most and work as implied. Spectroscopy measures the intensity of light emitted at wavelengths in the infra red, ultra violet or visible light spectrum by a gas. Hot wire anemometry is a technique using a thin wire stretched between the tips of two prongs to measure time resolved flow in the cylinder. Three color pyrometry looks at the relationship between the temperature of an object and the intensity of the radiation that it emits. Identifying an abnormal rise in the temperature of machinery that could result from problems such as bearing wear, lack of lubrication, poor electrical connections, etc.
Oil Analysis	The oil sample is taken and sent to a lab for analysis where it is subjected to physical testing, spectroanalysis, and particle testing. The physical testing looks at viscosity and water content greater than 1%. The oil is heated and run through the viscosity bath. The results are then compared to the new oil specification. This test is valuable in determining the condition of the oil and an indicator of water contamination and oxidation. Spectroanalysis examines metal content and additive package. This test checks around 19 elements and reports them in parts per million. The particle count is a critical part used to measure the efficiency of the system filtration. The particle count measures all particulate in the oil larger than 5 microns. Particulate include: dirt, carbon, metals, fiber, bug parts, etc. Using either laser or optical methods does this. The laser method reports the quantity, size and distribution of particulate, but not what they are. The optical method gives a quantity, size, distribution and identification.

Table 11-1. (*Continued*)

Oil Analysis (*continued*)	Controlling viscosity and other chemical impurities is the key to preventing premature failure of mechanical systems such as bearings, gears, etc.[1]
Motor Current Signature Analysis	Analyzing the unique signature patterns in the motor supply current created by fluctuations in load, high resistance joints in rotor bars and end rings or uneven air gap between the rotor and stator.
Operating State Dynamic Analysis	Correlation of machinery operating parameters such as load, pressure, speed, etc. to the dynamic characteristics of the machinery such as vibration and motor current signature.
Ultrasonic Leak Detection	Use of ultrasonic measurements to identify changes in sound patterns caused by high-pressure gas/steam leaks, bearing wear or other deterioration.
Balancing	The leading cause of imbalance in rotating machinery is vibration. The imbalance may vary with time and can be hard to correct. Without detection and adjustment, machine problems can occur through mechanical stress. In the case of power tool operators, long-term physical harm to humans called "White Finger" can occur. This is a condition in which the blood supply is interrupted to the hand. Vibration monitoring equipment can enable maintenance staff to detect, diagnose and correct misalignment and imbalance in rotating machinery.

[1]*http://www.oillab.com/oil.html* web site contains one of the best descriptions of the process.

cult challenge for any complex field of endeavor such as the facility maintenance function is sharing meaningful information with executive management. The reports generated in a first class CMMS are readable and easy to interpret.

One of the advantages of using a web-based system is that these reports can be shared independent of the platform. The data can be reviewed in real time via any internet-connected computer that has a browser. Previously, the system users created complex accounting reports manually by exporting data from MMS into Microsoft Excel, performing various calculations and then formatting as needed. The report was then ready to be distributed by email. Now they can simply press a button to get the same information in all the right formats—from basic reports to reports for the executives to government-mandated reports.

THE APPLICATION INTERIOR

Software Lifecycle Planning

A lifecycle is the procedure by which a piece of software is created, implemented, tested, and finally maintained. There are a variety of lifecycle systems in use today and the numbers continue to grow as internet programming derails the static models built in the last decade.

Every software-development effort goes through a "lifecycle," which consists of all the activities between the time that version 1.0 of a system begins life as a gleam in someone's eye and the time it is retired from production. A lifecycle model is a prescriptive model of what should happen between first glimmer and last breath. [8]

The waterfall method developed in the 1970s is one of the oldest lifecycle models still in use. In the waterfall method a project moves sequentially through non-overlapping phases. These transitions from one phase to another, often called "throwing it over the wall," reflect the assembly-line mentality of the method. Each phase is fully documented and must be completed before the beginning of the next phase. The waterfall method is considered too hard and too expensive to implement in its pure form. To capture all the system requirements and furnish a complete analysis before the design process can begin is extraordinarily costly in terms of time and manpower.

The sashimi model, a modification of the waterfall method, allows overlapping of the stages. This normally results in a cleaner product definition, but it makes project management difficult to chart and can lead to ambiguous milestones. The sashimi model is best suited for software projects that are well understood, but intricate to design. This is because the complexity can be tackled early on in the project lifecycle during the requirements gathering phase. CMMS programs are enormously complex, so for our project, we used the sashimi lifecycle model.

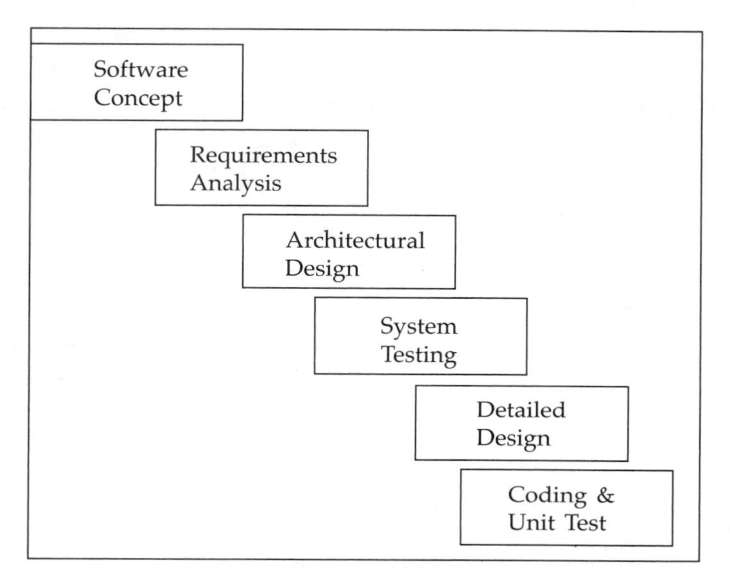

Figure 11-3. Sashimi Model—First implemented by a Japanese hardware manufacturer. Sashimi is the Japanese method of presenting fish sliced raw with the sides overlapping.[8]

Software Concept

Our MMS system had been operational since 1980 and was reflective of the command and control hierarchies of 20 years ago. Much of the work in the conceptual phase of the update was refinement and expansion of the original system. The application had to be enhanced to work with internet protocols and employ standard security and encryption technology. This revision recognized the reality that applications are beginning to migrate from a standalone executable sitting on a user's hard drive to a distributed application delivered by a Web server across the Internet. MMS needed to be redesigned to utilize Extensible Markup Language (XML) based documents. The Extensible Markup Language is the latest offering in the world of data access. XML is quickly becoming the universal protocol for transferring information from site to site via HTTP.

Requirements Analysis

In this phase of our project, we gather and itemize the high-level requirements of the system and identify those basic functions that the software must perform. These requirements are called **business rules**. According to the Object Management Group, business rules can be thought of as conditions that *must* be satisfied for business operations to be correct. [9] The business rule basically tells us what a system should do and what it should not. For example, a business rule would state that every piece of operating equipment must have a serial number or some identifying tag.

In capturing the business rules we used Ivar Jacobson's [10] approach to documenting system requirements. This object oriented approach called "use case modeling" has become the de facto standard in modern development methodology. Use cases define the system in short narratives. They help keep requirements gathering simple and understandable for all involved. Here is a short example of a use case:

Entering Update Information:

A service technician receives a work order for scheduled maintenance on compressor unit #4. He performs the 2000 hour maintenance. He completes the work order and submits it to the office. The data entry person logs into the system and enters the maintenance details, which the system validates and records. The system updates the resource data (labor and parts) and the schedule. The data entry person runs a new schedule for the upcoming week.

A formal use case is more involved than this, but the basic concept is the same: discover and write functional requirements by creating a narrative of the steps. The process begins with the domain expert. This is a person with extensive knowledge of the equipment or system. We were fortunate enough to have access to several domain experts at our facility, who could provide a rich and detailed accounting of the system.

ARCHITECTURAL DESIGN

A major part of any software project is the architectural design. Architectural design is the organization of data and other program components necessary to build a computer-based system. It encompasses the identification of all interfaces, relationships, and database layouts. It can also require the use of notation systems like the Unified Modeling Language (UML), which is a graphical shorthand for constructing and documenting software applications. The Unified Modeling Language (UML) is the successor to the wave of object-oriented analysis and design (OOA&D) methods that appeared in the late '80s and early '90s. [11]

From Monolithic to N-tier

The original DOS-based MMS was created in the monolithic configuration model. In computer programming, the term monolithic is used to describe applications where all the logic and any resources it may need to function are contained within the application. Monolithic software is confined to a single computer and cannot support multiple users on a networked drive. Some operating systems fall into this category. Monolithic software is simple, but it is stunted by the inflexible structure that defines it.

As the need for applications that could be shared across a network grew, a change in the way applications communicated with the data source was needed. The term client-server was first coined in the 1980s to describe the personal computer (PC) on a network. The client was the application on the user's PC. The server was a powerful computer with processes committed to running the drives (file server), printers (print server), or network traffic (network server).

The client-server concept was extended to mean breaking logic and data apart into modules and deploying the data across a network. Initially there were two layers: the application and a database server. The application consisted of a graphical user interface (GUI) and the code to control the data retrieval and updating. The GUI is exposed to a user; it is the screen or web page, a place to enter or read information. The design of the GUI determines a system's usability or user-friendliness. The data access layer is simply the database or a collection of files where the data is stored. Web applications introduced the three-tier model by default: the browser is the client tier, the database the back-end tier, and the web server and its extensions became the middle tier. [12]

We used a three-tier architecture for MMS. Popularized by Rational Software and Microsoft, three-tier architecture breaks the GUI and logic apart. The logic is moved into a third layer sandwiched between the GUI and persistent storage (databases). This is the business logic ("business rules") layer or the middle tier. The BusinessRules layer serves to implement all logic that is outlined in the system requirements. It validates data, performs calculations, and reads and writes data. When the middle tier itself is layered, the overall architecture is called "n-tier architecture."

Middleware Components

The software talks to the data source (server) with the help of Open Database Connectivity (ODBC) drivers, which *translate* the application's requests to the database. Because it resides between the two entities, the ODBC portion is called middleware. Transmission Control Protocol and Internet Protocol or TCP/IP[1] are middleware

[1]The higher layer, TCP, manages the assembling and reassembling of a message or file into smaller packets for transmission over the Internet. The lower layer, Internet Protocol, handles the address part of each packet so that it gets to the right destination.

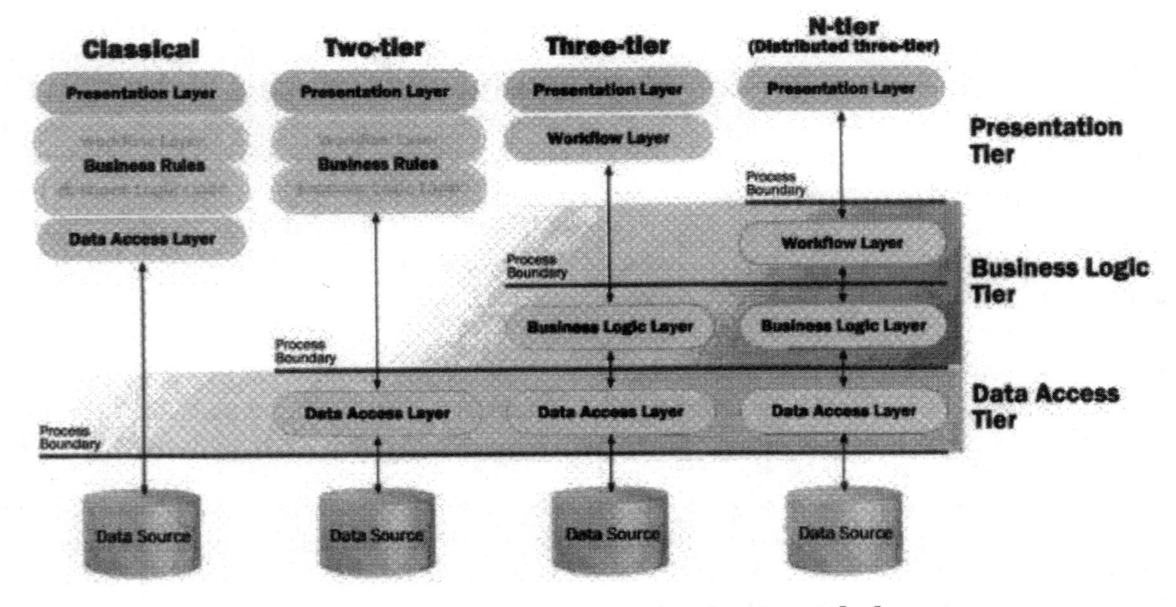

Figure 11-4. View of the Application Layers[13]

components and allow end-to-end communication from one point (computer) across the Network to another point or host computer. Without TCP/IP, dubbed "the cockroach of protocols" by Microsoft guru Don Box, the internet could not exist.

Detailed Design

Object-oriented programming (OOP) is a programming language model structured around "objects" rather than step-by-step routines. In top-down structured programming, the problem is broken into natural pieces and each piece is solved independently in sequence. Logic flows from the top to the bottom. Data is input, processed, and output.

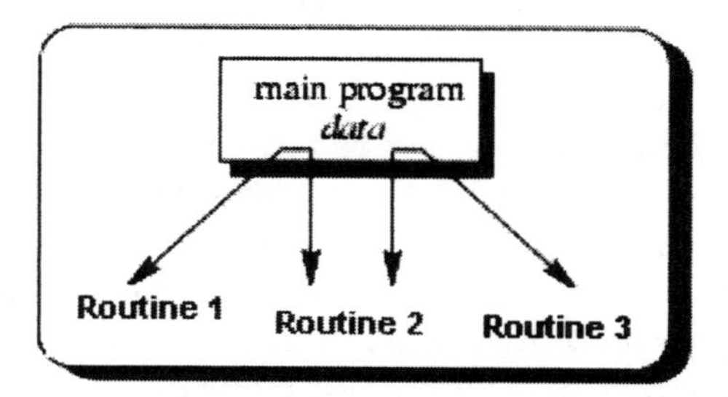

Figure 11-5. Program Constructs

In object-oriented programming we build islands of code and call them objects. These objects perform specific functions. They communicate with each other by sending and receiving messages. Each message contains a request for a particular set of instructions to be executed by the receiving object. We refer to a block of instructions as a method. An object can have as many methods as it needs to perform the tasks assigned during the design phase.. The term "class" refers to the static form of object. If you think of an object as a house, then the class would be the blueprint.

Data modeling is the first step in OOP. The purpose is to find and identify all the procedures that are used in a business and determine the relationships among these procedures. The next step is to refine these procedures into classes that have properties and methods. Properties define the kind of data the object contains and methods are the logic sequences that can manipulate it. The class is a nothing more than a blueprint.

In our system, we began with several base classes and used them to define the objects in our system. For example, we have a print service class to handle all external outputs from the system. There is also a data service class that talks directly to the tables and other files. Then there are a plethora of functional classes, each with its own specific identity and function. The customer class knows what tasks and what data to use when we need a customer process. The maintenance plan class understands maintenance planning functions. When these "classes are in use" they become objects. An object is an instance of a class. They talk among themselves by passing messages and share tasks and information. As in a team of players, each object is as a distinct "person" with a common goal.

Relational Database Management Systems (RDBMS)

To support our n-tier strategy we needed a relational database on the back-end. A relational database is a series of tables that represent entities related to one another. We began by mapping all fields within each

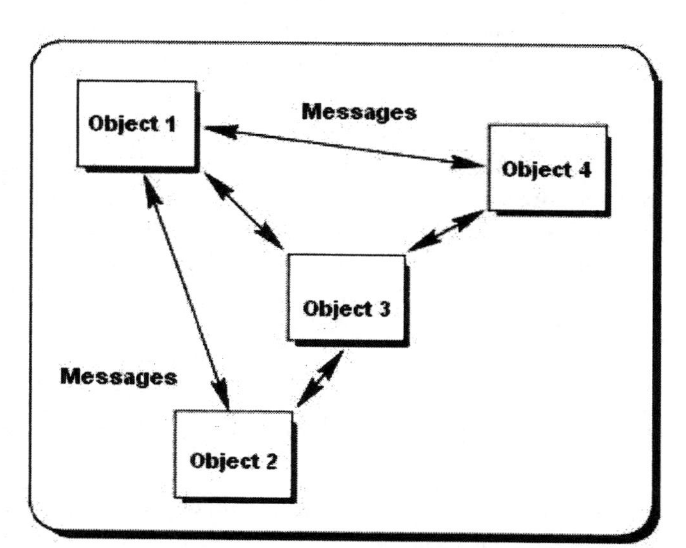

Figure 11-6 shows a simple interaction of several objects.

Object 1 has sent messages to objects 2, 3, and 4 telling them to execute one of their methods.

Object 2 and Object 4 receive the message execute the code and return a success message to Object 1.

To process the method on Object 3 additional information from Object 2 and Object 4 is needed. Object 3 sends and receives a message to the other objects. Once the information is received Object 3 continues processing the original request and returns a message back to Object 1.

Figure 11-6. Object Constructs

table and describing each field's function in the program. After each field was defined and categorized, we determined if the field existed in more than one table. If so, it was moved to a single table. The field's destination was based upon the dictates of the business rules and the relational model.

For example, our business rules for the customer told us that a customer can have many addresses. There is a shipping address, billing address, and the personal contact address. The address portion of the CUSTOMER table was removed and placed in a separate ADDRESS table. Why? Mathematics is at the heart of relational database theory. Based on predicate logic and set theory, the relational model tells that we must remove any unnecessary redundancy to be efficient. To store multiple customer addresses in the CUSTOMER table meant repeating the customer name and other details for each address. By moving the address to its own table and placing an identifying key on each record to tell us which customer the record belongs to, we saved space and increased our ability to maintain the integrity of the various relationships between tables.

To design the database, we used xCase from F1 Technologies and Visio. The xCase software allowed us to conceptualize the table structures. It also permitted us to relationally view and edit the physical data with integrity. For instance, if we decide to delete a customer from the database, it is not enough just to remove the customer from the CUSTOMER table, for this would leave dangling references to that customer in the ADDRESS table, among others.

Data flow diagramming shows business processes and the data that flows between them. Think of a process model as a formal way of representing how a business operates. To capture these details we used Visio, a Microsoft application for business and technical drawing and diagramming.

Coding & Unit Testing

Coding is the construction phase of the software. It is often the longest and most difficult part of the process. Earlier we defined a method as a set of coding instructions contained within an object. A good design will have many small methods in separate classes, and at times it can be hard to understand the over-all sequence of behavior. Unified Modeling Language or UML was conceived out of a need for a standard notation for modeling system architectures and behaviors. In the coding the new system we used UML models to sequence the movement of data through the system. These sequence diagrams were produced from a modeling software package called Visual UML made by Visual Object Modelers, Inc.

Sequence Diagrams

UML sequence diagrams model the flow of logic within the system in a visual manner. They enable the software designer to document and validate the program logic, and are commonly used for both analysis and design purposes. Sequence diagrams are a favorite tool for dynamic modeling.

Sequence Diagrams are about deciding and modeling. It details how the system will accomplish what the use case models have defined in the requirements gathering phase. Here is where we decide and describe how the system will create and store maintenance schedules or how adding a new equipment item will be handled.

Unit Testing

With the use cases, we can test each functional area of the program to make sure it is working. By daily testing, using a technique called "build and smoke" testing, the errors are caught and repaired quickly and the process moves forward. Once unit testing is complete, the software is assembled and made ready for system testing.

System Testing

This is the final phase before deployment of the software. The goal of system testing is to verify that the code satisfies all requirements. The objectives of system testing are as follows:

1. To measure the stability of the system and determine operating system specifications.
2. To measure the system's performance. (Issues such as speed and accuracy of the data retrieval process)
3. To measure the overall usability of the system.

Once the system has passed system testing it can be deployed. Our application has two user interfaces: a browser interface for our sites with web access and a standard application interface. Deployment can be done from the company's intranet or the software and support files can be distributed on media such as a CD or a flash memory USB drive.

SUMMARY AND FUTURE DIRECTIONS

This chapter has reviewed the development history of Cooper Compression's in-house CMMS system called Maintenance Management System (MMS). Computer-

Figure 11-7. Sample Sequence Diagram

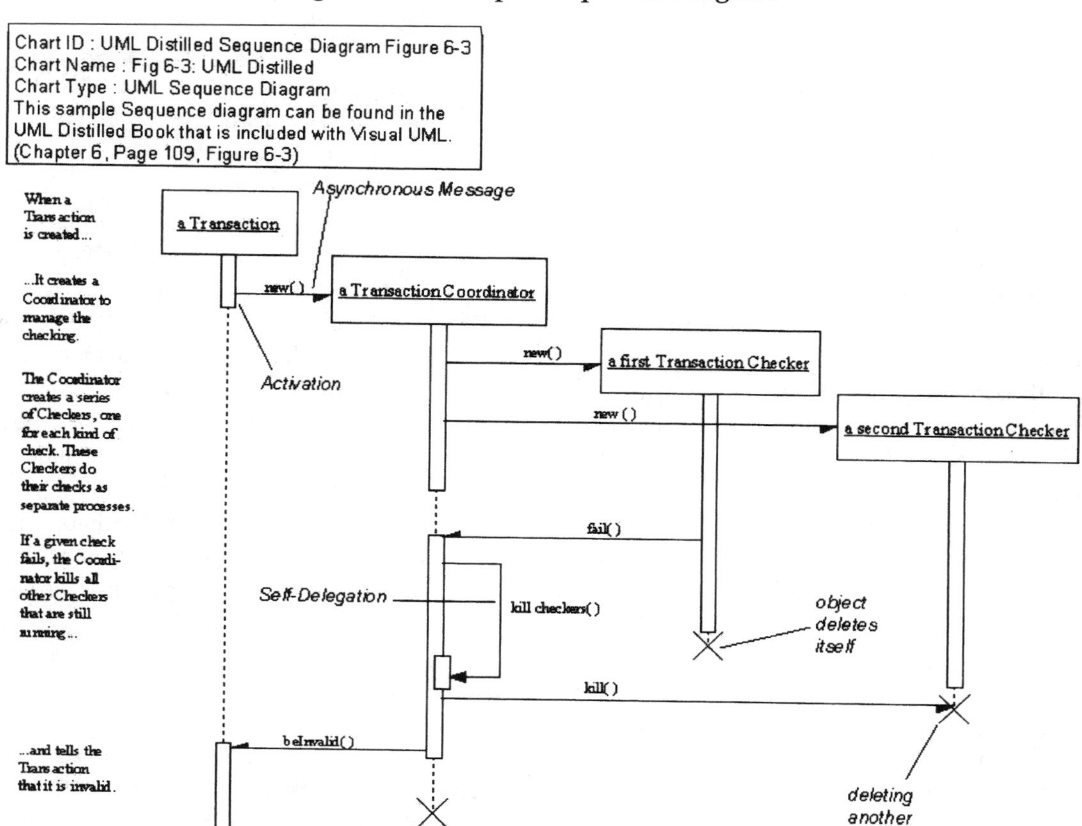

ized Maintenance Management System software maintains a schedule for preventive maintenance on operating equipment. The software can generate complex analytical reports and record the maintenance history of any machinery. Compared to periodic maintenance, predictive maintenance is what is known as condition-based maintenance. It records trend values, by measuring and analyzing data and uses a surveillance system, designed to monitor conditions via an on-line system.

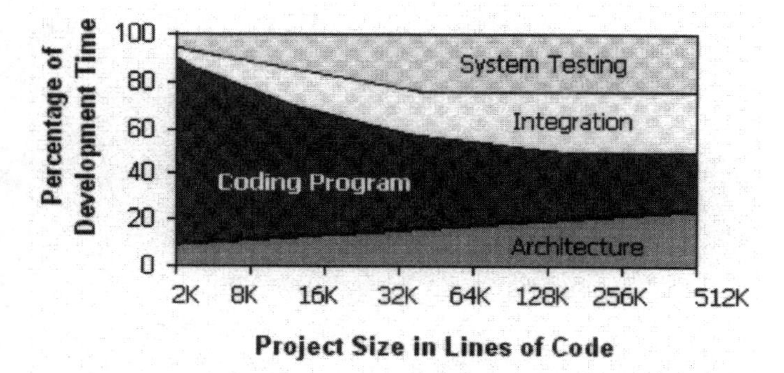

Figure 11-8. Determining Development Time by Software Size[8]

A good CMMS will increase the effectiveness and overall efficiency of any company's maintenance program. MMS is helping the Maintenance Technology Services department control costs, save administrative time, and increase customer service and satisfaction. And, because MMS was designed to be a scaleable system, it supports the client's needs today and will continue to do so well into the future.

As a rule, web-based systems can be rapidly deployed. To implement a standard CMMS system can be complex and normally requires the coordination of several departments. One of the chief benefits of using web-based software is that the software does not have to be installed on the computer, all support and maintenance is handled on the back end at the server. This means there are no computers to configure or costly hardware upgrades. Having the software run from a single source also simplifies troubleshooting and error detection efforts.

Another advantage of using web-based software is mobility. Web-based programs can be designed to operate in the Occasionally Connected Computing (OCC) environment. In the past, maintaining an accurate history of equipment in remote locations was an involved process for Cooper Compression field service personnel. Large volumes of data entered into the system from remote areas had to be manually uploaded to the LAN for planners to review and evaluate.

By enabling the program to work off-line, technicians can enter data readings into a hand-held or laptop for later synchronization into the database. Because the upload process is automated, the time to populate a data source is streamlined. Computer resources are used efficiently and "manual" steps in the data synchronization process are reduced.

Today, with competition in industry at an all time high, cost effective practices may be the only thing that stands between growth and failure for some organizations. Technology-driven predictive maintenance programs have been proven to work. These practices and tools can be adapted to work not only in industrial plants, but in building maintenance, transportation, construction, and in a variety of other industries.

References

[1] Sullivan, Pugh, Melendez, Hunt, "Operations and Maintenance (O&M) Best Practices Guide," US Department of Energy, 2002

[2] Labib, A. W., "Computerized Maintenance Management Systems—A Black Hole or a Black Box," Maintenance Journal, Vol. 17 No 1, 2004

[3] Oberg, P., "Managing Maintenance as a Business," EPAC Systems Handbook, 2004

[4] Universal Oil Products (UOP), "Reliability, Availability, Inspection, and Equipment Support Services," *http://www.uop.com/services/2060.html.*

[5] Weir, "An Impartial View of CMMS Functions, Selection and Implementation," The Plant Maintenance Resource Center, 2002

[6] US Department of Energy, *http://www.eere.energy.gov*

[7] *http://domino.automation.rockwell.com/applications/gs/region/gtswebst.nsf/pages/Integrated_Condition_Monitoring*

[8] McConnell, S., "Rapid Development," Redmond, Wash, Microsoft Press. 1996.

[9] Ulrich, W., "Legacy Business Rule Capture: Last Piece of the Transformation Puzzle," *http://www.systemtransformation.com/IT_Arch_Transformation_Articles/arch_legacybrc.htm*, 2004

[10] Jacobson, Ivar et al, "Object-Oriented Software Engineering: A Use-Case Driven Approach," Reading, MA, Addison-Wesley, 1992

[11] Fair-Wright, C. "UML Distilled: Interview with Martin Fowler," Component Developer Magazine, Issue 3. 2001. p34

[12] Komatineni, S., "Qualities of a Good Middle-Tier Architecture," *http://www.radford.edu/~wkovarik/design/ch1.html*, 2003

[13] "Microsoft Architecture Decisions for Dynamic Web Applications: Performance, Scalability, and Reliability," *http://msdn.microsoft.com/library/en-us/dnnile/html/docu2kbench.asp*

Bibliography

Automated Buildings website, *http://www.automatedbuildings.com*

Coleridge, R. (1999). "An Introduction to the Duwamish Books Sample Application," *http://www.aspxnet.de/XML_RK/visualstudio/duwamish/dw1intro.htm*

Edelstein, H. (May 1994). "Unraveling Client/Server Architecture," DBMS 7, 5: 34(7).

Federspiel, C.C. and L. Villafana. (2003). "A Tenant Interface for Energy and Maintenance Systems." CHI 2003, Florida, USA. ACM 1-58113-630-7/03/0004.

Levi, M., D. McBride, S. May, M.A. Piette, and S. Kinney. (2002). "GEMnet Status and Accomplishments: GSA's Energy and Maintenance Network." Proceedings of 2002 ACEEE Summer Study on Energy Efficiency in Buildings. LBNL-50733.

Norman, D. (1998). The Invisible Computer. Cambridge, MA: The MIT Press, 113-115.

Chapter 12

Using Virtual Metering to Enhance an Energy Information System

David C. Green
Green Management Services, Inc., Fort Myers, Florida
dcgreen@dcgreen.com

Paul J. Allen
Walt Disney World, Lake Buena Vista, Florida
paul.allen@disney.com

ABSTRACT

THIS CHAPTER explains how to enhance an Energy Information System (EIS) to do more than just report on energy consumption recorded at the various meters throughout a facility complex. Energy managers would like to see more details about their facilities. For instance, they would like to report consumption from some fraction of one meter. Or, perhaps they would like to report on a metered area represented by more than one meter (as if it was one meter). Defining *virtual meters* in terms of *real meters* makes this task much easier than it seems. An EIS administrator can define *virtual meters* in terms of percentages based on floor area (square footage) or some other criteria. A *virtual meter* might also be a combination of *real meters* represented by some arithmetic expression. We discuss how to create, configure and apply *virtual meters* in an Energy Information System.

INTRODUCTION

Automobile dashboards these days are reporting more and more details to us about the operation of our vehicle. We can see miles per gallon, miles to go before refueling, outside air temperature and many other useful details. This availability of information helps us drive safer and more comfortably. The commercial building controls industry is moving in the same direction. Centralizing the structure of an energy management and control system increases the volume of information available at any one place and time. Sensors hooked into a computer network can provide data continuously. Therefore, the possibility for improved efficiency is enhanced. Energy Information Systems also take advantage of this centralized structure to gather information about energy consumption in an effort to report energy consumption, evaluate and fine tune energy conservation programs or perform billing services. However, what if not all of the sensors or real meters are there? Many of the critical areas in facility complexes today are not monitored by sensors. This is where *virtual metering* may help to fill the void. [1]

"Virtual Metering" is a commonly used, but somewhat loosely defined term that needs some explanation. First of all, *virtual meters* do not completely replace the function of sensors or real meters. Virtual meters can typically provide an *estimate* of actual consumption for reporting or cost reimbursement purposes. Engineers use sub-meters or *information meters* to obtain actual data "downstream" of another meter. Sub-metering is common with apartment complexes that wish to apportion the actual amount of electric and water use to each tenant. *Virtual metering* would apportion the total electric or water consumption equally across all tenants for the purposes of cost reimbursement. E-MON, a sub-metering and energy monitoring system, uses virtual meters "...for (energy used in) common areas (such as security lighting, clubhouses and pools) to spread these energy costs across all tenants pro rata." [2]

However, engineers can use virtual meters to report *actual* consumption data in certain circumstances. Let's say a particular area "A" does not have its own meter or sensor but is part of a larger area tracked by a *master meter.* Information meters as well as the master meter track other adjacent areas "B" and "C." A virtual meter can accurately report on the un-metered area by subtracting the total values of all the information meters from the values for the master meter. Energy Saving Technologies, Inc. uses a virtual meter to "...calculate the remainder resulting from the sum of all tenant's meters being subtracted from the main meter." [3]

Northeast Utilities (NU) uses virtual meters to calculate energy use for areas where it would be very expensive to install a meter or to report on accounts with multiple meters. NU calculates values for virtual meters from actual meters using mathematical expressions. [4]

Virtual meters track energy use in terms other than just the metering technology used to collect the data. In their simplest form, virtual meters are defined in terms of floor area (square footage) for a particular living or work area, and they report energy consumption for that area. They report data just like any other electric, gas or water meter except that there is really no meter there and the values may only be an estimate of actual consumption. This certainly opens the door to other tracking opportunities. Areas defined in terms of square footage can be categorized as to how they are used; this allows us to estimate energy consumption at a more detailed level instead of simply by meter location. For instance, a school that adds a temporary classroom adjacent to a building to accommodate increased enrollment can create a virtual meter to report estimated consumption as part of the consumption for an entire building. This provides a meaningful record of the effect of the temporary facility on consumption. Reporting actual meter data alone does not always give us the detail needed to track energy use effectively.

WHY USE VIRTUAL METERING?

EIS data collection only goes so far toward the entire reporting needs of an energy management office. There may be a limit to how much sub-metering an organization can afford to install. Therefore, some areas may not have data available specific to their boundaries and so their energy use remains unreported. Also, as an organization evolves over time its infrastructure may not continue to correspond to the physical layout of utility supply lines and metering capabilities. Relocating those supply lines and meters may not be practical. Using virtual metering to *fractionalize* or *combine* real meter data for reporting purposes fulfills those extended needs of an energy management office.

In a typical EIS the meter data is the lowest level of reporting capability. So, if each building has only one meter, then buildings are the lowest reportable area. Large buildings today may contain multiple operational areas that require individual reporting capability. A virtual meter might define these areas as a percentage of the floor space of the real metered area. The EIS divides the consumption recorded among the areas in the same proportions as the floor area. However, the EIS administrator may base the formula for proportioning the consumption on *any* criteria. For instance, if multiple departments share the fiscal responsibility for the building, the percentages would be established accordingly and utility cost reimbursement made much easier. An EIS *fractionalizes* the data for a real meter into two virtual meters as shown in Figure 12-1 below. [5]

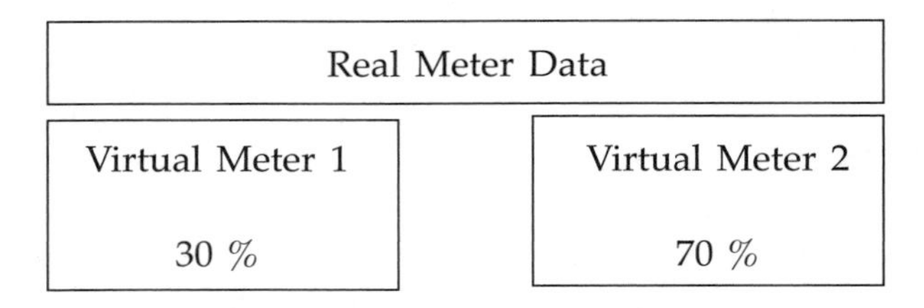

Figure 12-1. Fractionalized Meter Data

In other circumstances, combining real meters is more practical than fractionalizing them. Since the administrative layout of an organization may not match the infrastructure layout of the complex, it may become necessary to combine real meters to obtain more meaningful reports. For certain reasons it may be simpler to report an area served by multiple meters as if it were one meter. Below Figure 12-2 shows how an EIS *combines* the values from three real meters into values for one reportable virtual meter.

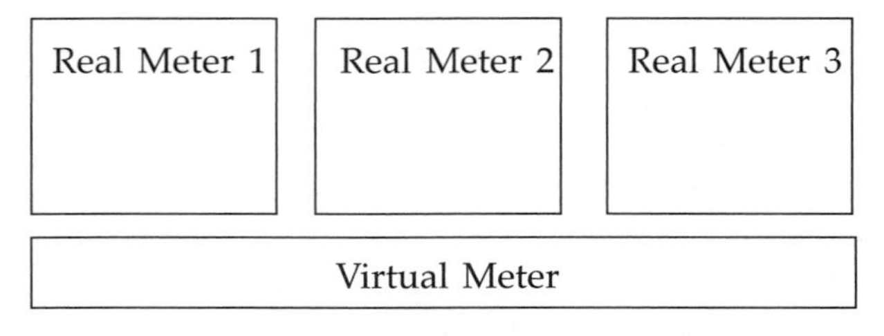

Figure 12-2. Combined Meter Data

Some areas may not be metered at all. In this case, an expression relating the sum of all metered areas to the values recorded for the substation feeding the areas (as well as the *un-metered area*) can produce reportable data for *the un-metered area.*

Figure 12-3 below shows how values for an un-metered area can be reported using a *combination* of real meter data. The recorded values for Real Meter 1 and Real Meter 2 are deducted (subtracted) from values for Substation A (master meter) to get a reportable value for the *un-metered area.*

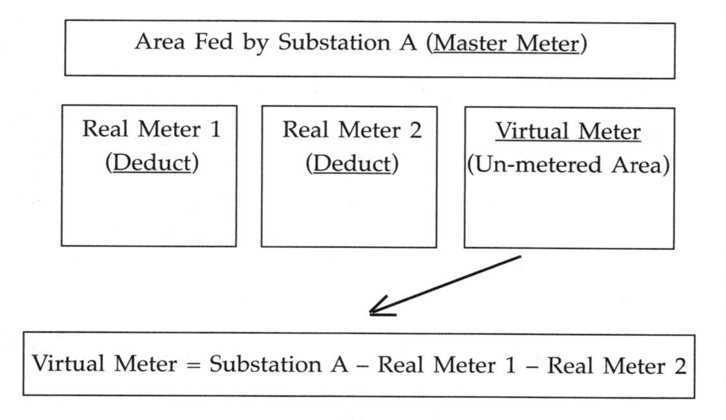

Figure 12-3. Un-Metered Data

In other cases, only portions of multiple meters combine to makeup a single reportable area. Let's say two buildings each contain a mix of office areas and storage areas and each building has its own meter. Technicians temporarily place separate informational meters on the office areas and storage areas. An energy audit of each building focuses on the percentages of the total consumption attributed to office space and storage space. Now, virtual meters can report the estimated consumption for each building in terms of its use, either for office space or storage space. Figure 12-4 shows how portions of areas served by real meters *combine* to produce data for one virtual meter.

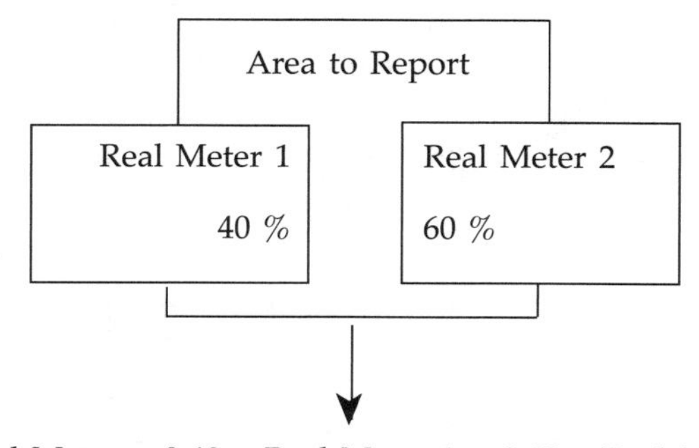

Virtual Meter = 0.40 × Real Meter 1 + 0.60 × Real Meter 2

Figure 12-4. Fractionalized and Combined Meter Data

An EIS administrator can fine tune virtual meter data to account for special situations. For example, suppose that a virtual meter uses the fractional meter data shown in Figure 12-1 to split the main chilled water meter used in a building. This split might be valid only during the day when all air handling units are operational. During unoccupied hours, one area might turn off all of their air handlers and not use any chilled water. During the unoccupied hours, we should reduce the virtual meter data for the area not using chilled water to zero to properly account for this occurrence.

An EIS administrator could create a virtual meter entirely from a pre-defined load profile. For example, suppose we need to deduct the electrical usage from an electronics room from a master meter, but it is difficult or expensive to meter. Since the load is virtually constant for an electronics room, we can multiply an instantaneous kW by –1 and use it as the hourly reading for this electronics room. The negative values effectively deduct these readings from the master meter when summed together. Another example of this technique deducts the electrical load on a chilled water air handler fan motor during unoccupied hours only. In this example, a virtual meter will multiply the fan motor kW load by –1 for the unoccupied hours and 0 kW for occupied hours. This would effectively deduct the fan motor kW load during the unoccupied hours.

Calculating values for virtual meters using values recorded for real meters gives an EIS more reporting power. It opens the door to an almost unlimited number of reporting opportunities beyond the already comprehensive reporting capabilities provided by the EIS. The challenge is to integrate this virtual metering approach into the EIS in a way that is easy to understand and use.

HOW DOES VIRTUAL METERING AFFECT THE EIS?

An EIS requires only a few enhancements to implement virtual metering. Since virtual meters have the same characteristics as real meters, the EIS defines them in the same meter definition table. The table requires some extra fields to accommodate the virtual meter definition. A small amount of extra disk space may be required to account for the additional virtual meter records. The interface to the meter definition table will have to provide the ability to define virtual meters. A calculation program will populate the database with calculated readings for the virtual meters. Once the calculations are complete the EIS reports the virtual meter data just like the real meter data. Since, in most cases, the values from real meters used in virtual meter definition expressions are no longer relevant, the EIS must be set up to prevent reporting of duplicate values for the same area.

The *expression field* is one of the additional fields required for virtual meter definition. It contains the formula for calculating virtual meter values from real meter values. The fields R1, R2, R3, etc. contain meter identification values for the real meters. Figure 12-5 below shows the additional fields required for a meter definition table to accommodate virtual meters. The *Meterid* field already exists in EIS meter definition tables.

METERID	EXPRESSION	R1	R2	R3	...R(N)
M506	R1 + R2	R693	R695		

Figure 12-5. Meter Definition Table Fields

A convenient interface is required to add virtual meter definitions based on real meter characteristics. Figure 12-6 below shows a simplified version of an interface to configure virtual meters. It allows the user to filter for the real meters required in the virtual meter expression and then provides a blank record to add the virtual meter to the table.

METERID	EXPRESSION	R1	R2	R3	R4
R693					
R695					
M506	*R1 + R2*	*R693*	*R695*		

Figure 12-6. Configuring Virtual Meters

A calculation program runs and replaces the real meter terms in the expression with the values read from the real meters, producing a virtual meter value for each virtual meter record in the table. That value is then stored in a *reading table* along with all of the other real meter readings. The EIS calculates virtual meter values as it collects the real meter data, probably daily. The calculation program should be able to calculate virtual meter data for any span of time, the most recent day, month or year if needed.

Once the virtual meter data is calculated, it can be reported by the EIS just as the real meter data is reported. The EIS knows which real data readings used in virtual meter expressions to eliminate from the report by their characteristics in the meter definition table. The meter definition table below in Figure 12-7 shows that the real meters used in virtual meter expressions have blank *EXPRESSION* and *R1* fields. It shows that a real meter *not* used in a virtual meter expression has values representing itself in the *EXPRESSION* and *R1* fields. The virtual meter definition is on the last line.

METERID	EXPRESSION	R1	R2	R3	R4
R693					
R695					
R697	R1	R697			
M506	R1 + R2	R693	R695		

Figure 12-7. Meter Definition Table

Most of the enhancements required for an EIS to accommodate virtual meters involve the meter definition table. Careful planning ensures that the meter definition format is complete enough to describe the virtual meters accurately. The meter definition table may require additional fields beyond what we have shown here. Adding the virtual meter calculation program to the data collection routine completes the programming. The EIS reports then include virtual meter data along with the real meter data.

INTEGRATING VIRTUAL METERING INTO AN EIS

Integrating virtual metering into an EIS involves modifying the structure of the meter definition table in a way to allow definition of virtual meter characteristics. Also, the tool used to maintain the meter definition table will need to accommodate the new structure. Using this tool, a virtual meter can be defined using real meter definitions already available in the database. The most important part of the virtual meter definition is the *expression*.

The EIS calculates virtual meter data using the *expression* entered in the meter definition table. A calculation routine evaluates the expression replacing terms with values from the real meter readings they represent. Once the calculation is complete for each virtual meter the program places the resulting values into the *readings table* along with the real meter readings. Then the EIS processes all readings and virtual meter values in the same manner. The characteristics for each meter defined in the meter definition table determine if the meter data is reportable directly or as a part of a virtual meter calculation.

METER DEFINITION TABLE

The meter definition table is the *backbone* of the whole EIS system. Most large complexes have hundreds of meters; a table to track them along with their characteristics is definitely required. Figure 12-8 shows an abbreviated meter definition table for some electric meters in a small school district. The table includes the fields required to use virtual metering. We will be creating virtual meters to obtain consistent electric consumption data at the *building* level. The EIS joins the meter definition table with the *readings table* to produce reports. Only meters with values in the EXPRESSION field are reportable. The EIS reports data from other meters as part of a virtual meter expression. Let's take a look at each of the schools in the table.

Note: For the sake of simplicity, the examples below involve schools with buildings that all have similar energy-consumption such as classrooms and offices. So, defining the virtual meters in terms of square footage is reasonable in these cases. Once EIS administrators understand the concept of virtual metering they will see opportunities to apply these principles to situations that are more realistic. In order to account for areas of high energy consumption such as a cafeteria, they can use temporary informational metering or estimates to create valid expressions for virtual meters.

1. *Alamo Elementary* has two buildings that are metered by only one meter. The EIS administrator reports consumption for the buildings by creating virtual meters V101 and V102 as a percentage of the total floor area (square footage) of both buildings. This is possible because the energy-consuming equipment in each building is similar and the student population is about the same. The total consumption for the school is the values for V101 plus V102.

2. *Aztec Elementary* has one building with two meters on the one building. The building is reportable using virtual meter V103. V103 reflects the combination of data for meters 1 and 2.

3. *Breakers Elementary* has three buildings. Two of them are individually metered. The third building has no meter. A substation provides electricity to the whole school. Therefore, V104 can report electric consumption for the un-metered building by subtracting the data readings for buildings 1 and 2 from the substation readings. Notice that buildings 1 and 2 are reportable as *real meters*. The sum of the two real meters and the one virtual meter then results in the total consumption for the school.

4. *Camden Elementary* has three buildings containing only classrooms with the same energy-consuming equipment. Two meters record electricity use for Building 1. One records use for part of Building 1 as well as all of Building 2 and one records use for the remaining part of Building 1 as well as all of Building 3. Technicians can evaluate which rooms in Building 1 are receiving electricity through meter R697 and which rooms are receiving electricity through meter R698. From this evaluation, they can apportion a percentage of floor area covered by each of the meters to Building 1. In this case, the virtual meter V105 can report on Building 1 electric consumption by adding together 40% of meter R697 consumption and 60% of meter R698 consumption. The expression for V105 is $0.4 \times R1 + 0.6 \times R2$. This represents the portion of each of the two electric meters floor area (square footage) that applies to Building 1. V106 and V107 report on the remaining portions of the meters that apply to the other buildings 2 and 3.

5. *Citrus Elementary* has one building with one meter. Virtual metering is not required for it.

VIRTUAL METER DEFINITION PROCESS

Usually adding a new meter to the EIS only requires adding a record to the meter definition table for that meter. The data collection routines will then start collecting data for the new meter and the reporting interface will report on it. The meter definition process involves using a tool to edit the meter definition table. The EIS administrator defines meters as they come on line or, in the case of virtual meters, as needed. The process is not complicated, but it is important to enter the correct information since the EIS depends heavily on the meter

METERID	DESCRIPTION	SCHOOL	SF	EXPRESSION	R1	R2	R3
R691	Meter 1	Alamo Elementary	8000				
R692	Meter 1	Aztec Elementary	9000				
R693	Meter 2	Aztec Elementary	10000				
R694	Breakers Substation	Breakers Elementary	40000				
R695	Building 1	Breakers Elementary	12000	R1	R695		
R696	Building 2	Breakers Elementary	7000	R1	R696		
R697	Meter 1	Camden Elementary	6000				
R698	Meter 2	Camden Elementary	6000				
R699	Building 1	Citris Elementary	4000	R1	R699		
V101	Building 1	Alamo Elementary	2000	0.25×R1	R691		
V102	Building 2	Alamo Elementary	6000	0.75×R1	R691		
V103	Building 1	Aztec Elementary	19000	R1 + R2	R692	R693	
V104	Building 3	Breakers Elementary	21000	R1-R2-R3	R694	R695	R696
V105	Building 1	Camden Elementary	6000	0.4×R1+0.6×R2	R697	R698	
V106	Building 2	Camden Elementary	3600	0.6×R1	R697		
V107	Building 3	Camden Elementary	2400	0.4×R1	R698		

Figure 12-8. Meter Definition Table

definitions. Figure 12-9 shows the meter definition table tool with a new virtual meter record at the bottom.

To create a virtual meter the process is as follows:

Step 1: Sort and filter the table so that the real meters used in the expression are visible.

Step 2: Enter the real meter identification values into the R1, R2, etc. fields of the new virtual meter record for the real meters needed.

Step 3: Enter the expression as a simple arithmetic formula. Do not group terms using parentheses.

Step 4: Transfer any other pertinent information from the real meter fields into the same fields in the new virtual meter record. In our example above this would be the SCHOOL and SF fields.

Step 5: Remove the values from the EXPRESSION and R1 fields of the real meters used in the expres-

METERID	DESCRIPTION	SCHOOL	SF	EXPRESSION	R1	R2	R3
R692	Meter 1	Aztec Elementary	9000				
R693	Meter 2	Aztec Elementary	10000				
V103	*Building 1*	*Aztec Elementary*	*19000*	*R1 + R2*	*R692*	*R693*	

Figure 12-9. Defining a Virtual Meter – V103

sion. This will insure there is no duplicate reporting of data.

Note: It is important not to distribute or combine more than 100% of any one particular real meter. Any one real meter only has so much floor area to distribute or combine. And, that floor area can only be distributed or combined once. It can be broken up as many ways as needed to accurately report the data. But, virtual meter definitions can only combine the total floor area with other real meters once. In the end, the total of all the floor area of any one real meter, reportable by itself or through a virtual meter expression, must be no more than the real meter is physically associated with. An automated procedure can check the accuracy of virtual meter expressions in this regard.

CALCULATING THE VIRTUAL METER EXPRESSIONS

In order to evaluate the expression in terms of the real meter readings, all of the pertinent data must be in one database record. Once it is in one record, it is easy to replace each term in the expression with the associated real meter reading values. Each record represents the lowest reading interval of the real data values: hour, day, month, etc. Therefore, when the EIS populates the *readings table* with the calculated values, it can summarize the data along with the real meter readings for reporting.

Let's walk through an example of how data for just one of these virtual electric meters is calculated. Keep in mind that normally the EIS calculates all of the virtual meters in a meter definition table at once. Figures 12-10 and 12-11 show extracts from the meter definition table and meter readings table. Note that before creating a virtual meter for Building 3 of Breakers Elementary and storing data for it in the meter readings table, the only way to report the total electric consumption for the entire school was to report the values associated with the Breakers Substation.

VIRTUAL METER CALCULATION PROCESS

Step 1: Read the value of R1 from the meter definition table for the virtual meter to calculate. In this case that is "R694."

Step 2: Look up the value of kWh from the meter readings table for the record where METERID = "R694" and the date is in the current month (May 2004). In this case that value is 80,000 kWh.

Step 3: Create a record in a temporary table for the current month. Store the date as the first day of the current month (May 1, 2004). Store the *meterid* and *expression* values from the meter definition for V104. Store the value retrieved in Step 2 (80,000 kWh) in the R1 field.

Step 4: Repeat steps 1 and 2 for R2 and R3 values and store them in the appropriate fields in the temporary table.

Temporary Table

DATE	*METERID*	*EXPRESSION*	*R1*	*R2*	*R3*	*kWh*
5-1-04	V104	1-R2-R3	80000	30000	40000	

METERID	DESCRIPTION	SCHOOL	SF	EXPRESSION	R1	R2	R3
R694	Breakers Substation	Breakers Elementary	40000				
R695	Building 1	Breakers Elementary	12000	R1	R695		
R696	Building 2	Breakers Elementary	7000	R1	R696		
V104	Building 3	Breakers Elementary	21000	R1-R2-R3	R694	R695	R696

Figure 12-10. Meter Definition Table

DATE	METERID	READING	kWh	DESCRIPTION	SCHOOL	REPORTABLE
5-1-04	R694	100000		Breakers Substation	Breakers Elementary	Yes
5-1-04	R694	180000	80000	Breakers Substation	Breakers Elementary	Yes
5-1-04	R694	260000	80000	Breakers Substation	Breakers Elementary	Yes
5-1-04	R695	200000		Building 1	Breakers Elementary	No
5-1-04	R695	220000	20000	Building 1	Breakers Elementary	No
5-1-04	R695	250000	30000	Building 1	Breakers Elementary	No
5-1-04	R696	210000		Building 2	Breakers Elementary	No
5-1-04	R696	240000	30000	Building 2	Breakers Elementary	No
5-1-04	R696	280000	40000	Building 2	Breakers Elementary	No
5-1-04	V104			Building 3	Breakers Elementary	

Figure 12-11. Meter Readings Table

Step 5: Replace each term in the expression with the appropriate value from the record for V104. Then evaluate the expression and store the result in the kWh field.

Temporary Table

DATE	*METERID*	*EXPRESSION*	*R1*	*R2*	*R3*	*kWh*
5-1-04	V104	80000-30000-40000	80000	30000	40000	10000

Note: If the database management software or programming language you use doesn't have the ability to evaluate an arithmetic expression, then an arithmetic calculator may be required. The JAVA Expressions Library has a good one to use to evaluate single line arithmetic expressions. It is available for free at *http://galaxy.fzu.cz/JEL/*. [6]

POPULATING THE DATABASE

Once virtual meter data values are calculated, they are stored with the real meter readings as if they were actual readings themselves. This process occurs on the same interval as the real data in order to keep the virtual meter values current with the other readings. However, a procedure must be available to go back and recalculate virtual meter values for any time period in case a virtual meter expression or real meter reading needs corrections. This insures accurate reporting at all times.

Updating the meter readings table is easy. Just store the kWh value from the temporary table above into the meter readings table for the appropriate date and *meterid*. In this case 10,000 kWh is stored into the kWh field of the record for DATE=5-1-04 and

METERID="V104." The REPORTABLE field is marked with "Yes."

Note in Figure 12-12 below that the REPORTABLE field is marked for each meter designating whether or not the meter data is reportable. This prevents the EIS from reporting data for one meter twice. For instance if meter values are summarized for the Breakers School, we want the electricity from the substation to be reported as the sum of Buildings 1, 2 and 3 only. Adding in the values for the substation as well would produce a result twice as large as the actual amount. The records must remain in the table even if they are not reportable so that the EIS can collect data for them. In this case, the Breakers Substation might remain reportable for some other purpose if we further qualify the records in the table. However, that discussion goes beyond the scope of what we are illustrating here.

REPORTING VIRTUAL METER VALUES

Reporting virtual meter values is exactly the same as reporting real meter readings. There is no need to provide additional formatting or functionality. That is the whole point of creating *virtual meters* is to make them behave like *real meters*. In fact, it may not even be necessary to distinguish between virtual meters from real meters in the reports. Figure 12-13 shows a typical report including values for a virtual meter V104. The first block shows information for the school. The second block breaks the information down by building. The third block is a breakdown of all of the reportable meters in one of the buildings (the one selected by the user) and the fourth block shows a breakdown of all real meters used in the virtual meter expression.

CONCLUSION

Energy management offices need to be able to track energy consumption at equivalent levels, such as at the building level. Typically engineers do this using informational meters. Fractionalizing and combining real meter readings into virtual meters is an alternative as well as complementary approach to providing that capability. The ability to easily integrate virtual metering into the EIS framework makes it a viable alternative to informational metering on a large scale. As long as the EIS design keeps virtual meter definitions and data closely related to their real counterparts, the EIS can easily report on virtual meters just as it reports on real meters. Virtual metering provides a means to report energy consumption along lines other than the physical boundaries represented by the metered area itself. However, any one virtual meter has no more ability to qualify data than a real meter. Each virtual meter represents only one utility type (electric, gas, water, etc.) and is defined by only one expression. That expression relates to the one unique physical, organizational or functional boundary that a real meter possesses. Also, remember that virtual meters may only represent estimated consumption values. Still, virtual metering opens the doors to some powerful analysis features. An example of using virtual metering is available at *http://www.utilityreporting.com/udvm.*

DATE	METERID	READING	kWh	DESCRIPTION	SCHOOL	REPORTABLE
5-1-04	R694	100000		Breakers Substation	Breakers Elementary	No
5-1-04	R694	180000	80000	Breakers Substation	Breakers Elementary	No
5-1-04	R694	260000	80000	Breakers Substation	Breakers Elementary	No
5-1-04	R695	200000		Building 1	Breakers Elementary	Yes
5-1-04	R695	220000	20000	Building 1	Breakers Elementary	Yes
5-1-04	R695	250000	30000	Building 1	Breakers Elementary	Yes
5-1-04	R696	210000		Building 2	Breakers Elementary	Yes
5-1-04	R696	240000	30000	Building 2	Breakers Elementary	Yes
5-1-04	R696	280000	40000	Building 2	Breakers Elementary	Yes
5-1-04	V104		10000	Building 3	Breakers Elementary	Yes

Figure 12-12. Meter Readings Table

GREEN COUNTY PUBLIC SCHOOLS									
ELECTRICITY JUNE, 2002	CONSUMPTION			COST			EFFICIENCY		
ELEMENTARY									
SCHOOL	Kwh	Previous Year Kwh	PCT Kwh Change	COST	Previous Year COST	PCT COST Change	BTU/SF	Previous Year BTU/SF	PCT BTU/SF Change
BREAKERS ELEMENTARY	112044	101636	10.24 %	8036.18	$ 8819.83	-8.89 %	956	1592	-39.95 %

BREAKERS ELEMENTARY									
BUILDING	Kwh	Previous Year Kwh	PCT Kwh Change	COST	Previous Year COST	PCT COST Change	BTU/SF	Previous Year BTU/SF	PCT BTU/SF Change
Building 1	37348	33879	10.24 %	2678.73	$ 2939.94	-8.89 %	956	1592	-39.95 %
Building 2	18674	16940	10.24 %	1339.37	$ 1469.97	-8.89 %	956	1592	-39.95 %
Building 3	56022	50820	10.24 %	4018.11	$ 4409.91	-8.89 %	956	1592	-39.95 %

Building 3									
METER	Kwh	Previous Year Kwh	PCT Kwh Change	COST	Previous Year COST	PCT COST Change	BTU/SF	Previous Year BTU/SF	PCT BTU/SF Change
V104	56022	50820	10.24 %	4018.11	$ 4409.91	-8.89 %	956	1592	-39.95 %

V104 = R694 - R695 - R696									
METER	Kwh	Previous Year Kwh	PCT Kwh Change	COST	Previous Year COST	PCT COST Change	BTU/SF	Previous Year BTU/SF	PCT BTU/SF Change
R694	112044	101636	10.24 %	8036.18	$ 8819.83	-8.89 %	956	1592	-39.95 %
R695	37348	33879	10.24 %	2678.73	$ 2939.94	-8.89 %	956	1592	-39.95 %
R696	18674	16940	10.24 %	1339.37	$ 1469.97	-8.89 %	956	1592	-39.95 %

Figure 12-13. Reporting Virtual Meter Data

References

[1] Roth, Kurt W., Ph.D., and Benekek, Karen; "Commercial Building Controls: Barriers & Opportunities"; Energy User News, February 2004, Vol 29, No 2, page 16; BNP Media, Troy, MI.

[2] "Submetering and Energy Montitoring"; Internet page, *http://www.emon.com/html/reading.html*, accessed 5/2/2004; E-MON, Langhorn, PA.

[3] "Meter Manager"; Internet page, *http://www.energysavingtechnologies.com/metermanager.htm*, accessed 5/2/2004; Energy Saving Technologies, Inc., Penticton, B.C.

[4] Demczuk, Cliff Northeast Utilities; Gowan, Geoff, ABB, Inc. and Fesmire, Bob, ABB Inc.; "One Database, Many Uses"; Internet page, *http://uaelp.pennnet.com/articles/_display.cfm?Section=ONART&Category-ONXTR&PUBLICATION_ID=22&ARTICLE_ID201950*, accessed 5/2/2004; Utility Automation and Engineering, T&D, April 2, 2004.

[5] Capehart, Barney; "How a Web-based Energy Information System Works"; Information Technology for Energy Managers; p. 25; Fairmont Press, Inc., Lilburn, Ga. 2004.

[6] Metlov, Konstantin L.; "JAVA Expressions Library," Internet page, *http://galaxy.fzu.cz/JEL/*, accessed 5/2/2004; JEL, 1998-2003.

Chapter 13

Providing EPA's Energy Performance Rating Through Commercial Third-party Hosts

Bill Von Neida, U.S. EPA
Sarah E. O'Connell, ICF Consulting

ABSTRACT

THIS CHAPTER presents the U.S. Environmental Protection Agency's (EPA) approach to providing the national energy performance rating system through commercial third-party hosts. The energy performance rating system was developed by the U.S. Department of Energy (DOE) and the EPA and provides building owners, operators, and energy managers with a way to benchmark commercial buildings to effectively and quickly communicate whole-building energy performance. Ratings are obtained by manually entering specific building data into EPA's Portfolio Manager, an on-line software application made freely available to the public through the EPA's Web site. While this energy performance rating system has been successfully introduced and accepted into the market, a broader application is now being explored. EPA is currently working with commercial energy information vendors to automate benchmarking. By automating the process, it becomes more realistic for organizations with large portfolios to assess their opportunities for improvement, prioritize upgrades across their portfolios, and realize significant financial and environmental savings.

EPA'S ENERGY PERFORMANCE RATING SYSTEM—INTRODUCTION

Systems Approach

Commercial building owners, operators, and energy managers consistently express a strong interest in understanding the energy performance of a building. While numerous methods exist for understanding the energy performance of commercial buildings (ranging from basic energy consumption benchmarking, to engineering audits and analysis, to more sophisticated computer modeling and simulation), all have significant shortcomings in their practical utility. EPA's energy performance rating system, as conceived by the DOE and the EPA, was designed to supplement these approaches by providing a benchmark for commercial buildings so they can cost-effectively measure and quickly communicate comparative whole-building energy performance for one or multiple buildings in a portfolio. This benchmark accounts for space use, weather, climate, occupancy and operational characteristics to: provide a method to compare the efficiency of a building relative to the national building stock; provide a simple 1-100 metric to communicate that relative performance; and establish a national performance target for excellence.

The Current Delivery Environment

From its inception in 1999, the primary delivery strategy for distributing the energy performance rating system into the marketplace has been through EPA's Portfolio Manager, a proprietary on-line energy accounting software application made freely available to the public through the EPA's Web site. To obtain a rating, building owners, operators, and occupants establish a Portfolio Manager user account, select their building space type and enter 12 months of energy consumption information along with basic building characteristics such as building type, size, and roughly half a dozen other basic building occupancy and operational characteristics. Having derived a rating for their building, users can then apply the rating to basic energy management functions, such as establishing baselines, setting energy and cost improvement targets, tracking performance over time, and aggregating individual building information to the portfolio level. Within Portfolio Manager, eligible buildings may also apply for ENERGY STAR certification and recognition, such as the Statement of Energy Performance and the ENERGY STAR label.

System Elements

The most critical initial system element for determining building performance to obtain a rating is space type. Space types included in the energy performance rating system were chosen primarily based on data availability and have been added to the system over time. For example, when the tool was first introduced to the market in 1999, only general office space could be benchmarked. Since then, the list of eligible space types has expanded to those listed below. Eligible space types, representing over 50% of US commercial floor space, include:

- Office (General, Bank Branch, Courthouse, Financial Center);
- K-12 Schools;
- Hotels (Upper Upscale, Upscale, Mid-scale with Food & Beverage, Mid-scale without Food & Beverage, Economy);
- Hospitals;
- Medical Office Buildings;
- Supermarkets/Grocery Stores;
- Residence Halls/Dormitories; and
- Warehouses (Refrigerated and Non-Refrigerated).

Data for these space types are, in most cases, derived from the Energy Information Agency's (part of DOE) Commercial Buildings Energy Consumption Survey (CBECS). CBECS is a national, statistical survey of building features, energy consumption, operations, and expenditures in US commercial buildings. If CBECS is inadequate to create a rating for a particular building type, other national data sets are used. In addition to space type, other building characteristic data from these datasets which is found to be a strong driver of energy consumption and that are determined to influence energy consumption (e.g. hours of operation, occupant density, etc.) are included as variables in the rating. In 2005, ENERGY STAR will use the newly released CBECS data to update the rating for existing space types, as well as extend it to retail and food service buildings.

To add validity to the rating, weather and climate normalization elements are included. Climate, specifically the typical annual values of cooling degree-days (CDD) and heating degree-days (HDD), affects building energy consumption. An office building located in Chicago, for example, will typically use significantly more energy on an annual basis than an office building of similar size and orientation located in San Diego. The underlying models in the ratings algorithms account for these typical climatic differences. Within a given location over any given year, buildings may also be exposed to relatively severe or mild weather as compared to historical averages. The rating factors out - or weather normalizes - this impact. Thus, the results one obtains using the rating account for both the climate conditions - such as the climate typically experienced in Chicago versus that of San Diego - and the year-to-year weather variations that a building in Chicago, for example, may experience in a given year. This ensures that the rating reflects comparative energy performance under typical conditions rather than climate or weather variations.

The process EPA uses to weather normalize building energy data as part of the energy performance rating system is based on E-Tracker, a software tool developed by Dr. Kelly Kissock of the University of Dayton. This process involves the following steps:

1. Based on its zip code, a building's monthly electricity consumption is regressed against the average daily temperatures for that area for the corresponding month to determine the building's response to the actual weather conditions experienced.

2. If one month's electricity consumption is significantly different from the building's average monthly consumption (i.e., at least 50% higher or lower than the mean), that month's value is considered an outlier error and not included in the regression.

3. Based on the results of the regression analysis, historical, 30-year average values for monthly average temperatures are then used to adjust the building's actual 12-month electricity consumption data up or down.

4. Steps 1 and 3 are repeated for natural gas and district steam to normalize the building's actual 12-month non-electric energy consumption up or down. The outlier test in Step 2 above is not performed for non-electric fuels since usage of these fuels more often varies widely over the course of a year. Weather normalization on non-electric fuels other than natural gas or steam is not attempted since actual monthly consumption is typically not precisely known. Nonetheless, consumption of non-electric fuels other than natural gas or steam (i.e. fuel oil) is collected and included as part of the building's total energy use within Portfolio Manager.

5. The normalized electric and non-electric energy consumption values are then converted to Btus and added together to determine the building's weather normalized annual energy consumption.

6. The resulting normalized energy use is used to determine the building's energy performance rating.

System Usership

The delivery mechanism has been successful at quickly and cost effectively introducing the energy performance rating system to the commercial buildings market. As of 2004, more than 4,000 Portfolio Manager users had evaluated the energy performance of 20,000 buildings totaling 3.0 billion square feet. For the 50% of the commercial buildings market that the rating is available for, this represents a 11% penetration rate (by square footage) of the market, with some building types (such as offices and supermarkets) having a 20% penetration rate. Furthermore, delivery of the energy performance rating system through EPA's Web site provided a one-stop shopping experience to a presold market of 950 ENERGY STAR Partner organizations that have committed 10 billion square feet of commercial buildings to be benchmarked and upgraded. Finally, delivery exclusively by EPA gave the program maximum control over algorithms, content placement, and brand usage that were all critical requirements to successfully introducing this powerful capability.

Growing Pains

While this delivery solution has been initially successful, its application as the sole option for obtaining ratings has been questioned. Success of the energy performance rating system is predicated on its use by a broad audience of end-users (owners, operators, and tenants), who would use it across their large portfolio of buildings to understand relative building performance and prioritize investment, and then continue to use the energy performance rating system to watch individual building and portfolio movement over time as building performance is upgraded. EPA has witnessed a flattening growth curve over the past two years, with continuous benchmarking over time lower than anticipated. Between 2002 and 2003, entry of new buildings has been flat at roughly 4,000 buildings per year, and rebenchmarking of those buildings at roughly half of that. Furthermore, the average portfolio size has remained consistent at five buildings per user. Given the burdens of manually entering and updating energy consumption information across all fuel types and meters in a building, users appear to be content with initial evaluations of their higher profile buildings, while not extending this practice across their portfolios over time.

Continually increasing market penetration for the energy performance rating system is a challenge for EPA given the competition Portfolio Manager faces for end-user attention. A wide variety of competitive energy management software is available to end-users, and a mature market exists for both in-house and third-party utility tracking solutions. The early attempts at restructuring the retail electricity market; the growth in advanced metering; and the opportunity to access real-time pricing through demand response are among the factors that have led to an increased interest in providing energy information services to commercial accounts. Esource surveys estimate that as many as two-thirds of the blue chip organizations and one-quarter of the broader commercial market already receive interval data regarding energy use. Convincing those users to enter that data again and manage it in another energy management software application is not realistic.

Finally, because the rating is only delivered through Portfolio Manager, it has a limited influence on energy decisions in the building. In its most potentially effective application, the rating should represent and clarify a heretofore-underrepresented financial aspect of the building in business transactions - such as buying, selling, leasing, insuring, retrofitting and buying energy for the building. Limiting the distribution of the rating to the basic energy management functionality of Portfolio Manager does not allow the rating adequate exposure to inform and promote energy efficiency within those transactions.

For these reasons, a new distribution mechanism for the rating was explored. Early on it was acknowledged that the required manual data entry was a significant barrier to widespread user adoption. While several projects have been undertaken to facilitate bulk data transfer into the tool, these solutions have done nothing to counter Portfolio Manager's competition from commercial energy information products, increase ENERGY STAR's ability to introduce the rating into decision making environments, or ease the growing information technology platform maintenance burden placed on EPA. To accomplish these objectives, ENERGY STAR would have to collaborate with commercial energy information vendors.

THE MARKETPLACE FOR ENERGY SERVICES

Flavors of Service

The ENERGY STAR program was founded on the tenet that "you can't manage what you can't measure." Energy information has long been the challenging and often missing element of building portfolio-wide energy savings initiatives. For companies struggling to understand the value of energy efficiency as an investment option for their facilities, the use of a commercial system to track, analyze, and present energy use and cost data over time may seem extravagant. However, systems that simply check for utility billing errors and find better rate schedules can save four times their annual cost for some commercial buildings, making energy information a potentially better energy-related investment than many more conventional upgrade projects.

Energy information tools have been available for decades, with Faser and Metrix being two notable examples from the pre-web era where single buildings were served by individual energy management systems. These single building applications morphed into web-based systems that can show data for many buildings in many locations, aggregated for viewing by managers from any location. Evolution of these portfolio-wide energy information tools has resulted in the formation of companies that provide services such as the following:

- Bill handling services use appropriate utility rate schedules to compare monthly bills against metered data and check bill estimates against actual bills. In addition, they can consider alternative rate schedules to help customers decide on the least expensive rate. Bill handling service companies which provide outsourcing of bill aggregation, verification, and payment of facility resource use are also able to aggregate facilities and organizations to develop supply-side procurement strategies to facilitate cost reduction for commodity resources.

- *Peak load management services* allow customers to analyze the history of peak loads in their buildings, compare these peaks to influences such as outdoor temperature, and monitor these loads in real time. In some cases, users can set alarm thresholds that allow them to shed load before some event occurs, such as their utility rate will ratchet due to rate demand thresholds being exceeded.

- *Energy analysis services* assist customers in understanding how their energy use varies over time and what end-uses or individual buildings are contributing most dramatically to their energy use. Many of these tools offer their own benchmarking approach, often a weighted approach to comparing buildings in a single portfolio.

- *Energy consulting and upgrade services* are often combined with or offered in conjunction with energy information services, since energy information can be used to identify opportunity and then to track success once an upgrade or operational modification has been completed.

- *Submetering and monitoring* is often combined with visual, web-based energy information to allow hardware applications to provide critical information. One example might be in-room controls for the hospitality industry, which serve an energy-saving function in each room where they are located, but can serve a building-wide function if their information is aggregated for centralized viewing.

Each of these flavors is represented in the marketplace, often with a sector-specific overlay and in any number of combinations. The actual product takes the form of web interfaces that facilitate energy-related decision-making, but also can include paper or electronic reports sent or emailed to customers.

Activity in the energy information arena has been driven by two factors:

- The advent of the internet: Combined with the dot com boom, the internet made it easy to connect building data collection to a central server and data delivery method, and fueled excitement about energy information. As a result, some start up energy information initiatives received venture capital to get started.

- The promise of deregulation: This caused a number of regulated and deregulating utilities to scramble for some kind of energy information product to keep their customer base intact and attract new customers. A number of energy information offerings were developed across the U.S. in a short timeframe to keep up with anticipated changes associated with deregulation.

Many enterprises underestimated the challenge of building, delivering, and profiting from energy information services. There are numerous stories of companies that lost their energy information provider and chose to do without rather than risk another relationship. This marketplace has stabilized somewhat around both established and newer players, and the past two years have proven to be a more reasonable time for ENERGY STAR to attempt to find win-win relationships built around the hosting of their energy performance rating.

Increasingly, full service, established energy service providers are finding that the addition of energy information is a good complement to their service set, especially where customers have portfolios of buildings. They can offer web-based energy information as part of their projects, making it easier for customers to buy into, participate in, and verify energy saving activities facilitated by the service provider. Because of the importance of good energy information for initiatives that attempt to save energy or energy costs over time, it is likely that energy information services will become a more critical component of large organizations' approach to portfolio-wide efficiency.

Energy Information—The Coin of the Realm

A partnership with service companies which hold energy information for commercial buildings seemed a logical extension of the ENERGY STAR commercial buildings program, which already has a history of partnering with energy efficiency product manufacturers, service providers, and retailers to promote the ENERGY STAR program. Before evaluating the market for potential partners to distribute the rating, EPA established several core requirements. First, in order to maintain quality, consistency, and currency for the rating, EPA would have to maintain control of the benchmarking algorithms and be able to maintain, update, and expand the scope of the energy performance rating system. Given the frequency with which refinements occur, the algorithms would have to be centrally controlled and distributed by EPA. Passing out software updates and relying on vendors to run the most current algorithm versions was not an acceptable option for maintaining consistency over time. This defined the need to house the algorithms on EPA servers, and to work only with web-enabled tools that could access those algorithms via the Internet. This eliminated stand-alone software applications from consideration (e.g. Faser, DOE2, etc.). Second, in order to evaluate the program's impact on the marketplace, EPA would need to monitor the use of the algorithms and be able to generically track the scoring of buildings over time. This also dictated that the algorithms be housed on EPA servers. Third, the rating would need to be complementary with the vendor's content. This eliminated information services that were wed to their own competing benchmarking services, and tools that were not focused on energy (e.g., tenant management and building services management software). Finally, as an ENERGY STAR branded product, proper application of the service mark would have to accompany the rating.

Actively engaging this market led to three potential targets. Energy utilities and bill processing bureaus, as the largest holders of building energy consumption data, were obviously considered first. Regulated utilities dropped out of consideration when it became apparent that no practical level of security could be provided to data transmission or off-site storage which would satisfy their state-mandated obligations to maintain privacy and confidentiality of their customer data. Deregulated operations within utilities were also an attractive target, given their closeness to larger customers and their growing interest in owning that customer's data to forge stronger service relationships. Over the past several years, however, this market, while receptive to this idea, has been difficult to pin down as the utilities undergo an identity and market positioning dilemma: do they package energy information services as a loss leader to sell energy commodities and other services, position it as a revenue generator, or simply provide it as a customer service? This, combined with the unanticipated startup expenses and long timeframes these utilities have incurred, have caused limited action in that market.

Having dismissed the idea of using utility partners, EPA approached the internet-enabled data service vendors (manual data entry and analysis), full service bureaus (utility accounting, analysis, bill verification, and payment), and energy management and maintenance service vendors with energy information services (e.g., ESCOs and upgrade contractors).

The fundamental business proposition of becoming a delivery agent or host of the rating content was immediately attractive to them for several reasons. First, although most had basic benchmarking capability, they recognized the significant investment ENERGY STAR had made in creating and maintaining a more robust benchmark than they currently offered. Those without whole building benchmarking typically focused on more granular interval data and system level analysis and welcomed the complementary and expanded content the rating represented. Second, in an increasingly competitive market where differentiation between vendors is difficult, the affiliation with EPA and association with the ENERGY STAR brand was also attractive as a method to differentiate themselves from their competition. Third, these vendors recognized a business opportunity to service the large ENERGY STAR partner base that had committed their portfolios to benchmarking, but had not begun to do so in earnest. When viewed as a preferred provider of ENERGY STAR benchmarking, the rating could become a new revenue service to their existing customers, or become a loss leader to a new revenue source from ENERGY STAR partners seeking to fulfill their program commitments. Having found a complementary fit between the rating and their business model, if a balance could be found to satisfy EPA's core requirements and the vendor's investment in creating the delivery mechanism, a new marketing channel for the rating would be created.

ESTABLISHING A CONNECTION

Creating a Third-party Hosting Model

As EPA began to consider providing the energy performance rating system through commercial energy information products, several key programmatic and functional issues surfaced that dictated the decision-making associated with the hosting relationship. The following issues made it essential that EPA maintain control of the actual calculation events rather than simply supply algorithms to host entities:

- EPA needs to own, host, and evolve the benchmarking algorithms used in the rating. It would not be feasible to provide all third-party hosts with algorithms and then periodic **updates to algorithms** since EPA's goals assume a single, consistent rating at all times. Sharing algorithms, although discussed, would have been a mistake, as end users would not accept a situation where some third-party hosts upgrade their algorithms immediately while others waited for the next upgrade to their software to be tested and released.

- EPA currently performs **weather normalization** using NOAA temperature data feeds run through an application developed by EPA. This process requires an ongoing relationship with the NOAA data source and could not reasonably be extended to all third-party hosts, who see EPA's balanced and consistent approach to weather normalization as one of the reasons to host the rating.

- Like most voluntary programs, ENERGY STAR has program assessment metrics, mostly based on the penetration of the benchmark in the U.S. marketplace. It would have been difficult to **count the total number of benchmarks** over time if these benchmarks were all being delivered in isolation through different commercial energy information products.

- Users of the energy performance rating system demand different levels of data **confidentiality and anonymity**; however, users that elect to apply for an ENERGY STAR Label need to identify themselves. Any third-party solution must preserve the ability of an energy information provider's end user to remain anonymous until they apply for a label.

This combination of factors resulted in the decision to process data for ratings in a single, consistent, EPA-hosted analytical environment. Ratings are supplied to the third-party hosts after being calculated by EPA on an EPA server. Third-party hosts receive the ratings and then display them as part of their product offering. Figure 13-1 diagrams this approach, where data are passed from the third-party host server to the EPA server, where calculations take place, and where the third-party host server retrieves ratings for display in their energy information system. This approach allows algorithms or the weather normalization approach to be updated by EPA for all end users simultaneously. It also allows EPA to count incidences of benchmarking and movement of benchmarking ratings over time. It does not require that EPA view the owner name or address, which meets some hosts' concerns about confidentiality.

In order to facilitate this approach to data transfer, EPA worked with early third-party host candidates to

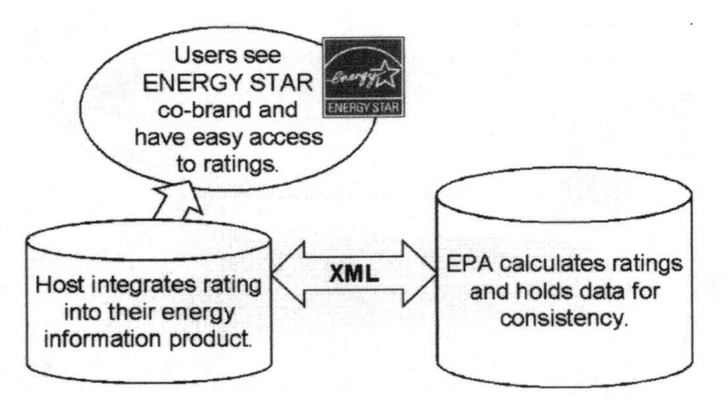

Figure 13-1. Data Flow Between EPA, Host, and Users

identify Extensible Markup Language (XML) as the best technology for third-party hosting. XML is a simple, easy-to-master approach that is already being used by the energy information community. This decision turned out to be a wise one as many of the third-party hosts already have in-house XML capability and it has taken as little as two weeks for hosts to develop and utilize the ability to transfer their utility data into the EPA XML data schema.

In partnering with ENERGY STAR to offer the rating, third-party hosts are committing to follow specific protocols associated with the XML data transfer approach, including updating this approach as new space types and other refinements are made to the rating. In addition, hosts commit to working with ENERGY STAR to ensure that messaging associated with the rating, its application, and associated program offerings is consistent. As these relationships evolve, it is anticipated that the third-party hosts will facilitate application for the ENERGY STAR Label, Statement of Energy Performance, and new program recognition offerings associated with exemplary, portfolio-wide energy performance.

EPA recognizes that business-to-business relationships, especially those that require consistent messaging and complex data transfer, are among the most challenging for extending the reach of voluntary programs and associated branded offerings. Therefore, EPA kept this initiative in a "pilot" mode that allowed interested hosts to help define the initiative. This prevented EPA from being overloaded with requests from potential hosts, or third-party hosts from being similarly overloaded with requests from potential end users. This approach has prevented a relatively common weakness of IT-driven program initiatives: attempts to anticipate, design for, and handle all possible user scenarios rather than phasing development to address issues as they are identified with pilot participants. The phased approach being used by EPA requires that buy-in is strong enough to allow flexibility over time. It can result in getting to the market more quickly and with a smaller investment.

With almost 250,000 sites represented by the current set of third-party hosts, the maximum throughput is promising and, importantly, EPA maintains the ability to ensure the quality of interaction with program partners and effectively monitor hosts to learn how to manage more relationships and expanded capabilities in the future. Hosts have offered to franchise the rating; take over the algorithms and their development; integrate other EPA tools into theirs; and/or train other potential hosts, so interest is high and varied regarding the future. Currently, however, EPA is working to fully build out and stabilize the existing host relationships around the rating and associated offerings.

SEVERAL MODELS FOR SUCCESS

Dramatic changes in the energy information industry during the deregulation era allowed EPA to experience all manner of business models, including start up and companies trying to add capabilities as their market share was contracting. Currently, there are several types of successes, each of which seems to be a viable avenue for moving forward:

- One of ENERGY STAR's award-winning energy service providers is integrating web-based energy information services for their customers. They are also a longtime user of the rating, having benchmarked over 1,000 buildings on behalf of their customers. Incorporating the rating into their information services offering seemed like a natural fit, and will significantly reduce the time required to provide their customers with benchmarks on a regular basis.

- Two of the leading billing services, which process, aggregate, and present utility billing information to identify errors and track expenditures over time, have signed on to host the rating. This is proving to be an easy way for them to expand their offering to include unbiased energy information, and, with over 150,000 sites between them, give ENERGY STAR access to an unprecedented number of buildings. Additionally, both companies offered to assist ENERGY STAR in collecting data required to develop ratings algorithms for space types not currently included in the ENERGY STAR algorithm set. This is an important opportunity, as census data for subsequent space types may not be sufficiently detailed to use for this purpose.

- A provider of operational data for industrial processes is expanding their reach to include commercial buildings. They have focused on the rating as an offering that provides entry to this marketplace. They involved their high level managerial and marketing staff early on, and have developed a sales pitch that is focused on the rating and large ENERGY STAR Partners in sectors where they have had some success. Their aggressive marketing approach has resulted in both contracts with key commercial building owners and added motivation on the part of their competition to host the rating.

- In 2003, ENERGY STAR participated in a very creative market relationship built around energy information and the rating. The California Hotel and Lodging Association used demand reduction funding from the California Energy Commission as a monetary incentive for association members to try energy information services for a one-year period. These energy information services include the rating. As a result, a number of California hotels will be getting the rating delivered to them via two companies that provide the rating, and, their hotel chains, many of which are national, will get some experience with both energy information and the rating.

There seems to an opportunity for all of these hosts to service more of the building portfolio for any of their customers. In other words, many large users of these types of services do not use them for their entire portfolio. As ENERGY STAR evolves, the program motivation may become associated with changing the performance of a participant's entire building portfolio, and ENERGY STAR participants will have good reasons to consolidate their energy information services under a single vendor. For hosts of the rating, the promise of expanding existing relationships seems to be more compelling than building new relationships, as the sales cycle is shorter and cost of sale lower.

TECHNICAL APPROACH

In order to facilitate data transfer associated with a third party, EPA has identified XML as the best technology for third-party hosting. The key reasons for this decision were:

- XML was repeatedly suggested by potential third-party hosts as their preferred data transfer approach, and many were already passing energy data using XML. This meant that they would have staff that could implement a data exchange schema in XML.

- XML was, and remains, a popular application in IT circles, which increases the chances that at least some staff on a third-party host's team will aggressively embrace the opportunity to work on the mechanical elements of the relationship. This has proven to be an advantage in several cases, compared to finding unmotivated or incapable IT staff, which is a huge potential impediment.

- XML is very well suited to the type of relationships EPA is pursuing. EPA's data requirements are well defined and critical to success, but present the most significant hurdle to developers. By choosing a straightforward data transfer language, IT staff can concentrate on getting the data right, rather than mastering a complex data transfer approach.

- XML is specifically designed for transferring complex information over the Web, and has an easy learning curve. See Figure 13-2, "What is XML?"

The process of engaging with third-party hosts involves their education regarding the nature of the energy performance rating system; the creation and submission of XML based on the EPA schema; and the modification of their commercial application to accommodate repetitive, high volume submissions and retrievals. There are two approaches to facilitating the mechanical aspects of the process:

1. Third-party hosts initially receive a Visual Basic desktop application based on the VB.Net specification that is designed to help them understand and test the XML data transfer process. This application provides a simple, manual interface for participants to use early in the process to submit and retrieve XML data. It is not intended to support long-term, high-volume data transfer over time.

2. Once hosts have become comfortable with the data transfer process and have proven that they can send and receive XML, they modify their commercial applications to send and receive via an instantiation of web service methods. This approach is intended to support the formal, automated data transfer relationship that is established by each participant through their hosting environment. Once a participant has successfully performed the data submission and retrieval steps manually using the VB application, the process should be automated using the web service.

In both cases, hosts send data to an EPA server and then retrieve results from that server. Users are notified by e-mail messages if the XML file is not in the proper format, including, to the extent reasonable, guidance regarding the nature of the problem. If data are incomplete or outside of acceptable ranges (both situations that require EPA processing to determine), the host is notified at the time results are retrieved. Again, there are a series of specific error messages designed to help the host correct the problem, primarily delivered as negative ratings, as follows:

If a benchmark rating cannot be calculated for a building, one of the negative values will be returned and an explanation from the list below will be provided in the description element.

1. The meter data for this facility did not meet the requirements for a benchmark rating.
2. Electrical meter data are required for a benchmark rating.
3. The space definition for this facility did not meet the requirements for a benchmark rating.
4. No energy use for the 12-month period.
5. The energy consumption for this facility is beyond the maximum amount for a benchmark rating.
6. There was an internal error calculating a benchmark rating for this facility.
7. Invalid Zip Code

Hosts can use an ID and password to access web views of ratings and facility information for their submissions. This allows them a second option for checking data that has been submitted to EPA and ratings associated with their submissions.

Both EPA and the third-party hosts are interested, first and foremost, in maintaining both the integrity of the rating and consistency of service to customers. As a result, the pilot program for third-party hosting of the rating has focused on setting and maintaining the expectations of all participants in this process. The following process for engaging third-party rating hosts was used to ensure that both EPA and the third-party host worked together to phase in this new service gradually and with a minimum of end user issues:

Test Phase

1. Register to participate in the pilot program to host the EPA energy performance rating system (this pilot program was closed by EPA in September of 2004, with an expected formal launch mid-2005).

2. Use the VB application for simple submissions (one or two buildings) to understand and master the mechanics of XML creation, submission, and retrieval.

3. Gain an understanding of EPA's program and goals by reviewing program materials and becoming familiar with the Portfolio Manager tool.

4. Work with EPA to compare the host's customer list with ENERGY STAR Partner list and identify customers who may be most interested in the rating.

5. Establish web service for automated submissions from the commercial application database.

6. Test XML submission and retrieval process using sample datasets (up to 100 buildings can be submitted during the test phase) and refine the automated process.

7. Work with EPA to design an interface for displaying rating information to customers.

8. Select a customer with whom to test the data exchange. Ideally, this should be an ENERGY STAR Partner since their commitment and understanding of the ENERGY STAR Program will facilitate motivation and results. Work with EPA and this pilot customer to get a co-signed commitment form completed and work with EPA to obtain all previous Portfolio Manager data.

9. Determine approach for submissions (e.g., frequency, quantity of data) and continue with regular submissions.

10. Determine schedule for moving the energy performance rating system to full production.

Move to Production

11. Notify EPA regarding commitment to a production rollout of the hosting capability.

12. Become ENERGY STAR Partner (if applicable).

13. Provide details on submission approach/delivery model number of customers, number of buildings, expected frequency and size of submissions, triggers for submissions (e.g., user data entry, receipt of utility data, and fixed schedule for processing).

14. Present interface and approach to ENERGY STAR management for approval.

What is XML

XML was designed specifically to address limitations in HTML that had become challenging as web applications demanded more sophistication. It allows structured information to be organized and presented with significant depth and flexibility and then used for a wide variety of applications. In laymen's terms, XML allows developers to provide or request many layers of information or data *about* information or data, where HTML is more oriented toward display and single hyperlink relationships. In addition, HTML provides a fixed tag set and associated tag semantics (The tag "**<P>**" refers, always, to a paragraph, but the tag "**<energy star rating>**" does not exist and cannot be created). XML allows the developer to define tags and relationships as needed to fit the application and subject matter. Some specific attributes of XML are notable and have resulted in wide acceptance:

- XML content is relatively simple to understand in its raw form. It does not require technical expertise or a translation program to look at an XML document and determine what it is, and where problems are occurring. This allows a wide variety of individuals to understand and participate directly in XML-based initiatives, not just highly skilled IT staff.

- XML combines simplicity of language, which allows it to be used for a variety of applications, with relatively broad capability. There are no hidden options that must be learned and the syntax is straightforward. This makes XML easy to learn regardless of what languages one used previously and makes XML applications fast to develop and share.

- Changes to an XML schema are easy to communicate and execute, making XML an ideal choice where a number of different entities are involved in using a single schema.

- It is easy to create different levels of educational or support information within an XML document. For example, a barebones schema could be shared among experienced team members, with heavily annotated and sample data versions of the schema used to initiate new team members.

While HTML is still the dominant approach for web page creation, XML has become a standard for moving complex data over the internet.

Figure 13-2. What is XML?

15. Receive approval of approach and confirmation of correct submission from EPA.

16. Begin to work with customers to determine if they have existing ENERGY STAR accounts in Portfolio Manager.

17. Work with EPA to determine which Partners will shift to automated rating, complete co-signed forms, have Portfolio Manager data transferred.

18. Work with EPA and customers to reconcile existing ENERGY STAR accounts, as applicable.

Production

19. Continue to automate and update services as more space types and other improvements become available.

20. Work toward ratings for all customers with applicable space types.

21. Support ENERGY STAR initiatives.

22. Provide information and support for customers to participate in ENERGY STAR initiatives, specifically, the ENERGY STAR Label, the Statement of Energy Performance, and the Leaders initiative.

Complete XML schemas for passing data between EPA and third-party hosts can be accessed via the following URLs:

- *http://ems-mx4.sradev.com/pmdxp/Benchmark-v.1.3.xsd* —provides the schema for sending data to EPA from the third-party host.

- *http://ems-mx4.sradev.com/pmdxp/EPAScore-v.1.3.xsd* —provides the schema for sending ratings and other information from EPA to the third-party host.

In addition, sample schemas with data included can be found at the following URLs:

- *http://ems-mx4.sradev.com/pmdxp/sampleXMLdocument.xml* — provides a sample of data sent to EPA from the third-party host.

- *http://ems-mx4.sradev.com/pmdxp/sampleXMLresult.xml* —provides a sample of data sent from EPA to the third-party host.

Appendices A and B provide Frequently Asked Questions and Common Errors in Passing Data, both current as of September 2004, with additions anticipated as this initiative moves from pilot to launch in 2005.

LESSONS LEARNED

Having worked for over two years to establish innovative and challenging business-to-business relationships that deliver voluntary program analytical benefits through a currently underutilized commercial resource, there are several lessons that should be passed on to others attempting to connect with energy information providers or attempt similar business-to-business program delivery channels:

- **Keep the value proposition simple**: Commercial enterprises do not want to be distracted from their core business, so this effort maintained focus on the ability of ratings hosts to sell services to large companies who are already committed to the rating. Now that we have established relationships, we are able to talk to our hosts about extending the relationships to include other programmatic offerings (i.e., financial analysis of energy saving opportunities).

- **Minimize the IT focus**: Information technology is often erroneously identified as the hardest and most immediate challenge associated with any relationship involving data exchange. In fact, the intrinsic value of the initial proposal and the ability to sell the initiative to key decision makers (not technicians), is more important. Each of the current third-party hosts has had little trouble mastering the mechanics of the XML data exchange relationship, but some have struggled to integrate the offering in their sales approach.

- **Get to vendors quickly** (before their business model or personnel changes): The energy information marketplace is still too dynamic for a relationship that requires too much patience or definition in the process. Our ability to quickly define both the value and the mechanics of participation was critical in getting companies to commit and move forward. On several occasions, a company champion dedicated to the success of the relationship left the company or got reassigned, resulting in a serious setback in the relationship.

- **Create a self-serve marketing environment**: Most publicly-funded programs like ENERGY STAR are obligated to serve all comers. For this reason, it was important to create, early on, a web resource to allow any potential host to understand and try out the data exchange using the ENERGY STAR Web site. This approach saved many contractor hours devoted to hand holding each company that expressed interest, and allowed focus on more likely success stories.

- **Focus on leaders**: While the ENERGY STAR web resource allowed all interested companies to participate, highly desirable and obviously aggressive companies were shepherded through the process. These included companies motivated enough to fly to EPA Headquarters for brainstorming and presentations. As a result, several recognized players were fast-tracked into the initiative. It was then extremely beneficial to be able to mention recognized players as hosts since this motivated other companies deciding whether or not to participate. Now that the rating is somewhat established among key market players, new hosts are approaching EPA and are greatly motivated.

- **Allow for creative application of content**: All of the current hosts make their living manipulating and displaying energy data. They have proven adept at integrating the rating into their offerings, in many cases offering their customers more creative and flexible viewing environments than those in which ENERGY STAR is likely to invest.

OUTLOOK FOR THE FUTURE

In 2005, it is anticipated that the third-party delivery channel will far exceed the original model in delivering ratings. Additional benefits associated with having a commercial delivery channel, both in educating program participants and soliciting new participants, will be tested.

APPENDIX A: FREQUENTLY ASKED QUESTIONS

GENERAL UPLOAD ISSUES

Are there any limits to the number of buildings for which I can submit data?

For the VB application, there is no limit to the number of buildings in the XML file that you submit just as long as the file is no larger than 3.5 MB.

If you choose to call the web service directly, there is no explicit limitation to the size of the file or number of buildings that can be submitted. A limitation may be imposed by the network, hardware, and/or software used to directly call the web service.

How frequently can I submit data?

For the VB application during the pilot phase, an organization can submit an XML file no larger than 3.5 MB once every two hours. There is no limit on the number of submissions allowed per day, as long as the file is no larger than 3.5 MB.

If you choose to call the web service directly, the web service can be called once every two hours. There is no limit on the number of submissions allowed per day.

How can I check the status of the data that I have submitted? How often can I do this?

You can check the status of your data and view benchmarking ratings anytime by going to *http://ems-mx4.sradev.com/pmdxp/xmldex/* and entering your Login Name and Password.

How often is the data I submitted processed?

Data processing begins immediately but the length of the process varies. It is based roughly on the amount of data submitted.

How long are values available to be retrieved? Once retrieved, how often can they be retrieved again and for how long?

Results are retrieved by the user via the FetchData method of the web service. After the results of a transaction are retrieved, the dataset will be marked as retrieved and will not be available to the user upon future calls to the web services via the FetchData method. To retrieve values that have been previously Fetched, the FetchBetween method of the web service should be used. It allows the user to retrieve the ratings for any building and any time period.

Where do the ID attributes come from?

The IDs for the organization, buildings, meters, and spaces are IDs that you provide and use to identify these items. We use our own internal IDs to map to those that you provide. The IDs can be text up to 30 characters long (as defined in the schema). The XMLParticipant ID is provided to you when you sign up to test the XML data transfer process and is unique to you.

Can I change my organization, building, meter, or space type IDs?

No, these IDs must remain static. In the case of the

building ID, changing the ID value will either create a new building in the system, if the ID hasn't already been used, or update the data if the ID already exists.

What is the degree of uniqueness for a given ID (building, meter, or space type)?

The ID must be unique within the appropriate scope of the element. A meter ID must be unique within a building, and a building ID must be unique within an organization, etc. Two separate buildings can have the same ID for a meter (different scope).

The following provides an example of allowable IDs:

XML Participant ID 1234
  Organization ID 234
    Building ID 1
      Meter ID 1
      Meter ID 2
    Building ID 2
      Meter ID 1
      Meter ID 2
  Organization ID 456
    Building ID 1
      Meter ID 1

What encoding is preferable?

Please limit submissions to US-ASCII (ISO 646).

If building data are submitted for testing purposes, is it possible for this data to be removed in order to clear out the account?

Test data can be removed upon request. Send an e-mail to energystarbuildings@epa.gov to request that your account data be cleared out.

What do you mean when you refer to this as a "pilot program" and when will this initiative leave the "pilot" stage?

During this pilot phase, we are working closely with our third-party hosts to solicit feedback regarding implementation. We will no longer be a pilot program when we have an offering that is proven over a variety of large ENERGY STAR Partner accounts and is comprehensive enough to be offered as a complete package to future hosts. All of our third-party hosts are currently working with one or more ENERGY STAR Partners to formally implement the delivery of ratings. We have just upgraded our IT infrastructure, which required some adjustments to the XML schema and web service. We are making a variety of improvements within that new infrastructure, many at the request of our third-party hosts, and plan to add new support for hosts in the form of messaging and technical guidance to offer customers. We are still getting important input from our third-party hosts, and a few of our hosts have finalized the delivery of the rating. In addition, there are capabilities that EPA is interested in adding as part of this third-party delivery channel that will require interaction with our hosts to select the most mutually advantageous approach. Our "pilot" status maximizes the interaction we can have with hosts as this offering is more fully defined.

Will we be able to help our customers obtain Labels and other ENERGY STAR recognition?

Yes, in fact, many service providers have already found ways to assist their customers with recognition. While we are currently finalizing the data exchange approach required for the third party to deliver the rating, we are also investigating options for integrating ENERGY STAR recognition functions into third-party environments. This may involve passing additional data, displaying forms generated by the EPA server, or both. Because we are still in "pilot" mode, we will have a chance to solicit your input regarding the most effective and desirable approach. We anticipate a proposed specification late this summer.

BUILDING ISSUES

Will existing data be overwritten if any of the building characteristics (building name, address, etc.) are different in an XML submission than in previously submitted data?

If building data are submitted with the same organization ID and building ID as an existing building, and the general building information (such as the name, address, or year built) has changed, the system will overwrite the existing data and replace it with the newly submitted data.

If a building is no longer active or in use, how can it be removed?

It is currently not possible to remove a building via the XML process, but this is on the list of future enhancements. However, if the building still exists, it is preferable to keep it in the system, even if energy data are no longer being submitted. If a building has been duplicated or created by mistake or for testing purposes, it can be removed upon request (see section 7 on "Removing Data").

Can I submit data for buildings located outside of the United States?

For the XML data transfer testing period, we are not accepting data for buildings located internationally. However, the on-line version of Portfolio Manager will provide benchmarking ratings for buildings located internationally.

Can I change the name of a building that has previously been submitted?

The building name can be changed at any time, as long as the building ID does not change. If you submit data with the same building ID, but change the basic building data, such as the name or address, this information will be updated. This is also true for the organization name, space name, and meter name.

SPACE TYPE ISSUES

Will existing data be overwritten if any of the space characteristics (gross floor area, occupancy hours, etc.) are different in an XML submission than in previously submitted data?

If data are submitted with the same organization ID, building ID, and space ID as an existing building in the system and the space information (such as space name, gross floor area, occupancy hours, number of occupants, etc.) has changed, the system will overwrite the existing data and replace it with the newly submitted data.

If a data element for a specific space type (i.e. office, etc.) is not included, what will happen?

Certain space-level data elements are required, such as the total floorspace and space type for all spaces, and the hotel type for hotels. If one of these required data elements is not provided, the data transfer will fail. However, there are other non-required space data elements (such as the number of occupants and PCs for offices) that do not need to be included in the data transfer. If no value is provided for a non-required data element, the system will provide a default value in order for the building to receive a rating.

How are the default values for each space type determined?

Each space characteristic has a default value that is filled by the system if no value is provided. Some are set values (such as the weekly operating hours for a given space type), while others are calculated (such as occupants based on a density per thousand square feet).

ENERGY METER ISSUES

How does the system determine how many months of energy data have been submitted?

The benchmarking calculations determine if there is sufficient data available based on the dates of the individual energy entries. All of the data submitted for a meter will be compiled, whether it was submitted in monthly, weekly, daily, or other increments. The system will then start at the end of the last full month (e.g., if data was submitted through February 12, the last full month would end on January 31) and count backwards to ensure that there are 12 full, consecutive calendar months of data. Then the system compares all submitted meters to determine if there are 12 full months of consecutive data that overlap for the required 12 month period. It does this by counting back from the last full month of the meter with the earliest end date. (i.e., if there are two meters, one ending March 15 and the other ending February 15, the system counts back from January 31, the last full month of the meter with the earliest end date).

If a building has 12 months of meter data for a gas meter but the last month of the electric bill has not come in, leaving only 11 months, what affect will that have on the rating?

Most likely, the building would not receive a rating for the latest gas meter date until the electric meter data are available for the same time period. The requirement is that there are at least 12 full calendar months of energy data across all meters in order to generate a rating for a year ending period.

If the final meter entries for a building do not run to the end of the month, is that last month excluded?

Yes, ratings are only calculated for full months of energy data, so any partial months would be excluded. For example, if you had data ending on

March 15, the rating would be calculated based on the energy data through the end of February. A rating for March would not be available until data are submitted through a time period that covers the end of March (such as April 15).

If we send 12 months of energy data for 2002, then a year later, we send another 12 months of new data, will the old data be overwritten?

The system maintains all energy data submitted over time. If data for 2003 is subsequently submitted, it will be added to the building data and used to calculate a new rating. If revised data for 2002 is submitted (using the same identifiers as the 2002 data), it will overwrite the existing data and the rating for that time period will be updated.

Is it necessary to send 12 months of energy data with each submission in order to receive a rating?

No. Once an initial 12 month period of energy data has been established for a building, data can be submitted incrementally in weekly, monthly, quarterly, or annual segments. As soon as an additional calendar month of energy data are available, a rating for the latest time period will be provided.

Does "energy cost" refer to the cost per unit of energy consumed or the overall cost for the total amount consumed?

"Energy cost" refers to the overall cost for the total amount of energy consumed.

Do the energy consumption and cost elements require 2 decimals?

No, these numbers do not require 2 decimals, but they will be accepted if they are provided.

What is the limit on how old the energy data can be and still be submitted via upload?

There is no limit on the timeframe covered by the meter data that can be benchmarked. If you submit data for January 1999 through December 1999, you can still receive a rating as long as the building meets the benchmarking requirements.

If a meter ID changes and subsequent energy data are submitted with a different ID, will that affect the ability for the building to receive a score?

There is currently no way to indicate that the old meter ID is no longer active or that a new meter has become active. Therefore, the benchmarking calculations will interpret this as two incomplete meters, because they do not both cover a full 12 month period, and the building will no longer receive a score. An approach for addressing this problem is on the list of recommended enhancements.

If a meter or meter reading is submitted by mistake, is there a way to remove them through the automated process?

There is currently no way to delete a meter or meter reading via the automated process. A special request to remove the data element could be submitted to *energystarbuildings@epa.gov*. Be sure to include the item that you want to remove, along with the Organization ID, Building ID, Meter ID, and/or meter reading period that you would like to have removed (see section 7 on "Removing Data"). An automated approach for removing meter data is on the list of recommended enhancements.

How is the rating calculation affected when zero usage records are inserted for Natural Gas or Electricity?

Meter readings with zero usage can be submitted for Natural Gas, Electricity or other fuel types. As long as the meter readings cover a sufficient period of time and the building is otherwise eligible, a rating will be provided. This is the recommended approach for indicating a period of non-use for a meter, rather than not submitting any usage for the time period at all.

RESULTS ISSUES

What is the definition of 'Energy Intensity' and what unit is it reported in?

Energy Intensity is the energy consumption per square foot in a building. This is an annual value, calculated by taking the total energy consumption for the building in kBtu for the annual analysis period and dividing it by the total floorspace of the building for the same period. The units are in kBtu/sf.

Is it possible to receive a rating of 'null' or greater than 100?

Valid benchmark ratings are in the range of 1 to 100. A negative rating could be received if the building data are not benchmarkable. Ratings of greater than 100 or Null will not be received.

APPENDIX B: COMMON ERRORS IN PASSING DATA

The following are common mistakes when transcribing data into the accepted XML schema:

1. **XML Participant element is missing the ID attribute.** The XML Participant element or node is the root of the XML document. The ID attribute of this element must be populated with the Participant ID given to you when you signed up with this program.

2. **XML Participant and Organization names are not assigned correctly.** The XML Participant who registers to participate in the data exchange process is the entity that is actually submitting data via the data exchange process. Most often, this is not the owner of the buildings for which data are being submitted, but rather an energy information provider who is providing a service to its customers. This entity is identified by the XML Participant ID in the schema, using the participant ID that was assigned at the time of registration. The Organization element in the XML schema is intended to identify the customers of the XML Participant. These are the owners and operators of the buildings for which energy and space data are being submitted. For example, if ABC Provider provides energy information services to customer XYZ Supermarkets and begins to provide automated data upload and rating services, their data submissions should indicate XYZ Supermarkets as the Organization name.

3. **Improper capitalization.** XML is a case sensitive standard. For example, there is a difference between "Building" and "building." Be sure that the proper case is used for the schema elements.

4. **Improper value for an enumerated Type.** Certain elements of the schema must only use values that have been enumerated. Values that are allowed for these enumerated Types can be found within the schema. An example is the "energyUnit" element under the "meter" element. This element is of Type "energyUnitType" and can only contain the values defined for "energyUnitType." These values are shown in the columns on the right.

5. **Incorrect date format.** The date format for the input follows the "mm-dd-yyyy" format.

6. **Overlapping meter periods.** Meter periods cannot overlap by more than one day.

For Electricity:
kWh (thousand Watt-hours)
kBtu (thousand Btu)
MWh (million Watt-hours)
MBtu (million Btu)

For Natural Gas:
ccf (hundred cubic feet)
therms
kBtu (thousand Btu)
kcf (thousand Btu)
cf (cubic feet)
MBtu (million Btu)
Mcf (million cubic feet)

For Steam:
lbs(pounds)
kLbs (thousand pounds)
MLbs (million pounds)
kBtu (thousand Btu)
MBtu (million Btu)

For Chilled Water:
ton hours
daily tons
gallons
kBtu (thousand Btu)
MBtu (million Btu)

For Fuel Oil (No. 2):
a. gallons
b. kBtu (thousand Btu)
c. MBtu (million Btu)

For Liquid Propane:
a. gallons
b. cf (cubic feet)
c. kcf (thousand cubic feet)
d. kBtu (thousand Btu)
e. MBtu (million Btu)

For Kerosene:
a. gallons
b. kBtu (thousand Btu)
c. MBtu (million Btu)

For Diesel (No. 2):
a. gallons
b. kBtu (thousand Btu)
c. MBtu (million Btu)

For Fuel Oil (No. 1):
a. gallons
b. kBtu (thousand Btu)
c. MBtu (million Btu)

For Fuel Oil (No. 5, 6):
a. gallons
b. kBtu (thousand Btu)
c. MBtu (million Btu)

For Coal (anthracite):
a. lbs (pounds)
b. tons
c. Ubs (thousand pounds)
d. MLbs (million pounds)
e. kBtu (thousand Btu)
f. MBtu (million Btu)

For Coal (bituminous):
a. lbs(pounds)
b. tons
c. Ubs (thousand pounds)
d. MLbs (million pounds)
e. kBtu (thousand Btu)
f. MBtu (million Btu)

For Coke:
a. lbs (pounds)
b. tons
c. Ubs (thousand pounds)
d. kBtu (thousand Btu)
e. MBtu (million Btu)

For Propane:
a. gallons
c. cf (cubic feet)
d. kcf (thousand cubic feet)
e. kBtu (thousand Btu)
f. MBtu (Million Btu)

For Wood:
a. tons
b. kBtu (thousand Btu)
c. MBtu (million Btu)

For Other:
a. kBtu (thousand Btu)

WHAT IS XML?

XML was designed specifically to address limitations in HTML that had become challenging as web applications demanded more sophistication. It allows structured information to be organized and presented with significant depth and flexibility and then used for a wide variety of applications. In laymen's terms, XML allows developers to provide or request many layers of information or data about information or data, where HTML is more oriented toward display and single hyperlink relationships. In addition, HTML provides a fixed tag set and associated tag semantics (The tag "<P>" refers, always, to a paragraph, but the tag "<energy star rating>" does not exist and cannot be created). XML allows the developer to define tags and relationships as needed to fit the application and subject matter. Some specific attributes of XML are notable and have resulted in wide acceptance:

- XML content is relatively simple to understand in its raw form. It does not require technical expertise or a translation program to look at an XML document and determine what it is, and where problems are occurring. This allows a wide variety of individuals to understand and participate directly in XML-based initiatives, not just highly skilled IT staff.
- XML combines simplicity of language, which allows it to be used for a variety of applications, with relatively broad capability. There are no hidden options that must be learned and the syntax is straightforward. This makes XML easy to learn regardless of what languages one used previously and makes XML applications fast to develop and share.
- Changes to an XML schema are easy to communicate and execute, making XML an ideal choice where a number of different entities are involved in using a single schema.
- It is easy to create different levels of educational or support information within an XML document. For example, a barebones schema could be shared among experienced team members, with heavily annotated and sample data versions of the schema used to initiate new team members.
- While HTML is still the dominant approach for web page creation, XML has become a standard for moving complex data over the internet.

Chapter 14

Web-Enabled GIS Platform with Open Architecture for Electric Power Utility Networks

Fangxing Li
ABB Inc.; 940 Main Campus Drive, Suite 300; Raleigh, NC 27606

ABSTRACT

THIS CHAPTER describes a flexible architecture of a Web-based GIS (Geographic Information System) application for utility network visualization. This Web-based application serves as a platform for high-level power system analysis. Different from regular applications, this GIS platform may be viewed as a semi-completed application. It implements fundamental features of a utility system application such as GIS-like drawing and visualization, built-in topology processor, and so on. Meanwhile, the platform defines open data structures for power system components with efficiency and flexibility being a major consideration. It also provides a mechanism to link itself with external engines, instead of implementing the engines directly. This architecture may achieve considerable flexibility. To maximize the benefits to users, the proposed platform is Web-enabled with a Java Applet. It is universally accessible, instantaneously upgradeable, and operating system independent.

INTRODUCTION

Web-based applications have attained great popularity in the past few years and are projected to become more popular in the future. Some of the benefits of Web-based applications to customers and application providers include:

- Minimal installation requirements
- Instant updates
- Mass customization
- Universal accessibility.

This chapter describes a flexible architecture of a Web-based GIS application for utility network visualization. This Web-based application serves as a platform for high-level power system analysis. Different from regular applications, this GIS platform may be viewed as a semi-completed application. It implements fundamental features of a utility system application such as GIS-like drawing and visualization, a built-in topology processor, and so on. Meanwhile, the platform defines efficient and flexible open data structures for power system components. It also provides a mechanism to link itself with external engines, instead of implementing the engines directly. Therefore, the architecture of this platform is considerably flexible and extensible. For instance, users may extend and customize data structures of power system components. Users may also develop their own analytic engines based on their specific needs and link them with the platform.

To maximize the benefits to users, the proposed platform is Web-enabled with Java client-side technology. It is universally accessible, instantaneously upgradeable, and operating system independent.

BASIC FUNCTIONS

The basic function of this GIS platform is fully developed. The core of the GIS platform is a graphic drawing tool that allows users to hand-draw a one-line diagram of a power system or import a large system through its data of (X,Y) coordinates. With the drawing tool, users can perform many common GIS functions. Typical functions are shown as below.

Editing

- Add one or more components

- Delete one or more components
- Move one or more components
- Resize one or more components

Zooming
- Zoom in
- Zoom out
- Zoom to a special window
- Zoom to display the entire system.

Layer management
- One background layer (roads, streets, and images)
- Up to 10 foreground layers.
- Each layer has three status: OFF (invisible and non-editable), GRAYED (visible but not editable), and ON (visible and editable).

Another important feature of the user interface is the ability to shade system components based on the value of a specific data field. This feature can be very helpful for utility analysts. For example, a utility's transmission and distribution (T and D) system can be shaded based on the voltage levels at each transmission or distribution line. Components with higher voltages may be drawn in darker color. Also, users may shade components based on loading percentage to obtain a system-wide view of regions or lines that are the bottlenecks of transmission capability. Or, as shown in Figure 14-1, users may shade components based on their upstream feeding substations, so it is visually clear about the service territory of each substation in a radial distribution system. Figure 14-1 also illustrates the zooming function that can display more details of a sub-area.

Other functions of the user interfaces include: open/save projects, edit components, redo/undo drawings, modify data fields, change the system scale and unit, system statistics, and many other necessary functions.

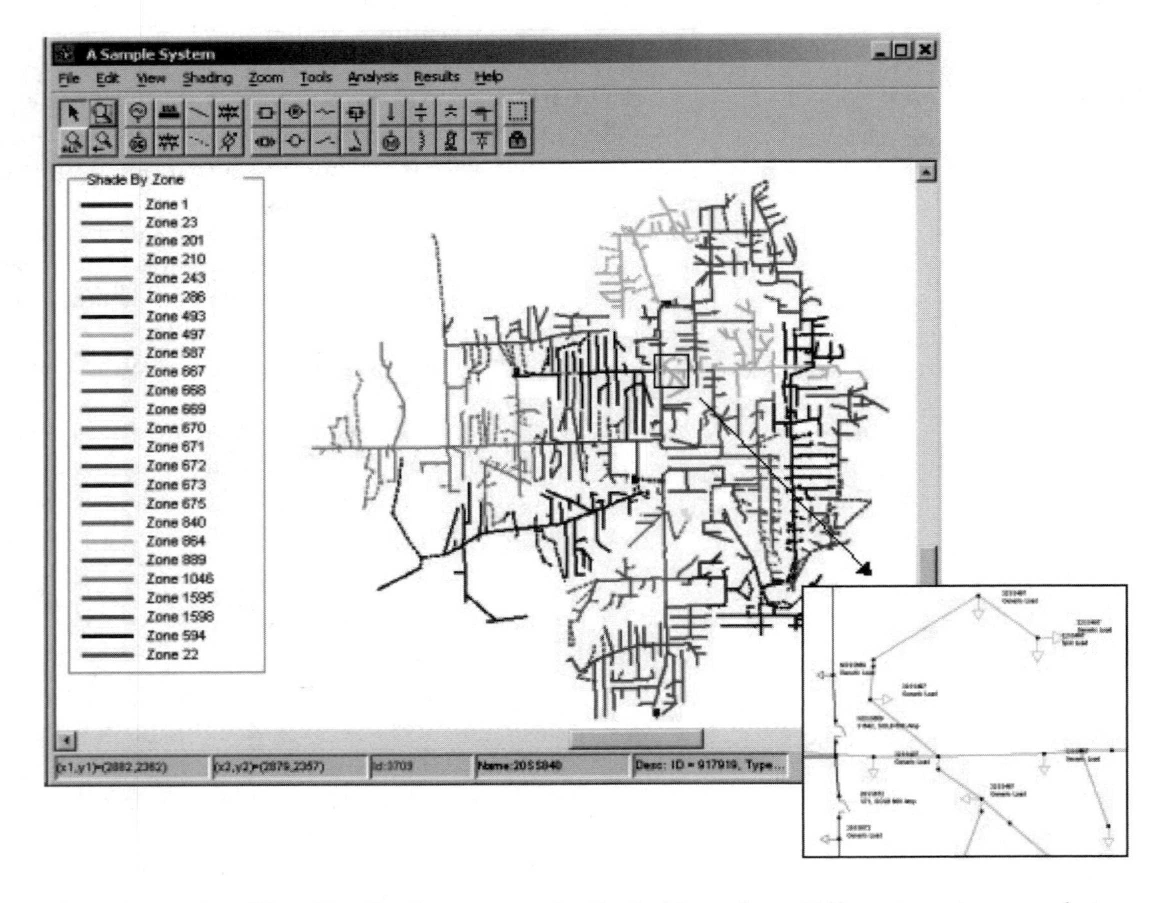

Figure 14-1. A utility distribution system is shaded based on different upstream substations. This is the main GIS window launched from the main Web page. This window is developed with Java Applet as a window floating on top of the main Web page. The small window illustrates the zoom-in function.

BUFFERED IMAGE FOR FAST GRAPHIC DISPLAY

The power utility network is bulk in nature. It is not unusual for a distribution system to contain tens of thousands or even hundreds of thousands components. The performance of the GIS system is therefore critical for usability, especially for a Web-enabled tool mainly based on Java Applet. This is because Java Applets do not operate with the local operating system. Instead they use a secure sandbox JVM (Java Virtual Machine) for operating system independence. (See Glossary for definition of "sandbox.") Visualization delay for the GIS system could therefore be longer than a regular application developed in other languages like C++.

A technique called Buffered Image is applied in the platform to achieve optimal performance in the graphical display. A straightforward way to display a geographic system is to draw every component one by one on the screen of the main window. This needs to be re-drawn whenever a new component is added, an existing component is removed, or the displayed area is moved. Efficiency is difficult to achieve because the re-draw for a big system may cause considerable delay. The Buffered Image technique is designed to reduce the visualization delay. Using this technique, the system initially draws a large background image. Then, a rectangular area of the image is displayed in the main window. The position of the rectangular area is indicated by the displacement of the main window's side and bottom sliders. When the user moves the side or bottom slider to view another part of the image, the origin of the displaying window will be determined in accordance with the displacement of the sliders. Then, a new rectangular area of the image will be displayed. This background image serves as a buffer between the screen and the user's drawing command, so it is called the "buffered image" technique.

Here are some underlying details when a user adds or deletes components. When the user adds a new component such as a line, the application simply draws a new line onto the background image and then displays the image on the screen. When the user deletes an existing line, the application draws a line on top of the existing line onto the background image using the background color. Therefore, when the updated image is displayed on the screen, the existing line is visually erased. Since there is no need to redraw the whole system after adding or deleting components, good performance will be achieved. It should be noted that the application updates the system's internal connectivity structure after the user adds or deletes a component.

Figure 14-2 shows the process of the Buffered Image technique.

BUILT-IN TOPOLOGY PROCESSOR

The GIS contains a built-in topological processor, which identifies the connectivity of system components. It can find islands of a system, de-energized components, all connected components of a specific source, etc. These functions are very useful for finding hard-to-detect human errors when a user tries to manually move or resize a component. For instance, after the user moves a line to connect it with another one, the two lines may appear to be connected but are actually unconnected due to a slight human eye error. Although the zoom-in function may be used to identify this drawing error, it will be a lot more efficient to identify many similar errors in a batch mode if the topological processor is run. Another important feature of this platform is that component connectivity is dynamically updated when users add, move or delete a component. This feature makes the topological processor very fast and efficient.

The topological processor considers drawing conventions of both transmission and distribution systems.

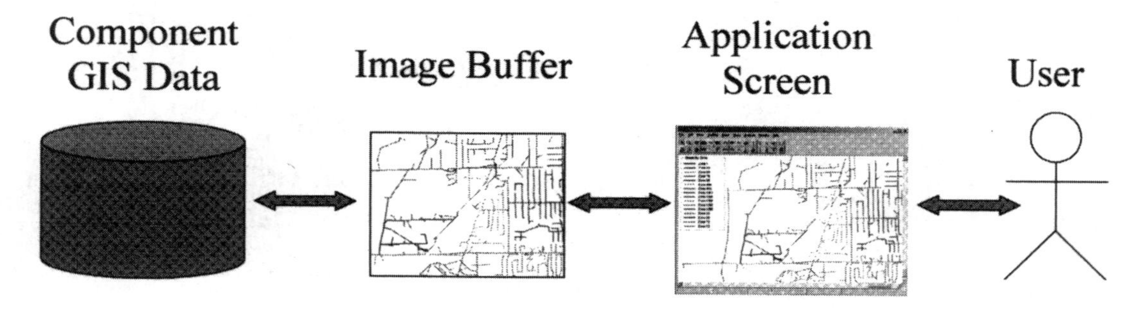

Figure 14-2. The technique of buffered image for fast graphic display

The common practice to identify connectivity of the two types of systems is illustrated in Figure 14-3. In a typical transmission system diagram, two components or lines are considered to be connected if each component has a node connected to the same bus, as shown in Figure 14-3(a). In a typical distribution system diagram, especially for distribution feeders and laterals, two components are considered to be connected if one component has a node directly connected to a node of the other component, as shown in Figure 14-3(b). The reason for this is that distribution systems may have a very large number of relatively short lines. In that case, it is neither feasible nor necessary to draw all the buses. This platform accepts both drawing conventions in order to support both transmission and distribution systems.

All topological searches in this platform are either upstream or downstream searches. The former identifies the path to power sources, usually substations, of a selected component. The latter identifies all power system components that must receive power through a selected component. This is particularly useful for power distribution systems since the majority of the US distribution systems are radial. Hence, the power source is unique for every component. Figure 14-4 shows the upstream and downstream search results for a radial utility distribution system.

It should be noted that this platform considers that each component may have up to four nodes. Hence, this design can directly model three-node components such as three-winding transformers and transfer switches, or even ad-hoc four-node components.

OPEN DATA SCHEMA

Each power system component is usually defined by various parameters (or attributes in object-oriented paradigm), which are employed to implement applications for power system analyses. The most common attributes usually include voltage, current, power, etc., which are essential for most power system analyses. But for some specific analyses, new attributes may be re-

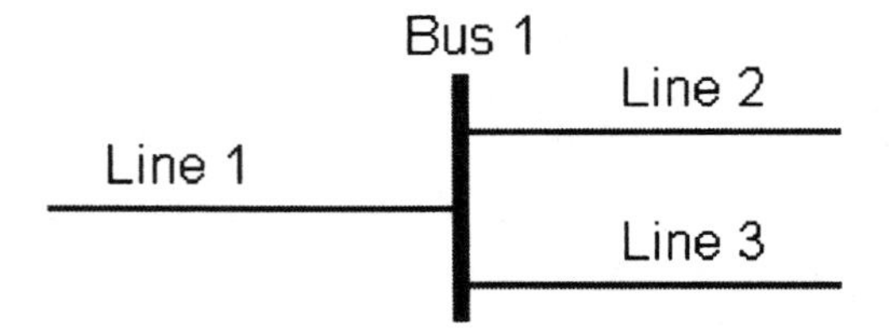

(A) Three lines connected through a bus

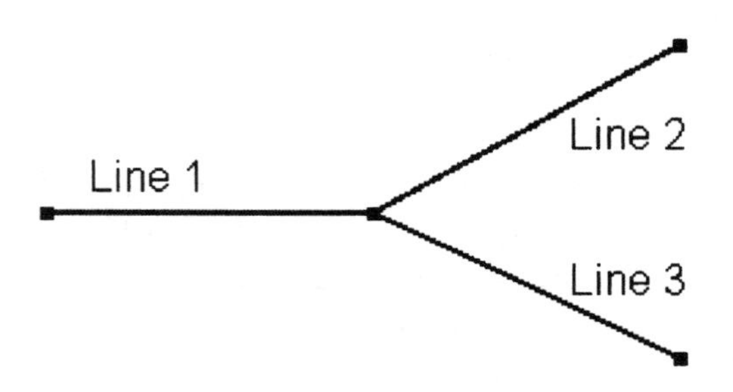

(B) Three lines connected through nodes

Figure 14-3. Identification of connected components for power transmission and distribution systems.

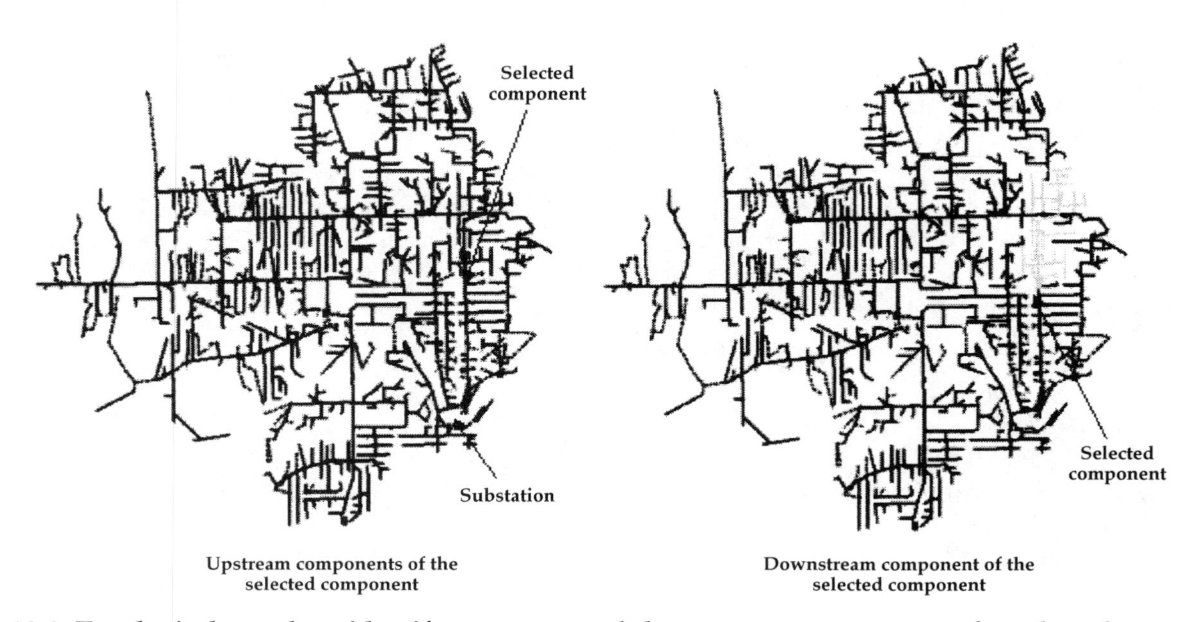

Figure 14-4. Topological search to identify upstream and downstream components of a selected component.

quired. As a generic tool for power system visualization, this platform only predefines the most common attributes but not all attributes that are potentially useful for a specific analysis. It is nearly impossible to predefine all attributes for future uses. However, this platform does allow users to extend component attributes whenever they are needed. Hence, this design balances efficiency and extensibility, because, generally, predefined attributes are more efficient but less flexible than user-defined attributes. The user-defined, extended attributes are briefly discussed as follows.

Each extended attribute is internally represented and stored as a string-like object. It may be costly to convert an extended attribute to its actual primitive data type like double, float, integer, etc. Hence, extended attributes are not as efficient as fixed, predefined attributes that are represented in primitive data types. However, extended attributes are more flexible from the viewpoint of user customization.

With the above design, each component has the following attributes:

- General information attributes: ID, Name, Type, Description, Layer
- GIS information attributes: X1, Y1, X2, Y2, X3, Y3, X4, Y4.
- Pre-defined, basic domain attributes (dependent on component types)
- User-defined, extended attributes (dependent on component types)

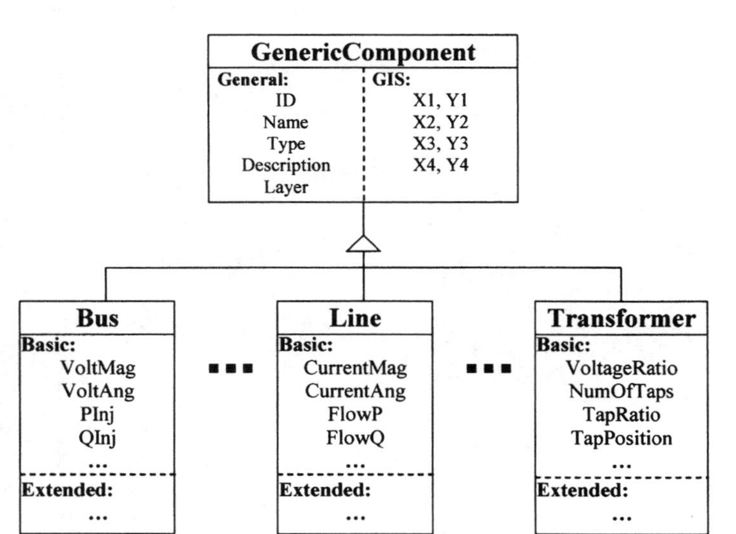

Figure 14-5. Hierarchy of the data structures of components. The most common component attributes are summarized in the GenericComponent class.

Figure 14-5 illustrates the above scheme of component attributes with a hierarchy structure. Inheriting all attributes from GenericComponent, each component has some pre-defined basic attributes as well as user-defined attributes, which essentially make this data structure open. While the pre-defined basic attributes are fixed, the extended attributes are flexible and editable. As Figure 14-6 shows, users may add, edit, or delete the extended attributes of components.

IMPORT AND EXPORT DATA TO SPREADSHEET

The previous discussion focused on the internal data structures of components. To make the internal structures more useful and eventually linkable with the external engines that are discussed in the next section, an external data presentation is necessary. This platform provides a flexible way to export or import component attributes to a tabular text file, which is in comma-separated-value (CSV) format. The CSV format is supported by many spreadsheet applications in various operating systems. The most popular one probably is Microsoft Excel. With the power of Excel, users can easily modify a large amount of component attributes by exporting data to a CSV file, opening the CSV file in Excel, manipulating data in Excel, and finally importing data back. If such a spreadsheet application is not available, users still can access, view and manipulate the data file, because the CSV file is basically a plain text and human-readable file regardless of the operating system.

This platform supports full or partial data export or import. That is, besides the key attribute, component ID, users may select a subset of the attributes to export or import as shown in Figure 14-7. This also implies the way to import the GIS information from an external data source to create a new system. Users simply need to arrange the external GIS data in CSV spreadsheet and then click the import button with the correct import options.

This platform also provides files in binary format in addition to CSV format for import or export. Usually this is not necessary; however, due to the high efficiency of the binary format, it might be desired under some scenarios that require, for example, many iterations of data import/export or manipulations of huge amounts of data. Basically, binary files are employed for this platform to permanently save project systems onto disk or to transfer them over the Internet for space and speed. For data export/import/manipulation at a local machine, CSV text files should be sufficient in most cases. In fact, it takes less than 15 seconds in a 700MHz machine to fully export a system with about 9,000 components, each of which on average has 35 attributes mainly in integers and double-precision numbers, and it takes only 10 seconds to import the updated data back.

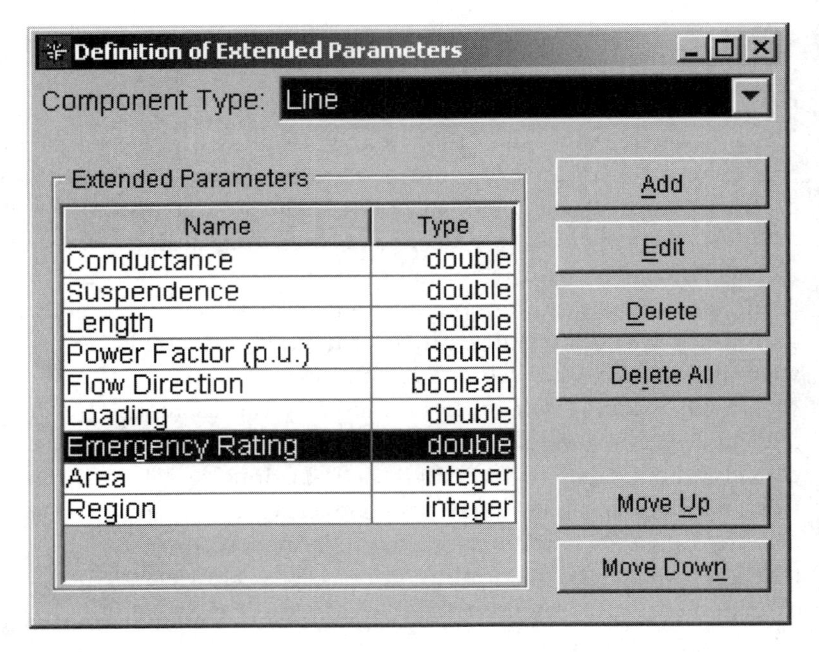

Figure 14-6. A dialog for users to add, delete, or change the extended attributes of a component.

UNIVERSAL ACCESSIBILITY

To maximize the benefits to users, this platform is designed to be deployable over the Web and accessible with any operating system. With this consideration, the platform is implemented in Java Applet. A Java applet is a Java program downloaded from a Web server into Web browsers and runs inside a Java Virtual Machine (JVM) that is integrated into the Web browser. Nearly as powerful as stand-alone desktop applications, applets are well able to locally process intensive user interactions such as drawings, which are often difficult or impossible to implement with other Web technologies like HTML, server-side scripts, or client-side scripts. In addition, applets have all the features of Java such as operating-system independence, built-in security mechanism, memory management, error handling, etc.

Like many other Web-enabled applications, this platform has the advantages of minimal installation requirements, instant updates, mass customization, universal accessibility, etc. Developed in Java Applet and embedded into Web-browsers, the platform is universally accessible from anyplace with an Internet connection, regardless of the local operating system.

LINKS TO EXTERNAL ANALYTIC ENGINES

As a generic GIS platform for power system visualization with open architecture, this platform is not designed to implement analytic engines to solve specific problems such as power flows, short circuits, optimizations, etc. From the standpoint of this platform, analytic

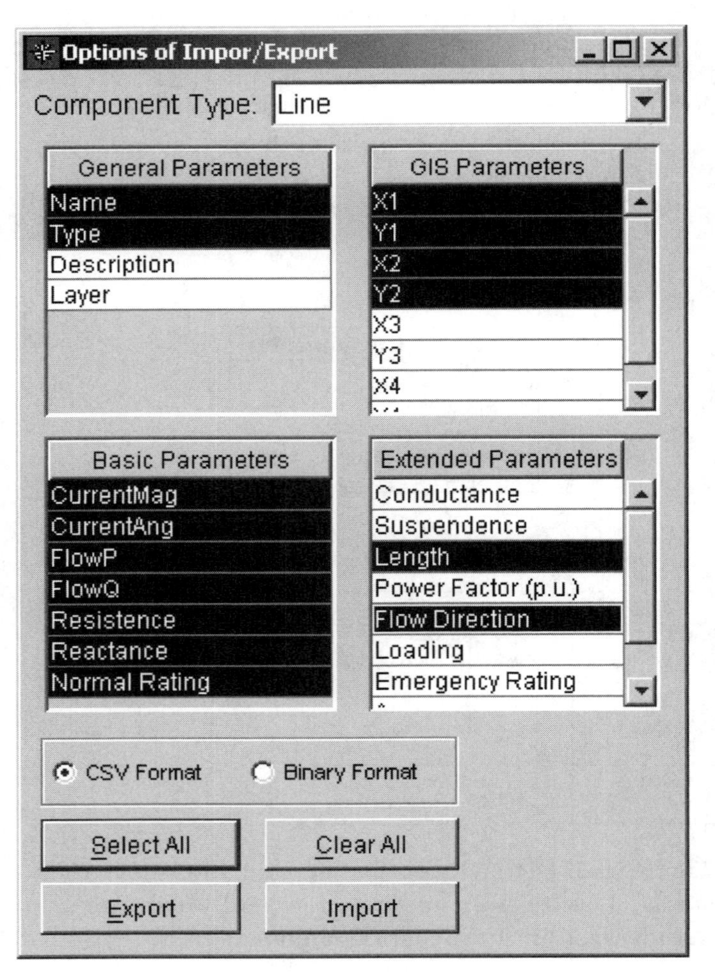

Figure 14-7. A dialog to set up the export or import options

engines are external or standalone modules that should be linked and integrated with the platform. The engines will be implemented by users of this platform or other third parties. The engines can be developed in any languages such as C/C++, Fortran, and Matlab, as long as they import or export data using the formats supported by this platform. With the flexible data schema, especially in the spreadsheet, it is easy for the external engines to work with the drawing tool. This platform also allows users to create and customize menu items that directly point to external engines that are standalone local applications or even as remote applications.

Figure 14-8 illustrates the overall architecture of the Web-enabled GIS system with an analytical engine for electric power system reliability analysis. In this figure there are three server programs that may or may not physically reside on the same server machine. The functions of these servers are described as follows:

- Web Server—GIS: a server containing Web pages related to the GIS platform
- Web Server—ENG: a server containing Web pages with embedded ActiveX control (C++) engines
- Remote Data Server: a data server hosting all data files.

A typical interaction of the Web-enabled GIS system with the analytical engine is described as follows.

1. The user logs into the system from a Web browser and the GIS platform is transparently downloaded from Web Server-GIS
2. The GIS platform is launched from the login web page
3. Data are loaded from the Remote Data Server (or local data files) and displayed on the GIS application
4. A Web page containing an analytical engine is downloaded from Web Server-ENG and invoked by the GIS application
5. The analytical engine communicates with the GIS system through a temporary file, performs computations, and returns results
6. Data are uploaded to the Remote Data Server or stored to local files.

DATA CONVERSION WHEN INTEGRATED WITH UTILITY GIS DATA

As previously described, this platform can work with utility GIS data through the spreadsheet. Given the format of the GIS data, a small query or macro applica-

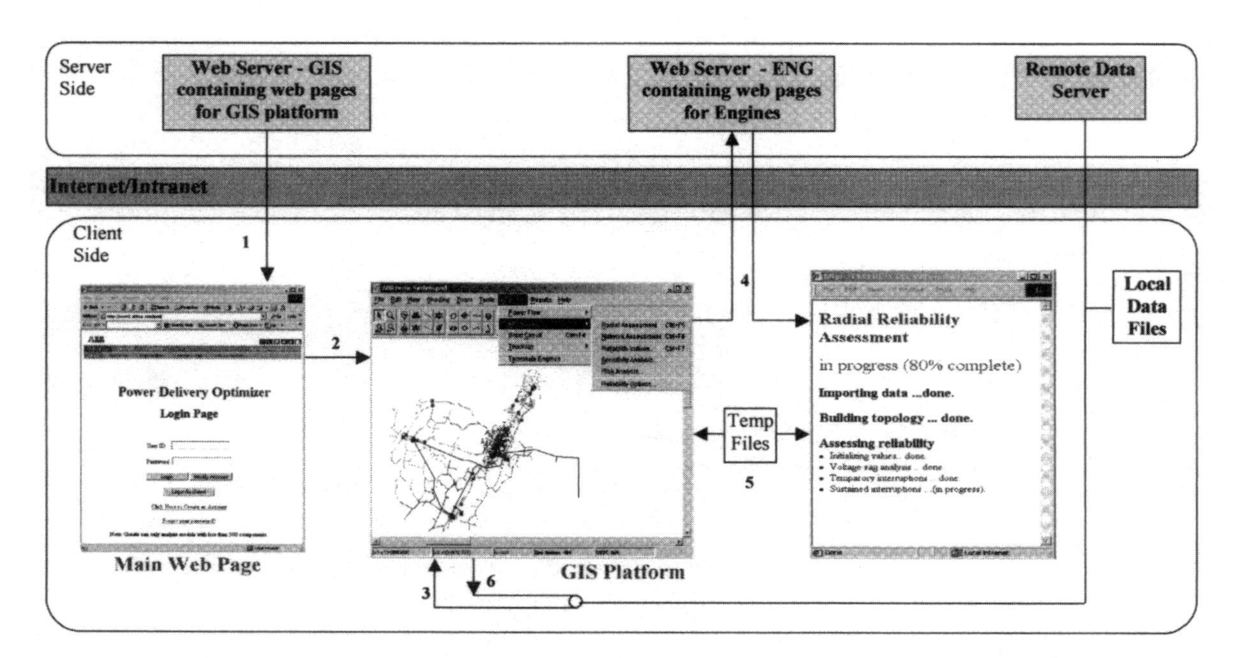

Figure 14-8. Architecture of the Web-based application showing interaction among the servers and the client

tion has been developed to translate various GIS data into the spreadsheet format. Figure 14-9 shows the process to extract the utility GIS data into spreadsheet format that can be imported into this platform for visualization. Then, with other inputs like component data, users may perform high-level analysis like load flow, reliability, short circuit through the external analytic engines, etc.

The utility GIS data usually consist of topological information of nodes and branches, the two essential components of a utility network from a topological viewpoint. Each node usually corresponds to a pole for overhead lines or a manhole for underground cables. Each branch usually corresponds to an overhead line section or an underground cable section, which is connected between two nodes. With this topological information, the converter can identify the (X1, Y1) and (X2, Y2) for all two-node lines and cables with cross-references and then dump it into a spreadsheet in the required format. This simple logic is the key part of the converter. When the conversion is done, the utility network can be imported and displayed on the Web through this platform.

SUMMARY

This chapter has presented an architecture and prototype of a Web-enabled GIS platform with open structures. The features of this platform are summarized as follows.

- It provides a Web-based GIS platform for power systems with drawing and zooming functions
- Systems can be shaded based on component attributes
- It has a built-in topological processor that supports topological search
- Data structures of components are flexible and extensible
- External analytical engines may be linked with this platform
- It is developed in Java Applet, which is browser independent, operating system independent and universal accessible.

Bibliography

1. F. Li, L.A.A. Freeman, and R.E. Brown, "Web-Enabling Applications for Outsourced Computing," IEEE Power and Energy, vol. 1, no. 1, January 2003, pp 53-57.
2. F. Li and L.A.A. Freeman, "Using Client-side Technologies to Develop Web-based Applications for Power Distribution Analysis," DistribuTECH 2003 Conference, February 4-6, 2003, Las Vegas, Nevada.
3. C. Horstmann and G. Cornell, Core Java, vol. 1 & vol. 2, The Sun Microsystems Press, 2001.
4. S. Ma, L. Qi, W. Liu, and W. Ma, "Power Station GIS Design and Implementation," IEEE Computer Applications in Power, vol. 15, no. 2, April 2002, pp 41-45.
5. B.R. Williams, C. Mansfield, R.E. Brown, and H. Kazemzadeh, "Engineering Tools Move Into Cyberspace," Transmission and Distribution World, March 2003, pp 27-36.
6. H. Tram, "The ASP Model for Energy Delivery Information Systems," Proceedings of IEEE PES Transmission and Distribution Conference and Expo 2001.
7. L. King, "Web Delivery of Engineering Software and Technical Reference Materials for Distributed Wind Generation Planning," Proceedings of IEEE PES Transmission and Distribution Conference and Expo 2003.

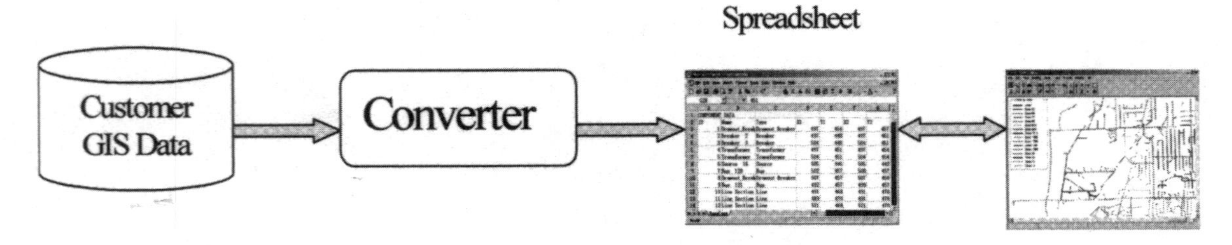

Figure 14-9. Utility GIS Data Conversion

Section Four

Web Based ECS Case Studies

Chapter 15

Intelligent Use of Energy at Work: A Detailed Account of Saving Energy and Cost at the Wellness Center of the University of Miami

Dirk E. Mahling, Ph.D.
Mark A. Noyes
William O'Connor
Richard Paradis

ABSTRACT:

THIS CHAPTER is a case study of a multi-use facility—the Wellness Center of the University of Miami. Savings greater than 11% are accomplished by equipping the existing building management system with an Enterprise Energy Management system called IUE (Intelligent Use of Energy). The case study shows how well-known energy management strategies, such as speed-reset or supply-air-temperature-reset can be leveraged by a rule-based expert system to enhance energy savings. It also shows how the automated strategies leverage the forecasting power of neural networks to predict and simulate building behavior as well as to reduce energy usage at the revenue meters. Detailed logs of two event sequences demonstrate how the savings are accomplished, step-by-step, without impacting comfort in the building.

INTRODUCTION: THE INTELLIGENT USE OF ENERGY

Purpose of the "Intelligent Use of Energy (IUE)" System

The best way to ensure that a building is operated in the most energy efficient manner would be to have a team of the worlds' best energy and building managers, each one an expert at what they do. One would be the world expert at "load rotation," another would understand the impact of weather on thermal storage strategies, another would be able to forecast energy usage, and so on.

Imagine this team taking care of the building for 24 hours a day, 7 days a week, without ever taking a break. Imagine they know all the industry best practices to conserve energy, to make sure energy spending stays within the budget, to keep hot/cold calls at a minimum, to ensure tenant comfort, and to operate equipment safely and efficiently.

In addition, wouldn't it be nice if this team would constantly monitor the weather forecast, including temperature, humidity, cloud cover, etc., and prepare for particularly hot or cold days, not needing to struggle when things get tight and energy prices are at a premium. Or if this team would actually monitor the price of energy and make informed decisions about energy cost savings and tenant comfort on a minute-by-minute basis? Or if this team continuously looked at the scheduling system to predict how many people are in each building zone in the next few hours? To do this, the team would need to have very detailed models of energy consumption patterns and degradation times for every air handler, every room, every compressor, and every other device that consumes energy. But it even gets more complicated. Even if such a model could be created, it would very soon be outdated, since equipment ages, seasons change, usage changes, etc. This complex and highly detailed model would have to be updated after a few short weeks or maybe even days.

So, is the idea of a "building engineering dream team" possible? Does such a team exist? If it could be assembled, would it be economic? Is it possible for the engineer to model devices and spaces at such a level of detail that they could provide meaningful forecasts [1,6]?

Of course, bringing together such a team to run just a few buildings in this manner would be impossible. Even the best building engineers in the world cannot achieve the level of modeling or information processing necessary to achieve all of the above objectives with the continual diligence necessary.

That's why WebGen has worked over the last three years with building managers and engineers to build a library of best practices in the area of building management and engineering, which accomplishes these goals. WebGen's Intelligent Use of Energy (IUE®) system is a collection of processes and algorithms, known as "intelligent agents," which incorporate the best practices in the world of facilities operation. IUE's agents employ such best practices as (to name just a few):

- Load rotation
- Supply air temperature reset
- Speed reset
- Lighting level adaptation
- Pressure reset
- Pre-cooling

All these strategies are well known to building managers and engineers. Through the IUE system, WebGen has methodically collected all these best practices and pulled them together into the world's largest automated library of best practices in energy engineering. This library is now available to building engineers and property managers. Building engineers can put these strategies into a "cruise control" mode, where the IUE agents act as experienced staff members, tirelessly implementing best practices.

In addition to the agents implementing the best energy management practices, IUE uses neural networks to continually learn the shifting behavior of the building. Since IUE is connected to the building in real-time, IUE's neural networks learn the changes in degradation ratios, ramp-up times for cooling per space, etc.

With the neural networks giving the most accurate prediction for temperatures, run-times, meter readings, etc. the agents can implement the most effective strategy for energy savings at a specific point in time, while maintaining comfort parameters.

Benefits for Building Engineers and Managers

In this way, WebGen's IUE system delivers the following benefits to building managers and engineers:

- Access to global best practices in energy management
- Additional "expert staff members" focused on energy management
- More bandwidth to provide physical customer service
- Early insight to equipment degradation

Benefits for Property Owners

Property owners are usually more focused on the business and financial performance of a building than on the day-to-day operations, which are the responsibility of building engineers. Property owners are interested in high-quality tenants and high customer satisfaction. They are keeping their eye on the overall performance of their properties. The financial performance, at a high level, consists of revenue and cost. Satisfied tenants tend to improve the revenue side. Intelligent controls are tools to keep the cost side under control.

The IUE helps with both sides. Utilizing various strategies, the IUE aids building owners by maximizing tenant comfort while minimizing energy consumption. This provides increased revenue and decreased costs.

ARCHITECTURE OF THE IUE

The landscape of systems in the domain of building and energy management is becoming very confusing. There are "Energy Information Systems," "Demand Response Systems," "Enterprise Energy Management Systems," and many more. We like to think of the IUE as the energy module of a building management system (BMS).

Conventionally, it is the job of a BMS to run the devices in a building according to a schedule or a set of given parameters (e.g. thermostat loop). The BMS does not care about weather forecasts, energy prices, or utility bills. The IUE brings these capabilities to the BMS, leveraging the control functions of the BMS to take smart action, while maintaining the integrity of devices and the safety of operations. In addition, the IUE manages energy across a portfolio of buildings, thus maximizing energy savings for many properties.

On the functional level, the IUE consists of 7 modules.

1. Web-I: this gateway and its drivers provide connectivity from the Internet to the BMS
2. TREND/Meter: remote metering and trending
3. TREND/BMS: archiving, trending, alarming and analysis of BMS data

4. FINANCE: computation of energy bills based on meter readings and tariffs; computation of savings reports against baselines
5. AGENT: forecasting and control using best practices in energy management
6. DEMAND RESPONSE: leveraging forecasting and best practices to accomplish demand reduction in response to a pre-established curtailment contract with an energy provider
7. DISTRIBUTED GENERATION: internet-based control of power generation equipment to support all previous strategies.

Each module leverages the capabilities of the previous modules. Figure 15-1 shows the connection to the building and the interaction between the modules.

The IUE System in Context

The IUE system works in conjunction with building management systems. It does not replace the building management system[1]. Figure 15-2 shows how the IUE works with the local building management system (BMS).

Meters and devices in the building are mapped from the BMS to the SCADA and then on into the IUE. A protocol driver in the SCADA unit (supervisory control and data acquisition) provides communication to the Internet. The SCADA unit actually has three functions:

- Working as a router to connect points in the BMS to the Internet
- Working as a store and forward device to provide a local buffer
- Working as a protocol translator, from the BMS to the IUE

[1]If the building has no BMS, the IUE can still be installed. In such a case advanced thermostats and DDCs, being a substitute for the BMS, need to be deployed.

In buildings without Energy Management and Control Systems where wiring would be a problem, e.g. multiple, distributed roof-top units, a wireless DDC (direct digital control) solution can be installed which connects directly to the Internet.

In locations where direct Internet access through existing LANs in buildings is a problem, DSL lines can be used.

On the other side of the Internet, the IUE's communication module, IUE COM, picks up the signals from the SCADA units. The control signals from the IUE take the same trip in return. IUE-COM puts the signals on the Internet and local SCADA units pick them up. Communications and addressing is handled via Internet IP addresses.

IUE's Control Layers

The IUE itself consists of three-tiered control layers:

- top level; Goals as triggers for agents,
- agents enacting energy management strategies based on goals,
- neuralnet providing forecast to aid in strategy decision
- the bottom layer in Figure 15-4 is a device layer executing decisions reached in the above 3 control layers.

Goals are at the top level; these goals set envelopes to the strategy agents to control. The strategy agents are

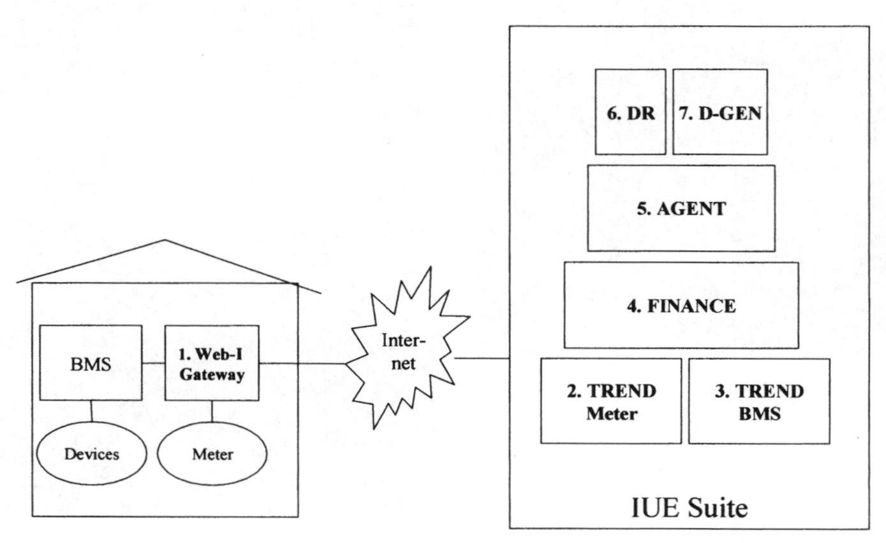

Figure 15-1. IUE Modules as an Energy Extension of the BMS.

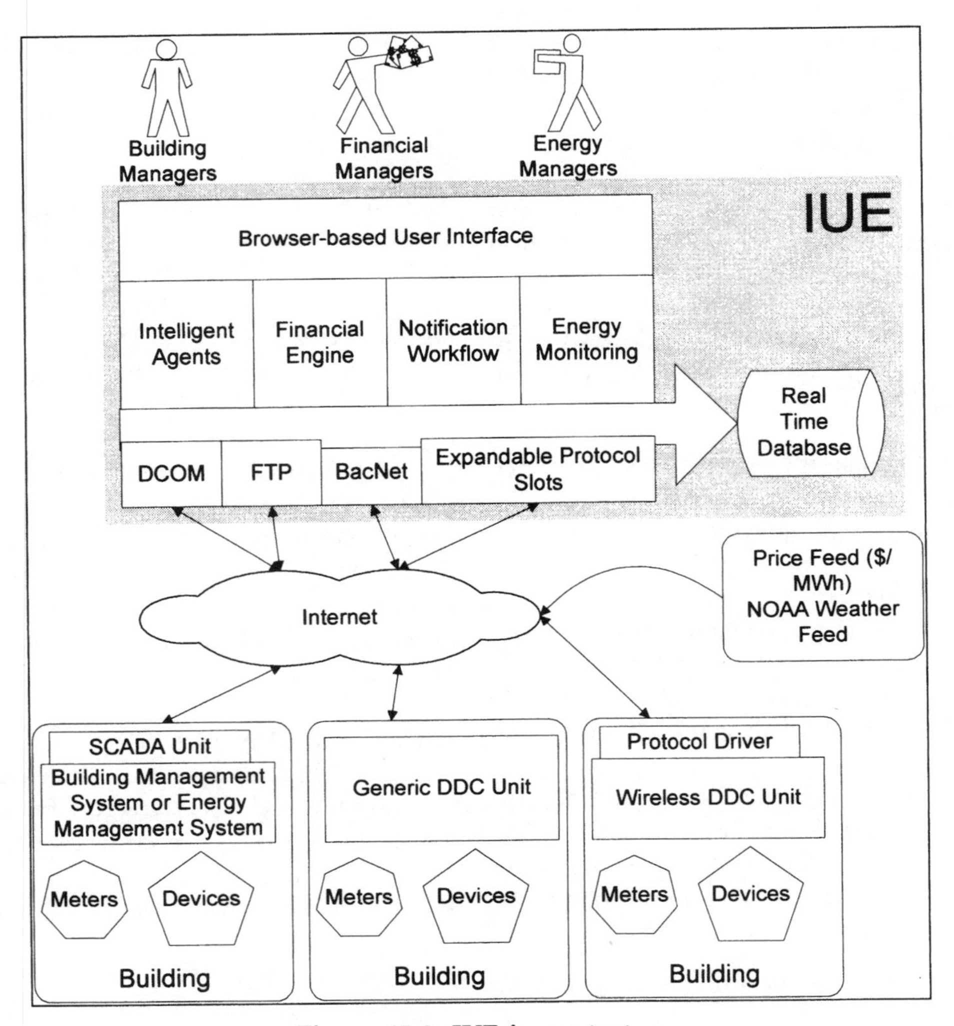

Figure 15-2. IUE in context.

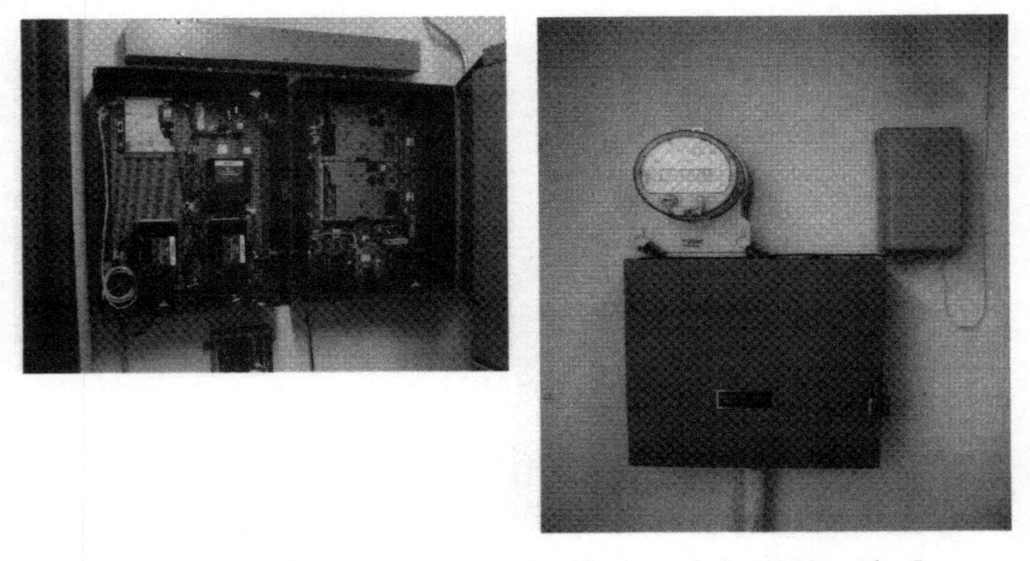

Figure 15-3. Connecting the meter, pulse block, and the BMS to the Internet

in the next layer down, they embody best practices in enterprise energy management, based on energy goals and neural network forecasts they device the currently optimal energy plan. The third layer down, houses the neural networks which continually learn and use weather and device data to make continual predictions on energy consumption and the related environmental quality. Each meter and each device has one dedicated neural network that learns its behavior.

Connectivity Protocols

IUE COM, the communications module, has a number of services that can communicate in numerous open and proprietary protocols, including BacNet, FTP, HTTP, XML, or any other format.

All the data are stored in the real-time database. This database is used by human users who request trending or graphing information as well as by computer-based intelligent agents, who use the information to improve their decision-making ability.

Triggers for Energy Strategies

At the top level are triggers that set the condition for the strategy agents [2,4]. Three such triggers exist:

- The default Business-as-usual trigger, which delineates normal operating temperatures, CO_2 envelopes, humidity envelopes etc. If one of these parameters does not exist, it is merely omitted.
- The Curtailment trigger, which delineates curtailment operating conditions. These parameters are the same as in business-as-usual, except they have more relaxed envelopes. The curtailment trigger becomes active when a human requests a curtailment for a certain number of kW for a certain duration under a specific curtailment contract; this is started by a human operator, navigating to "control, curtailment, new" in the system.
- The "Peak-Load-Avoidance (PLA)" trigger is based on the consumption forecasts of the neural networks which can predict when a meter will set a new peak, thus resulting in a new peak charge. The PLA trigger will start working to avoid this peak, first via gentle means (e.g. invoking a "precooling" agent or a "temperature reset agent"); if this does not sufficiently alleviate the situation, the PLA trigger will call upon "speed reset" agents and finally upon "load rotation agents" to curtail demand for short periods of time, thus keeping the meter from setting a new peak.

Energy Management Strategies as Intelligent Agents

The middle layer embodies the control strategies. These strategies operate within the limits set by the trigger agents and receive operating information from the neural networks. These are called the strategy agents. The strategies are well known in the field of energy management. The strategies are coded as expert systems

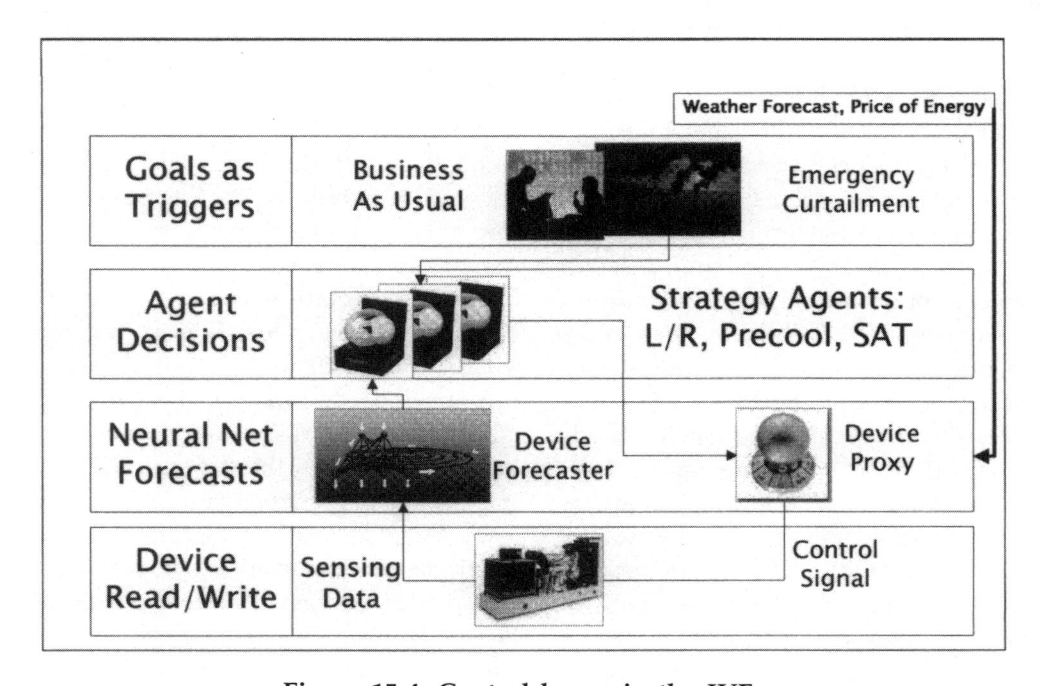

Figure 15-4. Control layers in the IUE

[4,5], i.e., each strategy is realized via agent technology. One, many or all agents can be deployed at a customer site, depending on the equipment and the BMS at the site. Some of the growing energy management strategies are:

- Supply air temperature reset.
- Price responsive supply air temperature reset.
- Speed reset.
- Static pressure reset.
- Pre-cooling.
- Load rotation.

The strategies are now described in detail. The **supply air temperature (SAT)** reset strategy for cooling is based on the predictions from the neural networks about temperature in the space related to the kW demand of the devices. This strategy agent determines if a slightly higher supply air temperature for the next 10 to 20 minutes is feasible without the space temperature increasing by more than one degree Fahrenheit. If the neural network associated with the respective airhandler shows that this is feasible, then SAT will write this higher supply air temperature register for the device in the BMS. Once the 10 to 20 minutes (depending on the predicted degradation curve learned by the neural net) are up, the strategy agent gradually returns the value of the register to its original setting. If during the override period any environmental parameter starts to go to the boundary of the envelope set by the trigger agent, then the original set-point for the register is immediately resumed.

Price Responsive Supply air temperature reset is a powerful variant of SAT. It affects the same control parameters that SAT reset influences. In addition, this strategy utilizes the fluctuating prices in deregulated markets. This strategy ensures maximum comfort, when the price of energy is low. On the other hand, this strategy avoids the consumption of a significant portion of the building's annual energy budget during a short time period, when the prices are skyrocketing. This strategy avoids being penny-wise and pound-foolish.

The **speed-reset** strategy is based on the predictions from the neural networks (one network per airhandler) about the speed of a VFD (variable frequency drive) and the related kW demand of the device; it determines if a slightly slower speed for a specified time interval (e.g., the next 10 to 20 minutes) is feasible without a noticeable loss of environmental quality. The neural networks can simulate the change in space air temperature that will result from the slower speed. If an opportunity exists, the speed reset strategy will give the BMS a new set-point and the BMS will write this lower speed to the device. Once the specified time (depending on the predicted degradation curve learned by the neural networks) is up, the strategy agent returns the set-point to its original value. If during the override period any environmental parameter starts to go to the boundary of the envelope set by the trigger agent, then set-points are immediately returned to their original values.

The **Static Pressure Reset** strategy is a combination of the supply air temperature and the speed reset strategies. Static Pressure Reset takes advantage of the fact that the BMS will maintain integrity of the overall device (airhandler speed, VAV setting, supply air temperature, duct work pressure, etc.). By resetting the static pressure to a lower value, the BMS will drop the fan speed and find an adequate temperature setting for SAT. The neural network that is attached to this airhandler's devices (speed, SAT, VAV opening, etc.) models this relationship and informs the decision of the SPR agent. If the static pressure reset strategy can find a lower static duct pressure than the current pressure that will result in kW savings it will leverage the BMS set-point and write this lower pressure (within the allowed limits) to the device. Once the specified time period (depending on the predicted degradation curve learned by the neural network) is up, the strategy agent returns the set-point to its original value.

The **Precooling** strategy employs the prediction of the neural networks concerning the capacity for cooling storage and the release gradient. If the peak load avoidance agent predicts a new peak for the day, the precooling agent will start precooling at the time and temperature derived from its observations of the thermal properties of the building. The precooling agent will use the BMS to run the precooling process until enough cooling energy is stored in the building to avoid setting a new peak later in the day.

Load Rotation is a well-known strategy among building managers [6]. If during a curtailment a large reduction in demand is required fast, the load rotation agent will determine which devices can potentially be shut off (or throttled down) for a short period. The L/R agent then groups these devices by similar environmental quality degradation times; in this fashion, the agent may create 2, 3, or more groups of devices with similar degradation characteristics. Once the agent has created the device groups, it starts shutting off the first group. The first group is switched back on either at the end of the degradation period or if any one device during that time interval starts to go outside of its operating range.

As the first group is switched on, the second group is switched off, etc. This process repeats until the end of the curtailment period is reached.

Neural Networks as Forecasters

The third layer in the IUE control stack, after the triggers and the strategies, is the neural network which functions using forecasting and simulation models. Forecasts and simulations are needed by all the strategies discussed above.

A simple example is the peak load avoidance strategy that leverages the kW forecast that was developed from a single meter. Figure 15-5 shows the kW and kWh curves for the electrical meter in the Wellness Center at the University of Miami.

While the kW (actual) and the kWh (actual) curves stop at the current hour (as of the writing of this chapter), the kW (forecast) curve continues into the future. The peak load avoidance agent is watching these forecasts and will spring into action if a new peak for the month is predicted.

Since the forecast demand is only for 560 kW while the old maximum is 650 kW, no action is currently necessary. Should the temperature rise and the forecast go to 660 kW, then peak load avoidance agent would take action to curb demand, keeping it from going over 650 kW.

Neural networks continually learn by observing data from the environment [3]. Just as a person keeps learning how objects move with every observation of a ball being thrown or a car braking, so do the neural networks keep learning about how the meters and the buildings will react. No permanent record of these learning instances is stored in a database. Instead, the learning is stored as changes to the network program.

Figure 15-7 shows the curve a network should learn and how it is actually getting there after a large numbers of iterations. Since the IUE gets a complete set of data points from the building every 2 minutes, its accuracy keeps growing. The neural networks keep learning, increasing their predictive power and changing with changes in the building.

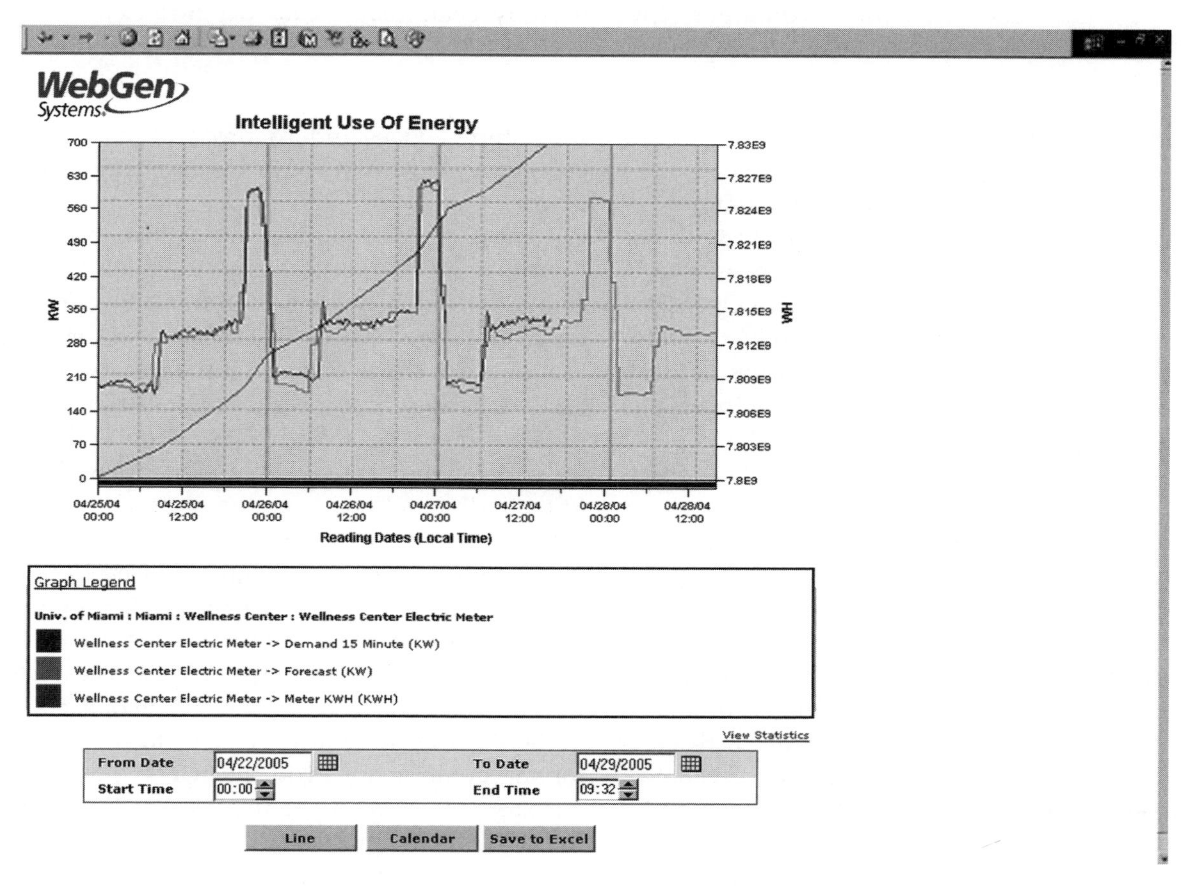

Figure 15-5. kW forecast on a meter for the next 24 hours.

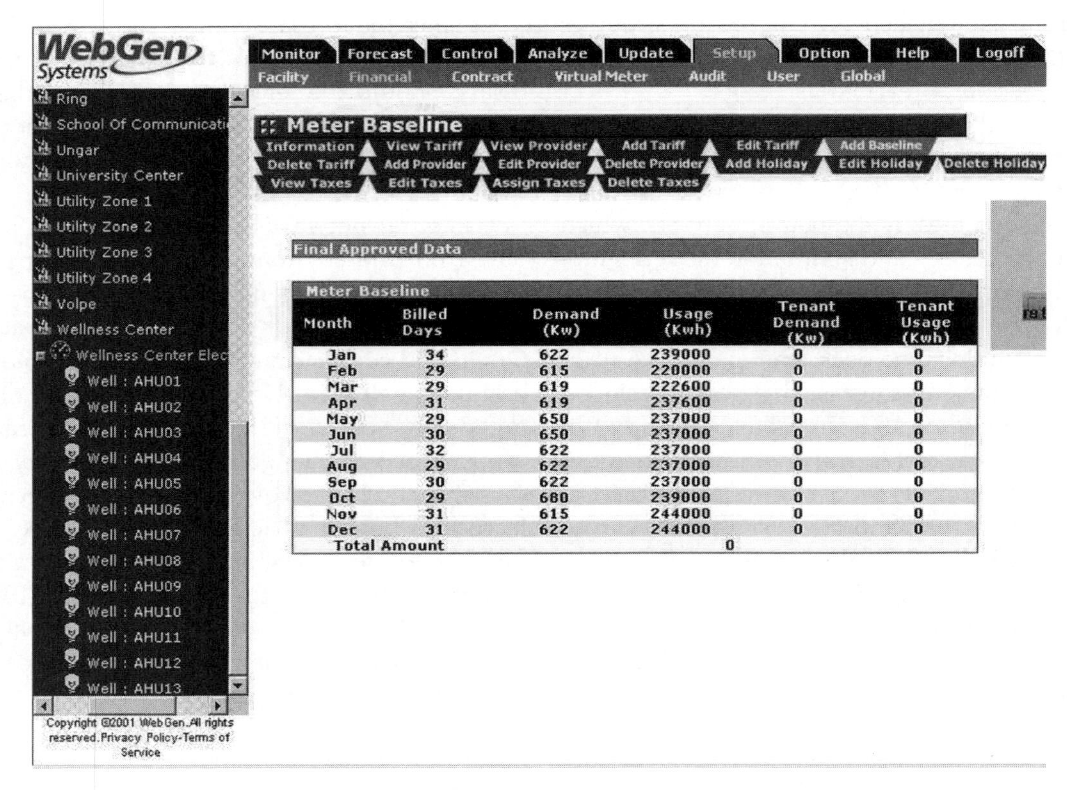

Month	Billed Days	Demand (Kw)	Usage (Kwh)	Tenant Demand (Kw)	Tenant Usage (Kwh)
Jan	34	622	239000	0	0
Feb	29	615	220000	0	0
Mar	29	619	222600	0	0
Apr	31	619	237600	0	0
May	29	650	237000	0	0
Jun	30	650	237000	0	0
Jul	32	622	237000	0	0
Aug	29	622	237000	0	0
Sep	30	622	237000	0	0
Oct	29	680	239000	0	0
Nov	31	615	244000	0	0
Dec	31	622	244000	0	0
Total Amount			0		

Figure 15-6. Baseline Maxima per Month used by Peak Load Avoidance Strategy.

IUE AT WORK IN MIAMI

The IUE installation at the University of Miami

There are many ways to communicate with the BMS. For the past several years, so called Open Protocol systems have become the industry standard. Open Protocol systems have some means of connectivity to a third party system. ASHRAE (the American Society of Heating, Refrigeration and Air-Conditioning Engineers), in an effort to allow this interoperability to flourish has created a standard protocol called BacNet. Systems adopting the BacNet standard are able to communicate with other systems using the BacNet standard. Another open protocol is called Lon Works.

In the case of a system that is neither BacNet nor Lon compatible, most are open to communications by either a "Gateway" or "Driver." Gateways are usually developed by the control system manufacturer as a means of having their system communicate with other systems or with BacNet. Drivers, not unlike drivers that facilitate communication between a computer and a printer, are often developed by third party device manufacturers to allow their system to talk to proprietary systems. These third party devices communicate with the Internet via standard protocols like BacNet, and communicate with the BMS via a control system specific driver.

In a typical application, a third party Interface Controller (IC) is physically installed in the building and connected to the existing BMS (usually) through a port on the BMS workstation. In the Wellness Center, we used the WebGen Web-I. When a software driver is installed and configured in the IC, the IC can transfer the point information to and from the BMS. This is done by creating software points in the IC and mapping the addresses to the individual points on the BMS. There are typically thousands of points in the control system, and knowing which points to map is where WebGen's building systems experts play a key role. There is no need to map every point that exists in the building.

An Internet Protocol (IP) address is established for the IC. When this process is completed, a point created in the WebGen system will be able to successfully monitor the points mapped to the IUE via the world wide web, as well as command the "virtual" points created in the BMS for optimization.

Another critical step is to connect the electric utility meter to the Internet. Some meters are capable of being

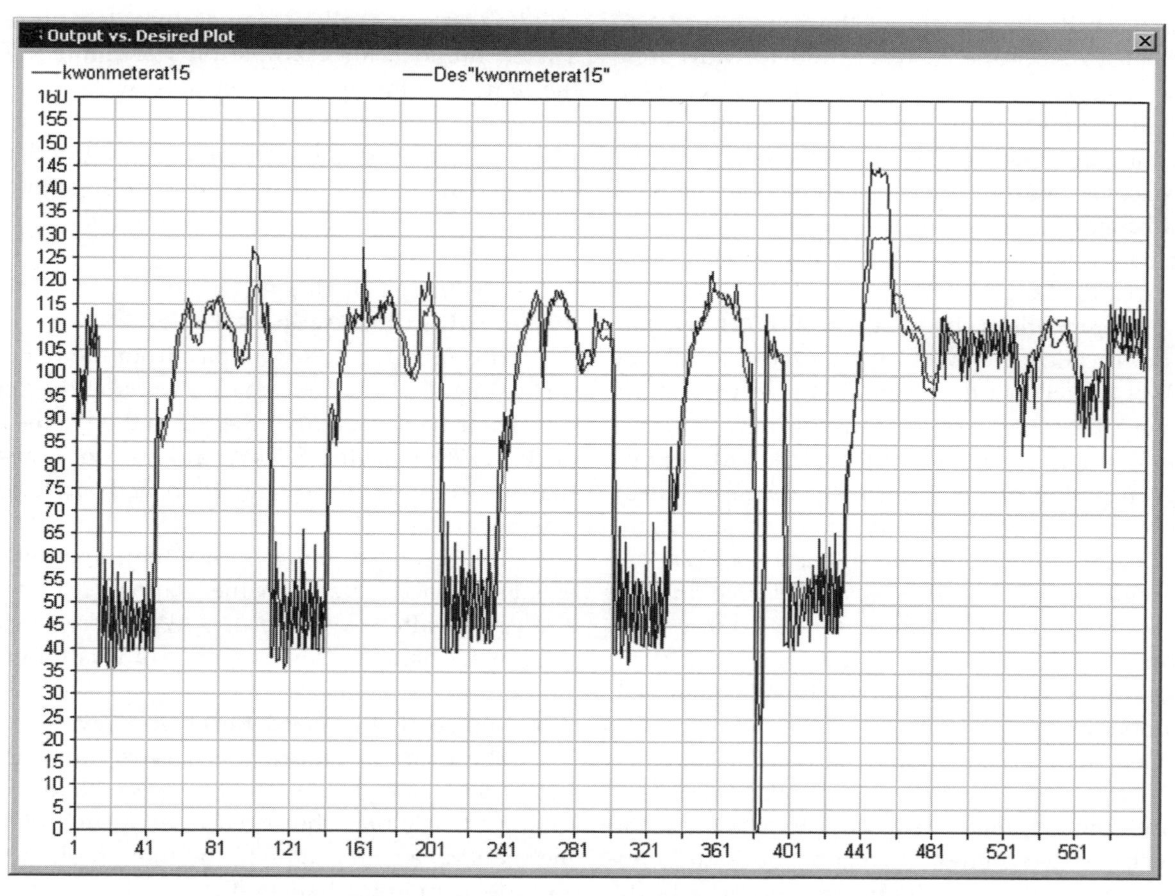

Figure 15-7. Neural Network learning from many events.

directly connected, others require a field device to take a pulse input from the meter, convert it to Kilowatts (kW) and Kilowatt-hours (kWh). In either case, a second IP address is assigned. Most larger locations already have a pulse generator connected between the BMS and the utility's revenue meter.

WebGen's IUE system can use the information in many ways. In addition to its agents using the information to determine when and how long they should run in normal operation (called Business As Usual), the IUE can predict when new peaks will be set and use load rotation to avoid setting new electrical peaks.

The following is a step-by-step breakdown of the installation process:

METERING

There are two options for providing electrical consumption data for WebGen's IUE. The lower cost option is to install a "pulse block" which is provided by the utility. Another option is to install an IP addressable electric meter. WebGen does not recommend any particular manufacturer, although any meter that is BacNet or Lon compliant should be suitable. We work with our clients to make sure they select the right hardware.

BMS CONNECTION

It is fairly common to need to add points to the BMS during the WebGen IUE installation process. The WebGen IUE needs to know the impact of its actions on the building environment. In buildings with a relatively open architecture, our first choice is to add wireless temperature sensors. Consisting of a transceiver base, which can support up to four field sensors, wireless signal repeaters, and wireless thermostats, these are generally more cost-effective than adding field wiring. The transceiver is wired to the BMS field controller and, as such, will require free space at those controllers. To be conservative, one day's worth of time for a controls technician should also be carried for every cluster of four sensors.

As a rule of thumb, at least one temperature sensor is required for sections of the building that are in one HVAC zone. The system must have the most accurate temperatures available throughout the zone in order to be able to best determine if, when and for how long agents may control without negatively impacting comfort levels. If there are many similar zones with consistent thermal and occupancy characteristics, a few sensors can cover multiple areas served by one unit. One additional benefit of wireless sensors is that they can be moved if necessary to make sure that the needs of a particular area are being met.

The only remaining change to the BMS is to create a virtual point called the WebGen Heartbeat, a fail-safe feature. WebGen will set this point to a new value from 001-100 every time the system scans for new information. A small program will be written in the BMS so that if for a given period of time (normally seven minutes), the value of the heartbeat point remains unchanged, the BMS will recognize that communications have been lost. The BMS will then return all points to Disable and the BMS will operate normally without the IUE. Once communications are re-established for three consecutive intervals, the BMS will again allow the WebGen IUE system to operate.

WebGen deployment cannot be considered to be successful if end-user comfort is compromised. The keys to occupant comfort and safety are the environmental limits, which are determined in conjunction with building management. These will dictate Agent activity. Ranges defined are directly proportional to energy savings—a wider range of limits will allow for greater savings.

The following are general environmental limit guidelines that will be set within the IUE. All limits are determined and agreed upon by WebGen staff and the building owner/operator to meet standard engineering practices that maintain comfort. If there are specific reasons why a particular zone should have different limits, that information is critical to a successful deployment.

For a typical Air Handling Unit, the limits are usually set as follows: return air humidity (approx. 30-65%RH), return CO_2 (approx. 800 ppm max), static pressure (approx. 1.0"-1.5") and space temperature (approx. 70-76 deg. F.)

Existing BMS inputs can be used as limits for the system. In the case of space temperatures, we prefer to utilize an average reading from multiple sensors, or even multiple averages of representative areas. The space temperature limits can be set to allow not only a minimum and maximum value, but also an offset from a set-point. This method lets WebGen IUE accommodate known variances in a particular space. Many buildings have varying individual space temperatures, and the building operators allow temperatures to drift relative to a fixed number, for example plus or minus 2°F. before the limit tells the agent to release.

All limits are determined and agreed upon by WebGen and the building owner to meet good engineering practice for comfort guidelines.

A log of Energy Savings Actions by the IUE in the Wellness Center

This section gives examples from the IUE database for two agents. The first one shows how the "speed reset" agent acted on air-handler 02 in the Wellness Center on 10/2/2003, saving 1.2 kW and 6 kWh. It also shows that the comfort from ventilation and air temperature was not decreased.

The second example shows the same detail for an event sequence initiated by the "supply air temperature" strategy. Again, 3 kW and 22.5 kWh were saved, while the impact on space air temperature was minimal.

A sequence of the Speed Reset Strategy

The chart in Figure 15-9 shows how the "Speed Reset" agent affects the actual fan speed and creates kW savings. Every time the speed reset agent enters by giving the BMS a better setpoint (bottom graph in the chart, 7 entries), an immediate and lasting impact on kW consumption in the fan can be seen. The most visible decrease comes after the 12: 14 event with a decrease in 1.2 kW of energy demand by the fan. The effects of that short 15 minute event lasted for about 1 hour and were reinforced through the later events that day, which kept fan speed "in check" all day long.

Figure 15-8. Wellness Center at the University of Miami.

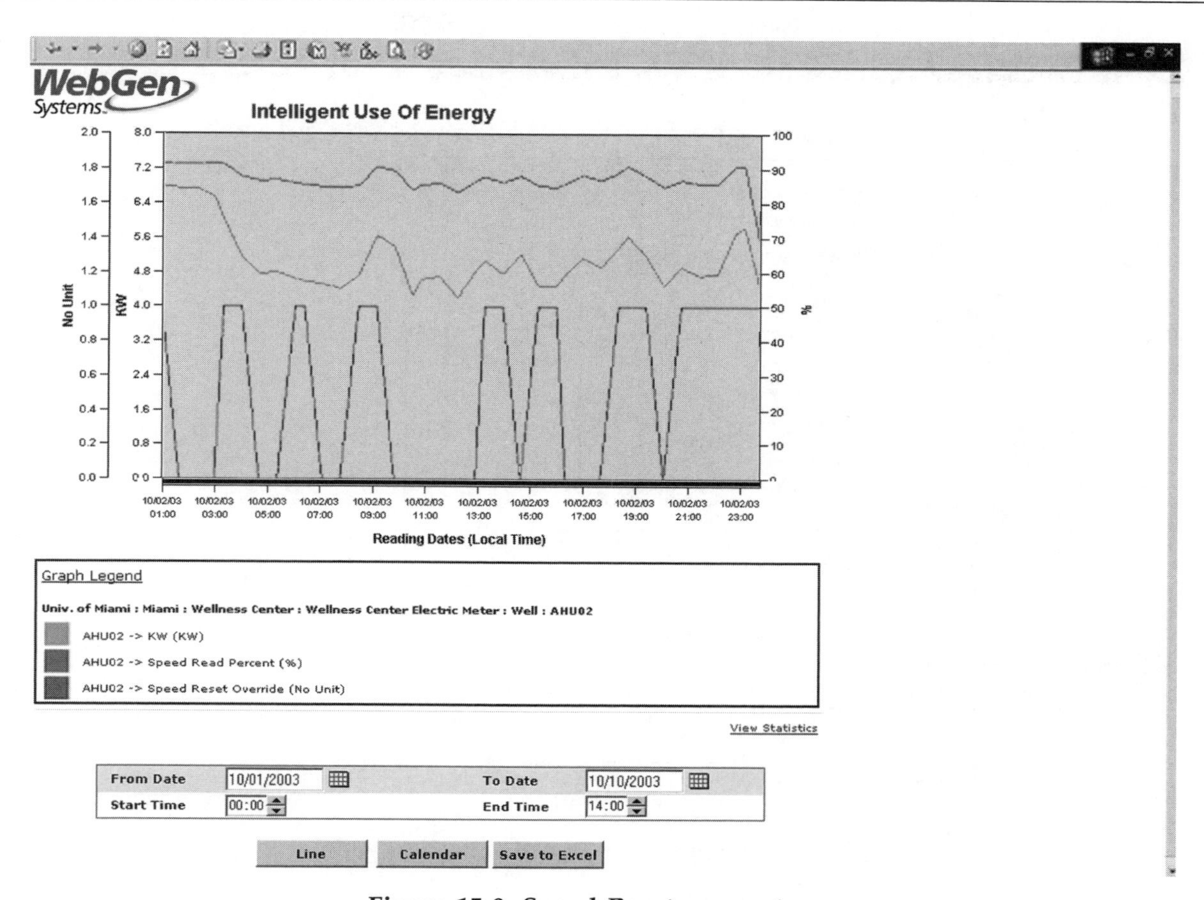

Figure 15-9. Speed Reset parameters

In six of the seven event sequences, the control action by IUE results in an absolute, immediately visible reduction in kW. At 12.14h, the fourth event of the seven, matters are a little more complex. IUE immediately notices the fan speed rising rapidly. The reason is the increasing outside air temperature. IUE's agents are not able to totally stop this trend without loss in comfort to the occupants. Yet the agents are able to decrease the slope of the kW rise long enough. Once "the worst is over," the agent actually sees the kW curve turn around. It stays in for a while and when the trend continues, it hands control back to BMS setpoint at about 13.50h. While the agent did not push the kW curve "under" the point where it started, it did keep it "under" the values it would have set without IUE-intervention.

The indoor air quality during the same time period did not decrease. As the top graph shows, the fan kept running well above the 80% mark during business hours.. Ventilation and meeting OSHA requirements was never a problem. The effect on space air temperature can be seen in the graph below.

The chart in Figure 15-10 shows the speed overrides as the bottom graph. "One" indicates the presence of the speed reset agent; "Zero" indicates the absence of it. While the presence of the speed reset agent had a large impact on speed and energy demand, very little effect was seen on the space air temperature. The temperature hovered around 73 degrees Fahrenheit after the AC turned on at about 6.30 am that day.

The Supply Air Temperature Strategy at Work

Table 15-1 shows the log entries from the supply air temperature agent for a fan/chiller combination for a space in the wellness center. At 11:10 am the building space temperature is 69°F and the fan plus the chiller are running at 40 kW. At 11:12 am the IUE reads the current set-point for the SAT, so it can return the building after the control event to the previous temperature setting. At 11:13 am the IUE changes the actual set-point and sets it to 60°F at 11:14 am. There was an immediate savings of 3 kW.

According to the predicted temperature degradation coefficient—which was calculated by the neural network attached to the airhandler for this space, before this control sequence was performed– the temperature actually rises in the space. Notice though that the tem-

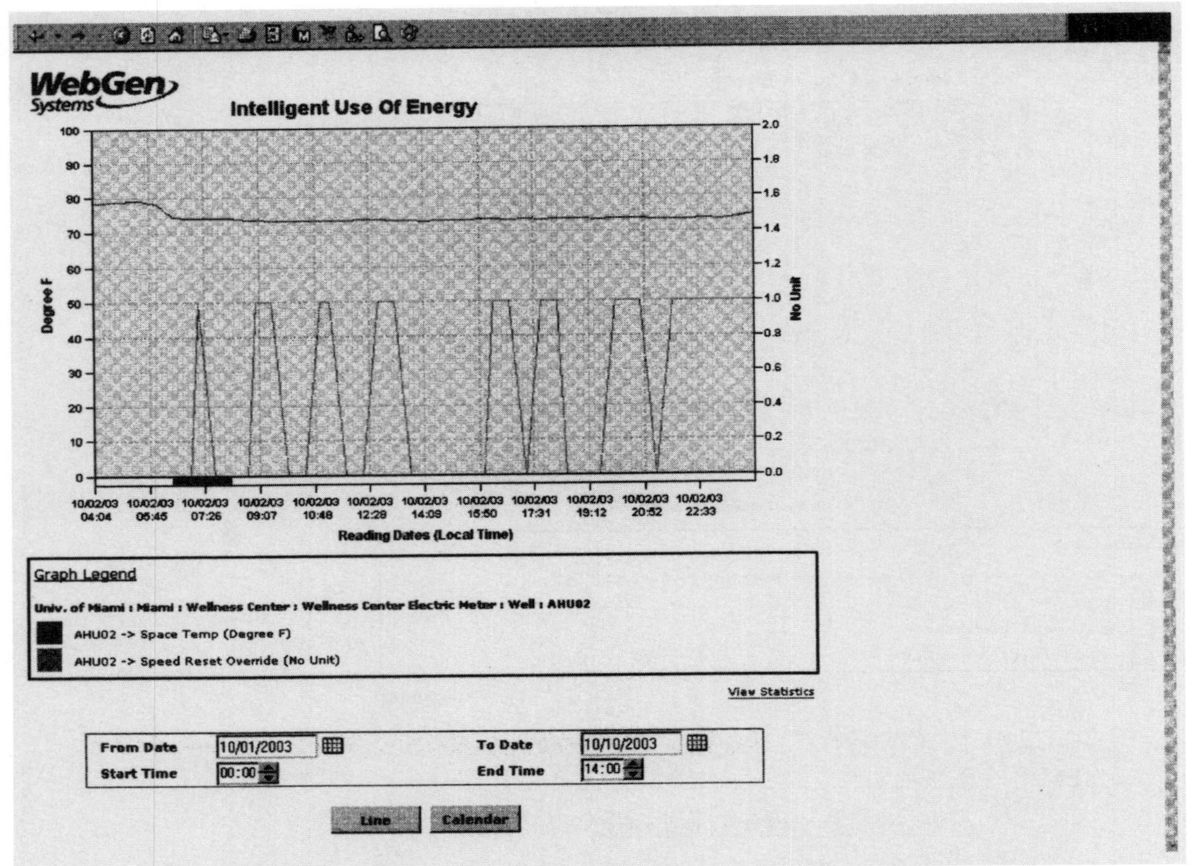

Figure 15-10. No noticeable effect of speed reset agent on space temperature

perature rise is: a) time delayed and b) not noticeable to the occupants, since it stays under the psycho-physical threshold that is perceivable by most people. The above sequence saved about 22.5 kWh. This may not seem like a lot, but since many of these events happen on all devices, all day long, large kWh savings are aggregating.

At 11:22 am the value in the SAT set-point register was returned to their original state from before the IUE sequence. Following the degradation gradient, space air returns to 69°F. The IUE has saved energy without compromising comfort or quality.

SAVINGS IN THE WELLNESS CENTER RELATED TO IUE ACTIVITIES

The previous section showed that IUE strategies could save energy—and thus money—without affecting comfort in the building. This section details how these energy savings are accounted for. First, the actual energy consumption is compared against baseline energy consumption. Then, kW and kWh are translated via the current tariff into actual dollars.

Savings: The Bottom Line

Figure 15-11 shows the savings report that was calculated on-line, on-demand by the IUE. The left column contains the various line items on the energy bill. Examples are "Full Day Base Demand" or "Full Day Capacity Charge." The next column shows the prorated and adjusted baseline numbers, e.g., 761.08 kW for the "Full Day Base Demand." This number is based on the raw baseline, adjustments (which will be explained later in this section), and proration (31 billing days in this report vs. 29 in the baseline).

The third column shows the actual energy bill. The actual demand for the month of October 2003 was 642.24 kW. The savings due to the IUE was 118.84 kW as shown in the last column. This amounts to 15.61% savings over the bill that would have been incurred without the IUE.

All in all, the University saved:

- 237.68 kW (base and capacity), or 15.61%, or $968.55
- 26,277.53 kWh, or 9.83% or $283.30
- 12.1% in related charges and taxes
- Total savings of $2,596.97 on a $22,175.59 bill, or 11.71%

Table 15-1. Database log of pertinent parameters for SAT strategy event.

Agent	Device	Time	Command	Space Temp	kW
	Wellness AHU08	11:10		69	40
SAT	Wellness AHU08	11:12	Presetting SAT Reset to 58	69	40
SAT	Wellness AHU08	11:13	Setting SAT Register to 1	69	40
SAT	Wellness AHU08	11:14	Writing SAT Reset 60	69	37
	Wellness AHU08	11:16		69	37
	Wellness AHU08	11:18		69	37
	Wellness AHU08	11:20		69.5	37
SAT	Wellness AHU08	11:22	Setting SAT Register to 0	69.5	40
	Wellness AHU08	11:24		69.5	40
	Wellness AHU08	11:26		69.5	40
	Wellness AHU08	11:28		69.5	40
	Wellness AHU08	11:30	69	40	

Determining the Baseline

A baseline is a reference point. The electric energy baseline reflects the amount of electricity that the school has historically used in a specific building. Florida Power and Light (FPL), the local utility company, provided us the electric bills for the 12 months prior to installation of the IUE so we could create our baseline.

How the Baseline is Made Comparable to the Actual Bills

In order to have a baseline we can compare with the actual bill, we had to take several steps. To begin with, we had to use a baseline for the same month of the year as the one we are considering. Second, we had to prorate that baseline so that it covered the same number of days; for example if last year's March bill was for 32 days, but this year's March bill is for 29 days, then we must take last year's kWh totals, divide by 32 and multiply by 29.

Example: Last year 88,080 kWh divided by 32-day billing period is equivalent to 2752.5 kWh/day
This year's March has a 29 day billing period, so the adjustment is:
29 x 2752.5 kWh/day is 79,822 kWh for a 29-day billing period (day adjusted baseline)

The third step is to normalize the consumption to discount any degree-day differential. The rationale behind this adjustment is obvious: the hotter it gets, the more energy is consumed to provide air conditioning. If the current month is hotter than the baseline month, then the baseline month must be adjusted up. If the current month is cooler than the baseline month, then the baseline must be adjusted down. ASHRAE has developed industry standard formulas, used extensively in the area of building management, to allow for these computations. WebGen is using those standard formulas to achieve fair comparability.

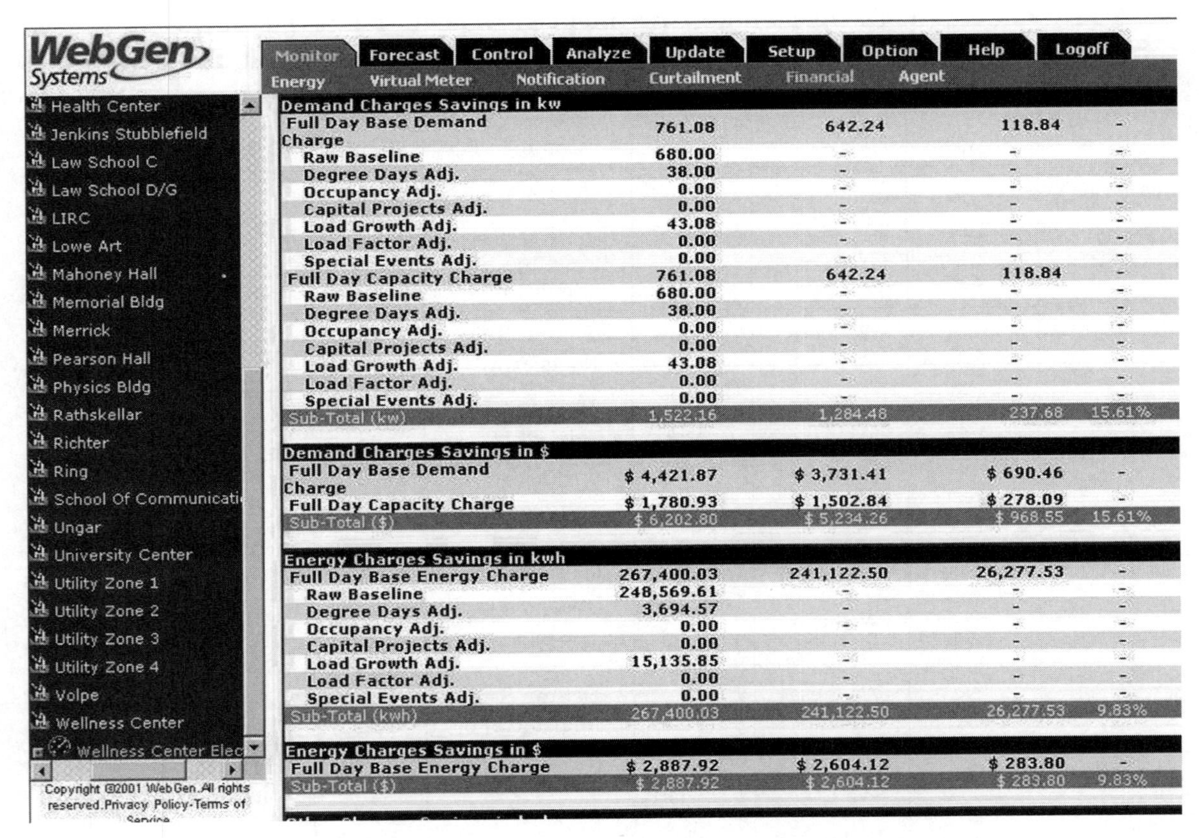

Demand Charges Savings in kw				
Full Day Base Demand Charge	761.08	642.24	118.84	-
Raw Baseline	680.00	-	-	-
Degree Days Adj.	38.00	-	-	-
Occupancy Adj.	0.00	-	-	-
Capital Projects Adj.	0.00	-	-	-
Load Growth Adj.	43.08	-	-	-
Load Factor Adj.	0.00	-	-	-
Special Events Adj.	0.00	-	-	-
Full Day Capacity Charge	761.08	642.24	118.84	-
Raw Baseline	680.00	-	-	-
Degree Days Adj.	38.00	-	-	-
Occupancy Adj.	0.00	-	-	-
Capital Projects Adj.	0.00	-	-	-
Load Growth Adj.	43.08	-	-	-
Load Factor Adj.	0.00	-	-	-
Special Events Adj.	0.00	-	-	-
Sub-Total (kw)	1,522.16	1,284.48	237.68	15.61%
Demand Charges Savings in $				
Full Day Base Demand Charge	$ 4,421.87	$ 3,731.41	$ 690.46	-
Full Day Capacity Charge	$ 1,780.93	$ 1,502.84	$ 278.09	-
Sub-Total ($)	$ 6,202.80	$ 5,234.26	$ 968.55	15.61%
Energy Charges Savings in kwh				
Full Day Base Energy Charge	267,400.03	241,122.50	26,277.53	-
Raw Baseline	248,569.61	-	-	-
Degree Days Adj.	3,694.57	-	-	-
Occupancy Adj.	0.00	-	-	-
Capital Projects Adj.	0.00	-	-	-
Load Growth Adj.	15,135.85	-	-	-
Load Factor Adj.	0.00	-	-	-
Special Events Adj.	0.00	-	-	-
Sub-Total (kwh)	267,400.03	241,122.50	26,277.53	9.83%
Energy Charges Savings in $				
Full Day Base Energy Charge	$ 2,887.92	$ 2,604.12	$ 283.80	-
Sub-Total ($)	$ 2,887.92	$ 2,604.12	$ 283.80	9.83%

Figure 15-11. October 2003 Savings Report for the Wellness Center

The degree-days used for these calculations are given by NOAA, and can be found in: (*http://www.srh.noaa.gov/*). It is necessary to match the degree days to the billing period established by the utility; therefore a continuous form allows us to add or subtract degree days so that the weather adjustment conforms to the utility billing period. Additionally, occupancy adjustments have to be made.

Example: In March of the base year there were 223 CDD (cooling degree days) for the 32-day billing period but only 202 in the same 29-day period as the current billing period. In this year, the 29-day billing period in March had 243 CDD or 41 more CDD than in the base year. Since it was a warmer year, the base year's consumption need to be adjusted accordingly to account for the increased use.

U × A × Delta CDD × 24hr/day × k ÷ 1000 = kWh thermal adjustment

0.19 (default when value is unknown) × 67,231 sf (walls and roof) × 41 × 24 × 0.06583 (k) ÷ 1000 = 827.5 kWh

Where:

U = Area average heat loss coefficient

A = Outside wall + roof area

k = Conversion factor (watts cooling per Btu)

k = kW/ton × (1 ÷ Btu/Ton) × Btu/kWh × 0.293 watt-h/Btu

0.06583 = .79 kW/ton × (1 ÷ 12,000Btu/Ton) × 3413 × 0.293

Cfm × Delta CDD × 1.08 × Peak Hours × k ÷ 1000 = kWh ventilation adjustment

10,000 × 41 × 1.08 × 20 × .06583 ÷ 1000 = 583 kWh

Where:

Cfm = cubic feet per minute of air flow (outside air)

1.08 = Sensible heat content of air

Peak hours = Coincident hours in average ventilation runs hours per day

k = Conversion factor (watts cooling per Btu)

k = kW/ton × (1 ÷ Btu/Ton) × Btu/kWh × 0.293 watt-h/Btu

0.06583 = .79 kW/ton × (1 ÷ 12,000 Btu/Ton) × 3413 × 0.293

Thermal + Ventilation = Total

Total CDD Adjustment = 1088 kWh + 583 kWh = 1671 kWh

Sometimes the 12-month floating sum of utility bills rises appreciably with time. This phenomenon is sometimes referred to in the industry as Plug Creep and can be caused by the gradual increase of installed appliance load in an office, or by gradual changes in usage patterns due to a change in management or ownership. Extrapolated over a three-year period, the effect generally amounts to a 2% to 3% plug creep per year. This increase is handled under the Occupancy load growth and is agreed to by both parties. In this case all parties agreed to adjust the baseline by a 2% occupancy increase after analyzing eight years of historical data.

SUMMARY

This chapter presented a case study that presents the energy savings impact of WebGen's IUE (Intelligent Use of Energy®) system. The goal of this chapter was to draw a clear connection between the financial savings accomplished and the energy saving actions that the IUE system was taking. The financial savings computed by the IUE system and were verified independently by comparing current utility bills with bills from a previous year. These financial savings were traced back to energy savings and individual control sequences. The energy savings in turn were correlated to agent activity and environmental readings. The agent activity in turn was based on standard practices in energy management and continuous learning by the IUE system. The chapter thus links the final goal, monetary savings, to the minute control actions of the automated energy strategies.

This case study focused on one billing period (October 2003) in one building (the Wellness Center at the University of Miami). A short overview of the IUE system and its architecture was provided to give the reader a basis for understanding the energy saving actions initiated by the IUE system.

In the end, the IUE system saved $2,596.97 of a $22,175.53 bill, which corresponds to 11.71%. Instead of 761 kW, the Wellness Center only used 642 kW, a reduction of 15.6%. Instead of 267,400 kWh it only used 241,122 kWh, a reduction of 9.8 %.

The IUE accomplished these results by forecasting new peaks on the meter, taking preemptive actions, and thus avoiding a higher peak demand charge (kW), while maintaining occupant comfort. The IUE also provided ongoing load reduction, which lowered the consumption part of the energy bill. Both permanent load reduction (reducing the kWh-part of the energy bill) and peak load avoidance (reducing the kW-part of the energy bill) are accomplished through a number of strategies well known to building engineers and energy engineers; examples are "supply air temperature reset," "load rotation," or "thermal storage."

Utilizing These Strategies

- at every possible opportunity,
- without fail,
- at the right times,
- for the optimal time interval,
- with the most optimal device settings,
- both from a local perspective (space air temperature, ventilation, etc.)
- and a global perspective (cost of energy, weather forecast),
- based on the most up-to-date forecasting models, gives building engineers a new tool to deliver savings to property owners and high level service to tenants. This truly provides a three-way-win situation: for the property owner, the operator, and the occupant.

Bibliography

[1] D. Edwards and D.E. Mahling: Toward Knowledge Management Systems in the Legal Domain; *Proceedings of ACM GROUP '97 Conference*, Phoenix, AZ 1997.

[2] D.E. Mahling and R. King: A goal-based workflow system for multi-agent task coordination. *Journal of Organizational Computing and Electronic Commerce.* 1998.

[3] B.D. Ripley: Pattern Recognition and Neural Networks. *Cambridge University Press.* 1996.

[4] B.G. Buchanan and E.H. Shortliffe: Rule-Based Expert Systems: The MYCIN Experiments of the Stanford Heuristic Programming Project, *Addison-*

Wesley, Reading, MA, 1985.

[5] D.E. Mahling: The Role of Technology in Knowledge Management; Panel Discussion, *Proceedings of ACM GROUP'97 Conference*, Phoenix, AZ 1997.

[6] D.E. Mahling: Knowledge Acquisition for Planners; in *Proceedings of the 3rd Workshop of Knowledge Acquisition*, Banff, Alberta 1988.

ACKNOWLEDGMENTS

The authors would like to thank Mr. Victor Atheron of the University of Miami for his support of this project. We would also like to thank Mr. Wayne Hart, a building manager extraordinaire. Further thanks go to the team at WebGen that made this project happen. Thanks also to many reviewers who added consistency and clarity to this chapter.

Chapter 16

Machine to Machine (M2M) Technology in Demand Responsive Commercial Buildings

David S. Watson, Mary Ann Piette, Osman Sezgen, and Naoya Motegi
Lawrence Berkeley National Laboratory

ABSTRACT

MACHINE-TO-MACHINE (M2M) is a term used to describe the technologies that enable computers, embedded processors, smart sensors, actuators and mobile devices to communicate with one another, take measurements and make decisions—often without human intervention.

M2M technology was applied to five commercial buildings in a test. The goal was to reduce electric demand when a remote price signal rose above a predetermined price. In this system, a variable price signal was generated from a single source on the Internet and distributed using the meta-language, XML (Extensible Markup Language). Each of five commercial building sites monitored the common price signal and automatically shed site-specific electric loads when the price increased above predetermined thresholds. Other than price signal scheduling, which was set up in advance by the project researchers, the system was designed to operate without human intervention during the two-week test period.

Although the buildings responded to the same price signal, the communication infrastructures used at each building were substantially different. This study provides an overview of the technologies used to enable automated demand response functionality at each building site, the price server and each link in between. Network architecture, security, data visualization and site-specific system features are characterized.

The results of the test are discussed, including system architecture and characteristics of each site. These findings are used to define attributes of state-of-the-art automated demand response systems.

INTRODUCTION

This chapter provides a summary of the control and communications systems evaluated and reported on as part of a larger research report. [1] The objective of the study was to evaluate the technological performance of Automated Demand Response hardware and software systems in large facilities. The concept in the evaluation was to conduct a test using a fictitious electricity price to trigger demand-response events over the *Internet*. Two related papers describe the measurement of the electric demand shedding and the decision making issues with the site energy managers. [2,3]

The two main drivers for widespread demand responsiveness are the prevention of future electricity crises and the reduction of average electricity prices. Demand response has been identified as an important element of the State of California's Energy Action Plan, which was developed by the California Energy Commission (CEC), California Public Utilities Commission (CPUC), and Consumer Power and Conservation Financing Authority (CPA). The CEC's 2003 Integrated Energy Policy Report also advocates Demand Response.

A demand responsive building responds to a remote signal to reduce electric demand. This is usually done by altering the behavior of building equipment such as heating ventilating and air conditioning (HVAC) systems and/or lighting systems so as to operate at reduced electrical loads. This reduction is known as "shedding" electric loads. Demand responsiveness and shedding can be accomplished by building operators manually turning off equipment in response to a phone call or other type of alert.

In this paper, the term "*Automated Demand Re-*

sponse" or "*Auto-DR*" is used to describe "fully-automated" demand response where electric loads are shed automatically based on a remote Internet based price signal. Although the facility operating staff can choose to manually override the Auto-DR system if desired, these systems normally operate without human intervention.

Previous Research

The California Energy Commission (CEC) and the New York State Energy Research and Development Agency (NYSERDA) have been leaders in the demonstration of demand response programs utilizing enabling technologies. Several studies associated with the California and New York efforts investigated the effectiveness of demand responsive technologies. In California, Nexant was charged with evaluating CEC's Peak Load Reduction Program. The Nexant reports document the performance of all the California funded technology projects including the magnitude of the response and the cost associated with it.[4,5]

In addition to research concerning utility programs, controls, and communications systems, several research studies have examined various topics concerning DR in commercial buildings, including how to operate buildings to maximize demand response and minimize loss of services. Kinney et al reported on weather sensitivity of peak load shedding and power savings from increasing the setpoint of temperatures in buildings to reduce cooling loads. [6] This research project also builds on previous LBNL work concerning the features and characteristics of Web-based *Energy Information Systems (EIS)* for energy efficiency and *Demand Response (DR)*. [7]

PROJECT DESCRIPTION

The Automated DR research project took approximately two years, beginning with a planning activity in summer, 2002, successful pilot tests in November 2003 and final reporting in March 14, 2004 (Piette, et al 2004). The building sites, including their use, floor area, and equipment loads shed during the Auto DR tests are listed in Table 16-1.

System Geography

Although all of the Auto-DR pilot sites were in California, the supporting communications infrastructure and several of the developers were distributed throughout North America (see Figure 16-1).

AUTOMATED DEMAND RESPONSE SYSTEM DESCRIPTION

The Automated Demand Response System published a fictional price for electricity on a single *server* that was accessible over the Internet (Figure 16-2). Each of five commercial building sites had *client* software that frequently checked the common price signal and automatically shed site-specific electric loads when the price increased beyond predetermined thresholds. Other than price signal scheduling, which was set up in advance by the project researchers, the system was designed to operate without human intervention during two one-week pilot periods. The test process followed these steps:

Table 16-1. Summary of Sites

	Albertsons	BofA	GSA	Roche	UCSB
Location	Oakland	Concord	Oakland	Palo Alto	Santa Barbara
Use	Supermarket	Office	Office	Pharmaceutical laboratory (Office & Cafeteria)	Library
Floor Area (ft^2)	50,000	211,000	978,000	192,000	289,000
Equipment loads shed during test	50% of overhead lighting, Anti-sweat heaters	Supply fan duct static pressure setpoint	Global zone setpoint setup and setback	Constant volume fan shut off	Fan speed reduction, Chilled water valves closed

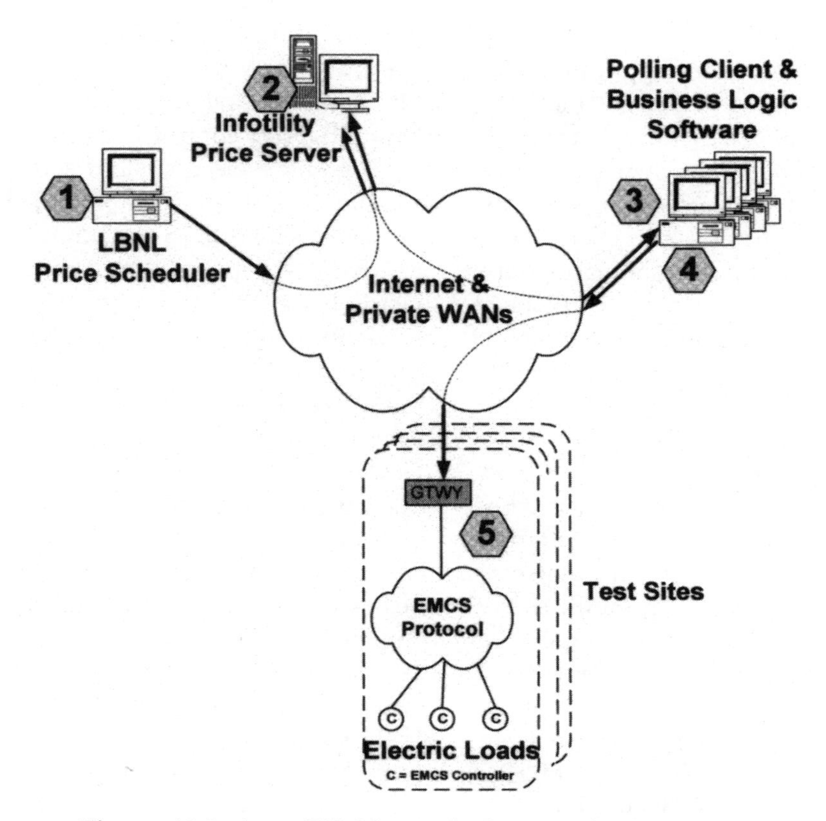

Figure 16-2. Auto-DR Network Communications Sequence

1. LBNL defined the price vs. time schedule and sent it to the *price server*.
2. The current price was published on the server.
3. Clients requested the latest price from the server every few minutes.
4. *Business logic* determined actions based on price.
5. *Energy Management Control System* (*EMCS*) carried out shed commands based on logic.

Web Services/XML

The infrastructure of the Auto-DR system is based on a set of technologies known as *Web services*. Web services have emerged as an important new type of application used in creating distributed computing solutions over the Internet. Properly designed Web services are completely independent of computer platform (i.e., Microsoft, Linux, Unix, Mac, etc.). The following analogy helps to describe Web services: Web pages are for people to view information on the Internet, Web services are for computers to share information on the Internet. Since human intervention is not required, this technology is sometimes referred to as *"Machine-to-Machine"* or *"M2M."* M2M is a superset of technologies that includes some XML/Web services-based systems.

XML is a "meta-language" (for describing other languages) that allows design of customized markup languages for different types of documents on the Web. [8] It allows designers to create their own customized tags, enabling the definition, transmission, validation, and interpretation of data between applications and between organizations. [9] Standard communication protocols (*TCP/IP, HTTP* and *SOAP*) are used on the Internet and *LAN/WANs* (Local Area Network/Wide Area Network) to transfer XML messages across the network.

Price Scheduling Software

Researchers at the LBNL used a software application to set-up the price vs. time profile published in the price server. The price profile could be set up hours, days or weeks in advance.

Price Server

The central Infotility server published the current price for electricity ($/kWh). Although the price used in the test was fictitious, it was designed to represent a price signal that could be used by utilities or Independent System Operators (*ISO*) in future programs that could be offered to ratepayers.

Web ServicesClients

The *polling client* is the software application that checks (polls) the Web services server to get the latest price data. The polling client resides on a computer managed by the building operators (or their representatives) for each site. In the pilot test, each client polled the server at a user-defined frequency of once every 1 to 5 minutes. The building operators were not given any prior knowledge of upcoming price increases planned by researchers. By checking their automatic price polling clients, operators could only see the current, most recently published price.

Polling-Client Price Verification

The price server included a feature that verified that each client received correct pricing information. This feature was implemented by requiring that each time the client requested the latest price from the server, it included its current price (from the client's perspective) and a *time stamp*. All pricing data were stored in a database. Although the intent of this feature was to verify client receipt of the latest pricing, there was another unforeseen benefit as well. When pre-testing began, researchers could see which sites were polling the server as each came on-line. After all systems were online, there were several cases where clients would stop polling for known or unknown reasons. When program managers observed these problems, they were able to manually make phone calls to the site system administrators, who restored proper communications.

Controls and Communications Upgrades

In order to add Auto-DR functionality to each pilot site, some upgrades and modification to the controls and communications systems were required. The upgrades were built to work in conjunction with the existing EMCS and Energy Information System (EIS) remote monitoring and control infrastructure in place at each site. For this project, custom software was written for each site, including: price polling client, business logic, and site-specific EMCS modifications.

Electric Price Signal and Test Description

Figure 16-3 shows the fictitious price signal that was in effect on the afternoon of November 19, 2003. During the rest of that day, the price remained at $0.10/kWh.

Auto-DR System Architecture Overview

Some Auto-DR facilities hosted the polling client software on-site and others hosted it at remote *co-location* sites (see Table 16-2). The geographic location of the computer that hosts the polling client is less important than the type of environment where it is hosted. Professional co-location hosting services, or "*co-los*" offer highly secure environments for hosting computers and servers. Co-los generally provide battery and generator backed electrical systems, controlled temperature and humidity, seismic upgrades and 24/7 guarded access control. For companies that don't have similarly equipped data centers, co-los fill an important need. For computer applications where high system *availability* is important, co-location facilities are often used.

Systems with a high level of integration between *enterprise* networks and EMCS networks tend to allow direct access to any or all control points in the EMCS without a need for excessive *point mapping*. Direct remote control of EMCS points from enterprise networks

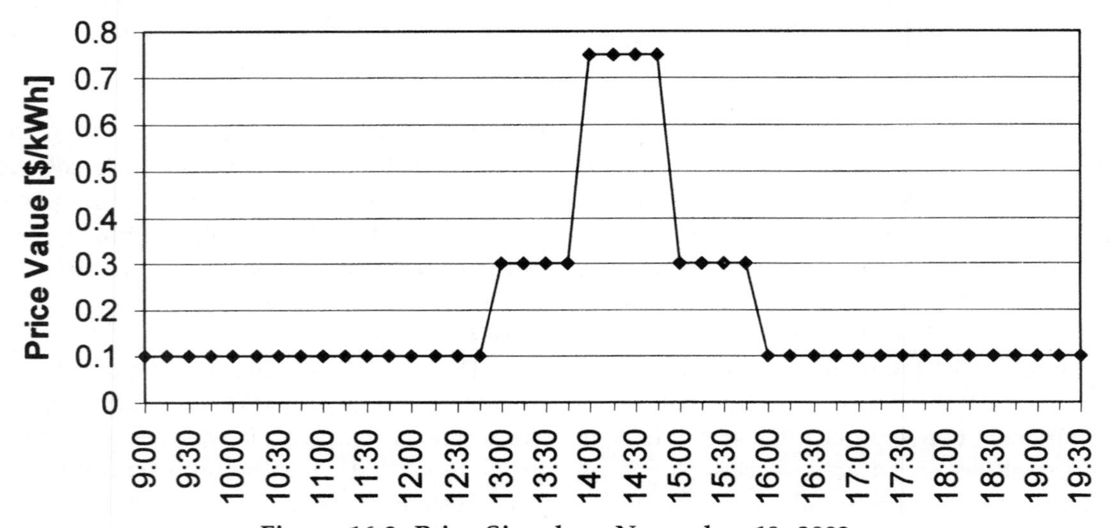

Figure 16-3. Price Signal on November 19, 2003

Table 16-2. Characteristics of Auto-DR Systems—Architecture

	Albertsons	B of A	GSA	Roche	UCSB
Client hosted at co-lo	Yes	Yes	No	No	No
Remote polling client	Yes	Yes	Yes	No	No
Remote control via Internet Gateway	No	Yes	No	Yes	No

allows the business logic computer to send commands over the network(s) directly to the EMCS *I/O controller* to shed HVAC or lighting equipment. In a highly integrated system, the EMCS becomes an extension of the enterprise. In these types of integrated systems, a *gateway* device is used to translate between the different protocols used in enterprise networks and EMCS networks.

Alternately, some systems used an Internet Protocol Relay (*IP Relay*) to interface between enterprise networks and EMCS networks. Relay contacts are commonly used in EMCS programming to define mode changes in HVAC equipment operation (e.g., smoke detector contacts). However, the use of relay contacts as an interface between networks is not as flexible as the gateway devices. Modifications to shed strategies would be more difficult with a relay interface system than with an integrated system with a translating gateway. However, when properly implemented, both gateway-based and relay-based interfaces between enterprise networks and EMCS networks can be effective for initiating shed strategies.

Gateway Type

Gateways used in building *telemetry* systems provide several functions. First, they connect two otherwise incompatible networks (i.e., networks with different protocols) and allow communications between them (see Figure 16-4). Second, they provide *translation* and *abstraction* of messages passed between two networks. Third, they often provide other features such as *data logging,* and control and monitoring of I/O points.

Of the five Auto-DR sites, two used *embedded* two-way communicating gateways to connect each site's EMCS networks to its enterprise networks (Table 16-3). *Embedded devices* are generally preferred over PC-based gateway solutions for scaleable, ongoing system deployments. Embedded devices have the following advantages:

- More physically robust. There are no hard drives or other moving parts.
- Less susceptible to viruses and other types of hacker attacks due to custom-designed operating systems and applications.
- Less susceptible to human error. Once they are set up to function, there is no reason for site personnel to interact with the device. Since they are not "general purpose" computers, there is no risk of memory overloads due to computer games, screen savers and other applications that may be inadvertently loaded onto them.
- Better form factor. Embedded devices are usually smaller than PCs and are designed to be mounted in secure server rooms with other *IT* equipment.
- Lower cost. Although volume dependent, application-specific embedded devices can be produced in volume for lower cost than PCs.

At Albertsons, an embedded *IP I/O device* (Engage EPIM™) was used for power monitoring and shed mode control. The EPIM provided power monitoring by directly counting *pulses* from power meters. The EPIM set various shed modes into operation by opening and closing *onboard* relay contacts. Although the EPIM IP I/O device effectively provides the interface between the EMCS and enterprise networks, it does not fit the most basic definition of a gateway because it does not connect the protocols of the two networks.

At UCSB, gateway functionality for monitoring was provided by software running on a PC. A previous version of gateway software also provided remote control functionality, but this feature was unavailable at the time of the Auto-DR test due to incompatibility issues that occurred after a software upgrade. To meet the remote control requirements of the Auto-DR test, an embedded *IP Relay* device was added. This device had onboard relay contacts similar to the EPIM, but direct measurement of I/O points (such as power meters) was not required.

The common source of electricity price and the communications protocol translations between the business logic and the final control element (relays, valves etc.) controllers that actually shed the electric loads is shown in Figure 16-4, "Network architecture overview of five combined Auto-DR sites,." Gateways or other devices are used to transfer necessary communications between dissimilar network protocols.

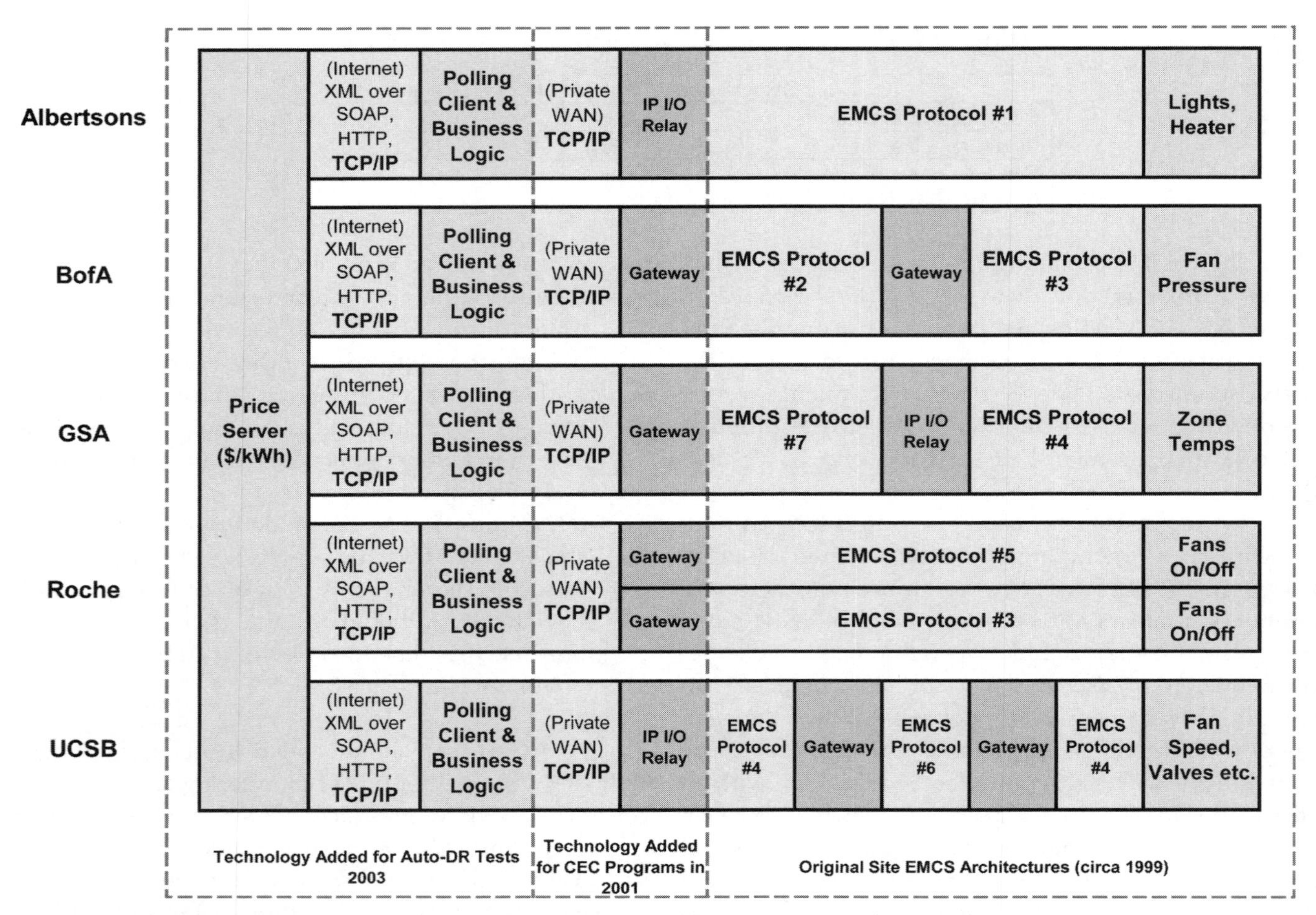

Figure 16-4. Network architecture overview of five combined Auto-DR sites

Table 16-3. Characteristics of Auto-DR Systems—Gateways

	Gateway type	Interface Device Description for Remote Shed Control	Remote Monitoring Description
Albertsons	Embedded IP I/O device	IP Relay (2 contacts)	Meter pulses monitored via EPIM™
B of A	Embedded Gateway	Control of EMCS via Internet gateway	Monitoring of EMCS via gateway
GSA	Embedded IP I/O device	IP Relay (2 contacts)	None – Local trending only
Roche	Embedded Gateway	Control of EMCS via Internet gateway	Monitoring of EMCS via gateway
UCSB	PC based Gateway for monitoring, Embedded IP relay for control	IP Relay (3 contacts)	Selected EMCS points monitored via gateway

Integration

For purposes of this study, integration between EMCS and EIS can be characterized by asking two key questions. First, can data from the EMCS and EIS be viewed and analyzed with one *Human Machine Interface (HMI)*? Second, do the EMCS devices such as energy meters reside on the same network as the EMCS devices? Table 16-4 summarizes the answers to these questions for each of the sites. Albertsons, B of A, and GSA either don't have EIS or else they are not integrated with the EMCSs at those sites. At Roche, the Tridium system integrates most of the EMCS points and a small percentage of the electric meters into a comprehensive HMI for viewing, archiving and analysis. UCSB has extensive monitoring of most of the electric meters and sub-meters throughout the campus. Data from these meters are available for visual representation, archiving and analysis through the SiE (Itron/Silicon Energy) server along with relevant EMCS points that have been mapped over to it.

One distinguishing characteristic of the Auto-DR sites was whether they leveraged the existing corporate or campus enterprise network to transmit EMCS and/or EIS data. Use of the existing enterprise network for this purpose has many advantages. System installation costs can be much lower if existing enterprise networks are used for communications instead of installing new, separate networks solely for EMCSs and EISs. In addition, the Information Technology department that manages the enterprise is often better equipped to assure network reliability and security than the facilities group that traditionally maintains the EMCS and EIS.

Each facility has different functional requirements and organizational structures that dictate how the enterprise, EMCS and EIS networks are designed, installed and maintained. Of the five sites in the Auto-DR test, three of them shared mission critical enterprise networks with EMCS/EIS/Auto-DR systems. Although *bandwidth* requirements for EMCS/EIS/Auto-DR systems are low, other organizational impediments may prevent the sharing of enterprise networks for non-standard purposes.

At GSA, a completely separate enterprise network was created for the GEMnet EMCS/EIS/Auto-DR system. This was the logical choice for this facility because it was not practical to share the existing enterprise networks with other tenants at the site: the Government Services Administration (GSA) and the Federal Bureau of Investigation (FBI). In such circumstances, it is easier to create a new enterprise network for local and remote access to EMCS and EIS data than it is to resolve complex security and maintenance issues associated sharing an enterprise network with another department or organization.

Shed Control Characteristics

Each Auto-DR site used different shed strategies. The control characteristics of these strategies also varied substantially. This section describes the characteristics of each shed strategy (Table 16-5). The number of shed control points that were adjusted or altered to invoke the shed strategy at each site is one characteristic of a given Auto-DR implementation. Shed control points include hardware control points (for example, valve position) and software points (for example, *setpoints*) that were altered during the shed. Software points other than setpoints were not included. Control granularity and closed loop shed control are additional characteristics that influence the likelihood and degree to which some occupants may be negatively affected by a given shed strategy.

Albertsons had only two control points (overhead lights and anti-sweat heaters). Because the size of the store is comparatively small, this was rated as "medium" control granularity. Switching off half of the overhead lights is an *open loop* type of control (i.e., there is no feedback to the system). The anti-sweat heater remained in closed loop control during the shed by operating with a reduced setpoint.

Table 16-4. Characteristics of Auto-DR Systems—Integration

	Albertsons	B of A	GSA	Roche	UCSB
Integrated EMCS & EIS	No	No	No	Partial	Partial
Primary enterprise network shared with EMCS/EIS/Auto-DR systems	Yes	Yes	No	Yes	Partial

Table 16-5. Characteristics of Auto-DR Systems—Shed Control

	Albertsons	B of A	GSA	Roche	UCSB
Number of Shed Control Points	2	1	~ 1,400	7	42
Shed Control Points per 10,000 ft.^{2}	0.4	0.05	14	0.4	1.4
Control Granularity	Medium	Coarse	Very Fine	Medium	Fine
Closed loop shed Control	Partial	Partial	Yes	No	No

B of A had just one control point (duct static pressure setpoint) for the entire 211,000 ft^2 building, hence the "coarse" control granularity rating. The shed strategy of resetting the duct static pressure setpoint while maintaining zone temperature is a type of closed loop control, but the dearth of temperature *sensors* in the zones reduced the closed loop rating to "partial."

In stark contrast to the rest of the sites, the GSA building used a fine granularity, closed-loop shed control strategy. The zone temperature setpoints for each of 1,050 VAV terminal boxes (1,400 including reheat side of dual duct boxes) were "relaxed" during the shed. In other words, the cooling setpoints were raised and the heating setpoints were lowered. This approach had an energy saving effect on the central HVAC systems while assuring a reasonable level of service modification to the occupants.

The Roche site used a coarse open loop shed strategy of shutting off fans during the shed.

UCSB used a variety of shed strategies of medium granularity. The shed strategies (including closing cooling valves, and reducing duct static pressure) were all open loop. The outside air dampers were opened to 100%, a strategy that could backfire in extremely hot conditions. The temperate climate in Santa Barbara made this scenario unlikely.

Open Standards

In the EMCS and EIS fields, protocols refer to the low-level communication languages that devices use to "talk" to one another on the network. Of course, one device can only talk to another if they are speaking the same language. Traditionally, each control system manufacturer built controllers and other devices that only spoke their own unpublished *proprietary protocol* (Table 16-6). Once a system is built using a proprietary protocol, the original manufacturer or their representatives are the only parties that can make substantial additions or changes to the system. Some control companies use proprietary protocols as a "lock" on their customers' systems so as to ensure future business and high profit margins.

Over the past fifteen years or so, there has been a movement toward "open" protocols in the EMCS and EIS industries. *Open protocols* are based on published standards open to the public. Interested companies can build products that communicate using open standards. In a truly open, interoperable system, products from a variety of open product vendors could be added at any time by skilled installers from independent companies. Several sites in the Auto-DR test use open EMCS and/or EIS products that include the *BACnet*, *LonTalk* (EIA-709) and *Modbus* open protocols.

Even with considerable interest from building owners few, if any, new or existing building EMCS or EIS systems are truly open and interoperable. Even when open protocols are used, they are often installed as part of a system that requires use of proprietary software or components at the higher levels of the system architecture. Another way "openness" is reduced is by designing products and systems that require proprietary software tools for installation.

In the IT marketplace, open protocols (e.g., TCP/IP), open database interface standards (e.g., *ODBC*) and open hardware standards (e.g., *SIMM*) have helped the industry thrive. This has allowed products from a wide variety of vendors to communicate with one another on internal LANs, WANs and the Internet. A service industry of independent *Systems Integrators* has grown to fill the need of integrating multiple vendor networks into cohesive systems.

Another important trend in the IT industry is the use of a new set of open standards, protocols and languages collectively known as XML/Web services. The use of XML/Web services in the building controls industry is increasing. This trend will help increase the ability to easily distribute, share and use data from disparate EMCS, EIS and other business systems. This will create opportunities for new products and services that will improve comfort and efficiency in buildings.

In the Auto-DR test, the use of XML/Web services over the Internet provided an overarching open-standards platform by which all of the proprietary and partially open EMCS and EIS systems could communicate. Although the number of commands transmitted between the systems in the 2003 test was minimal (e.g., price, shed mode, etc.), the implications of XML based "add-on" interoperability are very powerful.

RESULTS

Aggregated Whole Building Power and Savings

Figure 16-5 shows the aggregated whole building power and associated savings for all five sites during the shed. The shed period was from 1:00 pm until 4:00 pm on November 19, 2003.

The average savings (load shed) is shown in Figure 16-6. Each bar represents the average savings over one hour of the three hour elevated price test. The electricity price during the first, second and third hours was \$0.30/kWh, \$0.75/kWh and \$0.30/kWh, respectively. [2] The

Table 16-6. Characteristics of Auto-DR Systems—Open Standards

	Albertsons	B of A	GSA	Roche	UCSB
Open Protocol EMCS	No	Yes	No	Partial	No
Open Protocol EIS	Yes	Partial	NA	Partial	No
Open Protocol Auto-DR	Yes	Yes	Yes	Yes	Yes
Open Standards Auto-DR	No	No	No	No	No
Data Archiving in Open Database	NA	NA	Yes	NA	NA

NA = Not Available

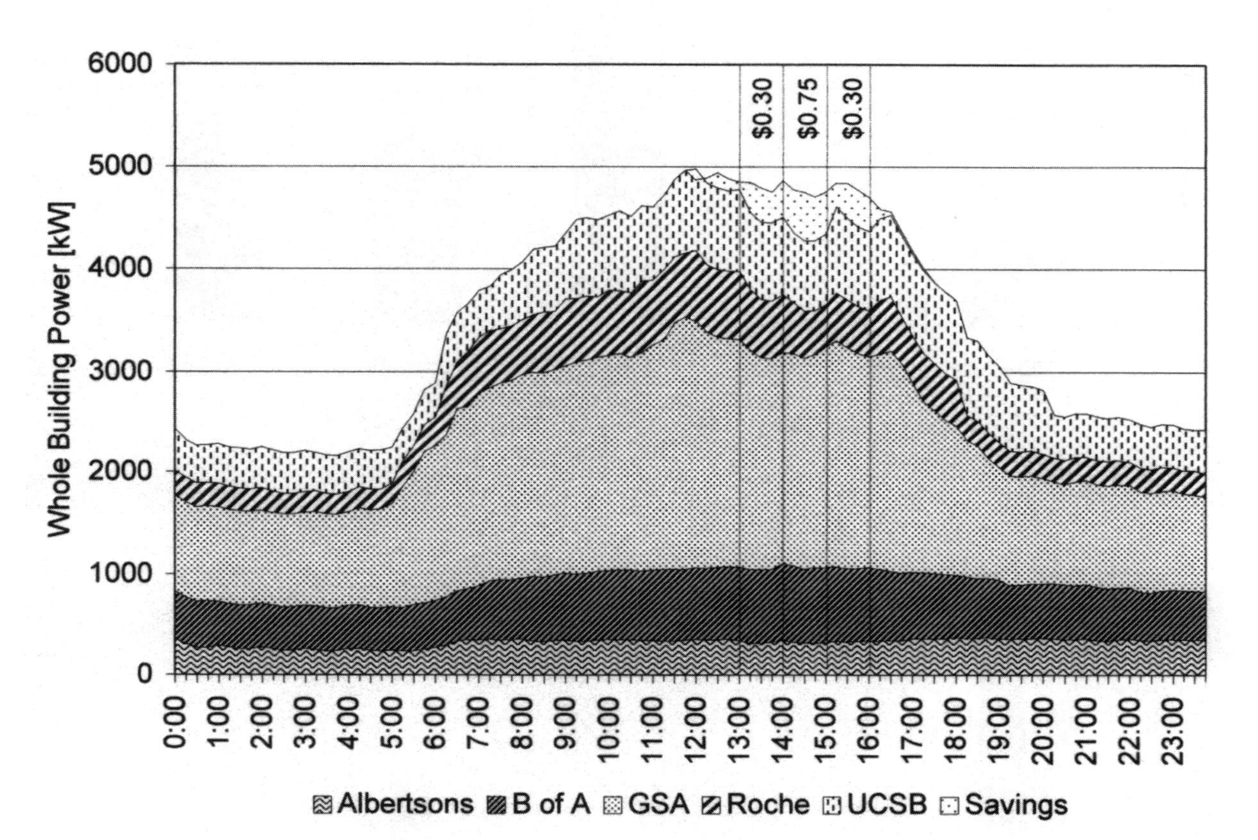

Figure 16-5. Aggregated Power and Savings of All Sites

left graph shows the total average power savings per site, while the right graph shows the savings normalized by floor area. This view presents a comparison of the aggressiveness and/or effectiveness of each shed strategy on an area normalized basis.

As an example, the results from an individual site (Roche) are shown below in Figure 16-7. Savings are determined by comparing actual metered power on the day of the shed with a calculated normal (non-shed) baseline. [2] The vertical lines show boundaries of the price range. The fan load component (cross-hatched) is superimposed on the profile of the whole building electric load (white). The savings due to the shed are shown both in the gray section above the whole building load profile and the inverted hat shape missing from the fan load profile.

SYSTEM CHARACTERISTICS OF EACH SITE

This section identifies the unique attributes of each participating Auto-DR facility. Controls and communications infrastructures and shed strategies are discussed for each system.

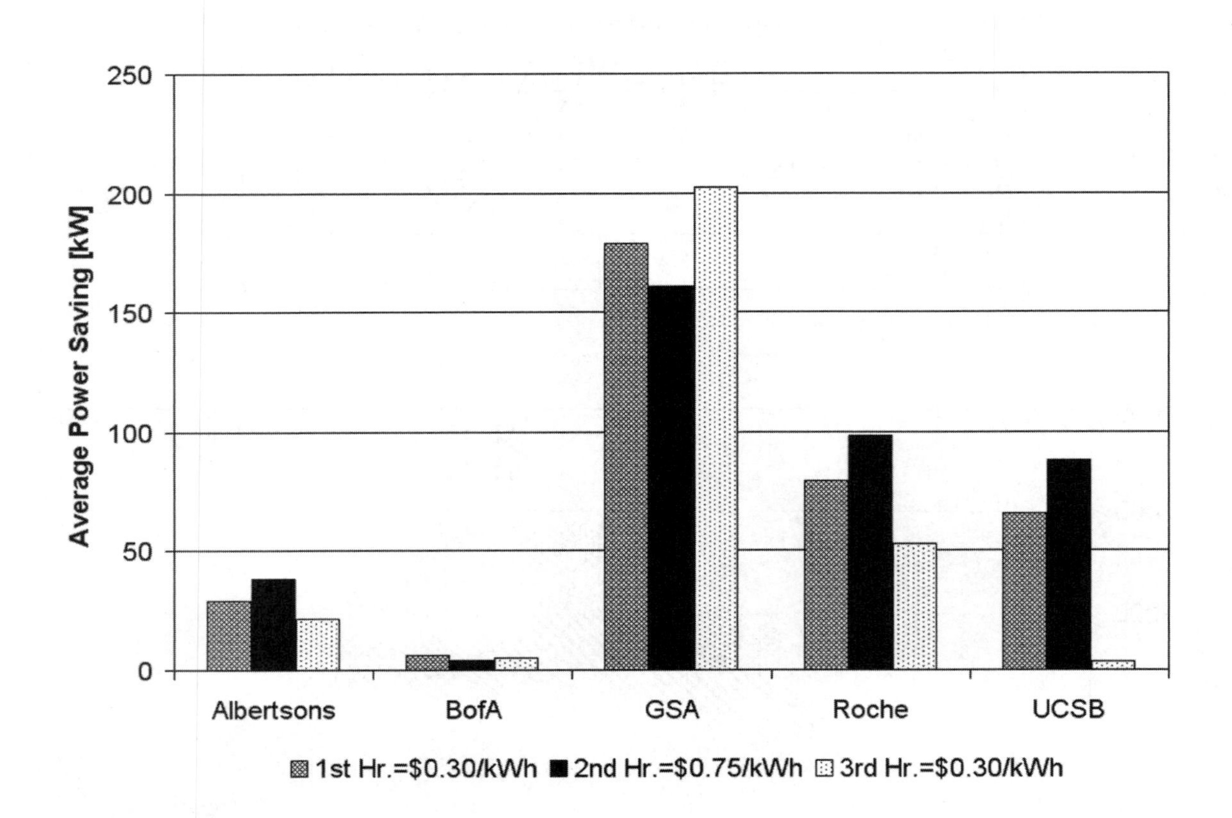

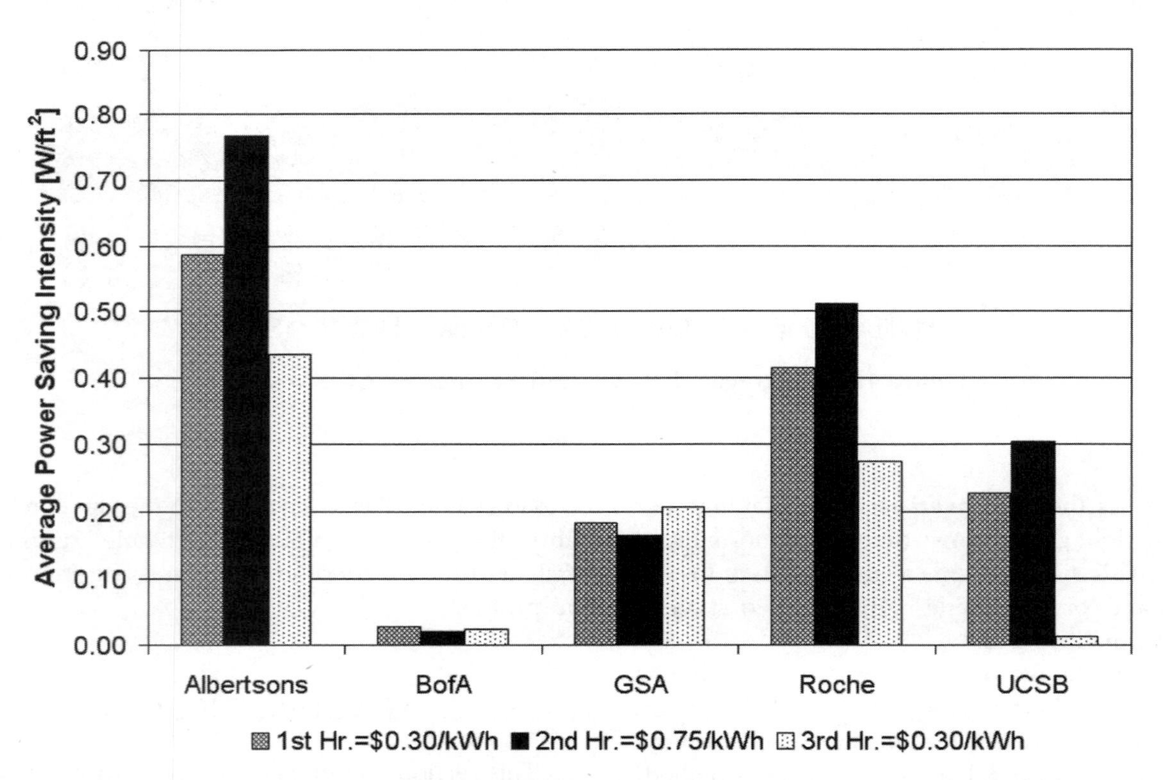

Figure 16-6. Average shed (kW and W/ft^2) during the 3-hour test on November 19, 2003

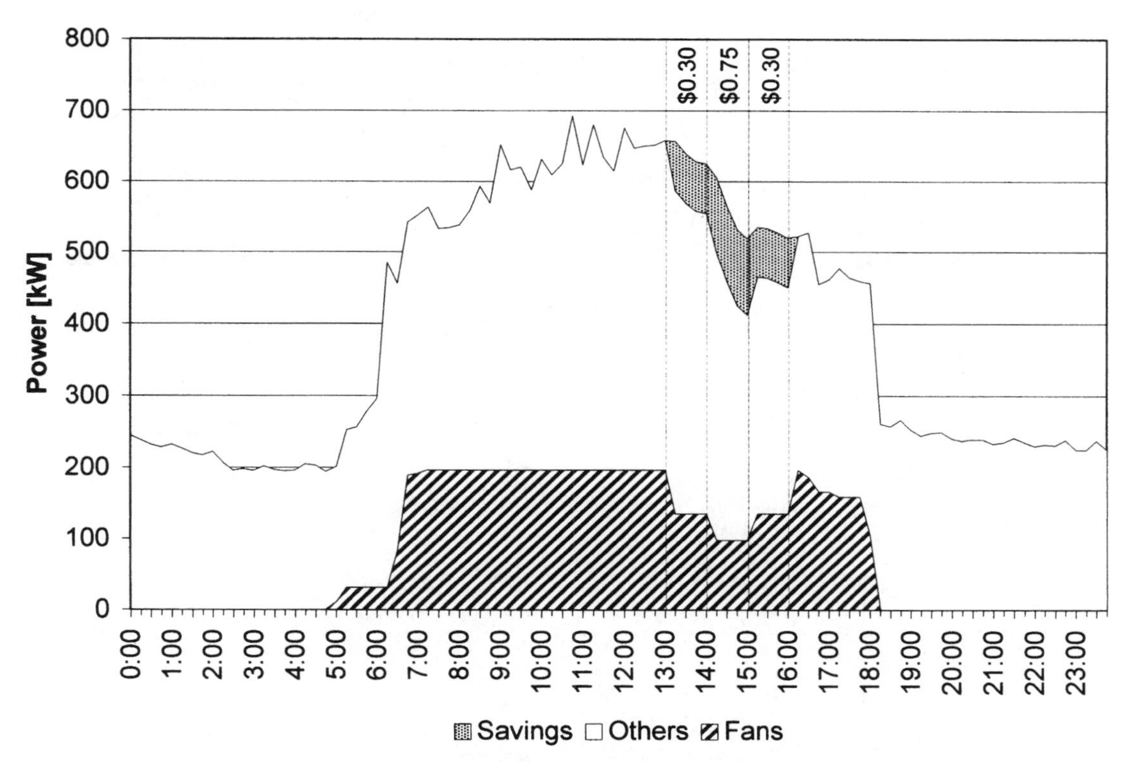

Figure 16-7. Roche Electricity Use, November 19, 2003

Albertsons (Supermarket)
System Overview (Figure 16-8)

The Albertsons building telemetry data system is split between two systems. The EMCS (excluding electric power monitoring) is accessed via a dial-up modem. EIS data is available via any Web browser through the EIS Web site. The segregated nature of the EMCS and EIS make it a burdensome task for the facility operator to change a temperature setpoint or lighting schedule and then observe the effect on electric usage.

However, the integration between the enterprise networks and *control networks* is rather tight. The corporate WAN (Wide Area Network) is used to communicate between the business logic/polling client and the on-site Internet Protocol Input/Output (IP I/O) relay device. The enterprise network is also used for mission critical *point of sale* data communications within the nationwide organization. The fact that the energy data are shared and communicated over the mission critical enterprise network indicates a high level of collaboration and trust between the Albertsons energy managers and other department managers involved with the core business of the organization.

The shed strategy was not objectionable to the store managers or patrons. Although the transition between 100% overhead lighting to 50% was noticeable, there were no complaints. The reduction of overhead lighting appeared to make the other light sources in the store, such as case lights, seem more intense. There is no evidence that the freezer doors fogged up during the shed, even though the setpoint of the anti-sweat heaters was reduced. If the transition of overhead lights to 50% were gradual (e.g., through use of dimmable ballasts) the entire shed would probably not be noticeable.

BofA (Bank Office) System
Overview (Figure 16-9)

Integration between enterprise networks and control networks at this site is tight. The BofA corporate WAN is used to communicate across the country to the on-site gateway. This network is also used for mission critical financial data communications within the BofA organization. Like Albertsons, the fact that energy and HVAC (heating, ventilation, and air conditioning) data are shared over the mission critical enterprise network indicates a high level of collaboration and trust between BofA's energy managers, IT (information technology) security managers and other department managers involved with the core business of the organization. The use of highly secure and reliable hardware *VPN* (virtual private network) routers and the use of a co-location site to host the polling client and business logic computers are indications that system availability and security are high priorities.

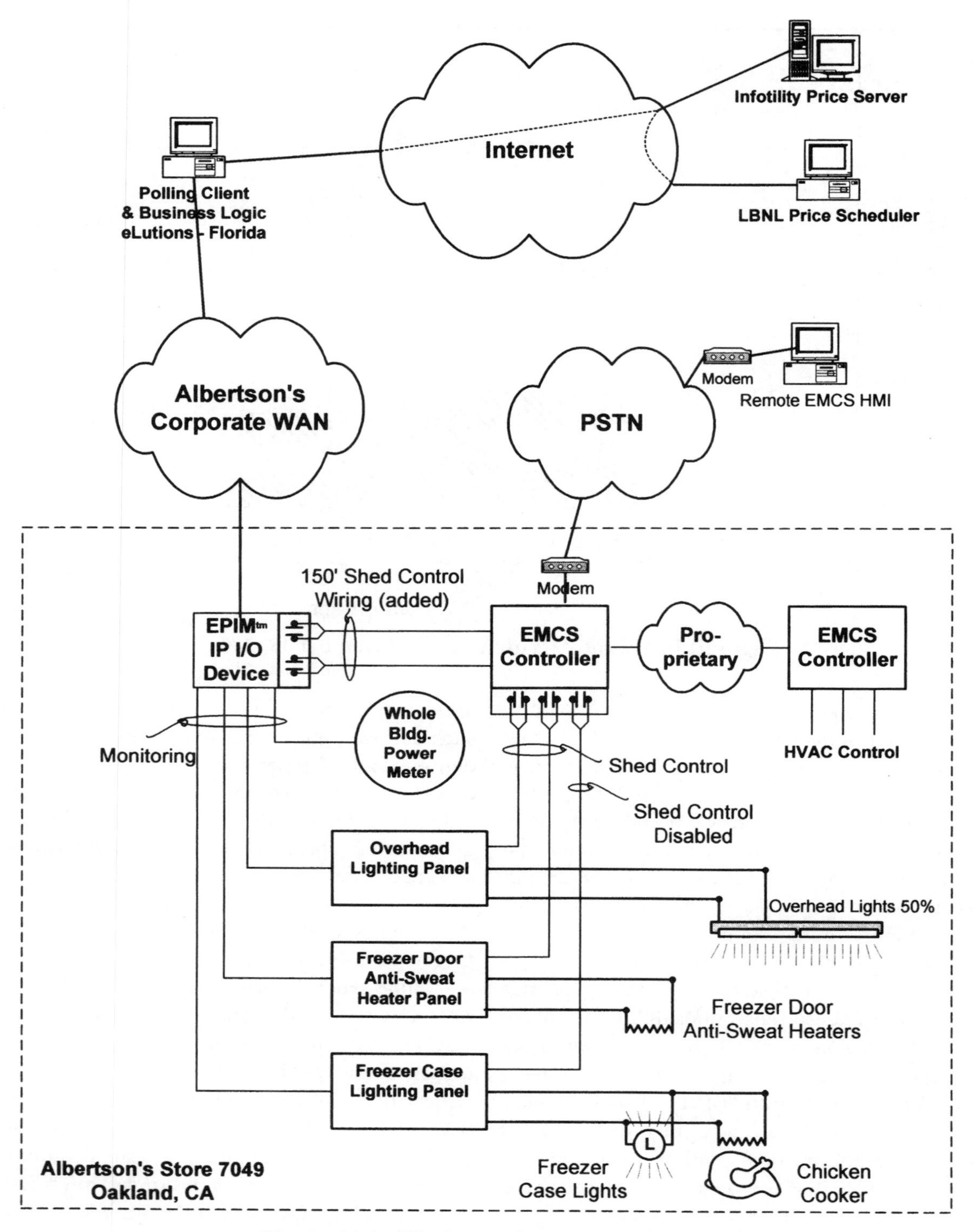

Figure 16-8. Albertsons System Overview

With regard to the shed strategy employed at this site, there is no evidence that a modest reduction in duct static pressure for short durations caused any negative comfort effects to the occupants during the test. However, as shown in the measured data, the extent of the electric demand shed is negligible. If this strategy were extended so as to produce significant sized electric sheds, the method may pose some fundamental drawbacks. When the duct static pressure is reduced below the minimum required by the terminal boxes in VAV (variable air volume) systems, airflow is reduced in the zones. But the reduction is not shared evenly between all

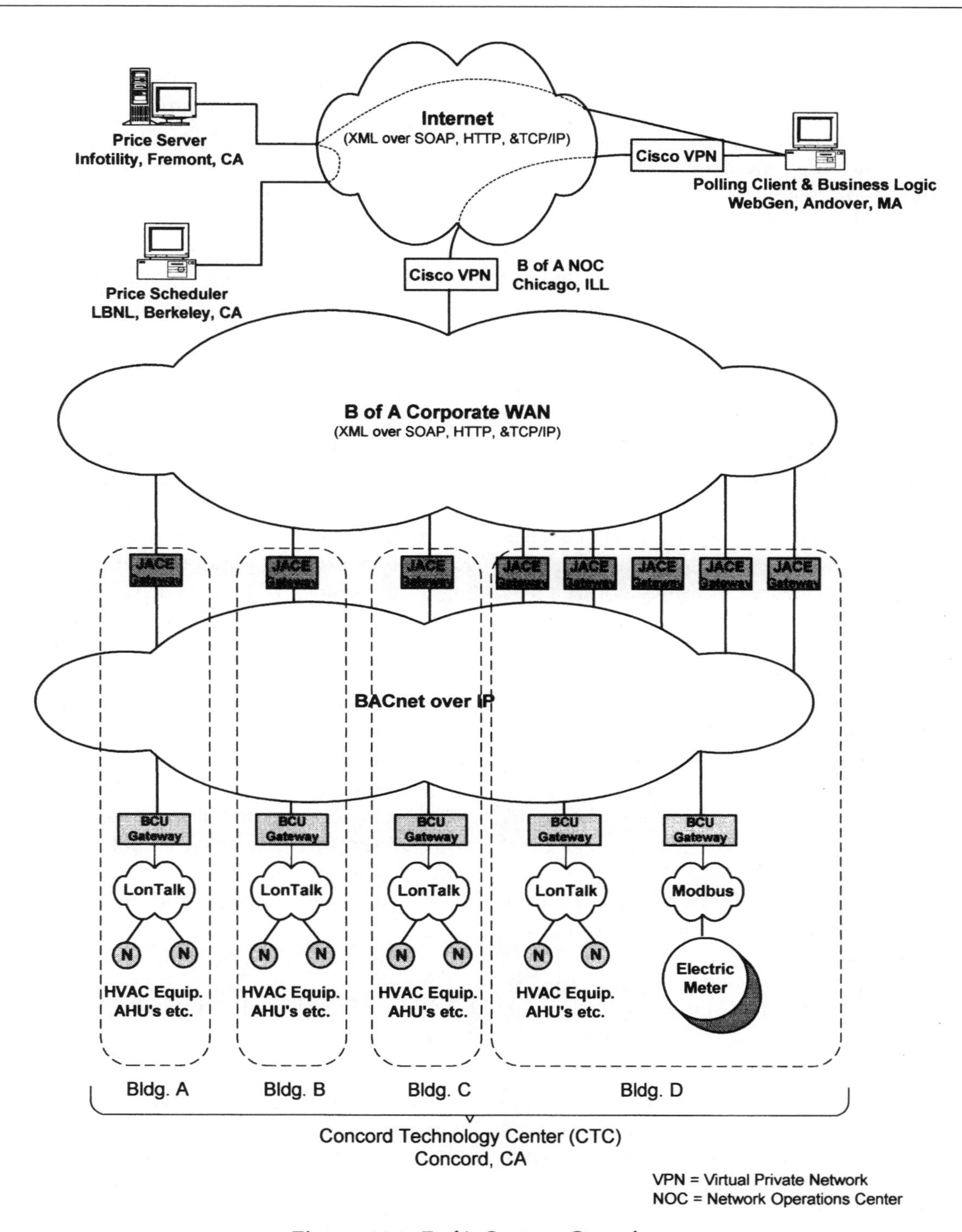

Figure 16-9. BofA System Overview

the boxes. The zones of greatest demand are the ones that are starved for air most immediately and most severely. In the building used in the Auto-DR pilot, the potential problem is exacerbated by the lack of sensors. Fan airflow is not measured and only nine "representative" zone temperature sensors are available for the entire 211,000 ft^2 building. There were not enough sensors to estimate the effect that reductions in airflow would have on occupants.

When the third party energy management company (WebGen) takes action to reduce energy at its connected sites, it uses a centralized control paradigm. While demand response systems are inherently centralized (signals to shed loads are generated in a one-to-many relationship), centralized control for day-to-day operation is less common.

In most control system markets (commercial buildings, industrial controls, etc.) there has been a trend for several decades toward decentralized control. In decentralized control, the control logic is moved (physically) as close to the sensors and final control elements (e.g., relays, valves, etc.) as possible. Decentralized control systems have traditionally been less costly, more flexible and more robust. However, in the IT community, there has been a movement in certain areas toward hosted solutions, application service providers and other centralized solutions. Ubiquitous Internet connectivity and other IT technology advances make these systems less costly, more flexible and more robust for certain applications.

The WebGen system alternates between centralized and decentralized paradigms on cycles as short as twenty minutes. At the end of one cycle, a fan system maintains a setpoint entered by on-site building operators. In the next minute, a neural network algorithm may define the setpoint from over 3,000 miles away.

GSA (Government Office) System Overview (Figure 16-10)

The enterprise and EMCS infrastructures used to enable Auto-DR at this site are linked together in a long series of serial components and communication links. The prototype system was assembled at low cost using spare parts. With so many links, it is not surprising that there were communication failures due to an unexplained equipment lock-up during the first test. To make the system more robust, a review of the components and architecture should be conducted.

The second test was quite successful, as communications were functional from end-to-end. The shed strategy produced an electric shed about as large as the other four sites combined. Because the temperature setpoint reset was at the zone level, comfort for each occupant could be maintained within the revised, relaxed constraints (Table 16-7). To implement this strategy, it was necessary to revise the software parameters and some logic in each of the 1,050 VAV terminal box controllers. For most EMCS systems, the labor required to make these revisions would be substantial (1-3 weeks). In this building, the process had been somewhat automated by previous system upgrades. This allowed EMCS reprogramming for Auto-DR to be conducted in about three hours.

Roche (Offices and Cafeteria) System Overview (Figure 16-11)

A third-party software framework (Itron/Silicon Energy) ties together three different EMCS protocols at Roche in a seamless fashion. The Web interface provides operators with compete monitoring and control capability from anywhere on the campus. It was relatively straightforward to interface the Auto-DR polling client and associated business logic to the system. The most challenging part of the project was setting up the "extra" computer outside of the Roche *firewall* and establishing communications to devices inside of the secure corporate network.

UCSB (University Library) System Overview (Figure 16-12)

Remote monitoring and control of the EMCS and EIS was available over the Internet prior to the Auto-DR pilot. However, at the time of the test, remote control of the EMCS was not available. The software gateway between the enterprise network and the EMCS network lost remote control functionality during an "upgrade" of Itron/Silicon Energy's third party server software. To meet the test schedule of the Auto-DR pilot, an IP I/O relay was added to allow the Auto-DR business logic to initiate the control functions such as initiating sheds. The shed strategy proved to be very effective. The books and other thermal mass in the library buildings acted as a thermal "flywheel" to help keep the space comfortable during the shed periods. In addition, the shed strategy reduced airflow without shutting off fans completely. The coastal climate of the site helped provide a temperate airflow even when the cooling and heating valves were closed.

STATE OF THE ART IN AUTOMATED DEMAND RESPONSIVE SYSTEMS

By evaluating the systems demonstrated in the November 2003 Auto-DR test, along with other existing technologies found in the EMCSs, EIS and the IT Industries, state-of-the-art Auto-DR systems can be envisioned. The five participating sites all successfully met the functionality criteria of the pilot (under tight schedules and limited budgets). However, a truly state-of-the-art system would use the "best of the best" components, systems and strategies from end to end. Such a system would be designed from scratch to meet a very specific set of requirements. The "best" system would meet or exceed the requirements at the lowest installed cost. State-of-the-art Auto-DR systems should have the following characteristics.

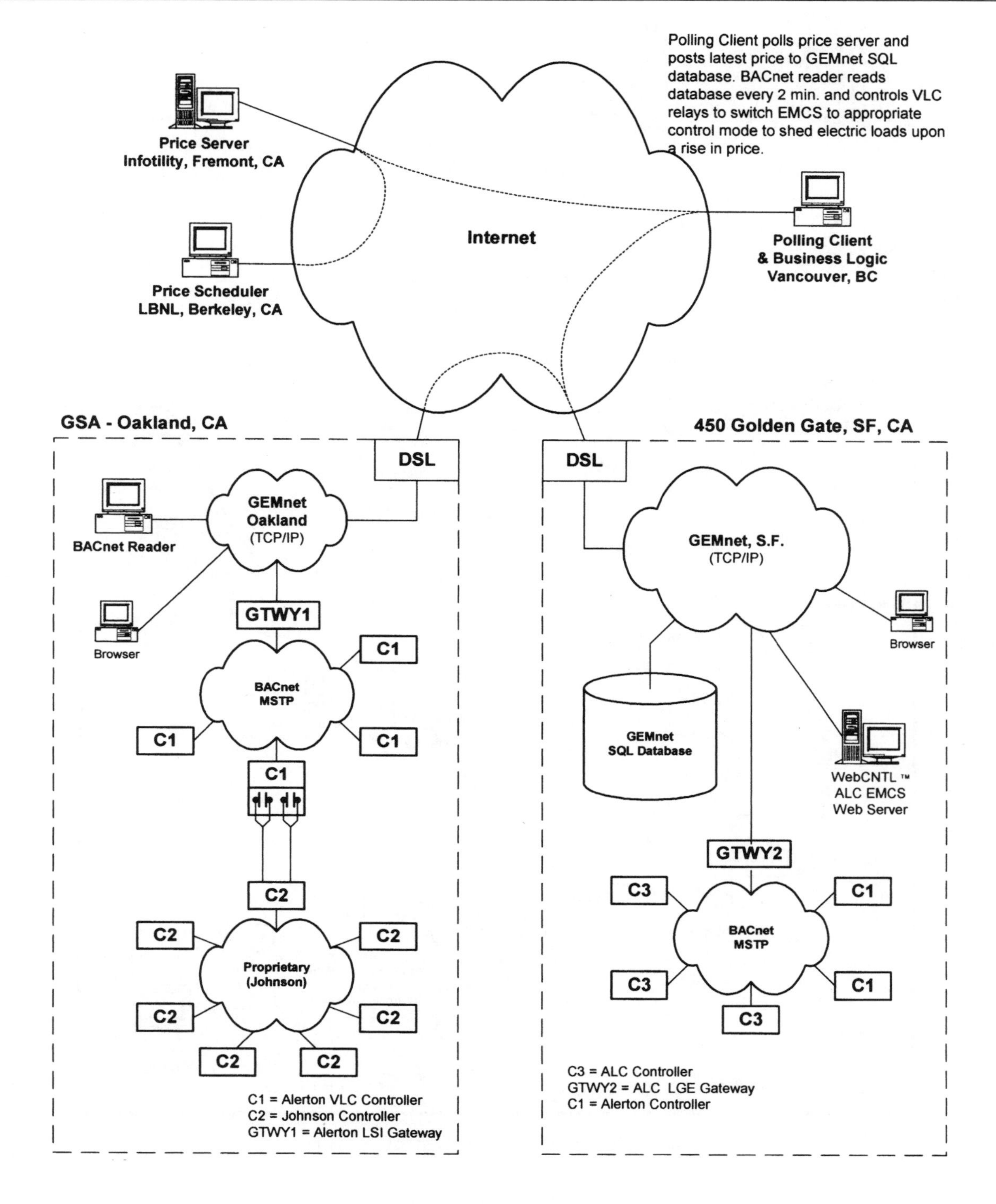

Figure 16-10. GSA System Overview

Table 16-7 Oakland GSA Zone Temperature Setpoints—Normal and Shed Modes

Shed Mode	Zone Heating Setpoint	Zone Cooling Setpoint
Normal ($0.10/kWh)	70°F	72°F
Level 1 ($0.30/kWh)	68°F	76°F
Level 2 ($0.75/kWh)	66°F	78°F

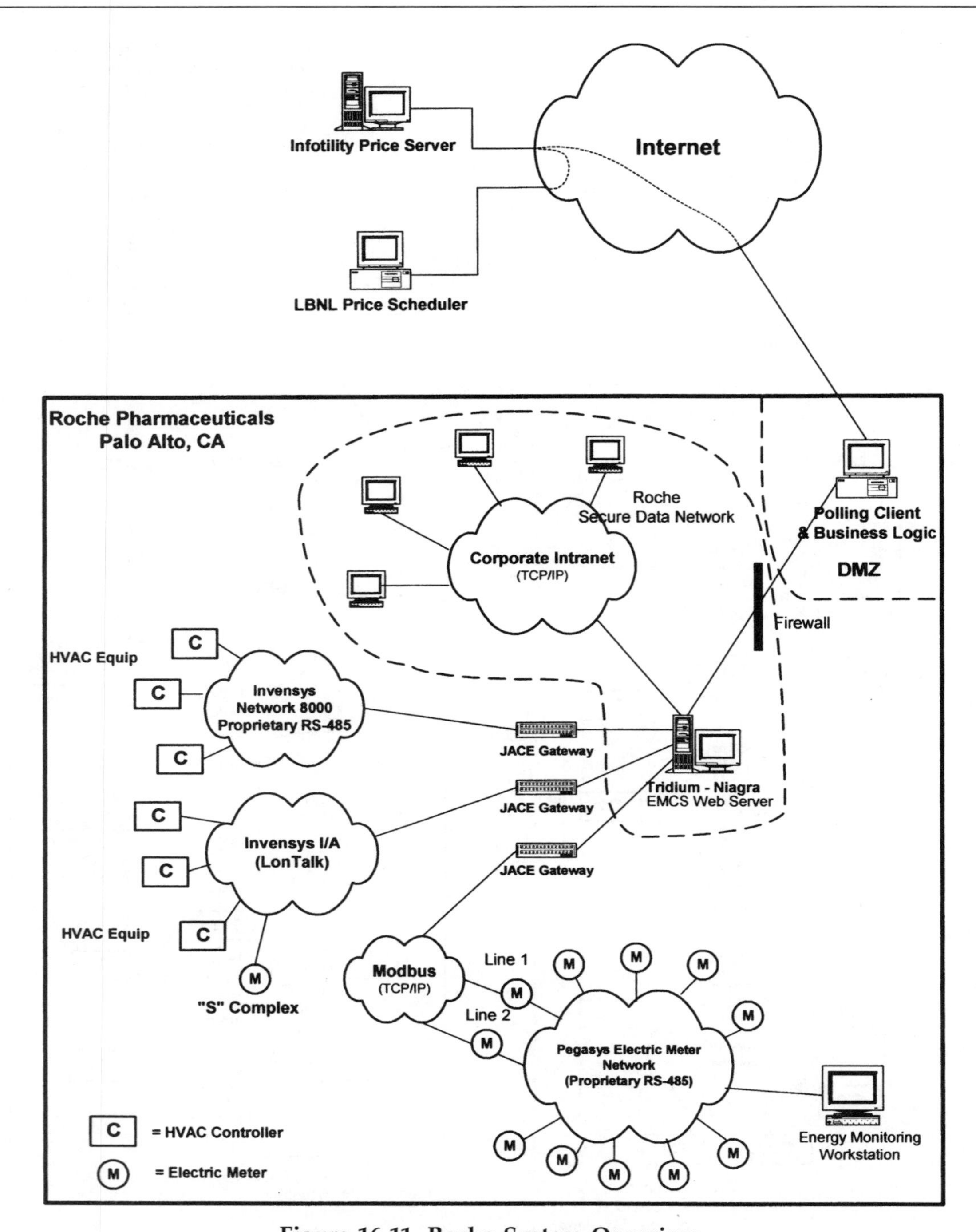

Figure 16-11. Roche System Overview

Flexible Designs for the Future

Today's state-of-the-art Auto-DR technology could be applied in many different ways, depending on the scenarios and applications that they are designed to satisfy. As the scenarios, applications, and driving forces behind Auto-DR become better defined, systems will be designed and deployed accordingly. Since these design criteria are likely to remain in flux, Auto-DR system flexibility and future-proofing have a very high priority.

Features

Customers should have numerous options about how they can participate in Auto-DR programs. For any given motivating force that drives customers to consider Auto-DR (i.e., price), each will have different circumstances under which they will want to participate. Any state-of-the-art Auto-DR system must have sufficient flexibility to meet the needs of a variety of customers. They should have the ability to use custom

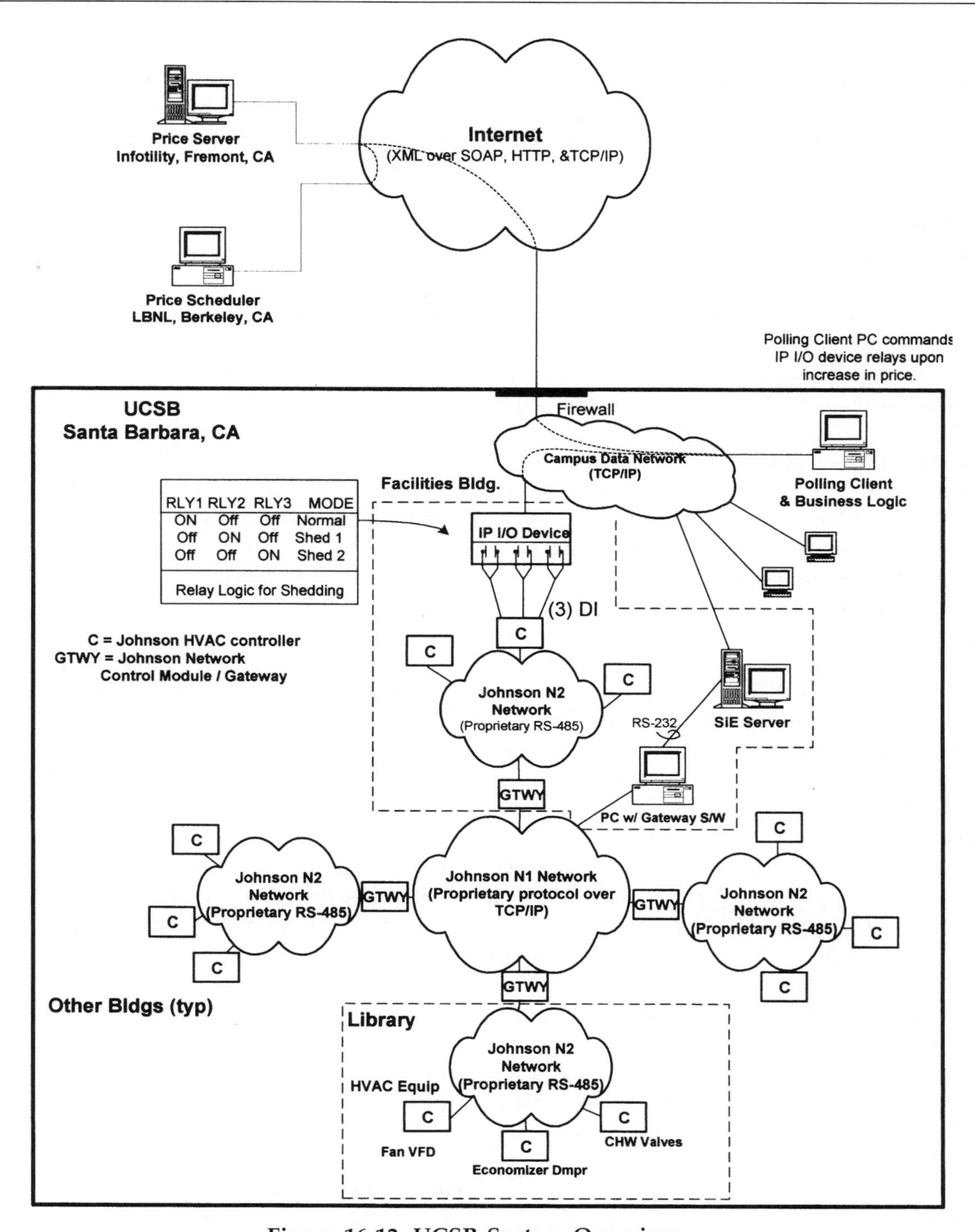

Figure 16-12. UCSB System Overview

business logic that is applicable to their own operations. Some may choose to allow remote *real-time* control (for its extra value) while others may want some advanced warning (via pagers, cell phones, etc.) and the ability to opt out, if desired. Other important features in state-of-the-art Auto-DR systems include real-time two-way communications to the final control and monitoring elements (e.g., sensors), high security, and high system availability.

Leveraging Trends in Technology

The lower the installed cost of state-of-the-art Auto-DR systems, the sooner they will find their way into mainstream use. One of the most important ways to keep costs low is to leverage existing trends in technology. For example, existing IT technology should be used in Auto-DR systems wherever possible. The public Internet and private corporate LAN/WANs are ideal platforms for Auto-DR controls and communications due to their ubiquity, especially in large commercial buildings. In addition to the availability of networks, the performance of IT equipment (e.g., routers, firewalls, etc.) continues to improve and prices for this equipment continue to drop.

Enterprise, EMCS & EIS Integration

Another way to obtain high system performance and keep the system costs low is through increased integration within the building. Since energy data from EISs is simply another type of measured data, EISs and EMCSs should share the same networks so as to maximize system performance and functionality and minimize cost. In addition to eliminating a redundant EIS network, other aspects of the system are also unified through this approach. Use of an integrated EMCS/EIS database and associated archiving and visualization tools increases user functionality while reducing cost. The ability to change setpoints for HVAC equipment and observe and analyze the effect on electric consumption from the same Human Machine Interface (HMI) is an important enhancement to both the EMCS and EIS.

State-of-the-art Auto-DR systems should also have tight integration between the EMCS/EISs network and enterprise networks within buildings. Once the integrated controls and communications infrastructures are in place, many applications in addition to Auto-DR are enabled. Some other telemetry applications include: energy management, aggregation, equipment maintenance, access control and regulatory record keeping.

The network architecture of a state-of-the-art Auto-DR system normally tends to be flatter than most of the sites in the November 2003 pilot. A flat architecture is one in which there are a minimum number of layers of control networks and protocols between the HMI and the final control and monitoring elements. The most robust and least costly systems should have no more than one enterprise network protocol and one control network protocol.

Open Standards

For flexibility and future-proofing as well as the option to choose "best of breed" products, state-of-the-art Auto-DR systems should use open standards wherever possible. Unlike proprietary systems, truly open systems are interoperable. In other words, a device from one company will easily and naturally reside on a network with products from other companies. Most products in enterprise networks are interoperable. They communicate using the TCP/IP protocol and can be set up and managed using common network management tools. TCP/IP is clearly the worldwide protocol of choice for LAN/WAN, Internet and enterprise networks.

There are several open standards control networks including BACnet (ASHRAE Standard 135-2001) and LonTalk (ANSI/EIA/CEA 709.1). Several database formats have become de facto open standards as well. Although the use of the meta-language XML is becoming a standard framework for communicating over enterprise networks and the Internet, XML alone does not define data formats that could be used to convey measured building or energy data. Standards of this type are being developed by OPC XML, *oBIX*, BACnet XML and other organizations.

With only two network protocols in the state-of-the-art Auto-DR system, only one type of gateway is required for translation/abstraction between them. An embedded gateway device that conforms with IT industry standards for reliability, security, and network management should be used.

Shed Strategies

State-of-the-art shed strategies should be designed to minimized discomfort, inconvenience and loss of revenue to the participating sites. Shed strategies should be devised by customers to meet their needs. In general, shed strategies that use fine granularity closed loop control are less likely to negatively impact building occupants. Ideally, sheds would vary, commensurate with a variable shed signal. Transitions would be fast enough to be effective, but slow enough to minimize attracting the attention of building occupants.

In addition to HVAC control strategies, lighting and switch-able plug loads should be considered for sheds as well. By increasing the controlled load to the point where it approaches the whole building load, each load type (HVAC, lighting, etc.) would need to shed a smaller amount in order to achieve a given shed target for the whole building.

Future Directions

In the industrial controls marketplace it is becoming more common for the TCP/IP based Industrial Ethernet to be used all the way down to the device level. In these systems, traditional open control protocols such as BACnet and LonTalk are eliminated all together. TCP/IP could be used in an end-to-end integrated enterprise, EMCS/EIS system. This trend is likely to gain momentum once the next *generation* of Internet Protocol, IPv6 is implemented. For greater flexibility, increased control granularity and lower costs, increased use of wireless devices in Auto-DR, EMCS and EISs is likely to occur.

ACKNOWLEDGMENTS

The authors are grateful for the extensive support from Ron Hofmann, a consultant to the California Energy Commission, Laurie ten Hope (CEC), Gaymond Yee (CIEE) and Karen Herter (LBNL), Joe Desmond and Nicolas Kardas (Infotility), and numerous individuals from the participating sites. This project was supported by the California Energy Commission's Public Interest Energy Research Program and by the Assistant Secretary for Energy Efficiency and Renewable Energy, Office of Building Technology, State and Community Programs of the U.S. Department of Energy under Contract No. DE-AC03-76SF00098.

Chapter 17

Mission Critical Web Based Building Monitoring and Control Systems

Travis Short

ABSTRACT

This chapter is devoted to the implementation of Web Based Building Monitoring and Control Systems (BMCS) in the Mission Critical building application. Mission Critical buildings include structures such as Data Centers, Financial Institutions, Telecommunications Facilities, and Hospitals. Items that will be discussed in this chapter include; web based monitoring and control, Human Machine Interface's (HMI's), benefits of Ethernet system architectures, Systems Integration, Fault Tolerant Design, advanced BMCS commissioning for Mission Critical systems, and real life examples of typical pitfalls in Mission Critical BMCS systems. This chapter will give the reader the ability to understand the complexities and solutions involved with a web based BMCS in Mission Critical buildings.

INTRODUCTION

Technology has finally caught up with the progression of Building Automation Systems towards the ultimate "Open" protocol, the Internet. Before we get into the implementation and proper installation of a web enabled Mission Critical BMCS we'll first go over a few definitions. One of the true secrets to understanding controls is interpreting the lingo associated with the systems. This is a well-hidden secret of the *Controls Contractor;* the basic ideas behind the system are only slightly more complex than the words associated with the business. Some of the terms that will be used throughout this chapter are:

- **Open Control System**—Open building systems are created using the products and systems from multiple vendors that in the end offer greater flexibility, easier management, higher levels of scalability, and lower life cycle costs.
- **LonWorks**—An "Open" protocol control technology developed by the Echelon Corporation to facilitate the use of multiple vendors at the device level. LonWorks communicates over the LonTalk communication protocol via a peer-to-peer communication between separate controllers. LonWorks does not require gateways to communicate with other vendor's products. All LonWorks devices utilize the neuron chip, which has the LonTalk protocol imbedded in it.
- **BACnet**—An "Open" protocol control technology developed by engineers to facilitate the use of multiple vendors at the network level. BACnet has several different medias by which it can communicate with disparate controllers. Due to the different levels of BACnet, gateways may be required for communication with other vendor's products. **BACnet** is defined by a specification developed by Engineers for ASHRAE. All vendors develop control products based upon an *interpretation* of this specification.
- **Systems Integration**—Systems Integration is the method by which separate systems utilizing diverse communication protocols are tied together and allowed to communicate and share information over one common system. These diverse protocols are converted into one common protocol. To accomplish the ideals of a truly "Open" System this protocol should be based upon one of the common Ethernet protocols, either XML or HTML.
- **Building Monitoring and Control System (BMCS)**—This system encompasses the typical DDC (Direct Digital Controls) system and is powered by Systems Integration with multiple vendors and systems.

The web enabled BMCS is the Cadillac version of controls while standard Electronic controls are comparable to the Yugo. When implementing a control scheme for a Mission Critical facility it is important to look at what the desired outcome of the application is. Data Centers for instance are classified by a Tier classification, which was developed by The Uptime Institute. Tier classifications range from a Tier I Data Center to a Tier IV Data Center. Tier I Data Centers incorporate the least amount of redundancy and therefore have the least amount of reliability, 99.671% of site uptime. Tier IV Data Centers on the other hand contain the highest amount of redundancy and encompass the ultimate in site uptime, 99.995% of site uptime. Control systems however are not directly included in the Data Center Tier Classifications. Nevertheless control systems do play an important factor in the buildings ability to stay operational 100% of the time. For instance, if the control system is not designed properly and a failure occurs, the Chilled Water Plant will become inoperable. The loss of the Chilled Water Plant will have a direct effect on the environmental conditions on the Data Center raised floor environment. Due to the critical nature of Mission Critical Facilities it is advised that the Cadillac of control systems be installed. Web enabled BMCS systems when properly installed will provide an added robustness to the system along with enhanced capabilities and reliability.

In regards to the robustness associated with the web enabled BMCS, key features include; the ability to access the system anytime and anywhere and receive real time data; additional information supplied from all building systems via the use of systems integration; and a fault tolerant fail safe design and installation. This added robustness is vital in the Mission Critical Facility where downtime is not acceptable and may cost the facility monetarily. Systems Integration is a key component to these systems with added information now available to maintenance personal at the touch of mouse. This added information could then be utilized to curb energy consumption during peak operating times, aid in fault analysis, diagnose a critical condition prior to an event, and much more.

When properly specified, designed, and installed the web enabled BMCS brings a high level of reliability to the facility. Additionally the system can respond to critical events and tailor control strategies to compensate for these events. When the design is incomplete or the system is installed incorrectly, the BMCS will become a problematic component of the building and not the aid that it was envisioned to be.

WEB BASED MONITORING AND CONTROL

System Components

Web based Building Monitoring and Control Systems are comprised of several components. These components if not properly specified and installed will not offer the robustness that is required by a Mission Critical Facility. The following section discusses the key components involved in the development of the web enabled BMCS.

- **Operator Workstation**: This component is not necessarily required for proper system operation. The BMCS graphics can be accessed from any computer via the use of a standard web browser and proper username/password access. However a dedicated operator workstation is beneficial when adds, moves, or changes to the system are required. Typical system configuration changes include the addition of a controller, graphics, or modifications, and dedicated data logging and access.
- **Web Based Integration Controller**: The web based integration controller is designed to convert the control protocol utilized by the BMCS into TCP/IP packets. A practical example of this type of conversion is found in any office building. When you have a document open on a computer and you send it to a printer, you are using a similar type of conversion. The information from the document that you are printing is converted into TCP/IP packets and is then sent over your office Local Area Network (LAN) and ends up at the printer. Another important feature of the web based integration controller is to house all Internet accessible graphics. Through the use of this configuration system, graphics are accessible over the Internet with or without an operator workstation acting as a web server. Under this arrangement the web-based integration controller is the web server.
- **Ancillary Devices**: Depending upon system configuration and size, Ethernet Routers and Switching Hubs will be used. These devices will be configured just as in any Ethernet network.

"Open" Web Based Systems:

The terms Open Architecture and Open Protocol are commonly used, but not always correctly. An "Open" Architecture is one that facilitates the use of multiple vendors and the ability to integrate diverse protocols. To truly have an "Open" System the architecture should utilize the Ethernet based TCP/IP protocol with all information converted to either HTML or XML. When discussing "Open" protocols two protocols immediately come to mind, **LonWorks** and **BACnet**. The use of either one of these protocols can produce the desired openness at the device or network level but not at a system level. If **LonWorks** or **BACnet** is utilized do you truly have an "Open" system? The answer in most instances is no. For instance **BACnet** has several levels of conformance while in order for **LonWorks** based system to be truly open all devices must at a minimum be LonMark certified. This is where the beauty of the Internet comes in. If the BMCS conforms to the standard Ethernet Protocols then you have a Truly "Open" Architecture. It is important that the system conform to one of the Ethernet protocols that support the ideals of an "Open" Architecture. For example the **BACnet/IP** protocol does utilize the Internet but it is not "Open" to multiple vendors.

There are several ways that an "Open" Architecture can be implemented in a BMCS. Like everything else, the more you spend the more you get. If high data transmission speeds are desired for alarm transmission and peer-to-peer communication, then an Ethernet Backbone must be used throughout the control system. Alarms, monitoring, trending, and control data will travel at high speeds over the Ethernet to either the Local or Remote operator workstation(s). If cost is a concern and ultra high speeds are not a necessity then the BMCS can be configured to utilize any standard protocol with a web-based integration controller, which will still provide the Openness of Ethernet protocols to the system. Data acquisition may not be as rapid but Local and Remote system access is still available over the Internet or Intranet at any number of Operator Workstations.

Web-enabled applications provide access to web-based graphics through any standard web-browser such Internet Explorer or Netscape. It is standard and recommended that the BMCS be provided a Static IP address in lieu of a Dynamic IP address. This ensures that connectivity to the system is easily accessible by all parties involved. Because security is a major concern in regard to access to the BMCS, the system being installed must have proper username/password protection.

ETHERNET SYSTEM ARCHITECTURE

System Configuration

To ensure proper system operation and reliability, the system architecture must be configured very carefully. The level of reliability required at the Mission Critical Facility determines which key functions must be applied.

The Ethernet network can be constructed in two ways. The first configuration is with the BMCS Ethernet network setup as a standalone network separate from the Facilities LAN or WAN. The second configuration is with the BMCS Ethernet network integrated with the Facilities LAN or WAN. Each configuration has its benefits. The standalone BMCS network affords a high degree of security and will not interfere with the bandwidth of the Facilities LAN or WAN. The integrated network provides secure access from any computer in the Facilities LAN or WAN, and with proper access through firewalls, the system is accessible from any Internet connection throughout the world. If an integrated solution is utilized, the BMCS contractor should consult directly with the Facilities IT department to ensure conformance to building standards.

No matter which of the above configurations is used, the Ethernet system architecture should be constructed in one of two ways. Proper configuration ensures Ethernet system fault tolerance and reliability. Option 1 provides a high degree of reliability while Option 2 is the ultimate arrangement for fault tolerant configurations.

- **Option 1 Dual Ethernet Backbones**—This is an A & B configuration, which uses two separate Ethernet Backbones. With this option if either one of the Ethernet Backbones fails, half of the system is still accessible through the Operator Workstation(s). When dual Ethernet Backbones are utilized, all controlled and monitored equipment should be carefully segregated to ensure that complete building monitoring of all managed and integrated equipment is not lost in the event of a failure. For example if the UPS (Uninterruptible Power Supply) systems are to be integrated and the modules are set up in an A & B configuration, the A side UPS modules must be integrated with a controller that resides on the A side Ethernet Backbone while the B side UPS modules are integrated with a controller on the B side Ethernet Backbone.

- **Option 2 The Spanning Tree Protocol**—The spanning tree protocol provides Ethernet Backbone redundancy at the Ethernet device level. This means the redundancy will reside at the Ethernet Routers and switching hubs. This is a very complex design, which requires a high level of expertise and special Routers and Switching Hubs.

WEB BASED HUMAN MACHINE INTERFACES

The Human Machine Interface (HMI) is the end user's window into the BMCS. If the HMI is not user-friendly, then its usefulness is extremely diminished. When specifying an HMI, great detail should be provided about what the end product should look like and what functionality is desired.

HMI Configuration

A true user interface is necessary to look at the system graphics for a Mission Critical BMCS installation. A large portion of the equipment monitored by the BMCS in the Mission Critical Facility now includes electrical equipment such as Electrical Switchgear and UPS systems. Although this equipment is vital to system monitoring, many of today's BMCS contractors do not understand how it operates or what the information that is gathered from systems integration means. This is where detailed system descriptions and graphical requirements become imperative.

In the truest sense the HMI graphics should provide monitoring of all critical equipment and manual software control of all BMCS controlled equipment. The manual override of controlled equipment through the HMI graphics becomes crucial during times of maintenance, testing, or failure of complex system.

A properly installed HMI gives the facility engineering team the ability to trend vital information, diagnose system or equipment problems prior to a failure event, troubleshoot critical events, and implement energy saving strategies. If the HMI is not configured correctly, it loses effectiveness for the end user and its essential functionality is missing.

BMCS SYSTEMS INTEGRATION

Systems Integration powers today's Mission Critical BMCS installations. The very first question is what is Systems Integration? Systems Integration is the means by which disparate systems utilizing diverse protocols are integrated into one common system. What does this mean to today's Mission Critical Facility? Simply put, Systems Integration provides Building Systems a common channel for communicating data. This will increase overall building performance as well as response time to critical events. Systems Integration is characterized by two unique performance features. This section will discuss each of these methods in detail and describe the situations for which each resource is best utilized. The first approach is equipment integration and the second is third party proprietary protocol integration.

Equipment Integration

This form of Systems Integration is becoming a staple in today's BMCS. Mission Critical Facilities are using equipment integration to give facility engineers the ability to gather previously unattainable information from Electrical and Mechanical equipment. The following list details some of the more noteworthy advantages of Equipment Integration:

- Facilities/Maintenance personnel need only learn one computer generated Graphical Interface commonly referred to as an HMI (Human Machine Interface) or an MMI (Man Machine Interface).
- Monitored systems that previously were not viewable over the Internet/Intranet are now web accessible.
- Diverse systems can now work together to curb energy consumption, respond to critical events, and perform tasks as if they were a single entity.
- Potential problems can be identified and diagnosed prior to equipment malfunction or failure, resulting in enhanced systems reliability.
- Cost savings can be realized when the troubleshooting of a system failure no longer requires extensive inspection and testing of multiple systems.

This is a condensed list of benefits, but if this form of integration is used, the advantages to the Mission Critical Facility are vast. The next issue in regards to equipment integration is how it works? Systems Integration is the methodology by which separate systems with either "Proprietary" or "Open" protocols are integrated into a common system. How exactly are these separate communication protocols imbedded in different systems able to communicate and share information? The key concept to understand in regards to separate protocols is the idea of ***Protocol Conversion***. Protocol conversion is the process by which one protocol is translated or converted into another protocol. This is a key component to today's Systems Integration.

Implementation of Systems Integration seems like a valuable tool, but what are the downsides? One of the main problems with Systems Integration is too much information. What this means is that some systems, for instance a PLC controller will have hundreds to thousands of points imbedded in its software. Out of those hundreds or thousands of points, perhaps 50 of those points are useful. The Systems Integrator will no doubt pose the following question to the Design Engineer, "What points do you want?" The common answer to this question is all of them, under the point of view that more information is a good thing. The problem here is which of these points are duplicated in the system in the form of Virtual (Software) Points and how are all of these points to be displayed. Another common problem is that in most cases the Systems Integrator is not familiar with all of the systems that are being integrated.

We ran into these two problems in a large Data Center/High-rise application. There were thousands of points to be integrated. First, integration with the PLC was assimilating hundreds of points into the system, but upon further review, less than 1/3 of these points were actually useful points. The second problem dealt with the Electrical equipment. The BMCS contractor knew in theory what a UPS was, but had no idea what one "looked" like or how it operated. The contractor was integrating around 40 points per UPS module. Furthermore, this system incorporated A&B Multi-module systems with System Control Cabinets. This lack of knowledge increased the project completion time exponentially; graphics and programming were affected in addition to the contractor's extra time to figure out the system. This is just one small example of the problems encountered at this jobsite.

Third Party Proprietary Protocol Integration

This method of Systems Integration is customarily employed in BMCS retrofit projects. Prior to discussing this type of integration it is important to answer the question "What is a proprietary protocol?" Several of today's installed systems are what the industry considers a ***proprietary*** protocol. To understand how a proprietary system is classified, let's look at an example.

- When computers were developed they were considered proprietary. Their proprietary nature was derived from the fact that in order to use a computer you could only use a single manufacturer's hardware and software. Combining one manufacturer's desktop with another manufacturer's monitor was not possible. Today computers are considered "Open" because a computer system can be developed through the use of multiple manufacturers' hardware and software.

- Similarly, when Direct Digital Control (DDC) systems were first developed they were also proprietary. The use of a multi-vendor system was not possible; for instance if your building utilized Andover Controls, all the controllers in the building had to be manufactured by Andover and use Andover's proprietary protocol in order to function. Today, much like computers, DDC systems are migrating toward "Open" protocols.

Until recently, a building was essentially locked into these proprietary systems unless the whole system was ripped out and a new system was installed. This was a costly endeavor and negated all the time and money dedicated to the existing system. Now with integration these once proprietary systems can be converted to an "Open" protocol.

When considering integrating these contrasting systems, you may wonder about the benefits to the End User. The most notable benefits are:

- The "Open" nature of the current DDC systems means a multi-vendor approach can be implemented for future upgrades.

- After the integration, failed proprietary controllers can be replaced with "Open" protocol based controllers from a multitude of vendors.

- Cost savings are realized when a system upgrade doesn't require a complete removal and replacement of the DDC but instead enables a system upgrade via integration.

The next question is "How does this type of integration work?" Integrating a proprietary system into an "Open" system requires that new hardware and software be applied to the system. This new hardware and software connects to the existing system, interrogates the system, captures all of the existing points in the system, and converts them to an "Open" protocol. This conversion then enables the new "Open" system to use the existing information for all graphics and programming.

The integration of an existing proprietary protocol is dependent on the existing system size and architecture. For instance, if the existing system has multiple controllers and VAV controllers then several Integration devices and a new architecture will be required to perform the integration. Compared to the cost of total system replacement this option is more viable and useful. However, it is important to understand that in order to have a truly "Open" architecture in place the existing proprietary controllers will have to be replaced over a period of time. The owner and the owner's budget will dictate this period of time. Thus, the beauty of integration is that the system can migrate to a truly "Open" system when existing controllers fail or money is allocated for replacement.

**Systems Integration—
Equipment Integration Project Example**

This example deals with a new Data Center construction project. In addition to Systems Integration, we will discuss proper equipment specification, submittal review, and system checkout. The new Data Center will house the following equipment, which will be integrated and controlled by the BMCS contractor.

1. Six (6) UPS Modules in an A&B plant configuration
2. Three (3) Chillers
3. Six (6) Variable Frequency Drives
4. Twenty-five (25) Computer Room Air Handling Units (CRAH's)
5. Two (2) Generators
6. Two (2) Automatic Transfer Switches
7. Four (4) Power Quality Monitors
8. A&B Switchgear

In considering this application, we start by reviewing the new equipment to be installed and how to ensure this equipment will be configured to facilitate Systems Integration. To accomplish this goal, the Engineer's equipment specifications must include certain accessories. This statement should be included in the equipment specifications to facilitate Systems Integration capabilities with the BMCS:

> "***Device***" shall be equipped with LonWorks, ModBus, or BACnet communication modules to enable integration with the BMCS. Integration with the "***Device***" shall incorporate read/write capabilities to enable the monitoring and control of the "***Device***" via this interface.

The next step that the Engineer must emphasize is the list of points for the BMCS. This is perhaps the most important part of the BMCS specification, yet it is usually given a quick once-over and is often incomplete or incorrect. When a BMCS contractor puts together a bid, the points list ultimately determines the type of system installed and the total cost for the project. Most projects list the physical hardwire points but rarely include the points required for the integration equipment. By leaving the integrated points list out of the specification the Engineer will either get all points for the systems (which may be far too many) or they may only get a handful of points chosen by the BMCS contractor (which may be far too few). (See Appendix "A" for sample integration point lists.)

After the specifications have been approved and a BMCS contractor has been awarded the job, the next step is the submittal review process. This is the engineer's chance to ensure that all specification requirements have been met. The following list details the key items to review in the submittal process:

- **BMCS Network Riser Diagram**—The network riser diagram if done properly will detail all equipment, which will be part of the BMCS network. This equipment should include all integrated equipment, DDC controllers, and operator workstations. In the above example if the riser diagram did not include one or more of the eight integrated systems then the Engineer would reject the submittal and the BMCS Contractor would have to revise and resubmit it.

- **Integration Hardware**—It is important to verify that the submitted integration device has all of the capabilities that are specified in the engineering specifications.

- **Point Lists**—Verify that the point lists included in the submittal have all specified points for each piece of equipment and integrated device to meet or exceed the design and the sequence of operations.

- **Point-to-Point Wiring Diagrams**—Verify that all physical points are assigned to inputs/outputs on the submitted drawings. This will ensure that the system meets the specified system requirements.

- **Sequences of Operations**—Confirm that the submitted sequences of operations have been written to reflect the engineer's specifications and have been developed to facilitate the system programming.

After the facility engineer approves the proposed system, the job is now officially released to the contractor for project completion. Upon project completion the final step in the engineering process is the control system checkout and commissioning. Due to the complex nature of this type of project a third party system commissioning agent is recommended. Before the commissioning agent validates the BMCS operation, the Engineer and the Controls Contractor should perform Level IV Commissioning. Level IV Commissioning is defined as the Controls Contractor's startup and validation of the system. This includes point-to-point checkout and sequence verification. The commissioning agent will perform Level V Integrated System Commissioning. Level V Commissioning performs complex system failures to determine system fault tolerance. Additional testing includes verification of the BMCS performance with all other building systems. Level IV Commissioning is fairly straightforward and is what is commonly referred to as system checkout. Level V Commissioning, however, is more complex and requires additional explanation.

The subsequent list details the minimum performance tests that should be implemented in Level V Commissioning:

- **Loss of power to control panel power at each standalone controller and integration controller**. This test will verify that the control system as a whole is not jeopardized by the not uncommon event of a loss of control power.

- **Loss of network communication between integration controllers and all integrated equipment.** This test will verify that integrated equipment is not adversely affected by a communication failure.

- **Loss of communication at all standalone controllers.** Similar to the above test, the loss of communication is used to verify that the standalone controller continues to function properly.

- **Initiate complex system failures to test all critical Mechanical Equipment directly controlled by the BMCS.** For example the Chillers in this example are configured in an N+1 configuration; therefore two Chillers are required to operate at all times with one Chiller in standby mode. Therefore one test would be to cut power to one of the Lead Chillers and verify that the standby Chiller immediately began operation.

- **Test all hardwired safety interlocks for proper operation.** For example all Air Handling Units with Supply Air Isolation Dampers should be checked. Two tests should be performed to verify the proper operation of the Supply Air Isolation Damper and the associated hardwire interlock. The first test would be to place the Supply Fan in "hand" at the motor control center. The Supply Air Isolation Damper should immediately spring open and the Supply Fan should only start after the End Switch on the Isolation Damper makes contact.. The second test is to verify that the Supply Fan does not operate when the Supply Air Isolation Damper fails to open. The failure of the Isolation Damper End Switch can be performed in a multitude of ways; they are:

  — Pull the relay cube out of the Isolation Damper End Switch relay base.

  — Remove one of the wires at the Isolation Damper End Switch on the actuator. Be especially careful if this is the chosen method; power will be live on one side of the End Switch Contacts.

  — Note that there should be no control point for the operation of this damper, or any other critical interlocked piece of equipment. If this is the case then the Isolation Damper may not operate correctly when the unit is placed in "hand."

- **Verify BMCS operation during a simulated loss of building power.** At this time UPS, Switchgear, and Generator integration points should be verified for proper operation. Since the BMCS should be fed off of UPS power in this application; it should continue to operate even with the power turned off.

The commissioning process is a very good opportunity for the owner's representatives to learn how the system functions in both normal and adverse applications. During both Level IV and Level V Commissioning, the owner's representatives should be given the ability to command the system, verify **H**uman **M**achine **I**nterface (HMI) operation, understand the points available via the BMCS, and ascertain how to operate the system in emergency situations. Time can be allotted to perform system training directly after the commissioning process while system operation is still fresh in the minds of the owner's representatives.

The Engineer should be available or on-site during the Level V Commissioning process. This will give the Engineer the chance to verify that the system meets the desired specifications. This also provides the Commissioning Agent the opportunity to ask the Engineer any questions about system operation. After the Engineer and owner are satisfied with the BMCS operation and training has been performed, the job is ready for the warranty period.

Full BMCS commissioning will be discussed further in this chapter. The above is a quick overview of what should be expected when commissioning a BMCS system in a Mission Critical Facility.

Systems Integration—Proprietary Protocol Integration Project Example

There are several buildings out there that have an existing proprietary system which they would like to upgrade but unfortunately do not have the money to accomplish this task. Integration of this existing system into an "Open" protocol based system is the answer to this problem. The following example gives an overview of the process involved when looking at migrating to an "Open" system.

The first step in this type of project is a site visit and review of the existing systems architecture and layout. The site visit encompasses the following tasks:

1. Review of existing DDC shop drawing which will determine exactly how the system architecture is laid out.
2. Review of system programming and graphics.
3. Documentation of existing points, which are to be integrated.

The second step is the development of a new system architecture detailing new integration hardware and connections to the existing system. The development of the new architecture includes:

1. Development of an integration strategy to determine the most feasible and cost effective solution.
2. Review of integration options with building owner.
3. New system architecture design.
4. All new hardware and software are ordered.

The third step is the preparation phase of the project. This includes installation of new hardware and development of software. This phase includes the following:

1. Integration panels are built and shipped to the site for installation after "in house" commissioning.
2. Integration programming is developed.
3. New system graphics are developed for the Human Machine Interface (HMI).
4. "In house" commissioning is performed on the system for verification of performance and operation.

The fourth step is the installation phase of construction. This phase includes the installation of the integration control panels, terminations, computer setup, and the loading of programs and graphics. This step includes the following:

1. Panel mounting, terminations, and power-up.
2. Connectivity verification with the existing system.
3. Computer setup and power-up.
4. Uploading of all programs and graphics into the system.
5. Graphics checkout for usefulness and correctness with the owner.

The final step in the project is the commissioning process. The commissioning process will verify that all integrated point values are correct and system sequences are functioning properly. All control system set points need to be verified for correctness and functionality at the HMI.

FAULT TOLERANT BMCS DESIGN

Although this facet of the Mission Critical BMCS is commonly overlooked, it is one of the more critical design features for a Mission Critical installation. Fault tolerant design includes several distinct features, including power distribution, relay configuration, and panel layout. Each of these attributes will be discussed in added detail below.

Power Distribution

In terms of fault tolerance this idea is the easiest to understand yet it is the most ignored concept. Power distribution modeled after fault tolerant ideals is broken up into two main areas, 120-volt power feeds and transformer configuration.

- **120-volt Power Feeds:** When laying out the power distribution to the control panels, large control valves, or application specific controllers, certain design concepts should be employed. Each control panel should have a dedicated power feed and if at all possible power feeds should be segmented between separate UPS-fed power sources. This will ensure that a breaker failure will not immobilize several control panels. Large control valves such as Chiller Isolation Valves should either have a dedicated power feed or be fed from the same power feed that is feeding the control panel; in this example, this would be the panel which has control over the Chiller. The power feed configuration for the valve will be based upon the amp draw that directly correlates with the power consumption of the valve. For application specific controllers, the power feeds need to be configured in a salt and pepper configuration. The salt and pepper configuration will guarantee that a breaker failure will not immobilize a whole control zone.

- **Transformer configuration:** If not properly configured, power distribution via control transformers will immobilize the control system in the event of a power?? failure. To follow fault tolerant design ideals, DDC controllers and control relays need to be fed from separate control transformers. In addition to the separate control transformers, the control panels should also be configured with HOA (Hands Off Auto) switches for each controlled device. This will ensure that if a controller malfunctions, the controlled system can be manually controlled.

Relay Configuration

All control relays should be configured to fail controlled equipment "On" if a controller malfunctions or loses control power. What this means in practical terms is that the control relays need to be wired up to the Normally Closed contacts on the control relay. Then if control power is lost to the relays, this will ensure that the equipment controlled by the relays will fail "On." To make sure that the equipment is failed "On" when a controller malfunctions, the control logic associated with the physical control point on the controller must be inverted. In other words, when the point is powered "On" through the control logic, the controlled equipment is actually "Off."

Panel Layout

When arranging point distribution and panel configuration, these important guidelines must be followed.

- **Position one system per BMCS control panel.** You may be wondering what this means? Here is an example of a correct configuration and an incorrect configuration of the same system:

— **Chilled Water System Example:**

1. The system in this example is a primary/secondary chilled water plant with (4) chillers, (4) primary chilled water pumps, (4) condenser water pumps, (4) cooling towers, and (3) secondary chilled water pumps.

✦ **Correct panel configuration**

1. BMCS Panel #1 – (1) chiller, (1) primary chilled water pump, (1) condenser water pump, (1) cooling tower, and (1) secondary chilled water pump.
2. BMCS Panel #2 – (1) chiller, (1) primary chilled water pump, (1) condenser water pump, (1) cooling tower, and (1) secondary chilled water pump.
3. BMCS Panel #3 – (1) chiller, (1) primary chilled water pump, (1) condenser water pump, (1) cooling tower, and (1) secondary chilled water pump.
4. BMCS Panel #4 – (1) chiller, (1) primary chilled water pump, (1) condenser water pump, and (1) cooling tower.

✦ **Incorrect panel configuration**

1. BMCS Panel #1 – (4) chiller and (4) primary chilled water pumps.
2. BMCS Panel #2 – (4) cooling towers and (4) condenser water pumps.
1. BMCS Panel #3 – (3) secondary chilled water pumps.

The major difference in the two configurations is that the correct configuration ensures if one of the BMCS panels malfunctions or loses power, control of the entire chilled water plant will not be lost. Using the incorrect configuration, if any one of the panels malfunctions or loses power, chilled water plant operation is compromised. As you can see distribution and control of equipment will determine the fault tolerance and redundancy of your complete control system. Note that there are several correct variations for segmenting your control panels; this is one of the most logical and straightforward methods.

BMCS COMMISSIONING FOR MISSION CRITICAL FACILITIES

BMCS Commissioning is broken into two different Levels of commissioning; they are Level IV and Level V commissioning. The practical definition of Level IV commissioning is the contractor's startup and commissioning of the system. This includes point to point checkout and Sequence Verification. Level V commissioning performs complex system failures to determine system fault tolerance. Additional testing includes verifying BMCS functionality with all other systems. The project plan determines whether the commissioning agent performs both Level IV and Level V commissioning or just Level V commissioning.

When considering a commissioning plan for your building a few questions should be asked to properly determine what kind of commissioning is required. Here are some of the basic questions that should be answered before formulating a commissioning plan:

1. What type of facility is the BMCS installed in?
   a. If the facility is a Mission Critical facility like a data center then both Level IV and Level V commissioning should be implemented.
   b. If the facility is a standard installation type then Level IV commissioning should be sufficient.
2. How important is fault tolerant design and reliability to the facility? Based upon this answer a plan can be formulated for how high a degree of commissioning is required.
3. If you have an existing system and are unsure how your system operates or if it is functioning correctly, then commissioning becomes a valuable tool in answering these questions. System type dictates the level of commissioning required.

Commissioning agent qualifications depend on system complexity and familiarity with the systems involved. Before selecting the commissioning agent, careful consideration should be given to a qualification vs. scope ratio.

Below is a sample outline for the steps followed in Level IV commissioning of a BMCS system. Most BMCS contractors do not truly perform Level IV commissioning. The typical BMCS contractor will only verify that points are displaying correctly and the graphics are working. System complexity will dictate how thoroughly the commissioning should delve into the BMCS.

The outline must be tailored to specific project requirements but it provides a good idea of what Level IV commissioning will entail.

On the following page is a sample outline for the Level V commissioning of a BMCS system. Level V commissioning is important when your BMCS is required to

---

I. Review Submittal and Shop drawings (provided by the BMCS contractor) to verify BMCS design conformance with the Engineer's specifications and design documents.

II. During the construction phase, check for proper system installation.
   A. Check for proper wire type
   B. Verify good workmanship for panel construction and installation
   C. Confirm proper end-device installation and validate that the correct device is used for the application.

III. After installation and checkout by the BMCS contractor, perform a point-to-point verification on all hardwire and integration points.
   A. Check point units and values for correctness

IV. Verify Sequence of Operations on all major systems. This is where the system is verified for proper operation.
   A. The Sequence of Operations should be broken down into different tests. For example the following tests can be performed on a Heat Pump Unit.
      1. Heat-Pump Enable Functionality
         a. Validate unit enable functionality in "Auto," "Hand," and Graphical Overrides.
         b. Check Occupied/Unoccupied modes of operation.
      2. Heat-Pump Proof of Operation
         a. Verify the unit status is functioning correctly.
      3. Heat-Pump Hardwire Safety Interlock Verification
         a. Check the proper operation of devices like Freeze Stats, High Pressure Cutout Switches, Smoke Detectors, and Damper Actuator End-switches.
      4. Heat-Pump Cooling/Heating Mode Operation
         a. Validate the Outside Air Temperature Sensor is used correctly in the control logic to switch between cooling and heating modes of operation.
         b. Verify that changes in the Room Temperature Set point affect the control logic correctly and that PID loops are properly tuned.

V. Sign-off all test procedures with the BMCS contractor, Engineer, and Building Owner.

I. Review submittal and shop drawings to validate that the BMCS design employs the proper level of fault tolerance and reliability.
II. If Systems Integration is used on the project then verify that proper integration techniques are used. Validate correct installation and device type prior to system turnover.
III. The ultimate goals of Level V commissioning are confirmation of the BMCS' fault tolerance and validation of BMCS operation with all other building systems.
   A. Sample fault tolerance tests are:
   1. Control Power Failures at critical controllers
   Verify that the loss of power at one controller does not jeopardize system operation
   2. Communication loss failures at critical controllers
   Verify that the loss of communication at one controller does not jeopardize system operation
   3. Perform complex equipment failures
      a. For example, manually fail a primary chilled water pump and verify that the system operates correctly and brings on the lag pump/system.
   4. If the system uses an Ethernet Backbone for communication, validate proper installation and check fault tolerance.
   5. Review the BMCS graphical interface for usability and execution.
      a. Sample Integrated procedures are:
         - Utility Power Failure test
            - ✦ The BMCS is watched to validate proper response and operation during this event.
         - Building operating on Generator
            - ✦ The BMCS is watched to validate proper response and operation during this event.
         - All equipment integrated with the BMCS should be verified during the manufacturer's on-site startup.
         - Chilled Water System equipment malfunctions
         - The BMCS is watched to validate proper response and operation during these events.

operate at all times and needs a high degree of fault tolerance. Level V commissioning will not be performed by the BMCS contractor and requires coordination between all trades in order to be performed correctly.

The above is just a sample of the critical tests that should be performed during Level V commissioning. Actual tests will reflect the BMCS installation and building configuration.

REAL LIFE EXAMPLES OF COMMON PITFALLS

If not properly installed, your BMCS could become an unwanted failure catalyst, which could potentially bring your facility to its knees. The following three examples, taken from actual commissioning documents, describe some potentially hazardous configurations that have been found while performing Level IV and Level V commissioning on multiple vendor systems, both proprietary and open. The examples are discussed and the findings and recommendations for each example should provide valuable information for other facilities.

Example 1: Fuel Monitoring and Control

In this example, fuel is distributed from two Underground Storage Tanks designated UST-1 and UST-2. Underground Storage Tank #3 is an overflow tank, which gathers fuel from the Day Tanks and distributes the fuel to UST-1 and 2 based upon tank fuel levels. Fuel Oil Pumps 1, 2, and 3 (FOP-1, FOP-2, and FOP-3) distribute the fuel from UST-3 to UST-1 and 2. Fuel Oil Pump's 4, 5, 6, and 7 distribute fuel from UST-1 and UST-2 to the Generator Day Tanks. The following findings and recommendations are from the commissioning process for the Fuel Monitoring and Control functionality which is controlled by an Andover control system, a proprietary protocol based system.

Control System Findings

- The Andover Controls wiring diagrams show all digital outputs to be powered from one power supply. Field verification showed this was not true. There were two separate power supplies for the various digital outputs.
- The Andover Controls system currently controls and monitors the Fuel Oil System delivery from

the basement of the facility to the Generator Day Tanks located on the roof of the facility. The system is monitored and controlled by an Andover CX-9400 controller located in the basement of the facility. Override capability of fuel oil pumps and valves is located in the Generator Control Room located on the roof of the facility adjacent to the generators. A Veeder Root Fuel monitoring panel provides the monitoring of the Fuel Oil Levels and alarms. The Andover Control System is integrated with the Veeder Root via an RS-232 integration. The integration of the Fuel Levels is an intimate part of the Andover Control Logic and is utilized for the Fuel Level Balancing functions of the Andover system.

- During the simulation of a loss of control power at the Andover Control panel, the following observations were noted:

  1. Fuel Oil Pumps #1 and #2 are configured to "Fail On" in the event control power is lost or the controller malfunctions. These are the pumps that pull fuel from Underground Storage Tank #3 and distribute fuel into Underground Storage Tank #1 and Underground Storage Tank #2.

  2. Fuel Oil Pump #3 is configured to "Fail Off." This pump is the Standby Pump for Fuel Oil Pump #1 or Fuel Oil Pump #2.

  3. Fuel Oil Pump's #4, #5, #6 and #7 are configured to "Fail Off." These pumps are configured to run in a lead-lag scenario and supply fuel from the Underground Storage Tanks to the Day Tanks.

  4. Underground Storage Tank Valves #1 and #2 and the System Drain Down Valve are not spring return valves. This means that when power is removed from the valve actuators the valves will "Fail in place."

  5. The HOA (Hand-Off-Auto) switches located on the roof of SBC-5 in the Generator Control Room derive their power from the basement control panel. Therefore when control power is lost at the control panel, the HOA switches become immobilized.

Control System Recommendations

- The fuel oil monitoring and control system should be arranged in the following fail-safe configuration.

1. Fuel Oil Pumps #1, #2, and #3 should remain in the current configuration.

2. Two lead pumps out of the set of four Fuel Oil Supply Pumps (Fuel Oil Pumps #4, #5, #6 and #7) should be reconfigured to "Fail On" if control power is lost or the system controller loses its memory or malfunctions. This will ensure that fuel will be delivered if the power fails and the Generators are started. The pumps will remain "On" until the problem is diagnosed and corrected. This feature will make sure that fuel oil supply is not lost if the control system fails.

3. The Underground Storage Tank Valves and the Drain Down Valve should be replaced with spring return type actuators. The Underground Storage Tank Valves should "Fail Open" and allow fuel to pump to the Underground Storage Tanks. The Drain Down Valve should "Fail Closed" thus ensuring Fuel Oil is delivered to the Generator Day Tanks.

4. The HOA control panel located in the Generator Control Room should be powered by a source separate from the CX-9400 control panel. This gives the user the true ability to override the Fuel Oil Pumps and Valves. This configuration can be accomplished by using control relays and a separate power supply fed from a source other than the CX-9400. However under the current configuration, the HOA's at the Motor Control Centers located in the basement of the facility are fully independent of the control system and can still be operated if the control system fails or loses power.

Example 2: Existing Control System Problems

The Mission Critical Facility has an existing Johnson Controls DSC-8500 Control System that was originally installed in 1985. This system was not commissioned due to its age. The commissioning agent did not want to take the risk of failing the system with adverse side effects. The DSC-8500 system is integrated into the new Johnson Controls Metasys System through the use of a separate Network Control Module. The age, configuration, and integration method contribute to the overall systems reliability and lack of fault tolerance.

Control System Findings:

Because of its age, the existing system should be replaced to ensure system reliability. Two reasons for this conclusion are: (1) replacement parts are not readily available, and (2) the current configuration is full of single points of failure.

Control System Recommendations

- The existing Johnson Controls DSC-8500 Control System should be upgraded to a newer system. This upgrade can be accomplished in several ways:

    1. Upgrade the existing system to a current version of the Johnson Controls Metasys system. Utilization of the DX-9100controller is recommended since this controller is currently installed on-site.
    2. Seek competitive bids on a complete Control System upgrade from any one of several vendors such as:
        a. Johnson Controls Inc. (Metasys – Proprietary)
        b. Invensys (LonWorks – Open)
        c. Staefa (LonWorks – Open)
        d. Andover (Continuum – Proprietary)
        e. Automated Logic (BACnet – Open)
        f. Alerton (BACnet – Open)
        g. Richards Zeta (BACnet or LonWorks – Open)
        h. Distech (LonWorks – Open)

    3. To fully maximize the newly installed Johnson Controls Metasys system the following strategy can be utilized. Replace the existing DSC-8500 Control System with a complete new system and integrate it with the new Metasys Control System. This option protects the owner's investment in the current project while still facilitating a competitive bid environment on the existing system replacement.

- If either option 2 or option 3 is chosen then the facility should consider using one of the "Open" protocols. The best option is option 3. This option will protect the capital investment in the current control system. The integration of the recently installed Johnson Controls equipment for the Chilled Water Plant now becomes part of the new system and can be systematically removed over a period of time based upon money allocated to control system upgrades.

Example 3 Human Machine Interface Issues

The system tested during the commissioning process was a Trane Tracer Summit control system. The subsequent findings and recommendations deal with problems involving the HMI and the Chilled Water Plant. The Human Machine Interface (HMI) is the end user's gateway into the control system. The HMI gives the end user the ability to override system controls, change set points, and view individual control systems such as Chillers, Air Handling Units, and Fan Terminal Units. Not all of this functionality is readily available at the existing HMI. The following breakdowns will detail the findings and recommendations for the HMI.

Control System Findings

- The following findings are designed to enhance the facility engineers' operation of the systems controlled via the BMCS. These enhancements are specifically critical to this facility since facility personnel will not be on-site 24/7.

    1. The HMI cannot be used to add or subtract a Chilled Water System. In order to add or subtract a Chilled Water System, the end user must leave the HMI graphics package and enter the operating system software.

    2. The Lead/Lag of the Chilled Water Systems cannot be changed by the HMI. In order to change the Lead/Lag of the Chilled Water System the end user must leave the graphics package and enter the operating system software.

    3. Primary Chilled Water Pump runtimes are not shown as requested by the specifications. The Primary Chilled Water Pump runtime should match the runtime of the Chiller to which the pump is dedicated.

    4. The HMI cannot add or subtract a Secondary Chilled Water Pump.

    5. Secondary Chilled Water Pump runtimes are not shown at the HMI as requested by the specifications.

    6. Equipment failures are not easily detected by using the HMI. Currently an alarm is generated if a chiller fails. However it is not easily discerned whether that failure is due to a pump failure or a chiller failure.

Control System Recommendations

1. A graphical Manual/Auto button should be added. When the button is in the "Auto" position the Chilled Water Systems will function under automatic control. When the button is in the "Manual" position, the end user will then be able to either add or subtract both Chilled Water Systems individually at the HMI. This type of functionality is extremely beneficial when trouble shooting the Chilled Water Systems or the control system.

2. In addition to the graphical Manual/Auto button, the following functionality is recommended. When the button is in the "Auto" position, the control system should control the Lead/Lag of the Chilled Water System based upon a predetermined schedule. When the button is in the "Manual" position the end user should have the ability to decide whether Chilled Water System #1 or Chilled Water System #2 is the Lead system.

3. If the Pump Runtime is not added to the graphics screens, it is recommended that language be added to the graphics screens showing that the pump runtime corresponds with the Chiller runtime.

4. A graphical Manual/Auto button should be added. When the button is in the "Auto" position the Secondary Chilled Water Pumps will function under automatic control. When the button is in the "Manual" position, the end user will then be able to either add or subtract both Secondary Chilled Water Pumps individually at the HMI. This type of functionality is extremely beneficial when trouble shooting the Secondary Chilled Water System or the control system.

5. Runtimes are helpful when determining proper pump usage and runtime distribution.

6. The ability to quickly determine where a failure occurred in a critical system of this nature is imperative. It is recommended that a color scheme be implemented at the HMI. The following is a suggestion:

   - Green outline around equipment that is running.
   - Red outline around equipment that has failed.
   - Flashing Yellow outline around equipment that is manually overridden in software.
   - Black outline around equipment that is off.

The above examples are just a small sample of common problems with BMCS installations in Mission Critical Facilities.

CONCLUSION

Web based Monitoring and Control systems in Mission Critical Facilities demand a high level of sophistication during the pre-design and design phases. Traditionally Mission Critical BMCS systems have NO redundancy and utilize NO fault tolerant design methods. In addition present day Mission Critical Facilities have evolved into a complex mesh of network equipment, mechanical and electrical systems, building controls, and site security and access. These multiple systems are required to interact and work together at all times.

Without a properly installed, web-based BMCS system, critical system information is only accessible locally at the equipment or through vendor supplied computers. The interaction of all systems is essential for reliable operation and control. When designing for a Mission Critical application, it is important to specify a first rate BMCS to ensure proper building monitoring, alarming, and control.

Bibliography

Integrated Building Solutions, Inc., *www.integrated-buildings.com*

Uptime Institute, *www.upsite.com*

Computer Site Engineering, *www.computersiteengineering.com*

APPENDIX A

The following includes sample point lists for typical systems found in Mission Critical Facilities. These point lists are not designed for generic reuse. The point lists give the reader a sample format and an idea of what points can be derived from integration.

Note: All register Numbers and point mapping addresses for integrated points are for example use only.

All values need to be correlated with each specific manufacturer's requirements and recommendations.

Point List Legend:
DI – Digital Input physical point
DO – Digital Output physical point
AI – Analog Input physical point
AO – Analog Output physical point
CI – Communication Interface integration point

Sample point list #1 Power Quality Meters

BMCS POINTS SCHEDULE - Power Quality Meters							PQM's
GENERAL				ENGINEERING			
Point No.	Register No.	Point Name	Point Decription	Point Type	Eng. Units	Device Range	Device Type
1	1001	FREQ-HZ-SI	System Frequency - .01 Hertz/Scale	CI	Hz	45-66 Hz	Modbus Interface
2	1003	CURR-A-SI	System Current Phase A - Amps/Scale	CI	Amps	0-32,767 Amps	Modbus Interface
3	1004	CURR-B-SI	System Current Phase B - Amps/Scale	CI	Amps	0-32,767 Amps	Modbus Interface
4	1005	CURR-C-SI	System Current Phase C - Amps/Scale	CI	Amps	0-32,767 Amps	Modbus Interface
5	1006	CURR-NEU-SI	System Neutral Current - Amps/Scale	CI	Amps	0-32,767 Amps	Modbus Interface
6	1014	VOLT-AB-SI	System Voltage Phase A to B - Volts/Scale	CI	Volts	0-32,767 Volts	Modbus Interface
7	1015	VOLT-BC-SI	System Voltage Phase B to C- Volts/Scale	CI	Volts	0-32,767 Volts	Modbus Interface
8	1016	VOLT-CA-SI	System Voltage Phase C to A- Volts/Scale	CI	Volts	0-32,767 Volts	Modbus Interface
9	1018	VOLT-AN-SI	System Voltage Phase A to N- Volts/Scale	CI	Volts	0-32,767 Volts	Modbus Interface
10	1019	VOLT-BN-SI	System Voltage Phase B to N- Volts/Scale	CI	Volts	0-32,767 Volts	Modbus Interface
11	1020	VOLT-CN-SI	System Voltage Phase C to N- Volts/Scale	CI	Volts	0-32,767 Volts	Modbus Interface
12	1031	TRPF-A-SI	True Power Factor Phase A-in 1000ths	CI	%	-100 to 1000 to 100	Modbus Interface
13	1032	TRPF-B-SI	True Power Factor Phase B-in 1000ths	CI	%	-100 to 1000 to 100	Modbus Interface
14	1033	TRPF-C-SI	True Power Factor Phase C-in 1000ths	CI	%	-100 to 1000 to 100	Modbus Interface
15	1034	TRPF-3PH-SI	True Power Factor 3 Phases-in 1000ths	CI	%	-100 to 1000 to 100	Modbus Interface
16	1039	RLPWR-A-SI	Real Power Phase A - kW/Scale	CI	kW	0-(+-32,767) kW	Modbus Interface
17	1040	RLPWR-B-SI	Real Power Phase B - kW/Scale	CI	kW	0-(+-32,767) kW	Modbus Interface
18	1041	RLPWR-C-SI	Real Power Phase C - kW/Scale	CI	kW	0-(+-32,767) kW	Modbus Interface
19	1042	RLPWR-3PH-SI	Real Power 3 Phase Total - kW/Scale	CI	kW	0-(+-32,767) kW	Modbus Interface
20	1047	APPWR-A-SI	Apparent Power Phase A - kVA/Scale	CI	kVA	0-32,767 kVA	Modbus Interface
21	1048	APPWR-B-SI	Apparent Power Phase B - kVA/Scale	CI	kVA	0-32,767 kVA	Modbus Interface
22	1049	APPWR-C-SI	Apparent Power Phase C - kVA/Scale	CI	kVA	0-32,767 kVA	Modbus Interface
23	1050	APPWR-3PH-SI	Apparent Power 3 Phase Total - kVA/Scale	CI	kVA	0-32,767 kVA	Modbus Interface
24	1051	THDCURR-A-SI	THD/thd Phase A Current - % in 10ths	CI	%	0-10000 thd	Modbus Interface
25	1052	THDCURR-A-SI	THD/thd Phase A Current - % in 10ths	CI	%	0-10000 thd	Modbus Interface
26	1053	THDCURR-A-SI	THD/thd Phase A Current - % in 10ths	CI	%	0-10000 thd	Modbus Interface
27	1055	THDVOLT-A-SI	THD/thd Phase A Voltage - % in 10ths	CI	%	0-10000 thd	Modbus Interface
28	1056	THDVOLT-B-SI	THD/thd Phase B Voltage - % in 10ths	CI	%	0-10000 thd	Modbus Interface
29	1057	THDVOLT-C-SI	THD/thd Phase C Voltage - % in 10ths	CI	%	0-10000 thd	Modbus Interface

Sample point list #2 Computer Room Air Conditioning Unit (CRAC):

BMCS POINTS SCHEDULE - CRAC Units							CRAC's
GENERAL				ENGINEERING			
Point No.	Register No.	Point Name	Point Decription	Point Type	Eng. Units	Device Range	Device Type
1	40001	CRAC-TEMP-SI	CRAC Unit Temperature	CI	Deg F	NA	Modbus Interface
2	40002	CRAC-HUM-SI	CRAC Unit Humidity	CI	%RH	NA	Modbus Interface
3	40003	COOL-MODE-SI	CRAC Unit in Cooling Mode	CI	On/Off	NA	Modbus Interface
4	40004	HEAT-MODE-SI	CRAC Unit in Heating Mode	CI	On/Off	NA	Modbus Interface
5	40005	HUM-MODE-SI	CRAC Unit in Humidification Mode	CI	On/Off	NA	Modbus Interface
6	40006	DEHUM-MODE-SI	CRAC Unit in De-Humidification Mode	CI	On/Off	NA	Modbus Interface
7	40008	CRAC-STAGES-SI	CRAC Unit Stages On	CI	On/Off	NA	Modbus Interface
8	40009	CRAC-CAP-SI	CRAC Unit Capacity Percentage	CI	%	NA	Modbus Interface
9	40289:0	COMM-FBK-SI	CRAC Unit Communications Alarm	CI	On/Off	NA	Modbus Interface
11	40289:1	HEAD-H1-SI	High Head Pressure 1	CI	On/Off	NA	Modbus Interface
12	40289:2	HEAD-H2-SI	High Head Pressure 2	CI	On/Off	NA	Modbus Interface
13	40289:3	CRAC-AFLOW-SI	CRAC Unit Loss of Airflow	CI	On/Off	NA	Modbus Interface
14	40289:5	CRAC-LD-SI	CRAC Unit Liquid Detected	CI	On/Off	NA	Modbus Interface
15	40289:6	CRAC-FILT-SI	CRAC Unit Filter Change notification	CI	On/Off	NA	Modbus Interface
17	40290:0	HUM-ALM-SI	CRAC Unit Humidifier Problem	CI	On/Off	NA	Modbus Interface
18	40290:1	HUMP-WAT-SI	No water in humidifier pan	CI	On/Off	NA	Modbus Interface
19	40290:2	COMP1-OVL-SI	Compressor #1 Overload	CI	On/Off	NA	Modbus Interface
20	40290:3	COMP2-OVL-SI	Compressor #2 Overload	CI	On/Off	NA	Modbus Interface
21	40290:4	FAN-OVL-SI	Main Fan Overload	CI	On/Off	NA	Modbus Interface
22	40290:5	MAN-OVR-SI	CRAC Unit Manual override	CI	On/Off	NA	Modbus Interface
23	40290:6	SMK-DET-SI	CRAC Unit Smoke Detected	CI	On/Off	NA	Modbus Interface
24	40290:9	LOW-SUC-SI	CRAC Unit Low Suction Pressure	CI	On/Off	NA	Modbus Interface
25	40291:0	LOSS-PWR-SI	CRAC Unit Loss of Power	CI	On/Off	NA	Modbus Interface
26	40291:2	STNDBY-FAN-SI	CRAC Unit Standby Fan enabled	CI	On/Off	NA	Modbus Interface
27	40291:3	LOSS-EPWR-SI	CRAC Unit Loss of Emergency Power	CI	On/Off	NA	Modbus Interface
28	40291:4	LOCAL-ALM1-SI	CRAC Unit Local Alarm #1 Present	CI	On/Off	NA	Modbus Interface
29	40291:5	LOCAL-ALM2-SI	CRAC Unit Local Alarm #2 Present	CI	On/Off	NA	Modbus Interface
30	40010	TSSETP-VIEW-SI	Temperature Setpoint View only	CI	Deg F	NA	Modbus Interface
31	40011	TTOLER-VIEW-SI	Temperature Tolerance View only	CI	Deg F	NA	Modbus Interface
32	40012	HSSETP-VIEW-SI	Humidity Setpoint View only	CI	%RH	NA	Modbus Interface
33	40013	HTOLER-VIEW-SI	Humidity Tolerance View only	CI	%RH	NA	Modbus Interface
34	40014	TSSETP-MOD-SI	Temperature Setpoint Modifiable	CI	Deg F	NA	Modbus Interface
35	40015	TTOLER-MOD-SI	Temperature Tolerance Modifiable	CI	Deg F	NA	Modbus Interface
36	40016	HSSETP-MOD-SI	Humidity Setpoint Modifiable	CI	%RH	NA	Modbus Interface
37	40017	HTOLER-MOD-SI	Humidity Tolerance Modifiable	CI	%RH	NA	Modbus Interface

Sample point list #3 Automatic Transfer Switch (ATS):

BMCS POINTS SCHEDULE - AUTOMATIC TRANSFER SWITCH								ATS
GENERAL					ENGINEERING			
Point No.	Register No.	Coil No.	Point Name	Point Description	Point Type	Eng. Units	Device Range/Information	Device Type
1	40001	1	ATS-RELAY-SI	Automatic Transfer Relay	CI	On/Off	1=ON, 0=OFF	Modbus Integration
2	40001	2	ATS-HAND-SI	ATS not in "Auto"	CI	On/Off	1=NOT IN AUTO	Modbus Integration
3	40001	3	ATS-FAULT-SI	ATS Fault Present	CI	On/Off	1=FAULT	Modbus Integration
4	40001	7	ATS-EMS-SI	Emergency Source Available at ATS	CI	On/Off	1=AVAILABLE	Modbus Integration
5	40001	8	ATS-NORM-SI	Normal Source Available at ATS	CI	On/Off	1=AVAILABLE	Modbus Integration
6	40004	34	ATS-NPOS-SI	ATS Normal Position Status	CI	On/Off	1=NORMAL POS	Modbus Integration
7	40004	35	ATS-EPOS-SI	ATS Emergency Position Status	CI	On/Off	1=EMERG POS	Modbus Integration
8	40004	36	ATS-COMM	ATS Modbus Card Comm Error	CI	On/Off	1=COMM ERROR	Modbus Integration
9	40006	NA	ATS-NVAB-SI	ATS Normal Pos Voltage Phase A to B	CI	Volts	NA	Modbus Integration
11	40007	NA	ATS-NVBC-SI	ATS Normal Pos Voltage Phase B to C	CI	Volts	NA	Modbus Integration
12	40008	NA	ATS-NVCA-SI	ATS Normal Pos Voltage Phase C to A	CI	Volts	NA	Modbus Integration
13	40009	NA	ATS-EVAB-SI	ATS Emergency Pos Voltage Phase A to B	CI	Volts	NA	Modbus Integration
14	40010	NA	ATS-EVBC-SI	ATS Emergency Pos Voltage Phase B to C	CI	Volts	NA	Modbus Integration
15	40011	NA	ATS-EVCA-SI	ATS Emergency Pos Voltage Phase C to A	CI	Volts	NA	Modbus Integration
17	40017	NA	ATS-NTRNS-SI	ATS Number of Transfers	CI	Number	NA	Modbus Integration

Sample point list #4 Standby Engine Generator (GEN):

BMCS POINTS SCHEDULE - Generator's							GEN's
GENERAL				ENGINEERING			
Point No.	Register No.	Point Name	Point Decription	Point Type	Eng. Units	Device Range	Device Type
1	40001	GEN-FBK-SI	Generator Status	CI	On/Off	NA	CCM Integration - Modbus
2	40002	GEN-HCTA-SI	Generator High Coolant Temp Alarm	CI	On/Off	NA	CCM Integration - Modbus
3	40003	GEN-HCTS-SI	Generator High Coolant Temp Shutdown	CI	On/Off	NA	CCM Integration - Modbus
4	40004	GEN-LCTA-SI	Generator Low Coolant Temp Alarm	CI	On/Off	NA	CCM Integration - Modbus
5	40005	GEN-LOPA-SI	Generator Low Oil Pressur Alarm	CI	On/Off	NA	CCM Integration - Modbus
6	40006	GEN-LOPS-SI	Generator Low Oil Pressur Shutdown	CI	On/Off	NA	CCM Integration - Modbus
7	40007	GEN-OVRS-SI	Generator Overspeed Shutdown	CI	On/Off	NA	CCM Integration - Modbus
8	40008	GEN-OVRC-SI	Generator Overcrank Shutdown	CI	On/Off	NA	CCM Integration - Modbus
9	40009	GEN-HAND-SI	Generator control switch not in "Auto"	CI	On/Off	NA	CCM Integration - Modbus
10	40010	GEN-EMERG-SI	Generator Emergency Shutdown	CI	On/Off	NA	CCM Integration - Modbus
11	40011	GEN-FRELY-SI	Generator Fault Relay	CI	On/Off	NA	CCM Integration - Modbus
12	40012	GEN-FRQ-SI	Generator Frequency	CI	HZ	0-65 HZ	CCM Integration - Modbus
13	40013	GEN-VAB-SI	Generator Voltage Phase A to B	CI	Volts	0-480 Volts	CCM Integration - Modbus
14	40014	GEN-CA-SI	Generator Current Phase A	CI	Amps	0-1000 Amps	CCM Integration - Modbus
15	40015	GEN-PWFA-SI	Generator Power Factor Phase A	CI	PF	0-1	CCM Integration - Modbus
16	40016	GEN-VBC-SI	Generator Voltage Phase B to C	CI	Volts	0-480 Volts	CCM Integration - Modbus
17	40017	GEN-CB-SI	Generator Current Phase B	CI	Amps	0-1000 Amps	CCM Integration - Modbus
18	40018	GEN-PWFB-SI	Generator Power Factor Phase B	CI	PF	0-1	CCM Integration - Modbus
19	40019	GEN-VCA-SI	Generator Voltage Phase C to A	CI	Volts	0-480 Volts	CCM Integration - Modbus
20	40020	GEN-CA-SI	Generator Current Phase C	CI	Amps	0-1000 Amps	CCM Integration - Modbus
21	40021	GEN-PWFA-SI	Generator Power Factor Phase C	CI	PF	0-1	CCM Integration - Modbus
22	40022	GEN-KWH-SI	Generator Total kWH	CI	kWH	NA	CCM Integration - Modbus
23	40023	GEN-RMSC-SI	Generator Total RMS Current	CI	Amps	0-2500	CCM Integration - Modbus
24	40024	GEN-PWR-SI	Generator % Rated Power	CI	%	0-100%	CCM Integration - Modbus
25	40025	GEN-RPM-SI	Generator RPM's	CI	RPM	0-1800 RPM	CCM Integration - Modbus
26	40026	GEN-CTEMP-SI	Generator Coolant Temperature	CI	Deg F	0-150 Deg F	CCM Integration - Modbus
27	40027	GEN-OILP-SI	Generator Oil Pressure	CI	PSI	NA	CCM Integration - Modbus
28	40028	GEN-HRUN-SI	Generator Hour Meter	CI	Hrs	NA	CCM Integration - Modbus
29	40029	GEN-BATTV-SI	Generator Battery Voltage	CI	Volts	0-30 VDC	CCM Integration - Modbus

Sample point list #5 Uninterruptible Power Supply (UPS):

BMCS POINTS SCHEDULE - UPS							UPS
GENERAL				ENGINEERING			
Point No.	Register No.	Point Name	Point Decription	Point Type	Eng. Units	Device Range	Device Type
1	30002	UPS-DFBK-SI	UPS Device Status	CI	NA	NA	I-Manager Modbus
2	30003	UPS-COMM-SI	UPS Comm Status	CI	NA	NA	I-Manager Modbus
3	30004	UPS-BPWRF-SI	UPS Power Failure (On Battery)	CI	NA	00 = True 01 = False	I-Manager Modbus
4	30006	UPS-BDISC-SI	UPS Battery Discharge	CI	NA	00 = True 01 = False	I-Manager Modbus
5	30007	UPS-LOWB-SI	UPS Low Battery Voltage	CI	NA	00 = True 01 = False	I-Manager Modbus
6	30008	UPS-HIGHB-SI	UPS High Battery Voltage	CI	NA	00 = True 01 = False	I-Manager Modbus
7	30009	UPS-BFAIL-SI	UPS Battery Failure	CI	NA	00 = True 01 = False	I-Manager Modbus
8	30010	UPS-BCHF-SI	UPS Battery Charger Failure	CI	NA	00 = True 01 = False	I-Manager Modbus
9	30012	UPS-BTEMPH-SI	UPS Battery Temp High	CI	NA	00 = True 01 = False	I-Manager Modbus
11	30019	UPS-BMFAN-SI	UPS Base Module Fan Failure	CI	NA	00 = True 01 = False	I-Manager Modbus
12	30020	UPS-SYFAN-SI	UPS System Mofule Fan Failure	CI	NA	00 = True 01 = False	I-Manager Modbus
13	30021	UPS-PMFAIL-SI	UPS Power Module Failure	CI	NA	00 = True 01 = False	I-Manager Modbus
14	30023	UPS-FBYP-SI	UPS In Forced Bypass	CI	NA	00 = True 01 = False	I-Manager Modbus
15	30024	UPS-SBYP-SI	UPS In Software Bypass	CI	NA	00 = True 01 = False	I-Manager Modbus
17	30025	UPS-HFBYP-SI	UPS Hardware Failure Bypass	CI	NA	00 = True 01 = False	I-Manager Modbus
18	30026	UPS-OVBYP-SI	UPS Overload Bypass	CI	NA	00 = True 01 = False	I-Manager Modbus
19	30027	UPS-SWBYP-SI	UPS Switch Bypass	CI	NA	00 = True 01 = False	I-Manager Modbus
20	30028	UPS-SHDWN-SI	UPS Shutdown from Bypass	CI	NA	00 = True 01 = False	I-Manager Modbus
21	30034	UPS-BFREL-SI	UPS Backfeed Relay Open	CI	NA	00 = True 01 = False	I-Manager Modbus
22	30036	UPS-EDCDIS-SI	UPS DC Disconnect Switch Open	CI	NA	00 = True 01 = False	I-Manager Modbus
23	30037	UPS-ICBRK-SI	UPS Input Circuit Breaker Open	CI	NA	00 = True 01 = False	I-Manager Modbus
24	30038	UPS-SYNC-SI	UPS Not Synchronized Fault	CI	NA	00 = True 01 = False	I-Manager Modbus
25	30047	UPS-REDL-SI	UPS Redundancy Lost	CI	NA	00 = True 01 = False	I-Manager Modbus
26	30070	UPS-NOBAT-SI	UPS Number of Batteries	CI	NA	1-99	I-Manager Modbus
27	30071	UPS-NOBADB-SI	UPS Number of Bad Batteries	CI	NA	1-99	I-Manager Modbus
28	30074	UPS-BATCA-SI	UPS Battery Capacity	CI	NA	0-1000 divide by 10 to yield 0-100	I-Manager Modbus
29	30076	UPS-NBTV-SI	UPS Nominal Battery Voltage	CI	NA	0-9999 divide by 10 to yield 0-999.9	I-Manager Modbus
30	30077	UPS-ABTV-SI	UPS Actual Battery Voltage	CI	NA	0-9999 divide by 10 to yield 0-999.9	I-Manager Modbus
31	30078	UPS-BCURR-SI	UPS Battery Current	CI	NA	0-9999 divide by 10 to yield 0-999.9	I-Manager Modbus
32	30085	UPS-IFRQ-SI	UPS Input Frequency	CI	HZ	0-9999 divide by 10 to yield 0-999.9	I-Manager Modbus
33	30090	UPS-PWMC-SI	UPS Power Module Count	CI	NA	0-5	I-Manager Modbus
34	30091	UPS-REDUN-SI	UPS Current Redundancy	CI	NA	0-4	I-Manager Modbus
35	30092	UPS-CLDCP-SI	UPS Current Load Capacity	CI	kVA	0-999 divide by 10 to yield 0-99.9	I-Manager Modbus
36	30097	UPS-IVA-SI	UPS Input Voltage Phase A	CI	Volts	0-9999 divide by 10 to yield 0-999.9	I-Manager Modbus
37	30098	UPS-ICA-SI	UPS Input Current Phase A	CI	Amps	0-9999 divide by 10 to yield 0-999.9	I-Manager Modbus
38	30099	UPS-IVB-SI	UPS Input Voltage Phase B	CI	Volts	0-9999 divide by 10 to yield 0-999.9	I-Manager Modbus
39	30100	UPS-ICB-SI	UPS Input Current Phase B	CI	Amps	0-9999 divide by 10 to yield 0-999.9	I-Manager Modbus
40	30101	UPS-IVC-SI	UPS Input Voltage Phase C	CI	Volts	0-9999 divide by 10 to yield 0-999.9	I-Manager Modbus
41	30102	UPS-ICC-SI	UPS Input Current Phase C	CI	Amps	0-9999 divide by 10 to yield 0-999.9	I-Manager Modbus
42	Verify	UPS-OVA-SI	UPS Output Voltage Phase A	CI	Volts	0-9999 divide by 10 to yield 0-999.9	I-Manager Modbus
43	Verify	UPS-OCA-SI	UPS Output Current Phase A	CI	Amps	0-9999 divide by 10 to yield 0-999.9	I-Manager Modbus
44	Verify	UPS-OMCA-SI	UPS Output Mac Current Phase A	CI	Amps	0-9999 divide by 10 to yield 0-999.9	I-Manager Modbus
45	Verify	UPS-OLA-SI	UPS Output Load Phase A	CI	VA	0-9999 divide by 10 to yield 0-999.9	I-Manager Modbus
46	Verify	UPS-OPLA-SI	UPS Output Percent Load Phase A	CI	%	0-100	I-Manager Modbus
47	Verify	UPS-OPPA-SI	UPS Output Percent Power Phase A	CI	%	0-100	I-Manager Modbus
48	Verify	UPS-OVA-SI	UPS Output Voltage Phase A	CI	Volts	0-9999 divide by 10 to yield 0-999.9	I-Manager Modbus
49	Verify	UPS-OCA-SI	UPS Output Current Phase A	CI	Amps	0-9999 divide by 10 to yield 0-999.9	I-Manager Modbus
50	Verify	UPS-OMCA-SI	UPS Output Mac Current Phase A	CI	Amps	0-9999 divide by 10 to yield 0-999.9	I-Manager Modbus
51	Verify	UPS-OLA-SI	UPS Output Load Phase A	CI	VA	0-9999 divide by 10 to yield 0-999.9	I-Manager Modbus
52	Verify	UPS-OPLA-SI	UPS Output Percent Load Phase A	CI	%	0-100	I-Manager Modbus
53	Verify	UPS-OPPA-SI	UPS Output Percent Power Phase A	CI	%	0-100	I-Manager Modbus
54	Verify	UPS-OVA-SI	UPS Output Voltage Phase A	CI	Volts	0-9999 divide by 10 to yield 0-999.9	I-Manager Modbus
55	Verify	UPS-OCA-SI	UPS Output Current Phase A	CI	Amps	0-9999 divide by 10 to yield 0-999.9	I-Manager Modbus
56	Verify	UPS-OMCA-SI	UPS Output Mac Current Phase A	CI	Amps	0-9999 divide by 10 to yield 0-999.9	I-Manager Modbus
57	Verify	UPS-OLA-SI	UPS Output Load Phase A	CI	VA	0-9999 divide by 10 to yield 0-999.9	I-Manager Modbus
58	Verify	UPS-OPLA-SI	UPS Output Percent Load Phase A	CI	%	0-100	I-Manager Modbus
59	Verify	UPS-OPPA-SI	UPS Output Percent Power Phase A	CI	%	0-100	I-Manager Modbus
60	Verify	UPS-PM1FBK-SI	UPS Power Module Number 1 Status	CI	NA	NA	I-Manager Modbus
61	Verify	UPS-PM2FBK-SI	UPS Power Module Number 2 Status	CI	NA	NA	I-Manager Modbus
62	Verify	UPS-PM3FBK-SI	UPS Power Module Number 3 Status	CI	NA	NA	I-Manager Modbus

Sample point list #6 Physical Control Points (DDC):

BMCS POINTS SCHEDULE - Pysical Control Points DDC Panel							DDC POINTS
GENERAL				ENGINEERING			
Point No.	System Served	Point Name	Point Decription	Point Type	Eng. Units	Device Range	Device Type
1	GEN	GEN-HCTA-HW	Generator High Coolant Temp Alarm	DI	On/Off	NA	CIM Hardwire Interface
2	GEN	GEN-HCTS-HW	Generator High Coolant Temp Shutdown	DI	On/Off	NA	CIM Hardwire Interface
3	GEN	GEN-LCTA-HW	Generator Low Coolant Temp Alarm	DI	On/Off	NA	CIM Hardwire Interface
4	GEN	GEN-LOPA-HW	Generator Low Oil Pressur Alarm	DI	On/Off	NA	CIM Hardwire Interface
5	GEN	GEN-LOPS-HW	Generator Low Oil Pressur Shutdown	DI	On/Off	NA	CIM Hardwire Interface
6	GEN	GEN-OVRS-HW	Generator Overspeed Shutdown	DI	On/Off	NA	CIM Hardwire Interface
7	GEN	GEN-OVRC-HW	Generator Overcrank Shutdown	DI	On/Off	NA	CIM Hardwire Interface
8	GEN	GEN-HAND-HW	Generator control switch not in "Auto"	DI	On/Off	NA	CIM Hardwire Interface
9	GEN	GEN-CFALT-HW	Generator Diagnostic Fault	DI	On/Off	NA	CIM Hardwire Interface
10	GEN	GEN-BCFAIL-HW	Generator Battery Charger Failure	DI	On/Off	NA	Dry Contacts at Batt. Charger
11	GEN	GEN-BLK-HW	Generator Fuel Basin Leak	DI	On/Off	NA	Dry Contacts at Basin Leak Sensor
12	GEN	GEN-LFL-HW	Generator Fuel Low Level	DI	On/Off	NA	Dry Contacts at Low Level Sensor
13	GEN	GEN-CBRK-HW	Generator Circuit Breaker Closed	DI	On/Off	NA	Aux Contacts at Generator CB
14	GLYCOL	GLY-PUMP1-HW	Glycol System Status Pump #1	DI	On/Off	NA	Current Sensor
15	GLYCOL	GLY-PUMP2-HW	Glycol System Status Pump #2	DI	On/Off	NA	Current Sensor
16	GLYCOL	GLY-FAN1-HW	Glycol System Status Dry Cooler Fan #1	DI	On/Off	NA	Current Sensor
17	GLYCOL	GLY-FAN2-HW	Glycol System Status Dry Cooler Fan #2	DI	On/Off	NA	Current Sensor
18	BLDG	ATS1-NORMF-HW	ATS-1 Normal Power Failure Indication	DI	On/Off	NA	A1-Auxillary Contact
19	BLDG	ATS1-EMERF-HW	ATS-1 Emergency Power Failure Indication	DI	On/Off	NA	A1E-Auxillary Contact
20	BLDG	ATS1-EMERP-HW	ATS-1 in the Emergency Position	DI	On/Off	NA	A3-Auxillary Contact
21	BLDG	ATS1-NORMP-HW	ATS-1 in the Normal Position	DI	On/Off	NA	A4-Auxillary Contact
22	BLDG	ATS2-NORMF-HW	ATS-2 Normal Power Failure Indication	DI	On/Off	NA	A1-Auxillary Contact
23	BLDG	ATS2-EMERF-HW	ATS-2 Emergency Power Failure Indication	DI	On/Off	NA	A1E-Auxillary Contact
24	BLDG	ATS2-EMERP-HW	ATS-2 in the Emergency Position	DI	On/Off	NA	A3-Auxillary Contact
25	BLDG	ATS2-NORMP-HW	ATS-2 in the Normal Position	DI	On/Off	NA	A4-Auxillary Contact
26	CRAC-1	CRAC1-FBK-HW	CRAC-1 Unit Status	DI	On/Off	NA	Current Sensor
27	CRAC-1	CRAC1-HTEMP-HW	CRAC-1 Unit High Temperature Alarm	DI	On/Off	NA	Terminal Blocks in CRAC
28	CRAC-1	CRAC1-LTEMP-HW	CRAC-1 Unit Low Temperature Alarm	DI	On/Off	NA	Terminal Blocks in CRAC
29	CRAC-1	CRAC1-HHUM-HW	CRAC-1 Unit High Humidity Alarm	DI	On/Off	NA	Terminal Blocks in CRAC
30	CRAC-1	CRAC1-LHUM-HW	CRAC-1 Unit Low Humidity Alarm	DI	On/Off	NA	Terminal Blocks in CRAC
31	CRAC-1	CRAC1-FALM-HW	CRAC-1 Unit Fan Status Alarm	DI	On/Off	NA	Terminal Blocks in CRAC
32	CRAC-1	CRAC1-DFILT-HW	CRAC-1 Unit Dirty Filter Alarm	DI	On/Off	NA	Terminal Blocks in CRAC
33	CRAC-1	CRAC1-HCAN-HW	CRAC-1 Unit Humidifier Canister Alarm	DI	On/Off	NA	Terminal Blocks in CRAC
34	CRAC-2	CRAC2-FBK-HW	CRAC-2 Unit Status	DI	On/Off	NA	Terminal Blocks in CRAC
35	CRAC-2	CRAC2-HTEMP-HW	CRAC-2 Unit High Temperature Alarm	DI	On/Off	NA	Terminal Blocks in CRAC
36	CRAC-2	CRAC2-LTEMP-HW	CRAC-2 Unit Low Temperature Alarm	DI	On/Off	NA	Terminal Blocks in CRAC
37	CRAC-2	CRAC2-HHUM-HW	CRAC-2 Unit High Humidity Alarm	DI	On/Off	NA	Terminal Blocks in CRAC
38	CRAC-2	CRAC2-LHUM-HW	CRAC-2 Unit Low Humidity Alarm	DI	On/Off	NA	Terminal Blocks in CRAC
39	CRAC-2	CRAC2-FALM-HW	CRAC-2 Unit Fan Status Alarm	DI	On/Off	NA	Terminal Blocks in CRAC
40	CRAC-2	CRAC2-DFILT-HW	CRAC-2 Unit Dirty Filter Alarm	DI	On/Off	NA	Terminal Blocks in CRAC
41	CRAC-2	CRAC2-HCAN-HW	CRAC-2 Unit Humidifier Canister Alarm	DI	On/Off	NA	Terminal Blocks in CRAC
42	CRAC-3	CRAC3-FBK-HW	CRAC-3 Unit Status	DI	On/Off	NA	Terminal Blocks in CRAC
43	CRAC-3	CRAC3-HTEMP-HW	CRAC-3 Unit High Temperature Alarm	DI	On/Off	NA	Terminal Blocks in CRAC
44	CRAC-3	CRAC3-LTEMP-HW	CRAC-3 Unit Low Temperature Alarm	DI	On/Off	NA	Terminal Blocks in CRAC
45	CRAC-3	CRAC3-HHUM-HW	CRAC-3 Unit High Humidity Alarm	DI	On/Off	NA	Terminal Blocks in CRAC
46	CRAC-3	CRAC3-LHUM-HW	CRAC-3 Unit Low Humidity Alarm	DI	On/Off	NA	Terminal Blocks in CRAC
47	CRAC-3	CRAC3-FALM-HW	CRAC-3 Unit Fan Status Alarm	DI	On/Off	NA	Terminal Blocks in CRAC
48	CRAC-3	CRAC3-DFILT-HW	CRAC-3 Unit Dirty Filter Alarm	DI	On/Off	NA	Terminal Blocks in CRAC
49	CRAC-3	CRAC3-HCAN-HW	CRAC-3 Unit Humidifier Canister Alarm	DI	On/Off	NA	Terminal Blocks in CRAC
50	CRAC-4	CRAC4-FBK-HW	CRAC-4 Unit Status	DI	On/Off	NA	Terminal Blocks in CRAC
51	CRAC-4	CRAC4-HTEMP-HW	CRAC-4 Unit High Temperature Alarm	DI	On/Off	NA	Terminal Blocks in CRAC
52	CRAC-4	CRAC4-LTEMP-HW	CRAC-4 Unit Low Temperature Alarm	DI	On/Off	NA	Terminal Blocks in CRAC
53	CRAC-4	CRAC4-HHUM-HW	CRAC-4 Unit High Humidity Alarm	DI	On/Off	NA	Terminal Blocks in CRAC
54	CRAC-4	CRAC4-LHUM-HW	CRAC-4 Unit Low Humidity Alarm	DI	On/Off	NA	Terminal Blocks in CRAC
55	CRAC-4	CRAC4-FALM-HW	CRAC-4 Unit Fan Status Alarm	DI	On/Off	NA	Terminal Blocks in CRAC
56	CRAC-4	CRAC4-DFILT-HW	CRAC-4 Unit Dirty Filter Alarm	DI	On/Off	NA	Terminal Blocks in CRAC
57	CRAC-4	CRAC4-HCAN-HW	CRAC-4 Unit Humidifier Canister Alarm	DI	On/Off	NA	Terminal Blocks in CRAC
58	DC	DC-TEMP2-HW	Temperature for new DC area	AI	Deg F	0-100 Deg F	Room Temperature Sensor
59	DC	DC-HUM2-HW	Humidity for new DC area	AI	%RH	0-100 %	Room Humidity Sensor
60	UPS	UPS-TEMP3-HW	Temperature for UPS Room	AI	Deg F	0-100 Deg F	Room Temperature Sensor
61	UPS	UPS-HUM3-HW	Humidity for UPS Room	AI	%RH	0-100 %	Room Humidity Sensor
62	LAB	UPS-TEMP3-HW	Temperature for new Lab	AI	Deg F	0-100 Deg F	Room Temperature Sensor
63	LAB	UPS-HUM3-HW	Humidity for new Lab	AI	%RH	0-100 %	Room Humidity Sensor

Chapter 18

Facility Energy Management Via a Commercial Web Service

Michael R. Brambley (michael.brambley@pnl.gov)
Srinivas Katipamula (Srinivas.Katipamula@pnl.gov)
Pacific Northwest National Laboratory,[1] Richland, Washington
Patrick O'Neill (poneill@northwrite.com)
NorthWrite Inc., Minneapolis, Minnesota

ABSTRACT

The World Wide Web (the Web) is rapidly transforming many business practices. The purpose of this chapter is to show readers how facility management tools provided by application service providers (ASPs) via the Web may represent a simple, cost-effective solution to their facility and energy management problems. All participants in the management of a facility, from building operations staff to tenants, can get convenient access to information and tools to meet their needs more effectively by simply using a web browser and an Internet connection. Further, energy tools that are integrated into these facility management software environments become readily accessible and can be linked to other tools for managing work orders for getting energy-wasting problems fixed. The chapter also describes how the Federal Energy Management Program (FEMP) is working to make such energy tools readily available.

INTRODUCTION

The World Wide Web is transforming the way business is done. Just a decade ago, most documents requiring immediate transfer between different geographic locations were sent via overnight delivery service. With the Web and its services (such as email), documents can be shared continents away in a matter of minutes or seconds. No longer do collaborators or business partners need to wait days, weeks or months to review, make revisions, or execute documents. Consumer-to-business and business-to-business purchase orders, once decided upon, can be placed over the Web in seconds. These are two of the many transactions that the Web has transformed and continues to transform.

Our use of software has been somewhat slower to change. Most of the software applications used in business are installed separately on individual computer workstations. There are some exceptions to this practice, but they still represent the small minority of software installations today. The consequences of this include collaborators with incompatible files, the need to disseminate and install bug and security patches on every computer (sometimes thousands or even millions) on which an application is installed, loss of data whenever an individual disk crashes that was not backed up regularly, and other difficulties associated with trying to manage interactions across software applications on many different machines.

Electronic access to software from dedicated providers, called application service providers (or ASPs), presents an alternative. This chapter describes the emerging availability of software tools for facility management (FM) via the Web, how energy management can benefit from this delivery mechanism and integration with other web-enabled FM tools, and some specific tools for managing energy use that are available now and others currently under development.

THE APPLICATION SERVICE PROVIDER (ASP) MODEL

Application service providers provide access to software via subscriptions. For the payment of a subscription fee, users obtain access to software on the World Wide Web using nothing more than a web

---

[1]Operated for the U.S. Department of Energy by Battelle Memorial Institute under contract DE-AC06-76RL01830.

browser to access it. An example of an ASP for facility management is shown in Figure 18-1. The software needs to be installed on only one computer, the web server, rather than on the individual workstation of every user. To provide reliability, the software is typically installed on several redundant servers to provide backup in case a server fails. Many users are then able to access a small number of installed copies of the software. User files are also maintained on the ASP's servers and backed-up in a similar manner.

This service delivery model enables a large set of software applications to be provided to many users at a site, without requiring a large on-site computer support organization. Keeping the applications up and running is the responsibility of the ASP, as are backing up data and installing patches and software upgrades. When software is upgraded, every user gets immediate access to the upgrades; there is no delay or time required for upgrades to get distributed, acquired and installed. These functions need not be provided by the end user organization. They are handled by the ASP.

Facility management ASPs can provide a complete suite of tools for facility management: asset tracking, communications, alarms and notifications, maintenance management, work order management, utility tracking, bill management, project management, and benchmarking tools. In addition, the environment provided by the ASP can provide convenient access to weather data, customized news, and on-line directories.

Under the appropriate licensing agreements, ASP tools can integrate work activities and communications of key participants within facilities (see Figure 18-2) through their computers, personal digital assistants, or pagers. Building occupants can enter requests for work directly into a work order management system. All work orders can be prioritized and assigned to building staff or contractors automatically, or by a dispatcher to whom that duty is assigned. Repair staff and contractors receive work orders for which they are responsible directly through the system. They can then report progress of the job and completion when it is done. If parts are required for repairs, vendors can also be integrated into the system so orders are placed, acknowledged, and status of delivery provided via the system directly to the parties affected. Such a system can effectively integrate all parties with an interest in the facility and keep them informed. Each type of user is given access to the tools necessary to carry out their job functions.

Many users at a site can have access to the software. Each user can have unique privileges to specific sets of tools. A financial manager, for example, would have privileges to a different set of tools than a building operator. For a campus of buildings or a company with buildings at many different sites, the management of diverse buildings can be monitored from a central location.

Many of these capabilities are available from software systems known as enterprise applications. Generally these are comprehensive tool suites installed on computers at the facilities (e.g., on a campus) and adapted to the unique needs of the site. These systems generally are licensed up-front for a fee and may be supported by a purchased maintenance contract. Often these systems are installed on a site server owned by the customer, and client software is installed on the personal workstations of each user of the system. These systems provide essentially the same capabilities as a suite of FM tools provided by an ASP. The primary differences are: 1) customers generally must provide their own computer support staff for maintaining the server and any backup systems, 2) the software license fee is incurred up front at the time of "purchase," 3) client software is generally installed on user workstations, and 4) software upgrades are made when purchased for the site (if not covered by a software maintenance or technical support contract) and installed on the site server. In some cases, enterprise FM systems can be set up as a web server. These systems can be accessed via a web browser, just like tools from an ASP, and can be installed on a site server or for a fee by a generic ASP service provider.

The key benefits of an FM tool suite provided by an ASP are:

- Low cost to subscribe, which can often be covered by an operating budget instead of a capital budget—no license fee—sometimes no long term agreement required
- Software maintenance and upgrades handled by the ASP
- Server hardware maintenance handled by the ASP—no on-site information technology support required
- Immediate access to software upgrades as they become available
- Compatibility among all users because everyone is operating on the same version of the software—eliminating file compatibility issues, organizational procedural variations, and training problems caused by version differences across an enterprise.

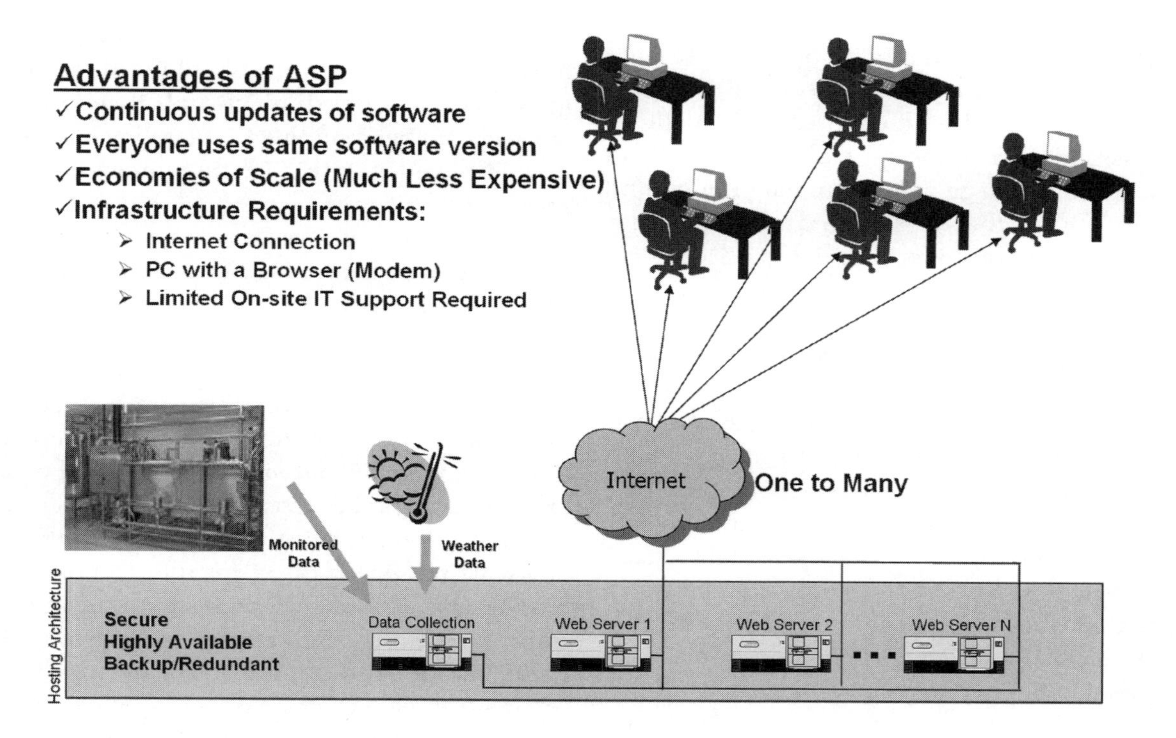

Figure 18-1. Diagram showing the basic ASP model and its advantages.

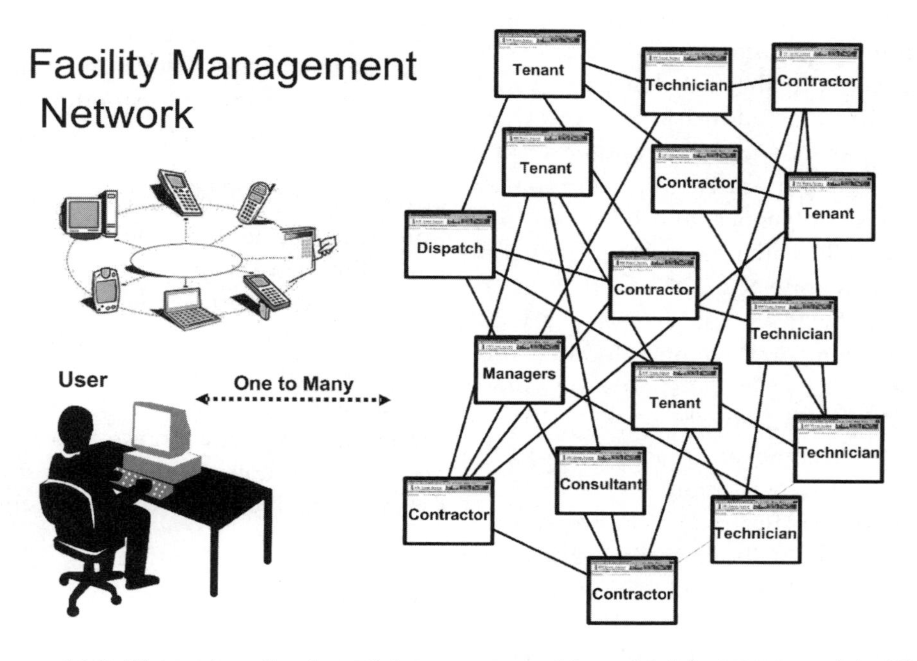

Figure 18-2. The network of participants created by a highly integrated facility management software suite where all stakeholders are connected.

WHY INTEGRATE ENERGY TOOLS INTO A FACILITY MANAGEMENT ENVIRONMENT?

A survey of 250 customers by NorthWrite Inc.[2] showed that facility managers devote 5 to 6% of their time to issues associated with utilities. This includes electricity, gas, and thermal energy (e.g., district heat) but also water, sewer, garbage collection, telephones, and others. They found that less than 1% of a facility manager's time on average is spent on electricity-related activities (Figure 18-3). In contrast to the small fraction of time devoted by facility mangers to utilities and energy, utility expenses represent about 34% of total facility purchases globally (see Figure 18-4).

Why is there such a discrepancy between the costs of utilities and the time facility managers spend on this building expense? If you have ever spent time in a facility manager's office you will quickly understand. Their phone is constantly ringing, they get emails, pages, and faxes every 5 minutes, and people are constantly stopping at their door to let them know about the latest issue, problem, or "emergency" that requires their immediate attention. They may have come to work that morning with the intention of looking at their latest utility bills to identify opportunities for savings, but that desire is quickly displaced by the daily demands of their job. When quitting time comes, they find that another day has gone by where they just didn't have time to think about their utility expenses.

Another issue has been the fact that energy tools are generally provided as stand-alone software. Although they can be part of a user's computer desktop environment, they still generally require separate purchase, installation, maintenance, and initiation to be used. Some innovative ASP's are now starting to provide energy tools that are seamlessly integrated with the other FM tools that the facility manager uses every day (e.g. Work Order Management, Preventive Maintenance, Project Management, etc.). By placing energy tools along side these other FM applications, these ASP's believe that energy tools will become more accessible to the facility manager and hence, used more.

Furthermore, many facility managers are still doing many of their tasks manually or are using software tools that are little more than bookkeeping aids. By employing the new generation of ASP-based FM software, facility managers should be able to perform their duties more efficiently, enabling them to reallocate more of their effort to energy and other utility issues, thus bringing both time spent and dollars spent closer to congruence. Because nearly all facilities have significant opportunities to improve the operating efficiency of their systems, energy tools should lead facility management to those opportunities and, therefore, improved efficiency in their buildings. Integrating energy tools into a web-based facility management system helps accomplish this.

For small buildings, a web-based approach may be especially important. Small commercial buildings often do not have a dedicated manager on site. Several of these buildings might be managed by a facility management firm. Tools are needed to keep facility-management staff informed of conditions in each building. This information can come from two sources: 1) occupants of the building and 2) automatically collected data. In both cases, data must be transmitted to a convenient central location where facility-management staff can easily access it and review the status of each building. A web-based facility-management system that integrates not only traditional stakeholders but also building occupants provides a mechanism by which occupants' reports of problems and needs can be readily identified and addressed. Other data from sensors in the building must come from building automation systems (BASs) or stand-alone monitoring devices or sensor networks.

Monitoring and diagnostic tools for building systems are available that can be integrated with web-based facility management software systems but very few are commercially offered. Monitoring tools hosted by an ASP have the advantage of economies of scale for set up and operation of data communication. Data from many distributed sites can all be sent to the ASP servers, processed by tools in the facility-management systems, and the results made available to users including automated email or paging notification for critical alarms. Economies for installation and operation exist for the ASP because it serves the needs of many customers with the same generic, essentially turn-key, monitoring systems. In the next section we provide two examples of how ASP's are providing a combination of energy and FM tools to their customers.

SOME EXAMPLES

Electric Utility Providing Energy Services as an ASP

Utilities have been providing energy management services to their customers for decades and have experi-

[2]An unpublished survey by NorthWrite Inc. of customers. NorthWrite Inc., Minneapolis, MN.

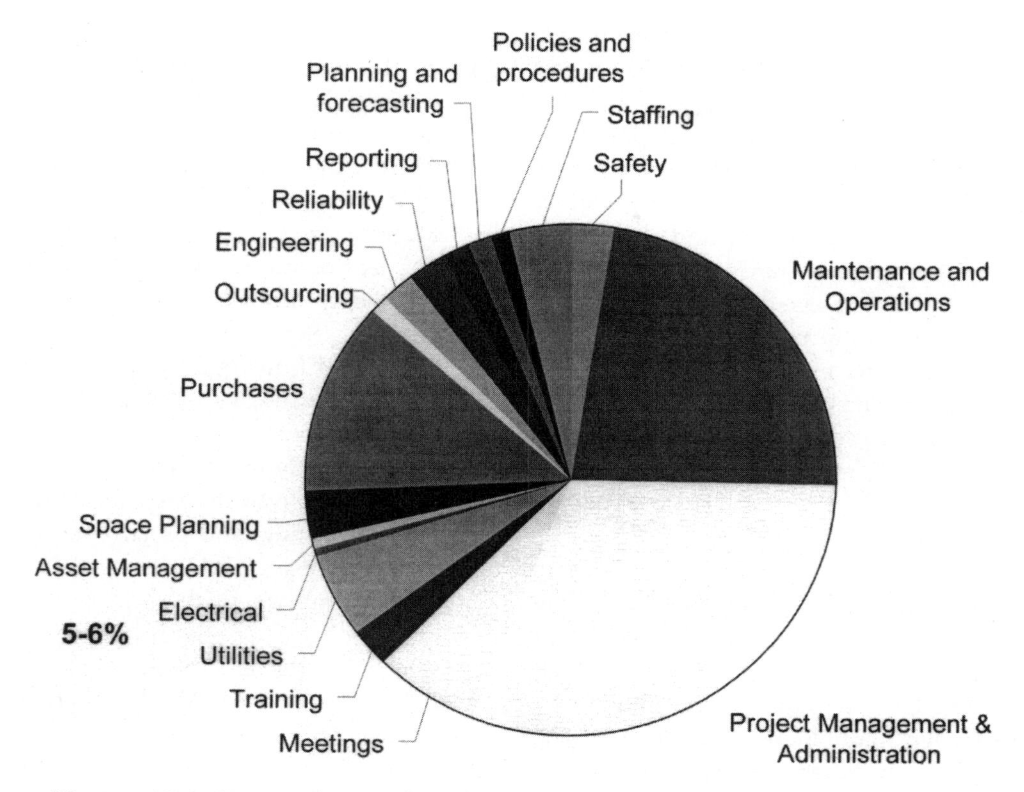

Figure 18-3. Proportions of time devoted to facility management tasks. Source: NorthWrite Inc. unpublished survey, 2002.

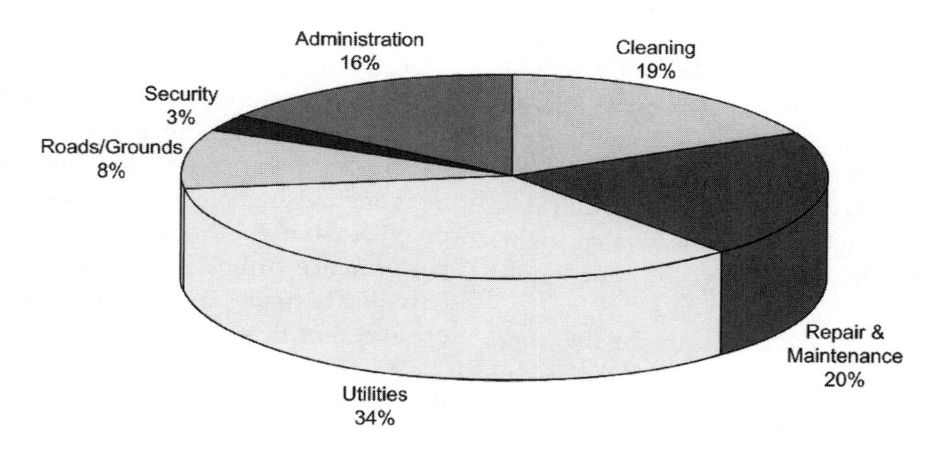

Figure 18-4. Fractions of $1.2 trillion of annual purchases by facilities globally. Data source: 1998 BOMA Experience Exchange Report. (1)

mented with many different delivery mechanisms. What has emerged as the preferred approach is via the Web. One of many examples is Portland General Electric's (PGE[3]) "E-Manager" energy information service. E-Manager enables their customers to view and analyze their 15-minute interval electric consumption data on a monthly, daily, or near real-time basis.

E-Manager replaced a legacy workstation-based software service that had been the source of significant problems for both PGE and its customers. With the old system, there had been continual difficulties in getting the meter data to the customers. Desktop software support was also something that the utility was not especially proficient at, and there were many problems with software versioning, hardware system incompat-

[3]*http://www.portlandgeneral.com/business/products/emanager/Default.asp?bhcp=1*

ibility, and training. PGE had to choose between finding a more sustainable delivery model and dropping the program altogether.

The utility decided to adopt a new energy information service delivered using an ASP model. Since moving to an ASP, about 1/3 of PGE's largest customers have subscribed to the E-Manager program, and the utility considers it to be a highly successful example of applying the right technology approach to real customer needs. PGE has also begun offering a more comprehensive set of FM services (e.g., work order management, preventive maintenance, project management, etc.) in an effort to address more of the facility manager's day-to-day needs.

Facility Management ASP

Another example of web-delivered services is NorthWrite's[4] ASP offering for facility management. One of their customers, St. Paul College manages a campus of buildings using NorthWrite's FM "WorkSite." This web service features standard FM software for managing work orders, preventive maintenance, documents, etc. However, it is unique in that there are also a number of energy management tools. St. Paul College tracks and assigns work orders, makes sure their equipment is properly maintained and manages their energy consumption all using the same set of software tools. They use their WorkSite to track utility bills, baseline their buildings against the U.S. Department of Energy's (DOE's) Commercial Building Energy Consumption Survey [2], identify up to 300 energy savings measures, and analyze 15-minute electric and gas consumption data. Having all these tools in one "place" means that sophisticated energy management tools are only one mouse-click away.

United Properties (UP), a large, progressive property/facility management company located in the Midwest, provides another useful example of value attained by use of facility management tools from an ASP. This company manages a 26 million square foot portfolio of commercial and industrial facilities and utilizes NorthWrite's WorkSite program to manage the day-to-day operations at many of its building locations. UP also uses several energy-related tools within its WorkSite to track the energy usage of its client's facilities. Recently, a United Properties facility manager noticed that one of his clients, a private college, had been receiving unusually large electric bills. Using the Energy Benchmarking Tool, he was able to spot that the facility was using significantly more energy than usual and discovered that his client was actually being charged for construction electricity that should have been paid by the new building's contractor. As a result of its NorthWrite usage, United Properties was able to recover more than $30,000 for its client.

Merging energy and facility management tools has a deeper, less obvious advantage. Often, energy savings are achieved through implementing a large number of operational changes within your facilities. Integrating energy and FM functionality means that many of these changes can be the result of automatically generated work orders that are created based on unusual consumption or demand characteristics. If larger capital improvements or retrofits are called for, project management, labor tracking, and procurement tools (also a part of a comprehensive FM ASP service) can be used to manage and implement these projects.

EMERGING WEB-BASED ENERGY TOOLS

Central collection and processing of data from many remote sites creates the opportunity to use a myriad of tools to extract information from the data. The cost of the software applications and maintaining the application servers can be spread over the facilities of all users, reducing costs per facility. Centralized monitoring also enables facilities and service providers to hire expert HVAC engineers and analysts to analyze several buildings rather than one or a few, and these experts can do their analyses remotely. A number of new tools that will provide better information about the energy performance of building systems and operating costs have just been or are now being developed. Discussions of several of these follow.

Remote Automated Diagnostics

For decades, building automation systems (also known as energy management and control systems, building management system, etc.) have provided the ability for building operators to log and graph data points over time. Unfortunately, these systems generally have required that users set up the trending of variables of interest and then view and interpret trends in the data to identify equipment faults, performance degradation, and opportunities for performance improvement. This requires that the building staff have time to set up and periodically review the trends as well as have the knowledge or experience to interpret them. Furthermore, understanding the condition of equipment

[4]*http://www.northwrite.com*

and systems often requires tracking changes in several variables simultaneously to detect a fault and then to diagnose it, making the task unwieldy for all but those most experienced at this type of analysis. Because of this, most operators tend to neglect much of the information provided by these systems. They just don't have enough time to do the urgent work and still analyze these data.

A new class of tools that is starting to emerge for practical use has been under development for a decade or more (see, for example, references [3] and [4]). These tools, known as automated fault detection and diagnostics, take data from equipment and systems, analyze it, and interpret it to provide easy-to-understand, actionable information. Most of these tools don't distract ushers with raw measurements, only providing them when requested by a user. Instead, they provide conclusions that can be acted on, such as "the actuator for the mixing damper in Air Handler 5 is not operating" or "Outdoor Air Sensor 8 has failed, is providing no reading, and should be replaced." This information is often presented in alarms to building staff, so only the alarms need to be monitored, not the raw data trends. Periodic reviews of data trends are not required; they are done continuously by the automated diagnostic tools.

Effective use of diagnostic tools can help facility managers and operators cut the cost of operations and consumption of resources while improving the comfort and the safety of occupants. Continuous diagnostics for building systems and equipment will help remedy many problems associated with inefficient operation by automatically and continuously detecting system performance problems and bringing them to the attention of building operators. [5] Some of these problems might otherwise go undetected. Advanced diagnostic tools can even suggest causes of problems, make recommendations for solving problems, and estimate the cost of not solving a problem. The value of this technology is evident in the findings of researchers at Pacific Northwest National Laboratory who found in applying an automated diagnostic tool for air-handling units on buildings at several locations that 32 out of 32 air handlers checked had existing faults. The estimated annual costs of these problems ranged from $130 to $16,000 (see Table 18-1). [6]

Access to data in real-time has been and continues to be a major obstacle to widespread deployment of remote automated diagnostic tools. The application service provider model described in this chapter helps overcome that problem and should make application of automated fault detection and diagnostics easier than it has been in the past.

Tracking Energy End-Uses

Because energy accounts for a significant portion of the operating cost in many facilities, facility managers, energy service providers, and owners alike will benefit from software tools that track energy end-use. For example, the benefits for an owner of a retail chain or a facility manager of a large campus with distributed facilities include:

- ability to generate reports in several different formats (e.g., by region, sales volume, or building type),
- ability to benchmark historical, normalized (e.g., with respect to weather, size, or sales) end-use consumption between similar facilities. Comparison with benchmarks can help identify operational inefficiencies.
- ability to forecast energy budgets and prepare energy purchasing plans.

An energy service provider who has signed a guaranteed savings (i.e., performance) contract with a facility can reduce risk and increase reliability by tracking end-use consumption and calculating savings continuously. From a central location, the energy service provider or facilities personnel can also identify problems associated with unscheduled operation of equipment (such as lights and HVAC equipment) because of control malfunctions or errant programming.

Table 18-1. Summary of air handling faults and their cost impacts for 32 air handlers monitored with an automated air-handler diagnostic tool. [6]

Fault	*Number of Air Handlers*	*Annual Cost Impact ($)*
Temperature sensor	7	*
Supply-air control	2	2000
Scheduling	2	130 to 700
Not fully economizing	10	115 to 16,000
Excess ventilation	9	250 to 12,000
Inadequate ventilation	2	**
Stuck damper	2	0 to 4000
Mis-calibrated sensor	1	*

*Improperly operating sensors prevent estimation of energy and cost impacts.

**Inadequate outdoor-air ventilation has not direct cost impacts but may affect occupants comfort, health and productivity.

The early developers of software tools for tracking savings often used special data logging equipment coupled with low bandwidth phone lines for communication—a cumbersome application. Application service providers can take this burden off individual building owners by collecting, analyzing, formatting, and displaying energy-use data more easily for a single building or several. Real savings from energy conservation measures can then be compared easily with estimates from engineers, contractors, and operators.

An example display from an energy tracking tool is shown in Figure 18-5. On the screen, the different major energy end-uses are shown by lines of differing colors. The interface for this tool quickly conveys excess consumption using enlarged data points. Additional information can be obtained simply by clicking the computer mouse with the cursor positioned over the point for which more information is desired. The system then provides the user additional information in a new display window (see Figure 18-6) including the amount of excess energy used, the normally expected range of consumption, and the cost of the excess consumption. The analysis of tracked data, normalization for factors such weather conditions and occupancy schedules, and interpretation of numerical results are all accomplished by the tracking tool itself, so that users can quickly access the information they need.

Load Aggregation

The electric power grid failure in the northeastern U.S. on August 14, 2003, made the limitations of the electric power grid painfully evident to the government, the electric power industry, and even the public. [7] The need to change how the grid is operated is acknowledged more today than ever before. In the future, customers are more likely to see electricity prices tied directly to the time-varying cost of producing and transmitting electricity. During times of peak consumption (often mid-to-late afternoon on a daily basis and during extremely hot or cold weather on an annual basis), customers are likely to experience much higher prices than during periods of low use (in the middle of the night and during mild weather).

In such a market, electric customers are likely to be able to negotiate more favorable utility rates, as well as control their electric demand, by aggregating loads across their facilities. This requires tracking of aggregated loads as well as the demand profiles of individual buildings whether they be at a single site or distributed across many locations. By centrally collecting and analyzing consumption data from meters and control networks, individual loads can be managed across many buildings to ensure aggregated compliance with electric-contract requirements. By aggregating real-time data across buildings, facility managers can identify where to curtail energy use when demand approaches a negotiated limit, insulating them from high peak prices.

Whole-Facility Cost Management

One of the greatest potential cost-savings opportunities for facility managers and operators lies in the ability to control and optimize whole-facility energy consumption. Utilities are beginning to offer rates that vary by hour-of-day and day-of-week using real-time pricing and time-of-use rates. To take advantage of time-varying rates, facilities need advanced control strategies. [8] Strategies include: HVAC load shedding (for chillers, thermal storage, supply and zone temperatures, fans, and pumps); load shifting (using precooling or thermal storage); and fuel shifting (gas, oil, and steam standby generators). [9] These strategies not only require centralized access to data from sensors and meters but also the ability to control equipment from a central location.

In addition to the energy charge, most of today's electric rate structures also have a demand charge that can be as high as $25 per kW or more. In some cases, the peak period is tied to the utility's system-wide peak. In such a case, predicting the system-wide peak and implementing the load-control measures during a 2- or 3-hour window surrounding the anticipated time of the peak can save a large facility a very large demand charge. Pacific Northwest National Laboratory (PNNL) developed a sophisticated software tool for naval bases in San Diego. The tool calculates the probability that any particular day's system peak will be the monthly system peak. The tool uses short-term (today), medium-term (1 week), and long-term (30-day) weather forecasts, as well as a baseline model of the utility's loads system-wide. All weather data are collected over the Internet for the analysis. Facility managers and energy service providers can optimize whole facilities using such sophisticated software tools, if the required data are available. Application service providers could provide access to these tools as well as the infrastructure for running them and storing data at a lower cost than each individual facility could.

Federal-Sector Leadership

The DOE's Federal Energy Management Program (FEMP) is currently working with a private contractor to web enable two energy-focused diagnostic tools previ-

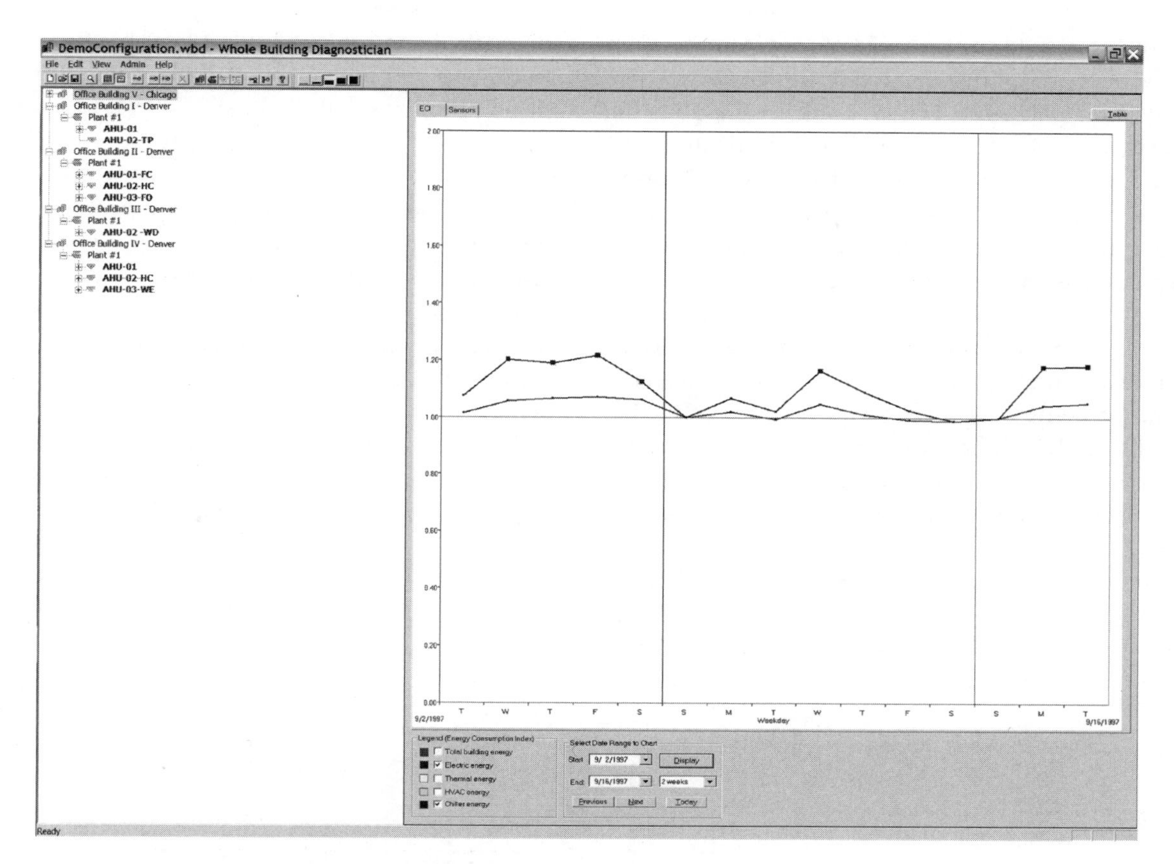

Figure 18-5. Main display for a representative energy tracking tool.

Problem Details

Office Building V - Chicago

Close

Date	Problem	ECI	Actual	Normal range	Daily cost	Weekly cost
9/10/1997	Actual Chiller energy is too high.	1.16	2413 kWh	1659 - 2242 kWh	$21	$150

Figure 18-6. Pop-up window that provides more detail about excess energy use on a particular day, obtained by clicking on an enlarged data point for the specific day of interest in Figure 18-5.

ously developed under DOE funding to make them available as part of ASP-offered web-based facility management systems. The first tool provides the capability to monitor energy use, automatically identify anomalies in energy use after adjusting for the influence of factors such as variations in weather or activity in a facility over time, and then alert facility staff with alarms. Essentially any energy use for which a meter is or can be installed can be monitored with this tool. This includes the electricity use of an entire building or campus, the electricity or fuel use for an entire HVAC system, or even the fuel use of an individual boiler or the electricity use of an individual packaged HVAC unit. By having this tool monitor energy end uses and automatically alert facility staff to high or low levels of energy use, problems with equipment scheduling or performance can be identified long before they might be otherwise. By accounting for weather and other factors that affect energy use, changes caused by underlying factors such as the performance degradation of equipment or override of a schedule are much more easily and reliably detected than by monitoring raw energy data. The tool also automatically analyzes energy use and detects anomalies with consumption so the time of building personnel is not needed to manually review data and visually detect anomalies.

The second tool automatically detects and diagnoses faults with air-handling equipment. This includes both built-up air handlers primarily found in large buildings and package HVAC units, which serve the many small commercial buildings. This tool detects and identifies potential causes of three generic types of operation problems with these systems: 1) excessive energy consumption, 2) inadequate ventilation, which causes deterioration of indoor air quality, and 3) sensor faults and other control problems. Energy problems can result from improper schedules (e.g., a temporary schedule change that is implemented but never reset to its initial values), improperly implemented controls (improperly-controlled outdoor-air dampers), or physical failure of equipment (e.g., a broken damper-actuation linkage that prevents modulation of the damper). These are a few of the more than 100 specific faults that this tool can detect. [10] Results are provided to the user graphically (Figure 18-7) so they can be reviewed quickly, using little of a building operator's time. If a fault persists, the user interface provides a simple mouse-driven interface to access additional details of the fault and potential remedies for it (see Figure 18-8). A key characteristic of this tool is that it does not simply provide plots of data, but

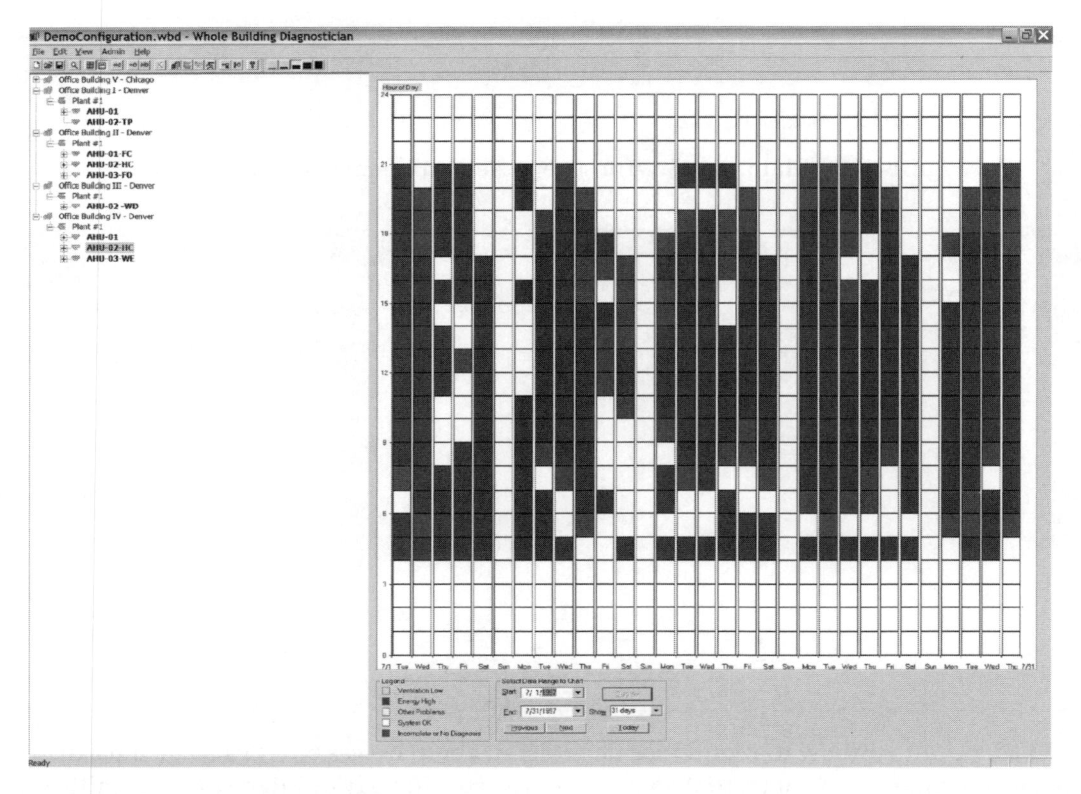

Figure 18-7. The user interface for the air-handling diagnostic tool. The actual user interface uses color to indicate whether a fault occurred and the type of fault for each hour of operation (white indicates no faults were detected). Each column represents a day with each cell representing a specific hour of day. Red cells, which identify hours during which an excessive use of energy occurred, appear as black cells in this figure. The slightly lighter gray cells identify hours during which the diagnosis was incomplete. The large number of red cells between 5 a.m. and 9 p.m. every day except Sunday indicates a persistent fault that is causing energy waste.

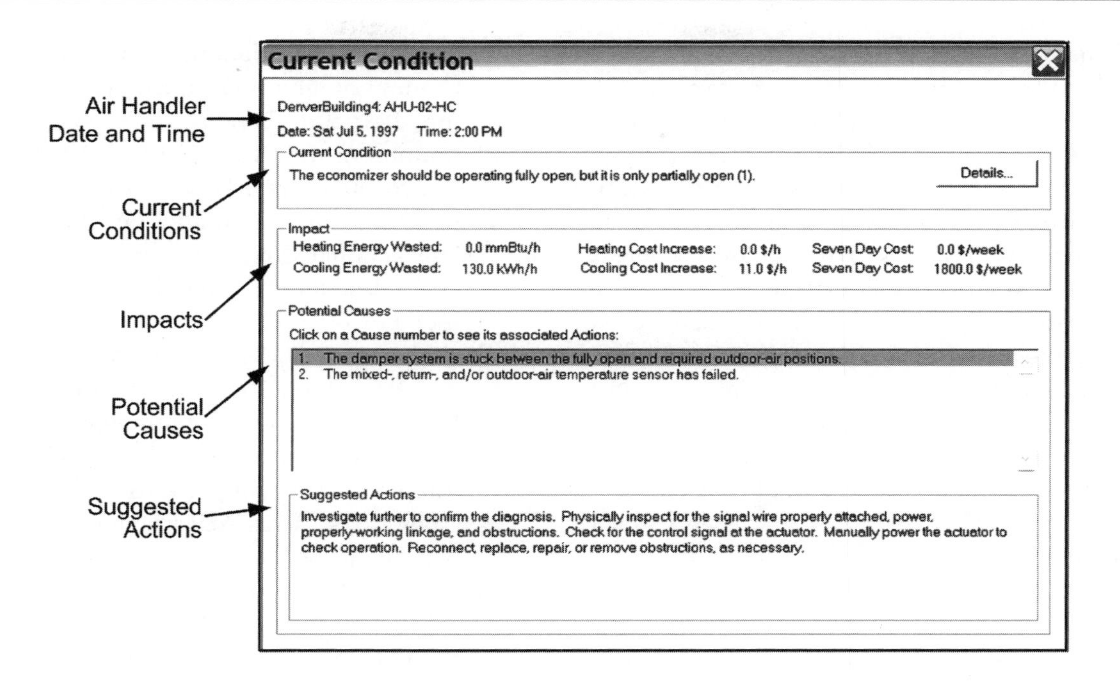

Figure 18-8. A pop-up window revealed by clicking on a cell in Figure 18-7 provides detailed information on the fault identified by the air-handling diagnostic tool.

interprets data to provide easy-to-understand information on which building staff can act. In web enabling this tool, the FEMP team seeks to further simplify the interface to streamline use and provide building operators and managers with key information for decision making literally at their fingertips.

By making results from these energy-focused tools conveniently available as part of a facility management environment, FEMP hopes to reduce the effort required of building staff to recognize and correct problems that increase energy use and might go undetected for months or even years without these tools. Although FEMP's focus is on helping Federal-sector facilities, once these tools are integrated into the web-based FM system, they will become available commercially to all potential users.

DATA GATHERING ISSUES

Software tools that perform energy analysis require monitored interval data (periodic data collected at a regular frequency—e.g., every 15 minutes). If the site has interval meters, access to utility end-use consumption data can come directly from the utilities,. Otherwise, end-use monitoring is necessary to collect the data, which then must be transmitted to the ASP. In addition, data from building automation systems (BASs) is needed for detailed diagnosis of equipment performance. Although the data can be collected using existing BASs, this is not always easy and the success of data collection varies widely depending on the type of BAS and its support of standard protocols. Furthermore, less than 10% of the commercial buildings have BASs. [2] Lack of adequate sensors to monitor performance of building systems is also another major stumbling block.

Widespread deployment of energy software tools hosted by ASPs will require low-cost sensors, reliable automated data collection, reliable low-cost communications to push the data from the building to the ASP server, and software tools that can be hosted by an ASP. Some ASPs are starting to market creative approaches to data collection that include the use of wireless networks to transmit the information back to their servers without requiring phone lines or connecting to their customers' intranets. We anticipate significant innovation in this area as increased customer demand drives ASPs to provide low-cost, reliable data monitoring services.

CONCLUSIONS

The World Wide Web provides a convenient and very cost-effective delivery mechanism for FM software. The ASP model shows promise for enabling well-managed facility management over a broad range of building sizes and business paradigms. It is generally cheaper than traditional software, reduces the information technology (IT) support burden on users, and brings additional features like the ability to network all the key participants in operating and maintaining buildings.

Integration of FM and energy tools shows promise as a way to make these tools more accessible to the facility manager. ASPs that provide a comprehensive suite of FM and energy tools enable users to more efficiently address a missing element of their cost of operations. Furthermore, by performing FM tasks more efficiently, facility managers will have more time to spend on proactively managing utility costs.

As more energy management and analysis software tools are developed and deployed by ASPs, building managers, facility operators, and energy service providers will gain access to more sophisticated and automated software tools that will enable them to manage distributed facilities more efficiently. These advances will provide better controls capability, enhance operation and maintenance by providing automated remote diagnostics, support predictive condition-based maintenance, help verify performance contracts, increase productivity, allow better integration and use of sensors, actuators and controllers, improve overall energy management, and lower facility management costs.

References

[1] Building Owners & Managers Association International (BOMA). 1998. 1998 Experience Exchange Report. BOMA, Washington, D.C.

[2] Energy Information Administration (EIA). 1995. 1995 Commercial Building Energy Consumption and Expenditures (CBECS), Public Use Data, Micro-data files are available on the EIA website: *http://www.eia.doe.gov/emeu/cbecs/microdat.html*.

[3] Honeywell. 2003. "The HVAC Service Assistant." Honeywell Home and Building Controls, Golden Valley, Minnesota. Available on the World Wide Web at *http://customer.honeywell.com/buildings/CBWPServiceAssistant.asp*.

[4] Friedman, H. and M.A. Piette. 2001. "Comparison of Emerging Diagnostic Tools for Large Commercial HVAC Systems." In Proceedings of the 9th National Conference on Building Commissioning, Portland Energy Conservation Inc., Portland, Oregon.

[5] Brambley, M.R., R.G. Pratt, D.P. Chassin, and S. Katipamula. 1999. "Use of Automated Tools for Building Commissioning." In Proceedings of the 7th National Conference on Building Commissioning, Portland Energy Conservation Inc., Portland, Oregon.

[6] Katipamula, S., M.R. Brambley, N.N. Bauman, and R.G. Pratt. 2003. "Enhancing Building Operations through Automated Diagnostics: Field Test Results." In Proceedings of the Third International Conference for Enhanced Building Operations. Texas A&M University, College Station, Texas.

[7] US-Canada Power System Task Force. 2004. Final Report on the August 14, 2003 Blackout in the United States and Canada: Causes and Recommendations. Government of the United States and Canadian Government. Available on the World Wide Web at *https://reports.energy.gov/BlackoutFinal-Web.pdf*.

[8] Kammerud, R.C., S.L. Blanc, and W.F. Kane. 1996. "The Impact of Real-Time Pricing of Electricity on Energy Use, Energy Cost, and Operation of a Major Hotel." In Proceedings of the ACEEE 1996 Summer Study on Energy Efficiency in Buildings, American Council for an Energy-Efficient Economy, Washington D.C.

[9] O'Neill P. 1998. "Opening up the Possibilities." Engineered Systems, Vol. 15, No. 6.

[10] Katipamula, S., M.R. Brambley and L. Luskay. 2002. "Automated Proactive Techniques for Commissioning Air-Handling Units," Journal of Solar Energy Engineering, 125:282-291.

ACKNOWLEDGMENTS

The authors wish to thank the Federal Energy Management Program (FEMP) for partially supporting preparation of this chapter and the research upon which it is based.

Chapter 19

Evolution to Web Based Energy Information and Control at Merck & Co., Inc. Rahway, NJ

Joseph LoCurcio
Central Engineering Division
Merck & Co., Inc.

ABSTRACT

The Merck Rahway site has grown substantially over the years, and has progressed through the various control technologies of pneumatic systems, electrical systems, and finally into DDC and web based energy information and control. As the site control requirements grew, and the maintenance and operational staff got smaller, there was only one solution—adopt a flexible BAS system that would allow remote monitoring and alarming of all HVAC and building control systems. An Andover Infinity BAS was installed at the site, and was modified and expanded over a period of several years to monitor and control the building systems in over 100 buildings at the 150 acre site. This web based system has allowed the small Rahway site staff to adequately monitor and control all of the important HVAC and lighting systems for the campus of buildings. The purpose of this chapter is to describe this web based system, and to describe its operation and its benefits to the Rahway site.

FACILITY BACKGROUND

The Merck & Co., Inc., site in Rahway, NJ has research and manufacturing facilities, as well as office buildings, warehouses, environmentally controlled areas. The site was formerly Merck's world headquarters, and currently, Rahway has a unique synergy due to research and development's close proximity to advanced manufacturing technologies. Manufacturing activities are directed toward the production of active ingredients for human and animal health products.

The site has a large infrastructure, and has designed and built its own centralized plants. These include a central powerhouse for steam and electrical cogeneration, specialty gas distribution, and large capacity (multiple 1500 ton electric and absorption) centralized chiller plants. The Merck Rahway site has been in operation for over 100 years, and has had every level of automatic controls for its HVAC and building control systems, from early pneumatic systems, to centralized Honeywell controls with remote reset, to today's use of a BAS with DDC control and a web based energy information and control systems structure. Figure 19-1 shows an early workstation BAS for HVAC control.

FACILITY PROBLEMS AND EARLY SOLUTIONS

With growth of the Rahway site has come an under-proportionate growth in the number of people on the maintenance and operations staff. This slower maintenance staff growth has resulted in an absolute reliance on the BAS for both remote trouble-shooting and alarm annunciation. There are simply not enough people to physically monitor all of the HVAC systems.

Since the campus had grown to more than 100 buildings across 150 acres, and the maintenance function had contracted to 2 shops, a complex backbone of short-haul modems and building telecommunications was required to bring operating and alarm information back to centralized areas, while leaving the controls decentralized. It was proven over time that local, terminal units with independent CPU, with local I/O, had the speed and independence necessary to perform the desired functions. Local, in-building networks started off as pro-

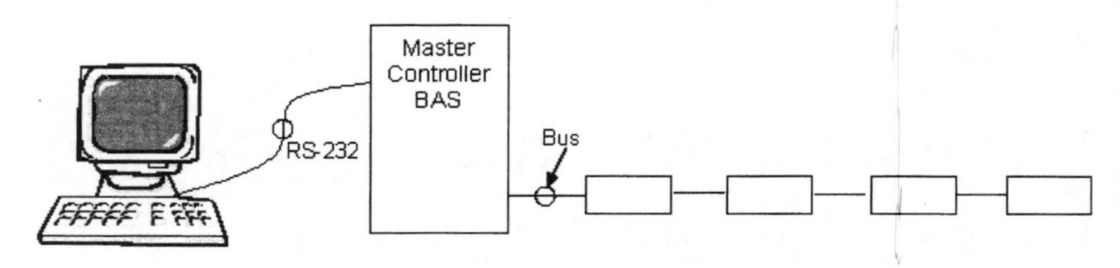

Figure 19-1. Early workstation

prietary 5 wire buss and serial type communications and evolved into RS-485 networks. This general system is shown in Figure 19-2.

A very flexible BAS was introduced on the site in the 1980's, when an Andover Controls Corporation AC256 was installed. The AC256 system was capable of complex data and I/O manipulation, serial communication with a generic PC front-end, modem communication for off-site monitoring and dial-out capabilities (pager alarms, phone calls to remote alarm monitoring stations, etc.). During this time a concerted effort to automate all HVAC and other building systems was accepted by the site as a standard, given its importance and value to the operation of the site.

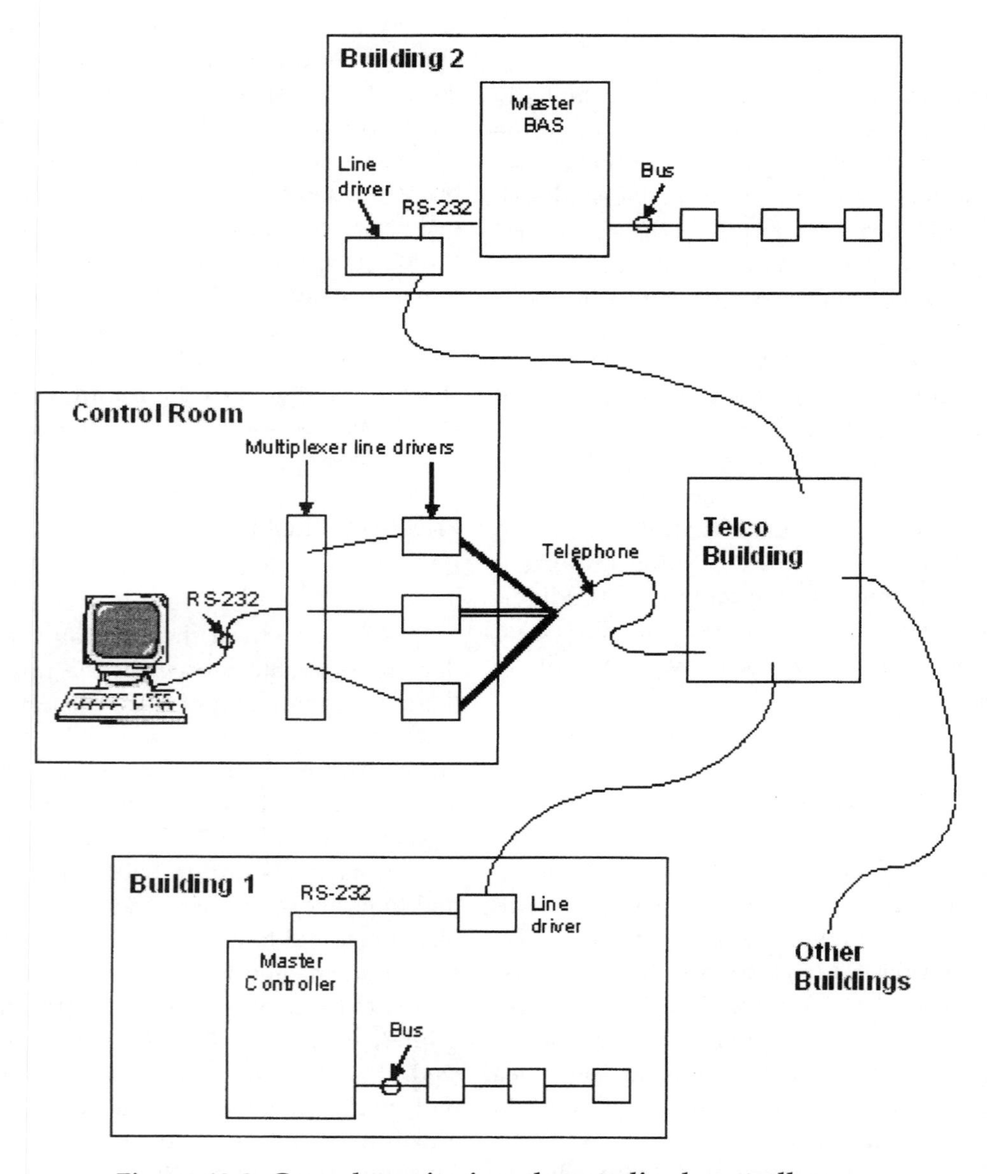

Figure 19-2. Central monitoring, decentralized controllers

EVOLUTION OF THE MODERN BAS AT THE SITE

Around this time the BAS evolved further with the purchase and installation of Andover Corporation's Andover Infinity system. This new system featured a centralized SQL database of each master and sub controller memory, TCP/IP enabled building master controllers, RS485 subnets (building level networking) extraordinary scalability and BAS equipment adaptability and performance.

The system is based on two philosophies that were of paramount importance to the site. The first was the existence of a building master controller with TCP/IP capabilities. This enabled us to install a master controller (or controllers if the facility was large enough) in each building that could communicate back to central monitoring via the company's standard computer communication network. The Ethernet distributed star configuration was adopted, and works very well when dealing with a multiple building campus. Each master controller supports two RS-485 sub networks for terminal devices such as air volume and temperature control boxes. It also interfaces with other manufacturer's equipment and motor control centers, in essence, handling large blocks of input/output devices for monitoring and control of discrete equipment assemblies.

The two sub networks can be fully configured with a total of about 254 devices, each having multiple discrete I/O capabilities for the equivalent of thousands of points on each master. Our distributed I/O and logic philosophy on this site ensures that control groups are on the same terminal hardware controller to minimize network traffic and maximize the continued operation of the facility in the event of a network failure. We strongly avoid having a sensor on one terminal controller and the controlled output on another.

The other philosophy is the use of a central database of the master and sub controller memory. Using Andover's system, when changes are made to the contents of a controller's memory on the network, the change is made to both the network device and the database. The database (and not the front-end workstation, laptop, or the PC connected via modem to the network) therefore always has the most current version of the operating software. This is important because in a large distributed facility, we have several thousand individual controllers. Attempting to keep an up-to-date, configuration controlled database of each controller's memory on one PC or server is impossible, unless it is somehow inseparable from the BAS infrastructure. For convenience, we have many more front-end workstations than we have programmers and operators. We also have contract programmers on our buildings currently under construction. There would be no possible way to have complete control over the software revisions without a central database and a built-in mechanism to keep the database updated no matter where or how the software changes are made.

The initial deployment of Andover Infinity was a few research buildings, 2 front end workstations, a PC-server, and dedicated pairs of fiber with a fiber hub located in the site main telephone distribution shed. The Building master controllers were initially interconnected with coaxial cables, in a thin net or bus configuration. This system configuration is shown in Figure 19-3.

The BAS LAN was initially designed to standalone on this site, with no support or interconnection to any other communication system. The HVAC department had to establish its own protocol and standards, and rely on its own internal people for all network engineering.

IMPLEMENTATION ISSUES FOR THE NEW WEB BASED SYSTEM

As the site BAS began to grow, so did the requests for dedicated fiber pairs for BAS, expensive fiber hub space, larger server hard drives. There was still a large installed base for the older AC256 system, a small base for the Honeywell remote reset, and a few dedicated all-pneumatic controls left on the site, but the site was positive that the Infinity system was the path forward.

Telecommunications and computer networks were also evolving, and new standards in computer networks developed. The site began to use fiber optic bundles, installed from the central telephone distribution service entrance to each building. At each building, the fiber was converted to copper and copper network distribution closets, to terminals in each area (office, lab etc.). The site chose TCP/IP Ethernet as the networking standard.

Cooperative ties had to be created between corporate IS standardization, telecommunication fiber planning, validation groups and the maintenance departments in order to gain acceptance for the BAS as a business critical entity, that would need to be incorporated into the growing site infrastructure master plans. The first step in acceptance was the acceptance of the HVAC SQL servers complying with the site standards. The site standards mandated minimum equipment that, on face value, increased the price of the equipment more

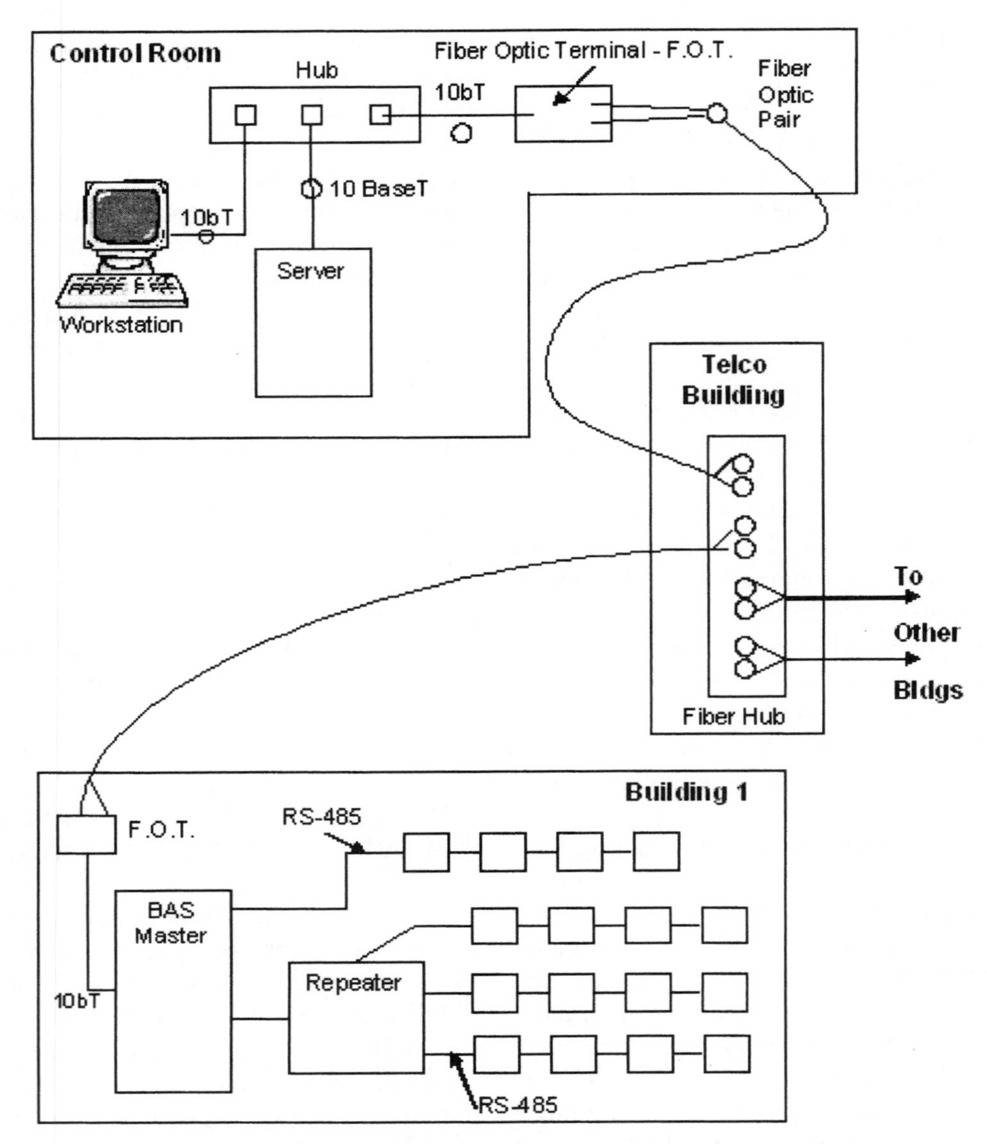

Figure 19-3. Development of BAS TCP/IP network

than five times. The price however included previously unused technology such as hot-swap RAID disk arrays, redundant power supplies, etc. The telecommunication devices, essentially the fiber optic transceivers and the required standard for switched, managed hubs actually lowered the cost of expansion of the BAS backbone, due to declining costs of Ethernet switches in comparison with the fiber optic hubs. Once these relationships were established, it became easier to handle such challenges as compliance with the site standards for TCP/IP static and dynamic addressing, and although we had some issues with identical IP addresses on the BAS backbone, these were relatively easy to clear up.

Andover had the forethought to include web instruction handling directly into the BAS Master Controller. Once the infrastructure relationships were established by HVAC, Information Services, and telecommunication departments, a BAS network to corporate network bridge was built, enabling instructions to be sent to the BAS controllers and received/displayed by a simple web browser equipped PC on the corporate LAN.

The site has since grown to over 47,000 hard wired I/O points. The coaxial cabling was replaced by CAT-5 UTP. All of the fiber optic hubs were replaced with switched, managed hubs and rack mounted fiber optic transceivers (FOTS). There are now 7 servers, and about 20 Infinity workstations. Figure 19-4 shows this system.

Security versus information availability, database change control, configuration management, data integrity responsibility requiring tight access control, activity logs, alarm logs, logon and password requirements, paperwork, signature responsibilities, SOP's training. On

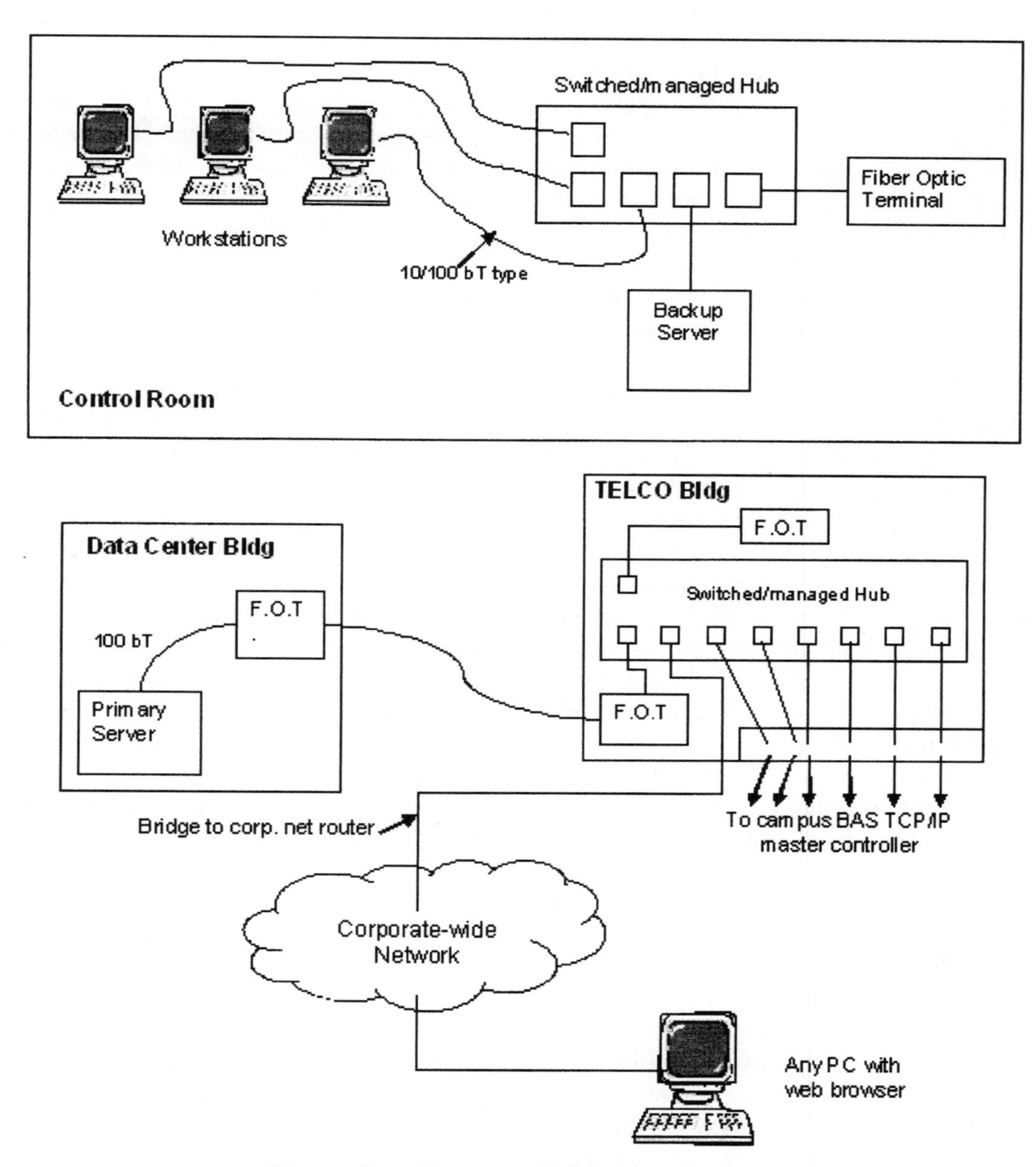

Figure 19-4. Modern BAS TCP/IP network

the other side, without the traditional building control room, how do you get local BAS controller operation information? This resulted in reasonable requirements for a one-way, read-only view of data on the BAS network.

SYSTEM DESCRIPTION AND INTERNAL WORKINGS OF THE ANDOVER INFINITY SYSTEM

The existence of the web-enabled instruction set of the Andover CX series controller chip, the site network infrastructure, complete with world wide web proxy servers and firewall, a well installed platform of Microsoft Windows based PC at every desktop across the site required only a small bridge between backbones. Since the HVAC BAS network had been following the network standards, it was easy to build a bridge between the corporate network and the BAS backbone.

Once installed, small report servers (PC's native to the BAS network) were set up to see the new bridges as a default gateway. This not only protected the BAS building master controllers from unsolicited network packet traffic, but also opened up a path such that BAS data could be retrieved from anywhere on the corporate LAN, or VPN extensions of the corporate LAN.

This bridge could also be used to send BAS data to centralized, controlled and secured data historians such as PI servers, for manageable long term history storage and retrieval. Previously, BAS long term data storage and retrieval was abysmally inadequate for the number of critical points on our control system. Terminal controllers have a mild amount of memory available for point value history. The point value history data however is typically stored in volatile RAM and can be erased very easily. In the best of cases there is enough memory to locally store values (at a very mild point/time resolution) for about 2-3 days. Longer term storage would ultimately include downloading the controller's memory at regular intervals to the building master controller memory. Once in the master controller, it can be stored in volatile RAM for weeks (at a very high risk) or a batch program can be created on a front-end workstation, which can automate the downloading of the master controller memory to a data file. Subsequent data files would either have to be appended to the original file, or a program can be written to change the data file name with respect to something descriptive, for instance, the time and date of the file creation.

These text files could then be batch copied to the server, where a DAT backup could be used to archive the data. Then, upon request or need, you could search the files for the data in question. This was the best solution in terms of data reliability, but it was bulky, manual, and there wasn't enough BAS technician manpower to manage the data, or page through it for that matter. The site's BAS strategy was to stick within our area of expertise, that is, HVAC control consistency and accuracy, and not data management. We would typically refuse to be responsible for long-term history storage; however, the demand for reliable storage was increasing every year.

The final disposition for long term data management was made at the manufacturing information management level, which ultimately decided on PI. Plant Information (PI), has the responsibility to make available and improve the use of process and manufacturing information. The PI system was chosen as a standard because it excels in collecting, storing, and providing historical process data. Readily available systems have been developed to use OPC Servers as the primary means to collect a complete copy, or a subset, of HVAC data for historical storage. Then OPC Clients, at desktop personal computers, has a quick and secure retrieval of data for review and analysis. HVAC data is kept current through an automatic updating system. As far as BAS HVAC was concerned this was perfect. They had to maintain their equipment; while other departments were responsible for the maintenance of the historian.

At current writing, we have some buildings that are still pneumatically controlled. In some cases we have Andover BAS monitoring points installed to see if air handling units are running, discharge temperatures, etc. We have some remnants of AC256 controllers, including some of our older laboratory buildings. We communicate to them on a stand-alone fiber optic backbone. Terminal devices (Hitchhiker Serial over Ethernet) convert serial communications to TCP/IP packets at the AC256 master controller panels in each building. A Hitchhiker device in the control room connects a simple PC to communicate with the remaining AC256 masters on the site.

The Andover Infinity BAS network is comprised of 3 separate fiber optic backbones in a distributed star configuration. We have 3 unmanned primary SQL servers in the site's World Wide Computing building. We have the 3 backup SQL servers in our HVAC control room office, located in a site maintenance building. Each server has 2 addressable hard drives, comprised of 4, RAID 0, hot-swappable, SCSI hard disks. Each unit has redundant power supplies. We have our own PDC server (all other SQL servers are BDC) in our maintenance building as well. Each of the 3 backbones has a connection to the corporate LAN, enabling the web interface and the connection to the PI server.

THE WEB BASED INTERFACE IS SIMPLE:

Utilizing Ethernet TCP/IP communications, building personnel can now control and monitor their buildings over the Internet. "Surf" your building with Andover Controls!

- Access real-time facility data through any internet browser
- Use familiar interface and monitor building systems
- Works with any popular web browser
- Generate real-time data, alarm status, and reports
- Possible future capabilities to create forms and menus to change setpoints, run programs & more
- Exchange information across multiple web sites
- Front-end workstation not required

BASIC WEB SERVER ACCESS

- The Web Server is built-into the CX firmware and the SX 8000 software as purchased. There is no specific hardware or software installation required other than normal setup for either of these products. When power is applied to the system, the Web Server starts-up and remains quietly running in the background transparent to other operations. While in the background, it is scanning for command requests that conform to a standard communications protocol used by Web Servers and Browsers. Commands sent to a CX controller or SX 8000 workstation can originate from many sources. These commands are sent in small chunks called "packets." CX controllers receive commands via their Ethernet network connection (Andover proprietary named EnergyNet). SX workstations can also receive commands via Ethernet or from their local keyboards. All Web Server command and request traffic is handled over the Ethernet using a protocol called TCP/IP. At present, it is the only method of communicating between the Server and the Browser. Web Server—Web Browser Connection

- Once you become familiar with HTML's simple formatting rules, you'll be creating screens in no time. For simple access to all the points in your system all you really need to know is how to create list views. Creating an HTML file takes spending a few minutes with a standard text-editor program

Designing Your Own HTML Pages

Designing your own HTML pages for use with the Infinity Web Server is a three-step process.

1. Create the HTML that defines the page.
2. Convert the HTML into Plain English print statements.
3. Load the Plain English function with the embedded HTML into the system.

There is one important concept to master before you construct your own Web Server pages. The Infinity Web Server sends these user-created HTML pages to your Browser in response to a call to a Plain English function. You create a custom function and include HTML codes that serve as your user interface within that function. When the function is called, the Server machine (CX or SX 8000) operates on the Plain English commands and sends any output it generates to the Browser. Output statements are simple Plain English print statements. To directly display a custom page without having your users access the built-in home page, you have to define a function including the HTML describing the page and save it under some descriptive name.

Advanced Design

Limited only by the imagination of the system administrator, many types of pages and many functions can be designed to operate over the Web Server.
Some other useful HTML pages can provide:

- Graphic panels with embedded values" Tables & Frames
- Alarm Displays with Interaction" Tables & Forms
- Selecting Points to Display" Forms
- Setting Points and Setpoints" Forms

Information Flow Diagram and Preparing Reports

Web report request - Corporate LAN - Gateway - Report running SX - CX- terminal device data value.

Web information—Request by link:
Department supported web page running on a corporate web server, with reports (hyperlinks to SX reports) organized by function or area. The person requesting information simply clicks the hyperlink to the report.

The hyperlink URL points to a report programmed at the SX workstation (BAS front-end workstation that is dedicated to the web report function). Typically such reports generate output on the local SX screen. The reports are programmed to organize data values into formatted pages. See example in Figure 19-5.

AHU Operations Report, Corporate Building 33
(Today's Date/Current time)

AHU	Command	Status	Supply Temperature	Failure
1	On	On	55.6	OK
2	On	On	68	High Discharge Temp.
3	On	Off	84	FAN FAILURE
4	On	On	54.9	OK

End of Report

Figure 19-5

The format for each report consists of static text and data values from each terminal device, and in some cases some time and date information.

Screen generated reports are programmed in a language very similar to BASIC, with simple commands such as "Print" and "Tab" and the ability to conjugate such as this excerpted example in Figure 19-6.

OPERATIONAL CONSIDERATIONS OF SYSTEM NETWORK SECURITY

Total system security is maintained due to the following:

- A front-end logon account is necessary to create the report.

```
Print  "          AHU Operations Report, Corporate Building 33"
Print  "                              "; b33c1 datetime
Print  " "
Print  "AHU    Command    Status  Supply Temperature   Failure"
Print  "1";b33c1 scxac1 sfc01;"       ";b33c1 scxac1 fan.status;"      ";b33c1
       scxac1 te__03;"        ";b33c1 scxac1 almtext
```

Figure 19-6

The report could be output to a line printer connected to the computer by changing the command to "LPRINT."

The report output could be to file by opening a new text file and adding formatted lines, then closing the text file.

The report can also be redirected to a computer internet browser by adding the appropriate HTML commands under the Print command (Figure 19-7).

- 2-way communications is limited to the programming; if the report is only a display of current data (as in the example) a simple read-only path is created, available to anyone on the company's network.
- Only data that is programmed in the report can be retrieved.
- Direct network pinging to the LAN master controller is restricted; the workstation however can see intranet traffic.

```
Print   "<body lang=EN-US style='tab-interval:.5in'>"
Print   "<div class=Section1>"
Print   "<p class=MsoNormal align=center style='text-align:center'>AHU
        Operations Report, Corporate Building 33</p>"
Print   "<p class=MsoNormal align=center style='text-align:center'>";b33c1
        datetime
Print   " </p>"
until all of the HTML codes, static text and dynamic data values are printed.
```

Figure 19-7

If the report with embedded HTML codes is triggered by a web request, the report is run from the workstation and sent back out to the internet browser requesting the view. The browser doesn't "see" the PRINT commands, just the HTML and plain text generated by the report server.

Since a company logon and password is required to see the intranet site, only traffic generated inside the corporate firewall can make it to the web server.

Other programmed methods for web information retrieval include the use of JAVA programming. In fact, most functions that can be executed on any generic web site can be replicated with this system. In order to run higher order programs, the cooperative use of both the corporate intranet web server and the BAS web interface would be implemented.

The web interface can therefore be programmed to allow setpoint changes to any system, and can be used to start and stop individual BAS programs or systems. Theoretically, corporate tenants could start and stop their own HVAC zones in their buildings, if they should come in during programmed unoccupied hours.

By allowing information flow through the company's firewall, this information could be available directly from the world-wide web, allowing for buildings to be controlled and monitored from anywhere in the world.

The use is restricted, however, due to the requirements of our unique site, to one-way (view only) type communications.

ECONOMIC AND TECHNICAL RESULTS OF USING THE ANDOVER INFINITY EIS/ECS SYSTEM

As I have described, there has been steady growth of connected I/O at the site, and a forecast that suggests continual growth over the next 5 years. There has also been very little appreciable increase in available mechanical or technical staff. The web-based interface to the BAS network has been an efficient way for mechanics and programmers to check systems status anywhere on the BAS network from anywhere within the campus. This efficiency will become critical as the areas where close monitoring is of vital importance will slowly outnumber the monitoring personnel.

This efficiency is enhanced by taking advantage of the large number and various locations of company computers, already equipped with internet browser software. Essentially, the HVAC mechanic has a valuable BAS front-end wherever they are on the campus.

Future applications can be easily integrated and maintained, and non-critical functions such as off-hour occupancy and uncontrolled, unclassified area space temperature set-points may be controlled by the end-user, or area occupant, thereby saving time for the BAS front-end operator to be used in more critical functions. With the company's adaptation of VPN, reports and status information can be retrieved and manipulated from anywhere there is a connection to the internet.

NOTES ON THE FUTURE OF WEB-BASED INTERFACES:

BAS systems all started out with highly proprietary front-ends. The most successful of the BAS systems were based on simple interfaces.

The decline in TCP/IP Ethernet enabled device costs is leading to an increase in products that are configured and programmed via network connections. I have also seen the price of switched hubs and routers plunge an order of magnitude in 5 years. I strongly believe that there will be a migration of TCP/IP devices including the lower level terminal controller devices and perhaps even the BAS input and output devices. It is plausible that an entire system based on Ethernet communications could be conceived. All devices in the BAS network could be viewed and configured via a web interface page. System graphic panels could be very simple (or very elaborate) web pages. With the proper planning for routers, I/O groups would be able to update fast enough to provide smooth PID control, and have the same ability to be interconnected that office work groups, computer workstations, and network printers have now. Using local hubs in a distributed star would limit the damage done by a singular network segment failure

CONCLUSION

DDC and web based energy information and control have allowed the Merck site at Rahway to monitor and control the HVAC and lighting systems in over 100 buildings with a small maintenance and operational staff. The heart of the system is an Andover Infinity BAS. This web browser technology is simple and robust, and the author believes that most BAS devices will have a web-enabled function similar to the Andover product in the near future.

Chapter 20

Interoperability of Manufacturing Control and Web Based Facility Management Systems: Trends, Technologies, & Case Studies

John Weber, President, Software Toolbox Inc

ABSTRACT

For years, facilities management and energy management systems have been primarily systems offered by major industry players seeking to provide the entire solution for the end user's application. There are excellent systems available; however, they all tend to share one thing in common, a general lack of openness. When users want to integrate or interface these systems with other plant systems, other vendor's hardware, or plant control systems, the cost of integration grows dramatically or users are told they just cannot do it.

This same challenge faced the manufacturing automation industry in the 1980's and early 1990's. Since then, open standards and open technologies have taken great leaps in decreasing the cost of integration. Technologies from the commercial information technology market, such as Ethernet, have made a significant impact on connectivity on the plant floor. There still is a long road to travel to reach the perhaps mythical "plug-and-play" control system. It may never be reached. In the manufacturing automation industry, the drive for decreased integration costs through open systems and standards will continue as manufacturers are forced by global competition to become more competitive.

The first half of this chapter will explore two trends in the manufacturing automation industry already having an impact on facilities and energy information systems: open connectivity standards and wireless connectivity. The second half of this chapter is dedicated to coverage of case studies where integration, interoperability, or utilization of technologies from the commercial information technology world or manufacturing automation world have made a significant impact on a facilities management system.

INTRODUCTION

Nearly every manufacturing operation involves some type of control system. Whether a single machine, a work cell, or an entire line, there is an electrical control system involved. That control system may be simple or complex, it may operate the entire machine without a human operator, or it may perform basic functions only upon the command of the operator. For purposes of this discussion, the term "Manufacturing Control" will be used to refer to techniques, technologies, and systems used in controlling the actual manufacturing process.

Of course, all manufacturing operations are housed in some form of a facility, and there are systems that manage the facility's energy usage, temperature, humidity, air quality, electrical system, and more. The needs of the facility's control systems are typically driven by the specific manufacturing and safety requirements of the actual manufacturing process, with consideration of any regulatory requirements included. For purposes of this discussion, the term "Facilities Management" will be used to refer to techniques, technologies, and systems used in managing the facility.

Without the proper facility and infrastructure, the manufacturing process cannot operate. Without a manufacturing process inside to generate products that people want to buy in sufficient volume to be profitable, there is no need for a facility or an energy management system.

Despite this obvious co-dependency, for years there has been a divide between Manufacturing Control and Facilities Management systems. Sometimes the systems are interconnected, often times they are not. Technologies used in each type of system sometimes overlap in use of hardware components or software, but often times they do not. In some cases, the lack of sharing of common technologies is justified as the specific needs of facilities management and manufacturing control in a specific business may not overlap. In other cases, there is great overlap. For example:

- If you break down each system to core components, they each are looking at analog or digital signals.
- Based on those signals, decisions about what to do are made by the system.
- The values of the signals are communicated to machine operators, facility managers, and other interested parties.

This gap between the systems results in missed opportunities for better coordination between manufacturing and facilities management and missed profitability improvement opportunities. Examples include:

- Use of open, standards based systems from Manufacturing Control to provide more flexible and open systems in Facilities Management.
- Use of similar hardware components allows for common spare parts, and common maintenance training
- Use of the same software systems allows for close and low-cost integration of facilities data into process operations displays and vice-versa
- Integrated systems allow for quicker and easier reporting and analysis to aid in drawing valuable correlations that can improve profitability, such as:
  - Product quality with facility conditions (air, humidity, temperature)
  - Energy usage with product production mix and time of day
  - Energy usage with machine operating conditions—i.e. machines that need maintenance may use more energy and thus provide an early indicator of machine maintenance needs.
  - And many others—take a moment and think about questions you might ask about your business and could get answers to, if only you could quickly get the data needed to answer the question.

As more energy management systems turn to the use of a group of technologies loosely grouped as "Web-based" or "Internet" or "Intranet," the availability of information is increasing for the facilities or energy manager. Manufacturing control systems are following a similar trend.

This chapter will explore the state of the use of open systems and web-based technologies in Manufacturing Control Systems. It will show how opportunities that can benefit the entire business exist for sharing of technologies, ideas, and techniques from the world of Manufacturing Control with Facilities Management systems.

TRENDS IN MANUFACTURING CONTROL SYSTEMS

There are two trends in the Manufacturing Control Systems domain to consider and discuss how they relate to the use of web-based systems in energy and facility management: Open Systems and the use of Web Based Technologies.

Open Systems

For most of the last decade, there has been a general trend towards openness in Manufacturing Control Systems. By openness, we mean the ability for the owner of the control system hardware or software to easily interface the product to that of another vendor, or easily modify the operation of the system to suit the user's particular needs.

You could in theory take the idea of open to a logical extreme where everything in the system was open for modification, as in open-source computer software, where you can obtain the source code and modify the system to suit your specific needs. There are even some examples of open-source software in the Manufacturing Control Systems market; however there are not signs of wide spread adoption of these systems.

What most users of Manufacturing Control systems seek are standardized interfaces at all physical and logical connection points to the particular piece of control system hardware. They also seek easily implemented ways to customize the software or hardware within reason. Some examples will help illustrate these points.

Example 1

In a Programmable Logic Controller (PLC), it is common today to have a TCP/IP Ethernet interface available either as an option module or built into the controller. Although the protocol used at the application layer of the TCP/IP packet is unique to that vendor (i.e. AB, GE, Modicon), the hardware required to connect to the controller is standard off-the-shelf commercial Ethernet hardware.

Example 2

There a variety of types of networks available to connect a PLC to field sensors, instruments and devices. These networks are typically called "fieldbusses." There are a number of fieldbus standards in the Manufacturing Control world, each suited to be strong in specific application types. Representative names include Profibus, Foundation Fieldbus, Hart, Devicenet, CAN, ASI, FIP, Interbus and more. Users purchasing a PLC expect to be able to obtain from the PLC manufacturer or a 3rd party a module that will allow them to be able to connect the fieldbus(es) of their choice to that PLC.

Example 3

Human Machine Interface (HMI) software vendors began to adopt the OPC standard for connectivity to communications drivers back in the late 1990's. Since then, it has become a near defacto requirement to offer OPC connectivity if a vendor plans to try and sell HMI software. By using OPC for device connectivity, the software vendor allows their user to choose from a wide range of vendors for the connectivity software required to connect to both Manufacturing Control and Facilities Management systems.

In Manufacturing Control Systems, users have found that often, by demanding the flexibility of open systems for connectivity to other systems and devices, they are able to choose best of breed components and integrate them to provide a system that meets their exact needs. There is a cost to such integration. The user should carefully weigh their integration costs versus the benefits they gain from choosing particular components from multiple vendors. They should also look carefully at any interoperability information available from a vendor before embarking down the path of a system based on multiple vendors, tied together with open system interfaces.

They key is the trend towards openness has resulted in greater freedom of choice for users. The trend towards more openness has reduced, but not eliminated, the possibility of getting locked into a single vendor's offering.

Web Based Technologies

Ethernet is the largest of the web or Internet technologies in use today in manufacturing. The use of Ethernet as a networking medium in Manufacturing Control Systems has grown remarkably in the last decade. As the internet boom and bust made commercial Ethernet hardware for PCs and setting up networks low cost and ubiquitous, providers of Manufacturing Control Systems have seized upon the opportunity to leverage the low cost, high volume production of these components. Nearly every control system now offers an Ethernet interface. Some control systems are beginning to offer wireless Ethernet interfaces as options. Providers of industrial grade network switches, wireless access points, and even cable components are proliferating in the Manufacturing Control Systems Market.

At this time, Ethernet has not all together replaced the many proprietary control networks that were used prior to Ethernet's rise in popularity. The reasons for this are multiple. First is the existing installed base of networks from a variety of vendors. Users have significant investments in cabling, support tools, and training. If the networks are working, doing their jobs, and can accept additional capacity, then there is no incentive to change. Second is the typical life cycle in control systems, which is about 10 years. That is the average. It is not uncommon to see 20+ year old control systems in use as companies seek to maximize their return on the capital invested in control systems. Over time, as older systems are replaced, the installation rate of Ethernet will increase. The third reason is there are applications where vendor proprietary control networks are needed. One example involves what is known as "deterministic response" or "determinism." Control systems engineers use the term deterministic response to mean that when a request is sent from one device to another on the control network, the time it takes for the transaction to complete is predictable, and within an application specific tolerance, repeatable from one trial to another. Different applications can accept different levels of variance in the two variables discussed that comprise a deterministic response. It has been shown in published studies that using 100baseT switched Ethernet, a level of determinism can be achieved that is the same as some popular proprietary control networks. That said, there will be cases though where the customer's definition of what constitutes acceptable "repeatability" and "predictability" will not be able to be met by even the best switched Ethernet networks. Thus, there will likely always be some need for vendor proprietary control networks to meet specific functionality required by specific systems and applications, either in security, capacity, level of determinism or other functional reasons.

The major providers of Human-Machine Interface software have also joined in and added web-browser based user interface options to their systems with the objective of providing information from the Manufacturing Control System anywhere. The implementation details of each offering are wide and varied, but the theme is the same—leverage the fact that every Windows based PC by default has a web browser installed.

CONNECTIVITY SOFTWARE AND WIRELESS TECHNOLOGIES IN MANUFACTURING CONTROL AND FACILITIES AUTOMATION

Connectivity Software for Manufacturing Controls and Networks

OPC Standards

Software for connectivity in the Manufacturing Control world can be handled in several ways, but the most popular way is using software that adheres to the OPC standard. OPC today stands for "Openness, Productivity and Connectivity." The OPC standard is an open standard for software-to-software connectivity that was first created in 1996 by a group of hardware and software vendors serving the manufacturing control business who came together to form the OPC Foundation. The OPC standard is managed by the independent OPC Foundation (*www.opcfoundation.org*). The OPC Foundation's primary mission is to foster open interoperability in automation software.

Today, the OPC standards comprise a wide range of specifications for exchange of different types of information between automation/manufacturing control software. When the OPC Foundation was first started in 1996, the first standard created was the OPC Data Access (OPC DA) standard. Prior to the OPC Data Access standard, every company that provided applications that needed data from the shop floor had to write their own communications software for the myriad of device types. (Figure 20-1). This resulted in a great deal of excess work done in the market and a wide variance of functionality and quality in the communications software offered by each vendor providing software that interfaced with the plant floor.

The purpose of the OPC DA standard is to create a way in which communications software ("OPC Servers") can be created that communicate with wide and varied types of control networks and expose that data through a standard software interfaces to applications that need to consume the data ("OPC Clients"). The primary objective is to allow software applications that support the OPC Client interfaces and need access to read/write data over manufacturing control networks to be agnostic relative to the device specific protocols. (Figure 20-2) The OPC client applications only have to be aware of tag names and have a standard way of reading and writing the data to the OPC server software applications. The OPC Server software applications handle all the specifics requirements of the device specific network topologies, hardware, and addressing.

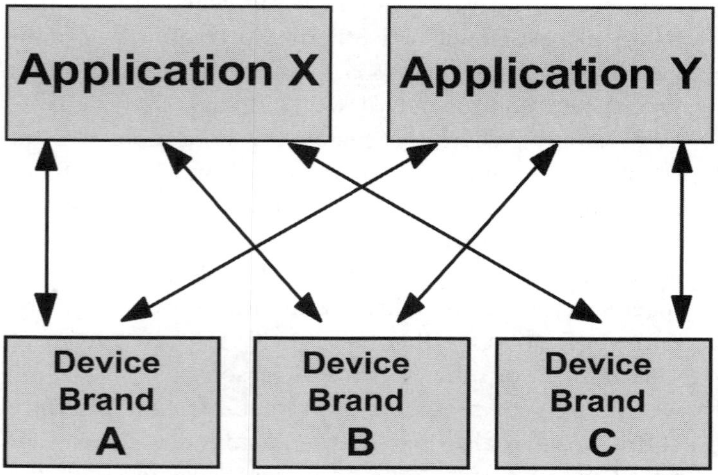

Figure 20-1. Device Connectivity Without OPC requires each Application Vendor to write their own drivers to talk to each device brand

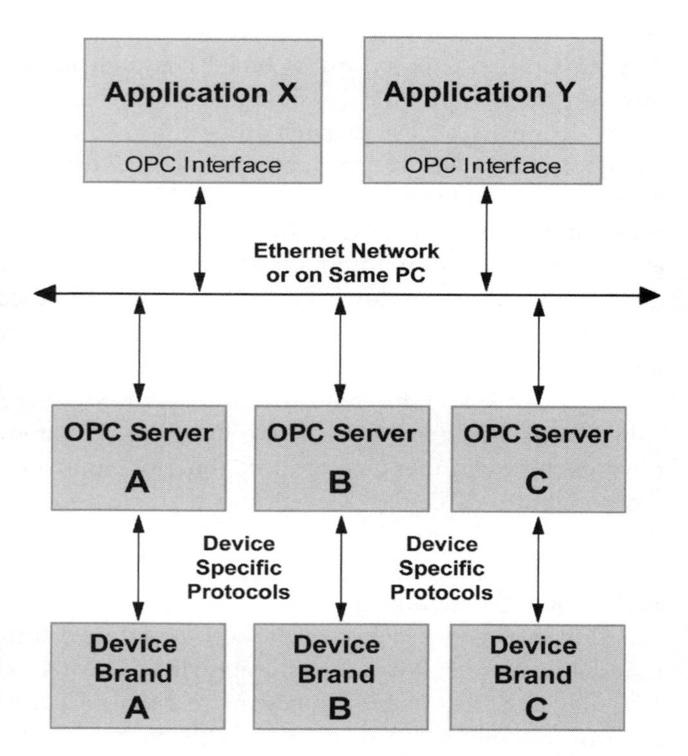

Figure 20-2. Utilization of OPC for Device Communications Separates Communications details from the Application

The result of this for users is they can obtain their connectivity software from any vendor that is able provide the solution that meets the user's requirements. For developers of OPC client applications, they can exit the business of writing connectivity software and let companies solely focused on connectivity provide the OPC server's software. For users, they are no longer limited to the connectivity offered by the OPC client application developer.

When talking about OPC compliant applications, the terms "server" and "client" are often heard. These are not hardware terms. These terms speak to the nature of the data handled by the OPC compliant software. In the world of OPC a "server" application provides data (Figure 20-3). It speaks a vendor specific protocol through one interface, and it serves the data to applications that need the data through it's OPC standard software interface.

An OPC client application (Figure 20-4) is one that wants data from a device or system. The OPC client gets its data from a server application. The same logic applies when writing data. An OPC client wants to write data to a device and it tells the server to perform the write on its behalf. It is the OPC server's job to perform the write of data to the device or system and report back a success or failure.

After OPC Data Access, the foundation moved on to address the exchange of complex data, historical data, batch process data, and alarms and events data between software applications. They have recently released OPC-XML specifications that provide means for exchanging OPC data over networks using standard web services and XML. The foundation is presently working in their next generations of technology that will build upon the existing standards with a unified architecture for access to all types of manufacturing control data.

Today there are over 300 member companies in the OPC Foundation offering OPC compliant products. Member companies are able to obtain OPC Certification for their products through a series of tests and interoperability workshops that strive to give users assurances that products carrying the OPC Certified logo can deliver on the promises of interoperability in automation. The OPC Foundation maintains a catalog of OPC Certified products on their website at *www.opcfoundation.org*.

For the building automation system manager, the relevance of OPC is that they can leverage the open connectivity options offered by the over 300 companies offering OPC solutions to provide connections to their building automation systems and to manufacturing control systems. By utilizing OPC standards based solutions in their systems, building automation managers can reduce their integration cost while expanding the scope of integration into areas that may have been closed to them in the past.

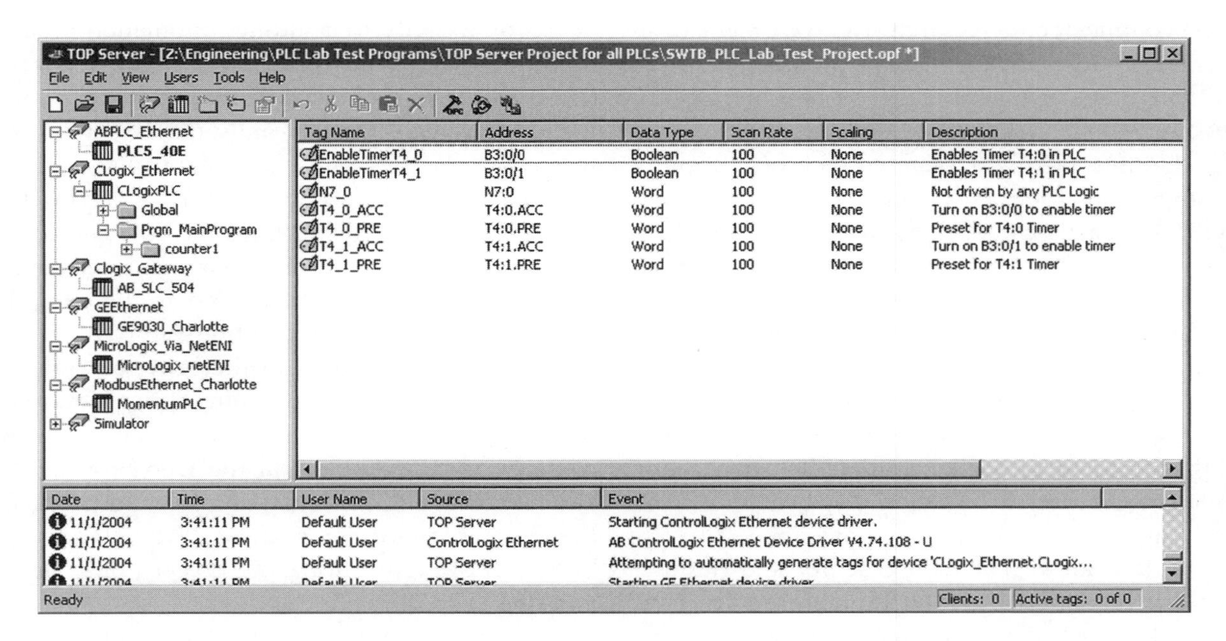

Figure 20-3. Example OPC Server Software User Interface with connections to 6 different control system types plus a simulation interface for testing.

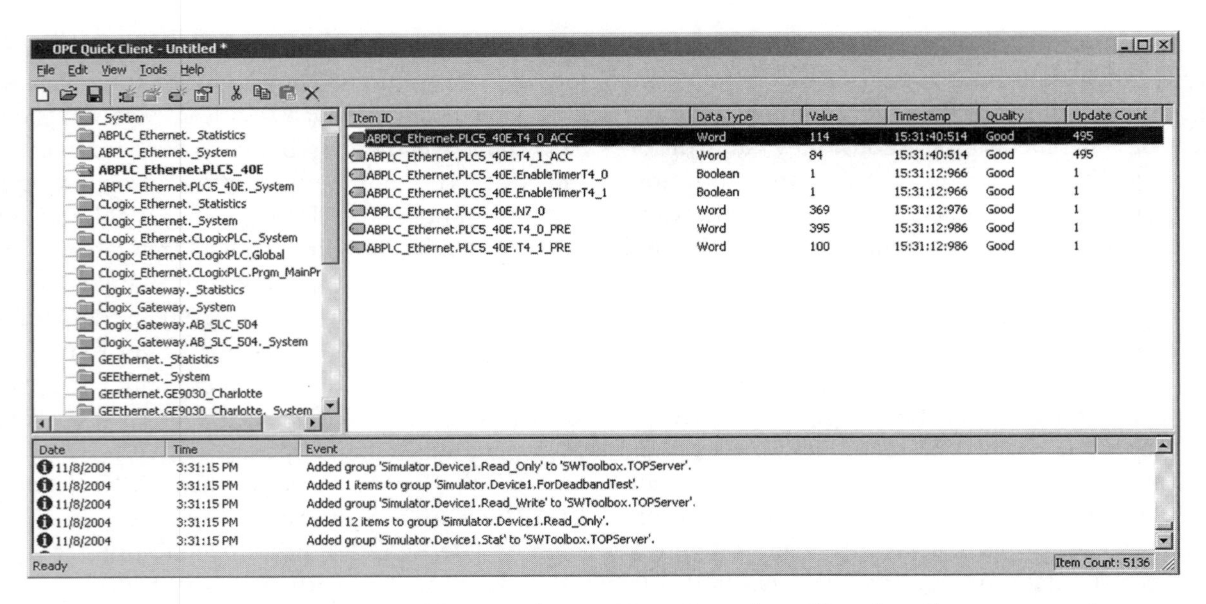

Figure 20-4. Example OPC Client Software User Interface Showing live process data

Other Connectivity Methods

There are other methods of software connectivity to control systems besides using OPC standards.

Within any given software application, the provider of the application may include their own drivers to connect to field devices either of their own manufacture or from third parties. In some applications, these built-in connectivity offerings will work just fine. Users should of course investigate these offerings with their own requirements in mind though to insure that they can get to the data that the user needs, at the data rates required, and in the volume of data required. There are some cases where connectivity software that is provided with an application is done to meet a basic set of requirements, but advanced functionality is not supported.

For the user looking to develop their own applications using tools such as Visual Basic (VB), Visual Basic.NET or Visual Studio, there are tools available in the form of ActiveX control software or .NET components that can handle communications to control system devices. Generally there are two classes of these control communications components: device specific or OPC connectors. The latter, OPC connectors are components in the form of ActiveX software or .NET components that plug into a VB or Visual Studio.NET application and can connect to any OPC server. The benefit of these types of solutions is that custom developed software only has to find an available OPC server and the user can stay out of the business of knowing device level protocols. This type of solution may have a higher per unit cost for deployment, but engineering costs can be lower. If you want to avoid all per unit deployment costs (i.e. runtime fees), then ActiveX or .NET components of the second class, device specific, may be a good fit for your applications. With these components you do have to know more about the device specific protocols, and you will need an ActiveX or .NET component for each device type that you wish to communicate with. For the savings of avoiding per system deployment costs for software licenses, you assume more responsibility for device level communications.

Web Connectivity

Connectivity to building automation systems and manufacturing control systems using web/browser based technologies has taken many leaps forwarding in the last few years. Until recently, displaying process data in a web browser could be done, but had its issues.

The most common method of displaying live data in a web browser was to have the web server read the data from process control systems and deliver it to the browser as a static value. To update the data, the user would have to refresh the entire web-page. Other solutions could deliver data to the browser and have the data update without a page refresh, but required you to install software on the client PC or the use of ActiveX controls or Java Applets in the web browser. Current security measures in companies could often prevent these small pieces of software from loading in the browser, thus preventing these types of systems from working.

In the last couple of years, the use of XML and Web Services has changed the way data can be delivered to a web browser. A Web Service is an intelligent application that runs on a webserver and the web browser can load a web page that will automatically pull new data from the Web Service without refreshing the entire page. As long as you have Internet Explorer 5.5 or higher on a Windows PC, you can consume a web service from a Web server.

There are now software products available that will run on a web server, connect to OPC server software on the web server, and allow you to publish process data in a way that users can see the live process data in a remote web browser, updating, without having to refresh the entire web page or load software on the client PC. One example is the OPC Web Client product (*www.opcwebclient.com*).

The OPC Web Client software is loaded on a webserver, talks to local or remote OPC servers, which in turn talk to building automation systems or manufacturing control systems. The system implementor builds web pages that access the OPC Web Services provided by the OPC Web Client to display the live data in the users web browser. The user simply points their browser to web pages on the server and they are able to read and write data from their browser. (Figure 20-6)

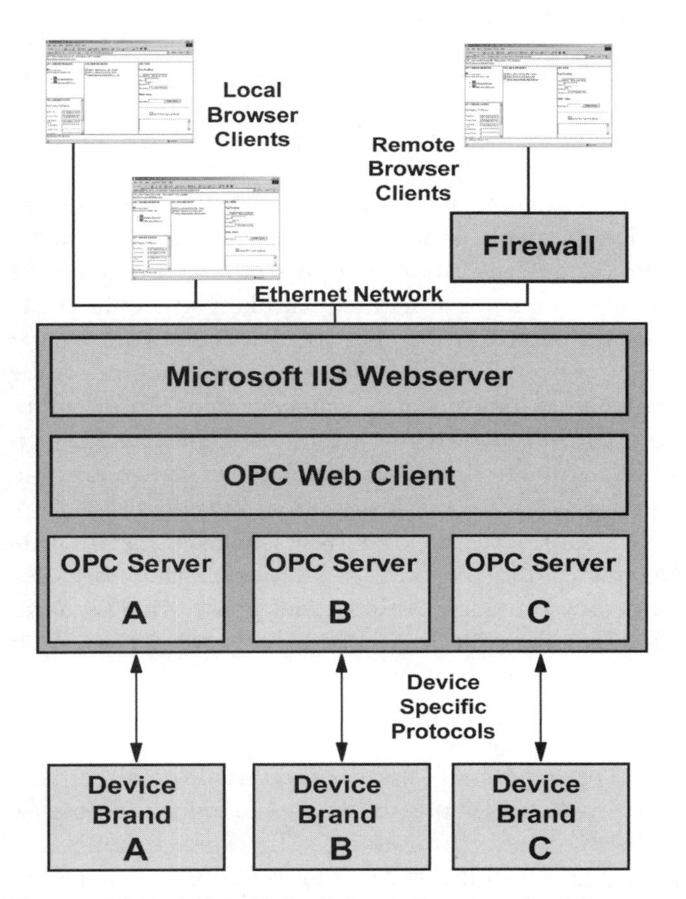

Figure 20-5. OPC Web Client System Architecture Overview

Wireless Technologies in Manufacturing Control Systems

There has been a growing trend towards the use of wireless technologies in Manufacturing Control Systems. There are two areas where wireless technologies are typically used in Manufacturing Control Systems: in-plant and remote site communications.

In-Plant Applications

With in-plant applications of wireless, the objective is typically to avoid the costs and time delays involved in running network cable to reach a remote device. In some cases, it may not even be practical to run additional cable. Using off-the-shelf wireless technologies used in commercial information technology applications, plants can quickly create Ethernet connections from one part of a plant to another. Typical wireless technologies employed are IEEE 802.11a, 802.11b, and 802.11g.

Typical ranges with in-plant wireless technologies can vary greatly depending on the physical site. Wireless routers and access points are available that advertise 108 Mbps speeds at up to 400 ft; however, the range can drop dramatically as signals have to pass through walls and other obstructions. The more obstructions that exist from the device to the nearest wireless access point, the weaker the signal and thus full speed may not be reliably achieved. Some manufacturers offer the ability to include repeaters or additional wireless access points to help with physical obstructions. Users planning to deploy wireless should carefully test signal strength by placing routers at proposed locations and then using a portable laptop computer with a wireless access adapter to identify the practical bounds of their wireless network.

When implementing wireless networks, one must also be careful to not assume infinite throughput. It is easy to think "oh this is 108 Mbps, I'll never use that much bandwidth." As previously discussed, the rated speed of wireless hardware is the maximum speed under proper conditions. Obstructions and interference can easily drop the practical speed on the wireless network. Also, wireless networks are not necessarily switched networks—which means in some cases with multiple devices talking concurrently, collisions can occur, just like on non-switched Ethernet, which means the more traffic on the network, the greater the risk of degradation of performance. Wireless can perform well if applied carefully and not used as if it is a connection of infinite bandwidth.

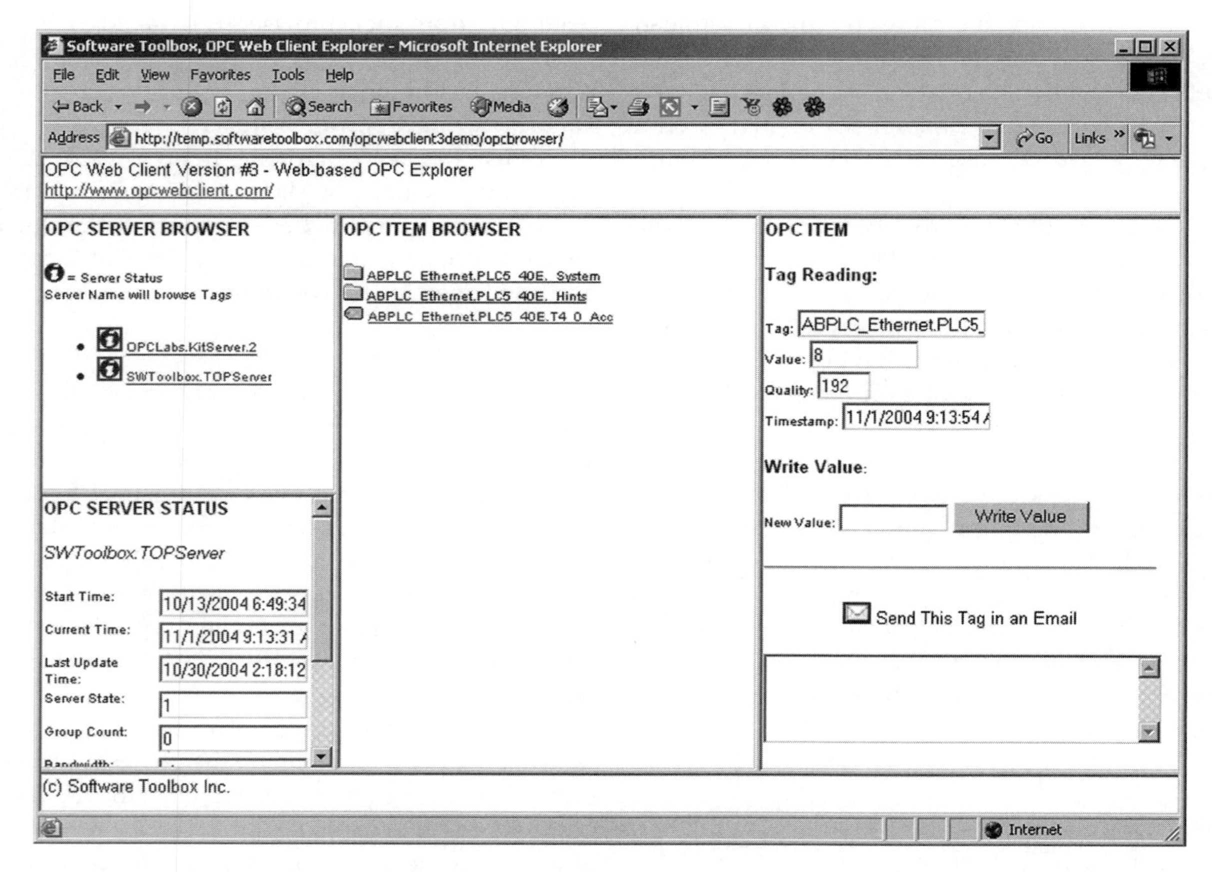

Figure 20-6: Web Browser displaying data from OPC servers using OPC Web Client

Security is a very important consideration in implementing commercial wireless technologies. Users should NEVER connect a wireless router or access point to their plant network without first at a minimum changing the default password for access to the router. Commercial wireless hardware typically ships in a non-secure state, with a default password. If a wireless router or access point is installed without changing the default password, an intruder can easily find the device on your network and access the login screen. Nearly all wireless hardware "advertises" the model and make on the login screen. (Figure 20-7) An intruder can go to the internet, download manuals from the manufacturer, obtain the default password, and take over control of the wireless network. By changing the default password, you prevent this.

An additional security measure that should be taken is to change the default wireless network name in the router and tell the router to not broadcast it's network name for discovery. This can help thwart an intruder finding the network. Next, you should consider only buying wireless hardware that supports encryption and use the highest level of encryption supported by your hardware. When configuring the encryption setup in the hardware, never use pass phrases that are included in manuals or tutorials in the hardware documentation. Create your own pass phrases that are long and alphanumeric in nature, just like you would create a secure password for your login to a network. Never write passkey codes or other such information on the hardware as a means of convenience under the assumption that intruders won't actually see the physical network hardware. Intruders many times can come from within the facility, not just random outsiders.

Manufacturers of wireless hardware include Netgear, D-Link, Belkin, and Linksys. There are also numerous providers of industrial grade wireless hardware including Black Box, B&B Electronics, Digi, Atop Technologies, and others.

Remote Site Wireless

For remote site applications, traditional IT technologies discussed previously are not a good fit because of their distance limitations. There are a number of emerging technologies for inter-building or inter-campus wireless connections that promise up to 30 miles line of site connectivity. Users needing remote site connectiv-

Figure 20-7. A NetGear Router Exposing It's Model Number through the Login Screen

ity should stay abreast of what new technologies are being adopted in widespread fashion in the commercial IT world that may be applied to facilities or manufacturing automation applications.

Presently, wireless technologies used for remote site applications are either private radio systems or public cellular based systems. Only recently have public cellular based technologies reached levels of performance and cost to make them practical alternatives to private radio networks.

Private radio networks have the advantage of being under total control of the user and can provide very high levels of security since everything is in the user's control. However, private networks can be costly to set up because towers must be built, licenses potentially obtained, and the entire network must be managed and maintained by the user. The water treatment industry has typically taken this route because they had large numbers of remote sites to manage and the economics of managing the network made sense for them.

Presently, the CDMA and GSM cellular technologies can offer up to about 105 kbps throughput on a public cellular network. In some major cities, public cellular data networks with 200-300 kbps throughput are becoming available. The advantage of public cellular networks are that someone else bears the capital costs of setting up the network and the ongoing maintenance costs. You simply pay a monthly fee for each device that has access to the network. Plans vary in cost based on whether you are using metered amounts of data or want unlimited data volume. Depending on your application, the unlimited volume plans can be the most economical solution as most plans are less than $100 per month and discounts can be obtained for multiple device contracts. Obviously local pricing will vary and it is important to work with local carriers to negotiate a solution that meets your needs. The downside of a public cellular network is that you do not control the network. This means that you are sharing the network capacity with others and there are greater security risks. The security risks can be mitigated with proper use of firewalling software and hardware. You should discuss capacity and security concerns with local carriers based on your requirements.

Regardless of the wireless technology chosen, it is important to insure that the communications software you choose can work well with the wireless technologies you choose. You should integrate the selection of communication software with your wireless technology search as the software provider may have vendors of wireless hardware that they have found work better than others. The communications software must also offer the flexibility to deal with the additional delays in transmission time that wireless technologies for remote locations can impose. If you are using public cellular radios, your radios will likely have network IP addresses that can change. The radios must be able to report their new IP address to the host computer automatically and the communications software (i.e. OPC server software for example) must be able to receive those reports and handle the situation appropriately.

Now let's look at some case studies that utilize the various topics that have been discussed in this chapter.

Case Studies

The case studies presented in this chapter provide examples of how Manufacturing Control systems and technologies have been integrated with and interoperate with Energy or Facilities Management systems. Some of the systems are web based, others are not. Regardless, the value to the reader is in generating creative thought about the many possibilities that may exist in their own business for integration and technology sharing.

All case studies are based on real-world user applications that are currently in operation. For privacy and company confidentiality reasons, the exact names of the subjects in some of the case studies are omitted.

Case Study: Gardening supplies manufacturer

Technologies Used:

- OPC Server Software
- Off-the-shelf webserver
- Programmable Logic Controllers (PLCs)

A major provider of supplies to plant nurseries in the United States has facilities in several locations around the country. Their maintenance manager needed a way to access operating conditions and predictive maintenance information from the sites quickly and effectively over the Internet. The systems they wished to connect to were all controlled by GE Fanuc Series 90 Programmable Controllers (PLCs). Connection options for each PLC were serial network or Ethernet.

A key design objective in building the system was the ability to access the system information anytime, anywhere, without having to load any software on a client PC. The system also needed to be cost effective and easy to implement.

The customer evaluated a number of off-the-shelf human machine interface solutions that could also offer Internet connectivity to view the operating data. Solution costs ranged from $10,000 to $20,000 total, with purchased software license costs of around $10,000 included in the total. The customer decided to also evaluate solutions that would require some engineering work on the part of the customer and utilize open standards and technologies to allow the user to build their monitoring application on their own.

The customer chose to utilize a solution using OPC Server software to connect to the GE Fanuc PLCs over an Ethernet network and an ActiveX component that would interoperate with Microsoft Internet Information Server (IIS) to display the operating data from the OPC server in custom web pages designed by the user. Using Microsoft Front Page as a web page editor, and utilizing example web pages provided with the products chosen, the maintenance manager was able to build their application in approximately 1 week. The total out-of-pocket software cost was $1500.

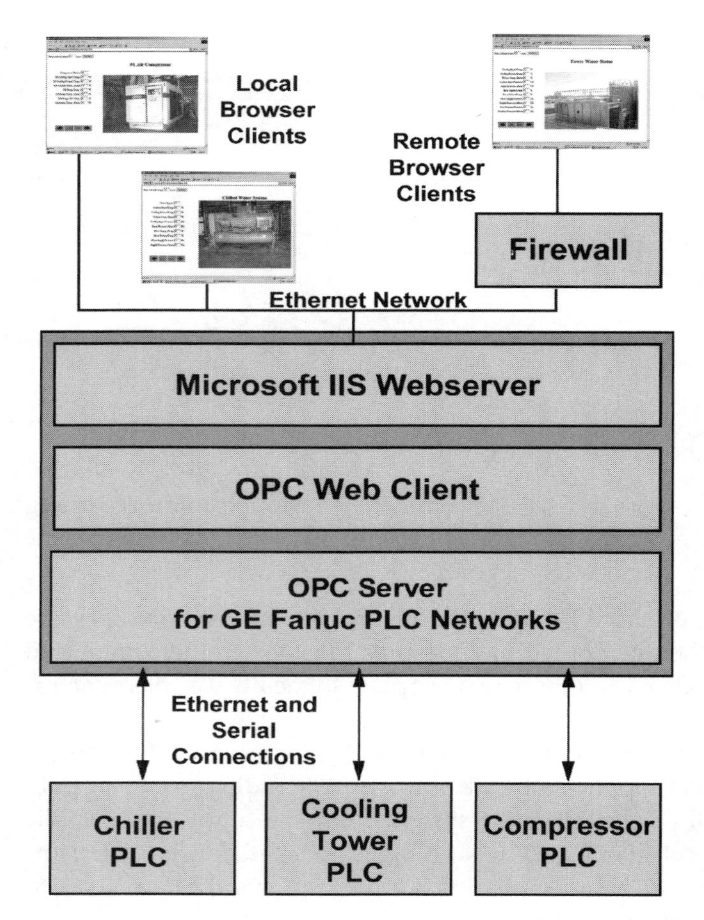

Figure 20-8. Web Based Facilities Monitoring Application Architecture

The system implemented monitors control systems on critical facilities equipment such as ventilation, chillers, air compressors, tower cooling units, pneumatic air dryers, and air reservoirs. As a result of installing this system, the customer has seen efficiency gains of nearly 15% and their annual maintenance budget is at an all time low. The customer attributes the improvements to "Knowing what's going on in our critical facilities systems at anytime from anywhere. With this system we literally feel like we can be in two places at one time. The knowledge we have obtained from our application allows us to make intelligent, informed decisions about what systems need maintenance and when. We are better serving our customers now with improved on-time delivery of products, reliability, and product quality."

The customer chose the OPC Web Client because it provided data in a browser that would update without refreshing the page, did not require any software to be loaded on client PCs, and was easy to implement. The OPC Web Client has allowed the customer to access equipment with the click of a button anywhere on site, corporate wide, or even from home. He no longer has to

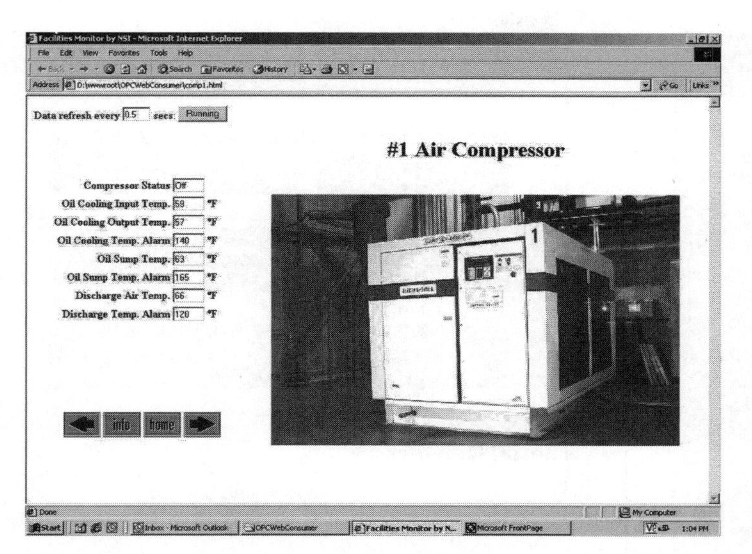

Figure 20-9. Example Screen Monitoring Air Compressor Status

wait on reports from facilities because he can see what's going on with critical equipment by clicking on our maintenance web site.

Products used in this application:

- TOP Server OPC Server—*www.toolboxopc.com*
- OPC Web Client—*www.opcwebclient.com*

Case Study: Supermarket Facilities Management and Haccp Compliance

Technologies Used:

- I/O Server Software
- Industrial Human Machine Interface Software
- Visual Basic
- Internet remote access

In this application, Ukrops, a major supermarket operator in central Virginia sought to implement a monitoring and management system in their stores that would provide a standardized user interface from one store to another, provide local and remote monitoring, and easily integrate data from all facilities infrastructure in the store. The new system needed to be intuitive, utilize open systems, and easily integrate with other facilities systems without 100% dependence on a single vendor.

The types of systems monitored in each store include cold storage systems in the stock room, generators, HVAC units, refrigerator and freezer cabinets on the store floor, lighting, and facility temperatures. Each type of system was often provided by a different company with their own means of connectivity or lack thereof. In some cases, the manufacturer of the equipment, for example refrigerator cases, would provide their own proprietary software applications for the monitoring of their equipment and their equipment only. Or they would provide a closed system that could only be expanded to integrate other systems by purchasing additional hardware and consulting services from that manufacturer.

Also, Ukrops needed to implement an automated means of tracking compliance with government Haccp regulations governing the tracking of handling of food products. Haccp stands for Hazard Analysis and Critical Control Point, and is pronounced "hassip." Haccp regulations require grocers to track when conditions exist that could cause hazards in food, log corrective actions taken, and the amount of time the hazard existed.

Over the years prior to this project, Steve Little of Ukrops had been working with Edward Stafford of Electronic Technologies Corporation (ETC) of Street, Maryland to implement a number of systems for management of the facilities. The existing systems utilized the latest technology at the time, and served their purpose, but over the years, Steve and Edward the team had identified shortcomings they wished to overcome in a new system.

The existing systems were text-based systems that required training for use and operation. When a problem occurred, only a trained maintenance person or operator could get details on the alarm condition and take action. The result of this was ongoing training costs as personnel moved from store to store, unknown impact on stock spoilage due to lack of response to alarms, and lack of management visibility into operating conditions. If a store manager or regional manager wanted to find out what current conditions were in a store, they had to go to a trained maintenance person who could gather the data from the monitoring system.

With these limitations in mind, Steve and Edward worked together to develop a solution was developed that would meet their requirements for their next generation monitoring system. The new system is based on an off-the-shelf Human Machine Interface software package named InTouch, provided by Invensys Wonderware and their local representative InSource Software Solutions. The InTouch software package provided ETC with the ability to build user friendly graphical screens to display operating data and open, standards based methods of gathering data from the various facilities systems. (Figures 20-11 and 20-12) InTouch provides an open OPC client interface for con-

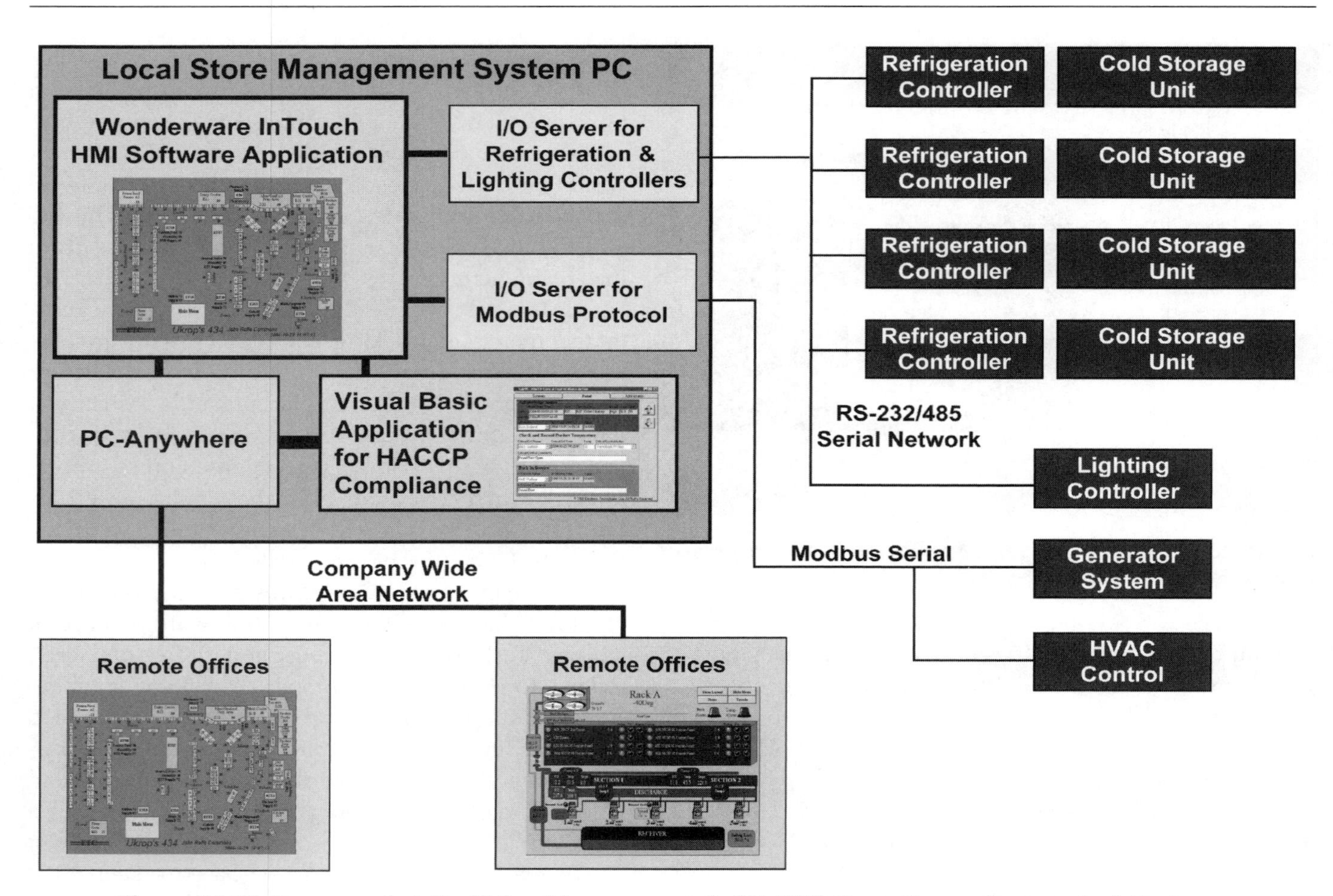

Figure 20-10. Supermarket Facilities Management & HACCP Compliance System Architecture

necting to OPC servers for data sources. InTouch also provides an open Modbus Serial and Ethernet connection that is used to connect to generators and some of the other systems in the store.

Part of the integration challenge in this application was connecting to the refrigeration controllers. The controllers in each refrigeration case on the store floor were networked connected over an RS-485 network using a protocol developed by the refrigeration controller manufacturer. The refrigeration controller manufacturer offered their own user interface software package for gathering data from their systems and to some degree, integration with other systems. There were no "open" communications software drivers available for connecting to the refrigeration controllers.

ETC worked with Software Toolbox Inc. of Charlotte, NC to develop an I/O server software application that would communicate to the refrigeration controllers. ETC was able to procure from the refrigeration controller provider the necessary protocol information for the controllers and test hardware for development of the I/O server. Software Toolbox provided a driver for their TOP Server OPC and I/O Server application that implemented the necessary protocol for communicating to the refrigeration controllers.

In this application, the system takes advantage of a native interface offered by the InTouch Industrial HMI system called Suitelink to connect to the I/O server instead of OPC because the Suitelink connection in InTouch is easier to configure than InTouch's OPC interface. However, by developing the communications software so that it could also connect to other applications using the OPC standard, ETC and Ukrops could be assured they could change the Human Machine Interface software application later or connect other applications to the OPC server software later without writing custom code. Essentially, Software Toolbox provided a means to take data from the closed system and expose it to any software application using an open standard.

Development of the connectivity to the refrigeration controllers was not without its challenges. Close coordination with the hardware manufacturer was required because during the design process, it was determined that the refrigeration controller's communications

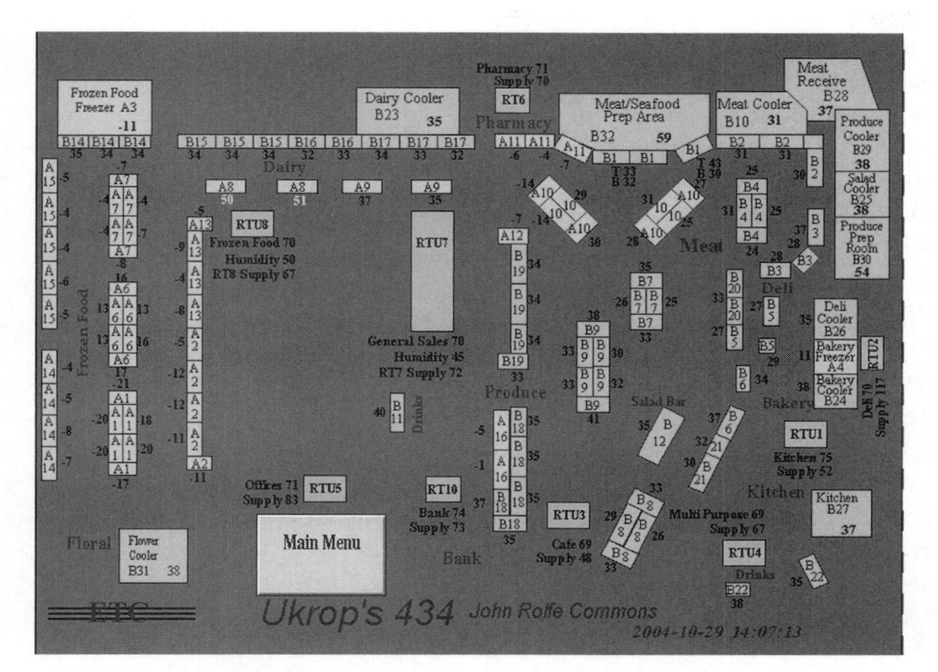

Figure 20-11. Store Overview Graphic in the InTouch HMI

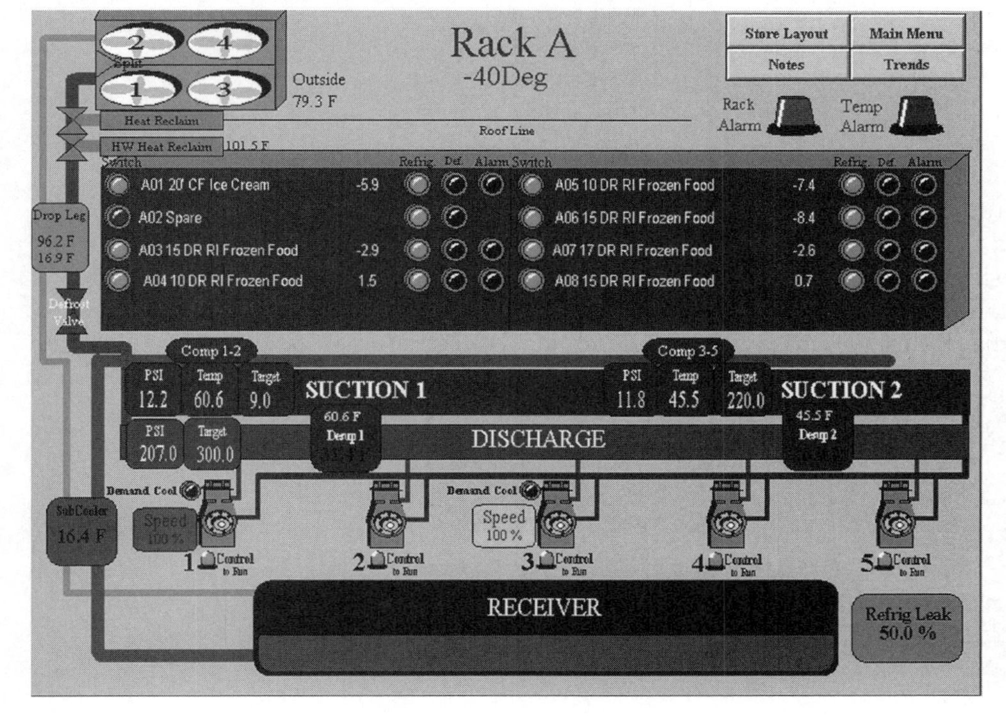

Figure 20-12. Refrigeration Rack Detail Screen in InTouch

protocol did not provide all the functionality needed to meet the supermarket operator's requirements. The refrigeration controller did not allow digital points in the controller to be turned on and off from a remote connection through the device's serial port unless you were using the vendor's proprietary software. Ukrops and ETC worked with the supplier of the hardware to persuade them to add the functionality to the communications protocol available for external connections to the controller. By upgrading the firmware in the refrigeration controller units, Ukrops and ETC were then able to perform the desired writes to the system from the InTouch HMI and I/O Server software. This application illustrates the importance of working with hardware companies who will be responsive and cooperative when the users of their hardware wish to interface to their systems with other applications. Without the protocol documentation and cooperation of the hardware vendor that was secured by ETC, Software Toolbox could not have provided the software required by the application.

To implement the Haccp tracking requirements, ETC developed an application in Visual Basic (VB). (Figure 20-13) The application connects to InTouch through an open DDE interface on the InTouch application and monitors for new alarms. When an alarm occurs, such as a high temperature in a freezer cabinet, InTouch displays the alarm on screen, and the VB application automatically creates a database record with the date/time/location of the alarm and current condition, and the currently logged in operator. When the operator acknowledges the alarm, he is required to input the corrective action taken, any temperature measurements taken from the food, when the problem was resolved, and if the food was moved to temporary cold storage to prevent spoilage. Operators cannot dismiss an alarm until the necessary Haccp information is entered into the system.

Remote access was implemented using PC-Anywhere operating over the supermarket operator's Local Area Network (LAN) in store and across their Wide Area Network (WAN) through the region. Management at the central office can connect to the systems in each store across the network using PC-Anywhere and see the operational screens the same way as if they were physically present in the store.

With the rich graphical user interface provided by Intouch and the connectivity provided by the TOP Server I/O server, Ukrops now has a system where anyone in the store who has had basic training can view alarms, react to them quickly to take corrective action, and notify maintenance. Store managers can go to any store in the regional chain and find the same system in place and know how to use it. Since store managers move about once a year from one store to another, this consistency of operation is important to the overall effectiveness of the managers. By reducing reaction times to problems, spoilage has been reduced, increasing the profitability of the stores.

Because the system is an open system, stores can easily add additional equipment to the displays and make changes to the displays. By using a device connectivity solution that offers both native interfaces to their current HMI and also an open standards based interface using OPC, stores are not locked into a single software application for the display of the data. This application is presently in use in 25 stores in central Virginia.

Products used in this application:

- TOP Server OPC and I/O Server— *www.toolboxopc.com*
- OPC and I/O Server development services— *www.softwaretoolbox.com*
- Invensys Wonderware HMI Software— *www.wonderware.com*
- Integration and Development by Electronic Technologies Corp, Street, MD, Edward Stafford.

Case Study: Integration of Chilled and Hot Water Production and Delivery Systems

Custom-Flo Incorporated of Cincinnati, Ohio is a maker of skid-mounted pumping systems used in chilled and hot water production and delivery. Custom-

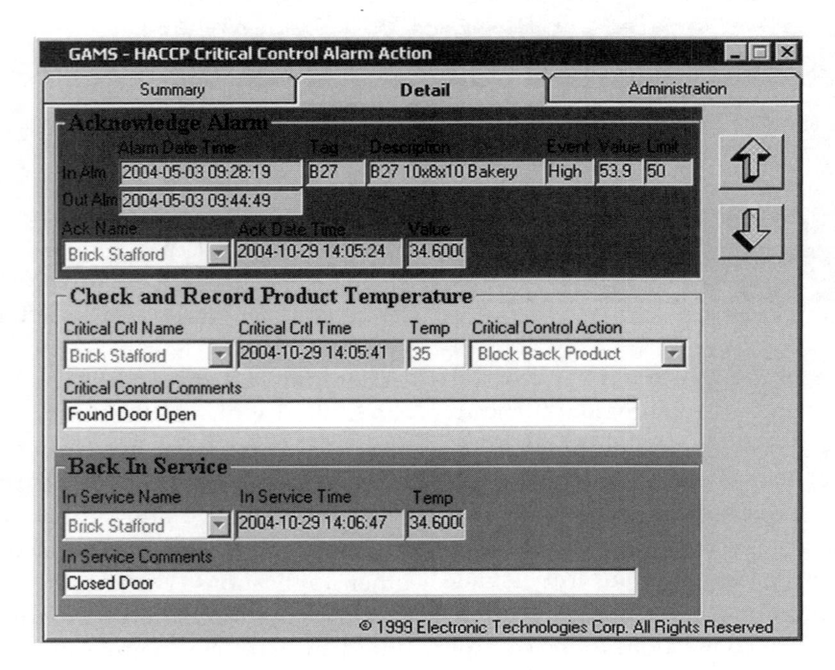

Figure 20-13. HAACP Compliance Application User Interface

Flo is often responsible for custody and delivery of chilled water and/or hot water in their customer's facilities. The facilities automation manager is responsible for the final end use of the resources. As a result, tight integration between the building automation system and systems provided by Custom-Flo is a standard requirement.

Custom-Flo has standardized their controls package for their skid-mounted systems on a PC-Based Control platform for over 4 years. Custom-Flo chose this platform because their customer ends up with a non-proprietary hardware and software solution. If a customer in the future changes direction or needs, he doesn't have to be tied to Custom Flo, he can go forth on his own. The customer can also upgrade to the next generation of PC technology and port the application up to the next revision without having to rewrite the application. The totally open solution offered by Custom-Flo also has saved their customers significant amounts of money whenever their requirements changed or new systems were added.

This case study will cover two areas:

1. Pumping System/Balance of Utility Plant Control System
2. Integration with building automation systems using manufacturing control technologies.

Pumping System Control System

The control system provided by Custom-Flo on their skid mounted systems consists of several parts.

Human Machine Interface (HMI)—Custom Flo uses Advantech Web Studio, and off-the-shelf, user configurable human machine interface application. With this application, Custom-Flo provides a rich graphical user interface in a package that is easily modified later without custom programming. All projects installed by Custom-Flo are web enabled for remote access. The remote access allows Custom Flo to connect through a secure port on the company's network or dialup to support their systems and users, and it allows the users a means for remote access to the systems provided by Custom Flo. In most installations the systems provided by Custom-Flo will be outside the company firewall but have their own router to secure them.

PC-Based Control—Custom Flo uses an IEC 1131 based PC control package, MultiProg Embedded Soft Logic Control, provided by Advantech. This system operates on Advantech TPC-1260 Industrial PCs with display built in so they can run the HMI and SoftLogic on the same PC. The PC-Based control package uses a real-time kernel for the control logic so the Windows operating system can crash, but control will continue to

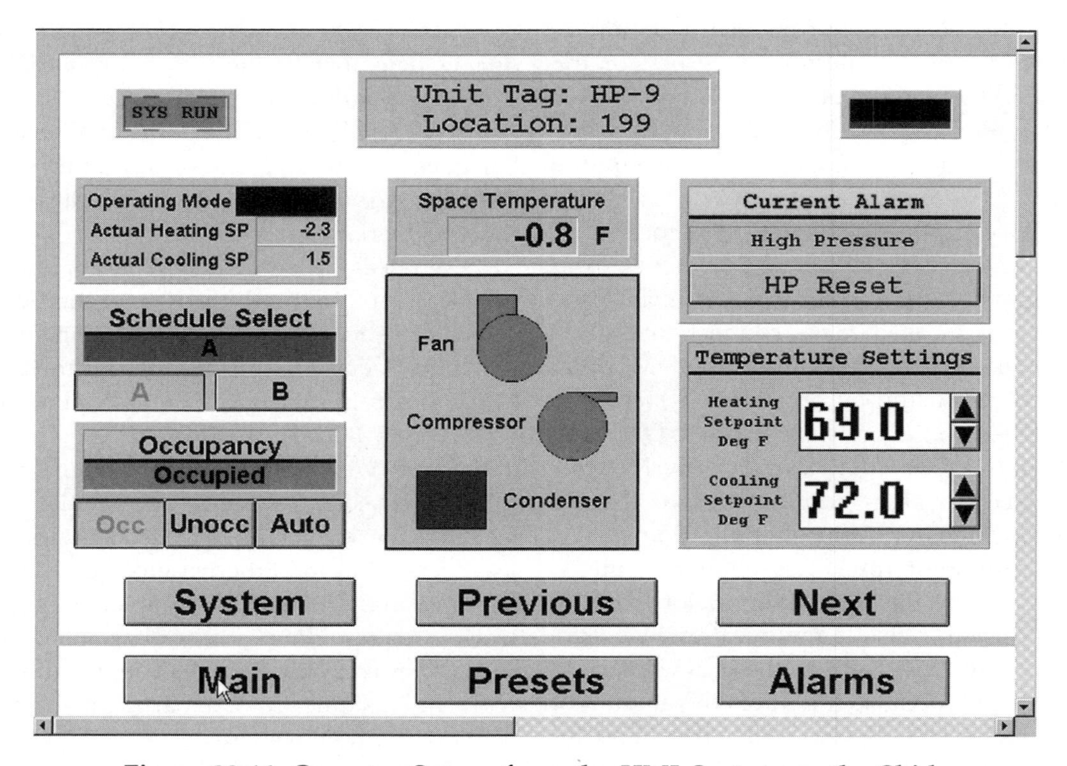

Figure 20-14. Operator Screen from the HMI System on the Skid

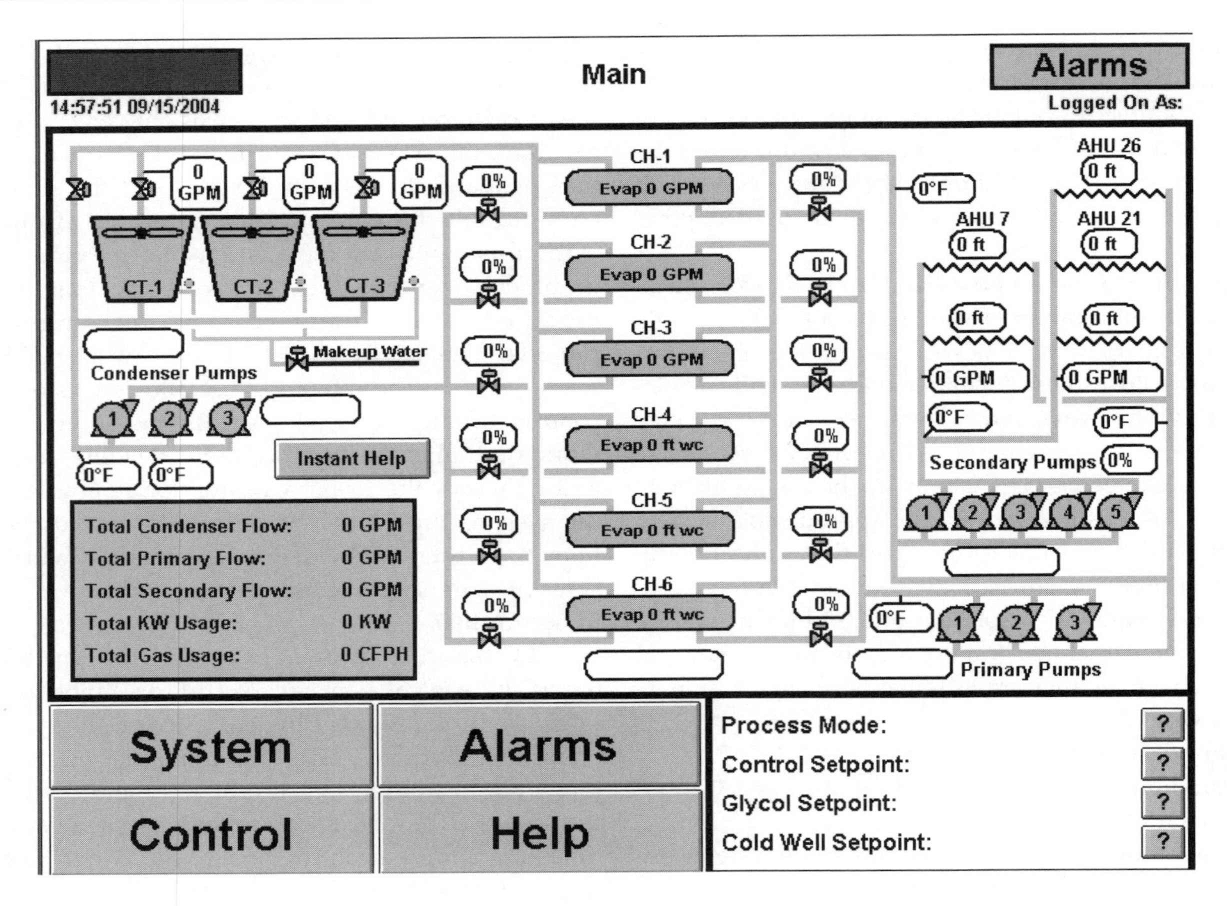

Figure 20-15. Overview Screen

function. Custom Flo believes strongly that users implementing PC based control should not cut corners on the hardware. They have found that by implementing robust, industrial grade PCs, they may pay more up front, but they see a return quickly in overall system uptime and reliability.

Ethernet Based I/O—For remote and local I/O, Custom-Flo uses Ethernet based I/O systems from Advantech, mainly the Adam 6000 and Adam 5000 systems. These Ethernet based racks connect to the PC based control system using Ethernet and Modbus Ethernet protocol. By using an open I/O system, the customer is not tied to Advantech for the I/O if they ever choose to replace it. They can add any Modbus Ethernet based I/O system and connect to it over Ethernet from the PC-Based Control application.

External Equipment Interfaces—Boilers, chillers, and variable frequency drives are the major external components the system must communicate with . Custom-Flo's choice is to try to talk to these devices via a digital/serial data link because it reduces the amount of physical I/O (analog and digital) and wiring and increases the system MTBF. This method also avoids a lot of conduits and wires where problems can be introduced during field installation. On many variable frequency drives, a Modbus serial interface is standard, and in many cases, Modbus Ethernet is available as an option. For devices with a Modbus serial interface, Custom Flo uses an Advantech Adam 5000 rack as a Modbus Ethernet/serial converter.

Wireless Ethernet—in some applications, it is not practical for Custom-Flo to run Ethernet cable in the customer's facility to connect to existing control systems and infrastructures. In these applications, Custom-Flo has successfully deployed commercial off the shelf IEEE 802.11b and 802.11g wireless Ethernet networks. They have used wireless Ethernet to connect to their own Ethernet based I/O systems and to other operator interface nodes in their HMI systems. Wireless has also been used to facilitate an Ethernet linkage to existing building automation systems. When implementing wireless networks, Custom-Flo takes advantage of all available security measures in the wireless routers and works closely with the customer's IT department to insure a secure network is implemented.

By using an open system for the control system,

Custom Flo provides their customer with investment protection and flexibility. Where other vendors charge significant amounts for system changes or adding additional monitoring points, Custom-Flo is able to make changes quickly, easily, and at low cost for their users.

<u>Interfacing to the Building Automation Systems—No Hardwired Point Panels</u>

When Custom-Flo connects to building automation systems at their customer locations, they first try to utilize the native network interface offered by the Building Automation System. Custom Flo prefers to connect to the Building Automation System using OPC because it provides their user with yet another open systems connection point should they ever need to connect another software application to the system.

Custom Flo maps all their data into an OPC Server that implements the native building automation system's protocol. For example, for one building automation system network, Custom-Flo needed an OPC server that would implement the slave side of the building automation network so that the building automation system could be the master. Custom Flo turned to Software Toolbox to provide the OPC server for that implementation.

By utilizing OPC for all their external software interfaces, Custom Flo can maintain the integrity of their systems and cope with changes quickly on site if needed. The OPC Server software has replaced the point panels that are commonly found at interface points in building automation systems, saving the customers thousands of dollars in up front hardware investment and installation costs, and creating significant flexibility for future system changes and expansion. Custom-Flo works to standardize a set of point names in the OPC server for each building automation system vendor, so that once they have implemented a few systems with an interface to a particular building automation system, future systems are quick and easy for both parties.

<u>Example—Airport Facilities Management:</u>

In a commercial airport, Custom Flo supplied 6 hot water systems in 6 concourses and interfaced them to a building automation system using an OPC server supplied by Software Toolbox Inc. After the systems had been running for a few years, the operator of the airport decided to replace their building automation system with one from another vendor. The change in the inter-

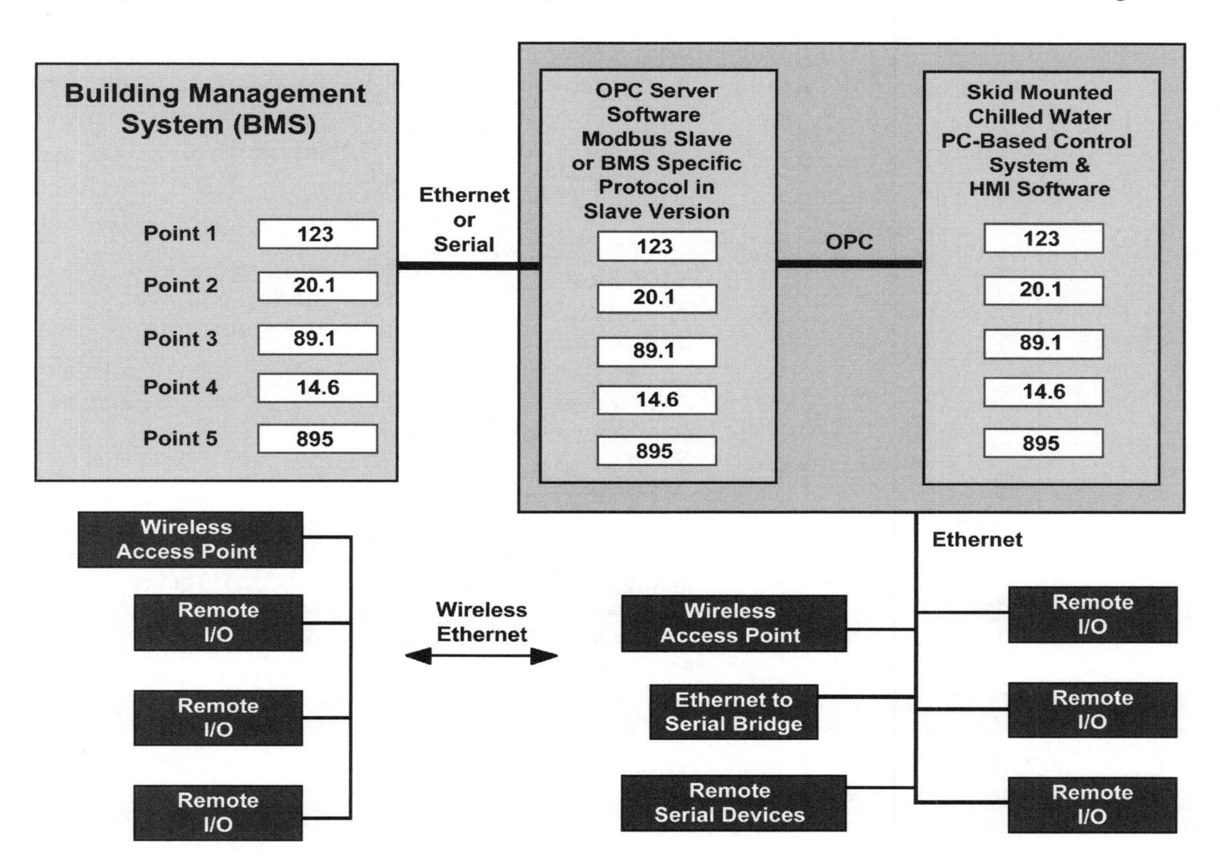

Figure 20-16. Building Management System to Skid Mounted System Interface Overview

face for Custom-Flo simply involved changing the OPC servers from one using a the native building automation system protocol to Modbus Slave OPC servers from Software Toolbox. The changeover of the interface was done in one day. There were no costs associated with replacing or rewiring point panels, or hardware involved in the interface change with the Custom Flo systems.

General advice to users implementing open systems:

During the research for this case study, Custom-Flo felt it important to emphasize some key points to users considering implementing open systems in facilities automation:

- Do not cut corners on PC hardware cost
- Anyone can purchase the pieces needed to implement an open system; however, it takes a competent, experienced controls engineer or integrator to implement the system and get the best return.
- The best way to evaluate the competency of an integrator is through references.
- If you choose to implement wireless Ethernet, you must realize that it too has a capacity limit, and that you should evaluate your data update rates and network utilization needs when implementing. 11 Mb/s or 54 Mb/s may sound like a lot of capacity, but if you take the "oh read everything that's there as fast as you can" approach, you can easily overload a wireless network.

Products used in this application:

- TOP Server OPC Server—*www.toolboxopc.com*
- OPC Server development services—*www.softwaretoolbox.com*
- Advantech Industrial Computers, Human Machine Interface Software, and PC Based Control Software—*www.advantech.com*
- Sample applications on-line—*http://www.custom-flo.com*

Case Study: Integration of Manufacturing Control System technology into a facilities management application

Technologies Used:

- OPC Server Software
- OPC to OPC Server Bridge Software
- Programmable Logic Controllers
- Modbus Ethernet Protocol
- Modbus Serial Protocol
- Windows CE based touchscreen panels
- DSL based Ethernet extenders

A major supplier of building automation software is often faced with customer requirements that they are not able to meet with the capabilities built into their facilities automation systems. Customers were frequently demanding to know "why can't we do this when I have a 3 Ghz, 1 Gb of RAM PC on my desk at home!" To solve a wide range of customer requirements including adding functionality, integration with other systems, and

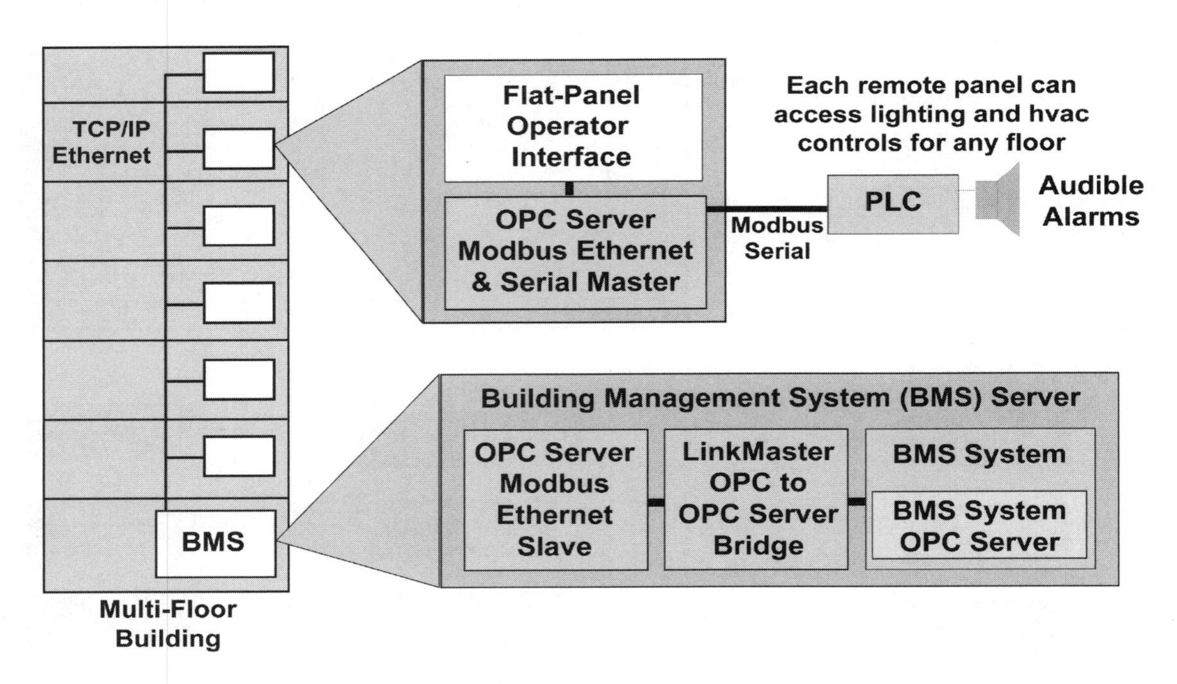

Figure 20-17. Medical Lab BMS to Remote Panel Interface System Architecture

more, the building automation supplier has turned to a wide range of manufacturing control technologies. This case study will explore a number of these applications in more detail. For the building automation supplier, they have been able to meet customer requirements their competitors could not meet, save their customers money, and provide their customers with more flexibility than they otherwise would have obtained from a traditional building automation system.

Example #1

The supplier was installing a system for a large medical lab. The system managed lighting control, HVAC control, security systems and more. The project required there be a touchscreen interface on each of floor in the facility. From each touchscreen, the user needed to be able to access data for not only their floor, but other floors. Budget constraints meant the system had to be done as economically as possible.

Because the building was already wired for Ethernet, it was easy to obtain an extra network jack at each panel. The building automation system provided an OPC server interface as a means of moving data into and out of the system.

To move data between the remote panels and the building automation system, the integrator chose to use open Modbus TCP Ethernet protocol as a fast and efficient means to communicate from the panels back to the Building Automation System over the existing Ethernet connections.

On each floor, a Siemens Windows CE Multipanel was installed running an application using Siemens ProTool HMI software and an INGEAR Modbus Ethernet OPC server. Back at the Building Automation System, the user installed a copy of the TOP Server Modbus Ethernet Slave driver and the Linkmaster OPC to OPC bridge. Linkmaster maps all necessary data from the Building Automation System to the Modbus Ethernet Slave. Each remote panel reads/writes data to/from the Modbus slave using the INGEAR Modbus Ethernet OPC server.

So how does it work? Lets look at some examples. Remember all these steps are happening in sub-second time over switched Ethernet.

Writing Data From the Touchscreen to the Building Automation System

When the user wants to write a value to the building automation system, the HMI software writes the data to the INGEAR Modbus Ethernet OPC server, which in turn writes the data to the TOP Server Modbus Ethernet Slave. The Modbus Ethernet Slave automatically notifies Linkmaster of the new value, and Linkmaster in turn writes the data into the Building Automation system's OPC server.

Reading Data from the Building Automation System and Displaying it on the Touchscreen & Sounding Alarms

When values change in the Building Automation System, the Building Automation System OPC server automatically notifies Linkmaster of the new value. Linkmaster notified the TOP Server Modbus Ethernet Slave of the new value. The INGEAR Modbus OPC server on the Touchscreen is periodically polling the slave for new data—as soon as the new data are received, the INGEAR OPC Server notifies the HMI application which updates the users screen.

In addition to the interface to the building automation system, the application also had a Siemens micro S7 connected to the touchscreen via its RS-232 port. The system uses the PLC to drive a local enunciator alarm/whistle for the customer. The Ingear Modbus OPC server on the panel is used to interface to the S7 PLC through an available Modbus interface on the PLC. The application sets PLC alarm tags from both Vbscript in the panel, as well as tags up in the building management system OPC interface. This solution was chosen because it overcomes the lack of an alarm device on the particular touchscreen chosen. Through use of the open Modbus protocol and a low cost, off-the-shelf PLC, the solution for an audible alarm was implemented for a fraction of the cost of a custom solution.

Example #2

The supplier needed to integrate several small, very low cost touch panels into a building automation system. The touchpanels offered a Modbus Serial protocol master communications interface. To make the connection to the building automation system, the supplier used the TOP Server Modbus Serial slave driver and LinkMaster OPC bridge.

On a PC, the supplier installed a multi-port serial port board, and connected all of the touch panels to these ports using RS-422 connections. On the same PC, the TOP Server Modbus Serial slave OPC server was installed and configured to listen for data requests from the touchpanels connected to each serial port.

The building automation system exposes an OPC server interface. To bridge from the TOP Server Modbus Slave OPC Server to the building automation system the supplier used the Linkmaster OPC to OPC bridge. Data

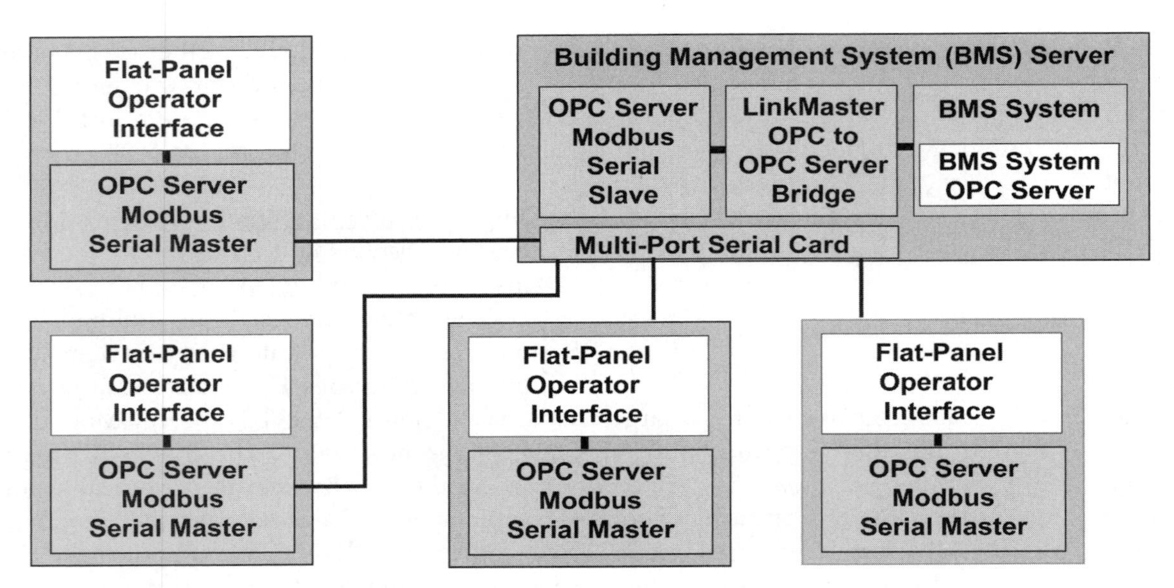

Figure 20-18. Application for Connecting Remote Panels to BMS via Serial Connections

points were mapped in software through a point-and-click interface between the modbus slave connections to each touch panel and the points in the building automation system. Once configured, data are able to flow bi-directionally between the touch panels and the building automation system. Changes are easily made through a point-and-click user interface.

Example #3

The supplier needed to integrate a large number of remote energy meters into the building automation system. Each energy meter exposes a modbus serial interface. The supplier used the TOP Server Modbus Serial Master OPC server along with Ethernet to serial convertors to connect to the remote meters over Ethernet. The connection from the TOP Server OPC server to the Building Automation System was made using the LinkMaster OPC to OPC bridge in the same fashion as other applications.

Products used in these applications:

- TOP Server OPC Server—*www.toolboxopc.com*
- Linkmaster OPC Bridge—*www.toolboxopc.com/linkmaster*
- Siemens MultiPanels and Siemens S7-200 PLCs—*www.sea.siemens.com*
- Patton Electronics DSL based Ethernet Extenders—*www.patton.com*

Case Study: DCS to Building Automation Application

Technologies Used:

- OPC Server Software
- Modbus Ethernet Protocol

A major petroleum producer needed to interface an Emerson DeltaV Distributed Control System (DCS) to a facilities automation system based on the Invensys Wonderware Intouch off-the-shelf industrial human machine interface software.

Design standards for the DCS system required that in all interfaces to external systems, the DCS system be the master. This means the DCS system would initiate all communications. The InTouch HMI system was designed to operate the same way, as it is typically the master when connecting to its data sources. The customer utilized OPC server software solutions from Software Toolbox to bridge the gap between these systems.

They installed the TOP Server Modbus Ethernet Slave OPC server on the same computer as the InTouch HMI based Building Management System (BMS). In the InTouch HMI system, they mapped all data points that were to be shared with the DeltaV DCS system (read or write) into points in the OPC server.

In the DCS system, they configured the Modbus Ethernet Master interface to connect to the OPC Server over Ethernet, using the open Modbus TCP protocol. When the DCS wants to read data from the InTouch HMI system, it reads the Modbus addresses in the OPC server that correspond to the data points mapped from InTouch into the OPC server. When the DCS wants to write data, it writes to the modbus addresses in the OPC server. The OPC Server automatically updates the InTouch HMI with the new values.

The use of an open system based on OPC and Modbus generated significant savings in integration time and engineering for the user of the application. In

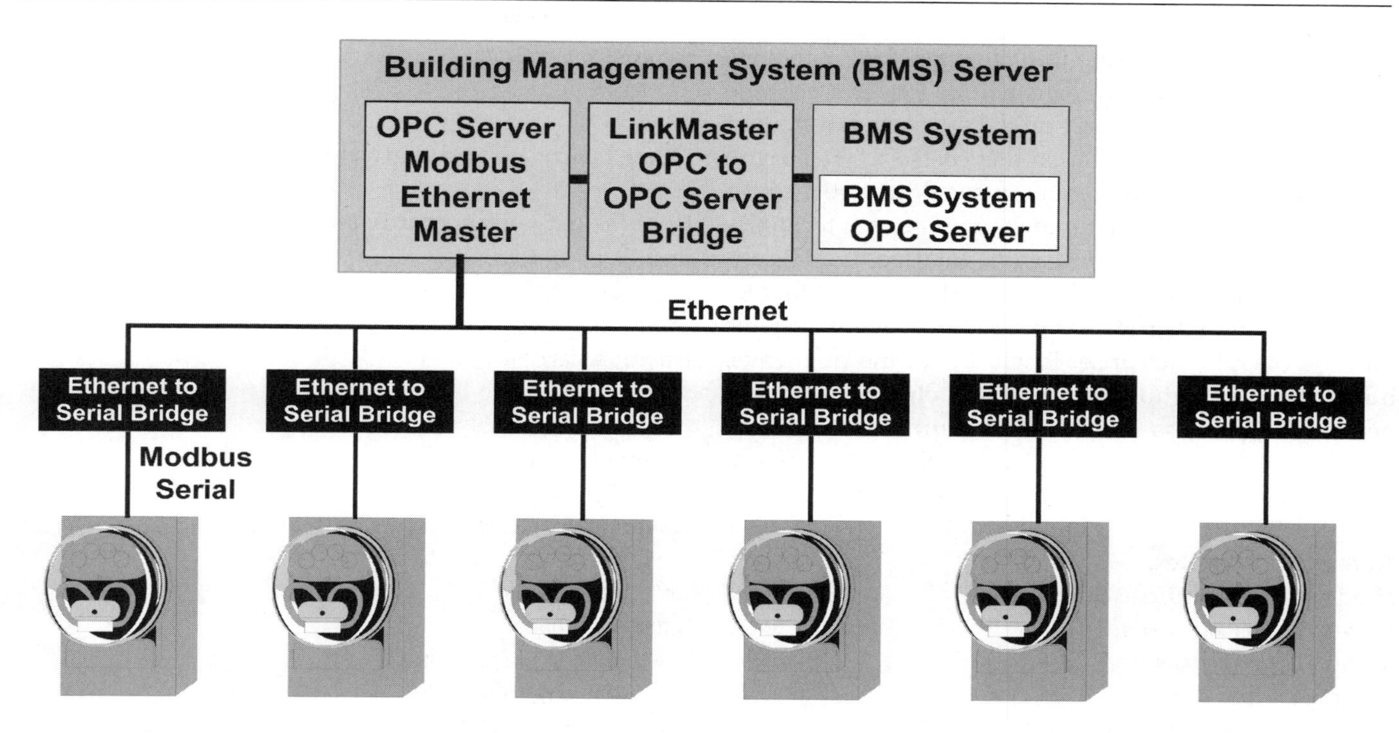

Figure 20-19. Connecting remote meters using modbus, serial to ethernet bridges, and OPC

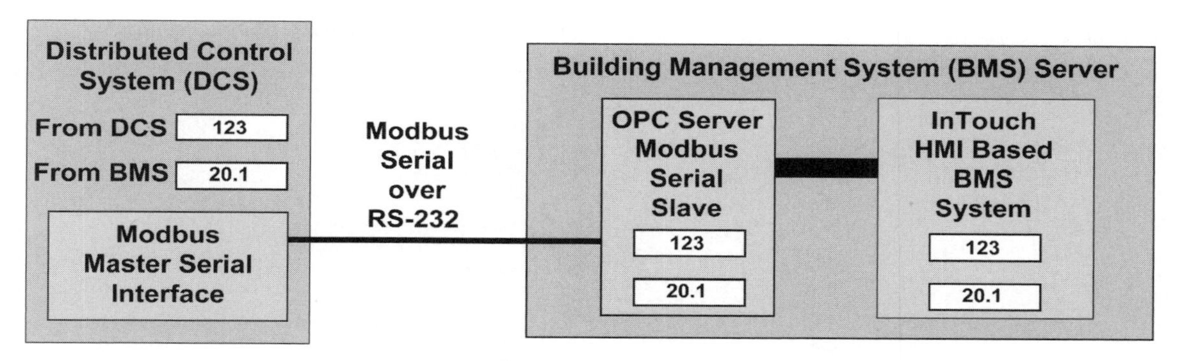

Figure 20-20. Distributed Control System (DCS) to Building Management System (BMS) Integration

addition to these savings, the open system made system checkout possible. The prime contractor for the job was in Europe, with an integrator in the Middle East, a planned installation in the Middle East, and a system checkout planned in the United States. Through the use of open systems, all interconnections were planned ahead of time via email correspondence between the parties. At checkout, a technician skilled in the use of OPC and Modbus was contracted in the United States to attend the system checkout on behalf of the integrator from the Middle East. In one day, all interface points where tested and checked, saving the integrator the expense and time of traveling to the United States for system checkout.

Products used in this application:

- TOP Server OPC Server—www.toolboxopc.com
- Consulting Services—www.softwaretoolbox.com
- Emerson DeltaV—www.easydeltav.com
- Invensys Wonderware InTouch—www.wonderware.com

CONCLUSION

The use of open systems and web based technologies from the commercial information technology world and from the manufacturing automation industry will become a major influence in the facilities automation market in the next 5 years. Implementors and managers of facilities automation systems will no longer accept the answers "well you just can't do that" or "we can do that but it will take a customization that costs $X." To be sure there will be cases where both answers will still apply. However, it will become more common to see technologies such as OPC, Human Machine Interface (HMI) software, Modbus, Modbus Ethernet, and Wireless Ethernet used to cleanly integrate Building Management Systems into the overall enterprise. It will be a time of change, but for those who embrace the change and the newfound openness possible, the returns to their businesses, whether they be suppliers, implementers of systems, or users of systems, will be significant.

Reference Websites:

Products Used in Case Studies

www.softwaretoolbox.com
www.paneldrivers.com
www.toolboxopc.com
www.toolboxopc.com/linkmaster
www.opcwebclient.com
www.ingearopc.com
www.siemens.com
www.advantech.com
www.patton.com

Other Hardware and Software Resources

Wireless networking hardware

www.netgear.com
www.dlink.com
www.belkin.com
www.linksys.com
www.airlink.com

Ethernet/Serial Convertors

www.digi.com
www.lantronix.com
www.bb-elec.com
www.sealevel.com
www.atop.com.tw

Industry Organizations

http://www.opcfoundation.org
http://www.isa.org

[1] *www.cfsan.fda.gov/~lrd/bghaccp.html*

Section Five

Web Based ECS Applications

Chapter 21

State of Practice of Energy Management, Control, and Information Systems[1]

Gaymond Yee
California Institute for Energy and Environment

Tom Webster, P.E.
UC Berkeley, Center for the Built Environment

Lawrence Berkeley National Laboratory
Berkeley, CA

ABSTRACT

It is important for energy managers to have a high level of knowledge and understanding of complex energy management systems. This chapter is intended to help energy practitioners with some basic informational and educational tools make decisions relative to energy management systems design, specification, procurement, and energy savings potential.

This chapter provides an evaluation of several products that exemplify the current state of practice of Energy Management, Control, and Information Systems (EMCIS). The available features for these products are summarized and analyzed with regard to emerging trends in EMCIS and potential benefits.

BACKGROUND

Traditionally, energy related subsystems used in building controls performed either energy management or combined energy management and control functions. These systems were usually called EMS or EMCS (see Glossary). However, current emerging energy related systems have expanded into information processing that includes data exchange and archiving, historical data visualization, and energy data analysis. In this chapter, the term EMCIS (energy management control and information system) is being used to refer to these types of systems.

With utility deregulation in the early 1990's and recent periods of energy crises, new products have emerged that augment or expand the basic functions covered by BCSs. These new energy information systems (EIS) include utility EIS, demand response systems (DRS), and enterprise energy management systems (EEMS) [1]. In many of the EMCIS products studied, some features of this new wave of EIS products are supported.

SELECTED VENDORS

Over twenty vendors were identified as possible candidates for study. Each of the candidates was categorized into three vendor types; OEM/Equipment, Building Controls, and Software. Although there are many more than twenty BCS vendors, those identified make up a representative sample of the industry.

For the study, a group of eight vendors were selected; one from OEM/Equipment vendor type, five from Building Controls vendor type and two from the Software vendor type. Building controls vendor types were further categorized into Large, "first-tier" and Small, "second-tier" vendors.

The selected vendors are:

<u>OEM/Equipment Vendor</u>
Trane

---

[1]This Chapter was originally published as a report for the Federal Energy Management Program, Trends in Energy Management Technology—Part 3, State of Practice of Energy Management, Control and Information Systems, Report No. LBNL—53545 (part 3)

Large Building Controls Vendors
Johnson Controls
Siemens Building Technologies

Small Building Controls Vendors
Automated Logic
Alerton Technologies
Teletrol Systems

Software Vendors
Tridium
Electric Eye

These vendors were chosen because their offerings represent current trends in the industry. In particular, these vendors support two key concepts that are becoming prevalent not only in the BCS industry, but in many other industries; open systems and Web enabled. Open systems support in the commercial BCS industry means support for the BACnet standard and LonTalk,[2] either natively or by the use of a protocol gateway. Web enabled means support for access to the EMCIS using Web-based technologies either internally on the corporate Intranet or externally over the Internet. Web-based access can mean the use of a Web browser as a user interface to access EMCIS functions or the use of XML (Extensible Markup Language) applications for data exchange with the EMCIS.

In general, the OEM/Equipment and Large BCS vendors sell and support their products directly. The Small BCS vendors mostly offer products and support through distributors/dealers or system integrators. Because Software vendors generally make no hardware, systems with Software vendor products are sold and supported through system integrators or are privately-branded under BCS or OEM offerings. For example, of the two Software vendors selected in this study, Electric Eye works almost exclusively with hardware products made by EnFlex Corporation. Products from the other software vendor, Tridium, are also available as third-party brands from both Honeywell and Invensys.

Organization of the Report

The remainder of this chapter is organized into three major sections. The first section is an EMCIS technology description with subsections on architecture, features and trends, and overall assessment. The second section discusses the potential of EMCIS products; their benefits, limitations, and cost effectiveness. The chapter ends with a summary and conclusions. There are two appendices to this chapter. The first contains a concise roadmap of EMCIS product features for the selected vendors and the second contains a discussion of XML, SOAP (Simple Object Access Protocol), and Web Services.

[2]LonTalk is a subset of the BACnet standard (ANSI/ASHRAE-135) and is a standard for home automation (ANSI/EIA-709).

TECHNOLOGY DESCRIPTION

EMCIS Architecture

The typical EMCIS system architecture consists of a 4-level hierarchical structure (see [2] for a diagram of this structure). Each level can be characterized by a network bus technology (a combined physical wiring or wireless specification and communications protocol specification).

(1) At the highest level is the *Enterprise/IT bus/network*. It consists of the day-to-day corporate desktop PC's, servers, printers, etc. The predominant Enterprise/IT network bus is TCP/IP over Ethernet (10 Mbps and 100 Mbps).

(2) The next level is the *building controls system backbone bus*. Here reside the large field panels or supervisory controllers, BCS workstations and other high bandwidth BCS hardware.

(3) Connecting to the supervisory controllers are terminal controllers. This level is called the *terminal bus* and typically has a lower bandwidth requirement than the backbone/equipment bus.

(4) Sensors and actuators can be interfaced to terminal controllers by the *sensor bus*. In most instances however, sensors and actuators are not interfaced via a sensor bus but are hardwired typically with 4-20mA, or 0-5/10 volt electrical signals. The level of the electrical signal (current or voltage) indicates a scaled analog value, an on/off digital status, or an on/off digital control. In the industrial process industry, there are emerging standards for a sensor bus [2]. One advantage of a sensor bus is the reduction of wiring cost. Instead of each sensor being wired directly to a controller, a single cable can be used to connect all the sensors. In this study, there was little evidence that BCS vendors support sensor bus technology.[3]

[3] A sensor bus is supported by at least one of the 20 vendors reviewed for this study, Delta Controls.

Although the four-level hierarchical architecture is typical, technologies are evolving such that the structure is being "flattened" [2]. For example, in some instances, the Enterprise/IT and BCS backbone/equipment bus co-exist on the same physical media. In another example, field panels and terminal controllers are on the same bus.

EMCIS Features and Trends

EMCIS systems, as represented by the systems in this study, are complex systems with a wealth of technologies and rich feature sets. The technologies and features for eight products in this study are summarized in the "roadmap" presented in Appendix B. These products exemplify the areas where advancements have been made in recent years. These areas include:

- Open Communications
- Applications Software
- *Web Access*
- Energy Information
- Integration to Other Systems

The first two items were covered in more detail in Chapters Eight and Twenty-seven of *Information Technology for Energy Managers*, [3], so only the last three will be emphasized in this chapter.

Open Communications

Clearly the BCS industry has embraced open communications. One of the main driving forces behind openness is the desire by customers not to be locked into any particular vendor's products. BCS customers want the ability to choose the best products from various vendors and be reasonably sure that the products inter-operate. Another driving force is the customers' desire to share information between different systems; for example between a process control system and the building control system.

A detailed assessment of the current state of practice of open communications is contained in Reference [2].

Applications Software

One of the most important elements of an EMCIS is the suite of software applications that users interact with on a day-to-day basis. This is where the intelligence of these systems is delivered. The most important applications from the user's perspective include data visualization, alarm monitoring/management, graphical programming, and support for reporting energy information. A comprehensive assessment of these applications is the subject of Chapter 21 of this book. Included in the tables of that chapter is a detailed listing of applications software capabilities for the products studied.

Web Access

There are two Web access trends evident among the vendors; (1) browser-based user interfaces that locally and remotely access the EMCIS via the World Wide Web, and (2) Web-based access into the EMCIS for data exchange.

The remainder of this section will discuss these Web access trends.

Browser-based User Interface

A browser-based user interface has many advantages:

- The user interface is simply a Web browser. There is no need for any additional (thin client) software besides standard browser plug-ins (Java).
- There is no need to keep track of where workstation software are installed and to keep the client software up to date.
- Access to the EMCIS can be from any device that has a Web browser and has access to the EMCIS Web server either internally over a corporate Intranet or externally over the Internet.
- The BCS industry can take advantage of emerging Web standards developed in the computer and information industries where the standards are usually defined and accepted much more quickly. Adoption of these standards will ensure compatibility within the corporate and business world.

The level of Web access support differs among the EMCIS vendors. Some use Web access exclusively as the workstation software for their product line. Others have standard Windows-based workstation software and offer Web access only as an option.

The features of each offering must be taken into consideration because of the impact on the cost, performance, user features, and reliability as well as any potential conflict with corporate IT policies. Things to consider are:

- **Browser support**. In general, this means the major browsers, Netscape and Internet Explorer. Almost all the vendors require Java as a plug-in because Java can provide real-time data access and other functions not possible on the browser.

- **User features**. For the case where Web access is optional, the feature set is normally not the same as the workstation software.

- **Web server platform**. The most commonly supported Web server platform is the Microsoft-Windows based server. A few vendors support Sun-Solaris and Linux platforms. One of the vendors, Alerton, offers its optional Web server via fully configured IBM hardware running Linux and Apache Web Server. Another vendor, Trane, offers its optional Web server as an embedded server. The embedded server may offer additional reliability, but may also lack the power to provide the features and the performance of standard server-based hardware.

- **Web page generation**. In most cases, tools are included with the Web server that facilitates the creation of Web pages. In other cases, standard third-party Web page authoring tools can be used. If the Web server is an optional feature, the Web pages are generated automatically from the standard workstation user interface. However if manual creation of the Web pages is required, considerable resources may be needed. This is a key issue that must be taken into consideration.

- **Internet connection security**. For secured Internet communications, a few of the vendors offer HTTPS (HTTP in secured mode) as the default connection method between the browser and the Web server. Other vendors allow the Web server administrator to configure the Web server for HTTPS connection as an option.

As an alternative to supporting browser based user interfaces via standard web servers, some vendors offer Web access via the Windows 2000/XP and Terminal Services (TS) feature. With TS, remote PC's can access the EMCIS Windows 2000/XP server over the Internet using Microsoft's Internet Explorer browser. The browser session is a remote login into the EMCIS system. The advantages are that all workstation features available locally are also available in the remote session[4] and the browser/server connection is encrypted (RSA RC4 cipher). The disadvantage is that the browser must be Internet Explorer which means that an ActiveX[5] control must be used. ActiveX controls do not work in other browsers.

Data Exchange using XML[6]

One of the emerging Web standards is XML. XML is becoming a standard language for exchanging data in many business sectors. A few of the vendors in this study offer XML as a Web access method to retrieve data from their EMCIS product line.

Whereas XML is a standardized language being developed by W3C (World Wide Web Consortium, *www.W3C.org*), the data defining schemas are not standardized. Here lies a weakness of XML. The dilemma is that businesses can begin defining schemas so that data can be exchanged, but these schemas may not be compatible with other businesses and sometime in the future, a standard set of schemas may make earlier definitions obsolete. In a recent GAO report [4], it was concluded, "key XML vocabularies, tailored to address specific industries and business activities are still in development and not yet ready for government wide adoption." Further, "using them at this time would mean taking the risk that future developments could diverge from these early standards and limit interoperability with them."

To begin the process of standardizing XML for use in the HVAC industry, to date, two committees have been tasked to begin XML standards definition for HVAC. The ASHRAE SSPC 135 BACnet XML Working Group is in the process of defining various application areas for XML technology. The ASHRAE GPC 20 (Guideline Project Committee) is establishing a common data exchange format for the description of commodity data and HVAC&R information using XML.

SOAP and Web Services[7]

Beyond XML is SOAP. SOAP is used to invoke applications running on another platform without regard to what the platform is or what language the application was written in. In fact, the client can be completely incompatible with the server.

SOAP is the basis for the emerging Web Services technology. Web Services perform functions, which can be anything from simple requests to complicated busi-

[4]Commercial software products that are similar to TS include Remote Desktop and PCAnywhere. There's also a free software product call VNC (Virtual Network Computing).

[5]ActiveX controls are the interactive objects in a Web page that provide interactive and user-controllable functions.

[6]A more detailed discussion of XML is presented in Appendix C of this chapter.

[7]A more detailed discussion of SOAP and Web services is presented in Appendix C of this chapter.

ness processes. Once a Web Service is deployed, other applications (and other Web Services) can discover and invoke the deployed service [5]. The few vendors that are offering XML support, are also planning to offer SOAP support and thereby also offer Web Services. One vendor, Automated Logic, is planning to provide limited support by permitting Web Services access to virtually anything in the WebCTRL database. However, exactly how this all plays out in the long term and what Web Services eventually will mean to the BCS industry remains to be seen.

Energy Information

In recent years, numerous Energy Information Systems (EIS) have been developed and deployed in a highly competitive market driven by energy deregulation and periods of energy crises [1]. These EIS are systems that typically: (1) collect and display metered usage data, (2) compare load profile data within and between facilities, and (3) correlate energy use with bill data and rate information [1]. The systems studied in [1] were Web-based EIS with some level of demand-response and monitoring/control capabilities. In contrast, the systems under study in this chapter are primarily monitoring/control systems with some level of Web-based, demand-response, and energy information capabilities.

Energy Monitoring

Main meters and sub-meters have communications capability, and can support one or more Modbus, LonTalk, BACnet or proprietary protocols. In practice, any EMCIS can perform energy monitoring if the EMCIS has support for the protocol. However, some EMCIS products have setups and configurations specifically dedicated to energy monitoring.

Energy Analysis

A few of the vendors are including more aspects of EIS as options in their systems. Customers with energy information needs can purchase the option from the EMCIS vendor instead of acquiring it from a third party. Graphs and reports available from these optional energy analysis add-ons can be used to:

- Adjust operations and scheduling to help minimize energy use.
- Negotiate energy contracts for more favorable terms.
- Benchmark against published statistical data to see if systems need optimization or upgrading.
- Cost allocate energy use.
- Validate utility bills.

Demand Limiting/Reduction

During peak electrical demand periods (e.g., midday during hot summer months), there is a high potential to reduce energy cost by limiting the electrical demand of the systems (HVAC, lighting, etc.) inside a building. Even if the building is not under a time-of-use or demand charge rate structure with the local electric utility, there are many curtailment programs with financial incentives sponsored by utilities and local and state governments to reduce energy use during peak demand periods.

Reducing electric demand by reducing HVAC load can be accomplished manually by creating new schedules or simply by manually turning off devices. A better approach is to have an automated system.[8] Many of the vendors have a demand limiting/reduction feature in their system. For example in WebCTRL (Automated Logic), devices can have different setpoints assigned to up to four demand levels. These demand levels can be triggered manually by an external input or automatically when a metering device reading exceeds the demand thresholds. In the automated case, the demand level trigger can be scheduled to be active or inactive depending on the time of the day, day of the week, or holiday.

Integration to Other Systems

Complementary to the trend towards open communications is the trend toward integration to other systems. These systems include various facility systems such as the lighting, security/access, fire/smoke, CMMS (Computerized Maintenance Management System), and CAFM (Computer Aided Facility Management) systems, as well as EMCIS from other vendors. The goal of the integration is to permit the use of a single system to monitor and manage multiple systems or to allow sharing of information between systems.

A few of the large vendors specifically offer non-EMCIS systems and support the integration to those systems natively. Other vendors specify support for integration to other systems by using gateways. However, all the vendors support either BACnet or LonTalk or both. By so doing, there is an implied support of integration to other systems if the integration mechanism is through the BACnet or LonTalk protocols.

[8]Even an automated system is not fully automated because the building operator must be notified of an impending curtailment by the utility and then must initiate the curtailment response manually.

In the case of integration to other EMCIS, the BACnet and LonTalk protocols again can be used as the integration mechanism. For example, the 450 Golden Gate Project [6] integrated systems from two of the vendors in this report using BACnet (the two vendors, Trane and Alerton were not specifically mentioned in the reference). For integration to EMCIS with legacy protocols (including the vendors' own legacy protocols), gateways are used.

Overall Assessment

The EMCIS products in this report are all complex systems with advanced hardware and software technologies. The products are rich in features and are indicative of trends in the industry. However, for each product, there are particular strong points that are summarized as follows:

- ***Tracer Summit (Trane).*** As an equipment vendor, *Trane* can provide complete turnkey systems that include chillers, air handlers, heat pumps, fans, VAV boxes, etc. in addition to a complete EMCIS.

- ***Metasys (Johnson) and APOGEE (Siemens).*** Both *Johnson Controls* (JCI) and *Siemens* have an extensive line of controls with numerous software and gateway options to third-party systems.

- ***WebCTRL(Automated Logic), BACtalk (Alerton), and eBuilding (Teletrol).*** These three products are completely native BACnet. BACnet is the direction the industry is heading and these products are positioned to take full advantage of the latest BACnet developments.

- ***WebCTRL (Automated Logic), eBuilding (Teletrol), and Vykon (Tridium).*** These three products have native Web-based user interfaces. These systems have one less component to worry about, the separate Web server. The Web is the future and these products are positioned to adapt the latest Web technology that is suitable to the industry.

- ***BACtalk (Alerton), Metasys (Johnson), APOGEE (Siemens), and Vykon (Tridium).*** These four products have integrated energy analysis tools or can be integrated with energy analysis tools. *BACtalk* has energy logs that trend energy demand and consumption as well as compute the averaged demand over a specified time interval. *Metasys* can be integrated into *Johnson Controls'* sophisticated *FX-TEM* product that has an energy bill rates database and can generate reports including usage, cost, and load profile. Similarly, *APOGEE* has the *Utility Cost Manager* option that provides reports for demand, consumption, and cost allocation. *Vykon's VES Energy Profiler* offers Web-based graphical reports that include energy trends, profiles, aggregate, summary, ranking, and relative contribution.

- ***Vykon (Tridium), and Electric Eye (Electric Eye).*** These two products are mainly software products. They are excellent candidates for retro-fit of systems for data gathering, archiving, and analysis. *Vykon* is particularly suited for systems with controls that use LonTalk. If needed, *Electric Eye* systems can also be implemented with controls from EnFlex.

POTENTIAL OF EMCIS PRODUCTS

Benefits

In general, the potential bottom line benefits of EMCIS are the reduction of operations and maintenance costs and the reduction of energy usage and cost without sacrificing occupant comfort. The specific benefits of EMCIS can be better summarized within the context of the categories of the emerging trends discussed in this report.

Web Access

- Reduce client software cost
- Allow system access from virtually any location
- Simplify data interchange between systems and across geographical locations

Energy Information

- Identify potential energy savings
- Energy usage data can be used to negotiate better utility rates

Integration to Other Systems

- Control and scheduling strategies can utilize data from multiple systems
- Integrate capabilities such as EIS with other business functions

Limitations

The numerous potential benefits of EMCIS are not totally free and automatic. The effectiveness of any system is compromised when:

- The system is not properly designed
- The system is not properly commissioned
- Features are not used or are used improperly
- Results of data analysis and reports are not properly interpreted

To mitigate these compromises there must be:

- Management structure, support, and direction
- Proper operator training
- Correspondence between product sophistication and operator/user expertise

Cost Effectiveness

Analysis of EIA/DOE statistics [7] indicate a greater fraction of federal buildings have BCSs than does the overall U.S. building stock and a high percentage of large buildings have BCSs.

There is still significant untapped potential in both large and small federal buildings; i.e., the buildings not yet fitted with BCSs amount to 28% of overall floor space for small and 33% for large buildings.

What is the cost effectiveness of an EMCIS? One needs to compare the O&M cost of a system before an EMCIS is installed and the O&M cost after an EMCIS installation. One must also factor in the cost of the EMCIS itself because an expensive system may not necessarily save the additional O&M cost to justify the additional capital expense.

It is beyond the scope of this report to analyze the cost effectiveness of EMCIS. The issue of cost effectiveness will be partially addressed in a future report that discusses case studies of actual EMCIS installations.

Summary and Conclusions

The EMCIS products reviewed in this report are some of the trend-setting products in the industry. These products have an impressive array of capabilities. Each of the EMCIS products supports several of the identified emerging trends. BACnet is clearly the most important open communications trend. Vendors have complete product lines that are totally BACnet compliant. This is a clear statement that BACnet is here and will remain in the future.

Another significant trend is Web-enabled. All vendors have or are introducing Web access with browser-based user interfaces. Others are adopting the IT industry trends of XML, SOAP, and Web Services for data exchange and remote applications over the Internet.

Utility deregulation and restructuring and the specter of high energy prices have brought about the trend towards energy information capabilities. Standalone EIS products are available from non-BCS vendors, but several of the EMCIS vendors in this report have included EIS capabilities in their products.

With all this capability and technology comes the question, have these systems become too complex? Can common operators use them effectively? Does all this technology work and is it useful? What are the actual benefits and savings that can be achieved? Chapter 21 will try to address these and other questions.

Appendix A
Acronyms and Definitions

Many acronyms used in this report can be found in the Glossary section of this book. Additional definitions are included here.

DRS	Demand Response System
EEMS	Enterprise Energy Management System
FDD	Fault Detection and Diagnosis
NMS	Network Management System
RF	Radio Frequency
SGML	Standard Generalized Markup Language
SNMP	Simple Network Management Protocol
SOAP	Simple Object Access Protocol
TP	Twisted Pair
TS	Terminal Service
WX	World Wide Web Consortium

Trademark Notices:

- LON, LonTalk, LonWorks, LonMark, SNVT, Echelon, and Neuron are trademarks of Echelon Corp.
- ARCNET is a trademark of ARCNET Trade Association
- BACnet is a trademark of ASHRAE

All other product, trademark, company or service names used are the property of their respective owner.

Appendix B
EMCIS Products Roadmap

Vendor	Trane	Automated Logic	Alerton	Teletrol	Johnson Control	Siemens Building Technologies	Tridium	Electric Eye - EnFlex
Product Name	Tracer Summit	WebCTRL	BACtalk	eBuilding	Metasys	APOGEE	Vykon/ Niagara Framework	Electric Eye - EnFlex
Estimated Market Share* (1=>15%, 2=10-15%, 3=5-10%, 4=<5%)	4	4	3	4	1	2	4	4
Vendor Type								
OEM/Equipment	•							
Building Controls		•	•	•	•	•		
Software							•	•
Sales Channel								
Direct from Vendor	•			•	•	•		
Distributors/Dealers		•	•	•	•			•
OEM							•	
System Integrator				•			•	•
Network Architecture & Protocols								
Enterprise/IT								
BACnet (Ethernet = BACnet over Ethernet, IP = BACnet/IP over Ethernet)	Ethernet IP ARCNET	Ethernet IP	Ethernet IP	IP	IP	IP		
JAVA RMI		•						
CORBA, IIOP		•						
TCP/IP - HTTP/HTML/XML				•			•	
COM/DCOM, DDE, OLE, OPC			COM/DCOM	OPC	OPC	OPC		
SNMP		•			•			
Proprietary					N1 Bus on ARCNET or Ethernet	•		•

*Market share estimates were extrapolated from "North American Building Control Systems Market", Frost & Sullivan, 1996.

Vendor	Trane	Automated Logic	Alerton	Teletrol	Johnson Control	Siemens Building Technologies	Tridium	Electric Eye - EnFlex
Backbone/ Equipment Bus								
BACnet (Ethernet = BACnet over Ethernet, IP = BACnet/IP over Ethernet)	Ethernet IP ARCNET	IP ARCNET MS/TP	Ethernet IP	IP	Ethernet IP			
TCP/IP - HTTP/HTML/XML				XML				
Proprietary					N1 bus on ARCNET or Ethernet	P2	•	•
Terminal Bus								
BACnet (IP = BACnet/IP over Ethernet)	PTP	MS/TP PTP	MS/TP PTP	MS/TP	PTP		IP MS/TP	
Lontalk	•				•	•	•	•
Modbus on RS-485							•	•
Proprietary	• (Legacy)	• (Legacy)		• (Legacy)	N2 Bus	P1		•
Gateways								
Gateway Offerings (Product name)	•	WebPRTL	BACtalk Ports	Integrator Controllers (legacy)	Metasys Integrator	APOGEE Open Processor	•	•
OPC	•			•		•	•	
Modbus	•	•	•	•	•	•	•	
Lontalk	•	•		•		•	•	
Profibus	•			•	•		•	
Allen Bradley (PLC)	•	•		•	•	•	•	
Simplex (Fire Panel)		•	•		•	•		
Vendor's legacy protocol(s)			•		•			
Various 3rd-Parties	•	•	•	•	•	•	•	•
Frontend / WS Software								
Standalone or Client/Server	Tracer Summit		Envision		M3, M5 Workstations M-Explorer	Insight		Electric Eye
Web-based (Thin-client)		WebCTRL		Envoy			Web Supervisor	

Vendor	Trane	Automated Logic	Alerton	Teletrol	Johnson Control	Siemens Building Technologies	Tridium	Electric Eye - EnFlex
Web-Enabled								
Web Server Product Name	Tracer Summit Web Server	WebCTRL	Webtalk Iport	Site Manager	M-Web Server & Web Access	APOGEE GO & Insight 3.4	Web Supervisor	Electric Eye
Web Server Platforms supported								
Windows		NT/2000/XP		NT/2000	• M-Web Server 98/NT • Web Access 2000 with Terminal Service	• APOGEE GO NT/2000 • Insight 3.4 2000 with Terminal Service	NT/2000	
Solaris		•		•			•	
Linux		•		•				•
Proprietary	Embedded Platform		IBM Hardware Linux OS Apache Web Server					
Web Page Builder Tool	Automatic	MS Frontpage with ALC extensions	Automatic	Edifice	Automatic	Automatic	Workplace Pro	
Internet Connection Security	HTTPS	HTTPS		HTTPS	• Web Access RSA RC4 cipher	• Insight 3.4 RSA RC4 cipher		
WS Features Supported by Server (F=Full or L=Limited)	L	F	L	F	F	• APOGEE GO L • Insight 3.4 F	F	L
Client Browsers Supported	Netscape & IE	Netscape & IE	Netscape & IE	Netscape & IE	• M-Web Server Netscape & IE • Web Access IE 5.5 or >	• APOGEE GO Netscape & IE • Insight 3.4 IE 5.5 or >	Netscape & IE	Unknown
Additional Client Software	Java	Java	Java		• Web Access MS TSAC ActiveX control	• Insight 3.4 MS TSAC ActiveX control	Java	Java

Vendor	Trane	Automated Logic	Alerton	Teletrol	Johnson Control	Siemens Building Technologies	Tridium	Electric Eye - EnFlex
Other Web Access Protocols		XML		XML			XML	
Database								
Relational	MS Access	MS Access MS SQL My SQL Oracle IBM DB2	MS Access	My SQL	MS Access	MS SQL Objectivity	SQL	
Flat File					InfoPlus.21		•	•
ODBC				•	•		•	
JDBC		•		•			•	
Product Offerings								
Field Panels	Building Control Units (BCU)	M-Line S-Line	Integrator (BTI) Lsi VLCP	Network Controller (eNC)	NCM N30	MBC MEC	JACE-NP JACE-5 JACE-4	EnFlex G-100 MG-200
Display Panels	Optional Touch Screen on BCU 320 x 240 pixel	BACview 2 x 16 char. 4 x 40 char.	Viewport 4 x 20 char.	2-Line LCD on eNC	• LDT on N30 4 x 20 char. • NT on NCM 16 x 40 char. 256 x 128 pixel	Local User Interface (LUI) 2 x 40 char.		
Terminal Units	Unit Control Modules (UCM)	U-Line	VLC series VAV series	eTRAC, TSC, iVAV	DX series VMA series AHU, UNT, VAV controllers	TEC		EnFlex LIO-21, IO-12
Sensors/Actuators (F=Full line, L=Limited, N=None)	L	L	L	L	F	F	N	N
Tools								
Programming								
Graphical	TGP	Eikon	VisualLogic (requires Visio)	Easel	GPL		Workplace Pro	
Menu/Table/Template			•				•	Site Manager

Vendor	Trane	Automated Logic	Alerton	Teletrol	Johnson Control	Siemens Building Technologies	Tridium	Electric Eye - EnFlex
Line	CPL			ECMScript	JC-Basic	PPCL	Java	
Program Debugging and Testing								
Simulator		•		•	•			
Real-time data monitoring	•	•	•				•	
System/Network Configuration		Site Builder	Device Manager	Edifice	M-Tool		Workplace Pro	Site Manager
Applications								
Monitoring,, Data Visualization, and Reporting								
Graphical equipment monitoring	•	•	•	•	M-Explore M-Graphics	•	•	•
Performance monitoring displays					Analog Profile Comfort Chart Color Spectrum River of Time Starfield System Analysis Tool			
Complex archive data plotting (xy, statistics, 2D/3D)		•		•	•		•	XY 3D Surface
Trends, (M= multivariable)		M	M	M	M	M	M	M Multi-day
Energy analysis graphs and reports			Energy Logs		FX-TEM	Utility Cost Manager	VES	
Energy monitoring (interval meter data displays)		•	•	•	•	•	•	•
Real-time data visualization	•	•	•	•	•	•	•	•
Historical Data Playback (VCR)				•	•			

Vendor	Trane	Automated Logic	Alerton	Teletrol	Johnson Control	Siemens Building Technologies	Tridium	Electric Eye - EnFlex
Report Generation	•	•	•	•	•	InfoCenter Suite	•	
Other								
Alarms (Additional notification methods)	•	•	•	•	•	•	•	•
SNMP		•			•		•	
E-mail		•		•	•	•	•	•
Instant Messaging					•		•	
Pagers		•	•	•	•	•	•	•
Telephone Call-Out					•	•	•	
Fax				•	•		•	
Scheduling	•	•	•	•	•	•	•	
Tenant Services (Override)	•	•	•	•			•	
Dial-up modem access	•	•	•	•	•	•	•	•
Demand Limiting/Reduction	•	•	•	•	•	•	•	
FDD (Fault Detection and Diagnosis)					•			
Integration (N=Native, G=Gateway)								
CMMS (computerized maint.)					N		G	
CAFM (facility documentation)							G	
Lighting	N			G	N	G	G	
Security/Access				G	G	G		
Fire/smoke	N		G	G	N	G		
Weather Data				G				
Utility Billing				G	N: FX-TEM			
Other			G: Hotel Check-In System				G: Tenant billing, maintenance & management	
Support & Costs								
Documentation								

Vendor	Trane	Automated Logic	Alerton	Teletrol	Johnson Control	Siemens Building Technologies	Tridium	Electric Eye - EnFlex
On-line Help	•	•	•	•	•	•	•	•
Printed Manuals	•		•	•	•	•	•	•
CD-ROM		•	•	•	•		•	
Download from Web/FTP site	•			•	•		•	
Support								
Direct from Vendor	•	•		•	•	•	•	•
Distributors/Dealers		•	•		•		•	•
System Integrator				•			•	•
Helpdesk Hours	Available to Trane personnel	Vendor 8:30 AM - 7:00 PM (EST)	Depends on Dealer		24 hours	24 hours	8:00 AM - 6:00 PM (EST)	Depends on Dealer or System Integrator
Cost								
Pricing model (HW=hardware, SW=software)	Pricing based on complete engineered system.	HW: by equipment SW: number of concurrent users and number of 3rd party points.	HW: by equipment SW: number of devices in system (1-50, 51-150 or 150+).	HW: capacity based SW: number of concurrent users.	Pricing based on equipment type, building type and size, and control specifications.	HW: by equipment SW: license fee Other: installation, coordination, Commissioning	Vendor price to channel partners. Channel partners adds charges for engineering, installation, commissioning, and markup.	Complete systems are priced by the number of points plus. Pricing also available for retrofits to existing systems.

Appendix C
XML, SOAP, and Web Services

XML

What is XML? XML is similar to HTML, both being based on SGML (Standard Generalized Markup Language). HTML is used to format and display information on a web page when viewed by a browser. While good at displaying information, HTML does not have the ability to structure data in a manner suitable for data exchange. XML, on the other hand, uses tags* to define the relationship between elements of the data. XML is being developed and defined by W3C.

For example, on an EMCIS web access page, an outside air temperature may be displayed as "OAT 68 °F." If the HTML code representing the displayed text was to be processed by an application, the application would need to parse the text to determine if the data being exchanged is 68 with a unit of degrees Fahrenheit and it is the outside air temperature. If the HTML code is slightly different from the above example (OAirT instead of OAT), then the data would not be parsed properly. With XML, there would be tags that identify the data value as 68, the data unit as degrees Fahrenheit, and the data represented as outside air temperature. The definition of these tags and their relationship is called an XML schema.

Example XML Schema:

```
<OutsideAirTemperature>
    <Units>Fahrenheit</Units>
    <Value>68</Value>
</OutsideAirTemperature>
```

In order for the proper exchange of data, the sender and the receiver must use the identical schema. In order for a whole business sector to exchange data between each individual business, a standard set of schemas must be agreed upon within the business sector. The dilemma is that businesses can begin defining schemas so that data can be exchanged, but these schemas may not be compatible with other businesses and sometime in the future, a standard set of schemas may make earlier definitions obsolete.

SOAP and Web Services

SOAP is used to invoke applications running on another platform without regard to what the platform is or what language the application was written in. In fact, the client can be completely incompatible with the

*An XML tag is descriptive text that is wrapped around content (i.e., data) and is used to identify the content within the structure of the XML document.

server. SOAP relies on HTTP to transport XML based messages that are encapsulated in a SOAP envelope. Like XML, SOAP is being developed and defined by W3C.

SOAP is the basis for the emerging Web Services technology. Many of the major Enterprise Software vendors have Web Services initiatives, including Microsoft's .NET, IBM's WebSphere, Sun Microsystem's SunOne, and Oracle's 9i among others [8]. An IBM tutorial describes Web Services as "self-contained, self-describing, modular applications that can be published, located, and invoked across the Web. Web Services perform functions, which can be anything from simple requests to complicated business processes. Once a Web Service is deployed, other applications (and other Web Services) can discover and invoke the deployed service" [5]. One early example is Microsoft Passport, an authentication Web Service hosted by Microsoft [9]. Many believe that Web Services technology will eventually replace some of the current object-oriented architectures such as CORBA, DCOM, IIOP, OLE (OPC) and RMI.

ACKNOWLEDGMENTS

The authors are grateful for assistance from the following people:

Kevin Caskey; Alerton Technologies Inc.; *www.alerton.com*

Erik Ahrens; Syserco Inc., (Alerton dealer); *www.syserco.com*

Steve Tom; Automated Logic Corporation; *www.automatedlogic.com*

Rob Vandergriff; Airco Mechanical Inc., (Automated Logic dealer); *www.aircomech.com*

Andrew Chiang; Electric Eye Pte. Ltd.; *www.eeye.com.sp,*

Kristopher Kinney; Engineered Web Information Systems, (Electric Eye system integrator); *www.enwise.com*

Fredric Smothers; F. Smothers & Associates, (Electric Eye consultant); *www.fsmothers.com*

Nicolina Guiliano; Johnson Controls Inc.; *wwwjohnsoncontrols.com*

Jason Daters; Siemens Building Technologies Inc.; *www.sbt.siemens.com*

Phil Weinberg; Teletrol Systems, Inc.; *www.teletrol.com*

David Rinehard; The Trane Company; *www.trane.com*
Kim Kam

Phil Bomrad; Tridium Inc.; *www.tridium.com*
Edward Kubiak

References

[1] Motegi, N. and Piette, M.A., "Web-based Energy Information Systems for Large Commercial Buildings," National Conference on Building Commissioning, May 2002.

[2] Webster, T.L., "BCS Integration Technologies—Open Communications Networking," LBNL-47358. See also Chapter 27, *Information Technology for Energy Managers*, Fairmont Press, Atlanta GA 2004.

[3] *Information Technology for Energy Managers*, Fairmont Press, Atlanta GA 2004.

[4] United States General Accounting Office, "Electronic Government: Challenges to Effective Adoption of the Extensible Markup Language," Report to the Chairman, Committee on Government Affairs, U.S. Senate, GAO-02-327, April 2002.

[5] Vasudevan, V., "A Web Services Primer," *www.XML.com*.

[6] Applebaum, M.A. and Bushby, S.T., "450 Golden Gate Project: BACnet's First Large-Scale Test," *ASHRAE Journal*, July 1998.

[7] Webster, T.L., "Trends in Energy Management Technology—Enabling Technologies," LBNL-47650 June 2002. See also Chapter 8, *Information Technology for Energy Managers*, Fairmont Press, Atlanta GA 2004.

[8] Craton, E. and Robin, D., "Information Model: Key to Integration," AutomatedBuildings.com, Article—January 2002.

[9] Glass, G., "The Web Services (R)evolution: Applying Web Services to Applications," *www-4.ibm.com*, November 2000.

Chapter 22

Review of Advanced Applications in Energy Management, Control, and Information Systems[1]

Gaymond Yee
California Institute for Energy and Environment

Tom Webster, P.E.
UC Berkeley, Center for the Built Environment

Lawrence Berkeley National Laboratory
Berkeley, CA

ABSTRACT

It is important for energy managers to have a high level of knowledge and understanding of complex energy management systems. Energy practitioners need some basic informational and educational tools to help make decisions relative to energy management systems design, specification, procurement, and energy savings potential. In this chapter, we review advanced applications in Energy Management, Control, and Information Systems (EMCIS). The available features for these products are summarized and analyzed with regard to emerging trends in EMCIS and potential benefits.

BACKGROUND

In this chapter, the term EMCIS (energy management control and information system) has been used to refer to emerging EMS or EMCS with expanded information processing capabilities beyond the traditional building control and energy management functions. These expanded capabilities include data exchange and archiving, historical data visualization, and energy data analysis [1].

Beyond the standard functions of EMCIS are advanced applications to help facility/energy managers operate and maintain their facilities in a manner that maximizes occupant comfort and energy efficiency, while minimizing costs. In most cases, the applications are stand-alone products that perform a specific function. Some are a component of an EMCIS while others integrate with and access data from an EMCIS.

ADVANCED APPLICATIONS CATEGORIES

There are multitudes of application types for EMCIS. In this chapter, the focus is on categories that are emerging and considered state-of-practice. Some application categories are mature while others are so advanced that many of the applications within the category are still at the research level. However, only applications that actually have working prototypes and have been field-tested were reviewed in this chapter.

The advanced applications categories we have examined are:

- Data Visualization Tools
- Portable User Interfaces
- Energy Analysis Tools
- Advanced Control Algorithms
- Fault Detection and Diagnosis Tools
- Occupant Interaction and Feedback Tools

[1]This Chapter was originally published as a report for the Federal Energy Management Program, Trends in Energy Management Technology, Part 4—Review of Advanced Applications in Energy Management, Control, and Information Systems, Report No. LBNL-53546 (part 4)

In all, a total of thirty-nine applications were reviewed. There are many others that were not reviewed due to resource and time limitations.

This chapter is organized into three major sections. The first section is a technology description of the different categories of advanced applications reviewed in this chapter. The second section discusses the potential of these applications; their benefits, limitations, and cost effectiveness. The chapter ends with a summary and conclusions section. There is also an appendix containing multiple tables that summarize some of the main features of each advanced application category.

TECHNOLOGY DESCRIPTION

This section describes the technology in each of the advanced application categories that were reviewed. However, a detailed features comparison for the applications product in each of the categories is provided as a series of tables in Appendix B.[2]

Data Visualization Tools

With direct digital control devices, intelligent supervisory panels, and multi-tiered networking, EMCIS have the potential to collect and archive a tremendous amount of data. The availability of the data allows operators, facility managers, and energy managers to keep an eye on the health of the BCS and to perform activities such as monitoring for abnormal conditions, verifying environmental comfort levels, and testing new strategies to save energy.

With large volumes of data, presentation of the data becomes an issue. Simple tabular data presentation may be sufficient for a small set of data. For a large set, it becomes difficult to identify trends, patterns, or irregularities. One option is to export the data to other software for analysis. For example, a spreadsheet program can easily be used to present exported data in a graphical form for visual analysis. However, the export process becomes tedious if data analysis is a routine day-to-day operational task, and the creation of such a spreadsheet requires considerable resources.

Data visualization software usually comes with standard tools such as graphical equipment monitoring and trend plots. The former displays real-time data on a graphical representation of the equipment. The latter are time-series graphs of the data contained in the trend logs.

[2]With the exception of the Occupant Interaction and Feedback Tools category—no features comparison table is provided.

More advanced data visualization tools improve on trend plots by trending either multiple data points on the same plot or multiple days of one data point on the same plot. Another variation on trend plotting is displaying separate trend plots for each of the days of the month on one screen. A third approach is to "stack" the daily trend plots onto a third axis. This creates a 3D surface profile plot where the 2D lines of each daily trend plot now form a surface. Any valleys or peaks can be easily identified as possible irregularities that may warrant further investigation.

Another advanced data visualization tool is XY scatter plot. This is a plot of data from two different physical points or variables. One point/variable is plotted on the x-axis and the other on the y-axis. A complete plot contains many dots (points on a graph); each represents the values of both points/variables at one moment in time. Directionality of clusters of dots indicates some sort of trend. Isolated dots indicate potential problems.

A new class of tools, called "performance" monitoring tools, is emerging from one vendor, Johnson Controls. These tools use "abstraction" and "metaphor" visualization techniques that can be used to help identify possible impending equipment faults. Johnson Controls has given names to these "performance" monitoring tools;[3] Analog Profile, Comfort Chart, Color Spectrum, River of Time, Starfield, and System Analysis Tool.

One of the performance monitoring tools uses the rate of change of the setpoint error to predict the future offset from the setpoint. This predictive feature is a type of fault detection and diagnosis (FDD) analysis. While a detailed description of the tools is given in the appendix, a detailed analysis of effectiveness of these tools is beyond the scope of this study.

Portable User Interfaces

The work environment of building operators and maintenance personnel has evolved in recent years. Building operators and maintenance personnel are constantly on the move, going from one service or maintenance area to another. Portable user interfaces are now available to these mobile personnel that provide access to system data when needed. This access was previously only available with stationary workstations or cumbersome laptops.

PocketControls from Abraxas Energy, and Sigma Compact Edition from Invensys Climate Controls (United Kingdom) are portable user interfaces implemented on Personal Digital Assistant (PDA) platforms

[3]Details of these tools are provided in Appendix B.

running either Microsoft CE or PocketPC operating systems. Both products provide the basic features of a standard workstation such as view/acknowledge alarms, view point values, view and graph trend logs, and change setpoints.

The two products differ in how they are used. The intent of the Invensys Sigma Compact Edition is to serve as a substitute for a standard workstation at equipment locations where no workstation is available. The Invensys product accesses the building control system either by Ethernet connection to the equipment bus or by direct serial or Ethernet connection to a nearby controller.

The PocketControl product, on the other hand, is intended to be a truly mobile device. Abraxas Energy provides the product as a turnkey system that includes a preconfigured server that interfaces to a building control system workstation and serves the data to the PocketControl device over a cellular data service. The service provider is the choice of the end user.

Energy Analysis Tools

A major concern of today's facility manager is the operation of the facility not only in the most energy efficient manner, but also at the least cost. Both are difficult tasks, but the latter is the most difficult. With electric and natural gas deregulation, there are many more choices for energy suppliers and additional choices for long-term energy supply contracts. These contracts normally have peak demand charges or the state's public utility commission may mandate peak demand charges. To make sure the facility's energy budget stays in the black, the facility must be operated not only in the most energy efficient, but also most power-demand efficient manner.

Energy Analysis software is commonly available as an option from building controls vendors. In recent years, many products specifically tailored for energy analysis have been developed and deployed. All provide graphs and reports that are used to:

- Adjust operations and scheduling to help minimize energy use and lower peak demand.
- Negotiate energy contracts for more favorable terms.
- Benchmark against published statistical data to see if systems need optimization or upgrading.
- Cost allocate energy use.
- Validate utility bills.

Most of the energy analysis products reviewed in this study follow the current software development trend by utilizing a web interface. The advantages of a web interface were thoroughly discussed in Chapter 20 of this book. [See also Reference 3.] A few of the other product vendors are also following the business model trend of being applications service providers (ASP). As an ASP, the vendor does not sell the software to the end user, but rather provides the software as a service. From the vendor's point of view, being an ASP potentially brings in a continuous and somewhat predictable source of revenue. From the end user's point of view, paying for the service reduces the risks of up-front capital expenditures for software and possibly hardware and recurring expenses for software operations and maintenance. This is especially true for small companies or for companies with limited budgets.

A few of the products offer comprehensive features that include:

- Energy analysis tools to benchmark energy usage by normalizing it with building parameters such as square footage and number of occupants and to compare energy usage between different buildings.
- Load management tools to monitor and optimize loads, and to manage strategies to shed loads and reduce peak demand.
- Billing analysis tools to do "what if" analysis by applying energy usage to different billing rates, to generate energy invoices for facility tenants, and to verify bills from the utility or energy provider.
- Forecasting tools to predict next day and next second-day energy usage for advance load shed strategy planning to avert peak demand charges.
- Reporting, real-time monitoring, alarming, and remote control tools.

Two emerging tools are tools for advanced data mining and distributed energy resource control. The advanced data mining tool utilizes Online Analytical Processing (OLAP) technology that can be used to interactively design custom reports that "slice, dice, and drill down" into large amounts of data. Distributed energy resource[4] control tools are used to monitor and control distributed energy resources. Controls can range from simple remote-control on/off, to predefined schedules, to automatic triggering by events, to control by sophisticated algorithms.

[4]This used to be called distributed generation, but distributed energy resource is a more general term that covers non-generation energy resources such as solar.

Advanced Control Algorithms

Back in the days of pneumatic controls, control algorithms included proportional (P) and proportional plus integral (PI) control techniques implemented in fluidic controllers. With the advent of microprocessor-based controls and direct digital control, P and PI were implemented in software. With the power and flexibility of the microprocessor, proportional plus integral plus derivative was also implemented (PID). However, even much more powerful algorithms can be implemented.

One innovation is the introduction of adaptive logic to the control algorithm. Adaptive logic is used to continuously adjust the control parameters (proportional band, integral times, and deadband) to maximize the responsiveness of the system while maintaining stability. The adjustments are made based on values of measured data or recognition of certain patterns in the measured data.

Adaptive logic is also implemented in the controllers for variable speed drives used on chillers. The optimal speeds for variable-load and water-temperature combinations are learned and remembered to maximize energy savings.

At the research level, more advanced algorithms including finite state machine control sequencing and neural networks are being developed. Finite state machine logic defines specific states, or modes of operation under various conditions. Within each state, control directives are issued. A neural network is a software system that has highly interconnected tasks, working in parallel like the human brain. Neural networks must be trained with data and expected output, and have the ability to learn. Neural networks can be used in highly nonlinear and dynamic processes where traditional control algorithms become overly complex.

Fault Detection and Diagnosis Tools

One of the important emerging EMCIS applications is software for fault detection and diagnosis (FDD). Many buildings do not perform as expected because of problems and errors that occur during the stages of the building life cycle: from design to commissioning to operation. FDD tools have the potential to quickly diagnose operational problems with the building HVAC equipment, and as a result, the equipment will operate as intended for a larger proportion of the total run time. The benefits of properly operating HVAC equipment include [4]:

- Improved occupant comfort and health
- Improved energy efficiency
- Longer equipment life
- Reduced maintenance costs
- Reduced unscheduled equipment down time

However, the impact of these benefits is difficult to quantify which is a barrier to widespread adoption of the diagnostic tools [4].

FDD tools are an emerging application for use in building commissioning and operation. Most tools are still in the research stage with functional prototypes. Many have undergone field trials and are being improved. Only a few of the tools are actually available commercially. However, none of the tools are components of current BCS/EMCIS offerings; i.e., all are stand-alone applications. Some operate in batch mode, taking data input from files. Others take data input on-line, either directly from the BCS/EMCIS or from a data gateway.

FDD tools employ either a top-down or a bottom-up approach. In a top-down approach, high-level performance measurements (such as whole building energy use) are used to detect possible problems in the lower level hierarchy of the BCS. The other, bottom-up approach, uses low-level performance measures to isolate the exact equipment fault and then propagate the problem up to higher levels to determine what the impact is on the whole building.

By its very name, FDD tools perform fault detection and fault diagnosis. The two are different operations. Fault detection is the determination that the building or equipment operation is incorrect or not within expected performance levels. Faults may occur over the whole operating range or be confined to a limited region and hence only occur at certain times. After a fault is detected, fault diagnosis identifies the cause of the fault as well as the location (which equipment) of the fault. This process involves determining which of the possible causes of faulty behavior are consistent with the observed behavior [5].

FDD tools utilize either manual or automatic methods to detect and diagnose faults. With manual methods, fault detection is performed by manually comparing the actual performance against some baseline. Manual fault diagnostic requires the knowledge of a human expert to infer the problem based on what the actual performance data show. Automated methods employ either automated comparison of actual performance data with a baseline or fault identification by recognizing patterns in the measured data. Automated diagnostics employ rule-based expert systems with decision trees to deduce the cause of the fault.

A fault is detected when the deviation of the measured performance from a baseline has exceeded a specified threshold. There are several methods to determine the comparison baseline.

- *Reference data.* Performance data from equipment manufacturers or from historical measurements can be used as the comparison baseline.

- *Modeled data.* Mathematical models based on principles of physics and thermodynamics are used to generate a predicted comparison baseline for proper performance. These quantitative models use measured data as input.

- *Neural Networks.* Neural networks are software systems that consist of large interconnected networks of simple processes or tasks. The processes or tasks execute in parallel (similar to the human brain). Similar to human learning, the neural network is trained with input data and the resultant output data. Neural networks are very good with non-linear systems where mathematical models are very difficult to derive. On the other hand, neural networks are only as good as the data that is used to train them.

  In FDD, a trained neural network can be used to predict output data, given a set of input data. The predicted output data is used as the comparison baseline. Neural networks fall into the category of empirical or "black box" models." (Note: No neural network based FDD tool was reviewed in this chapter)

- *Expert Systems.* Expert systems are qualitative models that use expert rules stored in a knowledge database to infer faults from the measured input. The expert rules can be based on human expert knowledge or past operational conditions. The rules are implemented as a decision tree, or in programming language using IF-THEN-ELSE statements.

- *Fuzzy Logic.* Fuzzy logic uses fuzzy set theory to take account of uncertainties in how a HVAC system is described. A fuzzy model is a set of fuzzy rules that describe the relationship between a set of inputs and a set of outputs in qualitative terms [6]. Data are not assigned with absolute values, but rather with levels of membership to qualitative sets. For example, a temperature of 25°C may be assigned a 50% membership to the qualitative set of "warm." The advantages of fuzzy logic are the ability to model uncertain, non-linear behavior, the simplicity of the rules, and low computational processing demands. The trade-off is that the results are also "fuzzy," or less precise. (Note: No fuzzy logic based FDD tool was reviewed in this chapter)

- *Electrical load disaggregation.* Electrical load disaggregation is not a baseline prediction method, but rather a method to deduce individual equipment power usage from one total system power usage profile. The advantage of this method is the elimination of the hardware and installation cost of submeters for the individual equipment.

Occupant Interaction and Feedback Tools

Building occupants have been an untapped source of information for facility managers trying to improve the performance of their facilities. Human occupants are mobile and intelligent and can provide sensory feedback that cannot be matched by the most sophisticated man-made sensors. Two tools, both developed by the Center for Built Environment (CBE) of the University of California at Berkeley, utilize occupant feedback to influence building system design and operation for better performance and comfort.

The first tool, *Web-Based Occupant Satisfaction Survey,* is an environmental quality survey that is more cost-effective and faster to complete than traditional paper-based surveys. The survey has a standardized set of core questions that is used to measure satisfaction with environmental factors such as office layout, office furnishings, air quality, thermal comfort, lighting, acoustics, and building cleanliness and maintenance. Branching questions are used to capture more details where the survey revealed perceived problems. Besides the standard core set of questions, there are optional sets that deal with wayfinding, safety and security, and underfloor air distribution systems. Core survey completion time varies, ranging from 5-12 minutes. Additional optional questions increased completion time up to 20 minutes. Responses to the survey are recorded in a secured SQL (Structured Query Language) server database. Results are reported using a web-based tool.

The survey has been extensively tested and refined. It has been applied in several research scenarios that include evaluation of the effectiveness of a design intervention, development of new building design guidelines, and benchmarking of facility performance. Over 40 surveys have been completed or scheduled.

Planned enhancements of the survey tool include reporting tool improvements with more graphics, the addition of a data mining tool, support for multi-languages, and the replacement of the General Services Administration's (GSA) paper-based Facility Management Performance survey for tenant satisfaction with a web-based survey.

The second tool, *Tenant Interface for Energy and Maintenance Systems* (TIEMS), is a web-based user interface for use by building occupants (or tenants). Tenants using such a user interface will provide useful occupant feedback that should improve thermal comfort, improve performance of energy management strategies, eliminate some redundant service requests, and improve the quality of data in maintenance databases. TIEMS is being designed as a component of the GSA's GEMNet, an integrated information technology infrastructure for energy and maintenance management. TIEMS is currently undergoing field-testing. Unlike the Web-Based Occupant Satisfaction Survey tool, TIEMS is an on-line system that has two-way communications with a computerized maintenance management system (CMMS).[5]

TIEMS has four key features: (1) the ability to submit and check the status of service requests, (2) the ability to see other service requests submitted during the last two hours from the same location as the current service request being submitted, (3) the ability to see a list of notices that maintenance personnel has conveyed to the tenants, and (4) the ability to check the indoor temperature.

Overall Assessment

Advanced applications are emerging because they service real needs in supporting facility operations and the maintenance group's function of providing the best environmental comfort to building occupants in the most efficient and cost effective manner.

Many of the applications would not exist without certain enabling technologies. These technologies include:

- High speed rendering of 3D graphics
- Personal Digital Assistants (PDA)
- Wireless cellular data communications
- The Internet
- Online Analytical Processing (OLAP)
- Neural Networks
- Expert Systems
- Web-based interfaces

As additional new technologies emerge, more advanced EMCIS applications will also emerge from the research community to eventually become standard tools for the facility/energy managers.

POTENTIAL OF THE EMCIS ADVANCED APPLICATIONS

Benefits

The potential benefits of the EMCIS advanced applications are the reduction of operation and maintenance costs and the reduction of energy usage and cost without sacrificing occupant comfort. The specific benefits of each category of advanced applications can be better summarized within the context of the category.

- *Data Visualization Tools*
- Help identify abnormal trends.
- Help identify and localize problems.
- Visual monitoring of system performance and system health.
- Reduce tedious work of raw data processing.

Portable User Interfaces

- Aid to mobile operators and facility/building managers.
- Substitute user interface when a workstation is not available.

Energy Analysis Tools

- Identify potential energy savings.
- Benchmarking.
- Use to negotiate better rates.
- Forecast loads for planning of load management strategies.
- Allocate energy use to tenants or corporate departments.
- Validate energy bills for errors.

Advanced Control Algorithms

- Optimize equipment control for better comfort, reduced energy usage, and longer equipment life.

Fault Detection and Diagnosis Tools

- Determine improper operations.
- Early detection of potential equipment failures.

[5]A CMMS software package helps organizations manage the maintenance of equipment and facilities. For example, it can keep track of preventive maintenance schedules, service contracts, problem reports, and material inventories. In the process, people and resources more managed more effectively.

- Provides better comfort with reduced maintenance and operating costs.

Occupant Interaction and Feedback Tools
- Occupant feedback to designers helps future designs.
- Used for benchmarking to identify under-performing buildings.
- Feedback to building operations helps to improve comfort, occupant satisfaction, and lower operations costs.

Limitations

As with any tool, the potential benefits of the EMCIS advanced application tools can only be realized if the tools are utilized properly. This means that proper operator training is very important.

Since these tools are in many cases additional to the cost of the EMCIS, one must determine whether the cost savings justifies the additional cost. For tools that are still being perfected by the researchers, one question is whether the research results can be applied to real world operating conditions.

Cost Effectiveness

It is beyond the scope of this chapter to analyze the cost effectiveness of EMCIS advanced applications. In many cases, the applications are still in the research stage and any cost effectiveness analysis is premature. In other cases, one can compare O&M cost before and after the advanced application tool is deployed as a means to determine cost effectiveness. And finally in some cases such as fault detection and diagnosis tools, a single use of the tool may justify the cost of the tool if the tool discovered a major fault and the correction of the fault resulted in a large savings in O&M cost.

SUMMARY AND CONCLUSIONS

The advanced EMCIS application products reviewed in this chapter cover a wide spectrum of functionality. Many, if not all, utilize the latest technologies. As even newer technologies emerge, more advanced applications with additional innovations will also emerge.

The standard functions of EMCIS do not serve all the needs of building, facilities, and energy managers. Advanced applications provide the tools for these professionals to operate their facilities with better comfort, higher or better performance, and at a lower cost.

Poor system commissioning and faulty equipment operations are common problems today. Re-commissioning is a costly process and identifying faulty equipment operations is an art form. As FDD tools emerge from research and mature, the FDD tools will likely have a significant and large impact.

APPENDIX A
ACRONYMS AND DEFINITIONS

Many acronyms used in this chapter can be found in the Glossary to this book. Additional definitions are included here.

ASP	Applications service provider
CBE	Center for Built Environment, University of California, Berkeley
CMMS	Computerized maintenance management system
FDD	Fault detection and diagnosis
GSA	General Services Administration
OLAP	Online analytical processing
P	Proportional
PDA	Personal digital assistant
PI	Proportional plus integral
PID	Proportional plus integral plus derivative
SQL	Structured Query Language
TIEMS	Tenant Interface for Energy and Maintenance Systems

APPENDIX B
APPLICATIONS TOOLS PRODUCTS AND FEATURES

The following tables list products and their features for each of the applications categories studied. Each table is customized to show the features of significance to the category. A detailed explanation of each feature is included below each table.

General Features

Applications Service Provider (ASP). An Applications Service Provider is a company that provides the use of applications software as a service to end-users. The ASP will host the software on its servers and access to the applications is via the Internet. Since the applications are not located on site, collected data is also transmitted to the ASP servers over the Internet.

Energy Analysis	Products (vendor)										
	Active Energy Management (Engage Networks)	Atrium (Honeywell)	EEM Suite (Silicon Energy)	Electric Eye (Electric Eye)	Energy Commander / Energy Analyzer (TruePricing)	Enerwise Energy Management (Enerwise Global Technologies)	EnterpiseOne (Circadian Information Systems)	ExcelSyus (Excel Energy)	FX-TEM (Johnson Controls)	InfoCenter Suite / Utility Cost Manager (Siemens)	Vykon / Vykon Energy Suite (Tridium)
General Features											
Applications Service Provider (ASP)		•			•	•					
Web Interface	•	•	•	•	•	•			•		•
Data Collection Gateway Hardware	•			•	•			•			•
Energy Analysis											
Benchmarking			•								
Multi-site Comparison	•		•				•				
Aggregation	•	•	•					•	•		•
Advanced Data Mining (OLAP)											
Load Analysis											
Load Management	•	•	•			•					•
Load Profiles			•	•		•	•	•	•	•	•
Load Duration	•		•								

Energy Analysis	Products (vendor)										
	Active Energy Management (Engage Networks)	Atrium (Honeywell)	EEM Suite (Silicon Energy)	Electric Eye (Electric Eye)	Energy Commander / Energy Analyzer (TruePricing)	Enerwise Energy Management (Enerwise Global Technologies)	EnterpiseOne (Circadian Information Systems)	ExcelSyus (Excel Energy)	FX-TEM (Johnson Controls)	InfoCenter Suite / Utility Cost Manager (Siemens)	Vykon / Vykon Energy Suite (Tridium)
Billing Analysis											
Rate Engine	•	•	•			•	•		•		
Energy Invoicing	•	•	•			•	•			•	
Bill Verification		•	•				•		•		
Forecasting											
Forecasting		•	•								
Weather Link	•	•	•			•	•				
Reporting											
Energy Reports	•	•	•		•	•	•		•	•	•
Exception Reports						•				•	•
Operations Support											
Real-time monitoring			•	•			•				•
Alarming	•		•		•	•				•	•
Remote Control		•	•		•			•			•
Distributed Energy Resource Control	•	•	•			•					

Web Interface. The user interface for the application is a standard Web browser. With a Web Interface, no user interface software is required for workstations accessing the energy analysis application. The use of a Web Interface reduces the cost of software maintenance and upgrade.

Data Collection Gateway Hardware. Some vendors provide a hardware gateway device to collect data from the control system. These gateway devices perform network bridge functions, protocol translations, and act as the Internet interface for those applications that are not located on site.

Energy Analysis

Benchmarking. To benchmark different buildings or sites, energy data can be normalized by square footage, number of occupants, outside air temperature, cooling/heating degree days, etc. and categorized by geographical region, building type, business type, etc.

Multi-site Comparison. This feature usually compares whole building energy consumption of each building in a bar chart. The feature is useful for multi-facility operators that need to target sites for energy efficiency retrofit.

Aggregation. Loads from several sources are aggregated (combined) into a total value. The aggregated data can then be analyzed like any other data point (for example load profiles). Using aggregated loads can save the cost of installing separate meters to measure total or subtotal loads. It also allows the flexibility to aggregate different sources and at different times if such analysis is required.

Advanced Data Mining (OLAP). Large amounts of data can be analyzed using custom reports that are designed interactively. Data can be "sliced, diced, and drilled down" to extract the information from different perspectives. The advanced data mining capabilities are provided by Online Analytical Processing (OLAP) technology.

Load Analysis

Load Management. This feature covers tools used to manage peak loads. These tools include tools for curtailment contract management, notification with pricing, bid submittal, event monitoring, and baseline comparison. Also included are tools for defining and activating curtailment and load shed strategies.

Load Profiles. Load profiles are trend plots of load over a period of time. Normally, the period is a 24-hour period. This common graph provides easy verification of operating schedules, identifies peak hours, and can also be used to develop baseline load profiles.

Load Duration. The load duration curve is a histogram that indicates the percentage of time the load was at a particular value. The graph has percentage as the x-axis and the load (kW) as the y-axis. If the curve spikes up at the low percentage area, it indicated that there is high peak demand at infrequent times.

Billing Analysis

Rate Engine. Past interval meter data can be passed to the rate engine software to compute a bill. Different rates can be used to determine the optimal one. The analysis aids in forecasting future energy costs and determining the lowest cost energy provider based on the energy usage pattern of the facility.

Energy Invoicing. Interval data from submeters are used to generate energy invoices for facility tenants or for energy cost allocation to corporate departments.

Bill Verification. Interval meter data and energy rate data can be used to compute a monthly energy cost for verification against the actual utility bill.

Forecasting

Forecasting. Forecasting tools use a combination of proprietary algorithms, historical load data and weather statistics, and temperature forecasts to generate predicted 24- and 48-hour energy usage. The forecast can be used to determine whether strategies should be deployed to avert peak demand charges.

Weather Link. Weather data is usually retrieved either from a commercial weather service or from one of the government-operated weather bureaus. Weather data can be used just for informational purposes, but in most cases, the weather data is used for more accurate forecasting.

Reporting

Energy Reports. Energy reports include both textual and graphical reports. Common reports include energy usage summary (by site, building, etc.), daily load profiles, aggregation reports, energy use ranking (vertical bar chart), and relative contribution (pie chart).

Exception Reports. Exception reports are generated for data that do not fall into a user-defined range during a specified time period. These reports can be used to identify problems.

Operations Support

Real-time monitoring. Data is collected and displayed in real time (or near real time).

Alarming. The alarming feature includes a set of tools that perform out of range detection, alarm notification (email, pager, fax, PDA, etc.), and alarm management (acknowledgments, resolution, logs, and audit trails).

Remote Control. Most applications perform only data collecting and monitoring functions. However, some applications permit remote control functions such as changing setpoint values and turning equipment on and off.

Distributed Energy Resource Control. Distributed energy resources such as internal combustion generators, fuel cells, micro-turbines, photovoltaic panels and wind power generators can be remotely controlled in a variety of ways. Control can be range from simple on/off control, to pre-defined schedules, to automatic triggering by events such as alarms, demand, temperature, and price thresholds, to sophisticated algorithms that factor in marginal cost of generation, marginal cost of grid energy, and forecast load.

Standard Data Visualization Tools

Graphical Equipment Monitoring. Instead of showing real-time equipment data in a tabular form, graphical equipment monitoring displays show the real-time data on a graphical representation of the equipment. One can even use a digital photograph of the equipment. The idea of graphical equipment monitoring is to give the operator a more intuitive sense of how the equipment is operating. Also, if there are any alarms, the alarms on the display can immediately point the problem to the operator. However, some operators find the graphical displays unnecessary and prefer lists and tables instead.

Trend Plots. A trend plot is a time series plot of a data point. Viewing trend data graphically is preferred over the traditional method of viewing a textual log. Graphical views allow easier identification of irregularities as well as abnormal trends.

Advanced Data Visualization Tools

Multi-Point/Variable Overlay Trends. Multiple variables or points are displayed on the same plot and with the same time period. In a multi-point/variable trend plot, one can see relationships between points/variables as time progresses.

Multi-Day Overlay Trends. Multiple days of data from the same point are displayed on the same plot (normally in a 24-hour period). Any trends that occur in a similar fashion from day-to-day, or only on weekdays or weekends can be easily identified.

Calendar Profiles. This plot is a continuous plot of trend data for a one-month period. The trend plot is segregated by boxes; each box representing a single day and the boxes organized like the days of the month on a calendar. The calendar profile can be used to identify any abnormal days, differences between weeks of the month, and differences between weekdays and weekends.

3D Surface Profiles. A variation of a multi-day trend plot but instead of plotting the days all on a 2D plane, the

Data Visualization	Active Energy Management (Engage Networks)	EEM Suite (Silicon Energy)	Electric Eye (Electric Eye)	EnterpriseOne (Circadian Information Systems)	Metasys (Johnson Controls)
Standard Data Visualization Tools					
Graphical Equipment Monitoring		•	•	•	•
Trend Plots	•	•	•	•	•
Advanced Data Visualization Tools					
Multi-Point/Variable Overlay Trends	•	•	•	•	•
Multi-Day Overlay Trends	•	•	•	•	
Calendar Profiles	•			•	
3D Surface Profiles	•	•	•		
X-Y Scatter		•	•	•	
Performance Monitoring Tools					
Analog Profile					•
Comfort Chart					•
Color Spectrum					•
River of Time					•
Starfield Display					•
System Analysis Tool					•

days are plotted on a third axis. The resultant plot is a 3D surface plot. For example, a 3D surface plot of system energy use will show valleys of low energy use and peaks of high energy use.

XY Scatter Plots. Scatter plots or XY plots graph one variable on the x-axis against another variable on the y-axis. Each point on the plot represents the values for both variables at an instance in time. Scatter plots normally contain points from a large number of samples. One can easily see the correlation between the two variables as well as potential abnormal conditions when a sample point lies far from all other samples. The ability to properly filter the data set for transient and spurious data is important to being able to see trends and spot the abnormalities.

Performance Monitoring Tools

Several advanced data visualization tools, classified as performance monitoring tools, use "abstraction" and "metaphor" visualization techniques to view several inter-related data at the same time. These advanced tools allow easy identification of system performance and can help identify impending equipment faults.

Analog Profile. The Analog Profile display consists of two horizontal bars. The left bar shows how much higher or lower the data point is from its setpoint. A small rectangle gives an indication of the rate of change for the data point. Several data points can be viewed by stacking the horizontal bars vertically. One can easily see normal conditions such as data point values over setpoints, but negative rate of changes. Abnormal conditions would be a positive rate of change. On the right of each bar is displayed another bar that represent the relative flow (0-100%) of a data point that's related to the first.

Comfort Chart. The Comfort Chart shows temperature on the x-axis and humidity on the y-axis. A visual pattern on the chart uses the ANSI-ASHRAE 55-1992 comfort standard to highlight the temperature and humidity range where 80% of people should feel comfortable. Temperature and humidity sensor values for rooms and zones of interest are plotted on the display in real time. Seeing where the data points lie and watching the movement of the points over time help in detecting possible heating and cooling problems that will lead to occupant discomfort.

Color Spectrum. In Color Spectrum, colors are assigned to particular historical values of a data point. The display shows daily, weekly, or monthly views of the data point as horizontal bars. The colors will bring out any daily/weekly/monthly patterns that can be identified for further investigation.

River of Time. River of Time is a dynamic display that uses a 24-hour bar to represent 12 hours of past data on the left, 12 hours of future scheduled state on the right and the current state in exactly the middle. As time passes, the display flows from right to the left. For example, the past 12 hours of a fan status can be displayed. Colors on the bar can indicate on/off/alarm status. The future 12-hour scheduled on/off state (setpoint) of the fan is shown on the right. Several parameters can be viewed with the same display. With this display one can quickly see what the state of various components in the system had been in the past 12 hours, and what the future states will be in the next 12 hours.

Starfield. The Starfield display consists of star clusters on a black background, with each cluster representing an area of interest in a building. For example, the central star in the cluster can represent a fan. The surrounding stars in the cluster can represent the zones controlled by the fan. The size and color of the central star show the fan's on/off status and normal/alarm state. The color of the surrounding stars show whether the zones are warm, cool, or normal, relative to the zone setpoints. The distance of the surrounding stars from the central star indicates how far the zone is away from the setpoint. By recognizing the star cluster pattern and colors, an operator can quickly identify problems.

System Analysis Tool. The System Analysis display can display data from many different types of points with different units and varying ranges. Each point value is represented as a bar and scaled individually. The rate of change is also displayed as well as little arrows to indicate the direction the point value is going. There are also minimum and maximum indicators that track the historical minimum and maximum values of the point. The System Analysis tool enables operators to quickly see the performance of the entire system in a single display.

Operating System

Windows CE. Windows CE is an embedded version of the Microsoft Windows operating system (OS) used mainly in PDA (Personal Digital Assistant) or handhand type PC's. Windows CE has been superseded by Microsoft's new embedded Windows OS, Pocket PC.

Pocket PC. This is the current Microsoft embedded Windows OS for PDAs.

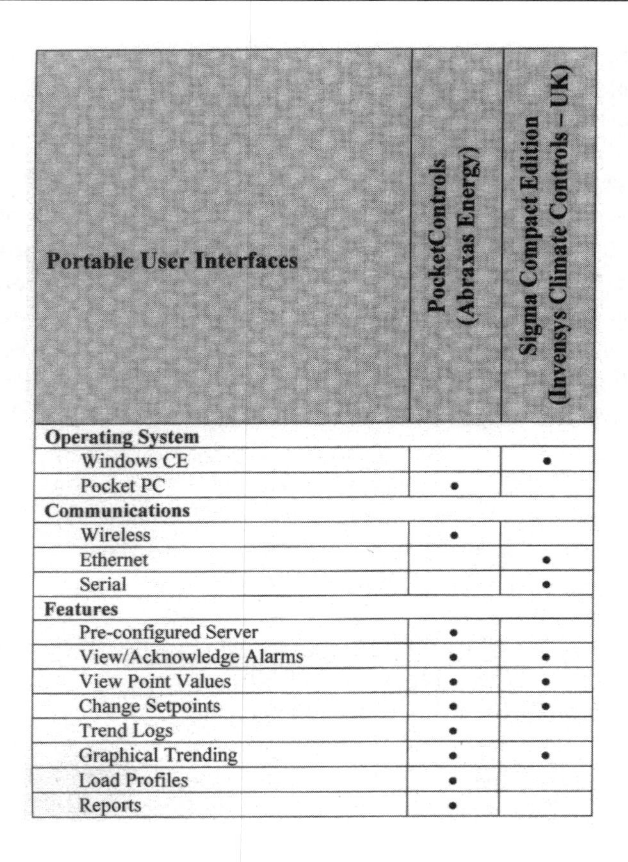

Portable User Interfaces	PocketControls (Abraxas Energy)	Sigma Compact Edition (Invensys Climate Controls - UK)
Operating System		
Windows CE		•
Pocket PC	•	
Communications		
Wireless	•	
Ethernet		•
Serial		•
Features		
Pre-configured Server	•	
View/Acknowledge Alarms	•	•
View Point Values	•	•
Change Setpoints	•	•
Trend Logs	•	
Graphical Trending	•	•
Load Profiles	•	
Reports	•	

Communications

Wireless. Wireless communication is provided by cellular modem technology.

Ethernet. Can connect directly to the Ethernet equipment bus.

Serial. Can connect directly to a local field panel using a serial cable.

Features

Pre-configured Server. A special server with pre-configured software is used to interface the BCS to the wireless service provider. The server communicates two-way with the BCS and wireless service.

Pre-configured Server. A special server with pre-configured software is used to interface the BCS to the wireless service provider. The server communicates two-way with the BCS and wireless service.

View/Acknowledge Alarms. Alarms can be viewed and acknowledged on the portable user interface.

View Point Values. View the current values of data points.

Change Setpoints. Setpoints can be changed from the portable user interface.

Trend Logs. Time series data (trend logs) can be viewed as tables.

Graphical Trending. Trend log data can be viewed graphically on the portable user interface.

Load Profiles. Trend plots of building kW and kWh usage.

Reports. The PDA or the server has a set of pre-configured reports.

Product Status

Some of the advanced control algorithms are still at the research level. Others are available commercially and already have an installed base.

Applied Method

Finite State Machine. Finite state machine is a new logic for program control sequence. The logic defines specific states, or modes of operation, for the process (such as heating, cooling, fan on, fan off). Each state also contains specific control logic. Finite state machine control sequence logic eliminates some of the common problems such as simultaneous heating and cooling and rapid cycling between heating and cooling modes. Finite state machine logic is easier to define, illustrate, and program than traditional control logic.

Neural Network. Neural networks distribute computational tasks onto many identical simple units that are highly interconnected and can work in parallel (some resemblance to the human brain). Neural networks deal well with arbitrary nonlinear problems with numerous simultaneous constraints, and they exhibit the ability to "learn." Neural networks are ideal for systems rich with data, but poor with theory, as well as systems that are non-stationary, nonlinear, and dynamic. In order for neural networks to operate, they must be "trained" with input data and expected output data.

Pattern Recognition. The control algorithm monitors specific input, output, and setpoint data for specific patterns. With each recognized pattern, the algorithm will issue the associated control directives.

Self-Adjusting Deadband. With this control algorithm, the deadband is dynamically adjusted based on the value of measured data. The result is a very responsive, but also a very stable control system.

Adaptive VSD Speed Control. Controllers on variable speed drives for chillers learn and remember the optimal

Advanced Control Algorithms	**Adaptive Capacity Control (York)**	**Enhanced Adaptive Control (Trane)**	**Finite State Machine (Johnson Controls / NIST)**	**HVAC-Pro (PRAC) (Johnson Controls)**	**Neural Networks (Colorado State University)**	**Neural Networks (U. of Colorado & U. of Nebraska)**	**P-Adaptive Control (Johnson Controls)**	**PRAC (Johnson Controls)**
Product Status								
Research			•		•	•		
Commercial	•	•		•			•	•
Applied Method								
Finite State Machine			•					
Neural Network					•	•		
Loop Tuning				•				
Pattern Recognition				•				•
Self-Adjusting Deadband							•	
Adaptive VSD Speed Control	•	•						
Adaptive Control Parameters				•			•	•
Equipment Application								
VAV				•			•	•
AHU			•	•				
Unit Controller				•				
Chiller	•	•				•		
Heating Coil					•			

speeds for variable-load and water-temperature combinations. The controllers instruct the drives to initiate the proper speed and pre-rotation vane position for maximum energy savings.

Adaptive Control Parameters. Control loop tuning parameters (proportional band and integral time) are continuously adjusted for optimal closed loop control.

Equipment Application

This section of the table lists the equipment application for each of the advanced control algorithms.

Product Status

Fault detection and diagnosis tools are emerging applications in building operations and maintenance. Many are still in R&D with functional prototypes at various universities and research laboratories across the country. A few are commercialized and available for purchase from private companies.

Data Collection Method

Data used by the fault detection and diagnosis applications can be provided either on a batch basis (electronic files containing the historical data) or on an on-line basis. In the on-line case, the application is either resident on the equipment or has communications access to the equipment.

Detection/Diagnosis Method

The detection and diagnosis of faults can be by manual or automatic means. Manual detection uses visual comparison of actual data against predefined data, reference data, statistical data, or historical baseline. Automatic detection either uses expert rules and decision trees to determine the presence of a fault or compares actual data against modeled data to determine the deviation that indicates a fault.

Manual diagnosis requires a human expert to analyze the state of the equipment at the time of the fault and diagnose the cause of the fault. Automatic diagnosis

Fault Detection and Diagnosis	ACRx (Hand Tool / Service Tool) (Field Diagnostic Services / Honeywell)	APAR/VPACC (NIST)	Enforma (Architectural Energy Corp.)	AHU Diagnostics Toolkit (Center for Built Environment)	FDD for Rooftop AC (Purdue University)	IMDS (LBNL)	NILM (MIT)	PACRAT (Facility Dynamics)	Universal Translator (PG&E / PEC)	MIFDD (PG&E)	WBD / OAE (PNNL)
Product Status											
Research		•		•	•	•	•		•	•	•
Commercial	•		•					•			
Data Collection Method											
Online	•				•	•	•				•
Batch		•	•	•				•	•	•	
Detection Method											
Automated	•	•			•		•	•		•	•
Partial Automated									•		
Manual			•	•		•					
Diagnosis Method											
Automated	•	•			•			•			•
Partial Automated							•				
Manual			•	•		•			•	•	
Applied Method											
Manual Data Visualization			•	•		•	•		•		
Expert System – Rule-Based	•	•			•			•			•
Compare w/Reference and Perf. Data			•	•					•		
Compare w/Modeled Baselines								•		•	•
Electrical Load Disaggregation							•				

Fault Detection and Diagnosis	ACRx (Hand Tool / Service Tool) (Field Diagnostic Services / Honeywell)	APAR/VPACC (NIST)	Enforma (Architectural Energy Corp.)	AHU Diagnostics Toolkit (Center for Built Environment)	FDD for Rooftop AC (Purdue University)	IMDS (LBNL)	NILM (MIT)	PACRAT (Facility Dynamics)	Universal Translator (PG&E / PEC)	MIFDD (PG&E)	WBD / OAE (PNNL)
Equipment Application											
Whole Building						•		•			•
Rooftop AC	•				•						
AHU (E-Economizer)		•	•	•				•	E		•
Cooling Towers			•			•					
Chillers			•					•			
Heating Plants			•								
Zone Distribution Systems (V-VAV)		V	•					•		V	
Hydronics Systems								•			
Fans and Pumps				•		•					
Elec. Equip. Cycling/Schedule Faults							•				

uses expert rules and decision trees to determine the cause of the fault.

Applied Method

Manual Data Visualization. This is the manual process of detecting faults. The process requires a human expert that can recognize abnormalities by viewing data logs either textually or graphically.

Expert System—Rule-Based. Equipment data are compared to a series of rules based on expert human knowledge of the operation of the equipment. If the equipment data point fails all rules for proper operation or fits a rule for improper operation, a fault is signaled.

Compare with Reference and Performance Data. Comparisons are made between actual data plots with pre-defined reference plots or from performance plots provided by the equipment manufacturer. Excessive deviations from the baseline indicate potential incorrect equipment operation

Compare with Modeled Baselines. Actual data plots are compared to baselines calculated from models. Excessive deviations from the baseline indicate potential incorrect equipment operation.

Electrical Load Disaggregation. The electrical load profile is monitored for patterns that signal equipment are being turned on or off. The equipment on-off profile can be compared with the equipment schedule to determine scheduling faults or excessive cycling faults.

Equipment Application

The diagnostic scope for the tools can be at the whole building, or high- level, top-down, approach. At this level, parameters such as energy consumption for the whole building are used to find causes for equipment-level problems. Alternatively, in a low-level, bottom-up approach, low-level parameters are used to detect equipment problems directly

APPLICATIONS PRODUCT REFERENCES

Energy Analysis

Motegi, N., Piette, M.A., Kinney, S. and Herter, K., "Web-based Energy Information Systems Energy Management and Demand Response in Commercial Buildings," Ernest Orlando Lawrence Berkeley National Laboratory, LBNL-52510, April 18, 2003.
http://eetd.lbl.gov/btp/buildings/hpcbs/pubs/E5P2T1b5_LBNL52510.pdf

Active Energy Management—Engage Networks, Inc.
http://www.engagenet.com/datasheets/products.html

Atrium—Honeywell International Inc.
http://atrium.honeywell.com/Atrium/Public/index.html

EEM Suite—Itron (formerly Silicon Energy Corporation)
http://www.siliconenergy.com/

Electric Eye—Electric Eye Pte. Ltd.
http://www.eeye.com.sg/main.htm

Energy Commander/Energy Analyzer—TruePricing, Inc.
http://www.truepricing.com/ec/overview.html

EnterpriseOne—Circadian Information Systems
http://www.circadianinfosystems.com/spec_sheet.html

ExcelSyus—Excel Energy Technologies, Ltd.
http://www.excel-energy.com/products.htm

FX-TEM—Johnson Controls
http://cgproducts.johnsoncontrols.com/met_pdf/120122.pdf

InfoCenter Suite/Utility Cost Manager—Siemens Building Technologies
http://www.sbt.siemens.com/sbttemplates/library/pdf/149201.pdf
http://www.sbt.siemens.com/sbttemplates/library/pdf/149197.pdf

Vykon/Vykon Energy Suite—Tridium, Inc.
http://www.tridium.com/products/ves.asp

DATA VISUALIZATION

Active Energy Management—Engage Networks, Inc.
http://www.engagenet.com/datasheets/products.html

EEM Suite—Itron (formerly Silicon Energy Corporation)
Silicon Energy Corp.
http://www.siliconenergy.com/

Electric Eye—Electric Eye Pte. Ltd.
http://www.eeye.com.sg/main.htm

EnterpriseOne—Circadian Information Systems
http://www.circadianinfosystems.com/spec_sheet.html

Metasys—Johnson Controls
Agrusa, R., and Singers, R.R., "Control Your World in a Glance—Data visualization simplifies the complex," Johnson Controls (On-line article).
http://www.johnsoncontrols.com/metasys/articles/article8.htm

Thompson, S., "Top 10 Technologies for 21st Century Buildings," Johnson Controls (On-line article).
http://www.johnsoncontrols.com/metasys/articles/article13.htm

PORTABLE USER INTERFACES

PocketControls—Abraxas Energy Consulting
http://www.abraxasenergy.com/pocketcontrols.php

Sigma Compact Edition—Invensys Climate Controls UK
http://www.satchwell.com/sigma/ce.htm

ADVANCED CONTROL ALGORITHMS

Adaptive Capacity Control—York International
He, X., "YORK Optispeed VSD CHILLERS (Principle and Application)" (Slides).
http://www.ashrae.org.hk/vsdchillers.pdf

"VSD-Equipped Chiller an Instant Hit at Compact-Disc Facility," Energy User News (On-line article).
http://www.energyusernews.com/CDA/ArticleInformation/features/BNP__Features__Item/0,2584,80971,00.html

Enhanced Adaptive Control—Trane
http://www.trane.com/download/equipmentpdfs/absds4.pdf

Finite State Machine—Johnson Controls/NIST
Seem, J.E., Park, C., and House, J.M., "A New Sequencing Control Strategy for Air-Handling Units," HVAC&R Research, Vol. 5, No. 1, January 1999.
http://fire.nist.gov/bfrlpubs/build99/PDF/b99018.pdf

HVAC-RPO—Johnson Controls
http://cgproducts.johnsoncontrols.com/met_pdf/63750414.pdf

Neural Networks—Colorado State University
Delnero, C., Hittle, D., Anderson, C., Young, P., and Anderson, M., "Neural Networks and PI Control using Steady State Prediction Applied to a Heating Coil," Proceedings of CLIMA 2000 World Conference on Indoor Climate and Comfort Science, Naples, Italy, Sept. 2001.

Neural Networks —University of Colorado and University of Nebraska
Henze, G.P., and Hindman, R.E., "Control of Air-Cooled Chiller Condense Fans using Clustering Neural Networks," ASHRAE Transactions, Vol. 108, Part 2, American Society of Heating, Refrigerating, and Air-Conditioning Engineers, Atlanta, Georgia, 2002.
http://www.ae.unomaha.edu/ghenze/Papers/ACC Control.pdf

P-Adaptive Control—Johnson Controls
http://cgproducts.johnsoncontrols.com/met_pdf/6375125.pdf

Turpin, J., "Making the Connection (May 2000)," Engineered Systems Magazine (On-line article).
http://www.esmagazine.com/CDA/ArticleInformation/features/BNP__Features__Item/0,2503,1451,00.html

Thompson, S., "Top 10 Technologies for 21st Century Buildings," Johnson Controls (On-line article).
http://www.johnsoncontrols.com/metasys/articles/article13.htm

PRAC—Johnson Controls
Seems, J., "Control and Fault Detection Research at Johnson Controls," Johnson Controls (Slides).
http://aes1.archenergy.com/cec-eeb/docs/DiagnosticsMtg/Seem_CEC/index.htm

Moult, R., "The New Role of Building Controls," Johnson Controls (On-line article).
http://www.johnsoncontrols.com/metasys/articles/article2.htm

Thompson, S., "Top 10 Technologies for 21st Century Buildings," Johnson Controls (On-line article).
http://www.johnsoncontrols.com/metasys/articles/article13.htm

FAULT DETECTION AND DIAGNOSIS

Comstock, M.C., and Braun, J.E., "Literature Review for Application of Fault Detection and Diagnostic Methods to Vapor Compression Cooling Equipment," Ray W. Herrick Laboratories, HL 99-19, Report #4036-2, December 1999.

http://www-library.lbl.gov/docs/LBNL/486/29/PDF/LBNL-48629.pdf

Friedman, H., and Piette, M.A., "Comparative Guide to Emerging Diagnostic Tools for Large Commercial HVAC Systems," Lawrence Berkeley National Laboratory, LBNL No. 48629, May 2001.
http://www.nist.gov/tc411/1043-RP_FDD_Literature_Review.pdf

Smith, V.A., and Scruton, C., "Progress Update on Automated Commissioning and Diagnostics under AEC PIER Research Program," National Conference on Building Commissioning, May 8-10, 2002.
http://www.peci.org/papers/smithv.pdf

ACRx—Field Diagnostic Services/Honeywell
http://www.customer.honeywell.com/catalog/pages/service_assistant.asp

APAR/VPACC—NIST
Castro, N., Schein, J., and House, J., "Project 2.3 AHU and VAV Diagnostics," PIER Diagnostics Meeting, Oakland Airport Hilton, April 16-17, 2002 (Slides).
http://aes1.archenergy.com/cec-eeb/docs/DiagnosticsMtg/CastroPAC5_NIST_FDDahu&vav/index.htm

Enforma—Architectural Energy Corporation
http://www.archenergy.com/enforma/default.htm

Fan Performance Tool—Center for the Built Environment, University of California, Berkeley
Webster, T., Ring, E., Zhang, Q., Huizenga, C., Bauman, F., and Arens, E., "Commercial Thermal Distribution Systems: Reducing Fan Energy in Built-up Fan Systems," Final Report: Phase IV, Prepared for the California Energy Commission, Fall 1999.

FDD for Rooftop AC—Purdue University
Braun, J., and Li, H., "Fault Detection and Diagnostics for Rooftop Air Conditioning," PIER Diagnostics Meeting, Oakland Airport Hilton, April 16-17, 2002 (Slides).
http://aes1.archenergy.com/cec-eeb/docs/DiagnosticsMtg/Braun-RooftopAC-Diagnostics/index.htm

Comstock, M.C., and Braun, J.E., "Development of Analysis Tools for the Evaluation of Fault Detection and Diagnostics for Chillers," Ray W. Herrick Laboratories, HL 99-20, Report #4036-3, December 1999.
http://www.nist.gov/tc411/1043-RP_FDD_Tools.pdf

IMDS—Lawrence Berkeley National Laboratory
http://eetd.lbl.gov/ea/iit/diag/index.html

NILM—Massachusetts Institute of Technology
Lee, K.D., Armstrong, P.R., Leeb, S.B., and Norford, L.K., "Non-Intrusive Electical Load Monitoring For Load Detection, Energy-Consumption Estimation and Fault Detection," PIER Diagnostics Meeting, Oakland Airport Hilton, April 16-17, 2002 (Slides).
http://aes1.archenergy.com/cec-eeb/docs/DiagnosticsMtg/MIT-PACTAG2/index.htm

Norford, L.K., Leeb, S.B., Luo, D., and Shaw, S.R., "Advanced Electrical Load Monitoring: A Wealth of Information at Low Cost," Diagnostics for Commercial Buildings: from Research to Practice, Pacific Energy Center, San Francisco, June 16 &17, 1999.
http://poet.lbl.gov/diagworkshop/proceedings/norford.pdf

PACRAT—Facility Dynamics
http://www.facilitydynamics.com/pacrat.html

Universal Translator—Pacific Gas and Electric/Pacific Energy Center
Universal Translator Help Files

MIFDD—Pacific Gas and Electric
"The Improving the Cost Effectiveness of Building Diagnostics, Measurement and Commissioning Using New Techniques for Measurement, Verification and Analysis," California Energy Commission, 600-00-024, Appendix IV, December 1999.
http://www.energy.ca.gov/pier/reports/600-00-024.html

WBD/OAE—Pacific Northwest National Laboratory
http://www.buildings.pnl.gov:2080/wbd/content.htm

OCCUPANT INTERACTION AND FEEDBACK

Web-Based Occupant Satisfaction Survey—Center for the Built Environment, University of California, Berkeley
Huizenga, C., Laeser, K., and Arens, E., "A Web-Based Occupant Satisfaction Survey for Benchmarking Building Quality," Center for the Built Environment, University of California, Berkeley, 2003.
http://www.cbe.berkeley.edu/RESEARCH/pdf_files/Laeser2002_indoorair.pdf

Tenant Interface for Energy and Maintenance Systems—Center for the Built Environment, University of California, Berkeley

Federspiel, C.C. and Villafana, L., "A Tenant Interface for Energy and Maintenance Systems," CHI 2003, Ft. Lauderdale, April 5-10, 2003.
http://eetd.lbl.gov/btp/buildings/hpcbs/pubs/E5P22T4c_public.pdf

References

[1] Webster, T.L., "Trends in Energy Management Technology—Enabling Technologies," LBNL-47650 June 2002. See also Chapter 8, *Information Technology for Energy Managers*, Fairmont Press, Atlanta GA 2004.

[2] Webster, T.L., "BCS Integration Technologies—Open Communications Networking," LBNL-47358. See also Chapter 27, *Information Technology for Energy Managers*, Fairmont Press, Atlanta GA 2004.

[3] Yee, G. and Webster, T.L., "State of Practice of Energy Management, Control, and Information Systems," LBNL-53545 (part 3). See also Chapter 20 of this book.

[4] House, J.M. and Kelly, G.E., "An Overview of Building Diagnostics," National Conference on Building Commissioning, Kansas City, MO, May 3-5, 2000.
http://fire.nist.gov/bfrlpubs/build00/PDF/b00022.pdf

[5] Haves, P., "Overview of Diagnostic Methods," Diagnostics for Commercial Buildings: from Research to Practice, Pacific Energy Center, San Francisco, June 16 &17, 1999.
http://poet.lbl.gov/diagworkshop/proceedings/haves.pdf

[6] Dexter, A., and Pakanen, J., "Demonstrating Automated Fault Detection and Diagnosis Methods in Real Buildings," Technical Research Center of Finland (VTT) Symposium 217, 2001.
http://www.inf.vtt.fi/pdf/symposiums/2001/S217.pdf

Chapter 23

Load Forecasting

Jim McNally PE
Siemens Building Technologies, Inc.

ABSTRACT

LOAD FORECASTING based on the popular linear regression technique is compared to a new technology—multi-variant, non-linear (MVNL) modeling. MVNL has a much smaller monthly error than linear regression analysis. Also discussed is the use of load forecasting for monitoring and verification (M&V) calculations, forecasting tomorrow's loads, spotting abnormalities, and comparing energy use for this year vs. last year.

INTRODUCTION

"How are we doing?" This is a question every facility manager asks about the facility's energy use. This question can take different forms, such as:

- "How are we doing so far this year with electricity and gas use?"
- "Was the electricity use yesterday 'normal'?"
- "Summer demand charges are sky high. I wonder if we should be concerned about setting a new peak this week?"
- "I know how much that conservation project cost us, but how much is it saving each month?"

What do all of these questions have in common? The answer to each requires the use of load forecasting technology. This chapter explores two techniques used to answer these questions, comparing a familiar load forecasting technology with a newer technique. The two methods are:

- Linear regression based on monthly utility bills and
- Multi-variant, non-linear [MVNL] modeling based on interval data.

USES OF LOAD FORECASTING INFORMATION

Load forecasting is the prediction of building energy use (i.e. electricity, gas, steam, etc.) based on external parameters. Load forecasting by itself is just an academic exercise. However, load forecasting is quite useful when applied to answering the operations and business questions posed in the introduction.

Expected vs. Actual Use

Load forecasts of daily expected consumption and peak can be compared against actual daily consumption and peak data to warn of near-term faulty operation.

Five-day Forecast

A five-day daily consumption and peak load forecast provides valuable operations information to warn against future excessive loads. The MVNL technology uses past energy use and historic weather data to develop a characteristic response. That response when played against forecast weather and operations schedules enables building energy loads to be forecast. If a facility manager knew that the facility was going to exceed a certain peak threshold, he could initiate a load-shedding program and maintain the load at a level that would not trigger rate penalties. See Figure 23-4 for an example of a five-day load forecast.

Comparative Energy Use

Many building managers like to compare this year's operation with last year's as one yardstick of operational efficiency. Some may wish to see 'continuous improvement' as suggested by ISO-14001 standards. However, things change from year-to-year. The weather isn't the same. The facility's use isn't always the same; internal changes may have been made, or building additions may have been constructed. The comparison is useful, but the energy use must first be adjusted for changes in weather, operations, internal connected load, and facility floor space for the comparisons to be valid.

Conservation Cost Savings

The cost savings associated with energy conservation projects is of interest to all involved in making such changes. The verification of energy cost savings involves all the parameters of the Comparative Energy Use discussion above. Additionally, energy rates must be considered. Rates can be simple 'cost per unit consumed' or complex involving time-of-use elements [i.e. on-peak, off-peak, shoulder periods], demand charges, rate holidays, real-time rates, ratchets, block structures, and taxes. The rates for the base period may be different from the rates for the present period. Within the base period, the rate may have sub-rates that change each month. Within the present on-going period, the rate may also have sub-rates that change each month. The changes might not be the same for corresponding months.

LOAD FORECASTING TECHNIQUES

In this chapter, the two techniques to be explored are:

- Linear regression based on monthly utility bills and
- Multi-variant, non-linear (MVNL) modeling using interval data.

They are briefly described below.

Linear Regression Using Monthly Utility Bills: How it Works

Input data

Monthly energy use is taken from utility bills. Daily high and low dry-bulb temperature data are obtained from a weather data service provider.

Calculation Model

The relationship between the monthly energy use and the average monthly outside air dry-bulb temperature is calculated by means of a linear regression model. In Figure 23-1, notice how the linear regression technique plots a best fit straight line through the data points.

Error

The monthly error can be rather large. However as larger time periods are selected, the error is reduced. It trends towards 0% when annual summaries are made.

Results

The results of the regression make it possible to weather-adjust the energy use and facilitate comparison of energy use for different time periods. This technique may be used to forecast annual, semi-annual, and perhaps quarterly energy use.

Multi-Variant Non-Linear (MVNL) Techniques Using Interval Meter Data: How It Works

Input data

Daily Weather data [dry-bulb temperature, dewpoint temperature, solar condition], operations data

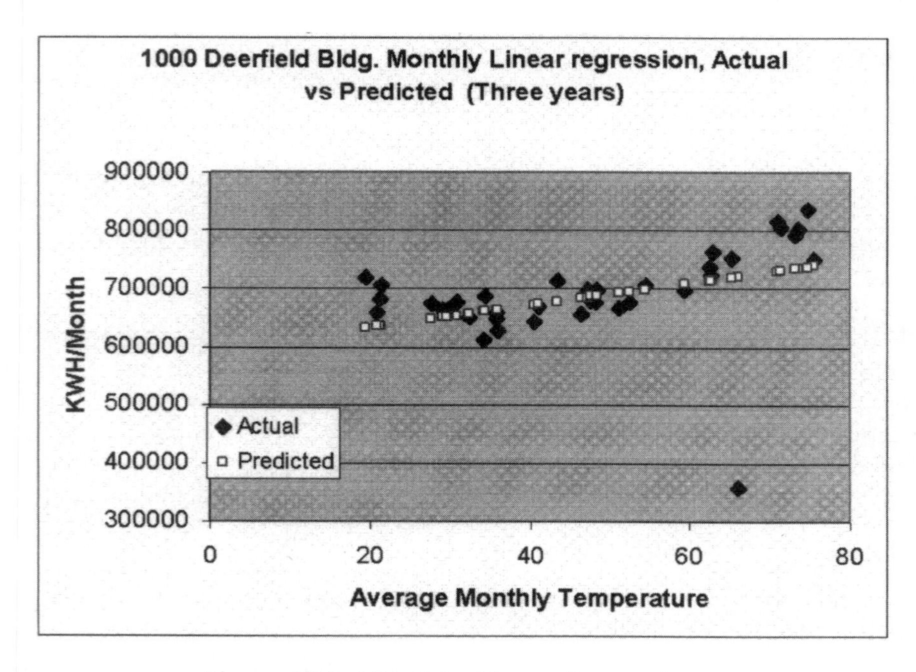

Figure 23-1. Monthly Linear Regression

[day type of use], calendar data [day of week, day of year], daily time-of-use energy consumption data, and time-of-use energy peak data.

Calculation Mode

The MVNL model solves simultaneous non-linear equations to best fit the data.

Error

The daily error can be modest-to-medium; the monthly error is rather small; the annual error tends toward 0%. In Figure 23-2, notice how the MVNL calculated data "blankets" the actual data points. The reason for this blanketing of the data is that the additional parameters used ensure a more accurate outcome.

Results

The result of this calibrated simulation is a forecast for energy consumption and peak load in each defined time-of-use period of each day using the inputs described above. This technique may be used to forecast daily, weekly, monthly, quarterly, semi-annual, and annual energy use.

COMPARISON OF TECHNIQUES

Both techniques work. Both have their uses, advantages, and limitations. The following table is a brief comparison of the two technologies.

Linear Regression

The Linear Regression technique is based on an existing utility bill. It has few data elements. Because there are few data elements, the result is a large error if computed on a monthly basis.

MVNL

The MVNL technique is based on interval data from meters on an on-line system which will have to be installed if they are not already present. It has many data elements which are automatically obtained from the metering system and updated every day providing a result having a small monthly error. In the example in this chapter, the Monthly Linear Regression monthly error is 32 times larger than the MVNL monthly error. [See Figures 23-1, 23-2, and 23-3.]

ISSUES WITH LOAD FORECASTING TECHNIQUES

Error Rates

The goal of using any load forecast simulation is to keep the variation between metered and simulated values sufficiently small so as not to incorrectly influence the application. For example, if one is predicting the peak electric load for tomorrow, an error of 1% is probably acceptable. However, if the load forecast with an error of 1% is part of a system to calculate energy savings, and the implemented measures themselves save

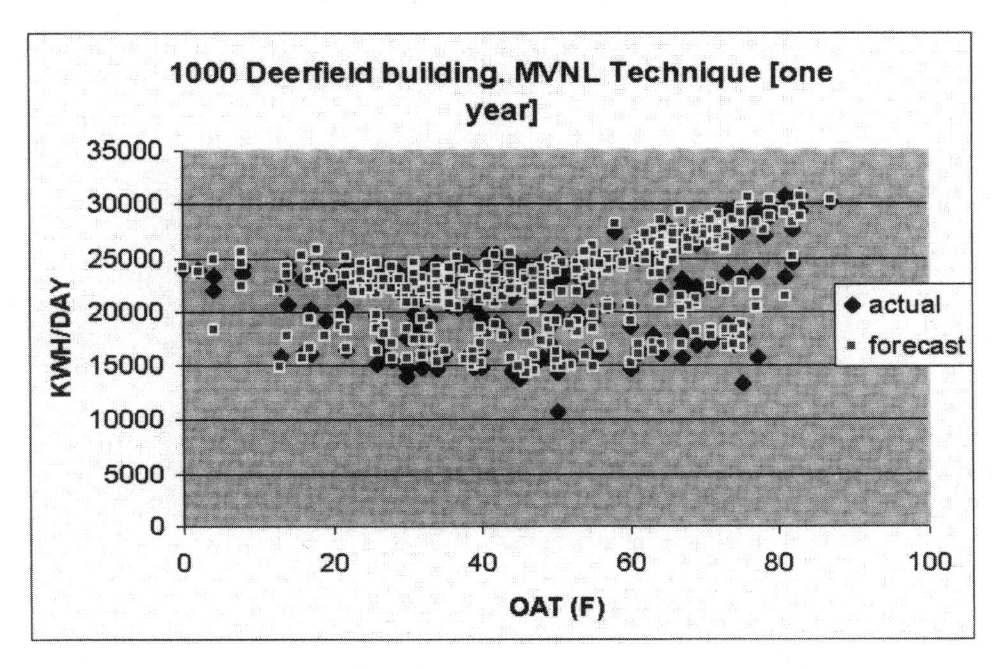

Figure 23-2. MVNL Technique [Daily Data]. Shows Actual vs. Forecast Values

Table 23-1. Comparison of Two Load Forecasting Techniques

	Linear Regression (Monthly Utility Bill-based)	MVNL (Interval Meter Data-based)
Data Points	One per month	2900 per month *
Update Frequency	Monthly	Daily
Time Lag	4-6- weeks (Time to receive the bill)	4-6 hours **
Load Adjustments Based on:	OA dry bulb temperature	OA dry bulb temperature OA dew point temperature Solar condition Time-of-Use periods Day Type (WD/WE/HOL) Day of Week Day of year Seasons
Model	Single variable, linear regression	Multi-variant, non-linear model that reflects how inputs affect facility energy use
Accuracy	Daily Error: Does Not Apply Monthly Error: ~5% -to-15% Annual Error: ~ 0% -to- 1%	Daily Error: 2% -to- 5% Monthly Error: 0.2% -to- 0.6% Annual Error: ~ 0%

* Based on 15-minute data.
**Based on daily processing and the wait for weather data to become available from National Weather Service.

1% to 2%, the simulation error will distort the results.

In the figure below, a comparison is made of the Daily, Weekly, Bi-weekly, Monthly, and Annual error rates of the MVNL (Interval data-based) and Linear Regression (Monthly utility bill-based) techniques. Note that for an annual calculation, both techniques trend toward 0% error. In general, as the time period is lengthened, more data points are used and the inaccuracies tend to cancel each other. The net effect is that the average error is reduced.

The minimum time period for the utility bill-based Linear Regression technique is one month. However, in the example below, the month-by-month error associated with utility bill-based linear regression is probably too large to make the results of any given month meaningful. The error is less for a three-month or a six-month period. However, in the example below, it is still too large for estimating energy savings. This means that reports based on monthly linear regression technology should not be given on a monthly basis. Quarterly and semi-annual reporting may still have errors which distort the results. Therefore, annual reporting is encouraged when using this method.

The MVNL technique, on the other hand, has daily updates based on 15-minute data. The minimum time period for the MVNL technique is one day. In Figure 23-3, the daily error is much less than the monthly error of the Monthly Linear regression technique. It may still be too large to use for calculating energy cost savings. However, it is probably acceptable to use in a 5-day peak load forecasting application. For comparative energy or cost savings applications, the monthly error is not large enough to distort monthly results. In some instances, the weekly or bi-weekly data may be used without severe distortion due to errors. Therefore, if bi-weekly or monthly updates are required for comparative energy or cost savings applications, the MVNL technique should be used.

Availability of Data and Cost

The MVNL technique requires interval meter data; the monthly linear regression technique is based on the utility bill. The availability of utility bill information and the expense of installing metering equipment to obtain interval data may lead one towards the Monthly Linear Regression approach where budgets are lower.

In cases where sub metering is required, there is no utility bill, and interval data are obtainable. This situation suggests that the MVNL technique is needed.

Cases where weekly, bi-weekly, or more accurate monthly updating is required, need the MVNL technique. In these cases, interval meters will have to be installed.

Updating

An energy reporting system that uses MVNL technology will need to automatically collect energy data from the submeters and weather data. The utility bill-based linear regression technology is most often manually updated each month when the utility bill is received. It is said that the most time-consuming of the manual bill-based process is the time (and effort) it takes to obtain the bills from customers.

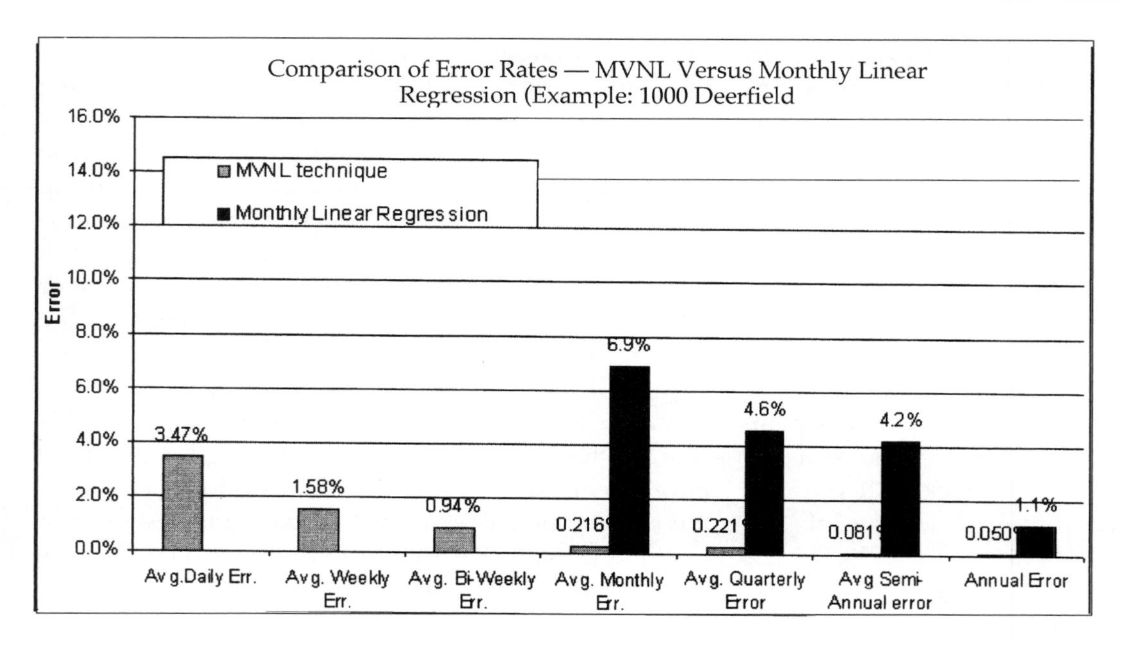

Figure 23-3. Comparison of Error Rates of Two Load Forecasting Techniques

USES OF LOAD FORECASTING TECHNOLOGY

Load forecasting plays a behind-the-scenes role in several applications. These include:

1. Quality control of data (compare expected with actual values)
2. Five-day forecast (demand and consumption)
3. Weather-adjusted comparisons (this year vs. last year)
4. Compare energy use to static, weather-adjusted baseline.
5. Calculate cost savings of conservation measures
6. Calculate cost of wasteful operations.

Quality Control—Compare Expected with Actual

A service which emails energy use data and reports to subscribers uses load forecasting technology to determine if the data it collects is 'reasonable'. A typical email sent by this service is shown in Figure 23-4. The energy use reports are included as attachments to the email. Note the "Actual" and "Expected" entries in Figure 23-4. The "Expected" entry is based on an MVNL Load Forecast of near-term historic data. It was originally used behind-the-scenes as a quality control test. It now is given to customers to be used as a daily benchmark of the previous days' operation. Large variations from "Expected" should be investigated.

Five-day Forecast—Demand and Consumption

The energy reporting service also uses the MVNL load forecast technique to predict the peak electric and daily energy consumption for the next five days. This is based on weather forecasts and operations schedules. Note the five-day weather and energy report as part of the email in Figure 23-4.

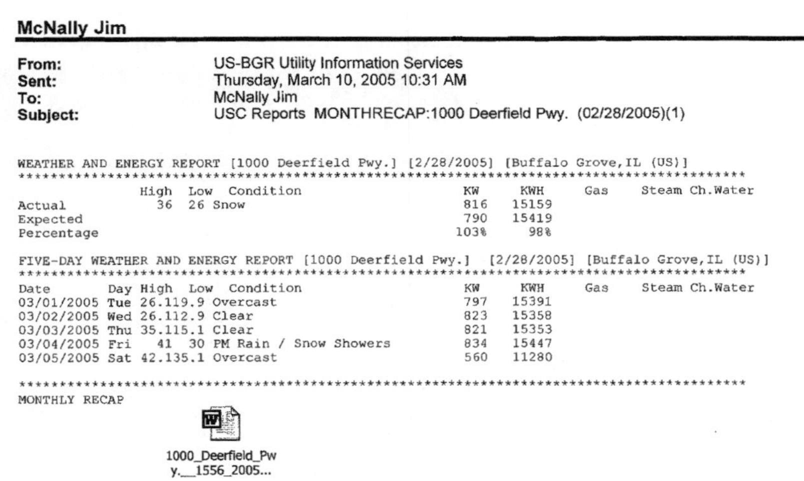

McNally Jim

From: US-BGR Utility Information Services
Sent: Thursday, March 10, 2005 10:31 AM
To: McNally Jim
Subject: USC Reports MONTHRECAP:1000 Deerfield Pwy. (02/28/2005)(1)

WEATHER AND ENERGY REPORT [1000 Deerfield Pwy.] [2/28/2005] [Buffalo Grove,IL (US)]
****

	High	Low	Condition	KW	KWH	Gas	Steam	Ch.Water
Actual	36	26	Snow	816	15159			
Expected				790	15419			
Percentage				103%	98%			

FIVE-DAY WEATHER AND ENERGY REPORT [1000 Deerfield Pwy.] [2/28/2005] [Buffalo Grove,IL (US)]
****

Date	Day	High	Low	Condition	KW	KWH	Gas	Steam	Ch.Water
03/01/2005	Tue	26.1	19.9	Overcast	797	15391			
03/02/2005	Wed	26.1	12.9	Clear	823	15358			
03/03/2005	Thu	35.1	15.1	Clear	821	15353			
03/04/2005	Fri	41	30	PM Rain / Snow Showers	834	15447			
03/05/2005	Sat	42.1	35.1	Overcast	560	11280			

****
MONTHLY RECAP

1000_Deerfield_Pwy._1556_2005...

****
Siemens Building Technologies, Inc. Building Automation Division is a building services provider with offices in major cities throughout North America.
The Utility Service Center is an energy data collection and analysis service of Siemens Building Technologies. Your local SBT account manager can help you with additional meter points; your account manager also has access to reports pertaining to your facility. Please contact your account manager for report interpretation.
For service requests, changes to your profile, or to resolve technical difficulties, contact the Service Center directly at 847-941-5204 during normal business hours 8:00 AM to 5:00 PM (Central Time). Voice mail may be left at any time day or night.
In Canada, contact Ludwig Galazka at 403-259-3404 (Calgary).
****

Figure 23-4. Typical Email Sent by Load Profile Service.

Weather-adjusted Comparisons (this year vs. last year)

Facility directors frequently compare operations from one year to the next. They often compare this year's energy use with last year's use. Because various factors may differ in the two years, adjustments must be made. The Comparative Energy Use Report shown in Figure 23-5 is an example of such a report. The MVNL load forecasting technique is used to make weather and operational adjustments. A second adjustment is then made to account for internal connected load changes. Because the monthly error rate of the MVNL technique is quite low, the monthly entries in the report have minimal distortion. The report below was used as the report mechanism to track energy use for ISO-14001 certification where "continuous improvement" is the goal.

Expected Range of Operational Changes

A Comparative Energy Use Report (containing MVNL technology) was run for 49 facilities comparing this year's power consumption against last year's weather-adjusted power consumption. The results in Figure 23-6 are shown in percent change in power use. Notice the continuum from –10% to +10%. Notice also the discontinuities at the extreme left (decrease) and the extreme right (increase). (The two facilities at the left had monitoring difficulties in 2003 (the meter was bypassed for several months during renovations). The two facilities at the right added to their load during 2003. (One had increased floor space; the other had experienced a major population increase.) The facilities in the middle had made essentially no equipment or structural changes. We also know that some of the building managers were actively trying to reduce energy use by making operational changes.

The informal conclusion to be drawn from this chart is that operational changes (both good and bad) have an annual effect between -10% to +10%. Looking at the conservation side, one might conclude: "Savings of up to 10% might be found with operational changes only. To save more, money will need to be spent!"

Compare Energy Use to Static, Weather-Adjusted Baseline

Many times, a static or fixed reference period is used for comparison. For example, a period before a major change was made is used as the baseline or refer-

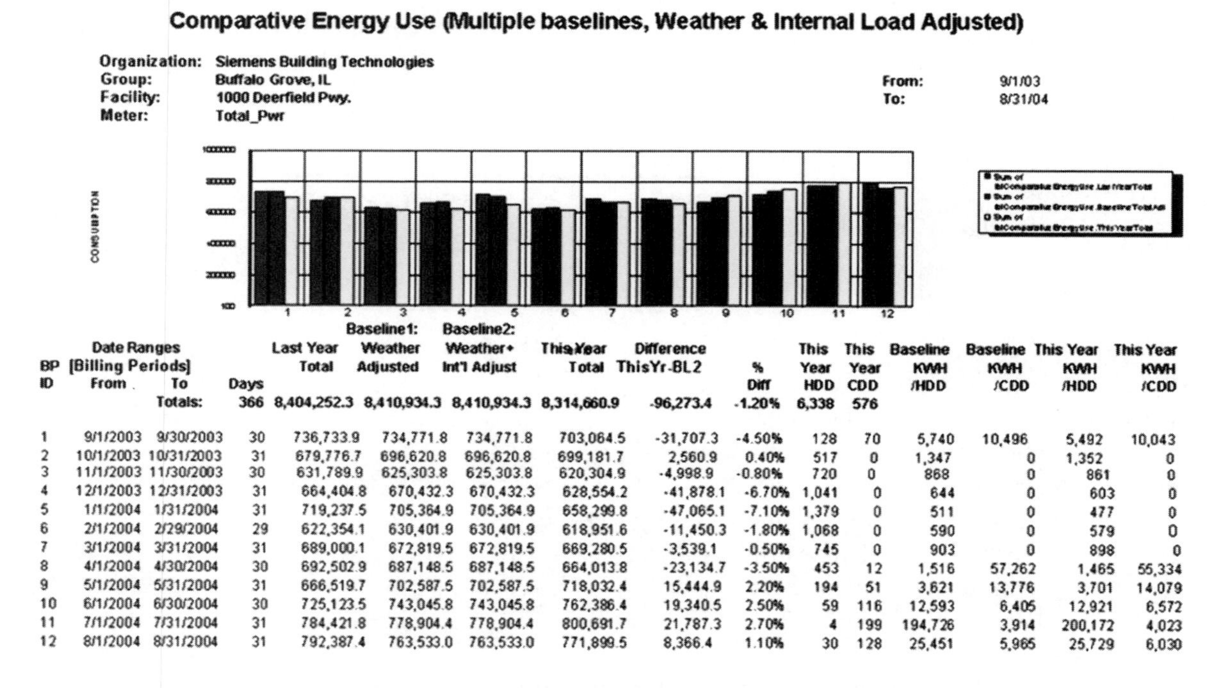

Comparative Energy Use (Multiple baselines, Weather & Internal Load Adjusted)

Organization: Siemens Building Technologies
Group: Buffalo Grove, IL
Facility: 1000 Deerfield Pwy.
Meter: Total_Pwr

From: 9/1/03
To: 8/31/04

BP ID	Date Ranges [Billing Periods] From	To	Days	Last Year Total	Baseline1: Weather Adjusted	Baseline2: Weather+ Int'l Adjust	This Year Total	Difference ThisYr-BL2	% Diff	This Year HDD	This Year CDD	Baseline KWH /HDD	Baseline KWH /CDD	This Year KWH /HDD	This Year KWH /CDD
		Totals:	366	8,404,252.3	8,410,934.3	8,410,934.3	8,314,660.9	-96,273.4	-1.20%	6,338	576				
1	9/1/2003	9/30/2003	30	736,733.9	734,771.8	734,771.8	703,064.5	-31,707.3	-4.50%	128	70	5,740	10,496	5,492	10,043
2	10/1/2003	10/31/2003	31	679,776.7	696,620.8	696,620.8	699,181.7	2,560.9	0.40%	517	0	1,347	0	1,352	0
3	11/1/2003	11/30/2003	30	631,789.9	625,303.8	625,303.8	620,304.9	-4,998.9	-0.80%	720	0	868	0	861	0
4	12/1/2003	12/31/2003	31	664,404.8	670,432.3	670,432.3	628,554.2	-41,878.1	-6.70%	1,041	0	644	0	603	0
5	1/1/2004	1/31/2004	31	719,237.5	705,364.9	705,364.9	658,299.8	-47,065.1	-7.10%	1,379	0	511	0	477	0
6	2/1/2004	2/29/2004	29	622,354.1	630,401.9	630,401.9	618,951.6	-11,450.3	-1.80%	1,068	0	590	0	579	0
7	3/1/2004	3/31/2004	31	689,000.1	672,819.5	672,819.5	669,280.5	-3,539.1	-0.50%	745	0	903	0	898	0
8	4/1/2004	4/30/2004	30	692,502.9	687,148.5	687,148.5	664,013.8	-23,134.7	-3.50%	453	12	1,516	57,262	1,465	55,334
9	5/1/2004	5/31/2004	31	666,519.7	702,587.5	702,587.5	718,032.4	15,444.9	2.20%	194	51	3,621	13,776	3,701	14,079
10	6/1/2004	6/30/2004	30	725,123.5	743,045.8	743,045.8	762,386.4	19,340.5	2.50%	59	116	12,593	6,405	12,921	6,572
11	7/1/2004	7/31/2004	31	784,421.8	778,904.4	778,904.4	800,691.7	21,787.3	2.70%	4	199	194,726	3,914	200,172	4,023
12	8/1/2004	8/31/2004	31	792,387.4	763,533.0	763,533.0	771,899.5	8,366.4	1.10%	30	128	25,451	5,965	25,729	6,030

Notes: 1. Baseline1 is weather-adjusted simulation based of actual energy use data from 8/1/02 through 9/30/03. It is adjusted to "This Year's" weather.
2. Baseline2 is weather-adjusted simulation further adjusted by internal changes such as floor area additions,lighting changes,etc.
3. "This Year" is the period from 9/1/03 to 8/31/04
4. "HDD" = Heating degree-days "CDD" = Cooling degree-days.
5. Values in Blue represent partially filled and empty Date Ranges

Figure 23-5. Comparative Energy Use Report

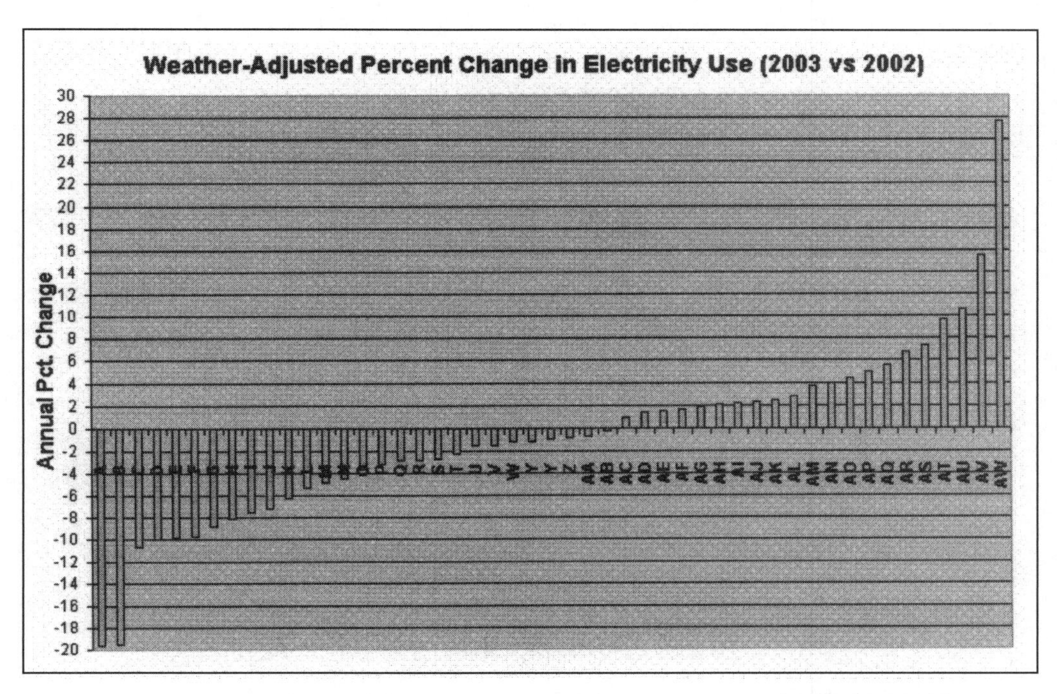

Figure 23-6. Weather-Adjusted Percent Change in Annual Electricity Use

ence against which future energy consumption is measured. In instances like this, the Comparative Energy Use Report in Figure 23-5 can be used also. However, instead of being "Last year," the reference period is a pre-defined static time period.

Calculate Cost Savings of Conservation Measures and Projects

Performance contracts often include provisions to verify promised energy and cost savings. This is usually done by establishing a base or reference period representing energy use patterns before the conservation measures of the performance contract were implemented. For several years after the changes, the energy use of the reference period is compared to the on-going energy use of the facility. Load forecasting plays an important role in this process. It is used to adjust loads for differences in weather and operations.

The MVNL technique with its small monthly error was used in generating the M&V Annual Meter Report shown in Figure 23-7 below. In the example, the energy use for calendar 2003 for an office building in Buffalo Grove, Illinois is compared to weather-adjusted energy use in calendar 2002. It shows a modest 1.3% reduction in power consumption but a 3.8% average increase in monthly peak loads. The On-Peak electric rate for the Actual On-going is $ 0.05599/kWh, whereas the On-Peak electric rate for the Baseline is $ 0.06000/kWh. All other rate elements were the same. The resulting annual *cost savings* is over $8,000. If the rates had been the same for the Baseline and the On-Going, the results would be a net annual *increase in costs* of $4,400.

Calculate Cost of Wasteful Operations

The same method that calculates savings can be used to calculate the cost of wasteful operations. The M&V Annual Meter Report shown in Figure 23-7 may also be used to quantify excessive cost practices. The MVNL technique is imbedded in the calculations. Its small monthly error means that wasteful operations may be accurately documented each month.

CONCLUSIONS

The new load forecasting technique (MVNL) is more accurate than the utility bill-based linear regression technique. MVNL technology enables monthly, weekly, even daily load forecasting to be made with acceptable accuracy.

The MVNL technique requires interval meter data which usually means the installation of a metering system. The downside of a metering system is its cost. The upside of a metering system is that it removes the limitation of monitoring only the main utility meter via monthly bills. With the availability of MVNL using interval meter data, meters may be installed wherever they are needed. For example, if a chiller plant is retrofit, submeters at the chiller plant will focus attention on the energy use of the conservation measure. The results will not be diluted by other parts of the facility as would be the case with utility bill-based monitoring of the main electric meter.

The MVNL approach is a paradigm shift to load forecasting. As this technology becomes accepted, the twin benefits of greater accuracy and forecasting loads over shorter periods of time will be reaped by the applications which employ it.

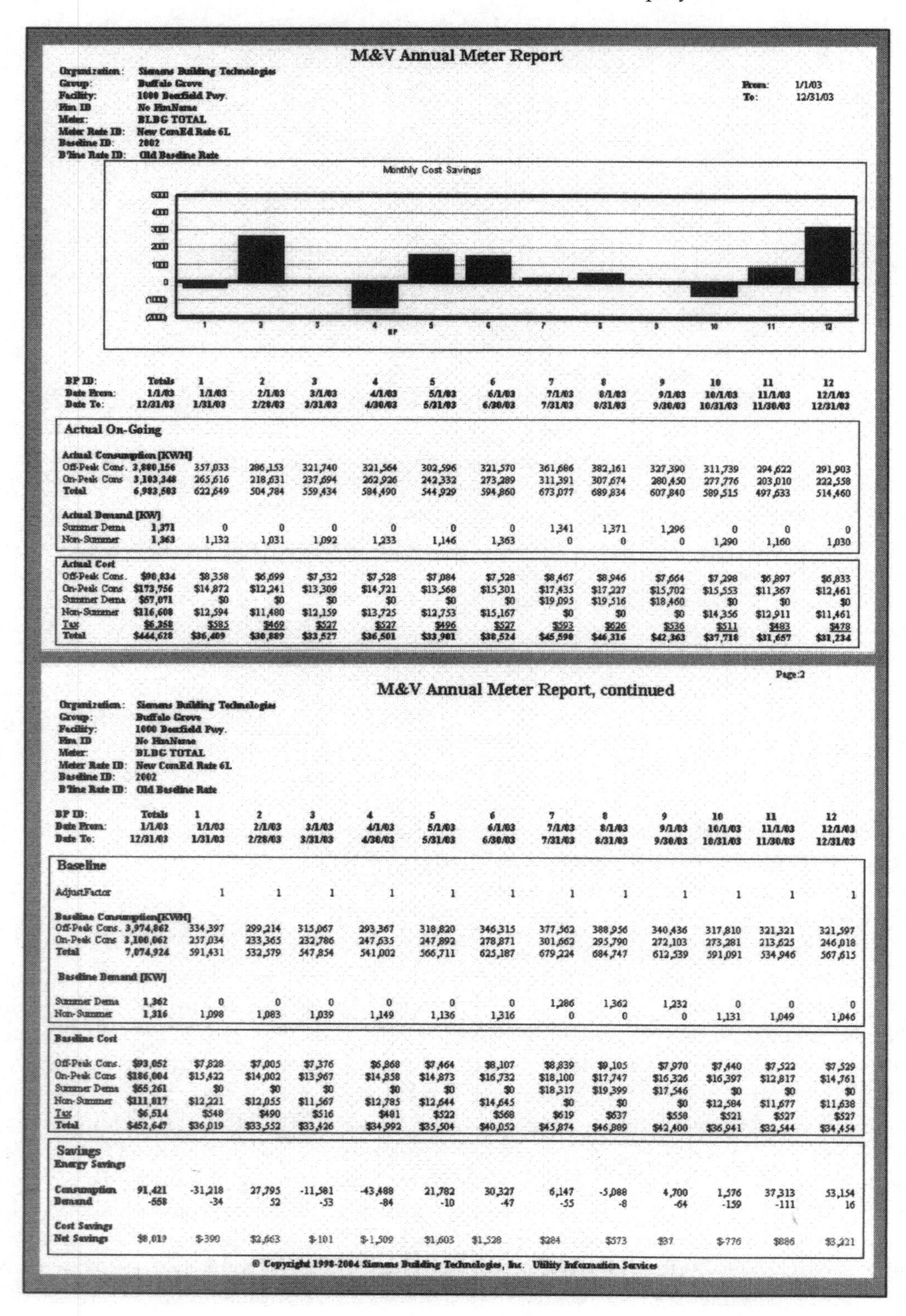

M&V Annual Meter Report

Organization: Siemens Building Technologies
Group: Buffalo Grove
Facility: 1000 Deerfield Pwy.
Fim ID: No FimName
Meter: BLDG TOTAL
Meter Rate ID: New ComEd Rate 6L
Baseline ID: 2002
B'line Rate ID: Old Baseline Rate

From: 1/1/03
To: 12/31/03

BP ID:	Totals	1	2	3	4	5	6	7	8	9	10	11	12
Date From:	1/1/03	1/1/03	2/1/03	3/1/03	4/1/03	5/1/03	6/1/03	7/1/03	8/1/03	9/1/03	10/1/03	11/1/03	12/1/03
Date To:	12/31/03	1/31/03	2/28/03	3/31/03	4/30/03	5/31/03	6/30/03	7/31/03	8/31/03	9/30/03	10/31/03	11/30/03	12/31/03
Actual On-Going													
Actual Consumption [KWH]													
Off-Peak Cons.	3,880,156	357,033	286,153	321,740	321,564	302,596	321,570	361,686	382,161	327,390	311,739	294,622	291,903
On-Peak Cons	3,103,348	265,616	218,631	237,694	262,926	242,332	273,289	311,391	307,674	280,450	277,776	203,010	222,558
Total	6,983,503	622,649	504,784	559,434	584,490	544,929	594,860	673,077	689,834	607,840	589,515	497,633	514,460
Actual Demand [KW]													
Summer Dema	1,371	0	0	0	0	0	0	1,341	1,371	1,296	0	0	0
Non-Summer	1,363	1,132	1,031	1,092	1,233	1,146	1,363	0	0	0	1,290	1,160	1,030
Actual Cost													
Off-Peak Cons.	$90,834	$8,358	$6,699	$7,532	$7,528	$7,084	$7,528	$8,467	$8,946	$7,664	$7,298	$6,897	$6,833
On-Peak Cons	$173,756	$14,872	$12,241	$13,309	$14,721	$13,568	$15,301	$17,435	$17,227	$15,702	$15,553	$11,367	$12,461
Summer Dema	$57,071	$0	$0	$0	$0	$0	$0	$19,095	$19,516	$18,460	$0	$0	$0
Non-Summer	$116,608	$12,594	$11,480	$12,159	$13,725	$12,753	$15,167	$0	$0	$0	$14,356	$12,911	$11,461
Tax	$6,358	$585	$469	$527	$527	$496	$527	$593	$626	$536	$511	$483	$478
Total	$444,628	$36,409	$30,889	$33,527	$36,501	$33,901	$38,524	$45,590	$46,316	$42,363	$37,718	$31,657	$31,234

Page:2

M&V Annual Meter Report, continued

Organization: Siemens Building Technologies
Group: Buffalo Grove
Facility: 1000 Deerfield Pwy.
Fim ID: No FimName
Meter: BLDG TOTAL
Meter Rate ID: New ComEd Rate 6L
Baseline ID: 2002
B'line Rate ID: Old Baseline Rate

BP ID:	Totals	1	2	3	4	5	6	7	8	9	10	11	12
Date From:	1/1/03	1/1/03	2/1/03	3/1/03	4/1/03	5/1/03	6/1/03	7/1/03	8/1/03	9/1/03	10/1/03	11/1/03	12/1/03
Date To:	12/31/03	1/31/03	2/28/03	3/31/03	4/30/03	5/31/03	6/30/03	7/31/03	8/31/03	9/30/03	10/31/03	11/30/03	12/31/03
Baseline													
AdjustFactor		1	1	1	1	1	1	1	1	1	1	1	1
Baseline Consumption[KWH]													
Off-Peak Cons.	3,974,862	334,397	299,214	315,067	293,367	318,820	346,315	377,562	388,956	340,436	317,810	321,321	321,597
On-Peak Cons	3,100,062	257,034	233,365	232,786	247,635	247,892	278,871	301,662	295,790	272,103	273,281	213,625	246,018
Total	7,074,924	591,431	532,579	547,854	541,002	566,711	625,187	679,224	684,747	612,539	591,091	534,946	567,615
Baseline Demand [KW]													
Summer Dema	1,362	0	0	0	0	0	0	1,286	1,362	1,232	0	0	0
Non-Summer	1,316	1,098	1,083	1,039	1,149	1,136	1,316	0	0	0	1,131	1,049	1,046
Baseline Cost													
Off-Peak Cons.	$93,052	$7,828	$7,005	$7,376	$6,868	$7,464	$8,107	$8,839	$9,105	$7,970	$7,440	$7,522	$7,529
On-Peak Cons	$186,004	$15,422	$14,002	$13,967	$14,858	$14,873	$16,732	$18,100	$17,747	$16,326	$16,397	$12,817	$14,761
Summer Dema	$55,261	$0	$0	$0	$0	$0	$0	$18,317	$19,399	$17,546	$0	$0	$0
Non-Summer	$111,817	$12,221	$12,055	$11,567	$12,785	$12,644	$14,645	$0	$0	$0	$12,584	$11,677	$11,638
Tax	$6,514	$548	$490	$516	$481	$522	$568	$619	$637	$558	$521	$527	$527
Total	$452,647	$36,019	$33,552	$33,426	$34,992	$35,504	$40,052	$45,874	$46,889	$42,400	$36,941	$32,544	$34,454
Savings													
Energy Savings													
Consumption	91,421	-31,218	27,795	-11,581	-43,488	21,782	30,327	6,147	-5,088	4,700	1,576	37,313	53,154
Demand	-558	-34	52	-53	-84	-10	-47	-55	-8	-64	-159	-111	16
Cost Savings													
Net Savings	$8,019	$-390	$2,663	$-101	$-1,509	$1,603	$1,528	$284	$573	$37	$-776	$886	$3,221

Figure 23-7. M&V Annual Meter Report

Chapter 24

Upgrade Options for Networking Energy Management Systems

Paul Allen, Walt Disney World
Rich Remke, Carrier Corporation
Steve Tom, Automated Logic Corporation

ABSTRACT

THIS CHAPTER examines the strategies and options involved in upgrading an Energy Management System (EMS) from a stand-alone PC-based workstation into a network-based client/server EMS. The benefits of upgrading are described to help users justify the additional costs required. The concepts presented should be able to help users understand what questions to ask and what options they might consider when moving their EMS to the next level.

WHY UPGRADE YOUR EMS?

There are several reasons why a user would consider upgrading their EMS. First, the rapid development of network technology has made EMS connectivity easier, cheaper and more reliable. Corporate IT departments have invested in network upgrades that have resulted in faster networks with larger bandwidth capability. Today, it is not uncommon to have a multitude of diverse systems using the Corporate IT network.

Another reason to move toward a network-based EMS is to provide greater visibility into how devices controlled by the EMS are working. Instead of just one stand-alone PC used for EMS access, a network-based EMS could potentially have many users accessing the EMS from various locations throughout the company. Authorized users from outside the company can also have access to the system. This means that small companies without an in-house engineering staff can obtain consulting services from experts who could potentially be located anywhere in the world. This also allows the EMS vendor to assist the owner with troubleshooting, maintenance, and upgrade work without needing to make costly site visits.

Traditionally, an EMS has always had the means to collect trend data on the points connected to the EMS. The challenge has been how to make this data more useful and available to a wider audience. A network-based EMS makes it easier to move data collected by an EMS into a centralized relational database for use in a web-based Energy Information System (EIS).

Web services are not built around any one specific relational database, so they are a more universal way to integrate an EMS with an EIS or any other enterprise integration. Web services also allow two-way communications (read and write) with the EMS itself, giving access to real-time data such as setpoints, current sensor readings, etc. which may not be available through the database. The write access allows the EMS to obtain real-time data on utility rates, weather forecasts, etc. which can be incorporated into the control logic. Web services also have the potential to become a universal method of exchanging data between different EMS protocols, like BACnet, LonWorks, and MODBUS. ASHRAE already has a draft standard to facilitate the use of Web services in EMS, and it's currently out for public review.

Finally, networking separate independent systems together allows the EMS to be managed globally. Uniform standards and operating practices can be developed and administered. Advantages to the user include:

1. Responsibility for all server and workstation hardware and software can belong to the IT department. These IT professionals are much better equipped to properly maintain the EMS front-end and have procedures in place for important routine functions for server backups and network security thus minimizing potential loss of EMS software and data.

2. The existing EMS can be expanded as new buildings are brought on-line. This can potentially mini-

mize additional server/client software costs.

3. A global EMS team can provide a consistent approach to EMS processes resulting in improved EMS operation and a more knowledgeable in-house maintenance staff. This can reduce or eliminate the need for EMS maintenance contracts.

EMS UPGRADE STRATEGIES

There are three main strategies available when upgrading an EMS from a stand-alone system to a network-based system:

1. Remove existing EMS and replace with new network-based EMS.

2. Update existing EMS with the same vendor's latest network-based system.

3. Install an EMS interface product that networks an existing EMS.

Each strategy has its advantages and disadvantages. Ultimately, the justification for the upgrade cost will be made based on the potential cost-saving features and improved operational processes that result from networking the EMS into a global system.

The first upgrade strategy is to simply replace the existing EMS with a newer network-based EMS that has been established as a standard within your company. The cost for this option is solely dependent on the size of the EMS that will be replaced. However, if the existing EMS requires high annual maintenance costs or has become functionally obsolete, then this approach might be justified.

The second upgrade strategy available is to contact the original EMS vendor and request a proposal for their upgrade options. Most EMS companies have developed some form of Ethernet network connectivity. Typically, some additional hardware and software is required to make the system work on an Ethernet network. The cost for this might be very reasonable, or it could be very expensive. It all depends on how much change is required and the associated hardware, software and labor cost to make it all work.

The third upgrade strategy involves the installation of a new network-based system that is specifically designed to interface to different EMS systems. These systems typically have dedicated hardware that connects to the EMS bus and software drivers that communicate to the existing EMS field panels. The new panels also have an Ethernet connection so they can communicate to an EMS server. Users view the EMS real-time data using web browser software on their PC. The advantage of this strategy is that a multitude of different EMS systems can be interfaced together into a common server. The disadvantage is that the existing EMS software must still be used to edit or add new control programs in the existing EMS field controllers.

The IT standards that have developed for the Internet in recent years have had a significant influence on the building controls industry. Few would question that what we see in the building controls industry today will most likely be much different in the future. Most building owners today already have a huge investment in one or more building control systems. Figuring out the most cost-effective EMS upgrade path is a challenge. The following section describes these upgrade options in more detail.

EMS UPGRADE OPTION 1: Put the EMS PC workstation on the corporate network.

The simplest and least expensive method to network your EMS system is to take the existing EMS PC workstation and put it on the corporate LAN. This might be as simple as installing an Ethernet card in one of the existing EMS PC workstations. However, IT departments are generally very particular about the PC hardware installed on the corporate LAN and will probably require that the existing EMS PC be replaced with one of their standard PC's. This might result in compatibility issues with older EMS software operating on a new PC's operating system (OS). The EMS vendor might have newer EMS software versions that are compatible with the newer PC OS.

This simple step can result in some significant benefits:

- By installing remote access software, like Symantec pcAnywhereTM or Microsoft NetMeeting, the EMS can be remotely operated from another PC on the corporate LAN.

- Data files created by the EMS can be copied and shared with other users and systems. As an example, EMS sensor trend reports that are stored on the EMS client PC could be shared with a wider audience by copying the files through the LAN to an energy information server. The report trend data could be pulled out of the report and written to a relational database that is accessible through a web-based energy information system program.

- The EMS PC Client hard disk can be backed up through the network on a routine basis by the IT Department.

This option is by far the simplest and least expensive to implement. However, this option really just provides remote access to an existing EMS. Additionally, there can only be one user on the EMS PC client at a time (either remotely or at the PC directly). This option might be used as an interim solution until funding became available to implement options 2 or 3 described below.

EMS UPGRADE OPTION 2: Ask your EMS Vendor for a proposal to network your EMS.

Asking the EMS vendor for solution(s) to upgrade your EMS to a network-based system is an obvious choice. Most EMS vendors are working on network-based solutions for their newer products and if you are lucky, their products are backward compatible to what you currently own.

These options will most likely require some type of connection to the corporate LAN. It is important to have participation from the IT Department in these discussions. The IT Department is generally very cautious with what they will allow on the corporate LAN. One thing that the IT Department will be concerned about is the amount of network communications (or bandwidth) that will result from connecting the EMS to the LAN. A second, and potentially more important, concern is security. Both concerns can be alleviated with proper planning and coordination, but you should plan on getting the EMS vendor together with the IT department early in the process. In general, the bandwidth consumed by the EMS is very small. On a 10 meg Ethernet we've seen brief "spikes" up to 0.5% bandwidth. Generally the EMS traffic is undetectable. The bandwidth consumed by the user interface web pages are much greater than that consumed by the EMS itself. The security issue generally involves convincing the IT department that this is just another web site. Most companies already have one or more web sites on the IT network, and the EMS should be treated the same. Put the EMS web server behind the IT department's firewall or inside its DMZ and use the same security (SSL, VPN) that is use for every other web site. Dedicated EMS web server applications can be stripped of all features not needed for the EMS (instead of being a general-purpose server with all the bells and whistles hackers like to exploit) but it should still be protected by appropriate IT security methods.

There are some general observations on what might be suggested for network upgrades that are not specific to any particular EMS vendor. EMS vendors generally use one of three options to provide an EMS networking solution: (1) connect the EMS master control panels to the Ethernet network; (2) install a dedicated Web Server that accesses the EMS network and is accessible through a web browser; and (3) install a dedicated EMS server on the corporate LAN and operate an EMS network as a client/server system. A user might consider asking the EMS Vendor for their solution to each of these options.

Connecting the EMS master control panels directly to the corporate LAN would have to be scrutinized by the IT Department. The main issue using this approach is the impact on the corporate LAN. Setting up a prototype and letting the IT department monitor the LAN communications between the master panels could resolve this issue. Assuming the IT Department approves, then it is a matter of getting LAN drops at each master control panel on the EMS network and upgrading the master panels to work on an Ethernet network. In this upgrade option, each PC EMS workstation would also be LAN connected and would communicate with the EMS master panels via the LAN. If you do this, you need to make certain you are using EMS workstation software that is designed to be used in a multi-user/multi workstation application. Running multiple copies of single-user workstation software on a network can be a recipe for disaster, as each workstation will use its own database and the databases may get out of sync.

Another option that EMS Vendors have started marketing is a web server device that connects the EMS network to the corporate LAN. Typically, the web server device has two connection ports, one to connect the EMS communications network and the other for connection to the corporate LAN. In this case only the web server device is on the corporate LAN, thus minimizing the IT Department's concerns over LAN bandwidth issues. Setting up a prototype using the EMS Vendor's web server device would let the IT Department fully evaluate this impact. The web server device allows users to access the EMS through their LAN-connected PC using a standard web browser (like Microsoft Internet Explorer). However, the web browser devices typically do not have full programming and configuration capability so a user typically must leave an EMS PC workstation directly connected to the EMS network for this purpose. This type of interface can be called a web-enabled system, which provides some functions through the web but which still requires a dedicated workstation for other functions. A web-based system, described later, will provide full functionality through the web interface. Web-based systems are built around web services and allow the users of the system to have a full-featured web browser interface to their system.

Finally, some EMS vendors offer EMS software that can be operated as a client/server. Using this approach, a dedicated server is installed on the corporate LAN to host the EMS server software. Typically, the IT Department would specify, purchase, install, and maintain this server according to corporate standards. The PC workstations would typically be the user's existing PC workstations that are already connected to the corporate LAN. The EMS client software is loaded on these PC's and works along with the other desktop applications they are using

Case Study: Carrier ComfortVIEW [1]

The EMS offered by Carrier Corporation is called the Carrier Comfort Network (CCN). ComfortVIEW (formerly called ComfortWORKS) is the Windows-based graphical user interface for the CCN and based on a client/server network-based system that uses an SQL database on a server and custom client software on the workstations. A diagram showing the components of a CCN is shown in Figure 24-1.

Carrier's previous front-end software was Building Supervisor, which was a stand-alone product that provided both the user interface as well as the programming tool for the field controllers. The Building Supervisor ran on a Windows 95/98 PC only and did not have any Client/Server capability. ComfortVIEW is backward compatible in that it can operate simultaneously with Building Supervisor on the same CCN, making the transition easier for users.

Since ComfortVIEW was designed as a Client/Server based system, it is already designed to accommodate multiple EMS networks using a common SQL server database. The process to add a stand-alone CCN to a ComfortVIEW server involves the installation of a Carrier CCN-to-Ethernet converter. This device provides the connectivity to the Ethernet Lan and the CCN RS-485 controller network. Up to 32 CCN networks can be connected to one ComfortVIEW server in this fashion.

Carrier has two other options that provide web access to a CCN network. These products use a web browser as the client software to view data from the CCN network. Although ComfortVIEW is not required for these products to work, functions like controller programming and configuration are not directly supported with the current versions of these products.

The first product is called CCNWeb. This device is installed similar to the CCN/Ethernet Converter described previously. The CCNWeb device can automatically upload all of the devices on a CCN and produce default web pages depending on the controller type found. Custom web pages can be created and transferred to the CCN using FTP. The older Carrier controllers do not have default web pages and require the user to create their own web page to display these points. The CCN web has the ability to display real-time point data, temporarily override point values and view alarm data. Access to equipment time schedules and setpoint tables are also available through the CCNWeb. The CCNWeb uses standard web ports (HTTP 80, FTP 21) for user access to its web pages. This standardized approach makes system setup on the LAN quick and easy.

The second product is called CarrierOne. Carrier private labels this product from Tridium. CarrierOne can share information and control between a Carrier system and non-Carrier systems that support open protocols such as Lon, BACnet or MODBUS. This type of third-party network interface product is explained further in the next section.

EMS UPGRADE OPTION 3: Look for a third-party network interface to your EMS.

If the options presented by the EMS Vendor are not acceptable, users might consider checking out the option of using third-party software to interface to their existing EMS network. Automated Logic (*www.automated-logic.com*), EnFlex (*www.enflex.net*), Richards-Zeta (*www.richards-zeta.com*), and Tridium (*www.tridium.com*) provide network connectivity and web-based options. These companies provide integration and browser-based user interface to many EMS manufacturers. The main problem with this approach is that the programming and setup for the existing EMS controllers still must be done using the original EMS manufacturer's software.

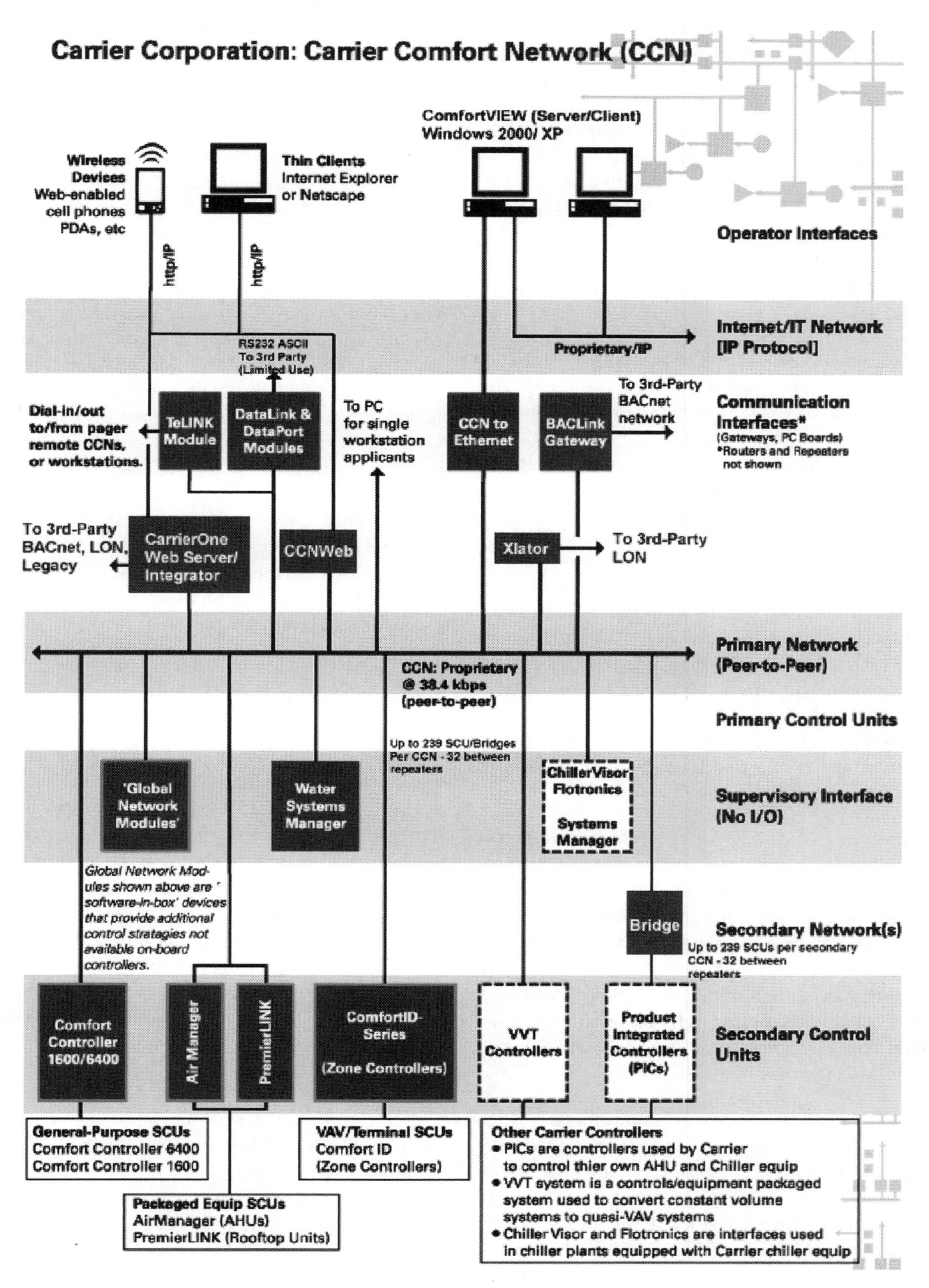

Figure 24-1. Carrier EMS Network Diagram

Case Study—Automated Logic [2]

Automated Logic Corporation (ALC) offers a complete web-based EMS called WebCTRL. ALC's hardware uses native BACnet communications from the IP network down to the ARCNET or MS/TP zone controllers. They also offer hardware and software interfaces to other protocols, including LonWorks, MODBUS, Johnson N2 Open, and Carrier CCN. This allows WebCTRL to connect to existing EMS hardware using multiple protocols, integrate it into a unified system, and provide a web-based interface to the entire system. Advantages include:

- Web browser based access to building systems for anytime, anywhere secure access. WebCTRL is a true web-based system, where all functions are available through the browser. No PC workstations running proprietary software are required because the browser *is* the workstation.

- Compatibility with open standards. WebCTRL is designed from the ground up to use Internet standards. IP addressing, HTML, GIF, Javascript, 3G phones—WebCTRL is fully compatible with existing wired and wireless networks. WebCTRL uses a Java-based server application, making it compatible with multiple operating systems such as Windows XP, Red Hat Linux, or Sun Solaris. It also supports industry standard and open source databases, such as Microsoft Access, Oracle, and MySQL.

- Web services (XML/SOAP) read and write access to any BACnet value, I/O point, or trend. This puts the EMS in the mainstream of IT technology, allowing easy integration with energy information systems, maintenance management systems, utility management systems, facility portals, tenant billing, and many other applications. Since WebCTRL allows third party points to be mapped to BACnet points, this effectively provides Web services connectivity to the existing EMS.

- Controller level integration to applications written in Eikon, ALC's legendary graphic programming language. This makes it easy to create custom applications that use advanced features such as learning adaptive optimal start, trim and respond setpoint optimization, and zone-based demand limiting. These applications can exercise direct control over the existing EMS applications for improved efficiency.

Figure 24-2 shows an architecture diagram of the WebCTRL system. At the top level, the Web services communications facilitate integration with enterprise applications and also provide a new route for exchanging data with EMS systems that do not support the BACnet protocol. ASHRAE has prepared a standard for the use of Web services in building automation systems, and this standard is currently out for public review. When approved, this will become an ISO standard that will provide a "roadmap" for integration of BACnet, LonWorks, and proprietary EMS protocols.

At the controller level, most communications are carried over EIA 485 twisted pair or similar field wiring. A WebCTRL router/controller transfers messages from the IP network to the ARCNET network used by ALC controllers or to the CCN, LonTalk, or other network used by third party controllers. This router/controller can also contain control logic designed to integrate control functions performed by the ALC and third party systems. It can also implement the high-level control features described previously.

CONCLUSION

There are many choices available to EMS users when upgrading from a stand-alone system to a network–based system. The process requires the Energy Manager to become familiar with their IT counterparts and have a basic understanding of IT fundamentals. The benefits of a network-based EMS appear as better standard operational practices and procedures, opportunities to share cost-savings programs and strategies, and wider access to building control processes. The key to justifying the costs associated with networking an EMS is that it be done at a reasonable cost and is relatively simple to implement and operate.

References

Carrier Comfort Network, Carrier ComfortVIEW, Carrier ComfortVIEW, CCNWeb and CarrierOne are registered trademarks of Carrier Corporation, Inc. Other products and company names herein may be trademarks of their respective owners.

WebCTRL and Eikon are registered trademarks of Automated Logic Corporation.

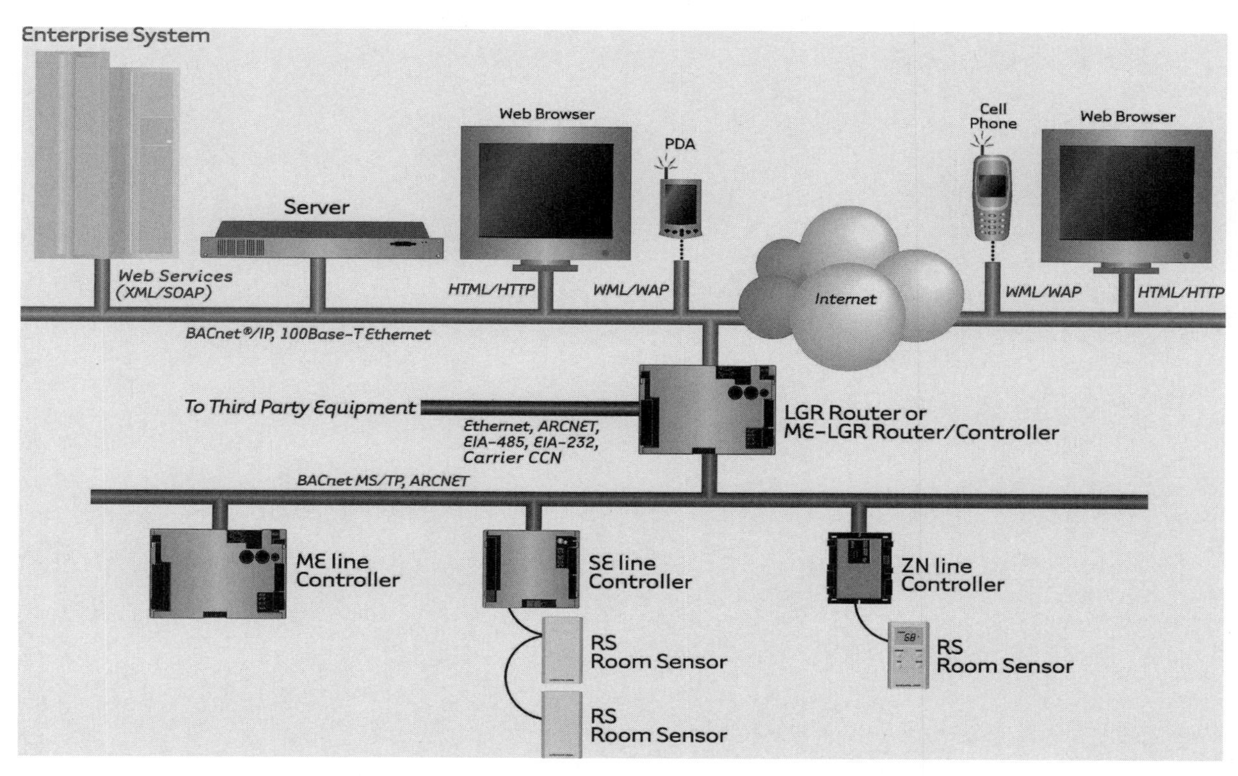

Figure 24-2. Automated Logic system with Web services and third party integration

Chapter 25

A Plan for Improved Integration and Web Based Facility Management Services For a Large University

David Brooks
Affiliated Engineers
Gainesville, FL

ABSTRACT

This chapter introduces the reader to the process of Enterprise Management design. The author takes the reader through the process of evaluation, planning, and implementation. The reader is taken through into the history of this particular university to get a better understanding of how they ended up with what has been described in the text as a melting pot of technologies.

The chapter gives specific examples of methods that were used to establish desired results and provides examples of what makes up a conceptual planning document. Examples of enterprise performance standards and a request for qualification (RFQ) used to identify qualified enterprise management system contractors. The chapter identifies the tools and concepts used to build and design an enterprise management system. It points out what makes this type of project successful and the recommended steps necessary to make it successful.

INTRODUCTION

The industry of facility management and energy management is undergoing a major transformation in this country. A new emphasis is being placed on the business of running an institution. Maintaining the institution's equipment and ensuring a comfortable and safe environment is no longer the acceptable minimum. Institutions, especially University and College campuses across the nation, must make the effort to be better stewards of the energy and dollars they use for operating their facilities. We must stop viewing these buildings and the systems that make up these institutions as a complicated series of parts and start looking at the entire complex as a single working entity. This entity needs energy to run, regular maintenance to work smoothly, and an intermittent tuning or on occasion a complete overhaul. The opportunity for energy conservation within these campuses are required and are critical for this and future generations. The building automation industry has focused on the automation and has lacked innovation.

One example came from a project at a large Ohio University. The consultant's goal was to reduce the $4 million/year energy bill by $1 million. The building had been built in 1971 and included ~700,000 sqft of net usable space. The facility was originally designed as a constant volume air system, with 50% of the building air volume being exhausted as laboratory exhaust. Air was being delivered to the space using a constant volume system sized for peak (when heating and cooling loads are at maximum conditions) load conditions. Energy recovery was non-existent and the retrofitted building automation system provided basic automatic control loop functions only. The consultant calculated almost $1.3 million in energy savings could be achieved with the addition of a variable air volume delivery system and a new energy recovery system. Additional strategies were presented that would improve the projected savings another $200,000 per year.

The above example represents only one building at one University in one state. Assuming a rather conservative estimate of 5% a year the potential for energy saving at the United State's Universities, amounts to some $12,761,664 for 40 Universities. Extrapolating to a total estimated population of ~ 1200 institutions nationwide the potential savings climbs to over $120 million. This analysis assumes 5% savings (rather conservative for the average building within a campus, considering the above example showed potential savings of over 25%). These savings estimates assume minimal investment and are attributed to simple changes in system control and system management strategies. Savings of approximately 10-15% could be realized by enhancing building management interface and real time information management increasing the potential dollar savings to a staggering $240-$360 million a year. Due to limited commitment from management and budgets, awareness of energy costs has not been a particular focus of facility managers within the University systems. To convince CEO's and/or CFO's of the value of these energy saving initiatives, the consultants and facility managers must be able to speak the language of these decision makers. This means showing a return on investment and developing a maintenance plan that confirms and monitors the projected savings.

This country, especially the building industry, must conserve and develop solutions to better manage this costly and limited resource we call energy. Information shared at the right time, for the right person, and in the right format is the essence of any Enterprise Management System (EMS). The rest of this chapter will focus on one particular project, requirements of that project, and a recommended process needed to build a new Enterprise Building Management System (EBMS). It is, however, important to note that an Enterprise Management system may be applied to other elements of a facility or campus to improve productivity, streamline processes within the organization, improve the human interface, etc.... The goal of this chapter is to focus on one element of the Enterprise Management System, the ability to manage energy at an enterprise level.

CASE STUDY INTRODUCTION:

The funding for this large university project came from an improvement bond which included funding for upgrades and expansion of the existing campus Energy Management and Control system (EMCS) that would be used to improve campus operations, especially the utilization of energy. The primary function of the current EMCS system was to monitor and control HVAC equipment to ensure environmental conditions are maintained at desired levels. This included room level temperature and humidity control, main equipment control, and life safety functions initiated by a fire alarm control system. Its secondary function was to monitor and manage campus alarms including equipment failure alarms, temperature high limit/low limit alarms, and other miscellaneous HVAC related alarms. Energy usage monitoring was being handled through a separate system interface with no integration to the EMCS. This separation of the energy monitoring management and building level control exemplifies the silos that had been created within the campus facility services department and that are so typical of other large campus operations.

Having experienced many different campus organizations it is utterly amazing how separate and divided the various operational entities become within these large organizations. The concept of enterprise management requires these management silos be torn down and rebuilt with common goals and objectives. The supply and demand side of building management must work as one if a large organization, such as the ones located within university environments, can achieve maximum operational efficiency. The tearing down of silos must begin at the highest level of an organization and be pushed down through the organization.

The goal of the project was to design a state-of-the-art Enterprise Building Management System (EBMS) with a system that leveraged the openness of Web Services and specifically the versatility of the XML/SOAP information transfer protocol. This EBMS was to be used as an energy management tool, a data normalization tool, an integrated alarm management tool, an integrated reporting tool, an integrated scheduling tool, and a depository for campus wide operational data. This system was to be designed in such a way that supported facility enterprise energy supply and demand optimization. Other goals of the project included optimization of the facility service business processes including, but not limited to: work order management, preventative maintenance processes, trouble call resolution, and improved staff utilization. This environment of integrated business processes, IT open standards, enterprise energy management and equipment control represents the vision of the total project initiative.

At the time funds were allocated to this project, the EMCS operation was a part of the University energy services division. Later, the EMCS operation was re-assigned to the University's facility services division bringing it under maintenance and the building operations group. This action affected the goals and scope of the project, and, in turn, forced the design goals for the project be scaled back to focus more on building level integration, process improvement and less on supply/demand integration. While some of the enterprise management strategies were removed, it was agree by all that the system must be flexible and powerful enough to implement enterprise management strategies in the future. Another impact on the goals and scope of this project has been the continuous and recent changes occurring in the building automation industry and the drive towards open communication standards.

The industry is searching for a common ground but is still divided in many ways. BACnet and LonWorks are the most common building automation standards but are currently being challenged by a new initiative, modeled after the IT industry. Web Services represent the latest initiative in the building automation industry. The primary driver of this initiative (Web Services) actually originated in the IT departments and has rapidly become the new buzz in the industry.

With the reassignment of the EMCS operation to the Facilities Services group, the focus of the project was then narrowed to upgrading and expanding the Building Automation Systems (BAS) to a new Enterprise Building Management System (EBMS). Eliminated from this project was the pursuit of technologies, management tools and business process tools related to the purchasing, generation and distribution of energy. The University realized the *vision* was too grandiose and not achievable with the budget that had been allocated for the project.

The revised focus of the project now included the procurement of EMCS associated infrastructure. This infrastructure includes both traditional and nontraditional information technology (IT) systems. The project's original vision of a working management system included, in part, the synthesis of specific and disparate IT elements. The project's current vision still required, to a lesser extent, the integration of some of these same elements. As a part of the lessons learned to date, it had become clear that the efforts made thus far had been undertaken in a marketplace that was still making rapid developments, but has not yet reached consensus on standards for interoperability, especially at the enterprise level. There were various issues that led to this conclusion:

1. Multiple IT communication standards that were still unresolved and uncoordinated.

2. Multiple proprietary vendor standards that preclude an integrated multi-vendor environment.

3. An extensive search of the marketplace has revealed that few companies have the necessary IT integration experience coupled with the business process knowledge to provide the university with a value added system.

Recent product releases by vendors of Building Automation Systems were supporting and/or integrating various levels of support for both new and existing IT standards. These recent developments supported the current intention to make use of the existing BAS vendors for the new EBMS environment, even when energy supply optimization were consciously removed from the project scope.

OBJECTIVES OF THE EBMS

The primary focus of the EBMS was to focus on increasing efficiency and diagnostic capabilities of the existing energy management systems.

The new EBMS goals included:

1. Reducing energy costs at the University as required by the Energy Conservation Measures (ECM). ECM requires Universities to undertake rigorous cost justification to earn funding, hence the need for upgrading the energy management system.

2. Fulfilling State requirement for saving energy, as resources dedicated to energy management are being cut.

3. Increasing predictability of energy consumption as the Campus expands and energy requirements increase.

The project followed a two phase approach. The first phase identified the technical requirements and business objectives of the project. This section provides the reader with one example of what was referred to as the Phase 1 or discovery phase.

The Phase 1 document was broken down into the following sections.

TABLE OF CONTENTS

Architecture
- Concept Diagram
- Existing Architecture
- Interim Architecture
- Final Proposed Architecture

Enterprise Building Management System

Security
- Introduction
- Security Guidelines
- Security Architecture (Logical)
- Security Architecture (Physical)

Project Implementation Guide

Introduction

Project Scope

Project Cost

Building List

Appendix

ILon 100 Web Services Pilot Project
- Scope of Project
- Project Architecture
- Set up Procedure

BAS/DDC Guide Specification
- Section 15900—Building Control System Integration
- Section 15901—Building Control Systems and Networks
- Section 15903—Control Instrumentation

Existing Systems Architecture

Integrator Consultant RFQ Draft

Potential Integrator Consultant List—with Qualifications

The above sections represent just one example of what an EBMS system and the scope of any discovery phase might include. Every technology project is different but the processes of design are similar.

EXISTING SYSTEM EVALUATIONS

The technology consultant and the University staff evaluated the various EMCS systems and associated local building Direct Digital Control (DDQ systems and the interface systems for the SCADA metering systems, chilled water and primary electrical facilities. The purpose was to provide configuration drawings highlighting the network hierarchy of the existing EMCS, chilled water and primary electrical facilities. They also analyzed existing IT capabilities to ensure that a proper infrastructure exists for system operation and upgrades.

Building control system

This particular client had DDC installed in many of their buildings with the majority of buildings containing the Invensys (Formerly Siebe Environmental Controls) system. Most of these buildings utilize the DMS product line, of which the DMS-3500 is the most common building-level (or supervisory) controller. In addition there are a few Network 8000 (NW8000) systems installed.

The Ultivist, a software package running on the OS/2 operating system, is currently the primary interface for the EMCS operators into the DMS and NW8000 systems. Invensys considers the Ultivist software package to be one of their legacy systems and no longer supports Ultivist upgrades. It was clear that Ultivist could not handle the integration needs of the EBMS.

Since 1997, the University had been in a dialog with Invensys about the future of their Ultivist product, and the migration from an OS/2 operating system to a Windows environment. Considering the lack of options available to upgrade the existing DMS system, the University decided to pursue other technologies and solutions.

While the decision was consistent with the goals of the University to pursue other technologies, it resulted in a technological melting pot of building automation systems. The following EMCS operator stations currently reside in the EMCS operations center and include the Invensys Ultavist, Invensys Sigma, JCI Metasys PMI, JCI NAE client interface, and the TAC Vista operating systems. This represents a common situation with many universities and exemplifies why our universities have been unable to maximize operations. While a sole source environment may help eliminate these variations in operating environments it is expensive and in most public universities almost never tolerated in today's competitive market. This melting pot is a symptom of poor planning and a lack of focus on the bottom line of facility management. The following history lesson exemplifies the industries need to standardize and give Owners innovation rather than partial automation.

In June 2000, Invensys delivered to the University an 1/0 driver for the InVue/InTouch to enable communication with the DMS ethernet devices. This was a major milestone for the University, since it allowed the primary building level controllers and the DMS-3500 to communicate with a Windows-based server over the campus ethernet.

The University has, however, been plagued with ethernet communication problems, centered around the DMS Ethernet Network Interface Module (NIM). Only recently had Invensys coded new firmware which they indicated would solve these problems. This firmware is currently in beta test at the University.

The Wonderware product InTouch is the basis for a user interface for the Invensys Intelligent Automation (I/A) product line. Invensys adds functionality to InTouch and issues their product with the name InVue. The University expected to be able to use InVue as an interface with the new I/A product line, while also enabling communication with the installed DMS product line. The Invensys product lines, DMS and I/A do not integrate directly with each other, without some "host" that is compatible with both. In addition to serving as host to both products, Invensys claimed the InVue would integrate well with other vendors' BACNET devices. The University control specification, however, called for LONMARK devices in the buildings and BACNET communication with the host. Different aspects of this specification have been problematic, with issues related to integration between both Invensys and Johnson Controls.

In August 2001, Invensys recommended a different direction for the University. They recommend Tridium's Java Application Control Engine (JACE box) with the Niagra Framework. Invensys signed an agreement with Tridiurn that year, in which Invensys would manufacture a JACE, and market it with the name Universal Network Controller (UNC). The JACE or UNC could then interface with a DMS panel via a serial cable connection and a custom 1/0 driver. This approach would require a JACE box to be installed at each DMS-3500 location. The University currently has approximately 70 DMS-3500 panels currently installed on campus.

In Winter of 2003, Invensys recommended yet another new building control system. The Sigma product line was introduced and is currently being installed at the University. Sigma can use Lon for controller level communications, but its head end communication to operator interface and database are based on proprietary protocol, something the University adamantly opposed.

Johnson Controls Metasys is the other system that has been used in several locations on campus. Metasys Extended Architecture is the current Johnson Controls product line and as been replacing/integrating the existing Network Control Module (NCM) technology. They have changed their front end interface, the field controllers and the communication protocols to this new extended architecture based on IT standards. A new database server was installed for supporting the new extended architecture. It was, however, the opinion of the University staff that the new Extended Architecture seemed a bit proprietary at this time. The SOAP messages and security applications are custom designed for JCI; thus not all the browsers can interpret the SOAP messages and the JCI server has a difficult time accepting XML messages. JCI has, however, assured the University that these difficulties can be overcome.

Vista, a building control system product from TAC Americas, is a new DDC based control system that is also being installed at the University. TAC and the University have been working towards a web services environment. Initially, TAC implemented a web based control system by constructing a server and a web server within the building location, and monitor and control the building processes via HTTP, from the EMCS control center, using custom designed web pages. The pages reside on a web server at the building level controller. The next phase included expanding the local control and establishing a Vista server at the EMCS control center. This allowed total monitoring and control of the building processes via the Vista proprietary protocol. The next phase of implementation for this system is an interface based on Web Services. It will start by creating ASP, AD security and SOAP consumption messages via an iLon 100 (an Echelon product).

Electrical Services

The University uses the ABB Power RICH System, as a Power Control and Monitoring System for their substations. The PCAMS monitors, through metering, all protective devices, analog (4-20ma) devices, and discrete devices (digital on/off). The protective and sensor devices monitored are in high voltage, and medium voltage switchgears. The PCAMS server communication protocol is based on Modbus. The system uses the campus backbone Ethernet for Modbus TCP. The field level communication is Modbus RTU over RS-485 and also wireless Modbus RTU.

The PCAMS is a Windows NT based SCADA system. The Database complies with standards for SQL (Structured Query Language) utilized with database acquisition and exchange of data (Client/Server) and ODBC driver support. The software provides network DDE servers that include data exchange data with other software applications, such as Microsoft Excel spreadsheet. The software also meets OLE and OPC (OLE for Process Control) as outlined by Microsoft and the OPC foundation. The dBase will allow connectivity with future EMCS servers via XML/SOAP custom configuration and adapters.

The system hardware includes;

- Two 10 Servers

- Network hardware and software for remote communications with other substations, remote points and users

- UPS and communication converters for wired and wireless devices

- One Engineering work station and support printer

Findings

Based on field visits and discussions with university staff, the technology consultant prepared Visio drawings of the existing DDC systems. The infrastructure deficiencies, the hierarchy drawings and the Input/Output points associated with these systems were presented and next step recommendations were made by the technology consultant.

In general, the following deficiencies were identified;

- The existing system documentation is poor. The contractors have not been supplying the required as built drawings to the University as required.

- Most of the DDC locations lack proper ventilation or cooling.

- Most of the DDC locations lack Campus backbone connectivity and a source for emergency power and proper grounding.

- Lon certified controllers have been the standard for the University for many years. The findings indicated that most of the controllers do not meet that requirement.

- Panel fabrication, wiring and field tagging needs standardization.

- Control system design including sequence of operations, alarm management, device quality lack consistency and standards.

- The various EMCS terminal units located at the EMCS operations center limit the ability of the operators to manage the campus as a single enterprise. The various operating systems and applications limit the operators to a reactive mode of operation.

This section focuses on the EBMS and the functionality required for a system operating on an enterprise level. Owners and Integrators, a term applied to firms specializing in the business of integration, typically focus on only the technology of integration and do not consider the goals of integration. Integration for the sake of integration will inevitably lead to a lack of direction and focus, ultimately leading to yet another investment with little or no documented return on investment.

Before an Owner ever discusses the option of integration it is critical they focus on the operation itself. In order to address the business aspect it is critical that the existing processes be dissected and documented. It is this process of identifying existing processes, identifying disconnects, identifying bottle necks that lead to real improvements in an operation. This is how to show a significant return on investment. Approaching any project, especially an integration project, this way will give credibility to the facility operations staff. This credibility will go a long way in convincing the CEO and CFO of these organizations to invest in other initiatives and upgrades.

Some of these functions may appear different from what is considered standard in the proprietary operating systems so familiar to the facilities management staff, but represent the essential components of an enterprise management system.

Specific Functions of the EBMS

In order to achieve the above objectives, the new enterprise energy management system was to include the following features:

1. Integrate all the enterprise information associated with:
   a. Building utility management—bill processing, bill validation, cost allocation and budget tracking.
   b. Facility operations—benchmarking, performance monitoring, demand management, price/use projections, and load shedding.
   c. Master Scheduling—Master schedule profiles with vendor transparent interfaces ad graphics.
   d. Master alarm management—Management and reporting capabilities that ensure vendor transparent messages, prioritization processing, and reporting methods.

2. In addition the EBMS must:
   a. Provide real-time monitoring of energy usage and consumption across the campus.
   b. Initiate system wide management actions using real-time information and rules.
   c. Enable easy connection of disparate data

sources using SOAP and other Web Service collection mechanisms.

d. Produce and consume SOAP messages. The messages are to expose all methods and properties required for monitoring status, real time values, historical values and operational control for all energy systems.
e. Distribute and gather data from all the relational Dbases within the EMCS.
f. Provide XML/SOAP interface for integration with Campus building control system.
g. Use Simple Vector Graphics (SVG) for display vendor transparent operator interface screens.

3. Software Requirements of the EBMS: The following software modules are to be implemented to perform the above functions:
   a. Alarm Management Module—The enterprise shall be capable of displaying all energy and facility alarms through an internet browser, deliver alarm messages, generate detailed audit trail reports for tracking current and historic alarms, as well as triggering real-time alarms. Alarms shall be triggered when energy costs, consumption and system operating conditions exceed set-points. The system shall also allow logging of comments made by personnel during the course of the alarm as well as alarm history details. The software shall provide a graphical representation of such alarm statistics as the total number of alarms by points, assigned operator alarm resolution, alarm severity and priority of a point.
      i. Alarms will be routed directly from building level controllers to the EBMS server. The alarm management portion of the software shall, at a minimum, provide the following functionality.
         1. Log date and time of alarm occurrence.
         2. Generate a "Pop-Up" window, with audible alarm, informing a user that an alarm has been received.
         3. Allow a user, with the appropriate security level, to acknowledge, temporarily silence, or discard an alarm.
         4. Provide an audit trail on hard drive for alarms by recording user acknowledgment, deletion, or disabling of an alarm. The audit trail shall include the name of the user, the alarm, the action taken on the alarm, and a time/date stamp.
         5. Provide the ability to direct alarms to an e-mail address or alpha-numeric pager. This will be provided in addition to the pop up window described above.
         6. Any attribute of any object in the system may be designated to report an alarm.
         7. The BAS/DDC will annunciate diagnostic alarms indicating system failures and non-normal operating conditions.

   b. Reporting—Reports shall be generated and directed to any of the following devices: User interface displays, printers, or designated archive server. As a minimum, the system shall provide the following reports:
      i. All points in the BAS.
      ii. All points in each BAS application.
      iii. All points in a specific controller.
      iv. All points in a user-defined group of points.
      v. All points currently in alarm for a specific building or group within the building.
      vi. All points locked out in a BAS application.
      vii. All BAS schedules.
      viii. All user defined and adjustable variables, schedules, interlocks and the like.
      ix. BAS diagnostic and system status reports.
      x. User defined reports using off-the-shelf data management software packages (e.g. EXCEL, ACCESS)

   c. Dynamic Systems Graphics—System graphics shall allow user to navigate through out the campuses and individual facilities without leaving the graphics mode. Graphics shall start at an enterprise level (all campuses) and allow navigation to campus, building and specific systems, equipment and points. The EMS shall include a "building home page" from which user's link to all buildings, building systems, and related information (e.g. CAD floor plans, O&M documents, work orders, etc.). A building systems and components tree structure shall be specified to show the relationships between systems and their geographic locations. Controller networks shall have a similar navigation "tree" showing all networks, and attached controls. An unlimited number of graphic displays may be generated and executed. Graphics will be based on Scalar Vector Graphic (SVG) technology.

i. Additional graphical user interface requirements
   1. Values of real time attributes displayed on the graphics will be dynamic and updated on the displays without a manual refresh.
   2. The graphic displays will display and provide animation based on real-time BAS data that is acquired, derived, or entered.
   3. The user will be able to change values (set-points) and states in system controlled equipment directly from the graphic display.
   4. System will include a graphic editing tool that allows for the creation and editing of graphic files. It will be possible to edit the graphics directly while they are on line, or at an off line location for later downloading to the system.
   5. The BAS system will be provided with a complete user expandable symbol library containing all of the basic symbols used to represent components of a typical BAS system. Implementing these symbols in a graphic shall involve dragging and dropping them from the library to the graphic.

d. Scheduling—The system shall provide multiple schedule input forms for automatic BAS time-of-day scheduling and override scheduling of BAS operations. At a minimum, the following spreadsheet types will be accommodated:
   i. Weekly schedules.
   ii. Temporary override schedules.
   iii. Special "Only Active If Today Is a Holiday" schedules.
   iv. Monthly schedules.
   v. Schedules will be provided for each system or sub-system in the EBMS. Each schedule shall include all commandable points residing within the system. Each point may have a unique schedule of operation relative to the system use schedule, allowing for sequential starting and control of equipment within the system. Scheduling and rescheduling of points will be accomplished easily via the system schedule spreadsheets.
   vi. Monthly calendars for a 12-month period will be provided that allow for simplified scheduling of holidays and special days in advance. Holidays and special days shall be user-selected with the pointing device or keyboard, and shall automatically reschedule equipment operation as previously defined on the weekly schedules.

e. Historical Trending and Data Collection—Trending and storage of point history data for all BAS points and values as selected by the user shall be possible. The trend data shall be stored in a manner that allows custom queries and reports using industry-standard software tools. At a minimum, the operating system will provide the following capabilities necessary to perform statistical functions on the historical database:
   i. Average.
   ii. Graphical Representation.
   iii. Arithmetic mean.
   iv. Maximum/minimum values.
   v. Range—difference between minimum and maximum values.
   vi. Standard deviation.
   vii. Sum of all values.
   viii. Variance.

f. Paging—The system will provide the means for automatic alphanumeric paging of personnel for user-defined BAS events.
   i. System will support both numeric and alpha-numeric pagers, using Alphanumeric, PET, or IXO Protocol at the owner's option.
   ii. Users will have the ability to modify the phone number or message to be displayed on the pager through the system software.
   iii. System will utilize pager schedules to send pages to the personnel that are "on-call."
   iv. If required the Contractor will be responsible for providing a modem for connection to the paging service.

g. Data Analysis Module—The enterprise system shall give the facility managers the ability to visualize and statistically analyze data using graphical and statistical tools. Data visualization will help managers to spot trends and abnormalities in energy demand and consumption, and take immediate corrective measures. Reports on data collected shall be customizable and transferable to other applications such as Excel, for further analysis.

h. Cost Analysis Module– The new system shall:
   i. Enable enterprise-level energy bill cost analysis
   ii. Enable comparison of monthly or annual energy costs across facilities.
   iii. Reproduce specific rate billing elements for bill validation
   iv. Index energy costs per area, weather degree days or user-defined production metrics.
   v. Produce the following enterprise reports per building:

      1. Summary Cost Report — Index costs by area, degree days, usage and production units.
      2. Monthly Cost Report — identifies the total expense by month for one or more commodity points or point groups. Index costs by area, degree days, usage and production units.
      3. Summary Rate Components Report — delivers total costs over a chosen time frame, identified by each rate component such as demand, energy, fixed charges, taxes, other.
      4. Monthly Rate Components Report — relates the total cost for each rate component by month for one commodity point or point group.
      5. Billing Charges Report — describes total costs, identified by each rate charge over a chosen time span for one commodity point.
      6. Hourly Cost Profile — reports the total cost of energy for each hour of any selected day.
      7. Rate Summary Report — Provides summary of the rate details for a selected rate.

i. Energy Analysis Module—The enterprise system shall provide for detailed reporting and analysis of real-time and historical interval energy data across all data sources within the Campus energy system. This software module shall enable:
   i. Benchmarking; for normalizing demand and usage data. This will help examine the average area and overall consumption of every site in the campus and identify the largest and most efficient energy consumers.
   ii. Load Duration Profiles—describes how many hours of a given period is load at or above any given demand level and compare predetermined load profiles for a given weather condition.
   iii. Peak Load Analysis—help identify what contributed to energy peak loads for investigative purposes.

j. Enterprise Navigation Module -The enterprise system shall provide SVG graphical based user interfaces for operators to view and control processes in the EBMS. Operators shall be able to drill down from the enterprise level summary down to an end-use device through an internet browser, and shall be able to perform the following:
   i. Interactive Navigation—The navigation module shall enable graphical quick movement between geographical sites, buildings, zones and system components and directory and hyperlinks shall provide instant access to all display screens in the system.
   ii. Build Custom Graphical Screens—The screen builder module shall allow dragging and dropping of graphic objects onto a screen using the Active Screen Builder program to rapidly build custom 2-D and 3-D scenes. SVG conversion tool shall be provided.
   iii. Real-Time Data Trending—The trending module shall allow the facility management staff to monitor the ongoing performance of buildings' energy-consuming systems
   iv. Live Data Visualization—The visualization module shall display visual cues of events such as alarm graphics to aid in making immediate corrections as potential cost-saving events occur in real time.

v. Remote Load Control—The control module interface shall be configured to have complete control of every energy consuming device in an enterprise using Internet Browser. It shall be capable of remotely initiating starts and stops of events and adjustment of control set-points.

vi. System Information—The Navigator shall provide an intuitive visual display of current operating conditions for every monitored device. This module shall also efficiently monitor facility performance at an enterprise level, rapidly spot cost or performance issues, and quickly locate and monitor a system that is operating below peak efficiency.

k. LonWorks/SOAP Gateway—Software gateways shall be installed as an interface to communicate with building and meter data collection systems, relational databases, and other data collection devices. These gateways shall run in real-time mode, obtain and translate data from these third party data sources and manage communication to and from the enterprise application servers.

l. An EBMS server to provide an aggregate overview of all the campus buildings and the respective local control systems shall be installed. Windows NET server running Active Directory shall be installed to provide a platform for secure web services applications

m. Internet browser software shall be used to enable operators to monitor and control the enterprise functions.

n. Data logging, storage and historical archiving applications shall use SQL based relational Dbase.

o. Security software/hardware shall be used. Secure data transactions shall be based on W3C's Web Services security standards.

The following RFQ sample was used by the University to establish a list of 3 pre-qualified vendors. The primary goals of the RFQ were to establish qualified vendors that could work in both the IT environment and the business processes typically encountered in the university campus system. When considering any enterprise level project it is critical that vendors be brought on board as early in the process as possible. The need to work within the vendor's capabilities and the hardware limitations of that vendor are critical for a successful project.

"RFQ SAMPLE"

SOLICITATION FOR: Strategic Consulting and Implementation Services

PROJECT NAME: Upgrade Campus Energy Enterprise Management and Control System (EMCS)

OWNER: (Project Name)

CLOSING DATE FOR SUBMISSION OF QUALIFICATIONS: Date

Project Description

The [Client Name or Facility Name] has recognized the importance of developing a strategic vision and detailed implementation plan for an Enterprise Energy Management System for the facilities on the campus at [Client Name or Facility Name]. [Client Name] has retained [Lead Consultant] to lead this effort and the basics of a plan have been formulated. [Lead Consultant] intends to retain an integration consultant to advance the details of the Concept phase to the point where specific implementation projects with not-to-exceed costs can be identified. Additionally, this Sub-Consultant may be retained to provide services for the implementation of the initial project having a total value of [$$$$].

Project Scope of Work

In general, the scope of work for this project shall include strategic planning for the development of an Energy Enterprise Management System (EEMS) that addresses the following issues:

a. Integration of systems to provide an EEMS to share system data and information among the EMCS Control Room, Energy Using entities (supply and demand side), the electric distribution system and other centrally distributed production facilities (Chilled water, Steam, Hot Water, Compressed Air, etc....).

b. Identify, review, detail and analyze appropriate processes and indicators for identification of improvement areas and measurement criteria.

c. Development of an enterprise solution to make use of energy data, provide analysis tools for making energy decisions, integrate relevant meter data along with other building level automation system data points. The EEMS shall be capable of integrating other enterprise wide technology solutions such as Asset management programs, Enterprise Resource Planning programs, Workforce Management programs and Inventory database tools.

d. Upgrade/Replace the automation system communications within buildings to enable central data acquisition.
e. Upgrade User Interface using graphics and other system tools available.
f. Specify a system that uses IT standards for integration and communication, making use of XML, SOAP, WSDL, and UDDI standards, as appropriate.
g. Expand upon DDC installed in buildings, to address control strategies enthalpy control, occupancy control, Indoor Air Quality control strategies, and other control strategies including demand side management.
h. Develop systems for billing and cost allocation for utilities using data from chilled water, steam and electric meters at each building.
i. Develop the web capabilities for customer service features and remote monitoring to its fullest.
j. Reducing energy costs at the University as required by the Energy Conservation Measures (ECM). ECM requires Universities to undertake rigorous cost justification to earn funding, hence the need for upgrading the energy management system.
k. Fulfilling State requirement for saving energy, as resources dedicated to energy management are being cut.
l. Increasing predictability of energy consumption as the Campus expands and energy requirements increase.

The final deliverables for this effort will include a detailed strategic master plan, including an executive summary, and follow –up presentations of the findings to [Client Name] personnel.

Submittal Criteria

Three copies of information should be provided to [Consultant Name] at the address noted below. Information must be submitted in 8.5″ x 11″ binders. The submittal information must be organized with separate tabs for each of the following items:

1. Corporate Information
2. Project organizational chart with team members identified, their roles clearly defined and resumes for all team members
3. Technical capabilities in the following areas:
   a. Similar project experience with strategic master plans
   b. Knowledge and experience of business processes
   c. Knowledge and experience of web Services
   d. Knowledge and experience of open systems (Lon, BACnet etc.)
   e. Knowledge and experience working in a campus environment
   f. Knowledge and experience with [Client Name] and their organizational structure.
   g. Specific technical expertise and relevant project experience in the following areas:
      i. Scenario/Energy modeling and tools used
      ii. Economic modeling and tools used
      iii. Solutions implementation
      iv. Utilities "fleet" management
      v. Design/layout of control rooms/operations centers
4. Customized project approach and schedule for Strategic Master Plan
5. Listing of any product affiliations or software preferences/agreements
6. References
   a. Project references
   b. References for key personnel

Selection Criteria Process

The following criteria will be used to evaluate the submittals:

1. Project approach and proposed schedule
2. Specialized or appropriate expertise in strategic planning for similar systems
3. Past performance on similar projects.
4. Proposed project team, experience and references
5. Proximity to and familiarity with the client and their organization.
6. Record of successfully completed projects without major legal or technical problems.

The submittal and the above criteria will be the basis for the selection. If there are several well qualified submitters and more detailed evaluation is necessary, interviews for selected firms may be held at AEI's offices the first week in June 2004. A pre-submittal review meeting will be held at AEI's office on May 19, 2004 to discuss the project scope in further detail and review the documentation developed to date. Attendance at this meeting is optional.

Proposed Schedule

Date 1 Pre-submittal Review Meeting at AEI
Date 2 Qualification submittal due by 5:00PM
Date 3 Submittal Review
Date 4 Interviews at AEI (If required)
Date 5 Notification of selected firm
Date 6 Proposal Submittal and Review
Date 7 Proposal Approved and Notice to Proceed

A report generated for the Oak Ridge National Laboratory and the United States Department of Energy (November 21, 2002) included 142 Universities and colleges. The report indicates a total steam load of 42,825,585 lb/hr (142 institutions reporting) and a chilled water load of 932,092 tons (107 institutions reporting). In addition the total fuel expenditures for these institutions totaled $255,233,285 (107 institutions reporting).

CONCLUSION

To conclude, there are three key elements to consider when designing an enterprise management system.

- The need to integrate must correlate with a need to improve the process of doing business. Integrating for the sake of integration will lead to nothing more than an expensive technological play toy.
- The design process must start with an evaluation and mapping of the existing processes. To fix a process one must document and analyze the process.
- Set goals and verify goals. Changing processes is an ongoing effort that must be managed and monitored to ensure changes are implemented and return on investment is realized.

It is this author's experience that Universities, in general, lack the commitment and funding to make significant facility management improvements. Other political factors often come into play including preferences given to specific departments, status in the organization, and student support. This lack of commitment and support is becoming less of a factor as energy budgets soar and operational budgets are stretched to their limits.

When considering an enterprise management system or any kind of integration initiative it is important that a consultant with no specific product alliances be considered. It is critical in any integration solution that a consultant with no bias towards a specific technology be used to identify the solution that makes the most sense for the situation.

Chapter 26

Overview of Digital Control and Integrated Building Automation Systems in K-12 Schools

Sandra C. Scanlon, P.E., Electrical Engineer
Scanlon Consulting Services, Inc.
https://home.att.net/~scanlonconsulting/

ABSTRACT

BUILDING ENVIRONMENTAL systems, such as heating, ventilation and air conditioning (HVAC), lighting, and water heating, account for approximately 85% of the energy consumed in school facilities. The ability to manage these systems and their energy use is critical and is dependent on their control, monitoring and communications systems. This particularly applies to HVAC systems, which can account for 50-60% of energy consumption within school buildings. Control of individual building systems is typically accomplished through separate control systems, including the use of direct digital control (DDC) systems for HVAC equipment.

This chapter addresses the use of an Integrated Building Automation System (IBAS) that integrates the DDC system with the other building systems (fire alarm, security, lighting, irrigation, etc.) to improve the level of monitoring, communications, and control of each individual building and group of buildings. It also discusses issues involved in the selection, operation and maintenance of an Integrated Building Automation System.

INTRODUCTION

The environmental control systems in buildings, such as heating, ventilation and air conditioning (HVAC), lighting, and water heating, account for approximately 85% of the energy consumed in school facilities. The ability to manage these systems and their energy use is critical and is dependent on their control, monitoring and communications systems. This particularly applies to HVAC systems, which can account for 50-60% of the energy consumption within school buildings. Control of individual building systems is typically accomplished through separate control systems, including the use of direct digital control (DDC) systems for HVAC equipment.

In this chapter, the term "direct digital control (DDC) system" can be used interchangeably with the terms building automation system (BAS), building management system (BMS), energy management system (EMS), and facility management system (FMS). These terms are *not* synonymous with IBAS. IBAS is the collective monitoring and control of many systems.

Typically, DDC systems control all the HVAC functions within a specific building. Separate, independent control systems provide functional control for other building systems: fire alarm, security, irrigation, lighting, etc. This level of control is commonly called the system level.

An Integrated Building Automation System (IBAS) is an extension of the DDC system, one that ties the DDC system together with the other building systems (fire alarm, security, irrigation, etc.) to improve the level of monitoring, communications, and control of each individual building and group of buildings. The IBAS concept also allows control and monitoring of multiple buildings from a single workstation. This level of control is commonly called the management or enterprise level.

Great care is needed in the selection, operation and maintenance of an Integrated Building Automation System. The three main issues to be considered in establishing an appropriate and feasible level of control and integration for the IBAS are:

1. The practical and cost implications of developing interfaces and/or introducing a common communications protocol;

2. Software complexity, reliability and ongoing support issues arising from the integration of multiple and/or large systems; and

3. The ability of a site-wide or District-wide software system to provide the functionality, performance, reliability, scalability and level of ongoing development available from single purpose systems, such as those developed specifically for environmental control (HVAC), access control and security, fire alarm, lighting, irrigation, elevators, and energy management.

The extensive capability offered by modern building automation systems can be a trap. It can be tempting to try to control or monitor too many functions; this can result in excessive costs and information overload. Systems that include constant polling of points can also have unacceptably slow response times over a network. To avoid high traffic over a District's network the IBAS design should include stand-alone intelligent controllers and therefore should only generate traffic on the network when there is an event or a user makes a request. These actions would have negligible impact on the IT infrastructure bandwidth, especially if non-critical communications are scheduled (programmed) for off-peak periods.

Another objective of the IBAS should be to avoid remotely controlling a function unless it is essential for operational or good facility management reasons. Monitoring should occur only on a genuine "need to know" basis. This approach will help ensure the IBAS stays simple, reliable, fast, and easy to manage, and will add value to the management of the District facilities. Also, the sheer volume of data and information that is available within the individual DDC systems can complicate and burden the IBAS. If systems are properly specified in the first place, extra control or monitoring points can always be added later once the system is well established and genuine needs can be defined.

Since the IBAS will integrate the operation of the various schools within the District, one key objective is to have *uniformity of DDC systems*. This will ensure that control strategies and approaches, as well as functional and performance requirements, are common to all sites.

Another main objective is to have systems based on *open system protocols* so they are flexible and scalable to meet changing needs. An open protocol will also allow the easiest integration with the existing non-DDC systems (fire alarm, security, irrigation, lighting, etc.). Systems should also have a *user-friendly front-end* that has consistent graphics and approach so as to minimize operator training and facilitate changes in personnel. These objectives are critical to optimizing facility operations, facilitating performance measurement, minimizing energy consumption, and providing timely and effective reporting.

SYSTEM PROTOCOL AND COMMUNICATION

Protocols are the digital language that a system uses to communicate between individual devices (such as a thermostat, air handler, or security camera) and the controlling or supervisory system. Protocols differentiate themselves from "hardwired" signals, as they are truly data streams of information. Hardwired signals only carry one signal at a time, whether it is a discrete signal such as "on" and "off," or an analog signal such as temperature. Protocol signals actually combine numerous pieces of data into a "stream" of information, encoded digitally.

Protocols may be open or proprietary. An open protocol is one where the programming language has been developed to allow multiple manufacturers to standardize communications. This allows equipment from one manufacturer to freely communicate with the equipment or control systems of another. With an open protocol, manufacturers can develop interoperable systems that are simple to interconnect. An example of this is the internet's standardization of hypertext transfer protocol (HTTP). Advantages of open protocols are:

- Simplification of the interconnection of equipment and systems, including easier setup and start-up of the communications between components.
- More generic communications parameters.
- Parallel development of competing or complementing systems and software with an agreed-upon design basis for communications.

A *proprietary* protocol is one that has been developed by a specific manufacturer or group of manufacturers. The code for the data is restricted so that it can only be interpreted by that particular manufacturer's series of products. Proprietary protocols are designed to be restricted in access for a number of reasons including:

- Protection of the system controls from unauthorized access, providing a more robust architecture.
- Possibly faster and more tailored control communications since proprietary protocols are generally application specific.
- Limitation of competition from other vendors not part of the consortium developing the protocol.

Often, proprietary protocols require specific hardware (such as a network chip), or specific software to allow communication within the system.

The initial DDC systems were mainly developed by HVAC equipment manufacturers in response to owner demands for more flexibility and centralized control and better system information. They were, and mainly still are, an extension of the original proprietary equipment control systems, typically focused on special purpose and intelligent DDC devices.

Early DDC systems were incompatible with each other and with field devices. These DDC systems all required proprietary software and a proprietary communications network. System field devices such as thermostats and actuators were incompatible between vendors. Many systems offered by the DDC market still suffer these same limitations, with owners tied to sole source suppliers, excessive pricing, and the need to compromise on functionality.

OPEN SYSTEM INTERCONNECTION

Original control systems relied on proprietary, single, complete package protocols for communications. This approach proved unworkable as the requirement for interoperability and wider connectivity increased. To address the interoperability issue, Open System Interconnection (OSI) evolved. This is an internationally accepted 7-layer reference model for communications systems. Protocols are now written as a series of layers in accordance with strict standards. The OSI plays an important role in facilitating communications between modern communication systems and hardware devices.

The seven OSI layers and their communications roles are as follows:

Layer 7 – Application:	Provides access to the communications system via services tailored to the system's use, e.g., file transfer, electronic mail, alarm reporting, etc.
Layer 6 – Presentation:	Performs translation necessary for presenting data to the application.
Layer 5 – Session:	Handles communications in which an on-going dialogue occurs.
Layer 4 – Transport:	Provides reliable end-to-end transfer of data regardless of the path.
Layer 3 – Network:	Provides a communications path when multiple networks are involved.
Layer 2 – Data-link:	Reliable transmission of data between two points on the network.
Layer 1 – Physical:	Transmission of bits over the physical medium.

For Layers 1 and 2, the standards adopted in the building industry are ethernet for high speed LANs, RS-485 for low speed local area networks (LANs), and RS-232 for point-to-point communication between field devices and the other system elements.

For Layer 3, Internet Protocol (IP) is the accepted protocol for communication between networks.

For Layer 4, the TCP (Transmission Control Protocol) is typically used along with IP (Layer 3) to send data in the form of message units between computers over the internet. While IP takes care of handling the actual delivery of the data, TCP takes care of keeping track of the individual units of data (called packets) that a message is divided into for efficient routing through the internet.

TCP is known as a connection-oriented protocol, which means that a connection is established and maintained until such time as the message or messages to be exchanged by the application programs at each end have been exchanged. TCP is responsible both for ensuring that a message is divided into the packets that IP manages and for reassembling the packets back into the complete message at the other end.

Layers 5, 6 and 7 are generally custom applications provided to suit each individual system. They can be either proprietary or open in terms of their hardware, software and programming.

PROTOCOL COMPARISONS

There are an increasing number of DDC and IBAS systems available with different protocols, interconnectivity and communications capabilities for management of multiple facilities. The main protocol standards for building systems are:

- TCP/IP-XML, HTML
- BACNet
- LonWorks
- Modbus

TCP/IP-XML/HTML

All IT (Information Technologies) systems worldwide utilize true standards. The standards of IT include TCP/IP, XML, HTML, and HTTP. The goal of integrated building systems will only be achieved using these 100% market share standards. The other communications protocols are only minor market share contenders (Lonworks 3%, BACnet 12%) and therefore do not represent true standardization. Technology and solutions that integrate with these and other third party non-standard protocols and that concurrently advance these systems and protocols into the IT world are required in order to migrate these building systems and subsystems into Integrated Building Automation Systems (IBAS).

BACNet

BACNet was developed by a consortium under the supervision of ASHRAE (American Society of Heating, Refrigeration and Air Conditioning Engineers) to specifically address interoperability and proprietary system concerns in the building automation industry. In general, the technology behind BACNet does not yet match that of the advance networking technology in other industries and standards. However, BACNet is generally accepted by independent experts to be the most complete protocol for the building industry and the only one without proprietary limitations. One of its main advantages is allowing the prioritization of information. BACNet will allow emergency or life safety signals to over-ride the control of a device by placing higher priority on the life-safety signal than the operational signal (e.g., fan control during the sensing of smoke within the ductwork versus temperature control).

The limitation of BACNet is that it does not address programming or device configuration. Fortunately for an IBAS application, these functions are contained within the building-level control systems.

BACNet is likely to become the official world standard for integration of building services. It is already established as a standard with ASHRAE (ASHRAE Standard 135) and CEN (European Committee for Standardization), and was adopted by the International Standards Organization (ISO) as Standard 16484-5.

A fundamental part of the design of BACNet is its planned continual evolution. While it is based on the OSI model, it only incorporates Layers 1, 2, 3 and 7. However, it defines multiple physical architectures necessary to handle both low and high-speed networks and emulates the structure of most automation systems. Its acceptance and use is now spreading to many intelligent building services especially lighting, security, access control and fire protection systems.

Development is also proceeding on the uses of Extensible Markup Language (XML) in BACnet systems. There are draft methods for using XML to exchange data between BACnet devices and the OSI Level 7 using the Simple Object Access Protocol (SOAP) and/or XML Remote Procedure Call (XML-RPC) technologies. The importance of this technology is to retrieve data from BACnet systems using standard desktop software instead of specialized drivers.

LonWorks

LonWorks is a protocol standard developed by the Echelon Corporation. While it generally achieves a higher degree of interoperability between multiple manufacturers than BACNet, in most cases access to the protocol requires access to an Echelon proprietary chip. The protocol, LON, is the simplest to implement because the majority of the development work has already been done. This makes it a lower-cost option for device manufacturers. However, while manufacturers provide users open access to LON protocols at the device and subnetwork levels (OSI Layers 1 and 2), at the higher levels (OSI Layers 3 to 7) involving information routing and management, the protocols are completely proprietary and closed.

Further, LonWorks' development potential is constrained by the limitations of the proprietary "Echelon" chip now wholly owned and manufactured by Motorola. In addition to the proprietary limitations, it is not flexible or scalable and so is only suitable for device-level networks. Another disadvantage is that the LonWorks wiring system is proprietary and therefore dedicated.

LonWorks does not allow for prioritizing information like BACnet. To compensate for this, a parallel signal would have to be configured from the life safety device (e.g., smoke detector) that would allow emergency or life safety signals to over-ride the control of a device.

In spite of these disadvantages, LonWorks has been adopted by many DDC and IBAS manufacturers, including Honeywell, Invensys, and TAC, mainly due to its low cost of development. Johnson Controls also uses LonWorks in some of its DDC applications.

MODBUS

The Modbus protocol was originally developed by Modicon in 1978 to exchange information between controllers within a factory. From the initial development, this protocol has become the de facto standard for exchanging data and communication information between PLC systems.

Modbus is an open, serial communications protocol based on the master/slave architecture. Since it is easy to implement on all types of serial interfaces, it has gained wide acceptance. The Modbus protocol was originally developed to link controllers into a network, but it has frequently been used to connect input/output modules. Because of the low transmission rate - a maximum of 38.4 kBaud, Modbus is best employed when the number of devices participating on the bus is small and low demands are placed on response time.

The bus consists of a master station and a number of slave stations, and the communication is controlled entirely by the master. Modbus offers two basic communication mechanisms:

- Question/answer (polling): The master sends an inquiry telegram to any one of the stations, and waits for the answer telegram.

- Broadcast: The master sends a command to all the stations on the network, and these execute the command without providing feedback.

Modbus is used on a variety of transmission media. An implementation based physically on the RS485 bus [twisted pair (two-wire) cable] is common.

Enhancements to Modbus include Modbus Plus and Modbus/TCP protocols, both of which allow Modbus information to be encapsulated in a network structure to support peer-to-peer communications. Modbus Plus communicates via a single twisted pair of wires and uses a token passing sequence for peer-to-peer communication sequences. Modbus/TCP is an open standard designed to facilitate Modbus message transfer using TCP/IP protocol and standard Ethernet networks.

Modbus is generally an open protocol at the device and sub-network levels (OSI Layers 1 and 2). At the higher levels (OSI Layers 3 to 7) involving information routing and management, additional protocols are required which are also generally open. The most critical limitation to developing a system using Modbus is the data transfer rate at Layer 1.

OTHER PROTOCOLS

Other custom protocols are used by some control system manufacturers. For example, Johnson Controls makes a DDC system that uses an N2 protocol, which is a proprietary protocol at the heart of the Metasys system.

Other examples of custom protocols are ASCII, DeviceNet, and Siemens FLN. Some of the custom protocols are proprietary (N2) and some are open protocols (ASCII).

IBAS FRONT END

The IBAS front end is the generic term for the workstation, software, configuration and other hardware and software that allows the IBAS to communicate with all of the other devices and systems within a school building. This is also known as an *Enterprise Interface*, since they interface between the owner's operations and the IBAS standard applications.

Each front end must provide communication for OSI Layers 3 through 7. Previously, interfaces were written on a manufacturer-by-manufacturer basis. This is still true for legacy systems. However, new devices and networks can be developed using standard protocols, so that they can easily interface with the IBAS. The protocols can be either open or proprietary systems, as described previously.

The interface can also use the internet (rather than the District Intranet) for communication Layers 3 and 4, utilizing TCP for transport and IP for network communications. This would minimize the impact and traffic on the District Intranet. Additionally, the system could also use an internet browser as the graphical interface (web-based). This would allow the graphics software to reside on a server, which would be universally accessible, while maintaining security. Also, this would allow access from any computer system rather than requiring a dedicated workstation. Finally, using a web-based front end during the development of system graphics can be more flexible than using standard software targeted toward HVAC systems because specific graphics (web pages) can be more readily created for unique systems.

Manufacturers take several approaches with web technology. Most are either web-based or web-enhanced. True web-based systems are designed from the ground up as a web application. Web-enhanced systems are those that achieve internet compatibility by adding software products that convert and export information to the web. In enhanced systems, the software itself is not internet compatible; only the data and information that has been passed through the conversion utility or gateway adheres to internet standards. Web-enhanced systems require both software and a web browser to be installed on each user's computer. Web-based systems have the advantage of having all system software reside on a single web server. A user simply logs into the server and uses the system through a standard web browser (e.g. Microsoft's Internet Explorer) thus eliminating the requirement for additional software and licenses for each machine. A web-based system is ideal for use in applications having multiple sites.

INTEGRATING EXISTING SYSTEMS

Integrated systems can be achieved by having one manufacturer provide multiple systems and by having all the intelligent systems in a building use BACnet as a high-level interface. This can be achieved using completely proprietary systems with BACnet gateways. This is not optimal, but is satisfactory. An optimal system would use *Native BACnet* as the standardized core of the installation. This will reduce integration costs.

Once the protocol and front end standards are determined, the IBAS supplier would need to develop specific graphical interfaces to bring data from the diverse systems (HVAC, fire alarm, irrigation, etc.) into the IBAS and tie it all together. Since many of the legacy systems are not BACNet, LonWorks or Modbus protocols, translator modules (gateways) would have to be developed to allow the information and system points to communicate with the IBAS.

It is critical to note that most existing DDC systems will need gateways to communicate with the new IBAS. This effort is likely to be more than just programming of gateways to translate data into the selected protocol. Some of the systems will need trending, scheduling and alarm features that are not currently resident within the existing controllers. Lastly, to improve the management of the building systems, some or all of the data communicated to the IBAS must be reported, alarmed, or archived.

IBAS COST CONSIDERATIONS

A number of interrelated costs must be quantified when considering implementation of a District-wide IBAS. Immediate personnel costs would include hiring a staff Energy Manager, training of maintenance and facility personnel in the use of the IBAS, and instituting cultural and procedural changes for the maintenance organization. Operational and maintenance costs would include ongoing maintenance of the IBAS and the system components, repairs to system components and equipment, and energy and demand costs associated with using the IBAS. Capital costs would include the costs of new equipment or upgrading existing equipment, and the cost of conforming to code changes when retrofits are performed (more stringent fire codes, elevator codes, ventilation codes, etc.). Finally, there may be other costs associated with service agreements for any of the new equipment or systems, and there will be costs associated with an annual review and fine tuning of the IBAS performance.

CONCLUSION

Numerous vendors can provide an integrated building automation system. The basics of selecting the correct technology for an IBAS depend on answers to the following questions:

- Is a building management system necessary or beneficial?
- If so, should open or proprietary protocol be used?
- If open protocol, should it be BACnet or LON or some other protocol?
- Are there emerging technologies that should be considered?

Selection of an IBAS supplier should be based on a life-cycle cost analysis, as well as the following:

- Training of maintenance and facility personnel
- Ongoing maintenance of the IBAS and system components
- Repairs to system components and equipment
- Upgrades and capital expenditures
- Energy consumption rates and demand charges
- Service agreements, if desired, for any of the above
- Conformance to code changes when retrofits are done (more stringent fire codes, elevator codes, ventilation codes, etc.)
- Yearly review of IBAS performance and fine tuning
- Utilization of an Energy Manager on staff
- Cultural and procedural changes for the maintenance organization

Prior to interviewing potential suppliers, an internal review should be performed to identify system needs as well as prepare a function list of the systems to be incorporated into an IBAS.

Section Six

Hardware and Software Tools and Systems for Data Input, Data Processing and Display, for EIS and ECS Systems

Chapter 27

Wireless Sensor Applications for Building Operation and Management

Michael R. Brambley, Michael Kintner-Meyer, and Srinivas Katipamula
Pacific Northwest National Laboratory[1] *Richland, Washington*
Patrick J. O'Neill, NorthWrite Inc., Minneapolis, Minnesota

ABSTRACT

THE EMERGING technology of wireless sensing shows promise for changing the way sensors are used in buildings. Lower cost, easier to install, sensing devices that require no connections by wires will potentially usher in an age in which ubiquitous sensors will provide the data required to cost-effectively operate, manage, and maintain commercial buildings at peak performance. This chapter provides an introduction to wireless sensing technology, its potential applications in buildings, three practical examples of tests in real buildings, estimates of impacts on energy consumption, discussions of costs and practical issues in implementation, and some ideas on applications likely in the near future.

INTRODUCTION

Wireless communication has been with us since the invention of the radio by Marconi around 1895. We have benefited from the broadcast of information for purposes of informing and entertaining. Radio technology has also enabled point-to-point communication, for example, for emergency response by police and fire protection, dispatch of various service providers, military communications, communication to remote parts of the world, and even communication into space.

We commonly think of communication between people by voice when thinking of radio frequency (RF) communication technology but need to look no further than a television set to realize that other forms of information, such as video, can also be transmitted. In fact, RF technology can be used to transfer data in a wide variety of forms between machines and people and even among machines without human intervention. This more generic wireless RF transfer of data and its application to operating and maintaining buildings is the focus of this chapter.

Wireless communication of data via WiFi (or IEEE 802.11 standards) is now routine in many homes, offices and even airports.[1,2] Rather than ripping walls open or fishing networking cable through them to install computer networks in existing homes and commercial buildings, many users opt to use wireless technology. These standards use license-free frequency bands and relatively low power to provide connections up to several hundred feet (although additional parts of IEEE 802.11 are currently under development for much longer ranges of up to 20 miles and higher data transfer rates). These standards are generally for relatively high bandwidth so that large files can be transported over reasonable time periods.

In contrast to the data rates required for general computer networking and communication, most sensor data collection can get by with much slower rates with as little as a few bits every second, every minute, 10 minutes, or even less frequently. Sensing generally imposes (or loosens) other constraints as well. For example, if the value of a single sensor point is low, its total installed cost must be very low as well. Furthermore, if power for sensing and communication is not conveniently available where sensor measurements are needed, an on-board power source may be needed. In general, we'd like to put sensors in place and then forget about them, so they should have long lives and require little attention. If a sensor requires frequent maintenance, the cost for its use increases rapidly, so power sources, like batteries, with lives of 10 years or more would be ideal. These requirements for sensors and sensor networks are leading to the evolution of wireless sensor network technology and

[1]Operated for the U.S. Department of Energy by Battelle Memorial Institute under contract DE-AC06-76RL-1830.

standards that provide specifically for convenient, widespread use of large numbers of sensors from which data are collected wirelessly.

The ideal wireless sensor would have very low installed cost, which would require that its hardware cost be very low and that it be installed quickly and easily using limited labor. One concept calls for wireless sensors that you "peel, stick and forget." The radio frequency identification (RFID) tag industry debatably has reached a cost as low as about $0.20 per tag and seeks to reach $0.05 per tag with a production of 30 billion tags per year for inventory tracking purposes. [3] Wireless sensors for active property measurements like those suitable for use in building operations still cost on average two to four orders of magnitude more than this.

To achieve easy and low-cost installation, wireless sensor networks, which provide the means for moving data from the collection points to where it can be used, will probably need to be *self-configuring*. This means that the sensors would assign themselves identifications, recognize their neighboring sensors, and establish communication paths to places where their data are used (e.g., on a personal workstation or a receiver connected to a building automation system). A self-configuring wireless sensor network would only require placing the sensors where the measurements need to be made and possibly providing a connection to a user interface or computer network.

To reduce the cost of maintenance, the sensors and sensor network would need to be *self-maintaining* and *self-healing*. For example, if a metal cabinet were moved into the communication path between two sensors, blocking communication between them, the network would automatically reroute the signal by another path with no human intervention. In addition, the sensors would need to maintain their own calibration reliably over their lifetimes (be *self-calibrating*), actively ensuring that they are within calibration periodically. These capabilities are critical to ensuring low cost and reliable sensor networks. If each sensor has to be maintained by technicians periodically during its life, the cost will be too high to justify its use in all but the most critical and high-value applications. To increase sensor use, lower life-cycle costs are essential.

Some wireless sensors may have access to hardwired power, but for many applications the sensor and its radio must be *self-powered*, using a battery that lasts for many years or harvesting power from the ambient environment. [4] In 2004, some manufacturers of wireless sensors claim battery lives as long as 7 years for some applications. Wireless sensors that use environmental vibrations as a source of power have also been developed for a limited set of applications, [5] but most ambient power harvesting schemes are still under development. Complementary developments are underway for a wide range of applications that reduce the power requirements of electronic circuits. Examples for sensor networks include: intelligent management of on-board power use by sensor radios to limit power requirements (and battery drainage), using sleep modes, transmitting only as frequently as absolutely required, and minimizing message size. Power requirements are also tied directly to the distance over which signals must be transmitted. By decreasing this distance and using multiple hops to span a long distance, power can be conserved. The mesh networking schemes described later in this chapter have the potential to significantly reduce the power requirements for wireless sensors.

These are some of the capabilities of the ideal wireless sensor. In the sections that follow, an introduction to wireless sensor technology is provided, potential applications for wireless sensors in buildings are described, potential benefits are discussed, a few real-world cases are presented, and the current state of wireless sensing and likely future developments are described. Three primary concerns are frequently raised in discussion of wireless sensing for building operation: cost, reliability, and security. This chapter addresses each of these, providing references for the reader interested in more detail. Some practical guidance for using wireless sensors in buildings today is also provided.

Why use Wireless Sensing in Buildings?

The cost of wiring for sensors and controls varies widely from about 20% to as much as 80% of the cost of a sensor or control point. The precise costs depend on the specific circumstances, e.g., whether the installation is in new construction or is a retrofit in an existing building, the type of construction, and the length of the wiring run. For situations where wiring costs are high, eliminating the wires may produce significant cost reductions.

Too often today operators are not able to effectively monitor the condition of the vast array of equipment in a large commercial building. Field studies and retrocommissioning of commercial buildings show that dirty filters, clogged coils, inoperable dampers, and incorrectly-implemented controls are all too common. [6, 7] Pressures to reduce operation and maintenance costs only exacerbate this problem. The problem can be even worse in small commercial buildings, which frequently don't even have an operator on site. Keeping apprised of the condition of equipment and systems in these buildings is nearly impossible for an off-site operator. If an equipment problem does not directly affect the occupants of a building (and this is quite common when the systems compensate by running harder and using more energy), it will usually continue undetected and uncorrected until conditions deteriorate and the occupants complain. This is often long after the problem started wasting energy and costing the bill payers money. Annual or semi-annual service visits by maintenance technicians, often catch only the most obvious problems. Incorrectly-implemented controls can go undetected for years unless major retro-commissioning of the building is undertaken.

More sensors to monitor the condition of equipment and systems, as well as conditions in the building, are needed along with software tools that automatically sort through data as it arrives and alert building operations and maintenance staff (or service providers) to problems. Building owners, however, often cite the need to keep costs down as the reason for not installing these sensors. By doing this, they are trading lower initial costs for higher expenditures on energy and lost revenue from tenant churn caused by poor environmental conditions in the building. This might be addressed by education and more evidence of the net value of good operation and maintenance over the building ownership life cycle, but lowering the cost of collecting data and obtaining useful results from it may be a more direct approach. This chapter focuses on the data collection issue by presenting information on wireless sensing; the need for tools that automatically process the data is a companion problem that is just as critical, but that is the subject of Chapter 18 in this book.

Better sensing in commercial buildings would lead to greater awareness of the condition of buildings and their systems. Operation and maintenance (O&M) staff would have the information to recognize degradation and faults in building equipment and systems and prioritize problems based on cost and other impacts. Today, most building staffs do not have this information. With it, the most costly and impactful problems could be identified, even those that are not usually recognized today.

The benefits of more data and tools that provide useful information from that data would be: lower energy and operating costs, longer-equipment lives, and better, more consistent conditions provided to building occupants. The value of these should all well exceed the cost of collecting and processing the information. With new, lower cost means such as wireless sensing for gathering data, first costs should also decrease making the financial decision to make this investment easier for building owners.

There are also some advantages directly attributable to the unique characteristics of wireless sensing beyond lower cost. Wireless sensors having their own power sources are mobile. Such a sensor can be readily moved from one location to another to investigate a problem. If a particular office, for example, were chronically reported as too hot, a wireless air-temperature sensor might be moved to that office or an additional one added to the wireless sensor network for that office to verify that the temperature was indeed unacceptably hot, then used to verify whether the corrective actions were successful. New sensors could be added to equipment for similar purposes without installing additional wiring. For example, if a pump motor were thought to be intermittently running hot, a wireless sensor might be installed on it to monitor its temperature and verify the need for repairs. If not wired, these sensors could be placed temporarily and then used at different locations as needed; no wiring costs would be necessary. One of the benefits of a wireless sensor network is that once it is in place in the building, sensors can be added or moved easily without installing new cables. As a result, wireless sensors have unique value for diagnostics.

Wireless Sensor Networks

Primary Components

Each wireless radio frequency (RF) sensor requires three critical components to sense a condition and communicate it to a point at which it can be used (whether by a human or directly by another machine): 1) a *sensor* that responds to a condition and converts it to a signal (usually electrical) that can be related to the value of the condition sensed, 2) a *radio transmitter* that transmits the signal, and 3) a *radio receiver* that receives the RF signal and converts it to a form (e.g., protocol) that can be recognized by another communication system, another device, or computer hardware/software. This is the simplest communication configuration for wireless sensing (see Figure 27-1).

At the sensor the device usually consists of signal processes circuitry as well as the sensor probe itself. This circuitry may transform the signal with filtering, analog to digital conversion, or amplification. The transmitter, in addition to modulating and sending a signal, may encode it using a protocol shared with the receiver. At the receiver, electronic circuits will perform similar operations, such as filtering, amplification, digital to analog conversion, embedding in another communication protocol (e.g. Ethernet or RS-232 serial), and transmission as

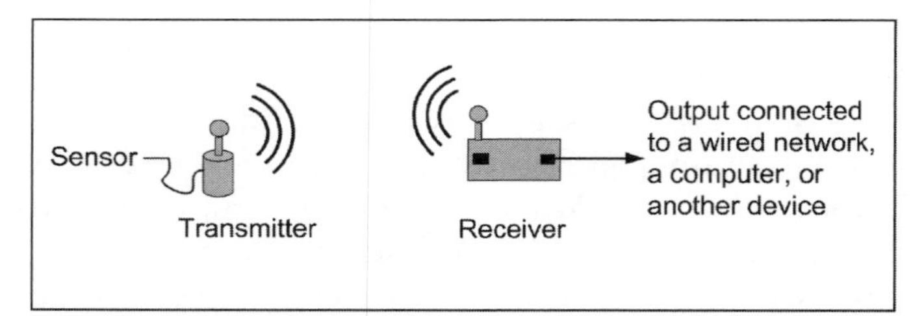

Figure 27-1.

output.

Many wireless networks replace the transmitter and receiver with radio transceivers (which have combined transmitting and receiving abilities). This permits 2-way communication so that the radio at the receiving point can send requests for data transmissions (poll the sensor transmitter) and send messages acknowledging receipt of both data and messages transmitted from the sensor's radio. The sensor's transceiver can receive requests and acknowledgments from the transceiver at the receiving point, as well as send the sensor data. In addition to these functions, both radios formulate packets of data that precede and follow the main data or messages sent that are specified as part of the protocol the radios use for communication purposes.

All of these components require electric power to operate and, therefore, a power supply, which is usually either wired power or a battery. The power supply then converts the source power to the form (e.g., direct current, DC) and voltage required by the device. Battery operated devices generally have sophisticated power management schemes implemented to conserve the battery's energy by powering the electronics down between transmissions. Another source of power for distributed devices under development is power-scavenging technology, which can extend battery lifetime or even fully substitute for a battery. Power-scavenging devices convert ambient energy forms such as vibrations, light, kinetic energy inflows, and temperature differentials into electric energy.

Networks of sensor nodes (the combination of a radio, other electronic circuitry, and the sensor) can be formed from the basic principle illustrated in Figure 27-1, but many sensor nodes transmit data to points of reception. Wireless sensor networks can have tens, hundreds, even thousands of nodes in the network, providing measurements from different kinds of sensors that might be located at many different positions. For example, a wireless network might measure many temperatures, humidities, and pressures throughout many HVAC systems, the electric power use of all major equipment, as well as the temperature and occupancy of rooms throughout a building, all reported to one receiver that sends the data to a computer for processing or display.

Network Topology

Wireless sensor networks have different requirements than computer networks and, thus, different network topologies and communication protocols have evolved for them. The simplest is the *point-to-point topology* (see Figure 27-2) in which two nodes communicate directly with each other. The *point-to-multipoint* or *star topology* is an extension of the point-to-point configuration in which many nodes communicate with a central receiving or gateway node. In the star and point-to-point network topologies, sensor nodes might have pure transmitters, which provide one-way communication only, or transceivers, which enable two-way communication and verification of the receipt of messages. Gateways provide a means to convert and pass data between one protocol and another (e.g., from a wireless sensor network protocol to the wired Ethernet protocol).

The communication range of the point-to-point and star topologies is limited by the maximum communication range between the sensor node at which the measured data originate and the receiver (or gateway) node. This range can be extended by using repeaters, which receive transmissions from sensor nodes and then retransmit them, usually at higher power than the original transmissions from the sensor nodes. By employing repeaters, several "stars" can communicate data to one central gateway node, thus expanding the coverage of star networks.

In the *mesh network topology* each sensor node includes a transceiver that can communicate directly with any other node within its communication range. These networks connect many devices to many other devices, thus, forming a mesh of nodes in which signals are transmitted between distant points via multiple hops. This approach decreases the distance over which each node must communicate and reduces the power use of each node substantially, making them more compatible with on-board power sources such as batteries. In addition to these basic topologies, hybrid network structures can be formed using a combination of the basic topologies. For example, a mesh network of star networks or star network of mesh networks could be used (see Figure 27-3).

Point-to-Point

In a point-to-point network configuration each single device (or sensor node) connects wirelessly to a

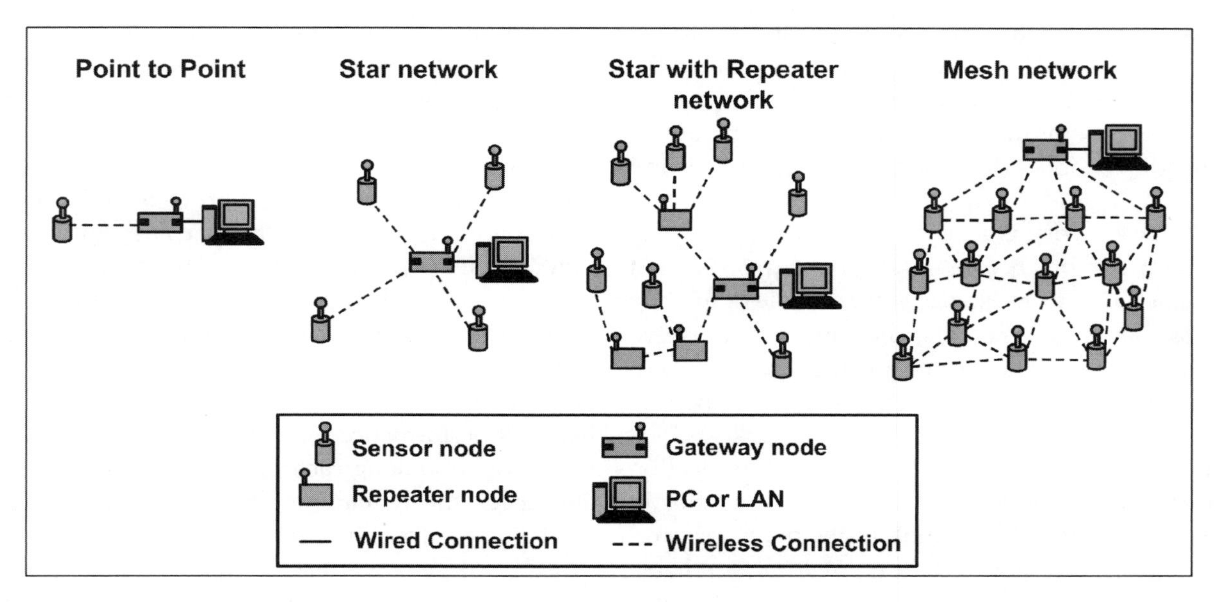

Figure 27-2.

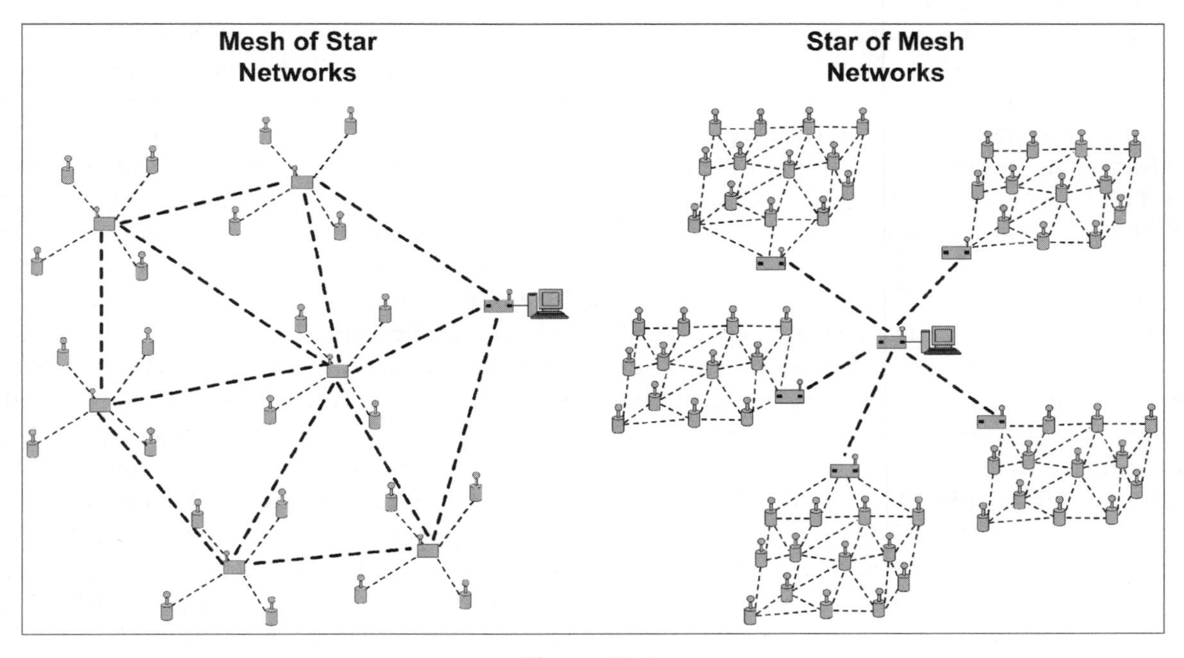

Figure 27-3.

receiver or gateway. An example would be a remote control for a TV, a garage door opener, or a wireless PLC (programmable logic controller) to turn on/off a remote pump or light. The communication can be kept simple with identification schemes that are either set up in the hardware with dip switches or by software during the initial configuration. Point-to-point wireless architectures apply a simple master/slave communication protocol whereby the master station issues a command for a single dedicated slave.

Star Networks

The star network is an extension to the point-to-point configuration. One central node broadcasts to many end nodes in the network (i.e. point to multipoint). Alternatively, the communication can originate from the end nodes, communicating to one single central point (i.e. multipoint to point). The latter is a typical architecture for currently available in-home and building security products. Remote sensors on doors and windows, when triggered, communicate to one central station, which then issues an alarm and performs other pre-programmed procedures such as dialing the police or fire department. A star topology can be used in building operation for monitoring zone-air temperatures with wireless sensors as described in References 8, 9 and 10.

The star network is a simple network topology to support many sensors. Before standard integrated-circuit (IC) manufacturing technologies were capable of making high performance RF chipsets, the only cost-effective wireless network was the star network because the sensor nodes often had only transmitters and not transceivers.

This topology provides only one communication path for each sensor node, so there is no redundancy in the network. As a result, each link in the network infrastructure is a single point of failure. Ensuring a reliable communication path for each sensor is critical, and a thorough RF site survey must be performed to determine the need and locations for repeaters to carry each sensor signal reliably to the receiver. Sufficient resilience should be built into the design of star networks so that reliable communications of all sensors can be maintained even if the interior layout of the building changes. Simply repositioning a bookcase into the path of a weak signal could add enough signal attenuation to stop communication between a sensor and the receiver.

Mesh Networks

With the significantly reduced cost of microprocessors and memory over the last decade, additional computational power at the device level can now be used to operate a more complex network that simplifies both the installation and commissioning of a sensor network while maximizing reliability. Mesh networks—where each device in the network acts both as a repeater and a sensor node—can achieve the long communication range of a star network with repeaters while also providing increased total network reliability through redundant communication paths. The nodes in a mesh network automatically determine which nearby neighbors can communicate effectively and route data through the network accordingly, changing the routing dynamically as conditions change. Having multiple links in a network provides built-in redundancy so data can be effectively routed around blocked links. This means that there are few single points of failure in the system, so the overall network is extremely reliable even if individual wireless links are not. Mesh networks also pass data from one node to another in the network, making the placement of additional sensors or controllers in the network akin to building out additional infrastructure. As additional devices are placed in a mesh network, the number of communication paths increases, thereby improving network reliability.

The most-used nodes in any sensor network use the most energy. So if the routing is static, even in a mesh network (when the "best" communication routes don't change with time), the energy demands will vary among nodes with those used most expending the most energy. For battery-powered nodes, this demand can rapidly drain the battery. Network protocols are being developed that are "energy aware." To help maximize network performance time, these protocols even account for energy use along each potential communication path and check the remaining charge of batteries along the paths in selecting the preferred route. [11, 12] This approach, however, works best where node density is high throughout the area covered by a network. In situations where node density is not high (as during initial adoption of wireless monitoring in buildings or other cases where sensor node deployment may be sparse), a single critical node or a small number of nodes that provide the path for all communication will be subject to excess power use and lower battery life (see Figure 27-4).

A disadvantage of mesh networking could be the use of the wireless data channels for network management and maintenance, which not only takes up part of the available radio bandwidth, but also uses power and drains batteries. For low-data-rate applications in facility monitoring and control as well as many other sensing applications, this limitation is likely manageable. The protocols under development for wireless sensor networks seek a balance between these factors. [11, 12, 13] Sophisticated network routing schemes, however, impose an overhead on hardware and firmware potentially adding a premium to the overall cost, but advances in electronics manufacturing should minimize the impact of this factor. Mesh sensor networking technology is in a nascent stage with early products just beginning to enter the building automation and monitoring market.

Frequency Bands

To minimize interference and provide adequately for the many uses of radio frequency communication, frequency bands are allocated internationally and by most countries. The International Telecommunication Union (ITU) is the organization within which governments coordinate global telecommunication networks and services. The United States is a member of the ITU through the Federal Communications Commission (FCC). The ITU maintains a Table of Frequency Alloca-

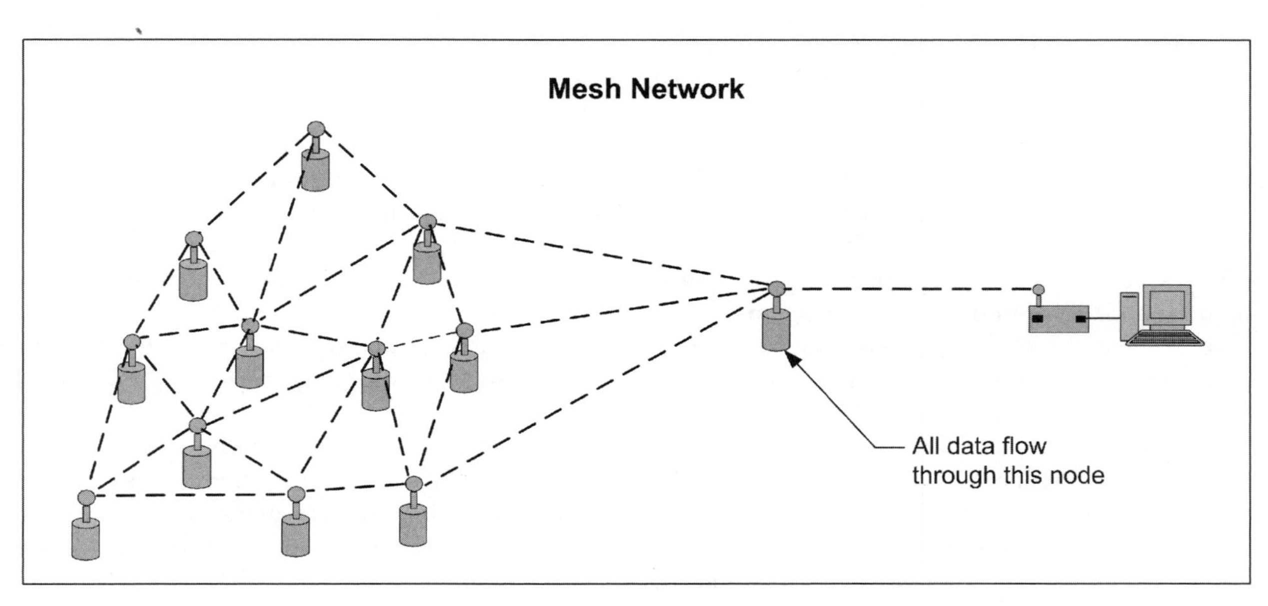

Figure 27-4.

Table 27-1. ISM Frequency Band Allocations and Applications. [13]

Frequency band	*Center Frequency*	*Band-width*	*Applications*
6,765-6,795 kHz	6,780 kHz	30 kHz	Personal radios
13,553-13,567 kHz	13,560 kHz	14 kHz	
26,957-27,283 kHz	27,120 kHz	326 kHz	
40.66-40.70 MHz	40.68 MHz	40 kHz	Mobile radios
902-928 MHz	915 MHz	26 MHz	In the US, applications includes Railcar and Toll road applications. The band has been divided into narrow band sources and wide band (spread spectrum type) sources. Europe uses this band for cellular telephony services (GSM)
2,400-2,500 MHz	2,450 MHz	100 MHz	A recognized ISM band in most parts of the world. IEEE 802.11, Bluetooth recognizes this band as acceptable for RF communications and both spread spectrum and narrow band systems are in use.
Cordless phones			
5,725-5,875 MHz	5,800 MHz	150 MHz	Cordless phones. The FCC have been requested to provide a spectrum allocation of 75 MHz in the 5.85-5.925 GHz band for Intelligent Transportation Services use.
24-24.25 GHz	24.125 GHz	250 MHz	Allocated for future use
61-61.5 GHz	61.25 GHz	500 MHz	
122-123 GHz	122.5 GHz	1 GHz	
244-246 GHz	245 GHz	2 GHz	

tion that specifies regionally and by country the allocations of radio spectrum. [14] The ISM (industrial, scientific, medical) bands provide frequencies for license-free radio communications given a set of power output constraints. The ISM frequencies and common applications are shown in Table 27-1.

Consumer products ranging from cordless telephones to wireless local area networks use the 2.4 GHz band. The trend for selecting higher frequencies is primarily driven by the need for higher data rates. As can be seen in Table 27-1, the bandwidth is greater at higher frequencies. Bandwidth is defined as the width of a particular frequency band. For instance, the 900 MHz band has a bandwidth of 26 MHz (928 MHz—902 MHz, see Table 27-1). Data rates and bandwidth of a frequency band are related. According to Nyquist, the maximum data rate in bits per second (bps) that can be achieved in a noiseless transmission system of bandwidth *B* is 2*B*. [15]Using the Nyquist theorem for the example of a bandwidth of 26 MHz, we would obtain a theoretical data rate limit of 52 Mbs. In practical applications where we encounter signal noise, the signal-to-noise ratio limits the actually achievable data rate to a value less than that determined by the Nyquist theorem. [16]

For wireless local area networks (LANs) higher bandwidth provides higher data rates, a generally desirable feature. Wireless sensor networks, on the contrary, are generally low-data-rate applications sending, for instance, a temperature measurement every 5 minutes. Hence, higher frequencies provide no bandwidth benefit for sensor network applications. In fact, higher frequency signals attenuate more rapidly in passing through media, thus shortening the range of the RF transmission as signals penetrate materials, e.g., in walls and furnishings. [17] To maximize transmission range, a low transmission frequency technology should be selected (see the discussion on signal attenuation in the section Designing and Installing a Wireless System Today: Practical Considerations).

Communication Protocols

There are a large number of wireless technologies on the market today, and "wireless networks" as a technology span applications from cellular phone networks to wireless temperature sensors. In building automation applications where line power is not available, power consumption is of critical importance. For example, battery-powered "peel-and-stick" temperature sensors will only be practical if they and their network use power at a very low rate. In general, a 3- to 5-year battery lifetime is believed to be a reasonable minimum. Although power is generally available in commercial buildings, it is often not conveniently available at the precise location at which a sensor is needed. Thus, for many wireless sensors, some kind of onboard power, such as a battery is necessary to keep the installed cost low. To maximize battery life, communication protocols for wireless sensor networks must minimize energy use.

Beyond power requirements, communication range is important. A radio that has a maximum line-of-sight range of 500 feet outdoors may be limited to 100 feet or even less indoors, the range depending on a number of factors including the radio's frequency, the materials used in construction of the building, and the layout of walls and spaces. Communication protocols for sensor networks installed indoors, therefore, must provide adequate communication ranges in less than ideal indoor environments.

Table 27-2 provides a summary of power consumption, data rate, and communication range for several wireless communication standards. The IEEE 802.11b and g standards (also referred to as "WiFi" for Wireless Fidelity), which were developed for mobile computing applications, are at the high end of data rate and have moderately high power consumption and moderate range. While these standards have proven very popular for wireless home and office networking and mobile web browsing, they are not suitable for most building sensor applications because of their high power consumption. Furthermore, in the long run, 802.11b and g are likely to see quite limited use for sensor networking because of their limits on the number of devices in a network and the cost and complexity of their radio chipsets, compared to simpler, ultimately lower cost, wireless sensor networking standards.

Bluetooth, another wireless communications standard, was developed for personal area networks (PANs) and has proven popular for wireless headsets, printers, and other computer peripherals. [18] The data rate and power consumption of Bluetooth radios are both lower

Table 27-2. Basic characteristics of some wireless networking standards.

Network Name/Standard	*Power Use (Watts)*	*Data Rate (kb/sec)*	*Line-of-site Range (meters)*
Mobile telecommunications			
GSM/GPRS/3G	1 to 10	5 to >100	>1000
Wi-Fi IEEE 802.11b	0.5 to 1	1000 to 11,000	1 to 100
Wi-Fi IEEE 802.11g	0.03 to 0.7	1000 to 54,000	>100
Bluetooth IEEE 802.15.1	0.05 to 0.1	100 to 1000	1 to 10
ZigBee with IEEE 802.15.4	0.01 to 0.03	20 to 250	1 to >100

than for WiFi, which puts them closer to the needs of the building automation applications, but the battery life of a Bluetooth-enabled temperature sensor is still only in the range of weeks to months, not the 3 to 5-years minimum requirement for building applications, and the communication range is limited to about 30 feet (100 feet in an extended form of Bluetooth). The number of devices in a Bluetooth network is also severely limited, making the technology applicable for only the smallest in-building deployments.

The IEEE 802.15.4 standard [19, 20] for the hardware layers together with the Zigbee standard [21] for the software layers provides a new standards-based solution for wireless sensor networks. IEEE 802.15.4, which was approved in 2003, is designed specifically for low data-rate, low power consumption applications including building automation as well as devices ranging from toys, wireless keyboards and mouses to industrial monitoring and control [19, 20]. For battery-powered devices, this technology is built to specifically address applications where a "trickle" of data is coming back from sensors or being sent out to actuators. The standard defines star and meshed network topologies, as well as a "hybrid" known as a cluster-tree network. The communication range of 802.15.4 radio devices is 100 to 300 feet for typical buildings, which, when coupled with an effective network architecture, should provide excellent functionality for typical building automation applications.

The industry group ZigBee Alliance developed the ZigBee specification that is built upon the physical radio specification of the IEEE 802.15.4 Standard [21]. ZigBee adds logical network, security, application interfaces, and application layers on top the IEEE 802.15.4 standard. It was created to address the market need for a cost-effective, standards-based wireless networking solution that supports low data rates, low power consumption, security, and reliability. ZigBee uses both star and meshed network topologies, and provides a variety of data security features and interoperable application profiles.

Non-standardized radios operating with proprietary communication protocols make up the majority of today's commercially available wireless sensors. They usually offer improved power consumption with optimized features for building automation applications. These radios operate in the unlicensed ISM frequency bands and offer a range of advanced features which depend on their target applications.

Technical Issues in Buildings

The primary issues of applying wireless sensor technologies in buildings are associated with 1) interference caused by signals from other radio transmitters (such as wireless LANs) and microwave ovens that leak electromagnetic energy, 2) attenuation as the RF signal travels from the transmitter through walls, furnishings, and even air to reach the receiver, and 3) security.

Interference generally stems from electromagnetic noise originating from other wireless devices or random thermal noise that may impact or overshadow a sensor signal. Spread spectrum techniques are used to increase immunity to interference from a single- frequency source by spreading the signal over a defined spectrum. Spread spectrum techniques utilize the available bandwidth such that many transmitters can operate in a common frequency band without interfering with one another. Spread spectrum, however, is not guaranteed to be completely immune to interference, particularly if the frequency band is heavily loaded, say with hundreds of wireless devices sending messages. Early technology demonstration projects with 30 to 100 wireless sensors in buildings have not revealed any problems with crosstalk or loss of data in the transmission; however, it remains unclear whether reliable communications can be maintained as the frequency band becomes crowded with hundreds or thousands of wireless devices. Experiences with the technology over time will reveal how wireless technology will perform under these conditions.

Signal attenuation is a weakening of the RF signal. It is a function of distance and the properties of the material through which the signal travels. Signal attenuation can be compensated by using repeaters that receive signals, amplify them, and then retransmit them to increase the transmission range.

With steadily increasing threats from hackers to the networking infrastructure, the *security* needs of modern facility automation systems have grown. The vulnerability of wireless networks is of particular concern because no direct "hard" physical link is required to connect. Data encryption techniques have been successfully applied to wireless LAN systems to combat intrusion and provide security. These techniques encode data in a format that is not readable except by someone with the "key" to decode the data. Encryption, however, requires additional computational power on each wireless device, which runs counter to the general attempt to simplify technology in order to reduce cost. These challenges are currently being addressed by researchers, technology vendors and standards committees to provide technology solutions with the necessary technical performance that the market demands.

Costs

Costs of commercially available sensor network components in 2004 are shown in Table 27-3. Excluded from the table are single point-to-single point systems based on RF modems. The table shows that costs vary widely, and as with many technologies, costs are expected to decrease with time.

According to a recent market assessment of the wireless sensor networks, the cost of the radio frequency (RF) modules for sensors is projected to drop below $12 per unit in 2005 and to $4 per unit by 2010. [22] While these costs reflect only one portion of a wireless sensor device, the cost of the sensor element itself is also expected to decrease with technology advancements. For instance, digital integrated humidity and temperature sensors at high volumes are currently commercially available for less then $3 per sensor probe.[2] The general trend toward greater use of solid state technology in sensors is likely to lead to lower cost sensors for mass markets.

To date, end users are caught between the enthusiastic reports of the benefits that wireless sensing and control can provide and skepticism regarding whether the technology will operate reliably compared to the wired solution. While advancements in wireless local area networks (LAN) have paved the road for wireless technology market adoption, it also has made end users aware of the inherent reliability challenges of wireless transmission in buildings and facilities.

Types of Wireless Sensing Applications for Buildings

Applications of wireless sensing in buildings can be placed into two broad categories that significantly affect requirements on the underlying wireless technology and its performance: 1) applications for which at least some (and often most) of the devices must be self-powered (e.g., with an on-board battery) and 2) applications for which line power is available for each device. In this section, we describe experiences in field testing both types of applications. The first (Building Condition Monitoring) is illustrated with wireless sensors used to measure the air temperature in buildings at a much higher resolution than possible with the wired thermostats usually installed. In the second (Equipment Condition Monitoring), data for continually monitoring the performance of rooftop packaged HVAC units is collected using a wireless sensor data acquisition system.

Building Condition Monitoring

As discussed above, eliminating the need for wiring makes wireless sensor technology particularly appealing and well suited for monitoring space and equipment conditions in buildings of all sizes. Without the wires though, some additional care must be exercised in engineering and installing the wireless network to ensure sufficient robustness of communication.

Starting in 2002, Pacific Northwest National Laboratory (PNNL) conducted some of the first demonstrations to assess the performance of commercially available wireless sensor technology in real buildings and to compare the cost of the wireless solution with that of a conventional wired system. The first demonstration building was an office building with 70,000 square feet of open office floor space on three floors and a mechanical room in the basement. The building is a heavy steel-concrete structure constructed in the early 1960s. The second demonstration building represents a more modern and structurally lighter building style with individual offices totaling 200,000 square feet of floor space in a laboratory building completed in 1997.

[2]Quote by SenSolution, Newberry Park, CA, February 2004.

Table 27-3. Cost ranges of commercially available wireless sensor network components in 2004.

Network Component	*Cost Range ($)*
Sensor transmitter unit	$50 - $270
Repeaters	$250 - $1050
Receivers	$200 - $900
BAS Integration units	$450*

*Only one is currently commercially available in 2004 specifically for connecting a wireless sensor network to a building automation system.

Demonstration 1:
In-Building Central Plant Retrofit Application

The building is located in Richland, Washington. The HVAC system consists of a central chiller, boiler, and air distribution system with 100 variable-air-volume (VAV) boxes with reheat distributed in the ceiling throughout the building. A central energy management and control system (EMCS) controls the central plant and the lighting system. Zone temperature control is provided by means of stand-alone and non-programmable thermostats controlling individual VAV boxes. The centralized control system receives no zone temperature information and cannot control the VAV boxes. The long-term goal of PNNL facility management is to network the 100 VAV boxes into the central control infrastructure to improve controllability of the indoor environment. As an intermediate step toward this, a wireless temperature sensor network with 30 temperature sensors was installed to provide zone air temperature information to the EMCS. The wireless sensor network consists of a series of Inovonics wireless products including an integration module that interfaces the sensor network to a Johnson Controls N2 network bus.[3] The zone air temperatures are then used as input for a chilled-water reset algorithm designed to improve the energy efficiency of the centrifugal chiller under part-load conditions and reduce the building's peak demand.

[3]N2 bus is the Johnson Controls network protocol.

The Wireless Temperature Sensor Network

The wireless network consists of a commercially available wireless temperature sensor system from Inovonics Wireless Corporation. It encompasses 30 temperature transmitters, 3 repeaters, 1 receiver, and an integration module to interface the sensor network to a Johnson Controls EMCS N2 network. The layout of the wireless temperature network is shown in Figure 27-5.

The operating frequency of the wireless network is 902 to 928 MHz, which requires no license per FCC Part 15 Certification [23]. The technology employs spread spectrum frequency hopping techniques to enhance the robustness and reliability of the transmission. The transmitter has an open field range of 2500 feet and is battery-powered with a standard 123 size 3-volt $LiMnO_2$ battery with a nominal capacity of 1400 mAh. The battery life depends on the rate of transmission, which can be specified in the transmitter. The manufacturer estimates a battery life of up to 5 years with a 10-minute time between transmissions. The transmitter has an automatic battery test procedure with a 'low-battery' notification via the wireless network. This feature will alert the facil-

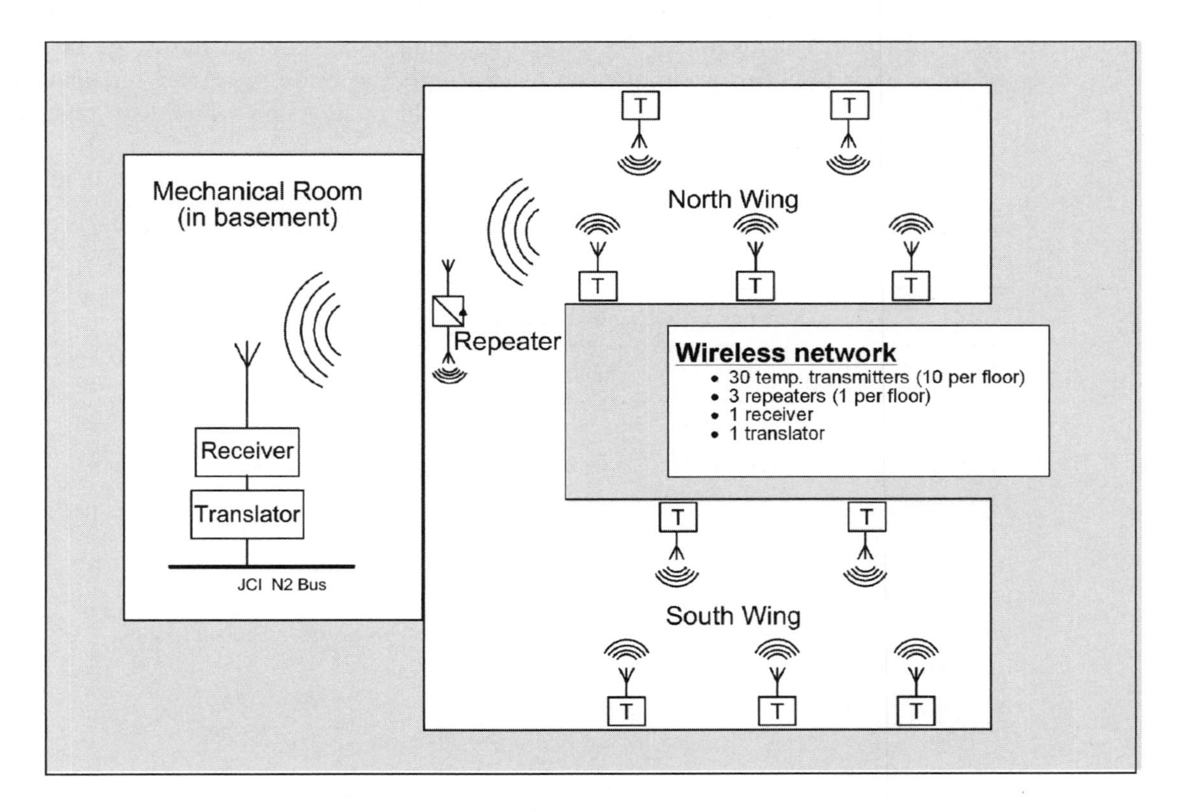

Figure 27-5.

ity operator through the EMCS that the useful life of the battery in a specific transmitter is approaching its end. The repeaters are powered from ordinary 120 volts alternating current (VAC) wall outlets and have a battery backup. Three repeaters were installed, one on each floor. Because the repeaters are line powered, the repeater operates at high power and provides up to 4 miles of open field range. The receiver and the translator are installed in the mechanical room in the basement. The translator connects the receiver with the Johnson EMCS system.

Design and Installation Considerations:

Installation of the wireless network requires a radio frequency (RF) survey to determine the proper locations for the repeaters to ensure that the received signal strength is sufficient for robust operation of the wireless network. RF surveying is an essential engineering task in the design of the wireless network topology. The signal attenuation in metal-rich indoor environments caused by metal bookshelves, filing cabinets, or structural elements such as metal studs or bundles of electric or communication wiring placed in the walls can pose a significant challenge to achieving robust wireless communication. Background RF noise emitted from cordless phones and other sources can also impair the transmission such that the receiver cannot distinguish noise from the real signal. There is no practical substitute for RF surveying a building because each building is unique with respect to its RF attenuation characteristics.

For the 70,000 square foot test building, an engineer performed the RF survey in about 4 hours while instructing others in survey procedures. This provided sufficient time for investigating several scenarios, whereby metal bookshelves were placed in the direct pathway between transmitters and a receiver. The result of the RF survey was a recommendation for three repeaters, one for each floor of the building (see Figure 27-5). An experienced surveyor should be able to perform this survey in about 2 hours, if not running special tests or instructing others.

The cost for the wireless system, including installation, was approximately $4000. See Table 27-4 for more details on the cost.

Operational Benefits

Operational improvements resulted from use of the wireless temperature sensor network. The wireless sensors enabled facility staff to respond to 'hot' and 'cold' complaints much more effectively. Because sensors can be easily moved and new ones readily introduced into the network, a spare sensor can be easily taped directly into a localized problem area for monitoring air temperature over a few hours or days. The much higher spatial resolution provided by the 30 zone air-temperature sensors enabled facility staff to identify individual VAV boxes that were causing uneven supply air. These malfunctioning boxes spread the range of air temperatures through the building. After repairing the faulty VAV boxes, the facility staff was able to raise the supply-

Table 27-4. Costs of wireless sensor systems in the two demonstration buildings.

	Cost per unit	Building 1		Building 2	
		Quantity	*Total*	*Quantity*	*Total*
Temperature sensors	$50	30	$1,500	120	$6,000
Repeaters	$250	3	$750	0	$0
Receivers	$200	1	$200	3	$600
Translators	$450	1	$450	3	$1,350
RF Surveying Labor	$80/hour	2 hours*	$160	2 hours	$160
Integrator configuration labor	$80/hour	4 hours	$320	8 hours	$640
Installation of Integrator labor	$80/hour	8 hours	$640	8 hours	$640
Total Cost			$4,020		$9,390
Cost per Sensor			$134		$78

*For an experienced surveyor.

air temperature by 2°F, alleviating the need for overcooling some zones in order to deliver enough cooling capacity through the faulty VAV boxes. Repair of VAV boxes improved the thermal comfort of occupants and eliminated the occasional use of space heaters during the early morning hours in both summer and winter months.

Energy Efficiency Benefits

The energy savings resulted directly from repairing several VAV box controllers, resetting the supply air temperature by 2°F during cooling periods, and reducing the use of small space heaters by occupants who were previously uncomfortably cool at times. In addition, a chilled-water reset strategy was implemented based on an average value of the 30 zone air temperatures. This allowed the chilled water set point to be reset between 45 and 55°F, the value depending upon the zone air temperature. Formerly, the chilled water temperature was fixed at 45°F. The average zone air temperature was used as an indicator for meeting the cooling loads. As a result the average coefficient of performance (COP) increased by about 7% due to the higher chilled water temperatures. The fan power for any given cooling load increased some but not nearly enough to offset the savings. The net result was an estimated cost savings of about $3500 over the cooling season (May through September). Additional energy savings were achieved by avoiding the use of space heaters and resetting the supply air temperature for a total estimated annual cost savings of about $6000. Based on the costs and estimated savings, the simple payback period for this wireless system was about 7 months.

Demonstration 2: Laboratory/Office Building

The second building, opened for occupancy in 1997, houses laboratories and offices. The gross floor space is about 200,000 square feet with three protruding office wings of about 49,000 square feet each. Only the office area was used for the demonstration. Each office wing has a separate air-handling unit and a variable-air-volume (VAV) ventilation system. Each VAV box supplies air to two offices controlled by a thermostat located in one of the two offices. The construction of the office area consists of metal studs with gypsum wall. The offices contain metal book shelves, and at a minimum, two computers with large screen monitors. The office space is relatively metal-rich, posing a challenge for wireless transmission from the sensors to the receivers.

Facility staff explored night setback options for the ventilation of the office space that would turn off the air-handling unit during the night hours after 6 p.m. The decision to implement such a strategy was suspended out of concern that those offices without a thermostat might be occupied during late hours and if so, that the air temperature in those offices could exceed the thermal comfort limits. Because of this concern, the ventilation system operated on a 7-day per week, 24-hour per day schedule. It was believed that if each office were equipped with one zone temperature sensor, the night setback could be implemented and then overridden if the zone temperature exceeded an upper threshold of 78 °F. A cursory cost estimate from a controls vendor for installing wired temperature sensors in the offices without thermostats yielded an installed cost per sensor of about $500, which exceeded acceptable costs.

After the initial positive experiences with wireless sensors in the other building, facility staff re-examined the viability of the ventilation night setback using a wireless solution and implemented wireless temperature sensors in early 2004. The same wireless temperature sensor network technology as deployed in Building 1 was used. Familiarity with the technology and experience gained from the first wireless demonstration greatly reduced the level of effort for a RF survey of the building and the wireless network setup.

The Wireless Temperature Sensor Network

Each office not previously equipped received a wireless temperature sensor. Forty wireless temperature sensors were deployed in each of the three office wings of Building 2, bringing the total to 120 sensors (see Figure 27-6). The temperature signals were read by three receivers, each located where the office wing meets the main hallway and connected via an integrator to the Johnson Controls network control module. The wireless network consisted of a total of 120 sensors, three receivers and three integrators. Facility staff tested the need for repeaters and found that with the use of one receiver for each wing, the communication was sufficiently robust. An alternative wireless network design was considered that would use one receiver in the middle wing and repeaters in each of the side wings to assure communication from the most distant transmitters in the exterior wings to the receiver. The integrator has a limit of 100 transmitters. Since this alternative used only one integrator, it could not support enough sensors for all the offices, and it therefore was rejected.

The temperature sensors are programmed to transmit a temperature measurement every 10 minutes. A sensor will transmit early when a temperature change is sensed that exceeds a pre-set limit. This is to enable

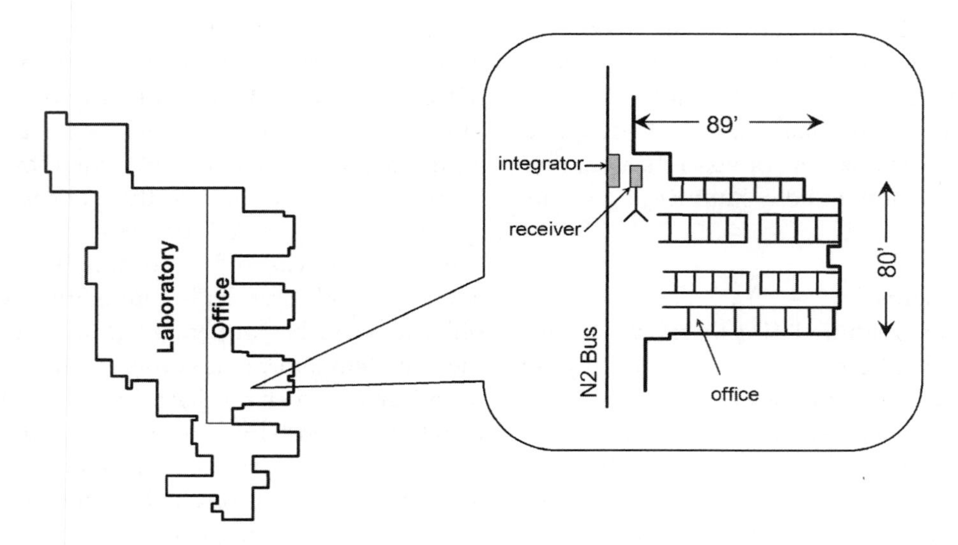

Figure 27-6.

detection of rapid temperature changes as quickly as possible.

Installation and Setup of Wireless Network

The installation costs for the wireless sensor network were minimal. They included a 2-hour RF survey, an initial setup of the integrator device to specify the number and ID numbers of the sensors, and the physical connection of the integrator and the Johnson Controls network control module. Configuration of the integrators was done in stages (each wing at a time) and the total time for setup of all 120 sensors was conservatively estimated afterward to be 8 hours. The integrator installation involves physically connecting the 24 VAC power supply provided in the Johnson Control network module and connecting the Johnson Controls N2 bus to the integrator using a 3-wire shielded cable. A short 4-wire cable connects the integrator and the receiver providing power supply and communication between the two devices. This work was performed by an instrument technician. The sensors were then attached to the office walls using double-sided tape.

Table 27-4 presents the cost components for the two demonstration buildings. The capital costs for the hardware represent the costs to PNNL and are representative of costs for a wholesaler. List prices would commonly be 75% to 100% higher than those shown.

Energy Savings

The supervisory control program was augmented to schedule night setback starting at 6 p.m. and suspending it if an office zone temperature exceeded a threshold temperature of 78°F during the cooling season or dropped below 55°F during the heating season, instead of maintaining the temperature continuously at a set point of 72°F. Initial estimates concluded that energy savings are largely attributable to the shut down of the supply and return fans and, to a lesser degree, to reduced thermal loss during the night as the temperature is allowed to float (rise in the cooling season and drop in the heating season). Trend-logs of run time using the new night setback strategy were used to estimate the electric energy savings. Preliminary estimates suggest that the night setback will achieve savings of approximately $5,000 annually. Verification of the savings is planned after one full year of night setback operation is completed. We attribute the cost savings to the wireless sensors because they enabled implementation of the ventilation night setback, something the facility operations staff was unwilling to do without the additional information provided by these sensors. Based on these energy savings, the wireless sensor system (which had an installed cost $9390) has a simple payback period of less than 2 years (22.5 months).

Other Impacts

Building operators also implemented a temperature averaging scheme for controlling the distribution system VAV boxes based on the average of the office temperatures in the zone served by each box. Although no energy savings resulted from this change, the building operators report that the number of occupant complaints about temperature has decreased significantly, saving building staff time and enabling them to devote that time to other improvements in operation.

Discussion on Costs for Demonstration Projects

Cost for the sensor and controls technology is a critical factor for the viability of any retrofit project or even in new construction. The wireless sensor solution was slightly more cost effective compared to an equivalent wired solution for Building 1. [9] For Building 2, the wireless sensor cost ($78/sensor) was significantly less than the estimated cost for the wired sensor retrofit (~$500/sensor). These examples tend to show that wireless sensor networks can compete with wired sensing on the basis of cost for retrofit projects. In both demonstration buildings, the wireless network infrastructure is sufficient to accommodate many more sensors at the cost of sensors alone. No additional infrastructure (repeaters, receivers, or translators) is needed to accommodate additional sensors. This enables facility staff to add sensors at the cost of the sensor itself plus a minimal setup time (a few minutes) for configuring the integrator.

Figure 27-7 shows cost curves for both demonstration buildings as a function of number of sensors installed. These curves are nearly identical. For 30 sensors, for example, the difference in cost is $22/sensor ($160-$138), and for 120 sensors, the difference is $6/sensor ($78-$72). This suggests that the cost of the wireless system per sensor might be nearly independent of the building itself but highly dependent on the number of sensors installed. The curves are actually dependent on the costs of the wireless components. The two curves shown are for the same brand and models of hardware. Average costs per sensor for systems built from components with substantially different costs will lie on other curves. Unless signal attenuation differs so significantly among buildings that it affects the number of sensors that can be served by each repeater or receiver, the curves for different buildings using the same wireless components should lie very close to one another. This observation proves useful in simplifying estimation of costs for wireless sensor systems.

The second insight from Figure 27-7 is that at high quantities of the sensors, the system cost on a per-sensor basis asymptotically approaches the cost of a sensor (in this case, $50/sensor). Therefore, for densely deployed sensors (high numbers of sensors per unit of building area), further cost reductions for wireless sensor networks must come from reducing the cost of the sensor modules (sensors plus transmitting radio) rather than decreasing the cost of infrastructure components—the receiver, repeaters and translators. In the short-term, however, while wireless sensing technology is just beginning to be deployed, sensor densities are likely to be relatively low, and as a result, all components will have a significant impact on cost. Users should realize, though, that once a wireless sensor network is installed in a building, additional sensors generally can be added to the network in the area covered by the network at the incremental cost of the additional sensors. The more uses the building staff can find for the wireless sensor network, the more cost-effective its installation becomes.

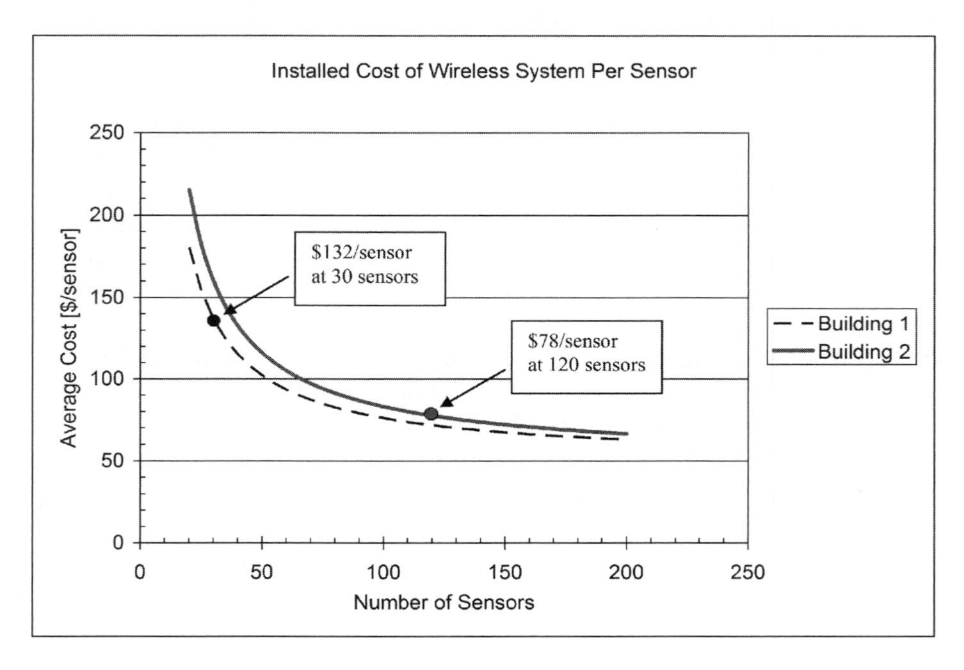

Figure 27-7.

Wireless Monitoring of Equipment Conditions

Heating, ventilating, and air-conditioning equipment is often run until it completely fails ("hard" faults), for example from a failed compressor, failed condenser fan, failed supply fan, or significant loss of refrigerant. Upon complete failure, the owner, operator, or building occupant calls a service company to repair the unit. Complete failure, though, is often preventable. Avoiding failures by properly maintaining the equipment would reduce repair costs, increase operating efficiency, extend equipment life, and ensure comfortable conditions, but this would require awareness of equipment condition and when the equipment needs servicing. Furthermore, several studies have noted that building systems operate under degraded conditions caused by insufficient refrigerant charge, broken dampers, stuck dampers, mis-calibrated and failed sensors, improperly implemented controls (e.g., incorrect schedules), electrical problems, and clogged heat exchangers [6, 24-29]. Many of these faults do not result in occupant discomfort because the system compensates by working harder (and expending more energy), and therefore, these faults are not reported nor are they corrected. Some of the faults require a service technician to correct, but many can be fixed with minor adjustments to controls or schedules; these faults are referred to as "soft" faults in this chapter.

With increasing pressure to reduce operation and maintenance (O&M) costs and with reduced staff in today's facilities, regular visual inspection by staff is out of the question. For small buildings without on-site operators, this was never a possibility. Service contracts providing scheduled but infrequent inspection and servicing alone are not likely the solution to this problem. Without a lower cost solution, package units are likely to continue to be maintained poorly and operated inefficiently.

Automated continuous condition monitoring provides a potential solution, but its cost is generally perceived as too high. Even installation of adequate sensors alone is usually viewed as too costly. Studies have shown, however, that automated monitoring and diagnostics implemented with wireless sensing and data acquisition can provide a cost effective solution [6, 8, 30]. In this section, we describe a wireless system for monitoring the condition and performance of packaged air conditioners and heat pumps, which are widely used on small commercial buildings.

Wireless System for Automated Fault Detection and Diagnostics

Functionally, packaged rooftop units can be divided into two primary systems: 1) air side and 2) refrigerant side. The air-side system consists of the indoor fan, the air side of the indoor coil, and the ventilation damper system (including its use for air-side economizing), while the refrigerant-side components include the compressor, the refrigerant side of indoor and outdoor heat exchangers, the condenser fan, the expansion valve, and the reversing valve (for heat pumps).

The choice of the fault detection and diagnostic (FDD) approach depends on the type of faults to be identified and the sensor measurements available. Many researchers have developed FDD algorithms to detect and diagnose faults in air-conditioning equipment. In this chapter we do not discuss the details of the diagnostic approaches, which can be found in other references [e.g., 31, 32, 33, 34] but instead describe the measurements needed, the faults that can be detected, and the system for collecting and processing the data. This system, which can be applied to both the air side and the refrigerant side of a heat pump is shown in Figure 27-8.

The minimum set of information required for monitoring the state of the air-side system with temperature-based economizer controls or no economizing includes: 1) outdoor-air dry-bulb temperature, 2) return-air dry-bulb temperature, 3) mixed-air dry-bulb temperature, 4) outdoor-air damper-position signal, 5) supply-fan status, and 6) heating/cooling mode. To identify whether the system is actually in heating or cooling mode, the status of the compressor (and the reversing valve for heat pumps) is required. If these measurements are available, economizer operations and ventilation requirements can be monitored and evaluated to verify their correct performance. If an enthalpy-based economizer control is used, then the outdoor-air relative humidity (or dew-point temperature) and return-air relative humidity (if differential enthalpy controls are used) are required in addition to the 6 measurements needed to monitor the performance of systems with temperature-based economizer controls. If supply-air temperature is also measured, additional faults relating to control of supply-air temperature can be detected and diagnosed. Details of the approach for detecting and diagnosing air-side faults are given in References 32 and 33.

Faults detected on the air side can be grouped into four categories: 1) inadequate ventilation, 2) energy waste, 3) temperature sensor and other miscellaneous problems including control problems, and 4) missing or out-of-range inputs. For more details on the faults that can be detected on the air-side, see References 6 and 32.

The minimum set of measurements required to monitor refrigerant-side performance include: 1) out-

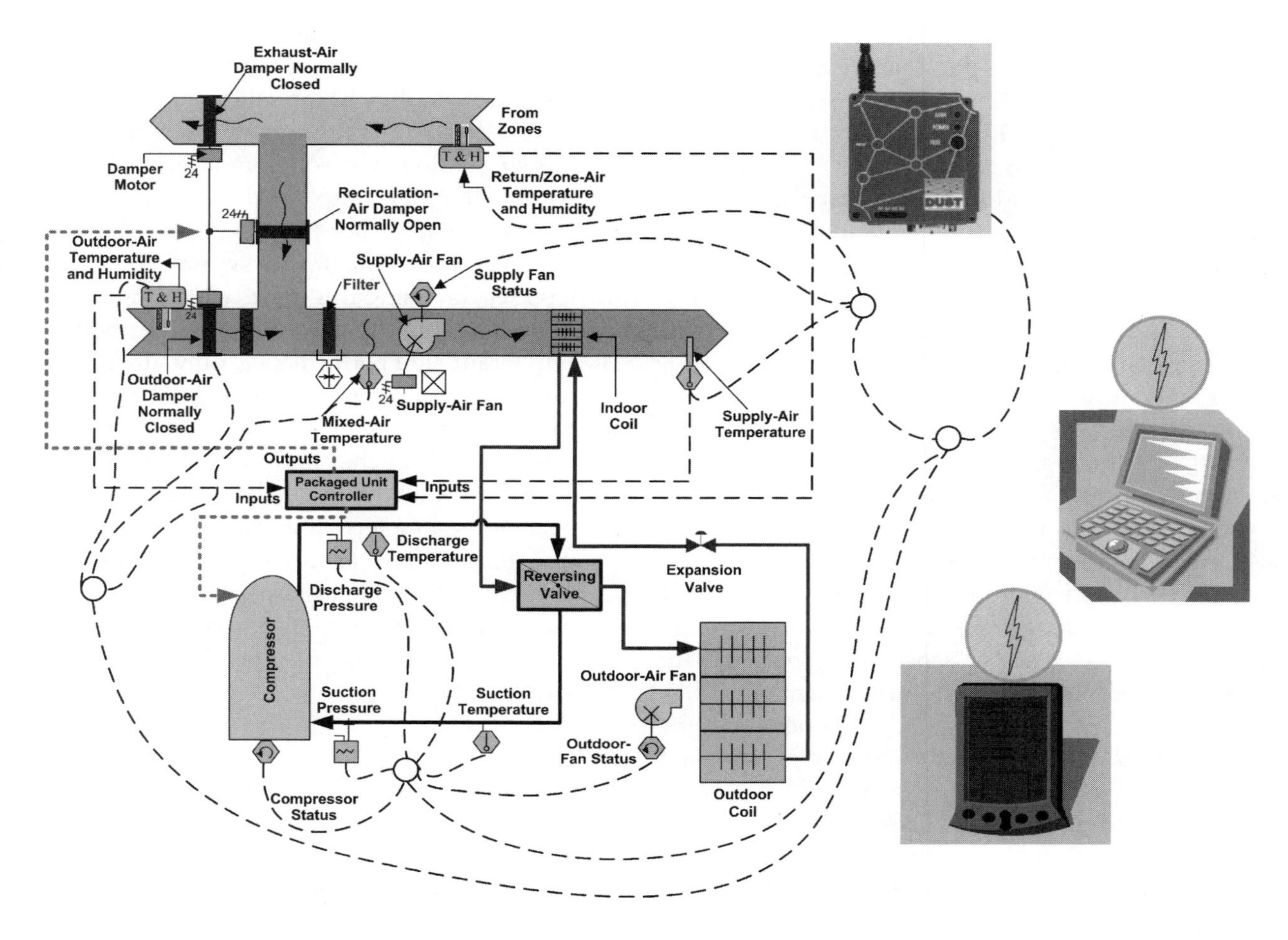

Figure 27-8.

door-air dry-bulb temperature, 2) liquid-line temperature (refrigerant temperature as it leaves the condenser), 3) liquid line pressure (as it leaves the condenser), 4) suction line temperature (refrigerant temperature at the compressor inlet), and 5) suction line pressure (refrigerant pressure at the compressor inlet). In addition to the five measured quantities, several derived quantities are used in monitoring the refrigerant-side performance: 1) liquid sub-cooling, which is estimated as a difference between the condensing temperature (calculated from liquid pressure and refrigerant properties) and the measured liquid line temperature, 2) the superheat, which is the difference between the evaporating temperature (calculated from the suction pressure and refrigerant properties) and the measured suction temperature, and 3) condensing temperature over ambient, which is the difference between the condensing temperature and the outdoor-air dry-bulb temperature.

The refrigerant-side faults that can be detected with these five measurements (two pressures and three temperatures) include: 1) evaporator (indoor coil) heat transfer problems, 2) compressor valve leakage (compressor fault), 3) condenser (outdoor coil) heat transfer problems, 4) improper supply-fan speed, 5) expansion device fault, 6) improper charge (too little or too much refrigerant), and 7) non-condensable substances in the refrigerant, such as air in the system. Details of diagnostics for the refrigerant side can be found in References 29 and 34.

Additional measurements that improve diagnostic capability and also increase the number of faults that can be detected include: 1) supply-air dry-bulb temperature, 2) mixed-air dry-bulb temperature, 3) mixed-air relative humidity (or dew point), 4) surface temperature of the condenser, 5) surface temperature of the evaporator, and 6) compressor power consumption. These measurements enable refinement of the diagnostics provided by the minimum set of sensors. In addition, cooling/heating capacity and efficiency degradation can be computed and tracked with these additional measurements. Although having pressure measurements makes diagnosis of the faults more reliable, pressure sensors are expensive compared to temperature and humidity sensors. The pressure sensors can be replaced with surface temperature sensors at the evaporator and condenser [31], and the temperature measurements can then be used as indicators of saturation temperature in the evaporator and condenser. Although the use of temperatures to estimate superheat and subcooling may lead to some error, their use will reduce the system cost and should still provide adequate diagnostics.

A wireless system providing data collection and diagnostics for only the air side of package HVAC units had a total installed cost per sensor of approximately half that of a wired system providing the same capabilities ($78 per point compared to $147-$193 per point for the wired system). [30] This wireless system uses one radio on each packaged unit, sending measurements from 4 thermocouples and a current switch used to measure the on/off status of the supply fan of the unit. Six units are monitored using one receiver unit, distributing its cost over the 30 sensors it serves. Power is tapped off the power supply for the packaged HVAC unit, so no batteries are used. Both the cost and benefits of a wireless condition monitoring system depend on several parameters, such as number of roof top units to be monitored, the size of the units, the size of building, the local climate, and potential savings from use of the monitoring and diagnostic tool. For a typical application on an 18,000 square foot 2-story building with six 7.5-ton units, the simple payback will be less than 3 years for most U.S. climates (assuming energy savings of 15% are achieved through better operation and maintenance) [30]. Paybacks will be shorter for larger units in more severe climates and longer for smaller units or units in milder climates.

Deploying Wireless Condition Monitoring

There are several ways to deploy wireless condition monitoring: 1) centralized data collection and processing at each building, 2) distributed or on-demand diagnostics and 3) centralized data collection and processing at a remote server—an application service provider model.

Method 1

The first approach is a conventional approach where all data from wireless monitors are collected by a wireless receiver that is directly connected to a computer. The data are continuously or periodically processed using automated software and results provided to the user through a simple and user-friendly graphical user interface. The authors have tested a prototype wireless monitoring and diagnostic system described in the previous section using this approach. Although the prototype system was capable of monitoring both the air- and refrigerant-side performance, only air-side diagnostics were tested. In this approach, data from packaged roof top units are automatically obtained at a user-specified sub-hourly frequency and averaged to create hourly values that are stored in a database. As new hourly values become available in the database, the diagnostic module automatically processes the data and produces diagnostic results that are also placed in the database. The user can then open the user interface at any time to see the latest diagnostic results, and can also browse historical results.

Method 2

Detailed diagnosis often requires historical data to isolate the primary cause of a fault or performance degradation; however, some faults can be detected with instantaneous or short-term measurements. The second deployment uses wireless data collected while servicing units along with simple rules-of-thumb to determine the condition of equipment. For example, data from rooftop packaged units might be accessed wirelessly by a technician visiting the site using a Personal Digital Assistant (PDA) with compatible wireless communication capabilities. This method can be effective in identifying incorrect refrigerant charge, blocked heat exchangers, and blocked refrigerant lines. The technician could get a report on each unit without even opening the units. Time at the site could then be devoted mostly to the units with faults or degraded performance. The authors have not yet demonstrated this approach, but a wired system with these sorts of diagnostic capabilities is available commercially. [35] The wired system requires physically connecting to previously installed sensors on each unit or connecting the instrument's sensors before use. Once the sensor system has been installed, the wireless approach is likely to save time and enable service technicians to identify units requiring the most attention immediately upon arriving at a site, improving the quality of service while decreasing cost.

Method 3

The third approach is similar to the first approach but all data are collected and sent to a central server possibly hosted by a third party—an application service provider (ASP). Ideally, the data are received at a central location at each building or site and then transferred to the central server. The transfer of data can be by phone line (wired or wireless) or through an existing wide area network (wired or wireless). The ASP provides access to software and data via subscriptions. For payment of a monthly subscription fee, users obtain access to software on the world wide web using nothing more than a web browser to access it. The software needs to be installed on only one computer, the web server, rather than on the individual work station of every user. To provide reliability, usually the software is installed by the ASP on several redundant servers to provide backup in case a computer fails. Many users are then able to access a small number of installed copies of the software. User files are also maintained on the ASP's servers and backed up in a similar manner. The wireless monitoring equipment can be purchased by the owner or can be leased from the ASP for a subscription fee. This type of approach is still in its infancy. The authors will soon be testing this delivery approach.

The three approaches may also be combined to provide information on equipment condition more flexibly. For example, once the wireless sensing and data acquisition infrastructure is installed on the equipment at a building, it can be connected for remote monitoring by building operations staff/management or at a service provider's office and also be accessed by service technicians when they visit the site. Availability of information on equipment condition and performance would provide the basis for a conditioned-based maintenance program that would help ensure that equipment gets serviced and repaired when needed rather than more frequently than needed or less frequently (which is all too common, especially for package equipment).

Long-distance Data Transmission

So far, this chapter has focused on short-range wireless data acquisition at a building for monitoring indoor conditions and equipment conditions and performance. Although not widely used yet, wireless communications have also proven effective in transmitting data between individual building sites and central monitoring systems. Deployment of this model by an ASP was discussed briefly in the preceding section. Central monitoring using wireless communication of data, however, can be implemented by any organization having geographically distributed facilities and the willingness to maintain the computer infrastructure necessary to implement and maintain such as system. This requires appropriate security and backup to ensure the system meets the necessary performance and reliability demands.

An example system is shown in Figure 27-9. Data collected from electric meters and sensors on equipment are transmitted by a wireless pager network to the operations center of a wireless carrier. Data are then sent through the Internet to the operations center of the ASP providing the service. There, the data are stored securely in databases and processed by the tools provided by the ASP. Customers can then securely access the processed results from their buildings from any computer with a web browser. The monitoring equipment for collecting and transmitting the data is provided by the ASP.

Designing and Installing a Wireless System Today: Practical Considerations

Laying out a wireless network indoors is probably as much art as it is science. Every building is unique, if not in its construction and floor plan, at least in the type and layout of its furnishings. Predicting wireless signal strength throughout a building would require characterizing the structure, its layout, and the furnishings and equipment in it and using that information to model RF signal propagation. No tools are available today for accurately doing this. Furthermore, when space use changes or furnishings are moved or change over time, radio signals encounter new obstacles in new positions. Despite these difficulties, there are several practical considerations for the design of a wireless network that are helpful for generating bills of materials and budget estimates and laying out wireless sensing networks.

Determining the Receiver Location

The decision with perhaps the most impact on the design of a wireless sensor network for in-building monitoring is determining the number and locations of the receivers. A stand-alone wireless network (not connected to a wired control network) may have some flexibility in choosing the location of the receiver. The best location from a communications perspective is one that is open and provides the best line-of-sight pathways between the most wireless sensors and the receiver. Convenient connection to a computer where data will be processed and viewed is another important consideration. These factors must be balanced. If the design requires integration of the wireless sensor network with an existing building automation system (BAS) infrastruc-

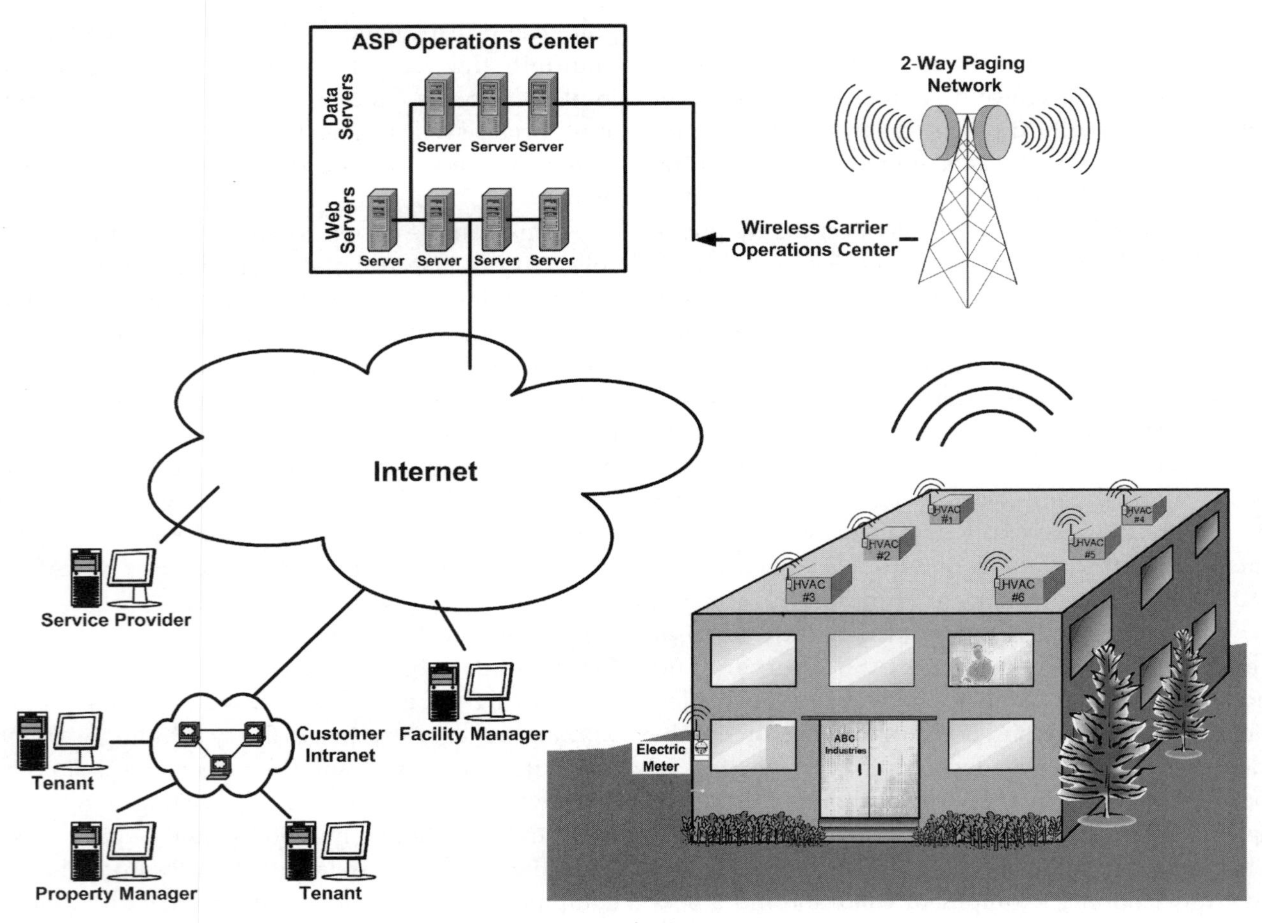

Figure 27-9.

ture, then receivers must be located near points of connection to the BAS. Locations are constrained somewhat in this case, but there are typically still many options. Frequently, a convenient integration point is a control panel that provides easy access to the communication cables as well as electricity to power the receiver and integration devices. In commercial buildings, the BAS network wires are often laid in cabling conduits (open or closed) above the ceiling panel and are relatively easily accessible. Often the lack of electric power in the ceiling space, however, renders this location less convenient than a control panel.

Signal Attenuation and Range of Transmitters

Estimating the range of the transmitting devices is important from a cost point of view. If the transmission range from a transmitting device to the ultimate end-node cannot be accomplished with a single transmission path, additional hardware is required for signal amplification adding to the total cost of the installation. The discussion below is designed to provide a general overview of this topic that may lead to generating some rough estimates of how many repeater or amplification devices an installation may need. It does not replace a thorough RF survey of a facility to determine the exact number and locations of receivers, repeaters, or intermediate nodes necessary to assure robust communication.

The range of a transmitter depends on the three key variables: 1) attenuation because of distance between wireless devices, 2) attenuation caused by the signals traveling through construction material along the signal pathways, and 3) overall electromagnetic noise levels in the facility.

The attenuation of the signal strength due to distance between the transmitter and receiver (free path loss) is governed by the relation of the electromagnetic energy per unit area of the transmitter to the distance of the receiving surface (see Figure 27-10). The energy per unit area at a distance d from the transmitter decreases proportionately to $1/d^2$. Therefore, for every doubling of the distance d, the energy density or signal strength received decreases to one-fourth of its previous strength.

This relationship accounts only for the dispersion of the signal across a larger area with distance from the source. In practice, other factors affect the strength of the signal received, even for an unobstructed path, including absorption by moisture in the air, absorption by the ground, partial signal cancellation by waves reflected by the ground, and other reflections. In general, this causes the signal strength at a distance d from the transmitter to decrease in practice in proportion to $1/d^m$, where $2 < m < 4$. [11]

The following example illustrates signal attenuation with distance from the transmitter in free air for a 900 MHz transmitter. This example shows how simple relations can be used to obtain an estimate of potential transmission range.

For this example, assume that the signal strength of a small transmitter has been measured to be 100 mW/cm^2 at a distance of 5 cm from the transmitter's antenna. The transmission path efficiency or transmission loss is customarily expressed in decibels, a logarithmic measure of a power ratio. It is defined as

$$dB = 10 \log_{10} (p_1/p_0),$$

where p_1 is the power density in W/cm^2 and p_0 is a reference power density (i.e., the power density at a reference point) in W/cm^2.

We choose the power density measured at 5 cm distance from the transmitter's antenna as the reference power density p_0. Table 27-5 shows the attenuation of the emitted signal as a function of distance from the transmitter for a signal traveling through air only. For every doubling of the distance, the signal strength decreases by 6 dB or, stated alternatively, the attenuation increases by 6 dB.

Further, assume that the ambient noise is measured to be -75 dB. For a signal to be detectable above the surrounding noise level, the strength of the signal should be at least 10 dB above the noise level (i.e., signal margin of 10 dB or greater is recommended) [36]. Using the results of Table 27-5, we can determine the transmission range of the wireless system in our example that meets the 10 dB signal margin requirements to be 80 meters, since –75 dB +10 dB = -65 dB, which is less than -64 dB at 80 meters.

Next, we extend this example to consider attenuation inside buildings. Suppose that the receiver is placed in a mechanical room of a building and that the signal from the furthest transmitter must go through two brick walls and two layers of drywall. Using signal attenuation estimates from Table 27-6, the combined attenuation of the brick and drywall is 14.6 dB [2×0.3 (for the 1/2″ drywall) + 2×7 (for 10.5″ brick wall) = 14.6], for practical purposes say 15 dB. Adding the material-related attenuation of 15 dB to the –65 dB signal strength requirement yields –50 dB as the new indoor signal strength requirement for the free air transmission segment. Using Table 27-5, we conclude that the transmission range is between 10 and 20 meters, only 1/8 to 1/4 of the range in open air. This example illustrates how significantly radio signals can be attenuated indoors compared to outdoors simply by the structure itself. Furniture further adds to attenuation and complicates prediction of the signal strength as a function of location in buildings. Therefore, to characterize indoor environments with respect to RF signal propagation, empirical surveying is recommended.

RF Surveying

The purpose of an RF facility survey is to determine the actual attenuation of RF signal strength

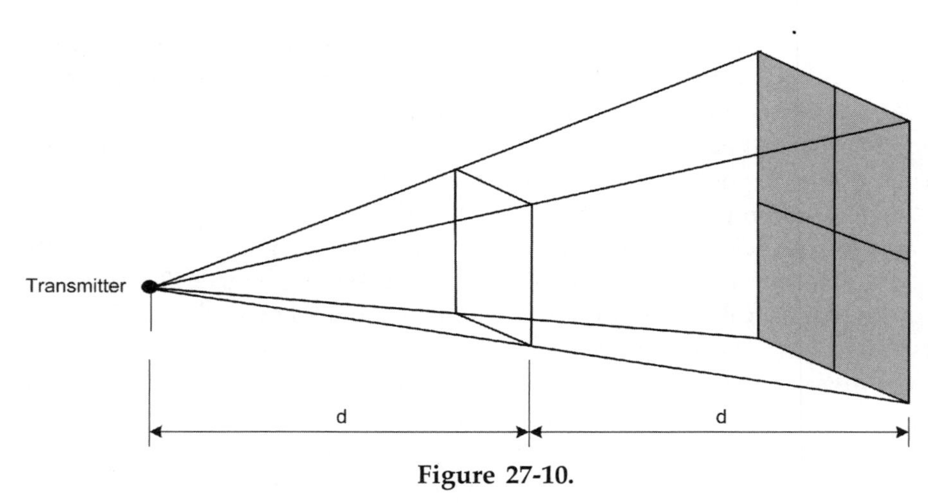

Figure 27-10.

Table 27-5. Attenuation of an RF signal in free air as a function of distance.

Distance in m (ft)	0.05 (0.2)	1 (3)	2.5 8)	5 (16)	10 (33)	20 (66)	40 (131)	80 (262)
Signal strength in dB	0	-26	-34	-40	-46	-52	-58	-64
Attenuation along line-of-sight in dB	0	26	34	40	46	52	58	64

Table 27-6. Signal attenuation for selected building materials for the 902-928 MHz band. [38]

Construction Material	*Attenuation (dB)*
1/4" Drywall	0.2
1/2" Drywall	0.3
3/4" Drywall	0.5
1/4" Plywood (dry)	0.5
1/2" Plywood (dry)	0.6
1/4" Plywood (wet)	1.7
1/2" Plywood (wet)	2
1/4" Glass	0.8
1/2" Glass	2
3/4" Glass	3
1.5" Lumber	3
3" Lumber	3
6.75" Lumber	6
3.5" Brick	4
10.5" Brick	7
8" Reinforced concrete with 1% ReBar mesh	27

throughout the facility. This information, together with knowledge of the locations at which sensors will be positioned, is used to lay out the wireless network. The layout will include the number of repeaters and receivers in the network and their locations. For instance, for a multi-story facility there may be good reasons for placing one receiver on each floor, provided the data are needed only on each floor (e.g., one user per floor for that floor) or there is another means to communicate the data between floors (such as a BAS connection on each floor). If the data are needed at a computer located on a specific floor (such as a control room in the basement), a repeater might be used on each floor to transmit signals to the location of a central receiver located close to where the data are needed. If communication between receivers on different floors is not sufficient, there may be opportunities to route signals inside an elevator shaft, stair case, or on the exterior of the building. The most cost-effective solution is in most cases determined by the difference in cost between repeaters and receivers and the cost of interfacing the receivers to pre-existing wired networks. The layout with the lowest total cost that provides sufficient (reliable) communication is generally optimal.

Most vendors of wireless sensor networks offer RF survey kits that are specific for the vendors' technologies. These kits consist of a transmitter and a receiver. The transmitter is often a modified sensor transmitter that is programmed to transmit at frequent time intervals. The receiver generally is connected to (or part of) an indicator of signal strength, together making a wireless signal-strength meter. These meters may simply give an indication whether the signal strength is adequate or provide numerical values of signal strength and background noise levels from which the adequacy of signal strength can be determined.

Before the RF facility survey is performed, potential receiver and sensor locations need to be known. The survey is then performed by placing the transmitter in anticipated locations for the receivers, then moving the signal-strength meter to locations where sensors will be positioned and taking measurements. By taking measurements throughout the facility, the limits of transmission range where the signal can no longer be detected (or is not of sufficient strength) can be identified. Repeaters will then need to be located in the layout within the transmission range to extend the range further.

The RF surveying is generally done by the wireless technology vendor or installer. Depending on the diversity of noise level in the facility and the complexity of its interior layout, an RF survey can be performed for office buildings with a floor space of 100,000 square feet in 2 to 4 hours.

Although RF surveys are critical for successfully designing and installing a wireless network that uses a star topology, systems using a mesh network topology with sufficient sensor density will ultimately not require RF surveys for installation. With sufficient densities of sensors (i.e., relatively short distances between sensors and multiple neighboring sensors within the communication range of each node), these networks will be self-configuring with the multiple potential transmission paths ensuring reliable, consistent communications. In the near term, care should be exercised in assuming that mesh networks will perform reliably for every application, especially in cases where high sensor density is not anticipated. For low sensor density installations, communication over long distances may require a higher-power repeater to connect a local mesh network to the point where the data are needed or a daisy-chain of nodes to communicate. In these cases, the advantages of mesh networking are lost in the region where individual devices carry all data communicated and those devices become potential single points of failure for the entire mesh that they connect to the point of data use.

Other Practical Considerations

Several other factors should be considered in deciding to use wireless sensing in buildings. Peter Stein [38] provides a nice summary of practical considerations for monitoring with wireless sensor networks. In addition to communication range, some of the key considerations that need to be assessed when selecting a wireless sensing network are:

- component prices
- availability of support
- compatibility with different types of sensors with different outputs
- battery backup for line powered devices
- low-battery indicators for battery-powered devices
- on-board memory
- proper packaging and technical specifications for the environment where devices will be located
- battery life and factors that affect it
- frequency of data collection and its relationship to battery life (where applicable)
- need for and availability of integration boxes or gateways to connect wireless sensor networks to BASs, other local area networks, or the Internet
- availability of software for viewing or processing the data for the intended purpose
- compatibility among products from different vendors—this is rare today but will improve with manufacturer adoption of new standards [e.g. IEEE 802.15.4 [19, 20] with Zigbee [21]]
- tools for configuring, commissioning, repairing, and adding nodes to the sensor network
- software to monitor network performance

Most important is ensuring the selected wireless network meets the requirements of the intended application. All factors need to be considered and assessed with respect to satisfying the requirements of the application and the specific facility. Each installation is unique.

The Future of Wireless Sensing in Buildings

The steadily growing number of technology companies offering products and services for monitoring and control applications fuels the expectation that the sub-$10 wireless sensor is likely to be available in the near future [22]. When we reach that point of technological advancement, the cost of the battery may then be the single largest cost item of a wireless module. Even the battery may be replaceable by ambient power scavenging devices that obviate the need for a battery as a power source. A self-powered sensor device creates fundamentally new measurement applications, unthinkable with battery- or line-powered technology. For instance, sensors could be fully embedded in building materials, such as structural members or wall components. They can measure properties in the host material that currently cannot be accessed easily or continuously by external measurement probes. In the energy efficiency domain, new diagnostic methods could be envisioned that use embedded sensors for early fault detection and diagnostics to prevent equipment failure and degradation of energy efficiency. Researchers are exploring different ambient sources for the extraction of electric power. Mechanical vibration emanating from rotary energy conversion equipment, such as internal combustion engines, pumps, compressors, and fans can be converted into electric power by induction driving a magnetic element inside a coil. Alternatively, piezo-electric materials can generate an electric potential when mechanically strained. Present research and technology development focuses on maximizing the energy extraction of mechanical energy by adaptive techniques that sense and adjust to a given vibration frequency and amplitude to maximize power extraction. [39] Thermo-electrical power generators utilize the Seabeck Effect, commonly used in thermocouple probes for temperature measurements. A temperature differential of a few degrees Celsius can, in cleverly designed probes, generate power in the micro-Watt range. [40] The small power generation from ambient power devices can be used to recharge a battery or stored in a super-capacitor to operate the wireless sensors when communication is required. Recent prototypes of ambient energy scavenging devices that generate sufficient electric power to operate a wireless sensor show promise for these revolutionary technologies to soon be commercially available. [41]

With an optimistic outlook on cost projections of wireless sensors and revolutionary self-powering devices, what are the likely impacts and opportunities of this technology for the building sector in general, and for energy efficiency improvement opportunities in buildings in particular? While the scenario of ubiquitous sensing by miniaturizing sensors to the size of paint pigments that can be painted on a wall may be in the realm of science fiction, there are real near-term opportunities for low-cost wireless devices providing value in the building sector now. Some of the applications where wireless sensing should have impact soon include:

HVAC fault detection, diagnostics, and control

- Higher spatial resolution of measurements of zone temperature and humidity to help assure better thermal comfort. Causes of localized hot and cold conditions can be detected and diagnosed. Each office or cubicle would be equipped with one or more temperature/humidity sensors.

- Expand terminal box control from a common single thermostat control point to multiple sensors located throughout the zone served. An average temperature that is more representative of the thermal needs could be used to control terminal boxes.

- Retrofit of terminal boxes for condition and performance monitoring. Because there are hundreds, sometimes thousands, of VAV boxes in commercial buildings, they receive very little inspection or maintenance except when suspected of causing a comfort problem. Wireless sensors placed on these units could be used to measure airflow rates, temperatures, and equipment status to enable central monitoring, performance-based alarms, and diagnostics that would support condition-based maintenance of this largely neglected equipment.

- Additional outdoor-air temperature sensors for improved economizer control. Ideally, place one or more air-temperature sensors near air intakes to air-handlers to minimize bias from radiative heat transfer and sensor failure.

- Equip packaged rooftop HVAC systems with sensors to continuously and automatically monitor performance.

Lighting control and monitoring

- In open-space office buildings, retrofit lighting controls for individual and localized control from the occupants' desks.

- Retrofit reconfigurable lighting systems with individually addressable dimmable ballasts.

- Retrofit light sensors at the work task location to turn off or dim lighting fixtures where daylight is adequate.

- Retrofit wireless occupancy sensors and control points on lighting panels to turn off lights during unoccupied periods.

Security and access control

- Motion sensors and door sensors for physical security systems.

- Environmental monitoring and physical security for IT systems and server rooms.

- Access control systems for retrofits and new construction.

Demand Responsiveness

- Retrofit wireless power meters for electricity end-use metering

- Retrofit wireless power meters and control on major loads to modulate or switch off power during grid emergencies or during periods of high power prices.

- Retrofit large appliances with wireless devices for receiving price signals or load control instructions from the power grid to respond to stress on the power grid.

CONCLUSION

Application of wireless communication for monitoring the conditions inside buildings and the performance of building equipment is feasible today. For retrofits, wireless sensing can be installed in many situations at lower cost than an equivalent wired system. Savings on energy, extended equipment life, lower total maintenance cost over equipment lifetimes, and maintenance of better conditions for occupants can even justify sensors using wireless communication where wired sensing has not been used previously.

Very few wireless products for building monitoring are available on the market today, but the technology is poised for rapid introduction soon. Generic hardware is available that can be adapted to building applications. Care should be exercised by those considering wireless technology for these purposes to ensure that wireless communication best matches the application requirements and that the specific system selected is the one best meeting needs. Every application is unique, and wireless technologies should be evaluated with respect to each project's unique requirements. Furthermore, special steps such as RF surveys of facilities in which wireless sensing is planned should be used to plan the proper layout of equipment to ensure reliable communication over the system life.

Data on the condition and performance of equipment can be used to implement condition-based maintenance for building equipment that may previously have been largely run until failure. Information collected from wireless sensor systems installed where no sensing previously existed can be used to improve control by adjusting set points, using sets of measurements throughout a zone rather than measurements at a single point in a zone as inputs for control, and diagnosing hot and cold spots. Control directly from wireless sensors is also possible but less developed and tested than monitoring applications, but today's wireless networks are not suitable for control requiring rapid response on the order of seconds or less. The network and its adaptation must be matched to the needs of the application.

Although wireless sensing can bring benefits not previously possible with wired systems, it is not a panacea for all monitoring and control applications in buildings. As pointed out recently by an author from a major building controls company:

> Part of the answer, at least for the near term, is that wireless networks can provide tangible benefits to engineers, consultants and clients alike. However, as we have witnessed with so many other fast-growing technologies coming of age, only time will tell if the technology will become an accepted and vital part of the HVAC industry. For now, all-wireless control of a facility is neither sensible nor realistic. Conversely, wireless technology cannot be ignored. Although every facility is unique with its own specific requirements, the most sensible building control solution could well be a balanced blend of wired and wireless devices that are strategically integrated for optimum performance and cost savings. [42]

Wireless technology for monitoring and control in buildings is emerging and can be used cost effectively today with care. In the next few years, new technology and products will make application of wireless easier and more reliable. Experience will build widespread support for this technology. Applications of sensors in buildings not fathomable yesterday will emerge based on wireless communication, bringing cost, comfort, safety, health, and productivity benefits.

ACKNOWLEDGMENTS

Work reported in this chapter was supported in part by the U.S Department of Energy Building Technologies Program of the Office of Energy Efficiency and Renewable Energy.

References

[1] *IEEE 802.11-1997.* Standard for Information Technology, Telecommunications and Information Exchange Between Systems, Local and Metropolitan Area Networks, Specific Requirements, Part 11: Wireless LAN Medium Access Control (MAC) and Physical Layer (PHY) Specifications: Higher Speed Physical Layer Extension in the 2.4 GHz Band. *Institute of Electrical and Electronic Engineers, New York.*

[2] *IEEE 802.11-1999.* Supplement to Standard for Information Technology—Telecommunications and Information Exchange Between Systems—Local and Metropolitan Area Networks, Specific

Requirements—Part 11: Wireless LAN Medium Access Control (MAC) and Physical Layer (PHY) Specifications: Higher Speed Physical Layer Extension in the 2.4 GHz Band. *Institute of Electrical and Electronic Engineers, New York.*

[3] *RFiD Journal. 2003. "The 5c RFID Tag."* RFiD Journal *1(1):30-34 (January 2004).*

[4] S. Roundy, P.K. Wright and J.M. Rabaey. 2003. *Energy Scavenging for Wireless Sensor Networks with Special Focus on Vibrations*. Kluwer Academic Publishers, Boston.

[5] Ferro Solutions. 2004. Energy Harvesters and Sensors (brochure). Ferro Solutions, Cambridge, Massachusetts. Available on the World Wide Web at *http://www.ferrosi.com/files/FS_product_sheet_wint04.pdf*.

[16] Katipamula, S., M.R. Brambley, N.N. Bauman, and R.G. Pratt. 2003. "Enhancing Building Operations through Automated Diagnostics: Field Test Results." In *Proceedings of the Third International Conference for Enhanced Building Operations*. Texas A&M University, College Station, Texas.

[7] Jacobs, P. 2003. *Small HVAC Problems and Potential Savings Reports*. Technical Report P500-03-082-A-25. California Energy Commission, Sacramento, California.

[8] Kintner-Meyer M., M.R. Brambley, T.A. Carlon, and N.N. Bauman. 2002. "Wireless Sensors: Technology and Cost-Savings for Commercial Buildings." In *Teaming for Efficiency: Proceedings, 2002 ACEEE Summer Study on Energy Efficiency in Buildings: Aug. 18-23, 2002, Vol. 7; Information and Electronic Technologies; Promises and Pitfalls*, pp. 7.121-7.134. American Council for Energy Efficient Economy, Washington, D.C.

[9] Kintner-Meyer M., and M.R. Brambley. 2002. "Pros & Cons of Wireless." *ASHRAE Journal* 44(11):54-61.

[10] Kintner-Meyer, M. and R. Conant. 2004. "Opportunities of Wireless Sensors and Controls for Building Operation." 2004 ACEEE Summer Study on Energy Efficiency in Buildings. American Council for an Energy-Efficient Economy. Washington, D.C. 2004.

[11] Su, W., O.B. Akun, and E. Cayirici. 2004. "Communication Protocols for Sensor Networks." In *Wireless Sensor Networks*, eds. C. S. Raghavendra, K.M. Sivalingam and T. Znati, pp. 21-50. Kluwer Academic Publishers, Boston, Massachusetts.

[12] Raghunathan, V., C. Schurgers, S. Park, and M.B. Srivastava. 2004. "Energy Efficient Design of Wireless Sensor Nodes." In *Wireless Sensor Networks*, eds. C.S. Raghavendra, K.M. Sivalingam and T. Znati, pp. 51-69. Kluwer Academic Publishers, Boston, Massachusetts.

[13] Ye, W. and J. Heideman. 2004. "Medium Access Control in Wireless Sensor Networks." In *Wireless Sensor Networks*, eds. C.S. Raghavendra, K.M. Sivalingam and T. Znati, pp. 73-91. Kluwer Academic Publishers, Boston, Massachusetts.

[14] FCC. 2004. The FCC's On-Line Table of Frequency Allocations. 47 C.F.R. § 2.106. Revised August 1, 2004. Federal Communications Commission. Office of Engineering and Technology Policy and Rules Division, Washington, D.C. Available on the world wide web at *http://www.fcc.gov/oet/spectrum/table/fcctable.pdf*.

[15] Nyquist, H. 1928. "Certain topics in telegraph transmission theory," *Trans. AIEE*, vol. 47, pp. 617-644, April 1928.

[16] Shannon, C.E. 1949. "Communication in the presence of noise," *Proc. Institute of Radio Engineers*, vol. 37, no. 1, pp. 10-21, January 1949.

[17] Pozar, D. 1997. *Microwave Engineering*, 2nd edition, Wiley, New York.

[18] Bluetooth SIG Inc. 2001. *Specification of the Bluetooth System-Core*. Version 1.1, February 22, 2001.

[19] IEEE 802.15.4. 2003. *Part 15.4: Wireless Medium Access Control (MAC) and Physical Layer (PHY) Specifications for Low-Rate Wireless Personal Area Networks (LR-WPANs)*. The Institute of Electrical and Electronics Engineers, Inc., New York.

[20] José A. Gutierrez, Ed Callaway and Raymond Barrett, eds. 2003. *Low-Rate Wireless Personal Area Networks. Enabling Wireless Sensors with IEEE 802.15.4*. ISBN 0-7381-3557-7; Product No.: SP1131-TBR. The Institute of Electrical and Electronics Engineers, Inc., New York.

[21] Kinney, P. 2003. "ZigBee Technology: Wireless Control that Simply Works." ZigBee Alliance, Inc. Available on the world wide web at *http://www.zigbee.org/resources/documents/ZigBee_Technology_Sept2003.doc*.

[22] Chi, C. and M. Hatler. 2004. "Wireless Sensor Network. Mass Market Opportunities." ON World, San Diego, California. Available on-line at: *http://www.onworld.com*. February 2004.

[23] FCC Part 15, 1998. *Part 15 Radio Frequency*

Devices. Code of Federal Regulation 47 CFR Ch. I (10–1–98 Edition), Federal Communications Commission, Washington, D.C.

[24] Ardehali, M.M. and T.F. Smith. 2002. *Literature Review to Identify Existing Case Studies of Controls-Related Energy-Inefficiencies in Buildings.* Technical Report: ME-TFS-01-007. Department of Mechanical and Industrial Engineering, The University of Iowa, Iowa City, Iowa.

[25] Ardehali, M.M., T.F. Smith, J.M. House, and C.J. Klaassen. 2003. "Building Energy Use and Control Problems: An Assessment of Case Studies." *ASHRAE Transactions*, Vol. 109, Pt. 2.

[26] Lunneberg, T. 1999. "When Good Economizers Go Bad." E Source Report ER-99-14, E Source, Boulder, Colorado.

[27] Portland Energy Conservation Inc. (PECI). 1997. *Commissioning for Better Buildings in Oregon*. Oregon Office of Energy, Salem, Oregon.

[28] Stouppe, D.E., and Y.S., Lau. 1989. "Air Conditioning and Refrigeration Equipment Failures." *National Engineer* 93(9): 14-17.

[29] Breuker, M.S., and J.E. Braun. 1998. "Common faults and their impacts for rooftop air conditioners." *International Journal of Heating, Ventilating, Air Conditioning and Refrigerating Research* 4(3): 303-318.

[30] Katipamula, S., and M.R. Brambley. 2004. "Wireless Condition Monitoring and Maintenance for Rooftop Packaged Heating, Ventilating and Air-Conditioning." *Proceedings, 2004 ACEEE Summer Study on Energy Efficiency in Buildings: Aug. 22-27, 2004*. American Council for Energy Efficient Economy, Washington, D.C.

[31] Breuker, M.S. and J.E. Braun. 1998. "Evaluating the Performance of a Fault Detection and Diagnostic System for Vapor Compression Equipment." International Journal of Heating, Ventilating, Air Conditioning and Refrigerating Research 4(4):401-425.

[32] Katipamula S., M.R. Brambley, and L. Luskay. 2003. "Automated Proactive Commissioning of Air-Handling Units." Report PNWD-3272, Battelle Pacific Northwest Division, Richland, WA. Also published by the Air-Conditioning & Refrigeration Technology Institute, Washington, DC. Available on the world wide web at *www.arti-21cr.org/research/completed/finalreports/30040-final.pdf.*

[33] Katipamula S., M.R. Brambley, and L. Luskay. 2003b. "Automated Proactive Techniques for Commissioning Air-Handling Units." Journal of Solar Energy Engineering—Transactions of the ASME 125(3):282-291.

[34] Rossi, T.M. and J.E. Braun. 1997. "A Statistical, Rule-Based Fault Detection and Diagnostic Method for Vapor Compression Air Conditioners." *International Journal of Heating, Ventilation, Air Conditioning and Refrigeration Research* 3(1):19-37.

[35] Honeywell. 2003. "The HVAC Service Assistant." Honeywell Home and Building Controls, Golden Valley, Minnesota. Available on the world wide web at *http://customer.honeywell.com/buildings/CBWPServiceAssistant.asp*

[36] Inovonics. 1997. *FA116 Executive Programmer, User Manual for FA416, FA426 and FA464 Frequency AgileTM Receivers*. Inovonics Corporation, Louisville, Colorado.

[37] Stone, William. 1997. *Electromagnetic Signal Attenuation in Construction Materials.* NIST Construction Automation Program Report No. 3. NISTIR 6055. Building and Fire Research Laboratory. National Institute of Standards and Technology, Gaithersburg, Maryland.

[38] Stein, Peter. 2004. "Practical Considerations for Environmental Monitoring with Wireless Sensor Networks." *Remote Site & Equipment Management,* June/July 2004 (*www.remotemagazine.com*).

[39] Roundy, S., P.K. Wright, and J.M. Rabaey. 2004. *Energy Scavenging for Wireless Sensor Networks with Special Focus on Vibrations*. Kluwer Academic Publishers, Norwell, Massachusetts.

[40] DeSteese, J.G., D.J. Hammerstrom, and L.A. Schienbein. 2000. *Electric Power from Ambient Energy Sources*. PNNL-13336. Pacific Northwest National Laboratory, Richland, Washington.

[41] Ferro Solutions. 2004. "Energy Harvesters and Sensors." Ferro Solutions. Roslindale, Massachusetts. Available at *http://www.ferrosi.com/files/FS_product_sheet_wint04.pdf.*

[142] Wills, Jeff. 2004. "Will HVAC Control Go Wireless?" ASHRAE Journal 46(7): 46-52 (July 2004).

Chapter 28

Net Centric Architectures

Klaus E. Pawlik
David C. Green
Paul J. Allen

ABSTRACT

The purpose of this chapter is to explain the terminology used when discussing different major types of net centric architectures. We classify the various architectures and technologies used to implement web sites. Additionally, we discuss some of the trade-offs considered when selecting various technologies. The emphasis here is to help one become a more informed client rather than an expert in web design.

INTRODUCTION

Have you ever found yourself lost in technical computer jargon? Is your ASP (Application Service Provider) providing a web site based on ASP (Active Server Page) technology? Hopefully, after reading this chapter, you will no longer get your TLAs (Two-Letter Acronyms) confused with your TLAs (Three-Letter Acronyms) when it comes to considering and reviewing web-based applications.

In this chapter, we classify various web or net centric technical architectures as well as describe some of the specific technologies available. We also define and discuss various terms and acronyms. Many of these terms are highlighted in a Definition box. This way, readers can view the definition as they read the discussion, but the explanation of the definitions does not interrupt the basic textual flow.

This chapter discusses various software, hardware, architectures, and methodologies. These discussions just provide an overview of the subject. The actual selection of software, hardware, architectures, and methodologies requires a careful consideration of the individual user's requirements including but not limited to the purpose, situation, expected use, available resources, expected number of users, infrastructure, and expected volume of transactions. Although the speed of computer technology innovation may render the specific example obsolete; many of the principles will always remain relevant.

NET CENTRIC ARCHITECTURES

In the context of computer architectures, this chapter will use the following terms somewhat interchangeably:

- Web Architectures
- Net Centric Architectures
- Internet/Intranet Architectures

Internet and intranet architectures do differ slightly when it comes to security. For example, intranet architectures may have a firewall that only allows specific users to view its content.

> **DEFINITIONS:**
> **Firewalls**: Security software or firmware that controls user access to information stored on computers or servers.
>
> **Firmware**: Software embedded into computer hardware.

Web architectures may be classified into two groups: static and dynamic architectures. Static web architectures are read-only, while dynamic web architectures are read-write. Dynamic architectures may be further classified as 2-tier or n-tier architectures (See Figure 28-1).

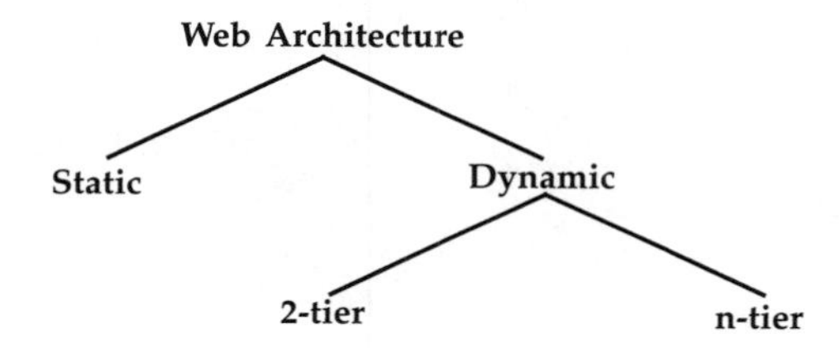

Figure 28-1. Net Centric Architecture Classifications.

The main difference between 2-tier and N-tier architectures is the number of layers between the client application and the data storage devices. To see the basic differences, compare the topology of Figures 28-2 and 28-3.

DEFINITIONS:

Client application: A client application is software that allows the end-user to view content. Common examples include web browsers like Microsoft Internet Explorer or Netscape. Other examples may include software on portable devices like Palm Pilots as well as other more functionally specific applications.

Server Topology: A server topology is a map or schematic indicating how computers or servers are connected. Figures 2 through 4 may be considered examples of simple server topologies. Various servers may reside on the same physical computer or across several computers. In other words, the relationship between servers and computers is not necessarily one-to-one. For example one physical computer server, sometimes referred to as a box, may contain the application server, web server, and database server. At other times, an application server, web server, and/or database, server may be spread across, many various physical boxes.

Server: The term server may cause some confusion because at times it refers to a physical computer or hardware, while at other times, it refers to an application or software. The terms web server, application server, or database server could refer to either depending upon the context.

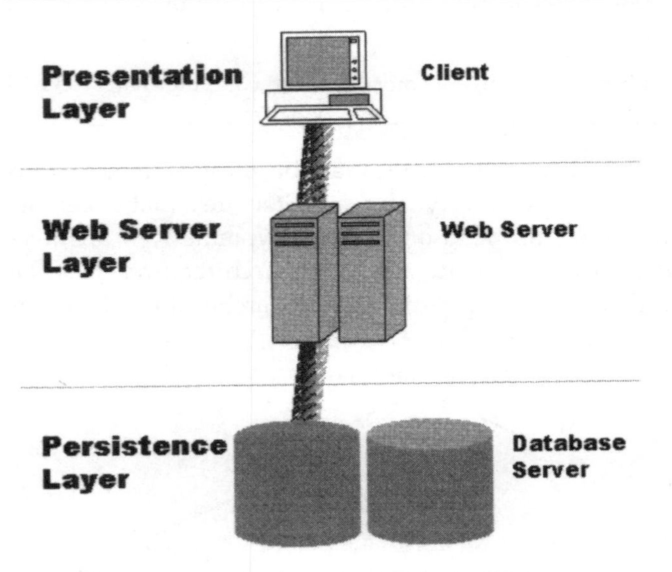

Figure 28-2. Basic 2-tier Web Architecture.

Two-tier web architectures have 3 layers:

- A layer to store data
- A layer to deliver the data
- A layer to display the data.

N-tier architectures have four layers:

- A layer to store data
- A layer that performs logic
- A layer to deliver the data
- A layer to display the data.

DEFINITION:

Data storage device: A data storage device is anything that allows a user to save data. Generally, a data storage device may be a database or a file system. Depending on the security rights or role assigned to a given user, users may perform the following operations to a data storage device:

- Read records or files—this operation may also be referred to as a select,
- Update records or files—this operation may also be referred to as a modify or change,
- Insert new records or files—this operation may also be referred to as a create,
- Delete records or files—this operation may also be referred to as a remove or erase.

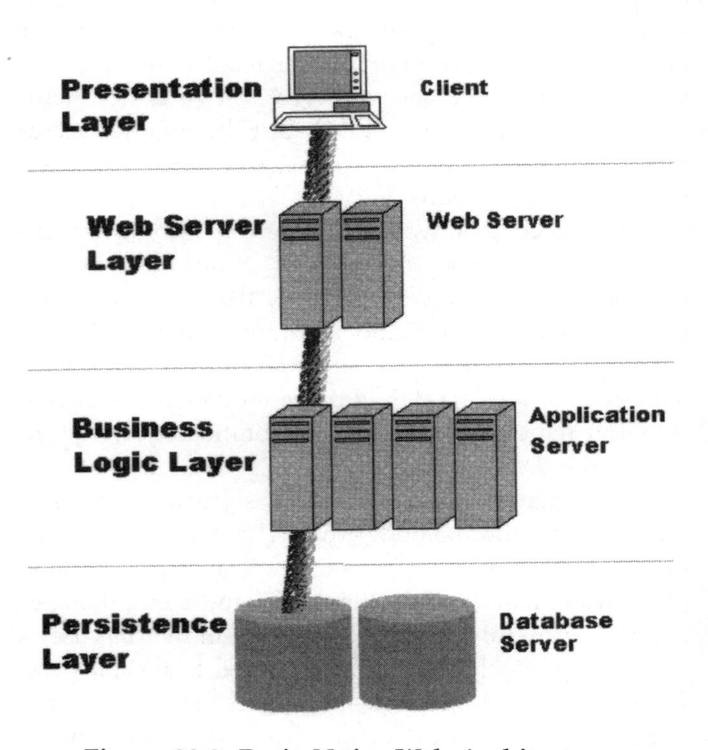

Figure 28-3. Basic N-tier Web Architecture.

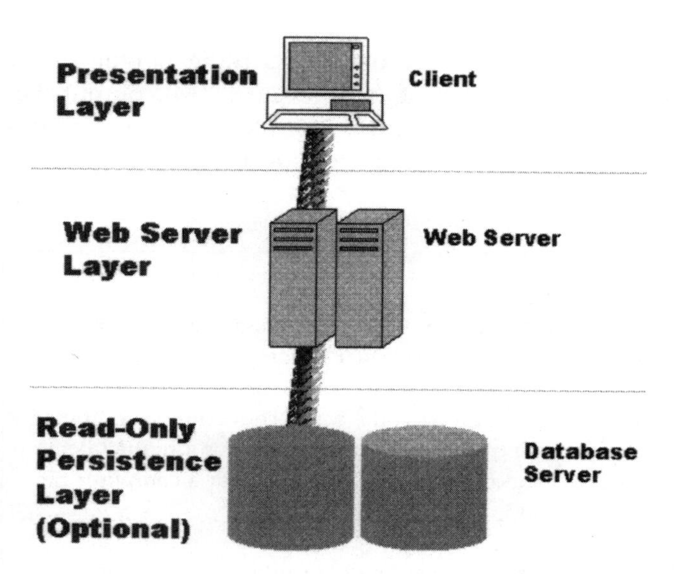

Figure 28-4. Static Web Architecture.

Static vs. Dynamic

The main difference between static and dynamic web architectures is how data storage devices are used—static web architectures have read-only storage devices while dynamic web architectures have read-write storage devices.

Static

Static web architectures only display data to the end-user. The page may have links to navigate throughout the web site. The web site may also have a data storage device like a database or a file system; however, the data storage device would be read-only. In other words, a user of the web site could not update, insert, or delete records or files. Static web architectures have the following layers (See Figure 28-4):

- Presentation layer,
- Web server layer,
- Optional read-only persistence (storage) layer.

For static web pages, these layers operate and serve a purpose similar to the 2-tier architecture except that the persistence (or storage) layer is optional and when it does exist, it is read only.

Dynamic

Dynamic web architectures allow the end-users to not only read from the data storage devices but also to update, insert, or delete records or files. Dynamic web architectures may be further sub-classified as 2-tier or N-tier architectures.

DEFINITIONS:
Throughout this chapter, we refer to different types of users: end-users and implementing-users. The type of user is relative to who is calling the person a user. A company who provides a packaged software solution may refer to the system administrator as the end-user, while the system administrator may consider the people using the web site the end user.

End-user: As used in this chapter, an end-user is the person viewing the web site.

Implementing-user. As used in this chapter, an implementing user may be a system administrator, a developer, a webmaster, some other person who maintains the web site or provides content to the end-user.

N-tier vs. 2-tier

Sometimes, N-tier net centric architectures are referred to as 3-tier or multi-tier architectures. All N-tier architectures have the following layers in common.

- Presentation layer
- Web server layer
- Business logic layer
- Persistence layer

All 2-tier architectures have the following layers in common.

- Presentation layer
- Web server layer
- Persistence layer

Some common examples of N-tier architectures include:

- J2EE—Java 2 Enterprise Edition
- Microsoft.NET
- Windows DNA—Distributed Networking Architecture
- COM based—Component Object Model
- CORBA based—Common Object Request Broker Architecture

DEFINITION:
Components or Common Objects: These terms refer to a grouping of programming code files that are packaged together to deliver specific or related functionality. It is sometimes referred to as a business component when the related functionality deals with a specific business function. For example, you may have a business component named billing verification that reads utility bills and looks for billing errors and compares the charges to the rate tariff structure. Other components provide for common technical servers like persistence or communication with a data storage device.

Some common technologies used to implement 2-tier architectures include:

- ASP—Active Server Page
- JSP—Java Server Page

These technologies may also be used in conjunction with some N-tier technologies in N-tier implementations.

Presentation Layer

The presentation layer displays information to the end-user through the client application. It provides two-way communication between the end-user and the web server layer. The presentation layer works the same way regardless of whether the content is delivered by N-tier or 2-tier web architecture. It may consist of a web browser, a smart device, or some other user-interface (UI) or graphical user-interface (GUI). Typical examples of client applications include Microsoft Internet Explorer (IE) or Netscape.

The presentation layer may be considered either a thin-client or a fat-client also referred to as a thick-client. In the thin-client presentation layer, the computer used by the end-user only needs one application to display its web content. For example, the end-user may have a laptop and the only application that needs to be installed on the laptop to display the web content is Internet Explorer. In the thick-client presentation layer, the computer used by the end-user must have more than one application installed to display its web content. For example, the website may require not only a browser to display the web content but also a specific graphics program.

Various static and dynamic web architecture technologies are used to generate the presentation layer. These include HTML, ASP, XML, XSL and OLAP.

Hypertext Mark-up Language (HTML)

No matter what technology is used, the end result sent to the presentation layer of a net centric application always accepts some type of HTML file as the final product. In other words, the presentation layer, which usually uses some type of web browser, interprets HTML files so that the information may be displayed to an end-user. Both static and dynamic web architectures use HTML. Static web pages may use HTML files that are stored on the web server, or the web server may redirect the request to a file server that contains the HTML files.

Active Server Page (ASP)

This architecture uses a scripting language like VBScript or JavaScript to build an HTML output. For dynamic web architectures, ASP files may contain business logic and may even access a database.

DEFINITIONS:

Scripting Language: A scripting language is a computer programming language that is compiled-at-run-time.

Compile: Compiling is the act of converting a computer programming language that is easier for a human being to understand to machine language that a computer can understand.

Compile-at-run-time: The file containing computer-programming code is compiled when an end-user makes a request that requires that file.

Extensible Mark-up Language (XML)

XML is used to organize data for transportation with a user-defined structure of tags. If an XML file were opened in text pad, it would look similar to an HTML file. Both have tags that can be identified with greater than and less than signs. For example, <some tag> is what a tag looks like in code. One difference between XML and HTML is that the XML tags are user defined where as HTML tags are predefined by convention. XML is to data as object oriented programming is to programming. In other words, it organizes data in a relational manner. For example, the user may have a tag named electric bill. Under the electric bill tag, the user may have a tag named dollar. Under the dollar tag, the user may have tags named energy use, peak demand, customer charge, and tax. The data would be inside or in between the beginning and end tags. The XML file for this example may look like the following:

```
<electric bill>
    <dollars>
        <energy unit=$>
            1000.00
        </energy>
        <demand unit=$>
            100.00
        </demand>
        <customer charge unit=$>
            25
```

```
        </customer charge>
        <tax unit=$>
            55.07
        </tax>
    </dollars>
    <Usage>
        ...
    </Usage>
    ...
</electric bill>
```

XML is only a way to transport data. XML will not create a web page by itself, but it can be used in combination with an extensible style-sheet (XSL) to create a web page.

Extensible Style-Sheet (XSL)

XSL is a style-sheet that takes an XML file as input and translates the data to an HTML output that a web browser can display to an end-user. XSL files typically reside on a web server and define the layout of how data should be displayed to the end-user taking an XML file as the data input. XML/XSL may be used by both static and dynamic web architectures; however, static architectures may only receive XML whereas dynamic architectures send and receive XML. Therefore, static architectures typically do not use the XML/XSL combination.

On-Line Analytical Processing (OLAP)

OLAP is a tool that creates HTML files based on templates, mappings, and the data in a database at a specific point in time. While OLAP tools may be used to assist in creating dynamic web sites, these tools normally are used to create static web sites. In other words, OLAP tools usually only read data from a database at a given point in time and display the data in a user-friendly format. It is quite useful to view static data by its various dimensions. For example, it could be used to view historical data such as various corporate financial measures after the closing of the books for a month, quarter, or year. In an energy management context, it could create reports to compare the energy use versus output for similar facilities.

DEFINITION:
Dimensions: As used in the discussion of OLAP tools, "dimensions" refers to the various ways that data can be dissected. For example, a dimension may be profitability by customer, location, or product. In an energy management context, dimensions may be energy usage, spend, or demand for various offices, manufacturing plants, warehouses, etc. Using OLAP, the data may easily be displayed to benchmark and compare facilities energy use by the purpose of the facility.

Web Server Layer

The web server layer compiles the information that will be delivered to the presentation layer and sends it in HTML. In addition, this layer may aid in security and web address (path) management similar to the file management system of an operating system like Windows Explorer. Web servers provide management of any resources needed to deliver content to the presentation layer. Examples of resources include the various files needed for making a web site operable like style-sheets, HTML files, graphic files, and server pages.

DEFINITION:
Style-sheets: Style-sheets are the application and files used in conjunction with a web server to define the format of content sent to the presentation layer. Examples include the following:
- Cascading Style Sheet (CSS)
- Extensible Style Sheet, (XSL)

Graphics files: A file containing a picture or drawing. Examples of various graphic files formats include the following (file extension is in the parenthesis):
- Graphics Interface Format (gif)
- Joint Photographic Experts Group (jpeg or jpg)
- Portable Network Graphics Format (png)
- Tag Image File Format (tiff or tif)
- Device Independent Format (bmp)
- Many others

Server page: A file containing a programming code used to compile content for delivery to the presentation layer. These files typically are stored on the web server. An example of a server page is Active Server Page (ASP).

N-tier

The web server layer of n-tier architectures provides two-way communication between the presentation and business logic layers. It receives and sends data from and to the business logic layer. Additionally, it receives data from the presentation layer and sends HTML to the presentation layer.

2-tier

The web server layer for 2-tier architectures provides two-way communication between the presentation and persistence layers. It receives and sends data from and to the persistence layer. Additionally, it receives data from the presentation layer and sends HTML to the presentation layer. Most business logic in 2-tier architectures is performed in the web server layer within various server pages. In addition, all database queries are defined in the web server layer of 2-tier architectures.

DEFINITIONS:
Business logic: Programming code that transforms input data to output data by performing various calculations as defined by the underlying functional or business design. For example, if the objective is to store in a database the current required to transport a certain amount of Direct Current (DC) power from one point to another, then perhaps the business logic would take as input the power needed, the resistance between the two stations, and the unit of measure of these inputs. Then, it would calculate the current by using the formula current = square root (power/resistance). After the calculation is completed, it would insert or update the appropriate record in the database.

Database query: Pieces of programming code used to define a select, update, insert, or delete for a database record.

Structure Query Language (SQL): SQL is a platform independent programming language that defines database requests like selecting, updating, deleting, and inserting records to or from a database. Platform independent means that for the most part SOL statements are the same regardless of the database being used. Please note that SQL is mostly independent of the type of database being used. For example, some SQL statements that work with Oracle will not work the same way or at all with Microsoft SQL server and vice-versa.

The following are examples of some web-server software packages.

Internet Information Services (IIS)

A product that comes standard with various versions of Microsoft Windows like Windows 2000 and Windows XP. Furthermore, it may be added to various Windows versions like Windows NT with a Service Pack (SP). This seems to be one of the most popular web servers for Windows platforms.

Personal Web Server (PWS)

A Microsoft product made available for home versions of Windows. This product is usually limited to home or very low volume use; however, it is relatively easy to implement and use.

Apache

Apache is a popular open-source code web server. This web server is especially popular with UNIX users. It is available from The Apache Software Foundation.

DEFINITION:
Open-source: This method of software development allows access to its source code. This access allows implementing users to modify the way the software operates through the modification of this source code. Usually, an organization manages the main code stream taking input from the implementing user community. For example, in the case of the Apache web server, The Apache Software Foundation acts as this coordinating organization.

Business Logic Layer

Often, the business logic layer is known as the middle-tier. This layer may use what is known as middleware or an application server; however, the term middleware is not exclusive to net centric architectures. The business logic layer performs all calculations and transforms data between the persistence layer and the web server layer. The business logic layer typically houses the application's compiled code in the form of binary files. Two-tier architectures do not have a business logic layer as these functions are performed in the web server layer. Only n-tier architectures have a business logic layer.

DEFINITIONS:
Middleware: Middleware is sometimes used in the web architecture context to mean an application used to register components with an application server. It aids in allowing applications to be distributed among multiple servers thereby allowing these applications to be scalable, portable, and transactional. For example, Microsoft Transaction Server (MTS) is an application that would be used to register Component Object Model (COM) components with Windows that maybe in the form of code packaged in Dynamic Linking Library (DLL). Another example of a middleware application is Common Object Request Broker Architecture (CORBA). These web middlewares seem to be a disappearing breed since the newer N-tier web architectures, like .NET or J24, can register their binaries

with the operating system of the application servers directly without the need for the middleware or middle-tier software, In addition, if the server is powerful enough to handle the load, then the need to distribute the application becomes less, which does reduce the complexity of the server topology. One example of how server hardware can be scalable is with blades, which are a processor and memory that can be plugged into a server that has blade, hardware and software.

Electronic Data Interchange (ED]): A standardized format for data transportation between computers. Sometimes, EDI uses XML files. Other times it uses flat files.

Flat File: A flat file is a text file that contains data separated by a delimiter. Common forms of delimitation include the following: comma-delimited, tab-delimited, bar-delimited, fixed-width, etc.

Enterprise Application Integration (eAI): Another meaning of middleware is Enterprise Application Integration (eAI) software. This software moves data from one computer server to another server. It may use a variety of data transportation mechanisms like XML, EDI, ODBC, flat files, or some other method. Some examples of eAI software include the following:

- SeeBeyond
- Microsoft BizTalk
- Vitra
- Tibco

DEFINITIONS:
Binary Files: Binaries are computer files that contain compiled code or collections of compiled code. You may see files of complied code for the following programming languages (with a file extensions in parenthesis) on your computer or server:

- Java (java)
- C (c)
- C+ + (CPP)
- C++ (cpp)
- C# (cs)
- Many others

Examples of collections of resources include (their file extension is inside the parenthesis):

- Web Application Archive (WAR)
- Dynamic Linking Libraries (DLL)
- Cabinet (CAB)
- Java Archive (JAR)
- An archive file (ZIP)
- Executables (exe)
- Others

Some of these collections or packaging of files may contain other resources than just the binaries. Some may contain HTML, graphics, or compile-at-run-time files.

N-tier advantages

The advantages of N-tier architectures are as follows:

- Security—Provides an extra layer of security by separating database access from the web server.
- Scalability—Allows the use of multiple application or middleware servers to handle requests from the web server.
- Reusability—The use of a business logic layer promotes the use of components that are callable from server pages.
- Extensibility—Because the business logic layer promotes the use of components it allows the use of built-in or third party components that can be extended or built upon to achieve the required objective.

2-tier limitations

Two-tier architectures have these limitations:

- Lower performance—Placing lots of code and business logic on the web sever and thus script (like Active Server Pages) is typically not very efficient.
- Lack of business logic encapsulation—Two-tier architectures does not encapsulate business logic. In other words, the end-user may be able to view propriety logic.
- Lower reusability—Since the business logic is wrapped within content delivery logic, if you wanted to reuse logic in another application, then you would have to strip the content delivery logic away from the business logic.
- Lower scalability—Because of the logic being wrap in script, it makes it harder to deploy code to multiple servers.

2-tier advantages (N-tier advantages)

Two-tier architectures are typically less expensive to develop, implement, operate, and maintain then n-tier architectures.

DEFINITIONS:
Portable: Portability refers to the ability to move code from one computer or server to another.

Scalable: Scalability refers to the ability to add capacity.

Transactional: In web architecture context, the ability to keep track of a transaction or request from beginning-to-end or data input to output or storage.

Extensibility: Extensibility refers to the ability to build upon or extend existing code.

Persistence Layer

The persistence layer provides two-way communication between a data storage device and the business logic or web server layer. Sometimes, the persistence layer is called the storage layer. It may use a database or file server to store its data.

N-tier

Uses database queries or file saves or retrievals as delivered from or to the business logic layer.

2-tier

Uses database queries or file saves or retrieval as delivered from or to the web server layer.

Some examples of databases that are typically used for web architectures include:

Microsoft Access

Access is a Microsoft Office product that is relatively inexpensive and often used to help organize data on a personal computer. While inexpensive, it has limited usefulness to distributed web architecture with multiple users.

Microsoft SQL Server

A Microsoft product that is relatively inexpensive to buy and operate. The latest version, Microsoft SQL Server 2000, has extensive functionality and good performance with lower load application.

Oracle Databases

Oracle Corporation database products provide higher performance and load capacities; however, they are typically more expensive to purchase, implement, operate, and maintain. Examples of Oracle database versions include 7, 8i, 9i or 10g.

Other

Other databases available include Paradox and Database2 (DB2). Corel makes Paradox, which is in a class similar to Microsoft Access. IBM offers DB2, and it is a high-end, mature database.

EXAMPLE

XYZ Real Estate Management Corporation—a fictitious company—leases buildings for professional use. Presently, it manages five buildings with five to 10 tenants each with a total of 35 tenants. Over the last five years, the number of accounts managed by the company has been growing at a rate of 20% per year, and they expect the same growth over the next five years. The corporate office has contracts to pass along utility costs to the tenants. To manage the utility cost pass through, XYZ has a manager, an accounts receivable person, and an administrative assistant.

XYZ Energy Management Website

To manage the accounts, the XYZ has set up a website that the tenants may access to manage and pay their utility costs.

Technology Selection Considerations

When deciding what overall architecture and technology to use, XYZ had the following constraints:

- Throughout the month there will be mostly three users: the account manager, the accounts receivable person, and the administrative assistant.
- At peak, 38 users will use the website: the 35 tenants and the 3 corporate people.
- Peak use only occurs the last week of the month and the first week of the month.
- They expect a 20% growth rate in the number of users over the next five years.
- Secure access is needed for payment.
- User security is needed, i.e. allow only the tenants to review their specific accounts.
- The website should provide for medium availability, but low fail-over protection.
- Costs should be minimized to achieve the business objectives.

To achieve these business objectives, XYZ chose a 2-tier architecture structured as follows:

DEFINITIONS:

Database Management System (DBMS): DBMS are programs that a database uses to access data.

Open Database Connectivity (ODBC): A communication method by Microsoft to access databases. It translates information from the application to the DBMS.

Java Database Connectivity (JDBC): JDBC is a communication method by Sun Microsystems to access databases. It translates information from the application to the DBMS.

- Limit the presentation layer to Internet Explorer.
- Use Internet Information Services (IIS) as the web server.
- Use Active Server Pages (ASP) to code the business logic.
- Use Microsoft SQL Server as the database.
- Use a high-end desktop computer as the production server that runs both the web server and the database.
- Use Secure Socket Layer (SSL) and digital certificates to maintain security.
- Code the individual account usernames and passwords in ASP to ensure that tenants may only view their own accounts.
- For fail-over protection, back up the database nightly using the desktop computer of the Administrative Assistant.
- Maintains the ASP code as well as the production server on the Administrative Assistant's desktop computer.
  - The Administrative Assistant has the skills necessary to maintain the website.

This setup gives XYZ the ability to keep implementation and maintenance costs low, while retaining the flexibility to expand in the near future by taking the following steps:

- XYZ may upgrade the production desktop to a newer more powerful desktop as prices decrease.
- XYZ may then use the former production desktop for fail over protection.
- XYZ may add a second desktop to run the database server as the tenant list grows.

XYZ does not expect to outgrow these system requirements with in the next five years. After five years, XYZ may add another web server computer with a load balancer to direct the user to one of the specific web servers.

NEXT STEPS

The world of computing changes rapidly, and the world of web computing changes even more rapidly. Definitions and meanings of acronyms are still evolving and changing. Sometimes, the same words or acronyms may have multiple meanings, which also requires context. In addition, you may find that the web address for information may often times change. Therefore, this chapter includes a bibliography of web sites with search engines that will allow you to find more detailed information on the topics discussed in this chapter. Use a web browser to go to the listed web sites, enter some key words into their search engines, and enjoy the detailed information about your chosen topic.

Bibliography

www.java.sun.com
www.microsoft.com
www.accenture.com
www.apache.org
www.ibm.com
www.oracle.com

Chapter 29

Enterprise Level Integration Using XML and Web Services

Steve Tom
Automated Logic Corporation
Atlanta, Georgia

ABSTRACT

JUST A FEW YEARS AGO, integrating a building automation system (BAS) with any other computer system was so difficult that successful integration projects were heralded in all the major trade publications. Today, integration is commonplace through the use of XML, or eXtensible Markup Language, an IT standard that is rapidly being adopted throughout many different industries. Today BAS manufacturers are adding XML support into their web servers and operator workstations because it is *the* standard for communicating with other systems at that level. XML is a logical extension to a BAS protocol.

INTRODUCTION

A college in Pennsylvania gathers data from multiple systems at a remote field station, compares utility consumption to a modeling program, and posts the results on a web site. Water consumption, electrical usage from multiple circuits, propane usage, temperature, and levels of CO_2, CO, and O_3 are all being monitored. An aeronautics research organization in Texas is pulling data from multiple chillers and AHUs into Excel for analysis and fault detection. They're also implementing a scheme to give over 2000 users a simple desktop application that will allow them to turn the air conditioning on when they're working late or on weekends. (Not surprisingly, this scheme will also keep a tally of the additional hours requested by each using organization.) An office complex in Melbourne, Australia, is taking this one step further, using similar data to generate a bill for the after-hours utilities used by each tenant. Should any tenant question the bill, the company can instantly produce a detailed report showing the exact start and stop time and date of every air conditioning request.

Figure 29-1. King Street Wharf in Sydney, Australia, uses Web services to analyze and manage multiple utility systems.

In Sydney, Australia, a building engineer is using data from his BAS to maintain his SEDA rating. (Sustainable Energy Development Authority.) Comparable to our LEEDS certification, Sydney requires all new buildings to track their energy usage and compare it to indexed values on a daily, weekly, and monthly basis. And finally, in the United States, a controls contractor is collecting performance data from every VAV box and reheat coil in a new building and compiling it into a commissioning report to submit to the contract manager.

XML FOR COMPUTER TO COMPUTER DATA EXCHANGE

So, if system integration is so commonplace, why did I bother to list all these examples? Because these examples were implemented using a new standard for data exchange called XML. XML, or eXtensible Markup Language is an IT standard that is rapidly being adopted throughout many different industries. Think of it as a logical successor to DDE, OLE, and OPC, but a successor that is much more powerful, much more widely accepted, and much more at home on the Internet than these previous standards. You think it's hard to find a common protocol to connect to HVAC vendors? XML has actually succeeded in getting Microsoft, Apple, and Linux to play nicely together!

Does this mean that BACnet, LonWorks™, and all those proprietary protocols are on their way out? Not at all. XML was designed as a tool to be used by computers to exchange data at a high level. The data files are actually composed of human readable text, and the data structure includes verbose (for a computer) descriptions of what data are contained within the files. This makes it relatively easy for human programmers to find the data they want and to quickly build links between computer databases, but it also means the resulting communication requires much greater bandwidth than any of the dedicated BAS protocols. Figure 29-2 shows a typical BAS architecture. XML is well suited to communications on the Ethernet IP network, but could strain the resources of the Field Controller network. Bandwidth issues mean it will be a long time, if ever, before you see XML used in a VAV controller.

By itself, XML does not provide standardized communications within a BAS, so in that regard it's a step backward to the days before BACnet and LonWorks™, when every vendor made up their own communication rules. Point data, trend logs, alarms—all can be presented in whatever format a vendor chooses and still fall within the XML umbrella. As discussed later in this chapter, ASHRAE and other groups have corrected this problem by developing a standard for presenting BAS data in XML.

Given that BACnet, or any other dedicated BAS protocol, is better defined and more efficient than XML, why is everyone so excited about XML? XML provides a standard method for a BAS to communicate with another computer, whether that computer is in another BAS or a completely different application. The initial version of BACnet did not support Internet Protocol (IP) addressing because when BACnet was created the Internet wasn't a big deal. That changed in a hurry! BACnet and other BAS protocols now support IP communications because it is *the* standard for high-end communications. Today BAS manufacturers are similarly adding XML support into their web servers and operator workstations because it is *the* standard for communicating with other systems at that level. XML is a logical extension to a BAS protocol.

OTHER FUNCTIONS FOR XML

XML also makes it possible to present a higher-level abstraction of building automation data, and to present it in a format that simplifies communications with other systems. A programmer writing a computer application for, say, a utility company might be interested in retrieving information about scheduled equipment start-up from a BAS, or he might want to send a utility curtailment command to the BAS, but he wouldn't want to learn an entire building automation protocol just for these few transactions. The same situa-

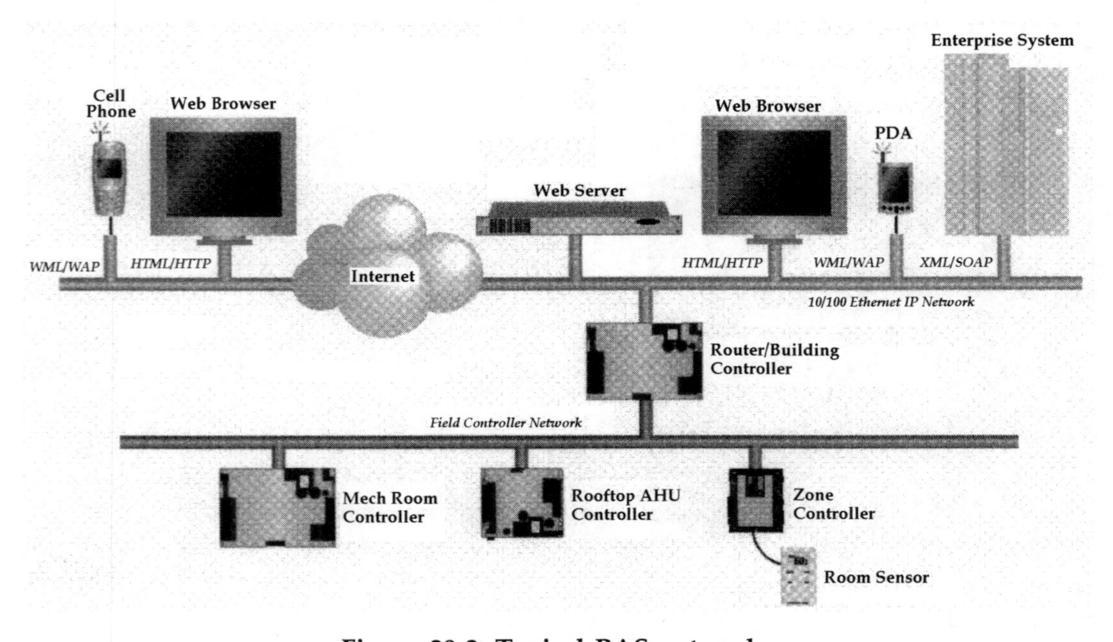

Figure 29-2. Typical BAS network

tion exists in computer programs used for accounting, classroom scheduling, hotel room management, hospital patient administration, maintenance management—there are many programs which a building owner may want to integrate with his BAS but which are not going to offer a BACnet interface, let alone a BACnet interface *and* a LonWorks™ interface *and* an interface to other proprietary protocols used in today's BAS.

Then of course there's the poor enterprise integration programmer who's trying to gather information from multiple computer systems throughout a college campus or a large corporation to prepare customized summaries for senior management. Is he going to want to learn the internal data structure of every application that contains information of interest? XML provides a way to present this data in a standard, self-documented interface that is independent of the data structure or the communication protocol used within the BAS. When combined with a standard service protocol like SOAP (Simple Object Access Protocol) it provides a universal interface that can be used to exchange data with thousands of individual computer applications.

XML Applications in the Facility-management Industry

XML has wide applications in two primary areas—as a data storage tool to be used within a building automation system and as a communication tool to exchange data with other high-level computer applications. The first area is very exciting to the programmers who work within the building automation industry, but of little interest to end-users. For example, Automated Logic already uses XML extensively throughout its product line and it helps them develop products faster, support multiple databases, and provide platform independent applications. While the end user may appreciate the end results, he doesn't much care whether the data storage was done using XML, a relational database, or flat files.

The second area should be very exciting to end users, as it opens up tremendous new possibilities for enterprise integration. XML, especially as used within Web services, is fast becoming the standard method of exchanging data within the facility management industry and within other industries as well. XML is probably already more widely supported than previous standards (DDE, OLE, OPC) ever were, and since XML is cross platform and web-friendly it can potentially be used to share data with virtually any computer, any place in the world.

The Migration Path for Implementing XML in the Facility-management Industry

Already XML data exchange features are being integrated into the front-end software of several vendors, and others should quickly follow. Initially every vendor will develop their own information model, or schema, for the data they make available, which means developing interfaces will require a fair amount of custom programming. XML will make this programming easier than other data structures because it is self-documented, but as long as each system uses its own unique schema, the interfaces will need to be custom programmed. ASHRAE and other groups are aware of this problem and are already working on establishing a standard XML schema for building automation, which will make the integrator's task much simpler in the future. ASHRAE has recently released a draft of their standard and as this book is written it is out for public review. More details on this are provided later in this chapter. The building automation industry is not alone in recognizing the need for standards. When XML first appeared, it was hailed as the key to successful B2B (business to business) data integration in many industries, but these industries soon realized that to eliminate the headaches of custom interfaces they needed a standard schema as well as a standard file structure.

JUST HOW UNIVERSAL IS THIS INTERFACE?

Dan Traill of U.E.S. Controls, Houston, Texas, gives two examples. A school district asked them to pull utility data from several individual schools, run a monthly energy audit, compare actual energy consumption to local weather data, and rank the schools based upon their energy efficiency. Since the BAS for these schools supported Web services, their in-house engineers had no trouble writing the report. "It was a very simple, straightforward process," Dan said. "They used the built-in XML support in Microsoft's Excel spreadsheet to gather the data and then wrote the report in Excel." A more complex situation arose when they were asked to develop a custom tenant override system that would allow 2000 building occupants to extend their office's hours of operation and would keep track of the resulting after-hours HVAC use. They hired a professional programming firm to develop this application, and the programmers presented U.E.S. with a rather hefty estimate of the cost to create the program using a custom data interface. When Dan told them the BAS system supported XML and SOAP, they cut their estimate in half.

WEB SERVICES

The combination of XML and SOAP has already gained wide acceptance in the IT world. Microsoft calls it .NET. IBM calls it WebSphere. Sun calls it SunOne. Other vendors have other names, but they all fall under the generic term of "Web services." Web services can be used to read or write data from one computer to another, but they can also be used to actually run applications on another computer. As a simple example, a weather system computer might provide access to hourly temperature and humidity data as well as routines to compute the average temperature, mean temperature, and heating degree days over any desired time period. If you ask for temperature readings over a given period the Web service will simply retrieve data for you. If you also ask for the mean temperature and heating degree days over the same period, the Web service could run a separate application on the weather computer to calculate this data. The weather computer might actually use Web services itself to gather data from other computers and combine it with its own data to respond to your request.

Web services have been used in B2B communications for several years and are fast becoming the accepted norm. As a few examples, Amazon.com uses Web services to allow partnering firms to integrate its products into their web sites. They sell the product, but the product information, availability, etc. is automatically updated by Amazon. Microsoft's Passport service uses Web services to provide a secure way for on-line vendors to obtain, with your permission, the billing and shipping information they need to process your order. And the State of New Mexico is using Web services to create a "portal," a single web site where users can access data and services from multiple government agencies.

XML and Web Services

XML is the syntax for data structures used by Web services. The structured data are transferred from one computer to another using a transport protocol such as the Simple Object Access Protocol, or SOAP. SOAP and XML form the backbone of Web services, just as HTTP (a transport protocol) and HTML (a file structure) form the backbone of the information we view on the Internet today. Web services can do more than just transfer data, however. Web services can include calls to applications that will generate, sort, filter, or otherwise prepare the data to meet specific needs. As a simple example, if you were gathering weather data from a number of different sources, a straight data transfer might provide a hodgepodge of data with different sampling rates and different units. A Web service, on the other hand, could let you specify the sampling rate and units, and each computer supplying data would then perform the conversion and interpolation required to supply data in the requested format. Defining what services should be available is one of the issues being tackled by the organizations establishing standards for Web services.

Webopedia™ defines Web services as "a standardized way of integrating Web-based applications using the *XML, SOAP, WSDL* and *UDDI open standards* over an *Internet protocol backbone*." It's easy to get lost in the alphabet soup of acronyms in this definition (see Glossary at end of book for details) so let's pare it down to what's important to HVAC engineers. Web services are a standard way of integrating applications over an IP network. XML, SOAP, WSDL, and UDDI are the essential "machinery" that makes Web services work, but you don't need to understand them to use Web services. The people who create Web services need to understand them, and many HVAC engineers will want to learn them so they can modify or create their own Web services, but trying to learn what Web services are by dissecting the acronyms is like trying to learn how to drive by studying an automatic transmission.

When HVAC systems first switched to digital controls, they used small stand-alone controllers and proprietary communications networks. This was not surprising, as people used them to replace small stand-alone pneumatic control systems. Digital systems were capable of much more, however, and it wasn't long until manufacturers were linking these digital controllers together into Building Automation Systems. These offered monitoring and control capabilities far beyond those of pneumatic systems. Unlike pneumatics, however, you couldn't easily connect digital controllers from two different manufacturers. Connecting dissimilar systems required an extensive knowledge of both protocols, a "gateway" or translator module, and expensive custom programming. The same was true when engineers tried to integrate their BAS with a lighting system, or a security system, or a fire alarm system. Frustrated users turned to ASHRAE for help, and ASHRAE created BACnet.

BACnet provides a solution to these problems by establishing a standard communication protocol for all building systems. This greatly simplifies integration between vendors using the BACnet protocol, but BACnet isn't the only game in town. LonWorks provides a different standard for integrating building systems. MODBUS offers yet another standard, and many vendors have opened up their own protocols to provide additional "standards" for the building industry. Having multiple standards is certainly better than having none, but multiple protocols means there is still a need for gateways and custom programming. The problem gets worse if you want to connect to a computer outside the BAS, say a computer at the power company that sets real-time utility pricing. There is very little chance this computer can connect to any of the standard building protocols. This problem is not unique to the building automation industry. It has long plagued business-to-business (B2B) transactions. To help solve this problem the Information Technology (IT) world established a standard called "Web services" for communications between dissimilar computers, and this standard enjoys wide support.

XML and BACnet and LonWorks

When combined with a transport protocol such as SOAP, XML is a way for computers to communicate, just as BACnet and LonWorks are. XML requires a lot more processing power and bandwidth than either of these two BAS protocols; so it's not likely to replace either of these protocols, particularly for unitary controllers, smart sensors, smart actuators, and other cost sensitive applications. BACnet and LonWorks also include functions that go well beyond the simple exchange of data and define the data structure, communication services, and similar system features. Both BACnet and LonWorks systems can use XML for high-level integration with other systems, and they can be used as the basis for a standard BAS schema. An XML schema that was based on just one protocol, however, would only partially ease the headaches of integrating today's building automation systems. What's needed is a high level information model that can provide an interface to any standard or proprietary BAS protocol.

How Could This Widespread Compatibility Benefit BAS Systems?

Imagine being able to incorporate weather forecasts into control algorithms. Ice storage systems, boiler start-ups, morning pre-cooling—all these control strategies could be more efficient if there was a way for the BAS to call out over the web and retrieve a weather forecast from a weather computer. Colleges and universities often have sophisticated computer systems to schedule classroom use. All too often, these schedules are then printed out and given to a BAS operator who has to type the schedules into the BAS by hand. Wouldn't it make more sense for the two computers to exchange schedules? Using Web services, they can. The accounting system computer would also be interested in accessing these schedules to see what department should be billed for each hour, especially if the system was tied in to utility meters that recorded the actual utilities used by each class.

Web services can also be used to read or write information on demand, making them very useful for constructing interactive web pages. As an example, there are many systems on a college campus which have information needed by a facilities engineer. Utility consumption, maintenance management, cost accounting, record drawings—a facilities engineer needs to interact with all of these systems but he shouldn't have to learn a new user interface for each system. For that matter, he shouldn't even have to log on to each individual system to gather data. Web services make it possible to create a "facilities portal," an interactive web page that collects data from all these systems, provides the engineer with a summary of key data, and allows him to "drill down" to more detailed information as required. (See Figure 29-3) These examples are not hypothetical flights of fancy. Several vendors have already incorporated Web services into their BAS products, and these examples of integration programs have already been demonstrated in projects around the world.

Interoperability Issues and XML

First, XML provides an ideal tool for communicating with systems outside the building automation industry. Many of these systems are already using XML in their business-to-business transactions, and the use of XML to integrate with BAS is a natural evolution. XML provides a high-level representation of HVAC data in a format that is useful to other industries, industries that have no need or desire to wade through the minutia of BACnet or LonWorks to find the information they need. Similarly, the HVAC industry can use XML to obtain utility rates, weather forecasts, occupancy schedules, or other useful data without having to learn the intricacies of the computer systems used in each of these other industries.

Second, XML has already become the standard protocol for data exchange within the IT world, so it is the natural choice for enterprise integration. A busy CEO who needs access to information from throughout his organization expects his IT staff to serve it all up on a single screen, with links that let him drill down for more details. These harried IT programmers, who are *not* going to want to learn BACnet or LonWorks just so they can extract a few tidbits of facility data, can pull that data out of an XML file in a heartbeat.

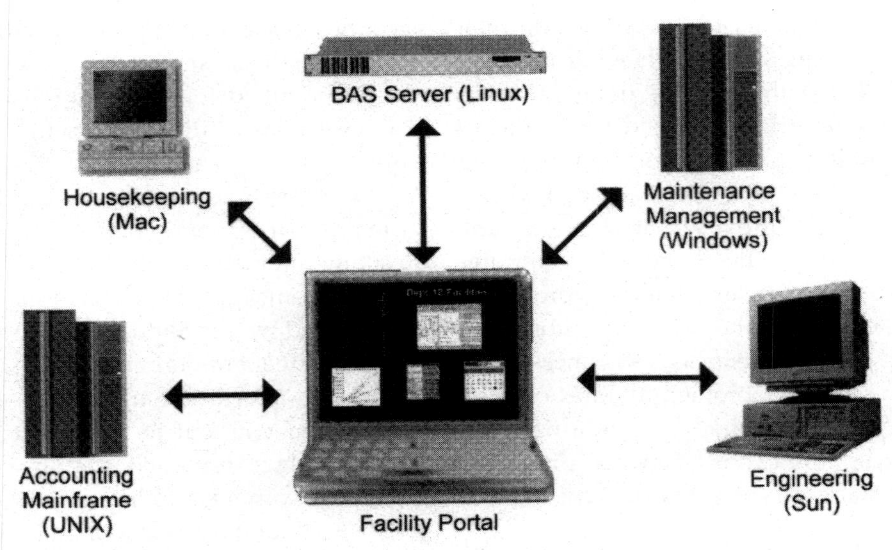

Figure 29-3. A Facility Portal uses Web services to integrate data from multiple systems.

Finally, XML could facilitate the integration of BACnet and LonWorks systems at the front-end computer level. This integration is possible now by using custom gateways, but these are primarily limited to a simple exchange of point data. If you want to exchange information that is not defined in both standards, such as alarms or trend logs, there are no off-the-shelf solutions. A standard XML schema for BAS systems would allow manufacturers of BACnet and LonWorks systems to provide an off-the-shelf interface that would work with any other manufacturer's system.

Physical and Organizational Infrastructure Needs of XML

In most locations, the physical infrastructure is already in place. XML could conceivably be carried on any kind of network, but to achieve its maximum interoperability potential it needs to be used in Web services with an IP connection to the Internet. Today most facilities have this Internet connection.

Standards are definitely needed. As mentioned previously, the fact that XML is internally documented makes it possible to hand-craft XML connections, but to fully realize the interoperability potential we need to get away from custom programming and automate these connections to the maximum extent possible. This requires standards, both for the communications media and for the data and services (the information model) to be communicated. The IT community is and should be responsible for developing a common standard for the communications technology. They are already doing a good job of this, as evidenced by the widespread acceptance of XML and Web services. The IT community should not, however, be the ones to develop an information model for any specific industry. Information models need to be developed by people who understand the needs of the industry, and the natural hosts of this development are the professional organizations that serve each industry.

XML and Existing Systems

XML will not, by itself, require replacement of an existing system. In the near term at least, XML connections will primarily be made at the top end, either from the operator workstation or from a server on the IP network. If an owner wants to add XML interoperability to an existing system, his first call should be to the vendor who supplied the system. That vendor may already have an XML compatible system available, or may have one in development. If the vendor has no plans to add XML capabilities to existing systems, it may be possible to purchase an XML server from another vendor and integrate it into the existing system. This would obviously be easier if the existing system used a standard protocol like BACnet, but it may be possible to use gateways to provide at least limited interoperability to a proprietary system.

Durability of XML

There are no guarantees in the digital world, but my guess is that XML will be around for a long time to come. It may be new to the HVAC world, but it has already been used in B2B connections for several years. This led to the development of Web services, which enjoy phenomenal support. Any time you have a communication standard that is supported by Microsoft, IBM,

Sun, Apple, Linux, and other major players in the IT world, it's going to be around for a while. It will evolve over the years, becoming faster and more powerful, but the installed base is already so large that each revision will of necessity include the backward compatibility needed to support existing systems.

ASHRAE SPC 135 ADDS WEB SERVICES TO BACnet

If BAS vendors are already providing Web services, where does ASHRAE fit in? ASHRAE is establishing a standard means of using Web services to integrate facility data from disparate sources. The IT world has established standards for the mechanism of Web services, but these say nothing about the actual data being exchanged. Without additional standards, vendors could claim support for Web services while providing as little or as much data as they wished through this interface, using whatever data structure and read/write interface they pleased. Even if every vendor tried to create useful Web services interfaces to their system, chances are no two interfaces would be alike and connecting two dissimilar systems would require hours and hours of custom programming. Some of the more farsighted members of our industry foresaw this problem years ago and began calling for a standard information model. ASHRAE answered this call by gathering input from facility engineers, equipment manufacturers, government agencies, and universities about the potential uses for Web services in facility automation and developing a standard information model. This model covers the types of data to be exchanged, the path used to locate the data, and attributes of commonly used data objects such as analog inputs or binary outputs. The services required to read or write values are defined, as well as services needed to obtain information about the available data or to return error messages if a service fails. The standard covers arrays as well as scalar data, making it particularly useful for handling trend logs.

Because this standard is designed for use with Building Automation Systems, it was developed by the technical committee that is in charge of standards for Building Automation Control networks, i.e. the BACnet committee. Once approved, it will become an addendum to the BACnet standard, which means it will also become an ANSI, CE, and ISO standard. Naturally the standard is compatible with the BACnet protocol, but it is not limited to BACnet. Indeed, one of its most useful applications may be to serve as a standard for exchanging data between building automation systems using different protocols. Web services could be an ideal way to make a "top end" connection between systems running BACnet, LonWorks, MODBUS, or any proprietary protocol. Engineers would not have to learn the details of each individual protocol to program the connections; they would only have to understand the Web services standard. A Web services connection would also avoid the problems with incompatible baud rates, wire types, proprietary communication chips, and all the other issues that can come into play when a gateway is used to connect dissimilar protocols. (See Figure 29-4)

Even though Web services have quickly become the standard for B2B communications, they are unlikely to replace BACnet, LonWorks, and other protocols within the BAS. The IT world has developed a standard that supports data exchange, but this Web services standard does not cover the content of the data or services to be exchanged. It would be nice to be able to find the data

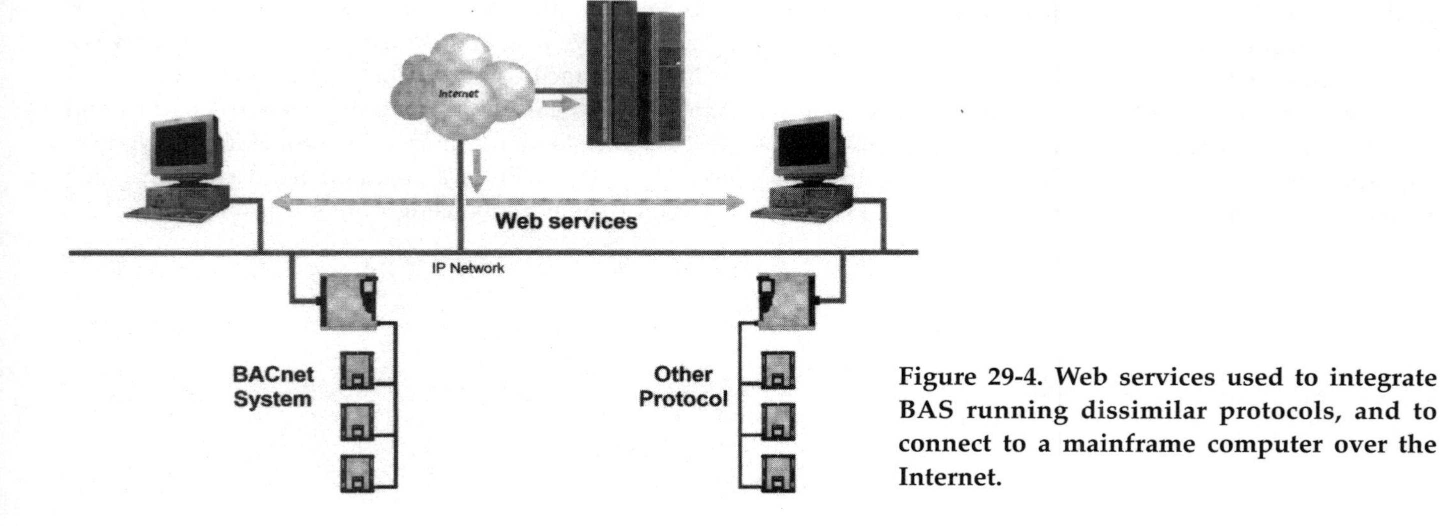

Figure 29-4. Web services used to integrate BAS running dissimilar protocols, and to connect to a mainframe computer over the Internet.

you need, and better still, to be able to retrieve it without having to hire a programmer to write a custom application. This requires a standard that goes beyond the packaging and transportation of data, a standard that covers the data itself. Since the range and variety of data available from building automation systems is huge, the standard itself has to be very flexible and all encompassing—an information model of the entire building system.

Furthermore, no one has developed a set of Web services that covers all the functions needed by a BAS. Broadcasts, alarms, time synchronization, backup and restore—there are a host of BAS functions that simply are not covered in the proposed Web service standard. Certainly such a standard could be developed, but it would in essence become one more BAS protocol fighting for acceptance in the marketplace. It would not be a protocol that was well suited for a BAS because the "overhead" required to implement Web services is beyond the capability of most BAS controllers.

By definition, Web services are using XML to communicate over an IP network. IP networks are great for connecting PCs, web servers, and other high-end computers, but it would be very expensive to run an IP network to every unit heater, VAV box, and exhaust fan in a building. Similarly, XML is a very "verbose" way to package data. It's designed to be human understandable, flexible, and self-documented. These characteristics also mean it needs to be processed by a powerful computer and transmitted over a high-speed network. This is beyond the capabilities of the price-sensitive controllers typically used for small HVAC equipment like VAV boxes. This may be a temporary limitation, as inexpensive microprocessors gain power and speed with each passing year, but since existing protocols like BACnet are already developed, are a more efficient way of integrating controllers, and are open for use by any equipment manufacturer there is very little incentive to switch these controllers to Web services.

Web services have been used in business-to-business (B2B) transactions for several years. As in the building automation industry, Web services were initially hailed as the ultimate solution to B2B needs for a universal tool for information exchange. It soon became apparent that Web services were an extremely valuable tool to expedite these transactions, but that the transactions still required much custom programming to locate and link the data to be shared. The problem could be greatly simplified if each industry adopted a standard information model, so programs could automatically find the data they needed without human intervention. The addition of a new ASHRAE standard to the Web services world promises even greater simplification, using IT technology and the foundation of BACnet to take building automation to the next level.

CONCLUSION

This chapter has provided a short, basic introduction to the application of XML in web based energy information and control systems, or BAS systems. XML provides a standard method for a BAS to communicate with another computer, whether that computer is in another BAS or a completely different application. XML also makes it possible to present a higher-level abstraction of building automation data, and to present it in a format that simplifies communications with other systems.

The combination of XML and SOAP has already gained wide acceptance in the IT world with products that all fall under the generic term of "Web services." Web services can be used to read or write data from one computer to another, but they can also be used to actually run applications on another computer. XML has already become the standard protocol for data exchange within the IT world, so it is the natural choice for enterprise integration.

Finally, the importance of an ASHRAE standard for Web Services for BAS cannot be stressed enough. It is critical to have a standard means of using Web Services to integrate facility data from many different sources. ASHRAE gathered data from all concerned parties and developed a standard information model for everyone to use. The result will be a standard for exchanging data between BAS systems using different protocols.

Chapter 30

Utility Data Web Page Design: Mining the Data

David C. Green
Green Management Services, Inc., Fort Myers, Florida
dcgreen@dcgreen.com

Paul J. Allen
Walt Disney World, Lake Buena Vista, Florida
paul.allen@disney.com

ABSTRACT

This chapter focuses on how to design a utility data web page interface that is both intuitive and easy to use. Users need to be able to quickly drill down through the data with minimal mouse clicks to investigate changes in meter data. Using the correct combination of options and links, hundreds of reports can be produced easily and quickly from one interface. Just like the dashboard of an automobile informs us of the inner workings of our cars, this interface can inform users of the inner workings of any facility or complex of facilities. This method combined with interactive reports as described in Utility Data Web Page Design—Presenting the Data from *Information Technology for Energy Managers—Volume I,* makes an intuitive and robust utility reporting web application as part of an *Energy Information System (EIS).* It helps us discover relationships, such as trends in our energy consumption over time or comparisons between similar facilities, which might otherwise go unnoticed.

INTRODUCTION

Mining information from a complex database of utility data can be a difficult and cumbersome task. WEBOPEDIA at *http://webopedia.internet.com/*defines *data mining* as:

> A class of database applications that look for hidden patterns in a group of data that can be used to predict future behavior. For example, data mining software can help retail companies find customers with common interests. The term is commonly misused to describe software that presents data in new ways. True data mining software doesn't just change the presentation, but actually discovers previously unknown relationships among the data.

What is needed is a data mining interface that suits both the novice user as well as the expert user. The *Utility Data Interface* described in this chapter provides a means to select specific data and create custom reports from it on demand. Novice users can create and modify reports quickly and easily using the options and links available. Expert users will quickly recognize additional strengths of the application and be able to compare and analyze data patterns that might not have been clear before. This tool helps to reduce the complexity and confusion involved in using utility data as part of a successful energy management program.

WHY CREATE AN ENERGY INFORMATION SYSTEM (EIS)?

If you examine successful energy management programs, you will generally find someone who champions the effort. However, there are several other pieces to the energy puzzle that must be in place in order for the energy management program to be sustainable. First, executive management must be committed and support the energy management program with resources and a budget. The next important piece is a facility-wide sense of ownership and accountability for energy usage. Everyone can help the energy program by identifying energy waste instead of assuming that it is someone else's responsibility. Once the desire and motivation to reduce energy costs is in place, there must be sufficient technical resources available, generally a combination of in-house, vendors and consultants, that come together to evaluate methods for reducing energy costs. Finally, there must be an energy reporting tool, a "report card" that tracks energy usage to budget to determine how well each area is doing. This is where the EIS comes in to the picture. The EIS must be easy to use and provide the answers regarding how well each area is doing relative to a budget, benchmark or to other areas. The EIS should rank the areas from best to worst and the resulting competition that results helps drive the process forward. The areas that are not performing are the source for further investigation by the energy management technical experts. The areas that are doing well should receive recognition from executive management. Examining utility data determines the degree of success of an energy management program.

UTILITY DATA AND REPORTING FORMAT

Utilities gather data on energy use by the minute, hour, month, year or any other increment desired. This data spans the realm of physical locations from individual meters to large complexes of buildings. For reporting purposes, the data may be organized into groups by physical location or some other criteria. Data will need to be reported in detail as well as summarized over these many time periods and groups. Comparisons need to be made between time periods and groups. It might be desirable to normalize the data by some other known value, such as floor area (square footage). "For most building types, floor area is traditionally the primary normalization variable for comparing building energy use and past work of this type has been based on it." [1] However, the data may also need to be normalized on a *per day* basis for utility billing data that has varying number of days from month to month.

Utility data are represented in many different units. Electricity alone may be recorded and reported in watts, kilowatts, megawatts, kilowatts hours, kilowatt hours per day or any other unit of electrical measure. Often, the data are recorded and stored in one unit of measure while the data reports use one or more other units. This requires conversion algorithms within the reporting application.

A wide variety of reports are needed to monitor all aspects of energy use across the many time periods, groups and units available. Users of an EIS may not be familiar with utility data and its units. These users may simply be trying to conserve energy by making informed decisions based on recorded data. They probably don't have a lot of time to spend on the task either. A few minutes a day, a few hours per week or one day per month may be all the time needed to make good informed decisions if the data are made available in a convenient manner. The amount of data collected for any one complex may become quite large so data will have to be filtered to create effective reports.

More and more data becomes available as automation takes hold of our facility infrastructure processes. Automated metering devices make collecting and storing energy data very easy. Thomas Jefferson University, near Philadelphia, is a good example of a large complex of facilities whose energy use is monitored using real-time automated metering technology and intranet reporting. [2] The problem for most is how to sort out all the meter data and present it in a way that is useful to many people. Many people need the information to make good, well-informed decisions about energy use and conservation.

UTILITY REPORTING GOALS

There have been many attempts at building customizable query tools that return user specified data from a database. They really don't provide much flexibility in reporting much less an interactive report with links to sort, filter and graph trends. For an example, take a look at **Advanced Query Tool** at *http://www.advancedquerytool.com/index.html*. Another method is to create a multitude of reports customized to a specific need and added to a long menu. This makes it difficult to decide which report to choose from and still doesn't open the door nearly wide enough to accommodate all user demands as quickly as needed. Another example is **InfoSurfer** by GanyMede Systems, Inc. at *http://www.infosurfer.net/infosurfer.htm* (registration required to view demo). So we have efforts at opposite ends of the spectrum of possibilities today.

Users of an EIS need to be able to reach into the massive database of utility data available and grab specific pieces of data and report the results quickly and effectively. A single point of entry into the database is a welcome change from a long menu of diverse *hard-coded* reports. The ability to search for data is important. Report templates can give novice users a start in designing reports to compare values by time periods and groups.

The reports must have the ability to summarize data by time period and group as well as provide detailed information. Users should be able to choose which units to display and specify the content of the report as much as possible. They should be able to quickly filter data by pertinent criteria. Users should be able to save their newly designed reports to a list of existing named reports for later use.

An ideal web page design would produce a printable report while keeping the selected options that drive the report visible to the user. Selection of options must be controlled so that the user cannot pick invalid items. An intelligent logic will present the user with only those options that are appropriate to previously selected options.

Any utility reporting interface should be easy to use but flexible enough to gain access to any and all data collected and stored in a utility database. It should be robust in design so that users can access the data in any number of ways to produce a clean report. And, it needs to be the focal point of the EIS, providing a starting point for exploring and reporting utility data.

UTILITY REPORTING SOLUTION—UTILITY DATA INTERFACE

The *Utility Data Interface* acts as the front-end to an Energy Information System (EIS). It could be compared to the dashboard of a car in that it provides information in a timely manner with little or no interaction on the part of the user (driver). Information on an automobile dashboard tells us the status of our car, its location, speed, temperature and many other bits of data important to our ability to drive safely and comfortably. A *Utility Data Interface*, working in conjunction with the processes of an EIS, can provide information about facility energy consumption. Even though EISs are more likely to provide information on a day-by-day basis rather then a minute-by-minute basis this allows facility managers to make well-informed decisions regarding energy conservation. Readers should note that the *Utility Data Interface* will likely require some customization for each different organization. There is also a great deal of potential for enhancement beyond what is presented in this chapter. The web page design of the *Utility Data Interface* is well suited to customization and enhancements.

OVERVIEW

The *Utility Data Interface* consists of HTML frames. One is the *title/options frame* across the top of the page. The other is the *display frame*. The *title/options frame* contains HTML elements, such as radio buttons and form fields, called *option selection elements*, to allow the user to select options that drive the database query and reporting. It also contains the application logo, title, time period option selections, a search field, other *option selection elements* and buttons to view or save the report.

The *display frame* is used for listing menu items, instructions and announcements, as well as displaying reports and graphs generated from the data. The *Utility Data Interface* design allows selected options to remain visible in the *title/options frame* while a printable report or graph is displayed. It represents a single entry point into the vastness of a utility database. The HTML frames allow users to configure and re-configure reports as needed without losing track of the selected *options*. The design does not require the user to generate a stack of data pages piled one on top of the other. The *option selection elements* used to query the database and design the report remain visible in the *title/options frame* (where?) at all times. The *display frame* contains only the information necessary to produce a printable report or graph without the option selection elements which would of course be quite a distraction. The *title/options frame* is really the best place to start designing a query and report since the user must select at least some options to produce the report desired.

TITLE/OPTIONS FRAME

The title/options frame spans the top of the web page for the complete width of the browser window. It contains the title of the application (here it is Utility Data Interface) which is probably the name of the EIS. In this case it is "Utility Data Interface." It may also contain a logo. The remaining space is for options used to filter the data such as time periods and locations. It also contains a search field.

The search feature allows searches of the database for any instance of the text entered into the search field. For example, the result of a search for the word "middle" is a page displaying all the middle schools with their associated types and districts along with any other pertinent information that may be used to quickly set options. Selecting links in the search list is a quick way to set multiple options in the *title/options frame*. This list is drawn from a table of valid option combinations called a *data catalog*. The *data catalog* is simply a list of all the possible option combinations without the values, or *readings* associated with them. The *data catalog* is used to populate other *option selection elements* with valid data. Valid options are important to designing a report quickly and easily.

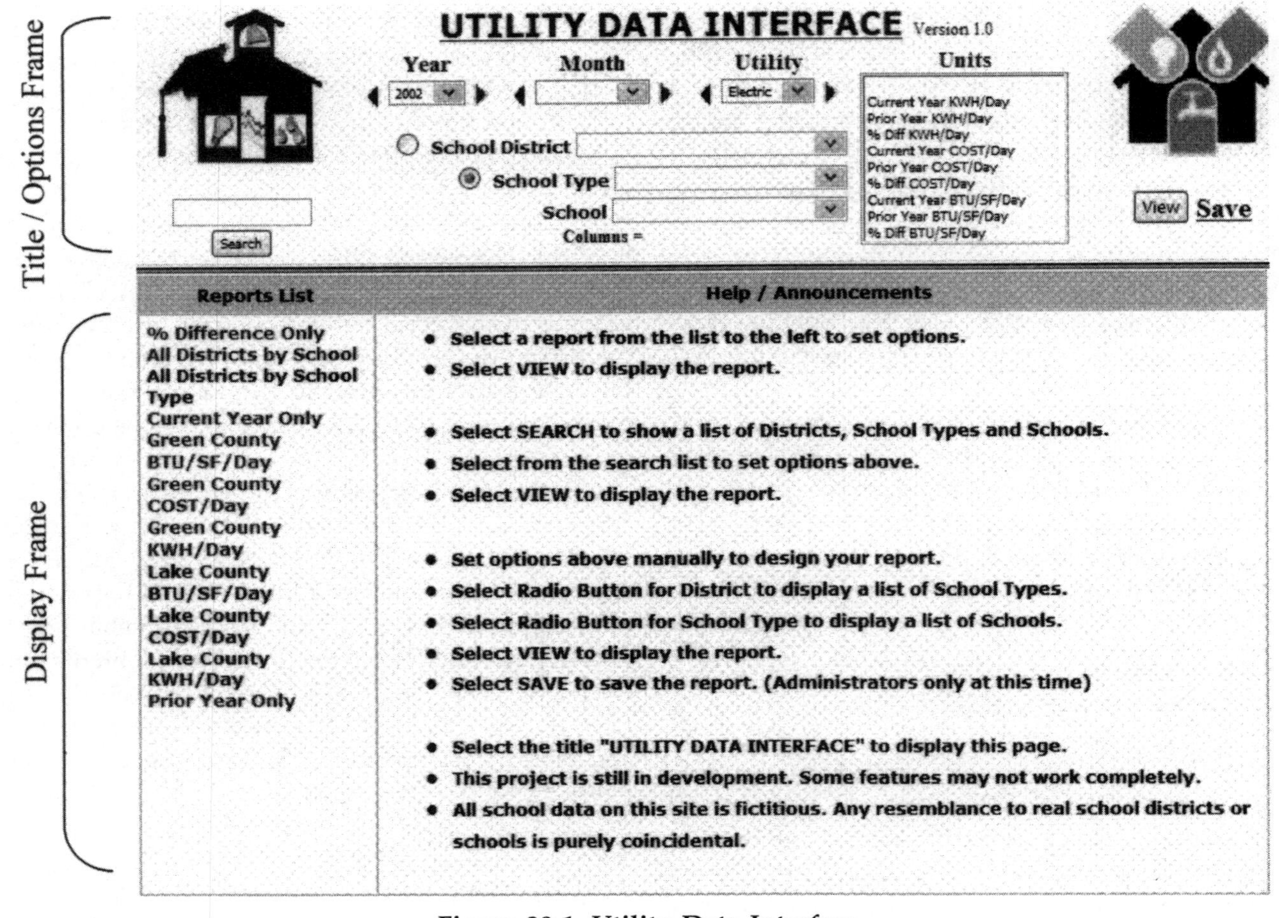

Figure 30-1. Utility Data Interface

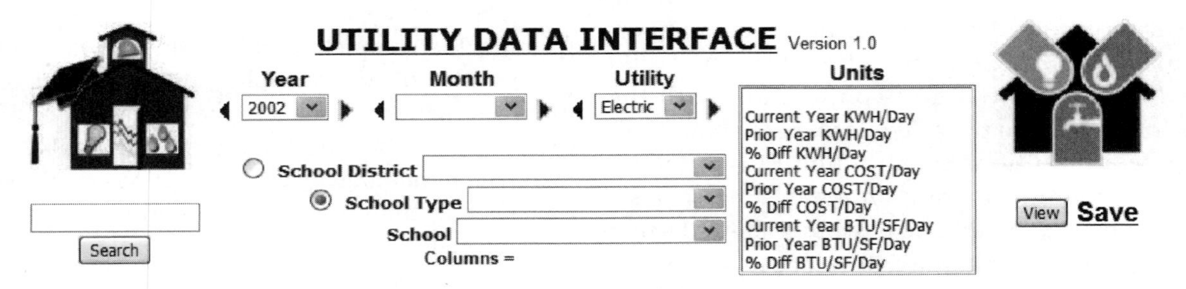

Figure 30-2. Title/Options Frame

As mentioned earlier, summarizing and filtering are very important aspects of utility reporting. The radio buttons on the *Utility Data Interface* represent control over the summaries. The form fields allow filtering control. Summarizing is performed on particular *groups*. A *group* might be a time period, a location or any other characteristic of the data. For instance, the radio buttons on the *title/options frame* in Figure 30-1 provide the ability to summarize by district, school type or school depending on which radio button is selected. Of course, only one of these three groups may be selected at a time. The form fields allow the user to select the district, school type or school by which to filter the data. Other form fields for year and month are used to further refine the reports. In our example, year and month fields both filter and summarize, since summarizing by time period is the focus of this example. The radio buttons could very well be placed above the time period fields as well for more flexibility in summarizing. EIS administrators may want to add date and hour fields also. For a *date* form field, a good calendar date picking tool called Tigra Calendar is available from SoftComplex at *http://www.softcomplex.com/products/tigra_calendar/*. This tool makes entering a date much easier. Other options further define the report.

The main purpose of the other options is to define which data to display. This is accomplished using the *utility* and *units* form fields. The utility and units form fields give the user a great deal of flexibility in creating the report. The *title/options frame* also contains the *view* button used to display the report. The report is displayed according to the options selected and more importantly the utility and units selected.

The utility list controls what data are shown on the report, electric, gas, water, etc. This typically affects the title of the report as well. The units list controls which units are shown as columns in the report: kWh, therms, gallons, cost, etc. Units for prior year data may also be listed such as prior year kWh, prior year cost, etc. Choices for baseline data can be included here as well. Also, normalization data can be added in this way such as *kWh/day or kWh/square foot*. Selecting these units determines which columns of data will be displayed. The order in which they are selected dictates the order in which the columns will appear on the report from left to right. Since the units list may be quite large, it might be helpful to select other options first since this will limit the possibilities for valid units. Using the *data catalog* described before, the options lists are filtered to contain only those items that apply to the previously selected options. For instance, if the year 2002 is selected and there is no electric data for 2002 then *Electric* would not show up in the utility list. If electric data are available for 2002, but not for 2001, then *prior year kWh* will not show up in the units list. If electric rates are not available then *cost* will not show up in the units list. This intelligent logic insures all of the radio buttons and form fields are compatible with each other.

The combination of the radio buttons and the other form fields make the *Utility Data Interface* very flexible in its ability to summarize or drill down into the data. For instance, selecting the radio button for *district* would produce a report showing all of the *school types* for the selected *district* and a grand total line at the bottom for the entire *district*. Similarly, if the radio button for *school type* is selected, data for each of the schools of the selected type will be displayed along with a summarization line at the bottom for the selected type. If no *school type* is selected from the list then data for all schools will be displayed along with summarization lines for each school type.

Let's look at an example. Selecting 2002 from the *year* list and picking May as the month sets those options for the report. Selecting the radio button for *district* and picking a district from the list will produce a report for all school types in the district. Select units from the units list as needed, say kWh/Day, Prior Year kWh/Day and % Diff kWh. Now if the *view* button is selected from the options frame, a report will display the data for May, 2002.

As you can see the number of reports available to the user is nearly limitless. Remember that the radio buttons relate to grouping and summarization and the form fields relate to filtering. Grouping controls the level of detail in the report. Filtering is fairly straight forward. The database is filtered by whatever items are selected in the form fields. Users can combine radio button and form field options to compile a report as needed.

Reports are created/configured and displayed in this manner at the users' discretion. Once a report is created that fits the users' needs, it can be saved by name and added to a list of saved reports. These reports can be retrieved immediately by selecting them from the report list initially shown on the *display frame*.

GREEN COUNTY PUBLIC SCHOOLS			Electric
≤ MAY, 2002 ≥			
	Current Year KWH/Day	Prior Year KWH/Day	% Diff KWH/Day
Total ALTERNATIVE EDUCATION	8659.95	9809.97	-11.72 %
Total ELEMENTARY	626565.67	699039.63	-10.37 %
Total GREEN EDUCATION CENTER	69595.64	78385.79	-11.21 %
Total HIGH SCHOOL	271382.15	313424.96	-13.41 %
Total MIDDLE SCHOOL	332007.22	364658.32	-8.95 %
Total NINTH GRADE CENTER	3138.08	3852.48	-18.54 %
Grand Total GREEN COUNTY PUBLIC SCHOOLS	1311348.71	1469171.15	-10.74 %

Figure 30-3. kWh/Day by School Type

DISPLAY FRAME

The *display frame* initially holds the list of saved reports, instructions and any program announcements. This acts as a kind of home page for the *Utility Data Interface* as well as the EIS. Customized report templates can be added to the reports list to help users get started in designing their own. The *display frame* is also used to display the reports and graphs. Links programmed into the reports with HTML elements simply refresh the same frame with a new report or graph. The links provide a method of instant sorting or graphing for trend analysis.

CONCLUSION

The difficult part of mining data from any database is doing it in a way that is comprehensive enough to get at all of the data, yet simple enough for the majority of users. Typical EIS users need to be able to access the complete data set from a *single entry point*. Preferably, this *single entry point* takes advantage of the common characteristics of all utility data in an effort to make querying the data easier for novice users. Hopefully, it is more than just an opportunity to return raw data from database tables. It needs to be a powerful analytical tool configurable at the click of a mouse button. It also needs to have the ability to store commonly used queries and report designs for later use. The ability to return instantly to a known configuration does have its advantages. Expert users will likely want to create new reports by modifying existing ones. Therefore, the *Utility Data Interface* approach to data mining is somewhat in line with the needs of EIS users.

The *Utility Data Interface* attempts to bridge the gap between other efforts by combining a database query tool with a report design tool. Utility data are unique in that they are always dependent on some date or time value and they are related to some physical location or characteristic. This is why the *Utility Data Interface* is a more descriptive and easier to use query tool. It has powerful configuration capabilities for summarizing and filtering data into a custom report that is *interactive*. The report itself contains powerful sorting and filtering features as well as trend analysis. Once reports are designed the users can save them for quick retrieval at a later date. So, the UDI provides a *single entry point* into the utility data database as well as a report design tool that suits both novice and expert users. This should be enough to provide many EIS users the opportunity they need to make well-informed decisions about energy use and conservation. A demo of the *Utility Data Interface* is available at *http://www.utilityreporting.com/udi.*

References

[1] Cary N. Bloyd, Asia-Pacific Sustainable Development Center East-West Center; William R. Mixon, TECH Support Services; Terry Sharp, Oak Ridge National Laboratory; "Institutionalization of a Benchmarking System for Data on the Energy Use in Commercial and Industrial Buildings," East-West Center for APEC, Nov., 1999, Honolulu, Hawaii.

[2] McGowan, Jack; "Selling Real-Time Metering up the Management Chain," *Energy Online*, February, 2004, Vol. 29 No. 2, Troy, Michigan.

Reports List	Help/Announcements
% Difference Only	• Select a report from the list to the left to set options.
All Districts by School School	• Select VIEW to display the report.
All Districts by School Type	
Current Year Only	• Select SEARCH to show a list of Districts, School Types and Schools.
Green County	• Select from the search list to set options above.
Btu/sf/Day	• Select VIEW to display the report.
Green County	
COST/Day	• Set options above manually to design your report.
Green County	• Select Radio Button for District to display a list of School Types.
kWh/Day	• Select Radio Button for School Type to display a list of Schools.
Lake County	• Select VIEW to display the report.
Btu/SF/Day	• Select SAVE to save the report. (Administrators only at this time)
COST/Day	
Lake County	• Select the title "UTILITY DATA INTERFACE" to display this page.
kWh/Day	• This project is still in development. Some features may not work completely.
Prior Year Only	
	• All school data on this site is fictitious. Any resemblance real school districts or schools is purely coincidental.

Figure 30-4. Display Frame

Chapter 31

Developing an Energy Information System: Rapid Requirements Analysis

David C. Green
Green Management Services, Inc.
Fort Myers, Florida—dcgreen@dcgreen.com

ABSTRACT

THIS CHAPTER describes a simple method of evaluating the importance of specific Energy Information System (EIS) requirements. This method helps organizations that have identified a need to develop an EIS but can't find the time to define the complex requirements. It is a method intended to speed up the process without sacrificing the quality of the process. Energy Management Teams typically know a lot about the need for each requirement but very little about the cost or effort to produce it. On the other hand, Information Technology (IT) Development Teams know more about costs and effort than the relevant importance of each of the requirements. Therefore the conditions are set for confusion and the likelihood of disagreements is high. Our goal in using this method of requirements definition is to prevent *cost* evaluation from tainting the *need* evaluation of any one of the proposed features of the EIS. This method breaks the requirements definition down into manageable tasks, evaluates the best balance between need and cost for each of the requirement tasks and then produces a priority of work document to guide the development team.

INTRODUCTION

With any EIS project one of the purposes of requirements analysis is to translate the needs and goals of the Energy Management Team into a document that the IT Development Team understands and can use to produce a cost estimate for the project. Requirements for an EIS, which may be more complex than many other software development projects, involve data collection (sometimes from remote locations), conversion formulas, input forms and reports for a wide variety of data. [1] In addition, utility data has unique characteristics. Meters produce the data in a wide variety of units of measurement and at many different intervals. The Energy Management Team understands these complexities very well. IT Development Teams however, often don't understand utility data well. Furthermore, the roles of the Energy Management Team and the IT Development Team are quite different. The Energy Management Team defines what the proposed application should do and how they want to use the application to do those things. The IT Development Team provides the tools to do it. But first, they need to work together to take advantage of both their strengths in order to work through the complex requirements of an EIS project.

EIS requirements fall into several categories. This may be one reason why the requirements analysis task is so difficult. There are collection tasks, input forms, reporting tasks, graphing, security issues and configuration requirements among others. The EIS must collect data from meters automatically or through input forms. The characteristics of the data vary from meter to meter so each meter must have a unique configuration description in the EIS. The EIS uses this configuration to collect and report data for each meter. This configuration description might include the interval at which the data are collected (minute, hour, day or month). Varying collection intervals sometimes present a problem to EIS processing tasks so they need to be part of the system configuration as well. Input forms have to validate data and make sure only certain users can change certain data at certain times. Security features may link users to input screens based on the time period or on which data are changed. The system will likely prevent users from changing previous month or year values. This is necessary to insure the ability to reproduce reports from the past exactly as they appeared at the time. Conversion formulas allow utility data to be presented in units other than those used to measure the utility data at the meters.

Utility data have unique characteristics that make the requirements for an EIS complex. First, the EIS usually collects utility data on an hourly basis, if not more often. This creates a huge amount of data. Systems collect data from many different locations and report it on different levels. The data can come from many types of utilities and is collected in a variety of units of measure. It may need to be divided into groups for reporting purposes. Looking at the utility usage for different "groups" is a reasonable expectation. The electric consumption of one building can be just as helpful as the consumption for a whole complex of buildings. It might also be necessary to combine data from several meters and report it as a single data point or fractionalize data from one meter and report it as more than one data point.

Utility data are usually date specific and require trending over time to be useful. Systems record all data with a time stamp. Hourly data for specific days and daily data spread over the period of a month provide the most meaningful reports and graphs. Comparisons to other data, such as outside temperature, are also helpful. Sometimes averaging data or summing it for each utility is required. Systems can average data or sum it together over periods of time. Data can also be summed by combining meters at various locations. These conditions obviously require the Energy Management Team to be deeply involved in the EIS development. [2]

The Energy Management Team understands the energy-related processes such as cost allocation, converting units of measurement, tracking meter reading schedules, to name only a few, that currently aid them in doing business. They also understand how the complex requirements mentioned above fit into those processes. So they know the priorities, formulas, problems, plans for improvements and potential benefits of an EIS. They know which tasks must absolutely be done and they know how to collect, input and store the data to produce the reports they need. They work around the problems that come up from day to day such as missing data or changes in price. They maintain formulas to calculate the desired results. They keep track of adjustments when needed. The Energy Management Team needs to express these requirements to the IT Development Team in a clear and concise manner. They need to specify tasks that are measurable and easily determined as being complete or incomplete.

The IT Development Team has unique capabilities that the Energy Management Team does not possess. They probably have developed applications similar to the desired EIS. They can develop applications that store, analyze and present data in a user friendly manner. They understand the maintenance requirements and all the hardware and software needed for the system. They understand the staffing requirements, strengths and weaknesses of the available personnel, and the amount of effort needed to do the job. Even though it is not necessary to put an *exact* cost estimate on each task, it should be easy for the IT Development Team to assign a *relative* cost to each individual task. They will still need to collaborate in some manner with the Energy Management Team to complete the relative cost assessment.

Bringing teams together to work on IT projects has been a problem for some time. Even well-defined Systems Development Life Cycle methods have failed to bridge the gap. Statistics show that more than half of large IT projects have significant cost overruns and require constant maintenance. Also, more than half the problems result from poor requirements gathering and functional specification definition. Surveys show that users and developers frequently fail to cooperate and coordinate with each other. Yet, organizations continue to attack the problem with more conversation and analysis rather than stepping back and trying alternative approaches. [3]

A case study concerning conversations among analysts and clients during requirements gathering shows results demonstrating that more conversation may not be the solution. The study showed that brief comments during the discussions were well informed and pertinent to the topics. However, there was a clear failure to put them all together toward common overall goals. This could be because of frequent topic changes and backtracking during the conversation. Interestingly, the study also shows that most participants gained satisfaction from the conversation based on how well the social interaction went rather than how many goals the group accomplished. [4]

It's clear that bringing two diverse teams together to deal with such complex data could be a challenge, and it is for many organizations. As long as the Energy Management Team can organize the requirements in such a way as to keep the IT development Team focused on one task at a time then the whole process becomes a stepwise effort. Merging the strengths of both teams together to produce a development plan in a reasonable amount of time is the goal of Rapid Requirements Analysis.

GOALS OF RAPID REQUIREMENTS ANALYSIS

Rapid Requirements Analysis provides a way to independently evaluate the tasks required to complete an EIS project yet combine the results of those independent evaluations into a balanced requirements definition and work plan. And, it seeks to do it in a short period of time so that those involved maintain interest and find that budgeting is easier. It provides a way for the Energy Management Team to present a clear picture of how important one task is relative to others by establishing a prioritized list of required tasks based on *need*. In other words, for each task required, the team asks the question: "how much is this needed relative to other tasks?" Ultimately, this tells the IT development Team which tasks are absolutely required and which tasks are not.

The IT Development Team then presents a clear picture of how *costly* each task is relative to others independent of the previous evaluation. They rank the requirements in some logical order based on cost and effort required to complete the task. Usually the *less costly* tasks get a higher priority just as the *most needed* tasks got a higher priority. This encourages everyone involved to examine individual tasks in-depth independently of each other. A facilitator can answer questions concerning the details of the tasks or act as a liaison between the teams to clarify the task details. The project manager or a consultant may fill this role. This is a more efficient manner of detailed analysis than conducting meetings with everyone involved since many times the details of any particular task only involves a few people. A later meeting would be a good way to combine the results of their investigations.

A final list of tasks is a combined evaluation of the requirements tasks with the most needed and least costly tasks near the top of the list and the least needed and most costly tasks near the bottom of the list. This list, prioritized by a combined need and cost rating, gives the teams a preliminary look at which tasks may be too costly to include in the project. The final list is the beginning of a detailed requirements document and priority of work that allows the teams to work together more efficiently.

Once the teams have evaluated the requirements list in this manner they can stay focused on one task at a time and be sure they are working in the correct priority. Evaluating the tasks independently prevents "cross contamination" of results due to the effects of one criterion on the other. For instance, if the teams meet together to discuss one particularly important task and they place a cost estimate on the task at that time, the effect of knowing the cost estimate might be to lower the task's importance in order to save time and money. In the same manner tasks that seem less costly might tend to get a rating of higher importance just because they appear to be easy.

So, this method of independently evaluating tasks using the strengths of each team and later combining the results of the evaluations into a comprehensive list of requirements helps to organize thoughts and processes while the project proceeds. Rapid Requirements Analysis encourages a well organized and accountable project life cycle.

CONDUCTING A RAPID REQUIREMENTS ANALYSIS

The following steps illustrate how to conduct a Rapid Requirements Analysis. The project manager may wish to add other tasks to suit an organization's system design rules. However, this framework is a proven method of requirements gathering. It has worked under the toughest of conditions and shows improvement over conventional methods in concept-to-prototype cycle times.

Step 1: **The Energy Management Team should identify who is involved.**

They should list everyone involved in the project and what their roles are to be. A contact list with phone numbers, email addresses, office locations and short description of their responsibility is helpful.

Step 2: **The Energy Management Team should list all primary and secondary tasks that the EIS must perform in order to completely replace the existing processes and accomplish the desired new processes.**

A mission statement can help to qualify which tasks the project absolutely needs to accomplish its goals. If tasks are categorized, they can be grouped together based on their dependence to one another. The team should list the tasks as clear and concise descriptions but not so detailed as to make them cumbersome to read and comprehend rapidly. Detailed analysis of the task later can clarify any questions that arise. An ID number will help keep track of individual tasks since the list may be quite large. See Figure 31-1 below for an example of how to list tasks.

Step 3: **The Energy Management Team and the IT Development Team must agree on criteria for evaluating the requirement tasks and a rating scheme for the criteria.**

It's likely the criteria will be "need" and "cost." Either team can add other criteria as desired. However, the more criteria the more difficult and time consuming the analysis becomes. Our examples will assume *need* and *cost* are the two criteria. The rating scheme determines how tasks compare to each other. For instance, the most common method is to assign the most needed tasks a lower number and the least needed tasks a higher number. Or, in other words, "lowest is best." The same method can apply to cost evaluations. The tasks with the lowest cost will have a low value assignment and tasks with a high cost will have a high cost value assignment.

Step 4: **The Energy Management Team independently evaluates the tasks based on *need* and at the same time, the IT Development Team independently evaluates the same tasks based on *cost*.**

The EM team puts a value in the "NEED" column associated with each task that represents its *need* relative to the other tasks, while the IT team fills in the "COST" column, ranking the tasks according to *relative cost* and not trying to estimate an exact cost for each task.. Each team should assign values in large increments, at least one hundred units apart, so that other tasks will fit between them later. The EM team can add new tasks that are required to complete other tasks but only if their *need* value is inherited from tasks that require the new task. These additional required tasks should be few in number and

ID	*Category*	*Task*
2	Electric	Maintain list of utility account numbers
4	Electric	Maintain electric billing days
5	Fuel	Generator gasoline
9	General	A security plan describing all security issues and procedures.
12	General	Data will be reported at various levels: by year, month, day, hour.
15	Network Access	Specific input screens will require security login procedures.
17	Square Footage	By Facility
18	Weather	Degree days
19	General	Data will be stored at various levels: by year, month, day, hour.
20	General	Data collected automatically will be available for reporting on an daily basis.
21	General	Graphs will be an image allowing them to be copied and pasted to other applications.
22	General	Distribution list
23	Distribution	Email notification of change to persons on distribution fist.
24	Email Subscription	Maintain list of name, email address, report URL, frequency of delivery.
25	Email Subscription	Email link to reports according to email subscriptions on a daffy basis.
26	General	Graphs will allow for multiple data sets, combined line and bar graphs and auxiliary date on the right axis
27	General	Graphs will allow for any specified data set as the auxiliary data on the right axis.

Figure 31-1. Task List

only added if the task enhances the EIS. Any other tasks required to complete the listed tasks should be part of the cost and effort involved in completing that task. Figures 32-2 and 31-3 show independent evaluations of tasks based on need and cost respectively.

Step 5: **The results of the two teams' work are combined.**

The project manager merges the task evaluation values together by adding the *need* values to the *cost* values as a "TOTAL" column. The project manager then sorts the list by "TOTAL" and calls a meeting to discuss the results. It could be that some tasks are not in the appropriate position in the list. Adjusting the *need* and *cost* values in the list produces a new prioritized list. This "adjusting" is repeated until the teams agree on a final requirements task list. Figure 31-4 (which is only a partial listing of the tasks) shows the "NEED" and ""COST" columns summed together as "TOTAL" and then sorted by "TOTAL."

Step 6: **The final analysis is performed.**

Once the teams agree on a final task list, the IT Development Team can draw a line or lines in the list and assign project costs to the tasks above or between the lines. This list becomes the basis for a detailed requirements document, cost estimate and priority of work.

CONCLUSION

EIS project requirements analysis is never going to be easy. As more and more technological innovations work their way into the energy management arena, it will only become more complex. As responsibilities grow and time becomes more and more valuable, it will be more difficult to get large groups of people to work together on such complex projects. This may be one rea-

ID	*Category*	*Task*	*Need*
2	Electric	Maintain List of Utility account numbers	100
4	Electric	Maintain Electric Billing Days	100
5	Fuel	Generator gasoline	100
12	General	Data will be reported at various levels: by year, month, day, hour.	100
9	General	A security plan describing all security issues and procedures.	100
15	Network Access	Specific input screens will require security login procedures.	100
17	Square Footage	By Facility	150
18	Weather	Degree days	150
19	General	Data will be stored at various levels: by year, month, day, hour.	200
20	General	Data collected automatically wig be available for reporting on an daily basis.	200
21	General	Graphs will be an image allowing them to be copied and pasted to other applications.	350
23	Distribution	Email notification of change to persons on distribution list.	400
22	General	Distribution fist	400
24	Email Subscription	Maintain list of name, email address, report URL, frequency of delivery	600
25	Email Subscription	Email link to reports according to email subscriptions on a daily basis.	600
26	General	Graphs will allow for multiple data sets, combined line and bar graphs and auxiliary data on the right axis.	600
27	General	Graphs will allow for any specified data set as the auxiliary data on the right axis.	

Figure 31-2. Independent Evaluation by Need

ID	Category	Task	Cost
2	Electric	Maintain List of Utility account numbers	100
4	Electric	Maintain Electric Wing Days	100
22	General	Distribution list	100
9	General	A security plan describing all security issues and procedures.	100
17	Square Footage	By Facility	100
18	Weather	Degree days	100
5	Fuel	Generator gasoline	200
21	General	Graphs will be an image allowing them to be copied and pasted to other applications.	300
12	General	Data will be reported at various levels: by year, month, day, hour.	400
19	General	Data will be stored at various levels: by year, month, day, hour.	400
20	General	Data collected automatically will be available for reporting on an dally basis.	400
15	Network Access	Specific input screens will require security login procedures.	400
23	Distribution	Email notification of change to persons on distribution list.	500
24	Email Subscription	Maintain fist of Name, email address, Report URL, Frequency of delivery	500
25	Email Subscription	Email flak to reports according to email subscriptions on a daily basis.	500
26	General	Graphs will allow for multiple data sets, combined line and bar graphs and auxiliary	600
27	General	Graphs will allow for any specified data set as the auxiliary data on the right axis.	600

Figure 31-3. Independent Evaluation by Cost

son why outsourcing is so popular these days. This may also explain why Energy Information Systems have been slow to prevail. Yet the advantages of EIS projects continue to multiply rapidly. Political, environmental and weather trends among others affect our energy consumption and cost. For the most part these factors are uncontrollable. Monitoring consumption regularly is the most productive way to cut costs. This justifies an effort to streamline the requirements gathering process and help produce Energy Information Systems specific to the needs of an organization.

This Rapid Requirements Analysis method attempts to bridge the gap between two diverse professional teams in an effort to take advantage of each of their respective strengths toward developing an EIS. It avoids the well known pitfalls of large lengthy meetings hammering out details of an EIS project without losing the detailed analysis needed to get the job done. It combines needs and costs of tasks so that the tasks can be compared and prioritized more effectively. It brings everything together into a well organized document that reflects the goals and capabilities of both teams.

As with any other idea, this method is open to adaptation. Sub-categories assigned to each task may be helpful. Also, the project manager may want to add a reference column to keep track of data or other information for each task. However, the project manager should be cautious to avoid over-complicating the process since simplicity is one of its virtues. A demonstration of using Rapid Requirements Analysis is available at *http://www.utilityreporting.com/rra*.

ID	*Category*	*Task*	*Need*	*Cost*	*Total*
2	Electric	Maintain List of Utility account numbers	100	100	200
4	Electric	Maintain Electric Billing Days	100	100	200
9	General	A security plan describing all security issues and procedures.	100	100	200
17	Square Footage	By Facility	150	100	250
18	Weather	Degree days	150	100	250
5	Fuel	Generator gasoline	100	200	300
12	General	Data will be reported at various levels: by year, month, day, hour.	100	400	500
22	General	Distribution list	400	100	500
15	Network Access	Specific input screens will require security login procedures.			
19	General	Data will be stored at various levels: by year, month, day, hour.	200	400	600
20	General	Data collected automatically will be available for reporting on a daily basis.	200	400	600
21	General	Graphs will be an image allowing them to be copied and pasted to other applications.	350	300	650
23	Distribution	Email notification of change to persons on distribution list.	400	500	900
24	Email Subscription	Maintain fist of Name, email address, Report URL, Frequency of delivery.	600	500	1100
25	Email Subscription	Email link to reports according to email subscriptions on a daily basis.	600	500	1100
26	General	Graphs will allow for multiple data sets, combined fine and bar graphs and auxiliary data on the right axis.	600	600	1200

Figure 31-4. Merge Evaluations

ID	Category	Task	Need	Cost	Total
2	Electric	Maintain a list of utility account numbers	100	100	200
4	Electric	Maintain Electric Billing Days	100	100	200
9	General	A security plan describing all security issues and procedures.	100	100	200
17	Square Footage	By Facility	150	100	250
18	Weather	Degree days	150	100	150
5	Fuel	Generator gasoline	100	200	300
12	General	Data will be reported at various levels: by year, month, day, hour.	100	400	500
22	General	Distribution list	400	100	500
15	Network Access	Specific input screens will require security login procedures.	100	400	500
19	General	Data will be stored at various levels: by year, month, day, hour.	200	400	600
18	General	Data collected automatically will be available for reporting on a daily basis.	200	400	600
21	General	Graphs will be an image allowing them to be copied and pasted to other applications	400	500	900
23	Distribution	Email notification of change to persons on distribution list.	400	500	900
24	Email Subscription	Maintain list of name, email address, report URL, frequency of delivery	600	500	1100
25	Email Subscription	Email link to reports according to email subscription on a daily basis	600	500	1100
26	General	Graphs will allow for multiple data sets, combined line and bar graphs and auxiliary data on the right axis.	600	600	1200

Figure 31-5. Cost Lines

References

[1] Burns, Kathleen; "Energy Information Systems: Knowledge About Power"; *Energy User News*, August 2000, *http://energyusernews.com.*

[2] Capehart, Barney Ph.D., C.E.M.; "Utility Data Web Page Design: An Introduction"; *Information Technology for Energy Managers*; Chapter 23; Fairmont Press, Inc., Lilburn, Ga. 2004.

[3] Jennerich, Bill; "Joint Application Design"; Internet page, *http://www.bee.net/bluebird/jaddoc.htm*, accessed 8/2/2004; Bluebird Enterprises Inc., Berwyn, PA

[4] Urquhart, Cathy Ph.D.; "Strategies for Conversation and Systems Analysis in Requirements Gathering: A Qualitative View of Analyst-Client Communication"; *The Qualitative Report*, Volume 4, Number 1/2, January, 2000; (*http://www.nova.edu/ssss/QR/QR4-1/urquhart.html*).

Section Seven

Enterprise Energy Management Systems

Chapter 32

Defining the Next Generation Enterprise Energy Management System

Bill Gnerre
Gregory Cmar
Interval Data Systems, Inc.

ABSTRACT

THIS CHAPTER highlights the key functional requirements of an Enterprise Energy Management System (EEMS) and describes how this functionality can be used to reduce costs, increase efficiency, and improve energy planning and cost allocation, all while improving or maintaining building comfort. Appendices to this chapter contain case studies showing the value of using interval data from the building automation system (BAS) as a tool to diagnose and monitor facility operations.

INTRODUCTION

The challenge is simple: how do facility organizations find new and innovative ways to ensure maximum operational efficiency, reduce deferred maintenance budgets by extending the life of systems and equipment, be good stewards of the building assets, forecast energy needs more accurately, and achieve the lowest energy purchase?

The answer is an Enterprise Energy Management System.

Today, large campuses and facilities typically have one or more building automation and control systems, campus metering systems (automated and/or manual read), a lighting management system, and some form of space management system. In addition, they deal with several utility companies, each of which has changing rates and rate structures. These systems generate an enormous amount of valuable operational information—which is nearly all thrown away without even being looked at because it is difficult to capture and access data.

While historically organizations have attempted to control energy and building maintenance costs by managing each individual energy source and energy consumer (e.g. building automation system), without a comprehensive 360° view of the facility's current energy consumption true energy optimization cannot be achieved.

An EEMS provides actionable insight through the consolidation of data from all of the institution's disparate energy and building management systems and the interactive access to that data, providing the facility's operations and engineering departments with an accurate picture of operations. With facts in hand, they can steward their assets, lower total energy consumption and operational costs quickly and effectively, and have the ability to verify and measure results.

EEMS DEFINED

At the most simplified level, an Enterprise Energy Management System consolidates *all* energy related data (sources, costs, control and monitoring points) into a data warehouse and provides tools to access and truly *interact* with the data. Conceptually straightforward, but today's energy management systems just do not do it.

It is worth noting what an EEMS is not. It is not a control system and should not be confused with building automation systems (BAS). An EEMS is much broader in scope than control systems, reaching well beyond the BAS. It provides data collection, data access, diagnostic and monitoring capabilities, a historical data warehouse, and a lot more as detailed throughout this paper. Similarly, an EEMS should not be confused with utility billing systems. It encompasses billing and meter data, but extends far beyond and connects billing information directly to the related operational data.

The EEMS makes data available so that the end user is able to perform in-depth diagnostics, analysis, and monitoring in a small fraction of the time it took with earlier methods. This, in turn, provides facilities' staffs with actionable information; i.e. information that enables them to make informed decisions to reduce energy consumption, accurately identify energy costs by cost center, or forecast energy costs in the future.

A true Enterprise Energy Management System is based upon five simple, but crucial, principles:

1. All energy related data must be consolidated into a centralized data warehouse.

2. The collected data must be 'normalized' and 'structured' to be usable.

3. Access to data must be 'interactive' and the information presented must be 'actionable.'

4. The system must measure and verify results.

5. The system must provide a platform that embraces industry standards for data collection, management, analysis, and publication.

EEMS DESIGN PRINCIPLE 1: CONSOLIDATE ALL ENERGY RELATED DATA INTO A DATA WAREHOUSE

Energy data come from purchased utilities, generated utilities, building automation systems (BAS), metering systems (both advanced and manually read), weather, and space planning systems. (There are also calculated data, but that will be covered later.) Additionally the EEMS must manage rate and billing data, users, and organizational information.

In order to be able to utilize energy data, the first step is to identify and collect the right data into a data warehouse so that accurate and actionable information can be available. The EEMS needs to collect *all* data, as one cannot optimize the whole by optimizing each component.

This section identifies the different data sources and attributes that define an EEMS and populate its data warehouse.

Purchased Utilities

For each utility within a campus or hospital, consumption data and billing information is generated.

Consumption information is typically time-based, regardless of the type of utility. For example, electric utilities use 15-minute intervals, natural gas utilities use a daily time interval, oil uses time between tank fill-ups, and water uses monthly or quarterly intervals. Other utilities such as steam or chilled water often use 15-minute to hourly interval data. Eventually these different time series need to be "normalized" (an issue that is discussed later in this white paper) so that information can be presented in consistent intervals.

Billing information is equally complex: for large campuses, there are often multiple vendors for each utility type, each with differing rates, billing cycles, and pricing structures.

An EEMS must manage both consumption and billing information, and present this data in an intelligible, clear, and actionable format. The diagram below is a simplified view of the issues related to collecting utility data.

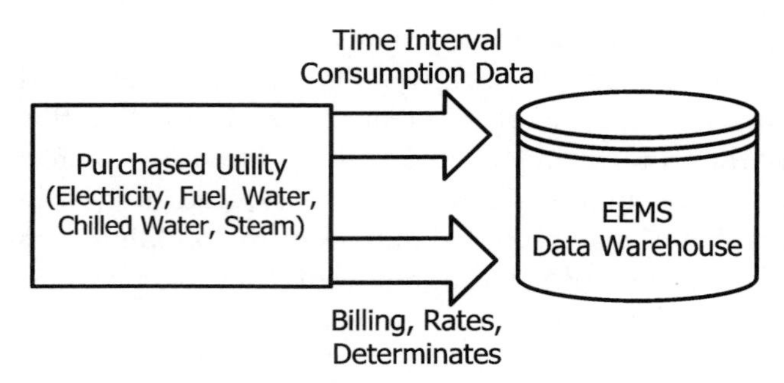

Figure 32-1. Billing & consumption data must be collected in tandem.

Time Interval Consumption Data

Many utility companies do not offer a way to track interval consumption data as it happens. The interval data are made available at the end of the billing cycle through reports or spreadsheets, which leaves the EEMS without current consumption data for a month at a time (with water being much worse). Other companies provide interval data through Web-based reports, which although more current, are far from ideal for populating the EEMS data warehouse.

The best option for collecting consumption interval data is for the meter to provide the data directly at regular intervals, or to attach a reading device that can provide the consumption data to the EEMS.

Bills, Rates, Determinants

The EEMS must understand and track the hierarchy of meter data that comes from a purchased utility. The data hierarchy goes from utility type, to supplier, to account, to meter and rate. A single rate is typically used for multiple meters spanning multiple accounts.

A large facility will have a significant amount of billing data to collect and manage. For example a university may have 1,000-3,000 utility bills per year. Today this is often captured in spreadsheets—limiting the accessibility and usability of that information. The EEMS should consolidate all the billing information, including the bill's underlying determinants.

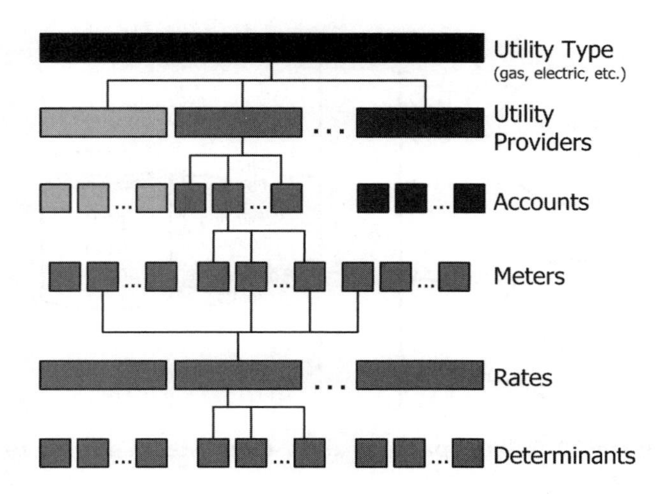

Figure 32-2. The EEMS must understand the hierarchy of utility meter data—from utility type, down to meter, and then to rates and determinants.

Billing Determinants				
Determinant	**Data Type**	**Rate 1**	**Rate 2**	**Rate 3**
Start Date	Date	■	■	■
End Date	Date	■	■	■
Total kWh	Integer	■	■	■
Billed kWh	Integer	■		
Total kW	Number	■	■	■
Billed kW	Number	■		
Rate Billing	Money	■	■	■
Customer Charge	Money		■	
Fuel Charge	Money	■	■	■
Sales Tax	Money	■	■	■
Municipal Franchise Adj.	Money	■	■	
Total Current Bill	Money	■	■	■

Figure 32-3. The EEMS must have the flexibility to handle multiple rates with varying determinants.

The bill is made up of a varying number of determinants. It is common for a large campus to have a dozen or more different rates from a single utility company. As an example, Figure 32-3 shows the billing determinants for three different rates in use at a university from a single electric company.

Electricity

Today most electric utilities quantify consumption by averaging the demand over a 15-minute period (standard interval). The majority of electric utilities make the interval data available electronically to the customers, although again, not always in convenient ways to collect it.

Both 15-minute average demand and month-to-date consumption are required for the EEMS. Similarly, electric bills with determinants must be stored in the database too (for reasons discussed later) and, because billing rates change over time, it is important that the EEMS can accommodate this dynamic data and propagate these adjustments.

Natural Gas

Due to the fact that natural gas utility companies rarely bill based upon readily obtainable standard time intervals (a fact which has led many institutions to install their own gas meters to validate billing), it is important that the natural gas meters installed throughout the site are connected to the automated metering or building automation system for data collection. It is also critical that the meter configuration and BAS point configuration collect running totals of consumption flow, etc. as well as instantaneous readings. Running total data is required to make it possible to reconstruct the inevitable gaps and missed readings.

Chilled Water

While many organizations generate their own chilled water for air conditioning, etc., some chilled water is purchased from third-party utility companies. Similar problems exist concerning metering, again leading some organizations to purchase their own meters to validate bills. Because chilled water generation is tied to electric consumption, suppliers are increasingly moving towards more accurate time-series billing.

Steam

Like chilled water, steam is often produced by an organization itself, but, when purchased, it is typically billed based upon time intervals ranging from 15 minutes to one hour.

Generated Utilities

Many large campuses/facilities have their own power plants that generate chilled water or steam that is subsequently distributed to the buildings. The EEMS must collect data from the control system(s) of the power plant as well as from the distribution network and end users of the energy.

Billing information must also be collected, much the same as it is for purchased utilities. Facilities that generate their own chilled water, steam, or even electricity will have their own rate structure and determinants and bill internally based on consumption.

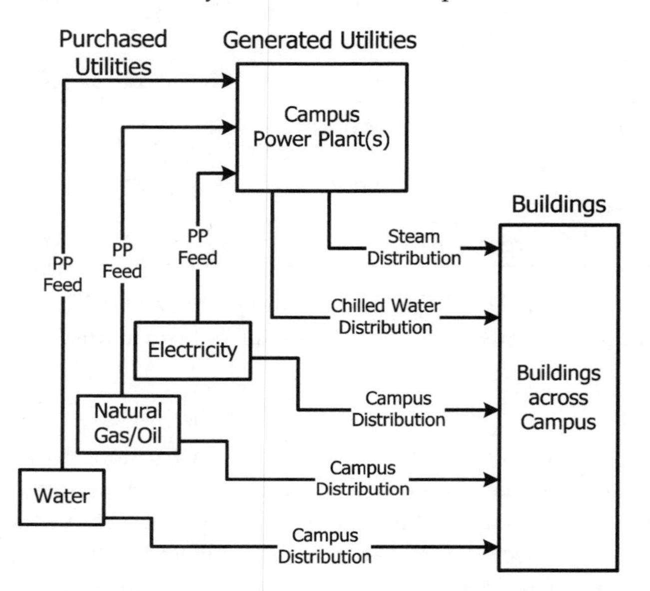

Figure 32-4. Institutions typically deliver purchased and generated utilities to the campus.

Chilled Water

Chilled water is distributed throughout the campus to different buildings. Typically, a building automation system will control the chiller plant (chillers, cooling towers, primary distribution loops, etc.).

The EEMS system must collect data from both the generating plant and the distribution network (buildings). In some cases the chiller plant will be run by a different BAS than the building and an EEMS must be able to display chiller plant efficiency as well as allow the user to determine how the chilled water is being used in the distribution system. Operating the chiller plant at maximum efficiency does not necessarily mean that the distribution system can benefit from a high delta T chiller operation, for example.

Steam

Like chilled water, steam is distributed throughout the site to different buildings. In order to determine the steam usage of a building, both a steam flow meter and condensate return meter are required: where the steam meter measures the energy delivered to the building while the condensate meter measures that which is returned. An EEMS then accesses the steam and condensate return meters via the building automation system that is normally used to monitor and control this equipment.

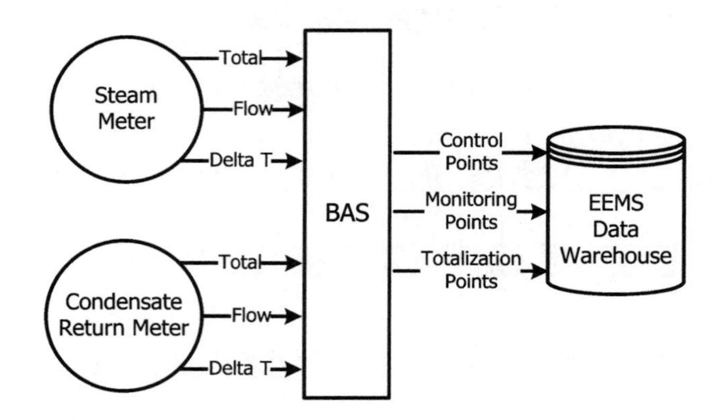

Figure 32-5. The path of data from steam meter to EEMS.

Building Automation Systems

First, an important point of distinction between a Building Automation System (BAS) and an Enterprise Energy Management System (EEMS) should be noted—BAS controls the HVAC equipment, lighting, security systems, etc. whereas the EEMS provides management information derived from the BAS, as well as all other energy systems across the site. An EEMS is not a control system.

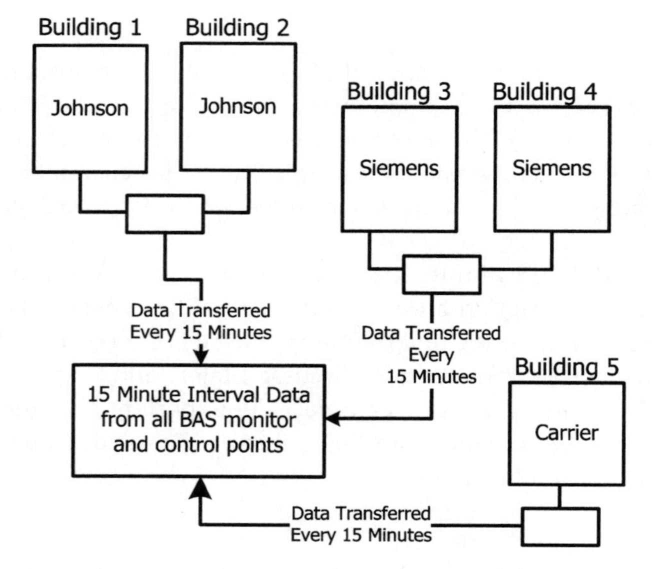

Figure 32-6. EEMS collecting data from multiple building automation systems.

Many institutions operate more than one BAS and so an EEMS provides a complete, holistic view by collecting data from all systems and overcoming the limitations of relying on the BAS for data. It is this comprehensive 360° view that enables organizations to more quickly and effectively diagnose energy usage.

As important as data collection is, it is equally important for the EEMS to provide fast access to the data, and to present a holistic view in a comprehensible form. This is covered in greater detail in "EEMS Design Principle 3."

Extracting Data from the BAS

Virtually all BASs (particularly modern systems) allow external applications to collect data without negatively impacting performance. To provide a comprehensive picture, an EEMS requires data at 15-minute intervals from control and monitoring points.

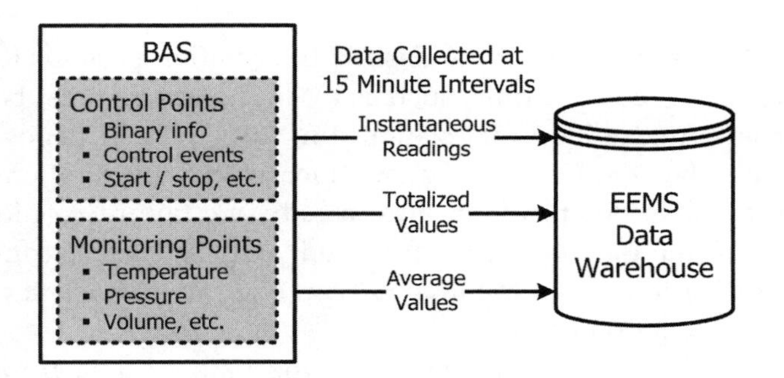

Figure 32-7. The EEMS collects various kinds of data from control and monitoring points every 15 minutes.

It is important to recognize that gathering data only from monitoring points means that, while you may identify something to improve, it is unlikely that you will gather sufficient information to know with certainty how to improve it. (This is what happens when advanced metering systems are installed instead of an EEMS.) Control points must also be gathered to be able to monitor any changes and understand how control changes affect behavior throughout the systems.

For operations at a fairly large facility or campus, it is not unusual to have 20,000 points or more. With such a large number of data collection points, each generating one record every 15 minutes for 365 days a year, the total data set comprises over 700 million records per year. From a user's data-handling perspective, 700 million records are overwhelming, and without an EEMS, much, if not all, of this information was thrown away.

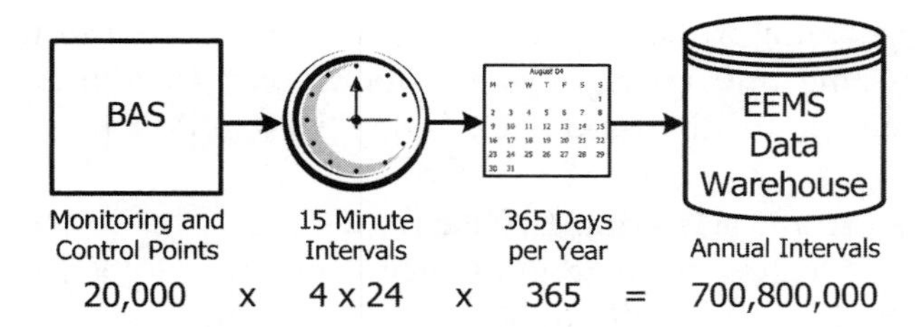

Figure 32-8. The ability of an EEMS to store, manipulate, and display very large volumes of data in an efficient manner is mission critical.

The Importance of Trend Data

Trend data are the foundation of diagnostics, monitoring, measurement and verification, and building a historical record of facility operations. While BASs typically contain a function called Trend Logs, they are a poor mechanism to collect trend data for the EEMS to operate against.

Trend Logs provide an excellent medium for viewing real-time BAS data, particularly where a known problem exists. The operator simply identifies the points required for the trend log and initiates data collection. However, there are several issues that severely limit their ability to serve as a useful data collection device or as a diagnostic or monitoring tool:

- All data are not collected. At any point in time, Trend Logs are typically only active for a few hundred points out of the thousands within the BAS.

- Trend Logs cannot collect all data. They were not designed to be active for all data points, and attempting to do so severely impacts the performance of the BAS server, affecting its ability to properly execute control functions.

- Data are limited to the BAS. Trend Logs have no ability to combine data from multiple BASs or other energy data sources critical to facilities operations.

- There is no meaningful historical data. With no constant data collection, there is no ability to look back in time at points of interest. Logs are turned on and then facilities staff must wait until enough data are collected. Earlier data are lost forever.

Fortunately there are approaches to collecting trend data without relying on Trend Logs. The EEMS needs to access the BAS monitoring and control points directly or through an independent server that does not impact the primary function of the BAS—to control the building and energy system operations. How this is done will be discussed later, under "Design Principle 5."

Meters and Metering Systems

Meters and metering systems are typically located at buildings and throughout the distribution system to measure usage of electricity, steam, chilled water, fuel, etc. Different energy types will require different approaches to move the data from the meter into the EEMS warehouse.

Most electric meter manufacturers have software applications that consolidate the data and display consumption demand and power quality data. In effect, metering systems are designed to simply collect and display data needed to understand usage. The EEMS requires this data to also be present, necessitating a connection between the metering systems' databases and the EEMS data warehouse.

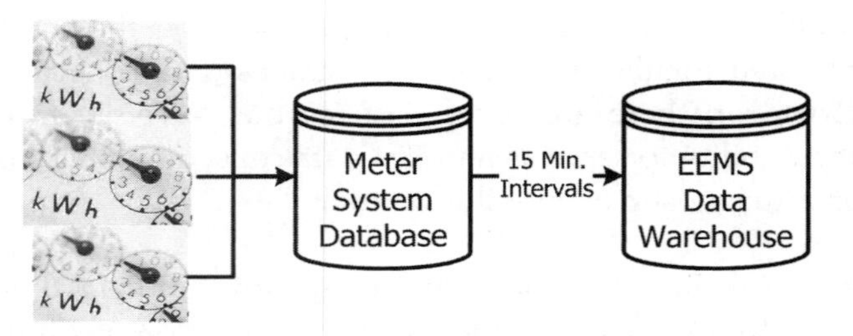

Figure 32-9. A connection between the metering databases and EEMS must exist.

For electricity, access to metering data via the metering database is a relatively straightforward process, but for those utilities that do not leverage advanced metering systems, data can be obtained and transferred through the building automation system; i.e., the BAS collects the data from the meter and the EEMS acquires the data from the BAS system. In reality, however, this connectivity is not a simple task. Meters must be reconfigured to collect and total the data, and the BAS points must be configured to access and deliver the required data to the EEMS.

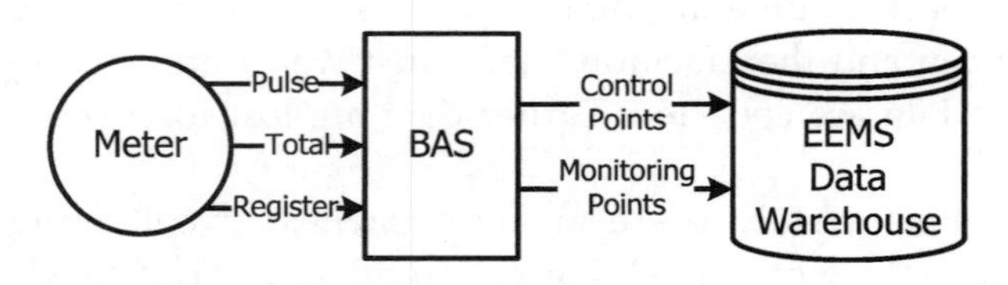

Figure 32-10. Data can be obtained & transferred through a building automation system.

In addition to the automated data collection described above, there is also the information from manually read meters that must be added to the EEMS warehouse in order to provide a complete picture for error diagnosis and identification. Many organizations already do this at some level, but the data are typically put into spreadsheets where there is limited access and no connection to other operational data. For manually read meters, like other meters, a configuration is required; i.e., the data being collected by the meter reader must be the right data for the EEMS.

Considerations for Meter Data Collection

Whenever meters are used for EEMS data collection there are a number of important considerations to make certain the right data are collected, even in the event of network failure. A decision must be made concerning the type of data needed. This can then be used as the basis for selecting the most appropriate collection device.

For example, if the desired information pertains to energy used over time, such as kWh, the device must be configured with a totalization function. While it is possible that the building automation system can total energy, it is not prudent since a network failure would mean the data was gone for good. With the meter configured to contain the total energy, one can go back and "fill in the data."

The diagram above shows the path of data from the device to the EEMS database. It is important to recognize that there are a number of potential points of failure—at each hardware device and each network link. One must plan for a loss of data and ensure that the data collection process incorporates the appropriate safeguards to minimize data loss.

As the diagram shows, data can be collected at the

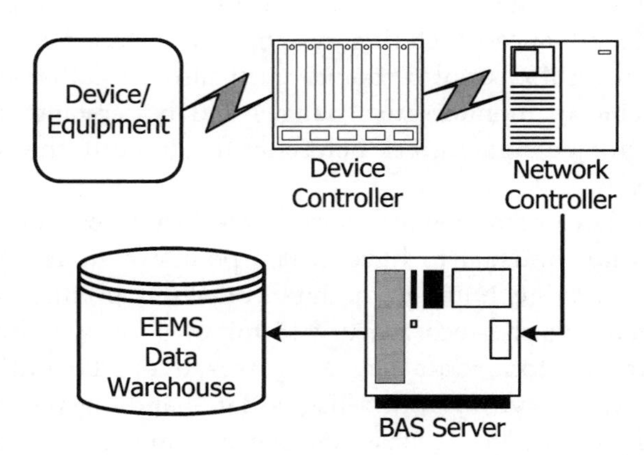

Figure 32-11. The communication path of data from device to EEMS.

device, device controller, network controller (sometimes referred to as the panel board), or the BAS server (via OPC). In selecting and specifying data collection points it is crucial to consider the goals and objectives for the project. For example, if you need to know the total amount of energy consumed you must configure the meter for register data *and* pulse data.

One other consideration when collecting meter data is deduct meters. This occurs when there is a main meter and then a series of sub-meters that cover part, but not all, of the energy consumption through the main meter. In this case the EEMS will want to collect interval data directly from the sub-meters, but then must deduct their consumption from the main meter before apportioning energy usage to other spaces.

Weather Data

To enable accurate energy forecasting, it is important to understand the context within which energy usage occurs. Weather is possibly the single biggest factor in this equation. For the most accurate results, weather data from a local airport (official METAR—Meteorological Terminal Aviation Routine Weather Report) should be used in the EEMS. While BASs typically have temperature and humidity sensors, invariably they do not provide quality reference information necessary for a number of reasons: temperature sensors may be located too close to the building, in the sun, on a roof, on the south side of the building; humidity sensors may be broken, etc.

An EEMS therefore should be configured to receive a weather feed from a local airport so that actual weather data can be used in energy forecasting applications and in assessing building performance characteristics. At a minimum, weather feeds should include the following data:

- Hourly temperature and dew point
- Wind speed and direction
- Barometric pressure
- Sky conditions (clear to cloudy)

Space Planning Data

Space planning data are required by the EEMS in order to identify energy costs at the space level. Space planning systems (SPS) contain information concerning the use and allocation of all areas within a campus or facility. They map the hierarchy of the campus by site, zone, building, floor and room.

Space planning systems also understand the relationships between space and cost centers. Both SPS and

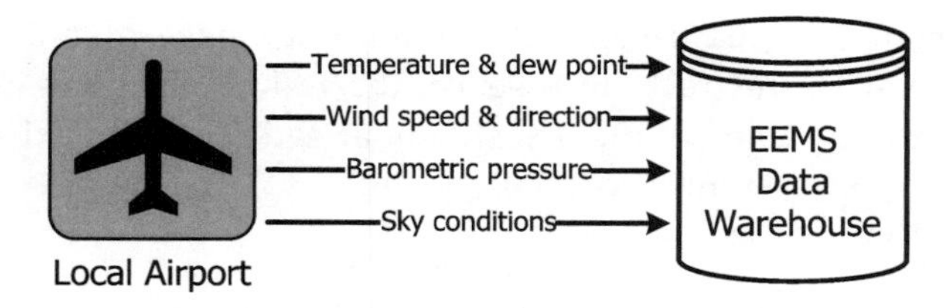

Figure 32-12. Weather data from a local airport should be used within an EEMS.

EEMS have distinct, complementary roles. Space Planning's role is to maintain the space relationships (since occupancy and cost centers change) and to transfer cost center information into the general ledger. The role of the EEMS system is to deliver accurate energy costs down to the space level where the SPS can roll up the costs by cost center.

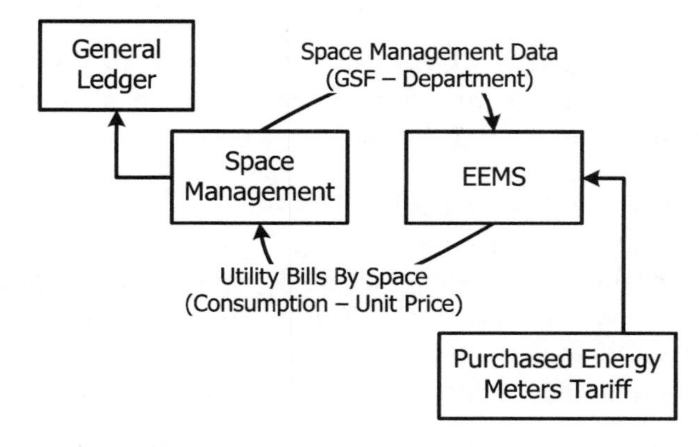

Figure 32-13. The EEMS and SPS work to combine energy and space data and write costs to the General Ledger.

Maintaining space planning data is an important discipline so that campus utility organizations, facility operations, planning and engineering functions can access common reference information. Today each of these distinct groups creates their own variation of a standard building name (for example) which makes their information useable only to themselves rather than many other departments.

Organizational Information

Organizational information is required for the EEMS so that it can roll up cost center information. Once the EEMS allocates costs to each space, organizational information is required that can relate space to department (or cost center). It is incumbent upon the EEMS to adapt to any hierarchical structure and to the constantly changing organizational structure of the institution. The EEMS system should not burden the space management system nor the personnel maintaining space planning data with this task. In essence, the EEMS functions increase the value of the investment already made in existing space planning systems.

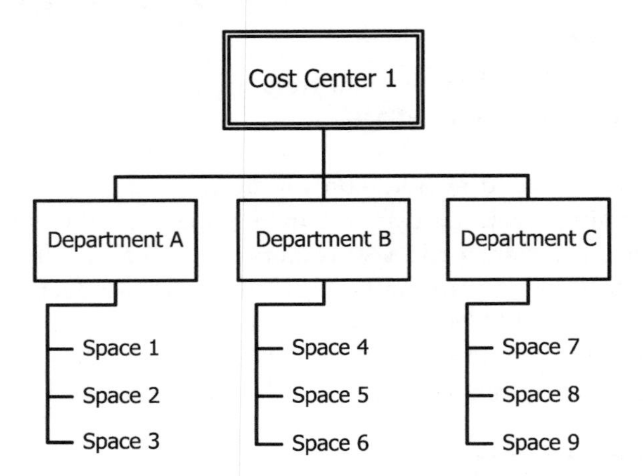

Figure 32-14. The EEMS needs to understand the dynamic nature of the relationships between organizational hierarchy and space.

Market and Pricing Data

For any purchaser of large amounts of energy, purchasing at favorable rates can make a significant impact on the bottom line. This is difficult to do since prices, pricing structures, and regulations are constantly changing. The role of the EEMS in purchasing energy is threefold:

- To display past and future energy usage patterns
- To convert and present utility billing and consumption usage into an equivalent real-time price
- To present real-time pricing information

All of this needs to be accomplished within the same interface.

Easy access to actual and predicted usage patterns enables the organization to make more informed utility purchasing decisions than ever before. For organizations that have secured the services of a third party to assist in making utility purchasing decisions, access to the EEMS must be made available to those individuals.

Don't Throw Data Away

Most large facilities and campuses spend millions of dollars annually on energy, millions on systems like building automation systems, and tens of millions for HVAC equipment. Yet the data that they generate is thrown away largely because there is so much of it, and it is difficult to access, manipulate, and interpret.

An EEMS captures this data and leverages it to provide a complete picture of utility consumption and an organization's energy infrastructure over time so that the most expeditious analysis can take place to reduce the total cost of ownership and operation.

EEMS DESIGN PRINCIPLE 2: NORMALIZE AND STRUCTURE DATA

EEMS Design Principle 1 focused on the importance and requirements of getting data into the system. EEMS Design Principle 2 constructs the data warehouse of historical trend information based upon the least common data denominator, a standard time interval. Once all data are normalized to this common standard, one can utilize it for a variety of applications.

Optimal Time Interval

There are a number of considerations in selecting a standard time series interval used by the EEMS, the most significant of which is ensuring that it is able to display sufficient data to identify transitions—this means that there should be enough data points gathered to discern performance fluctuations across transition time periods such as between day and night, 'office hours' and non-working hours, etc.

With this information, behavior patterns and problems become apparent quickly. For this reason, and because electricity is frequently metered within the same time interval, 15 minutes is an appropriate time series upon which an EEMS can be normalized. Longer intervals do not provide sufficient data granularity to always see behavioral changes. Shorter intervals can increase the data storage and processing requirements for the EEMS by 300% (or more) while increasing the information value very little.

Additionally, because capturing data at 15-minute intervals does not require a great deal of storage space to house the data records, data can be stored for the lifetime of a piece of equipment—20 years or more.

Normalized Time Series Data

Normalizing the data means that data from all sources are stored in the warehouse in the same time series interval. As discussed earlier, not all data sources provide data in the same time series, so an EEMS system must reconcile these differences. In some cases the actual time interval of the data source is one per day so the EEMS system must automatically convert this data into the normalized times series data. This problem isn't always due to the time series specified by the utility provider; it may occur for a number of reasons, including equipment failure.

The problem of time series inconsistencies is a fact that must be acknowledged and addressed. A well-implemented EEMS accepts these inconsistencies and "fills the gaps" with estimates that, when totaled, account for the total energy consumed and represent the pattern of that energy usage in a precise and accurate manner.

Data Calculations

Part of the function of any data warehouse is to provide access to the data in the most efficient means possible. This includes calculating and storing certain commonly needed values.

Once the data have been normalized, you can calculate additional trend data at each interval. For example, a Delta T calculation, simply the difference between supply and return temperatures, can be computed and stored as another monitoring point at each interval, making it available to the user as a Trend Line.

Storing calculated values is important to the overall data warehouse structure because it dramatically speeds up access by the user. The small amount of storage space used by the calculated data is more than made up for by the performance gains. (See more on calculations in a later section.)

Naming Conventions

There is a complete lack of uniformity in how buildings and systems are labeled within an institution. Today's facility organizations use building automation systems, store utility bills and meter readings in Excel or some utility system, generate their own utility bills for steam and chilled water, create campus maps and engineering drawings, maintain space planning systems, and work with outside engineering and construction firms. Each group has their own systems for specific tasks, each with a different nuance to the same information.

This is unlikely to be prevented or be brought under control—hence the EEMS should be able to present to each specific user group the naming convention they are familiar with, while providing the cross reference information required.

Data Warehouse Structure and Hierarchy

It is clear that an EEMS handles and stores an enormous amount of data from many disparate systems. In order to derive value from such a large data set, a defined structure and hierarchy must be implemented to make the data readily consumable. The structure of the data must be flexible too, since physical configurations are constantly changing. Buildings may be added, equipment may fail unexpectedly, and space may be modified to accommodate organizational changes. The design of the EEMS system must be flexible to adapt and keep pace with this dynamic environment.

Warehouse Objects

Structuring of data should mirror the manner in which those data are to be used to gain insight. For example, it should parallel the physical facility so that data can be viewed either in aggregate or in isolation when focusing on an individual building or piece of equipment.

The EEMS warehouse needs to support logical objects—sets of information and relationship hierarchies—that allow for this structure. The list on the following page shows six warehouse objects and the hierarchy that needs to exist within them. The warehouse structure also must support the way different elements within the object interrelate, i.e., the way meter interval data must connect to billing rate data and physical building data.

EEMS DESIGN PRINCIPLE 3: PROVIDE INTERACTIVE ACCESS TO ACTIONABLE INFORMATION

Defining Interactive and Actionable

Interactive access allows the users to work with the data in a dynamic fashion, moving seamlessly through the data with tools that provide near instantaneous response. This allows users to "work the way they think" rather than being limited to a series of static queries and reports. The EEMS must address both usability and performance issues to successfully provide interactive access to the data.

Actionable information can be used as the basis and rationale for effective decision-making, as opposed to merely indicating "status." To create actionable data the EEMS must collect *both* monitoring and control points. With monitoring data alone (without control information) it is difficult to verify or quantify the savings opportunity. Monitoring data and control data should be viewed in tandem to become "qualified" as actionable information.

Time Matters

It is crucial when collecting tens of millions of records per month that insight can be gained within minutes rather than days. The use and operation of an EEMS cannot be an arduous and time-consuming task.

Facility Objects

- Site
  - Zones
    - Buildings
      - Floors
        - Rooms

- HEGIS group
- HEGIS classification

Issue Objects

- Issues

Organizational Objects

- Departments

User Objects

- User privileges
  - ➢ Users

Interval Data Collection Objects

- BAS & OPC point data

- Calculations
  - ➢ Balance efficiency calculation
  - ➢ Cooling efficiency
  - ➢ Cost calculations
  - ➢ Delta T
  - ➢ Theoretical water loss
  - ➢ Etc.
- Weather data

Meter Objects

- Utility type
  - Utility information
    - Account information
      - Meter information
      - Rate information

- Billing information

Users must be able to derive insight about the data (useful, actionable information) in a very short period of time if the system is to become embedded within the facility's management operations.

To accomplish this, data must be accessed and presented to the user at the *speed of thought*—able to view hundreds of thousands of data intervals in 15 or 20 minutes. In essence, if the user has to wait for information, their thoughts will wander. If they continually have to wait—if they are not constantly engaged—then their reaction will be that the EEMS is wasting their time. Once that point is reached, the EEMS will not be used and any potential value will be lost.

Currently, most facility personnel waste an enormous amount of time collecting and distributing data. Due to the historical lack of availability, this wasted time is culturally accepted as "part of the job," when in fact, instead of spending days manually gathering and piecing together data, a well-implemented EEMS can deliver it in seconds. When you have immediate access to the data, staff is freed up to devote time to the real engineering work of diagnostics, analysis, and planning.

Usability Matters

The amount of time required to gain insight is directly related to the usability of the system. With such vast amounts of data stored in the EEMS warehouse, it is essential that data can be assembled dynamically by the user via a simple, intuitive interface. The elements of the interface include the data organization and presentation, data display, and the program interface itself. The EEMS also needs to support other aspects of the facility staff's workflow, such as tracking identified issues or interacting with external analysis tools.

Data Organization and Presentation

As discussed in "Design Principle 2," the data must be structured in a manner that is inherently useful. This needs to happen not only at the data structure level, but also at the user presentation level.

With such large volumes of data, the EEMS must allow them to be categorized in meaningful ways, such as organizing within the facility hierarchy (zones, buildings, floors, rooms), organization (departments, rooms), or systems (chiller system, air handling system, etc.). In many cases data must be accessible through multiple views so that, for example, a building manager has access to the building information while a plant engineer can look at chiller operations facility-wide.

The EEMS must make the presentation simple by providing a master organizational structure that offers a mechanism for users to select which data they see. Users must be able to "drive through the campus" from their desktop viewing thousands of data intervals per minute. The EEMS must also provide a method for users to define their own organized views.

Trend Lines

An earlier section discussed the importance of collecting trend data as the basis of diagnostics, monitoring, M&V, and more. The EEMS must use the trend data to provide *Trend Lines* as the primary mechanism to display and interact with the EEMS data. These are very different from BAS Trend Logs—a difference that elevates, by orders of magnitude, the effectiveness of an EEMS for diagnostics, monitoring and other applications.

Trend Lines provide insight into operational data over extended time periods and permit the expeditious identification of problems—problems spanning multiple BASs and inefficiencies that arise suddenly due to changing circumstances such as weather, equipment degradation over time, or system configuration adjustments. In contrast to Trend Logs, Trend Lines:

- Contain all the data from every monitoring and control point
- Do not impact BAS control performance at all
- Combine data from multiple BASs and other data sources such as meters, utilities, and weather data
- Capture all data from the moment the EEMS is turned on, so it is constantly available
- Create a historical record of building operations—both monitoring points and control settings

BAS Trend Logs do have their place in providing needed information—where their ability to collect real-time data is useful—as shown in Figure 32-15.

Calculated Data

Calculations are often done ahead of time and stored in the data warehouse. The efficiency gains in having commonly desired calculations available for monitoring and diagnostics are tremendous-having a dramatic impact on usability. A visual display of an ongoing trend built on a calculation can provide insight instantly that would otherwise take hours of number crunching and charting in Excel.

A Delta T is a simple calculation example commonly viewed as a basic performance measure of a chiller system. A Delta T is even more effective when the EEMS displays the supply and return temperatures used in the calculation at the same time. This way, if the Delta T fluctuates, the user can immediately see if the change was affected by a supply or return temperature rise or drop.

More complicated calculations can provide users with an overall operational efficiency rating, or, applying billing rate data, even a Trend Line that shows at every 15-minute interval the total energy cost for that 15-minute period. Calculations of this complexity rely on the EEMS's ability to fully integrate data from all sources and present it in a normalized fashion.

The EEMS should support calculations of any complexity, although it is appropriate to restrict the creation of calculated points to a system administrator who understands how calculations fit into the underlying data warehouse.

A sample of desired calculations—calculated every 15 minutes—include:

- Balance efficiency
- Chiller efficiency
- Chiller total cost
- Chiller plant total hourly cost of operations
- Cooling tower cost & efficiency
- Cooling tower make-up water cost & efficiency
- Delta T (primary & secondary)
- Pump brake horsepower
- Pump efficiency
- Pump kW
- Theoretical water loss
- Tons output
- Tower total cost

Application Interface

All of the usability factors mentioned must come together in the software user interface (UI). It is how end users interact with the data-through mouse clicks, menu selections, etc.

The EEMS should use interface conventions already familiar to its users, such as expandable data trees to display available points and hierarchies, tabs for organization, contextual menus, drag and drop, etc.

Users should be able to take an iterative approach with each action building on the last for diagnostic purposes. Monitoring should be fast and efficient, allowing the user to quickly cycle through hundreds of Trend Lines—hundreds of thousands of intervals—in minutes. Trend Lines are the source of actionable information, but it is the UI and system performance of the EEMS that allow the user to work at the speed of thought.

Tracking Issues

As an expert reviews the data within the Trend Lines, they are able to identify areas of concern or high-

Application of BAS Trend Logs and EEMS Trend Lines		
Criteria	**BAS Trend Logs**	**EEMS Trend Lines**
When to Use		
Diagnosing operations	After problem has been identified as under the control of the BAS and further data is needed for final diagnosis	Always—far superior tool for nearly al diagnostics and all cases where historical data or data outside the BAS must be considered
Monitoring operations	When real-time data for a small number of BAS points needs to be watched	Always—provides the ability to monitor hundreds of trends in minutes, combining data from any and all sources
Typically used by	BAS control engineers and technicians	BAS control engineers & technicians, energy engineers, facility managers, performance monitoring contractors, commissioning agents, HVAC design engineers
Technology Perspective		
Data storage	Stores point data for trends defined	Stores data for all points from all systems
Time interval of data	Captures data in increments from milliseconds to minutes	Captures data from all systems every 15 minutes
Displays data from	Native BAS	Multiple BASs, metering systems, utilities, weather, billing
Display time period	Typically a few days or weeks	Between a day and a year, with historical data going back years
Data storage	Up to a few months and data is discarded	Up to 20 years

Figure 32-15. Some of the uses, users, and differences between BAS Trend Logs and EEMS Trend Lines.

light known problems. For each problem discovered, the EEMS should take a snapshot of the Trend Line(s) at that point in time, which can be annotated by the expert and sent to the appropriate building control technician. That technician is then able to take action and modify settings within the BAS as appropriate. The technician can then log the fix while the expert can verify that the change is working appropriately before closing out the issue.

Issues must be accessible by a variety of personnel that need the information to put together the action plan and ultimately resolve the issue. It is appropriate, however, to have controls that limit some users' scope of what they can see or modify.

Interact with External Analysis Tools

As good of a diagnostics tool as an EEMS is, there are many types of data analysis that are best performed by other tools created for that purpose. For example, Excel can do curve fitting, regression analysis, and many other calculations that would be wasteful to duplicate within the EEMS.

To make using external applications an easy process, the data from the EEMS should be easily exported into Excel or other analysis software. This allows engineers who have built their own analysis routines in Excel (or other packages) to continue to use them.

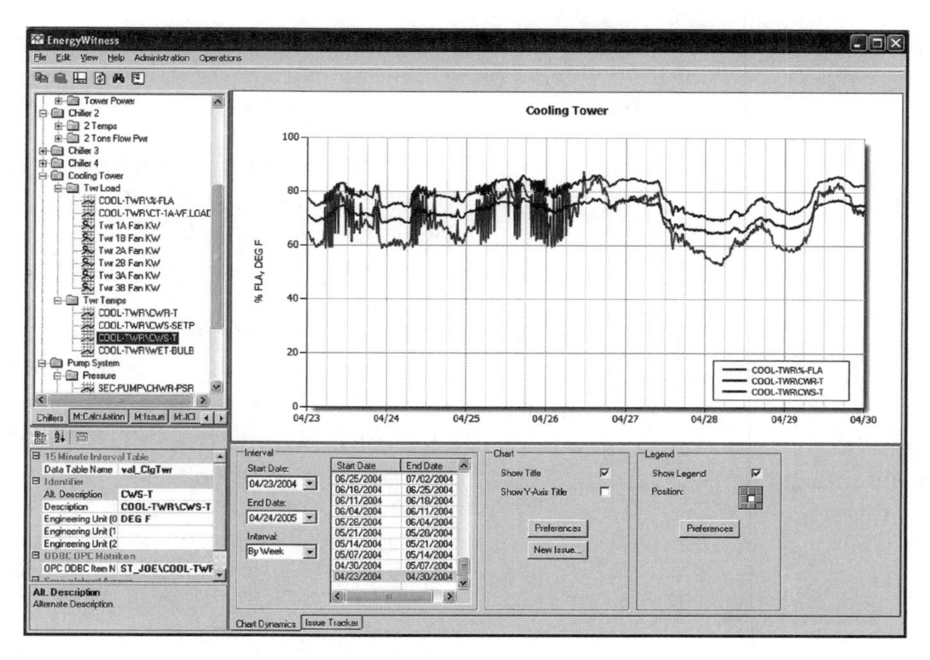

Figure 32-16. An EEMS should allow interaction with data, such as drag & drop capabilities to add more Trend Lines onto an existing chart, one-click scanning back through time, and zooming in or out.

Present User Specific Information

An EEMS must support a variety of applications and users. Data must be pertinent for each user while providing the ability to conduct additional investigation and access data not typically made available via other systems. Users of an EEMS may be BAS engineers and technicians, energy engineers, HVAC design engineers, facilities engineers and managers, commissioning agents, energy purchasers, and/or performance contractors.

Users of information from an EEMS extend well beyond that group to building managers, zone maintenance groups, department heads, finance personnel, and anyone who has occasional needs for some facilities information.

Information Publishing

In many cases, this second tier of users do not need interactive access to the data at the individual point level. Their interests are mostly static and may be better served by periodic reports.

An EEMS needs to provide a variety of output options that enable everything from detailed reports with data tables and charts of multiple Trend Lines, to summary reports that roll up information into cost breakdowns and overall operating efficiency ratings. The output needs the flexibility to be delivered via paper, electronic documents, or the Web.

EEMS DESIGN PRINCIPLE 4: MEASURE & VERIFY RESULTS

All too often performance measurement and verification (M&V) ends up neither measuring nor verifying performance. It is a simple case of not having access to data to do M&V properly[1]—a problem an EEMS solves.

In measuring and verifying results it is first important to define terms often used to justify and quantify the impact of investments in utility operations.

Real Savings versus Stipulated Savings

Real savings is proving cost savings via actual dollar savings, while stipulated savings is based upon savings of equipment. For example, new lighting fixtures may consume 40% less electricity while generating the same amount of lumens and, while the savings per lumen are real, the heating costs may have increased because the new lighting now produces less heat. An EEMS is required to view both the data for the electricity consumed by equipment type and for the utility consumed to heat (or cool) the area, all within the context of a specific space if a real savings assessment is to be made.

[1]IPMVP Volume I: Concepts and Options for Determining Energy and Water Savings, 2001, Sect. 5.6

Energy Savings versus Dollar Savings

The EEMS must be able to account for both energy savings and cost savings (actual dollars). For example, just because the utility bill has dropped does not necessarily indicate that money/energy has been saved; rather it confirms only that less was spent. It confirms nothing about reduced energy consumption.

Factors like price, weather, a new construction coming on line, and consumption rates are required to understand whether actual dollar savings have occurred. Consumption savings occurs because of the way energy usage is being controlled and this can only be validated through monitoring and control points that highlight energy consumption changes.

To understand the dollar savings realized, one must account for variations in the price of energy, the actual weather versus the planned weather for the time period, and the change in the amount of energy used.

To realize true energy savings, which in turn lead to dollar savings, consumption must be reduced independent of the factors above and this can only be achieved when energy usage is being controlled more efficiently.

Use Life Cycle Costing

An important philosophical concept to adopt is "Life Cycle Costs." It is the most appropriate way to assess equipment and building costs. The initial purchase of HVAC equipment is a significant capital investment, but its true costs lie in this number plus the cost of its operation, service, maintenance, and total life span. The dramatic impact of an EEMS on life cycle costs is discussed later, but it is important to highlight this "real cost" assessment as a true measure of the total cost of equipment ownership.

EEMS DESIGN PRINCIPLE 5: A PLATFORM THAT EMBRACES INDUSTRY STANDARDS

Using industry standards and an open architecture is the right way to build an enterprise-class application. This has been proven repeatedly at all levels of technology and business where broad support and interoperability are significant benefits. A platform architecture and the use of standards protects the organization by minimizing dependency on any single vendor, even allowing functionality to be added outside the vendor's development cycle.

There are standards in several areas that an EEMS should adhere to:

Operating System

There are three platforms, sufficiently open and standardized, that an EEMS could run on. The first, and by far the most popular, is Microsoft Windows. It offers the greatest availability of tools and options, and is already installed and supported nearly everywhere. The Microsoft .NET platform is excellent for developing and integrating application components. Other options include Linux, which has strong server support and tools such as J2EE, but is a limited end-user platform; or a Web-based platform, which is typically made up of Windows or Linux servers using the Web for network communications and a browser for the application front end (which instills limitations on the UI).

Database Management

Equally important is that the data stored within an EEMS is housed in a manner that is open, accessible, and interoperable with other systems such as a standard relational database (i.e. SQL Server, Oracle, etc.). Proprietary data managers will handcuff users to rely on the vendor for everything.

Data Collection

As discussed within "Design Principle 1," data are the lifeblood of an EEMS, and access to this data is a complex and arduous task. The mechanism for ensuring that this data extraction/transfer takes place can be both cost prohibitive and extremely difficult without adherence to standards such as OLE for Process Control (OPC), BACnet, LonWorks, and Modbus.

Analysis

Data analysis tools range from the most general purpose and broadly available, Microsoft Excel, to highly specialized analytics. Ideally the EEMS will provide direct support for Excel, allowing users to take advantage of analysis routines already developed. Minimally the EEMS will allow data to be exported into any analysis program through cut and paste or by using an intermediate file.

Space Planning and Classification

The EEMS should support HEGIS groups and classifications for space planning. It should also interface with the leading space planning system, FAMIS.

Publishing

Making information available throughout the organization is an important function that saves significant time. There are four document formats an EEMS can consider publishing to—Microsoft Office, PDF, HTML, and XML. Office, which includes Word, Excel, and PowerPoint, is ubiquitous in business settings. The other formats are also nearly universally readable, although they require less common tools or special skills to edit.

BUSINESS APPLICATIONS OF AN EEMS

The EEMS has many business applications, each with different benefits to the campus. Some of these applications deliver benefits that are operational savings (reduced energy consumption) and some of the applications enhance the infrastructure. The most common applications of an EEMS are:

- Operational diagnostics & monitoring
- Empowering efficient building control strategies
- Enabling continuous commissioning
- Chiller plant efficiency calculations
- Controlling building comfort
- Accurate energy cost allocation
- Capital request justification
- Information publishing
- Providing more accurate budgeting & forecasting
- Purchased utility accounting
- Vastly improved performance measurement & verification

Operational Diagnostics & Monitoring

Operational diagnostics is the process of reviewing operational energy data and identifying targets for energy savings, highlighting engineering design deficiencies, and alerting staff to malfunctioning equipment, to name just a few. While these tasks sound familiar, it is important to contrast the manner in which an EEMS performs this function with that of today's technology and processes.

In short, an EEMS streamlines the process. It provides a complete picture that can be explored quickly and easily to identify problems, validate improvements, and test hypotheses—all without the technical literacy and timescales required by today's approaches. With an EEMS, this process simply involves a straightforward visual analysis of the data, quickly digging deeper into the information where anomalies are evident to uncover problems and gain real insight.

An engineer diagnosing a problem using the data in an EEMS can typically resolve the issue to its root cause in one fifth to one tenth the person-hours than with standard approaches. Also, the elapsed time from the first indication that something is wrong to resolution can be reduced by a factor of 100 or even more. Appendix A shows a case study where this was the situation.

Once an issue has been resolved, the results should be verified and measured. This is to ensure that any actions taken had the intended result, and also to be sure that no unintended side effects occurred. Trend Lines provide the ability to start verifying results as soon as 15 minutes after the change has been made.

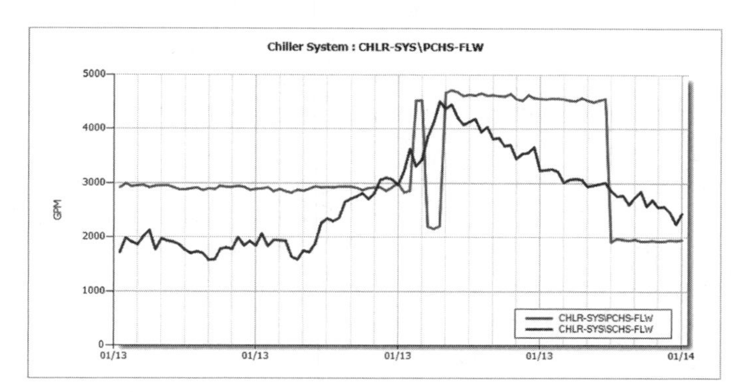

Figure 32-17. An example of a chart used in the process of identifying a problem. Here the diagnostician has identified significant differences between the Btu output of different chillers even though they should be producing equal amounts.

When not diagnosing specific issues, an EEMS is the ideal tool for ongoing monitoring of operations. The data organization, performance, and Trend Lines allow engineers to monitor as many as a half million interval data points in 20 minutes. Instead of watching unconnected real-time monitors and waiting for BAS alarms to sound, an EEMS's monitoring capability provides far more in-depth information about how systems are operating and reveals issues at their earliest stage.

Efficient Building Control Strategies

A great deal of effort is expended programming building automation systems to meet the disparate comfort needs of building occupants. Since seasons and weather change constantly, the strategy used to control the building will need to change accordingly.

A key role in the development of efficient building control strategies is the EEMS Trend Lines function. They show how the change in the control strategy is working, or highlight areas where the strategy should be adjusted. Importantly, by using a data collection interval of just 15 minutes, control strategy change requirements and the results of implemented changes will be apparent almost immediately.

Continuous Commissioning^{SM}

Continuous Commissioning[2] is an ongoing process for monitoring systems, diagnosing and resolving issues, and making energy consumption as efficient as possible while maintaining or improving building comfort. It includes anything from physical maintenance, to control strategies, to prioritizing and implementing retrofits. While other forms of commissioning on existing buildings have initial design specifications as their goal, continuous commissioning seeks to optimize the current operations—how the building is occupied and used today, taking into account changes since the original design.

An EEMS provides a continuous commissioning engineer the information to perform many of the steps and meet the objectives of a successful continuous commissioning program:

- Document the existing state of operations, space comfort, and energy consumption
- Locate system issues, diagnose, and develop a plan to resolve them
- Create new recommendations for control settings, setpoints, etc.
- Identify and prioritize retrofit projects that will have the greatest impact
- Measure and document improvements in system performance and energy consumption
- Monitor ongoing operations to ensure all benefits are sustained

Chiller Plant Efficiency Calculations

A true EEMS must enable holistic utility performance measurement and management, including the frequently neglected chiller plant data. Chiller plant efficiency calculations provide simple to understand graphs that show the cost of running the chiller plant; accounting for energy usage by each chiller, cooling tower, and primary and secondary distribution loops. As the control strategies change, operators can assess the true overall impact on costs.

Building Comfort

As well as operational efficiency and energy cost related applications of an EEMS, it is important to remember that the real purpose of HVAC systems is building comfort. Well-maintained systems improve comfort and health of building occupants through properly controlled temperature, humidity, and ventilation.

Improving comfort levels and maintaining their consistency can be accomplished while lowering energy consumption and its associated costs. This was proven by the Energy Systems Laboratory at Texas A&M University where their Continuous Commissioning efforts resolved major comfort issues while reducing energy consumption by 15-30%[3].

An EEMS provides the data to see exactly how building or room conditions change and what the control settings are to try to maintain comfort. For example, a large meeting room filled to capacity will typically experience a rise in temperature during a long meeting. The solution of "cranking up the AC" halfway through does not work very well, yet is often what is done in response to occupant complaints. With an EEMS, engineers can use the data to determine the proper control settings to manage temperature, relative humidity, supply air static pressure, terminal damper position, etc. This is a more complex control, but maintains comfort better and is often more energy efficient.

More Accurate Energy Costs Allocation

By integrating all energy sources and consumers across the campus along with space planning information, organizations can identify specific energy consumption by space. Once space consumption is determined, the rates can be applied to generate accurate costs for each space. These can then be rolled up to assign costs by department or other organization structure.

Today, energy costs per space are typically allocated by a combination of approximation and averaging, producing an inaccurate cost allocation. In essence, the high energy consumers are being subsidized by the low energy consumers, resulting in cost centers being billed inaccurately.

Capital Request Justification

Large capital requests tend towards two categories: replacing old worn-out or out-dated equipment, and doing major overhauls resulting from design flaws or

[2]Energy Systems Laboratory, Texas A&M University, Continuous Commissioning In Energy Conservation Programs, *www-esl.tamu.edu/cc*

[3]U.S. Department of Energy, 2002, "continuous Commissioning Guidebook, Maximizing Building Energy Efficiency and Comfort," ch. 2, pp. 1.

changing requirements that render the design obsolete. Replacing old equipment is part of a normal cycle, even if only done every 20 years. Proving the need for an overhaul is much more complicated and requires proof to show that the investment is needed.

Today, engineers design systems and specify how they should be operated. However, they receive very little feedback as to how a particular piece of equipment performs in the field, under load, supporting usage patterns that may be different from when the original design was done. EEMS Trend Lines can be used to "instrument" the equipment and provide verification of air handling system performance, distribution loop efficiency, and overall system performance under various loading conditions, to name just a few.

Many design engineering firms build various computer models to simulate operations; but now, Trend Lines provide real feedback, based upon live data for the first time. The information is based on actual operations, not original design intent.

While a building automation system cannot provide this feedback due to the large volume of data created, and the inability to look outside itself, an effectively implemented EEMS can perform this function with ease.

In short, an EEMS can be used to identify and verify design flaws, whether from bad initial design or due to changing requirements or equipment upgrades. This capability gives facilities personnel the information needed to justify the capital budget requests needed to fund system overhauls.

Information Publishing

The current approaches to publishing operational data fall into two categories: the ad hoc, "that information should be here somewhere" approach, and the comprehensive reporting approach. Both consume vast amounts of time from engineers, technicians, and other facilities personnel because the data are not readily available, and what is available is not organized. It is a giant exercise in manually compiling and distributing spreadsheets.

The EEMS provides the starting point where all data are present. It is organized so that building-specific, departmental, or system-centric views of data can easily be reported. Reports can be financial or consumption based, or correlate the two with ease. The EEMS makes more information available at a fraction of the effort and associated cost.

More Accurate Budgeting and Forecasting

An EEMS should also show deviations from budget in terms of causes (consumption, weather, price and new construction).

Through the EEMS, individuals can view daily updates versus budget on a graph, thereby preventing month-end budget shortfalls. By providing this insight concerning current status on an ongoing basis, the EEMS provides opportunities to prevent budgetary overages before they blossom, while at the same time providing the vehicle through which the most effective cost savings can be realized should a budget overrun actually occur.

Purchased Utility Accounting

Today, at their own significant expense, many organizations install their own meters in parallel with those of the utility company for the purpose of verifying the utility bills. However, there are other techniques to verify utility billing without requiring additional meters.

An EEMS can summarize energy usage from all of the consumers of energy, identify losses, and provide supporting documentation to reconcile discrepancies with the utility. Furthermore, an EEMS can do this in a fraction of the time and cost.

Vastly Improved Performance Measurement & Verification

In the past, the first phase of a performance contract required that the contractor collect baseline information regarding the status of the building. This labor intensive and expensive task invariably only provided small samples of data from locations across the campus. Attempts to collect more data through BAS Trend Logs will have a negative impact on the BAS's ability to control operations. With such a random data collection process, it is simple to understand why this form of baseline information is inadequate for use to make significant investment decisions.

EEMS Trend Lines, discussed earlier, can significantly improve performance contracting for both the customer and contractor. With an EEMS, all the point data can be collected (with zero impact on the BAS) to provide a complete and accurate baseline as a matter of course, rather than as an expensive manual task, and the subjectivity can be removed from performance contracting. As improvements are implemented, their impact can be measured and verified quickly and easily.

FINANCIAL BENEFITS OF EEMS

It is clear that an EEMS is able to provide the infrastructure to support many different business purposes. Importantly, these applications of EEMS technology are also able to deliver a rapid return on investment in a number of areas including:

- Lowering operational costs
- Positive cash flow
- More effective staff deployment
- Greater indirect cost recovery
- Reducing equipment maintenance costs and increasing equipment life
- Improving the efficiency of energy purchasing

Lowering Operational Costs

EEMS Trend Lines provide the insight necessary to identify and achieve operational savings. Today, organizations typically identify operational savings targets by engaging a seasoned HVAC engineer to tour the campus, every building, and each piece of equipment to determine what is happening. Decisions are made based on a tiny fraction (often just one or two percent) of the data and many problems are not found until long after they appeared or are completely missed.

With an EEMS, this process can be accomplished within days instead of months, without leaving the office, through the rapid visual analysis of data. Unlike a manual campus tour, an EEMS uses all of the operational data to quickly identify common operational problems that, when addressed, lead to a reduction in operational costs. Below are samples:

- Fans running constantly when they do not need to
- Chillers and air handling systems unable to keep up with setpoints
- Over controlling of the building
- Operation of the equipment that deviates from the instructions of supervisors

Furthermore, using an EEMS, the operational changes and the financial benefits derived from them are instantly measurable and verifiable through simple monitoring of the data.

Positive Cash Flow

A typical retrofit project involves replacing dated components with new, efficient equipment. Such projects require a huge up-front investment to fund a few months of design and planning, followed by 6-24 months of installation/construction. It can be multiple years of negative cash flow before the project is complete and stipulated savings begin, and much longer before the return on investment is fully realized.

With an EEMS, savings start after just a few weeks. The initial investment is typically recovered in 5-6 months, and thereafter generates a positive cash flow. The realized savings and additional cash could even fund other projects once the operational issues are resolved.

More Effective Staff Deployment

Most facilities staffs are operating on guesswork due to the lack of data. Energy and control engineers may study real-time monitors and run BAS Trend Logs, often taking weeks to correct an operational defect.

With the data provided through the EEMS, facilities staff prioritizes the right problems and resolves them in far less time. Less time is spent on fire fighting and more on improvements that upgrade campus/facility physical assets in line with the master plan.

Equipment Maintenance and Life

Access to Trend Lines and comparison data between the weather, like-equipment, and facilities can serve as a powerful tool to reduce maintenance costs and prolong the life of equipment.

For example, cooling tower fans are often run at 100% capacity unnecessarily when the dew point is too high for the environment to accept the heat that the tow-

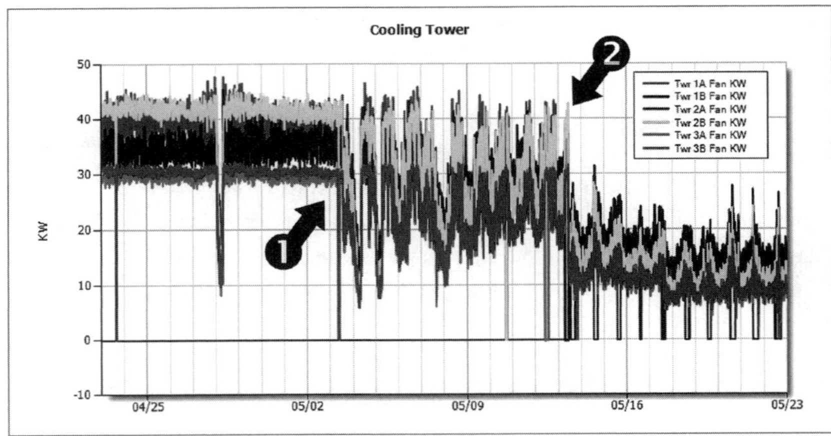

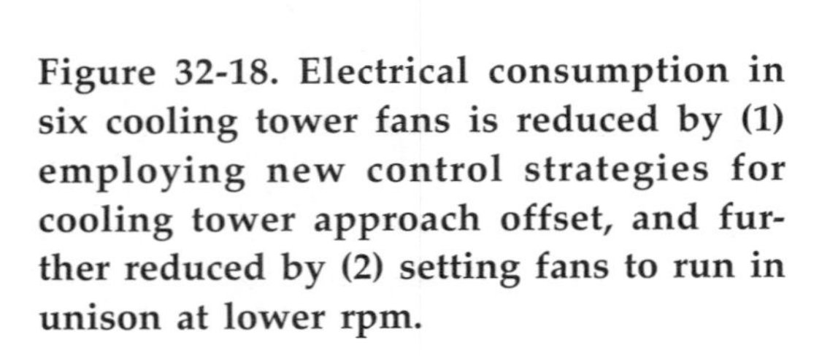

Figure 32-18. Electrical consumption in six cooling tower fans is reduced by (1) employing new control strategies for cooling tower approach offset, and further reduced by (2) setting fans to run in unison at lower rpm.

ers are attempting to transfer. In this example, running the fans at 50% or less would be equally beneficial as running at 100%. Access to this data via an EEMS can extend the life of the fans, reduce the service call frequency, and reduce overall costs dramatically. The same approach can be applied across other equipment types.

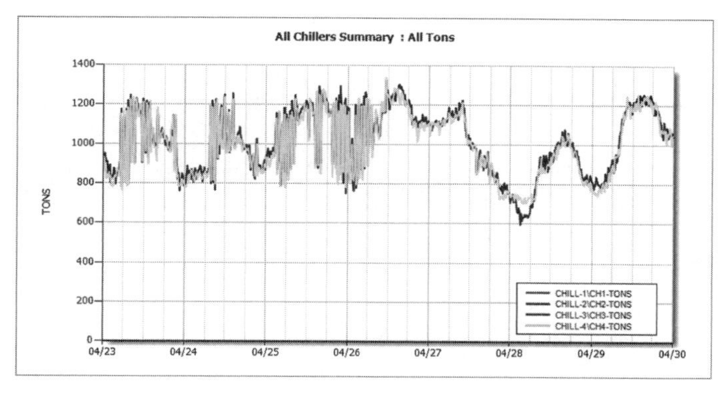

Figure 32-19. Equipment trashing was stopped when the chiller setpoint was set to a fixed value, resulting in an expected increase of life and reduction in repair/maintenance.

More Efficient Energy Purchasing

Whether an organization makes energy purchasing decisions alone, or with the help of consultants, an EEMS can have a positive and profound impact on the energy purchase process.

Purchasing energy is a complex task that requires a thorough understanding of demand. In addition, if energy costs are to be minimized, it requires an insight into how to manipulate the timing and height of peaks in demand.

Also, because energy pricing structures change regularly, effective purchasing requires a sound knowledge of the energy market's price drivers so as to "lock in" the best price. Organizations frequently buy extra energy, locking in prices via futures contracts. When decisions concerning contracts of this magnitude are made, organizations, like people, make better choices when presented with all of the facts. Without a robust knowledge of an institution's consumption patterns, the drivers of these patterns, and the impact of the weather on operations, an energy purchase is being made without visibility of the whole story.

Indirect Cost Recovery

Allocation

By accurately identifying energy cost for research space, rather than using an estimated allocation approach that invariably falls below the actual energy cost of the space, institutions are neglecting access to government funds.

Through the use of an EEMS, energy cost for space can be accurately and verifiably attributed to ensure that all of the institution's revenue opportunities are maximized and an institution's low energy users do not necessarily have to subsidize the high energy consuming departments.

Productivity

When examining financial benefits and focusing on cost savings, it is easy to overlook the impact of having space comfort. Studies have shown that maintaining a consistent, comfortable environment has a positive impact on attention span and productivity, and lowers time lost due to illness. Also, fewer comfort complaints mean that facilities engineers and technicians are not forced to spend time servicing comfort complaints.

CONCLUSIONS

When implementing a true EEMS, the following criteria must be specified:

- The EEMS must collect data from all monitoring and control points from all sources—building automation systems, utilities, metering systems, weather, and space planning systems. It must gather consumption data (in instantaneous, totalized, and average forms as appropriate), control settings, billing and rate information to provide a holistic view.

- It should structure the data within a data warehouse in a manner that provides the flexibility to handle complex relationships, hierarchies and calculations, and adjust to meet evolving requirements. Data should be normalized so that it may be more easily compared and contrasted, and so that problems can be pinpointed more precisely.

- It must present actionable information and Trend Lines so that data are represented in an informative manner. Performance and user interface must combine to provide interactive access to the data that can then be manipulated, supporting detailed investigation and efficient monitoring.

- The EEMS must deliver real savings that can be measured and verified, and demonstrate these returns over time.

- The EEMS should be more than an application—it should be a platform for energy management and other facilities operations that is expandable. It must conform to open standards so that integration and interoperability between related systems is a seamless and straightforward process.

An effectively implemented EEMS provides unparalleled insight into the day-to-day, week-to-week, and month-to-month operation of an institution's utilities. It nearly eliminates the time wasted by facilities staff gathering needed data. It provides individuals with the ability to rapidly find and address inefficiencies (fixing the root cause, not just treating symptoms) that can result in immediate cost savings and an ongoing financial return—often when these problems have gone undetected for many months or even years.

An EEMS also reduces the total cost of ownership of equipment by reducing maintenance costs and extending its life, all while providing the most actionable data to effectively secure lower utility rates and both improve and expedite master planning. It makes it easy to publish operational data to constituents within and outside the facilities staff.

An Enterprise Energy Management System presents the opportunity to span all existing building automation systems and energy related data so that assessments can be made in the context of the whole facility, environment, and billing climate. Complete information leads to better decisions—decisions that address building comfort, energy consumption, operational costs, capital investments, and stewardship of the assets.

APPENDIX A: SHORT CASE STUDY DIAGNOSTIC PROCESS USING EEMS

This short case study is from a hospital in the southeastern United States. The data shown is real, taken from a live EEMS.

Problem

The water temperature leaving the cooling tower was not meeting setpoint.

Based on traditional data and investigation, all that was known was:

- The first indication of the issue from the BAS was an intermittent alarm
- Alarms came from cells 2 and 3 of the 3-cell system
- The problem had been going on for over a week without being able to identify the cause

With no prior knowledge of the systems or operations at this facility, the diagnostician using the EEMS set out to find the root cause of the problem.

Process

The first step was to look at the water temperatures leaving the cooling towers for each cell. Tower 3 had the most difficult time achieving setpoint (80°F).

A correlation with weather was investigated by overlaying wetbulb with Tower 3's water temperature.

Scanning back through time verified that the behavior was related to weather. This was the first time in several months that wetbulb reached that high.

The next area of interest was the power consumption of the two fans in the tower. The EEMS immediately showed that two identical fans were using very different amounts of power. Adding control information to the monitoring data showed the exact same signals being sent to each variable speed drive (VSD).

Another scan back in time shows the fan behavior starting about the same time the BAS alarms started. Earlier, the fans run in unison (although wetbulb was not as high).

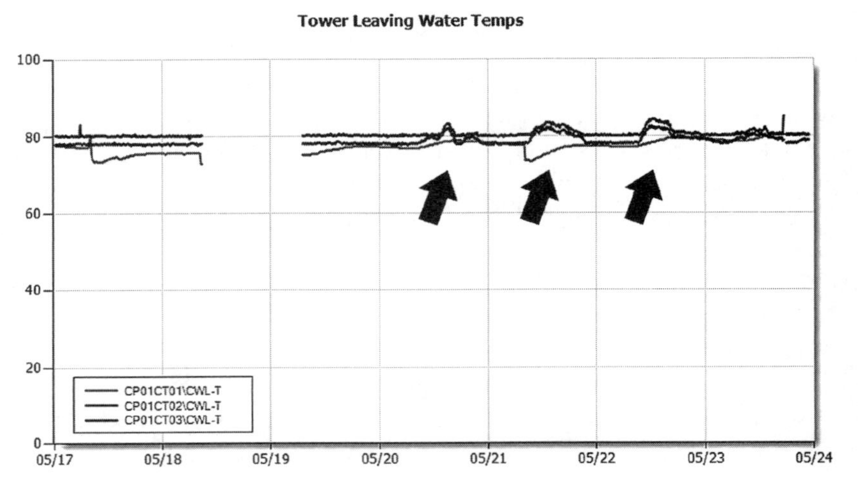

Figure A-1. Bumps in trends show where setpoint is not held, causing BAS alarm to sound.

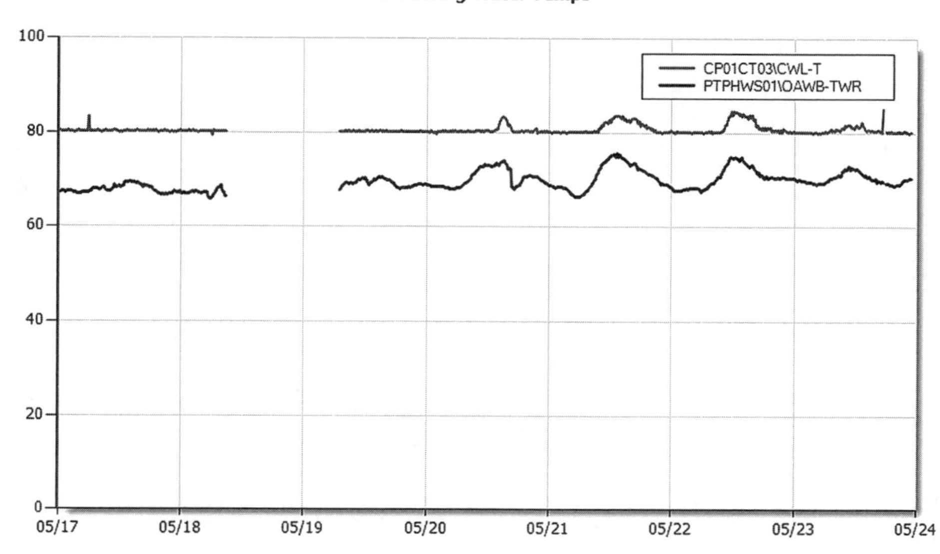

Figure A-2. Each time the wetbulb exceeded 73.4°F, the tower could not maintain set-point.

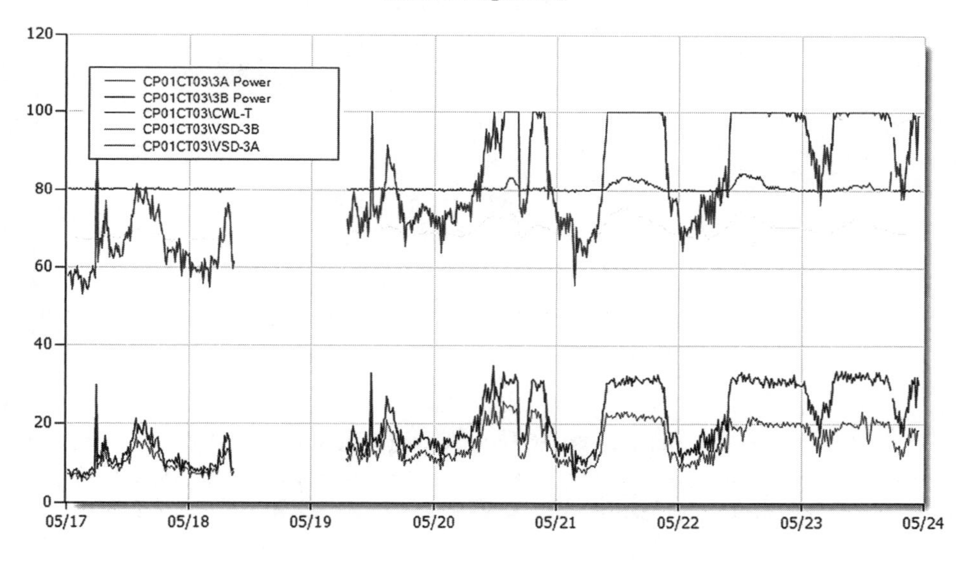

Figure A-3. The bottom two lines show the separation in power when wetbulb rises. The upper line that follows the same shape is actually two lines—the control signals.

Diagnostics Conclusion

The conclusion was that a fan belt was likely slipping because it was a VSD, causing the problem. It only took about 30 minutes to complete the diagnosis.

The fan belt was checked and found to be loose. It was tightened and the EEMS verified the result.

Although the problem was properly resolved, its return suggests that the equipment ultimately needs replacing, not adjusting. The EEMS provided the facilities manager with the data to prove that this was a necessary expense.

APPENDIX B: HOSPITAL CASE STUDY

Introduction

This case study shows the value of using interval data from the building automation system (BAS) as a tool to diagnose and monitor facility operations. You will get a close-up view of the first six months of using EnergyWitness™—an Enterprise Energy Management System (EEMS) from Interval Data Systems—from installation and initial data collection to discovering, fixing, and validating a variety of issues, including operations, comfort, and engineering design. The case

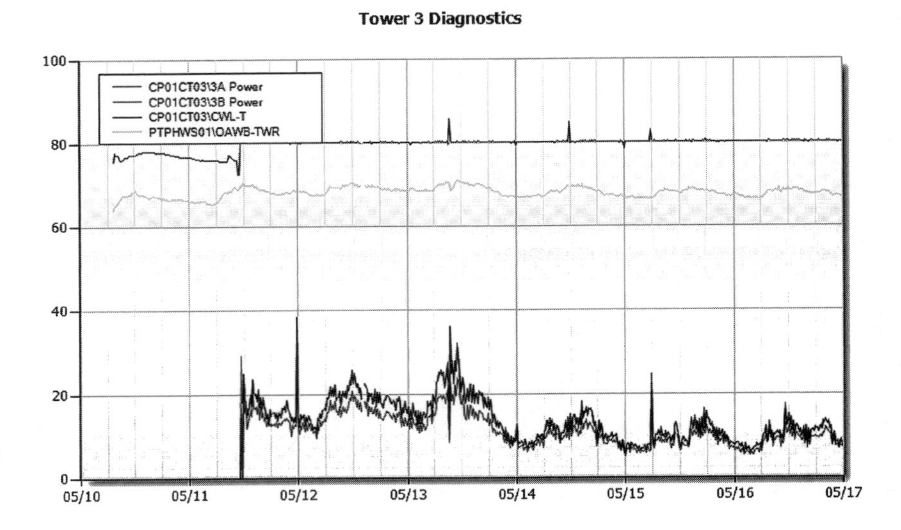

Figure A-4. The fans first started to split a week earlier.

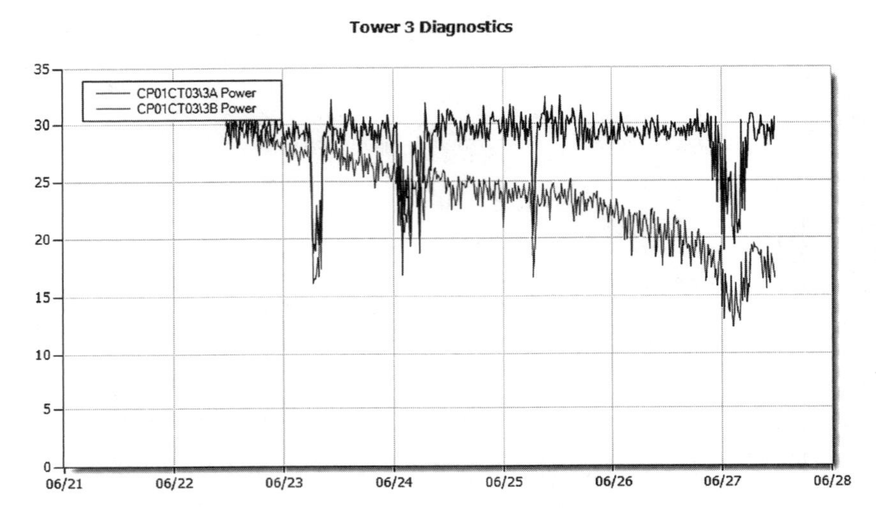

Figure A-5. After the belt tightening, the fans ran properly again, but monitoring showed slippage again a month later.

study traces issues starting in the chiller plant, then out through the distribution system, into the air handlers, and finally to the terminal boxes. In all, we'll see more than $260,000 of annual cost savings quantified and verified through the data.

The site is a large hospital complex in the southeastern U.S. with nearly 1,000 beds. To meet the demanding comfort requirements of a southern hospital, there are four chillers and three cooling towers capable of producing over 5,000 tons of output at the heart of their HVAC system, which contribute to a multi-million dollar annual energy cost.

Situation

The hospital had come to suspect that there were significant issues within its chiller plant. They believed it was operating far from optimally, and that there were perhaps engineering design issues at the root of the problems. If true, a significant investment would be needed to redesign, re-engineer, and repair the system. Additionally it was known that there were space comfort issues in several areas of the hospital, including the operating rooms.

Before seeking funding, the facilities staff knew they needed to prove that engineering design problems did indeed exist. Further, if they could quantify the ongoing wasted cost of current operations, it would provide the justification to proceed with the anticipated redesign.

Getting Started

Hospital staff found that they were unable to extract enough information from their BAS to analyze the problem. Engineering studies were also inconclusive. They needed a system that could help diagnose the issues beyond the obvious symptoms they were already

aware of and find the source of the problems.

Selecting an EEMS

It was their BAS vendor, Johnson Controls, who knew of a company that provided a system with the level of comprehensive data gathering and analysis tools needed. Interval Data System (IDS) was brought in and put EnergyWitness to work. IDS also provided expert diagnostic services to work with the hospital staff interpreting the data, identifying the issues and making recommendations for an action plan.

Installing the System

Installing EnergyWitness required addressing server hardware, software, networking, security, and remote access components. Once installed, BAS monitoring and control points had to be mapped into a relational database. The challenges associated with the installation fell into two categories—technical and project management. It took four to five days of labor time over a three-week period to get to a point where data was being collected 24x7.

Technical Installation

There were three main components to the hardware and software installation:

- An OPC (OLE for Process Control) server: a software component that resides on the BAS server hardware, typically supplied by the BAS manufacturer, used to collect BAS point data.
- OPC client: the software that EnergyWitness uses to communicate with the OPC server to collect data.
- The EnergyWitness server: a Windows PC running the EnergyWitness server application and Microsoft SQL Server for database management.

In order to provide the necessary diagnostic services, as well as providing software support (technical support, software updates, backup of data), IDS staff needs remote access to the EnergyWitness server. As with any site, it was critical for the installation and ongoing support to follow the policies set forth by the IT department. In this case, the hospital set up a VPN (virtual private network) line to provide secured access.

The final phase of installation was data collection. Since the initial focus was the chiller plant, all of the BAS point data for the chillers, cooling tower, and chilled water loop were defined in EnergyWitness as collection points. This was the most time-consuming part of installation, as it is important to not just identify the points, but also determine which need instantaneous, averaged, and/or totalized readings. All of the interval data are collected every 15 minutes. At this point the system was monitored, and given a couple weeks to collect data before data configuration and diagnostics would begin.

Project Management

The project management related activities (coordination with IT, BAS managers, and project leaders) required far more time and effort than did the technical issues. Most times the management issues were the critical path to *allow* the technical work to happen. This is normal, but often overlooked, for any enterprise-scale software installation—especially when it involves network connections and a need for remote access. Communication between the vendor, facilities team, and IT department is the foundation to a smooth, successful installation.

Data Configuration

Data configuration is an ongoing process. EnergyWitness uses a combination of a tabbed interface for major grouping of information, and a tree structure to organize and group the points to be meaningful and easily accessible to the user. Points were gathered from the BAS into a master list—the raw data coming into the system. The tree in back shows the start of the master list, including all of the chiller plant points and points for some of the air handling systems. The master tab categorizes approximately 3,000 BAS points.

Phase two of data configuration organized points from the BAS into more navigable structures. The front list shows a Chillers tab that further organized the data into a tree structure. The structure makes locating the right data easier and trend lines for all points within a folder can be displayed as a group for fast monitoring. When configuring the data within the Chillers tab, weather data was also included for convenient access. You can also see two types of trend lines: the first, as can be seen in the CHW Tons folder, are BAS points; the second, seen in the Tower Power folder, are calculations.

Over 60 calculations, based on the chiller plant data, were added to the structure to provide a higher level of information. These range in complexity from Delta Ts, to balance efficiencies and theoretical water loss, to overall chiller and plant efficiency and operational cost calculations. All calculations are updated at every 15-minute interval.

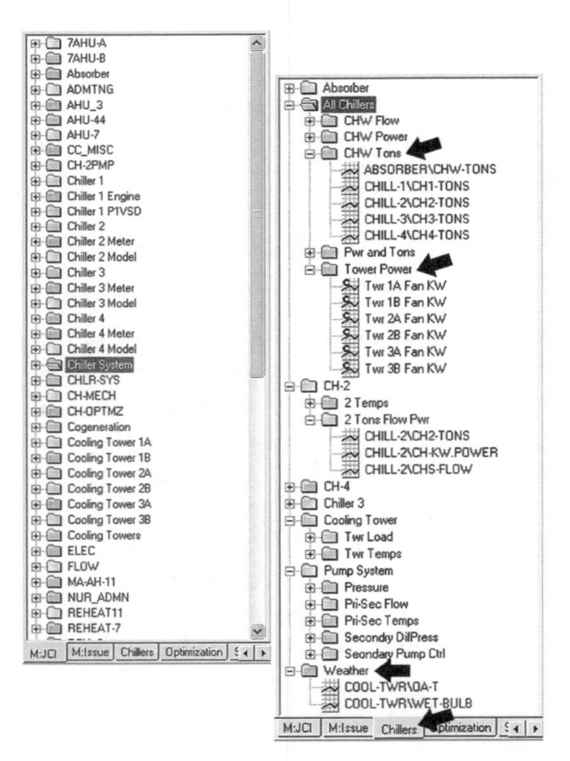

Figure B-1. Data trees organize the interval data for instant access.

Diagnostics Phase I: Operations and Control Issues

The data capture and configuration was phased in, as mentioned above. At the start, with only 700 BAS points from the chiller system and weather data being collected, the IDS diagnosticians started to review the data. In just three days of diagnostics, 45 preliminary issues were identified and documented (Figure B-2). The issues spanned control problems, sensors that needed calibration, and the suspected engineering design issues began to surface.

Using the collected data, IDS and the hospital staff were able to categorize each issue, define the implications to the overall operation, and identify (quantification comes later) the likely financial benefit of correction. This classification effort helped prioritize the plan of action.

Diagnostics Reveal Instability

Before the analysis on the engineering design problem could begin, it became clear that there were operational problems that could be resolved first. The system was showing a significant amount of instability that needed to be addressed to identify the root causes of problems and inefficiencies. Much of the instability was a result of hospital staff addressing symptoms—they simply did not have the data to get to the source of the issues.

The next three charts show some of the initial concerns found by studying the data for literally just a few hours. The issues shown revealed sensor problems, out-

Hospital Chiller Plant Preliminary Investigation Summary

Issue Num	Issue Description	Category	Implications	Comment	Financial Benefit of Correcting Issue
1	CW set point adjust every 10 minutes causing CW & CHW systems to hunt	Over Control	Cost of Operations, Equipment Life		Straight forward change to BAS will pay for itself very quickly
2	Cooling tower fans (all) controlled in unison	Inadequate number control points and control sequence	Cost of Operations, Equipment Life	Each pump will need own control points added so that flow can be controlled	Eligible for rebate with significant energy cost savings
3	Although systems may be new, they do not seem to take advantage of high CW delta T for increased chiller efficiencies and reduced pumping rate.	Engineering design	Cost of operations (significant)	More effort required to determine savings benefit as costs could be high	TBD
4	Secondary pumping systems seem to operate from the same pressure controller thus causing increased pump hp operating costs.	Inadequate number control points and control sequence	Cost of Operations, Equipment Life		Straight forward change to BAS will pay for itself very quickly
5	Current CHW operation causes flow in path of least resistance through idle chillers/absorber.	Suspect of engineering design	Large cost of operations, Equipment Life	Needs further investigation to determine course of action. Possible piping change req.	Candidate for rebate
7	Current CHW operation causes mixing of 42° F water to higher temperature which will reduce design cooling load because control valves will be open trying to dehumidify.	Engineering design	Cost of operations (significant)		
8	Current CW operation causes flow in path of least resistance through idle chillers	Engineering design	Large cost of operations	Likely addition of smaller in series valve will fix situation	TBD
9	Current CW low ΔT causes chillers to work less efficiently and increases hp energy.	Chiller Control	Cost of operations		
10	Primary CHW pumps are controlled through circuit setters instead of VFDs which can cause high annual operating costs. i.e.. 10ft head at 3000 gpm x \$0.08/kWh ≅ \$5K/year pumping hp	Engineering design	Cost of operation		Eligible for rebate with significant energy cost savings

Figure B-2. The beginning of the initial diagnostics report—the first ten of 45 preliminary findings.

puts cycling by as much as 500 tons every 15 minutes (Figure B-3), temperatures within the primary and secondary chilled water loops that were not as they should be (Figure B-4), and bypass flow issues (Figure B-5).

Disable Chiller Reset Program—Fix Setpoint at 42°F

In order to stabilize the overall chiller plant operations, the first recommendation was to discontinue using the reset program that was in place and fix the setpoint at 42°F for all chillers. This suggestion was made to the hospital during meeting on 4/26 at about noon. The control change was made immedi-

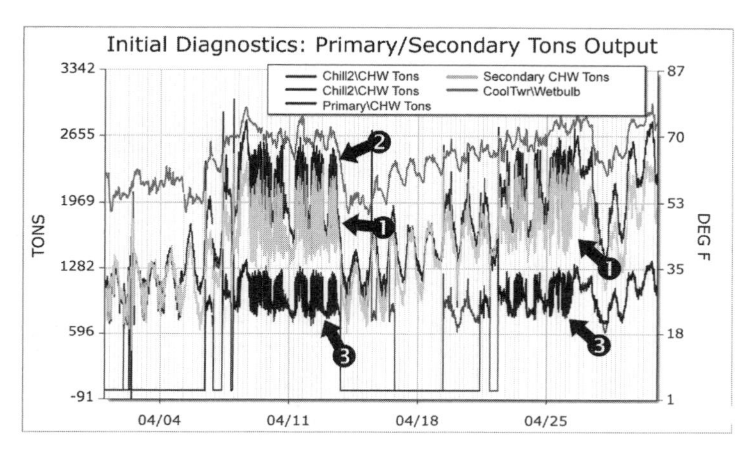

Figure B-3. The chiller operations were unstable. Tons output measured from the primary and secondary sensors (❶-green and ❷-dark purple lines) and the chillers (❸-blue & dark green lines) showed cycling by as much as 500 tons every 15 minutes.

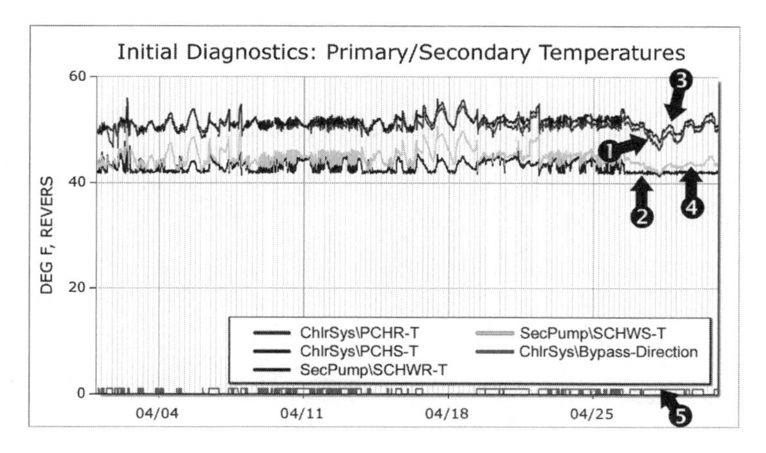

Figure B-4. The primary supply and return temperatures (❶-dark green and ❷-blue lines) should not be lower than the secondary temperatures (❸-purple and ❹-bright green lines) when the flow is reversed (❺-magenta line at bottom).

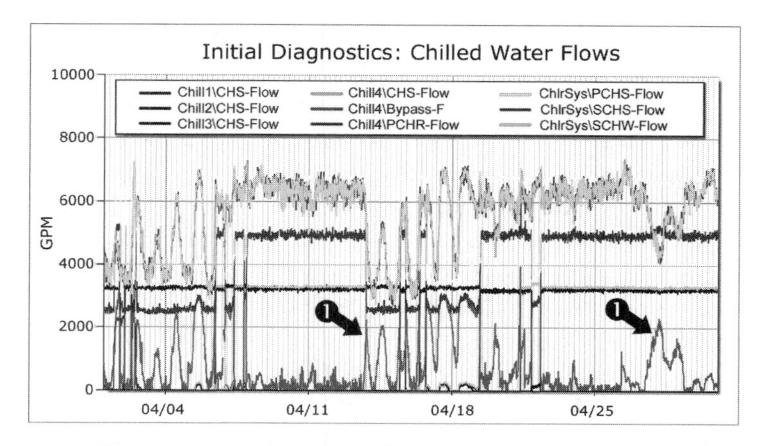

Figure B-5. There is often too much flow through the bypass (❶-magenta line across bottom). This causes the secondary supply temperature to rise, reducing cooling capacity.

ately afterwards, and you can see in Figure B-6 the cycling of output that was happening immediately settled down. The stabilization that occurred was immediately verifiable because the interval data are constantly collected, showing the impact of the control change in setpoint.

Looking at the impact of chilled water and condenser water temperatures, the change had a similar positive impact. The chilled water supply temperature, which had been oscillating between 42° and 45°, stabilized at the 42° setpoint (Figure B-7). The vacillating chilled water return temperature and the condenser supply and return temperatures also smoothed out with the fixed setpoint.

After any operational change it is important to verify that the intended impact occurred, as we saw above. It is also critical to check that there are no unexpected side effects. Because all of the chiller plant BAS points are available, it was trivial to look at cooling tower data and check the impact of the setpoint change. Similar positive results had occurred. As seen in Figure B-8 the condenser water temperatures smooth out and the full load amps percentage (which had been cycling 20 percentage points) stabilized significantly.

Change Approach Offset of Cooling Towers

After stabilizing operations, the goal was to improve efficiency of the chiller plant operations. Attempting to improve the efficiency of the chillers, the hospital had set the cooling towers to achieve the lowest possible condenser water temperature. However, this led to the cooling tower fans running at full load all the time, negating any savings the chillers were achieving.

A series of adjustments were made to the approach offset for the cooling towers. A change was first made to a 3°F offset (on 5/3) and then to 3.5° (on 5/7). Figure B-9 shows the impact, where the first change immediately allowed the fans to draw less power at night—20 to 30 kW less power for each fan—especially between midnight and 7:00 a.m. The additional half-degree approach resulted in the tapering off of kilowatts used shortly after noon.

Tower Fan Change

The final operational change was to the control strategy for running the six cooling tower fans. Instead of cycling each one up to full speed and adding additional fans as needed, the change was made to run them all at once, all the time, but at lower speeds. Also, the sixth tower fan that had been offline was fixed on 5/13 and added to the mix. The impact was another significant drop in electricity used by the fans, and a significant cost savings for the hospital as a result.

Verification of Results

The reason for the drop is that the power used by the fans is related to the cube of the fan speed. So, as fans were pushed to 100%, the power consumption was going up exponentially. Keeping the fan speeds down keeps the kW down.

Further analysis was done in Excel, plotting the kilowatts used versus cooling tons output. In Figure B-11 you can see that running only five fans uses more power at all load levels. As the load increases, the penalty of only running five fans increases. At 1,600 tons output, the difference is about 35 kW, but at 2,400 tons the difference is nearly 60 kW.

Savings Estimate

The final stage of this operational change was to determine the annual cost savings resulting from the improvements. The bottom line was a calculated $40,000 annual savings. The calculation is summarized in Table B-1.

Ongoing monitoring in the weeks and months following this change have further confirmed that the fans continue to run at a lower rate, validating that the projected cost savings are being realized.

Also during this initial phase, because the data from the initial diagnostics showed that many pressure and temperature sensors needed calibration, the

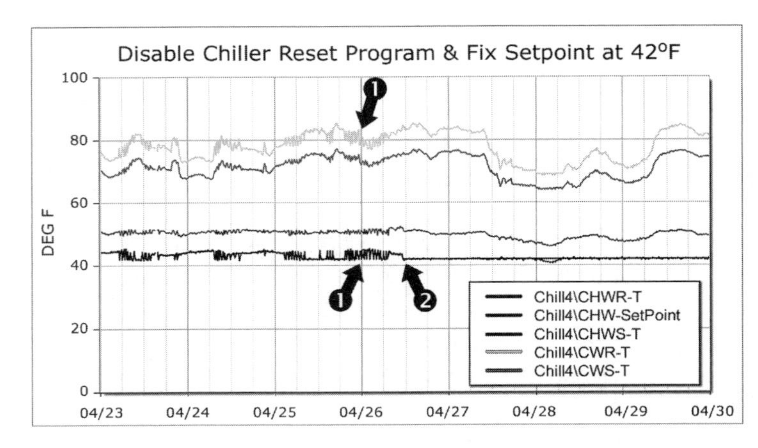

Figure B-6. The tons output load was cycling widely (❶-green line) until the setpoint (❷-magenta line) was fixed at 42°F on 4/26.

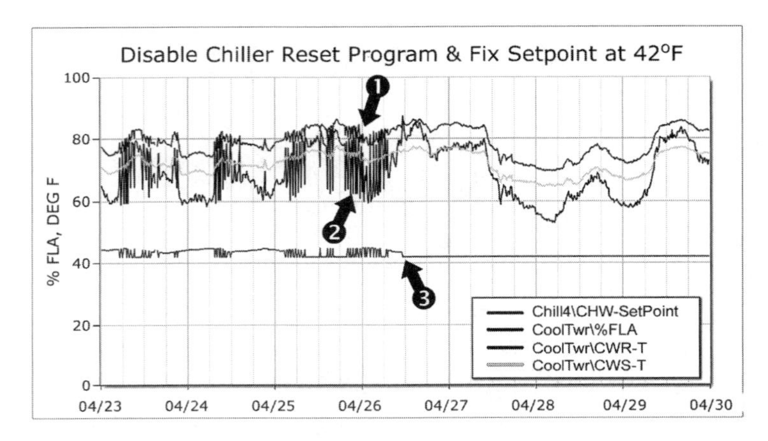

Figure B-7. Temperatures of chilled and condenser water stabilized (❶) immediately when the chiller setpoints were set to 42° (❷).

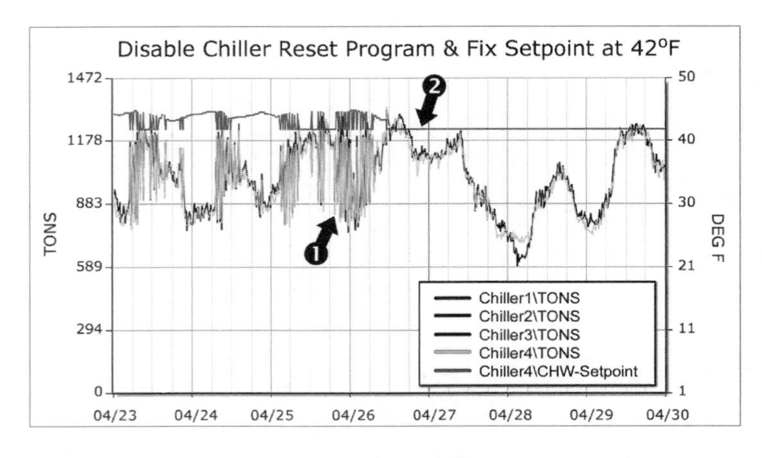

Figure B-8. The effects of the chiller setpoint change on the cooling tower show condenser water temperature (❶-purple line at top) and % full load amps (❷-blue line) smooth out considerably with setpoint (❸-green line at bottom) change.

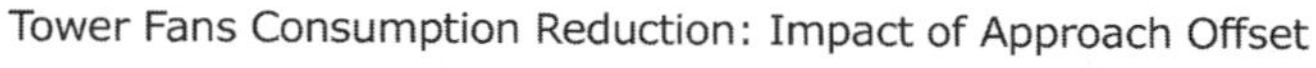

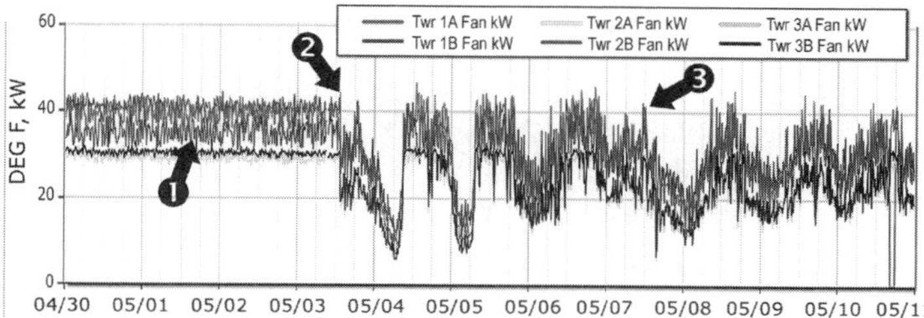

Figure B-9. Electrical consumption of the six cooling tower fans (❶-fan kW lines across the bottom) was reduced by setting the approach offset to 3° on 5/3 (❷) and then 3.5° on 5/7 (❸).

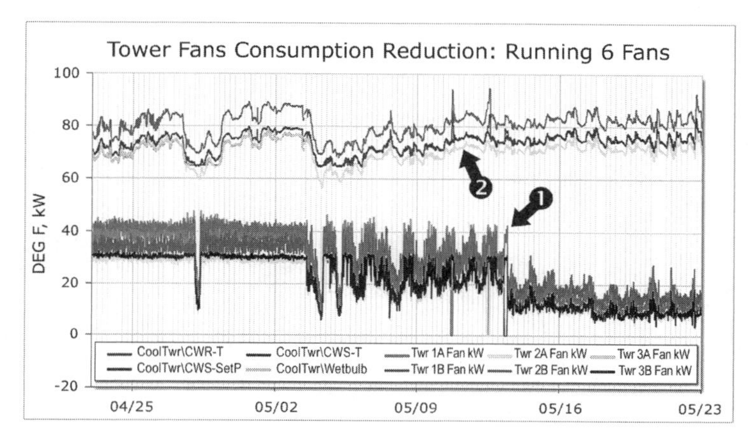

Figure B-10. In this monthly view, you see the impact of the approach offset changes also seen in Figure B-9, then an equally dramatic impact of the fan repair and control change on 5/13 (❶) to run all six fans together at lower speeds. The control change benefit occurs despite higher wetbulb temperatures (❷-bright green line).

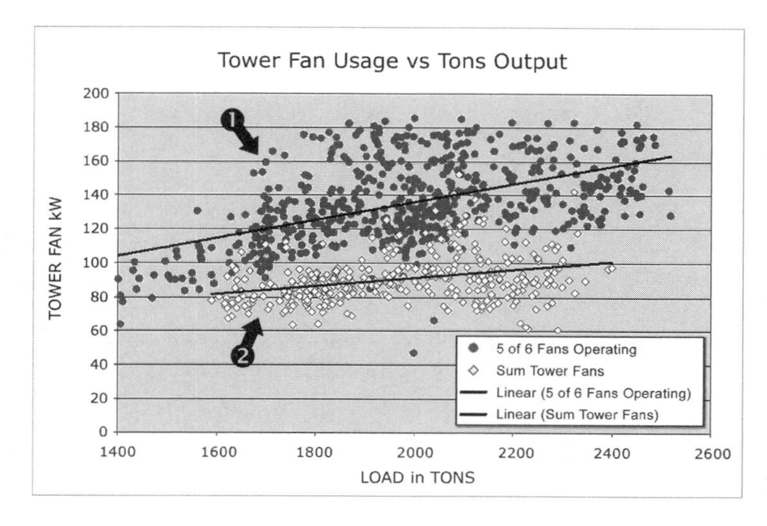

Table B-1. The data show an average savings of 78 kW for all six fans. The calculation assumes that there are 100 days/year where (in the south) the wetbulb isn't high enough to cause this condition.

A.	Fan energy savings (all fans):	78 kW
B.	Electricity rate:	$0.08/kWh
C.	Daily savings (A*B*24):	$150
D.	Days/year affected by change:	265
E.	Annual savings (C*D):	$39,800

hospital reprioritized their work plan and fixed them to ensure the accuracy of incoming data.

Diagnostics Phase II: Engineering Design Issues

With the chiller plant operating in a much more stable manner, the focus could finally turn to the larger set of issues due to the engineering design of the chilled water piping. The belief was that a significant amount of money was being wasted and staff resources were spending considerable time controlling around the deficiencies—treating the symptoms, as they didn't have enough information to identify and tackle the root problems.

Given that the expected cost of a piping redesign is in the millions of dollars, the hospital needed quantitative proof of the issues to justify the investment in a redesign.

In order to assist the diagnostics process of determining the redesign requirements needed to optimize the chiller performance, a series of calculations were put into the system. These calculations are applied at each 15-minute interval and made available to the system users as trend lines, exactly in the same way the individual point data are available.

Underutilized Low-Cost Chiller

The chiller plant consists of three electric chillers, each with a 1,500-ton capacity, and a gas-fired chiller with an 800-ton capacity. The electric chillers need to maintain minimum loads to prevent surging and resulting damage. Because of that, the gas-fired chiller is only used during peak hours. Despite its

Figure B-11. Using only 5 of 6 fans (∂-purple dots) uses more kW at all load levels than using all 6 (❷-green diamonds), and the difference gets greater the higher the output load.

800-ton capacity, the gas chiller rarely outputs more then 600 tons, and its output trails off during the course of the day. That leaves, on average, 236 tons of unused capacity.

The unused capacity is so important because this chiller operates at a much lower cost per ton than the electric chillers do. As the analysis in Figure B-13 shows, the gas chiller (chiller 1) costs an average of $0.07 per ton to operate, while during on-peak hours the electric chillers cost $0.22 per hour. By running below 75% of capacity, there is an unrealized savings of $83,000 per year (Table B-2).

Table B-2. The gas chiller is 14.73 cents/ton cheaper to operate during the 9-10 peak hours per day.

A.	Electric chiller on-peak:	$0.2182
B.	Gas chiller on-peak:	$0.0709
C.	Savings/ton-hour (A-B):	$0.1473
D.	Unused gas tons:	236
E.	Hours per day:	9
F.	Days per year:	265
G.	Unused ton-hours (D*E*F):	562,860
H.	Lost savings (C*U):	$82,900

Reduced Chiller Efficiency

Issues in the chilled water flow of the electric chiller systems (chillers 2, 3, and 4) were discovered that caused additional inefficiencies. At times the primary chilled water flow was higher than the secondary flow, resulting in secondary chilled water mixing with primary chilled water supply (note the bypass loop in Figure B-14). This, in turn, caused the temperature of the water returning to the chiller to be lower than it should (a low Delta T). The result was a decrease in chiller efficiency and a corresponding increase in operating costs. Figure B-14 shows the increased primary flow during peak hours, and Figure B-15 shows the matching drop in return temperature. The picture is completed in Figure B-16, which shows the kilowatts per ton increasing each time the flow problem occurs.

The operational cost associated with the increase in power used is $38,500 per year, summarized in Table B-3.

Reduced Primary Pump Efficiency

The three electric chillers also have pumps working at less than peak efficiency. The chillers were run at a reduced capacity—approximately 300

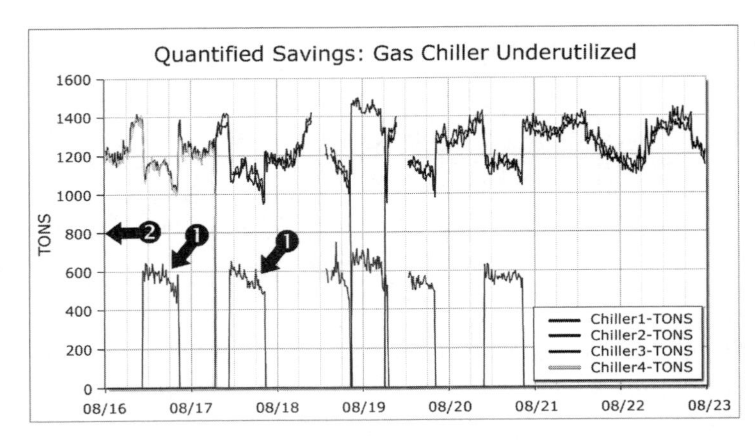

Figure B-12. Chiller 1, the gas-fired chiller (❶-dark green line), operates well below its 800-ton capacity (❷).

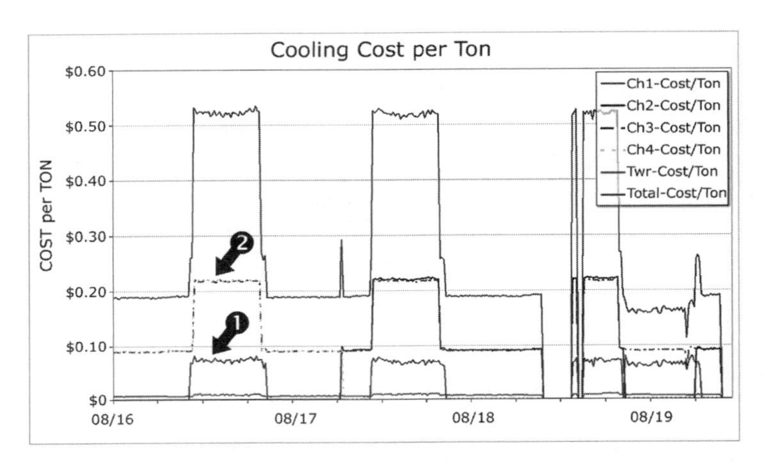

Figure B-13. Chiller 1, the gas chiller (❶-light blue line), operates at a third of the cost of the electric chillers (❷) during on-peak hours.

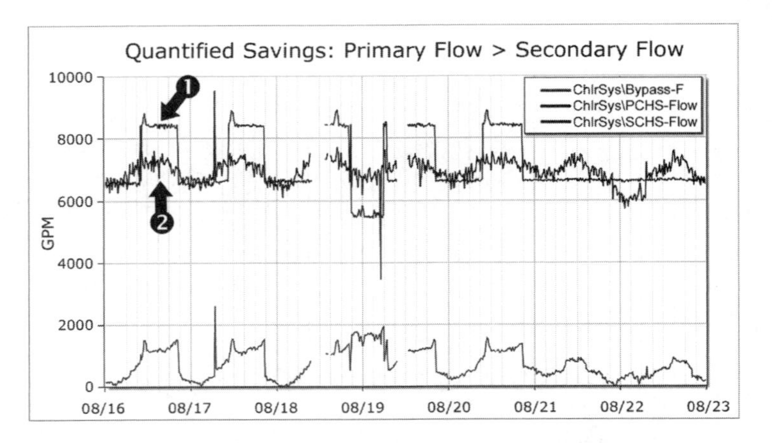

Figure B-14. Primary chilled water flow (❶-blue upper line) often exceeds secondary flow (❷-purple line) during peak hours.

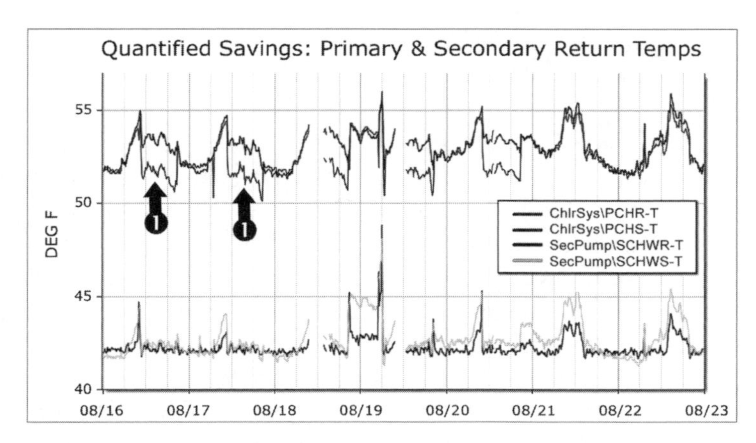

Figure B-15. Lower chiller water return temperatures (➊) result when the primary flow exceeds secondary flow (B-14).

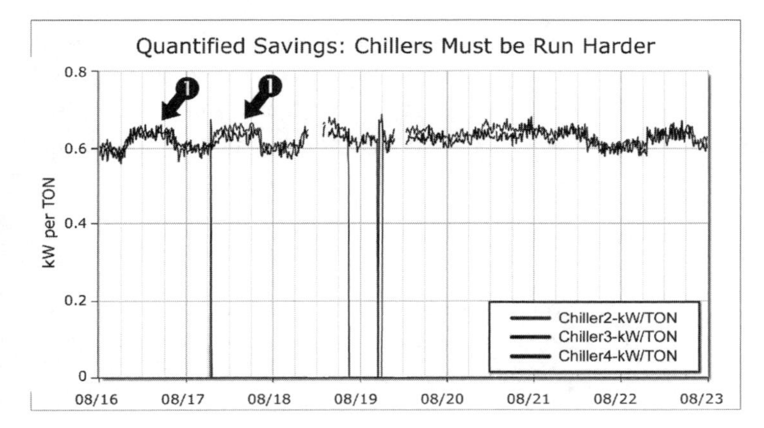

Figure B-16. Power consumption (➊) increases when the primary flow exceeds secondary as well, resulting from a higher supply temperature and corresponding lower Delta T.

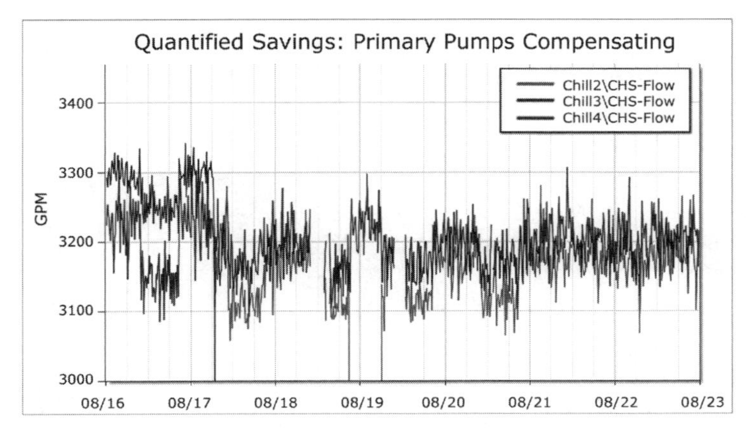

Figure B-17. The three electric chillers running below capacity.

Table B-3. Monthly savings calculated based on a lost Delta T average of 1.73°F for 197 hours/month, differential kW/ton of 0.027, on-peak rate of $0.124/kW, and off-peak rate of $0.053/kW.

A.	Total monthly savings:	$5,500
B.	Months savings will occur:	7
C.	Lost savings (A*B):	$38,500

gallons per minute flow reduction—to minimize differential flow problems and help balance the system. The way the flow was reduced was by increasing the pump head pressure, resulting in a 15 kilowatt power consumption increase per pump.

The annual cost of the primary pump inefficiencies is nearly $23,000.

Table B-4. Daily savings based on a 6" water column differential pressure, on-peak rate of $0.124/kW, and off-peak rate of $0.053/kW.

A.	Total daily savings for all three electric chillers:	$85.50
B.	Days/year savings will occur:	265
C.	Lost savings (A*B):	$22,700

Reduced Secondary Pump Efficiency

A much bigger problem, as it turns out, were efficiency issues with the secondary pumps. There are seven zones served by the secondary header with four 75 hp variable speed pumps. The control for the header is managed by the worst-case pressure across all seven zones. The system was working the pumps too hard, providing too much pressure to five zones in order to meet the needs of zones 4 and 6—the worst-case zones. In Figure B-18 you can see that the pressure in five of the seven pumps is well above setpoint nearly all of the time.

The average demand placed on all of the pumps causes an hourly cost excess of $9.35 per hour. Since this condition is constant, the annual savings possible is nearly $82,000.

Summary of Quantified Lost Savings

Using the interval data collected by the BAS, EnergyWitness provided the means to diagnose several problems with the chilled water system. The data enabled IDS and the hospital staff to quantify the excess expenditure these issues are currently costing

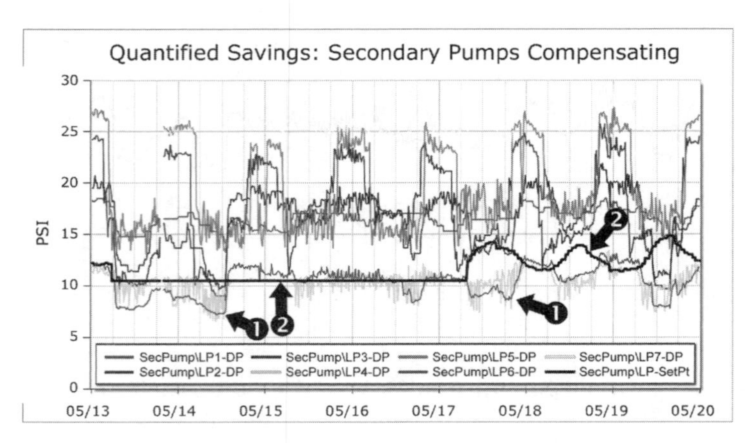

Figure B-18. Controlling all seven zones to meet the worst-case pressure demands of zones 4 & 6 (❶-bright green & teal lines) causes the other five zones to exceed setpoint (❷-gray line), forcing the pumps to work harder.

Table B-5. Average demand cost related to over-pumping is based on demand HP values ranging from 39.9-177.8, demand kW values ranging from 42.7-190.2, and demand costs (rates of $0.053 off-peak, $0.124 on-peak) ranging from $3.77-$22.54.

A.	Average demand of secondary pumping system:	$9.35
B.	Hours per day affected:	24
C.	Days per year affected:	365
D.	Lost savings (A*B*C):	$81,900

on an annual basis. All of this analysis was done by a small team of diagnosticians and facility engineers, requiring just a week of time from each team member, spread over about two months. The final compilation of results speaks for itself:

Diagnostics Phase III: Space Comfort Issues

A most important goal of the pending engineering redesign effort is to correct space comfort deficiencies. Although not quantifiable in the same manner as the operational cost issues in the previous section, it is mandatory that the hospital be able to reach required comfort levels throughout the facility as a completely separate concern from the cost of doing so.

The current design was simply making it impossible to deliver the desired air quality to some areas. Appropriate cooling levels were not achievable or could not be maintained throughout the day. Priority areas to examine included certain floors of the main building, floors in the north wing, and the children's and cancer care facilities. Plus, a dozen operating rooms were critically affected.

At this point there is data from approximately 3,000 BAS points being collected by EnergyWitness for continuing diagnostics and ongoing monitoring. It includes the original 700 chiller plant points, plus an additional 2,300 points throughout the distribution loop, air handlers, and terminal boxes.

Unable to Provide Sufficient Cooling due to Uneven Flow

The data examined in the Reduced Chiller Efficiency section show how that flow problem directly impacts cooling performance in the spaces. Earlier, we quantified the costs associated when primary flow exceeded secondary flow. Cooling issues also existed when secondary flow exceeded primary flow (Figure B-19). This resulted in higher chilled water supply temperatures delivered to the air handling units, reducing cooling efficiency and space comfort. Figure B-20 shows the rise in supply temperature caused by the flow problem.

Table B-6. Summary of quantified lost savings.

Quantifiable Cost of Chilled Water Problems		
Affect	**Condition**	**Cost per Year**
Underutilized low-cost chiller	Gas chiller unable to reach full capacity	$82,900
Reduced chiller efficiency	Primary flow greater than secondary flow	38,500
Reduced primary pumping system efficiency	Operating water pumps to compensate	22,700
Reduced secondary pumping system efficiency	Operating water pumps to compensate	81,900
Total		$226,000

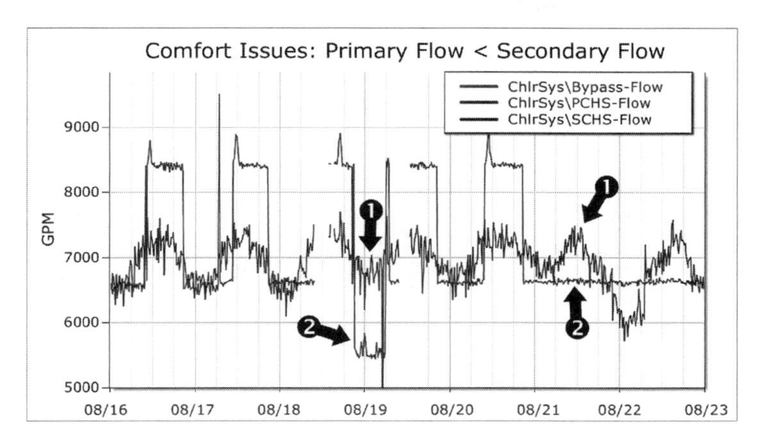

Figure B-19. Secondary chilled water flow (❶-purple line) exceeds primary flow (❷-blue line).

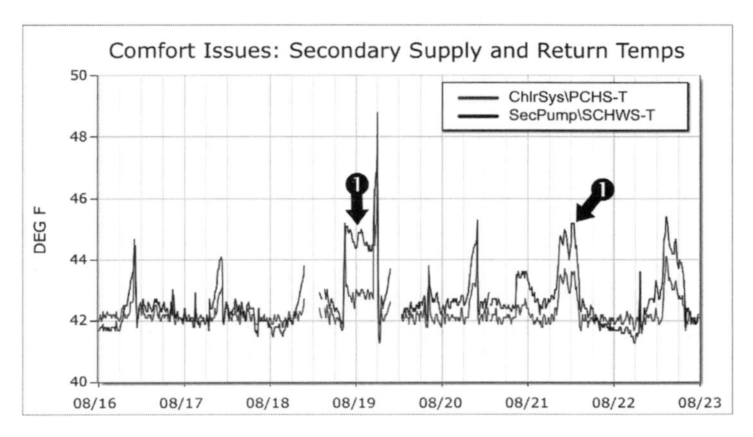

Figure B-20. Supply water temperatures (❶-blue line) rose when secondary flow exceeded primary flow (Figure B-19).

Inadequate Pressure for Cooling to Reach Some Zones

During peak cooling periods (e.g. summer) some zones are unable to meet their pressure setpoint. This indicates inadequate flow and insufficient cooling capacity, allowing increased humidity and temperature in those areas. Although the cooling capacity exists, the pump differential pressure issues keep it from being delivered to all of the hospital zones that need it.

Doctors "Requesting" Colder Temperatures

Additional evidence of space comfort problems were found by examining the operating room controls. The temperatures in the operating rooms were seldom as low as the doctors wanted them, as evidenced by looking at the warm-cold slider control data. Doctors most often push the sliders down to get the rooms colder, as seen in Figure B-22.

Space Comfort Summary

The information gathered through the interval data supplied the hospital facilities director with ability to:

- Document (for management) the affects on comfort and air quality resulting from the current HVAC design
- Define the requirements for the engineering design work to ensure that all known comfort problems get resolved

SUMMARY

This case study represents only the beginning of the impact that using an EEMS, EnergyWitness, has had on the hospital. The information that has been extracted from the interval data, and the quality of the decisions and planning that have come from it, was not possible before.

Earlier efforts relying on the BAS trend logs took substantially more time to garner much less information even though most of the data used in this case came from the hospital's BAS. In short, the hospital personnel determined that he BAS is unequipped to supply the level of diagnostic capabilities the EEMS has provided, as are the tools used by the engineering company to determine the redesign. This is why even the BAS vendor suggested EnergyWitness.

So far EnergyWitness has been used to:

- Improve chiller plant operations that are saving the hospital $40,000 a year in electric costs, plus anticipated maintenance reductions and extended equip-

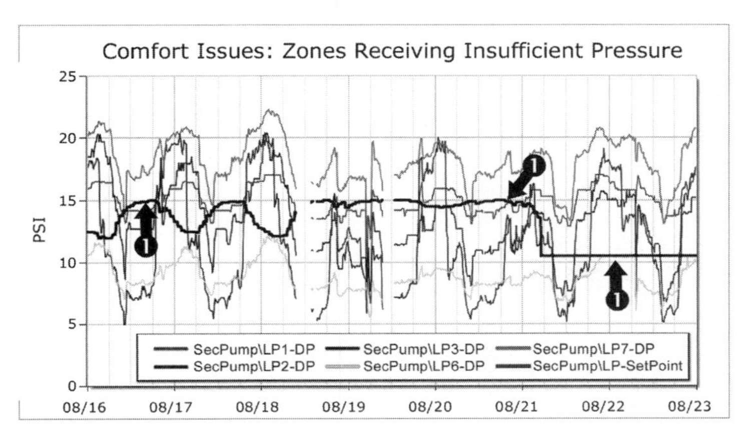

Figure B-21. Pump pressures are unable to meet setpoint (❶-thick teal line) in many of the zones.

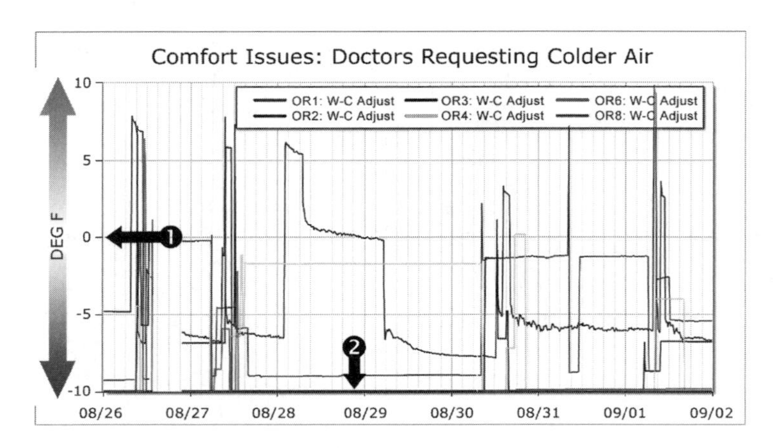

Figure B-22. Colder temperatures are requested most of the time in each OR as shown by the sliders in the negative position (❶-below the zero setting). Note that three of these six ORs sat with the sliders at -10 all weekend (❷).

ment life due to smoothing out operations, which are not quantified. (More operational savings to come.)

- Identify and document piping design issues that cost the hospital nearly a quarter-million dollars ($226,000) per year.
- Identify and document space comfort problems, showing the root cause, which should disappear with the right engineering redesign.
- Reprioritize staff assignments to address the most important issues.
- Focus the right resources on the right problems—stop addressing symptoms with technician time and fix the real problems with engineering work.
- Provide the facilities staff with a tool to monitor operations, continue driving down operational costs, and document needed equipment and services.
- Prove the necessity of instant access to all the data in a visual form in order to achieve goals.

This information has become the foundation of the proposal to upper management justifying the need for a multi-million dollar investment to correct the design problems. It also has provided the hospital with data that can be used by the engineers hired to do the piping redesign, and then enable the hospital to validate or question their recommendations.

Data have proven to be a new and viable tool to provide operational information to help manage the facility—one that has never been available before. The hospital realized that when more people have access to the data in a manner they can understand, more people can participate in discussions and strategy decisions. An Enterprise Energy Management System such as EnergyWitness is the only tool capable of delivering this level of operational diagnostics. Facilities directors now have the information—and the credibility that goes along with documented facts—to make the right decisions, fund the right projects, and fix the root problems.

Chapter 33

Data Quality Issues and Solutions for Enterprise Energy Management Applications

Greg Thompson, Jeff Yeo, and Terrence Tobin
Power Measurement

ABSTRACT

WEB-BASED ENTERPRISE energy management ("EEM") systems are delivering the information and control capabilities businesses need to effectively lower energy costs and increase productivity by avoiding power-related disruptions. However, the quality of energy decisions is directly affected by the quality of the data they are based on. Just as with CRM, ERP and other business intelligence systems, EEM systems have data quality issues, issues that can seriously limit the return on investment made in energy management initiatives.

Data quality problems result from a number of conditions, including the reliability and accuracy of the input method or device, the robustness of the data collection and communication topology, and the challenges with integrating large amounts of energy-related data of different types from different sources. This chapter describes how dedicated data quality tools now available for EEM applications can be used to help ensure that the intelligence on which an enterprise is basing its important energy decisions is as sound, accurate, and timely as possible.

THE IMPORTANCE OF MANAGING ENERGY WITH RELIABLE DATA

Under growing competitive pressures, and spurred by recently introduced energy policies and mandates, businesses are becoming increasingly aware of the need for, and advantages of, proactively managing the energy they consume. Industrial plants, commercial facilities, universities and government institutions are looking for ways to lower energy costs. For some operations, the quality and reliability of power is also critical, as it can negatively affect sensitive computer or automation equipment, product quality or research results, provision of 24/7 service, and ultimately revenues.

In response, facility management and engineering groups have been tasked with finding the latest available technology capable of delivering and managing the energy commodities that have such a significant affect on their bottom line. A number of options have emerged in recent years, the most comprehensive of which are enterprise energy management ("EEM") systems.

An EEM system typically comprises a network of web-enabled software and intelligent metering and control devices, as well as other inputs (Figure 33-1). A system can track all forms of utilities consumed or generated, including electricity and gas, as well as water, compressed air and steam. Data can be gathered from the utility billing meters or other meters positioned at each service entrance, from tenant or departmental submeters, and from instruments that are monitoring the conditions of equipment such as generators, transformers, breakers, and power quality mitigation equipment. Other inputs can include weather information, real-time pricing information, occupancy rates, emissions data, consumption and condition data from building automation systems, production data from enterprise resource planning (ERP) systems, and other energy-related data.

Based on these diverse inputs, EEM systems deliver analysis and reporting for energy, power quality and reliability information. Armed with this intelligence, managers can verify utility billing, sub-bill tenants, aggregate energy use for multiple locations, compare the affect of utility rate choices by running "what if" scenarios, procure energy or manage loads and generators in real-time based on pricing inputs, identify power conditions that could potentially cause downtime, and perform a variety of other tasks that can improve energy performance.

Though EEM technology effectively gathers, stores, processes and delivers customized information from key

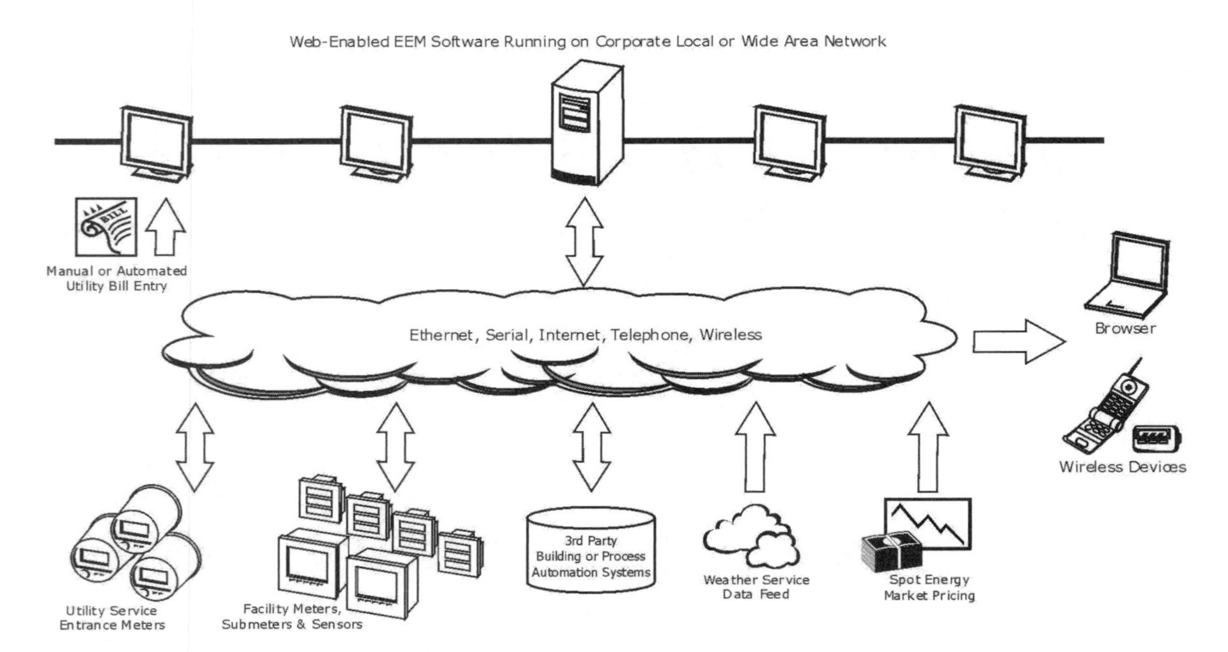

Figure 33-1. Enterprise energy management ("EEM") system inputs

points across an enterprise to the people that need it, the important economic and operational decisions based on that information can be negatively impacted by poor data quality. EEM systems, like other business intelligence systems such as customer relationship management (CRM) and enterprise resource management (ERP), are susceptible to the effects of poor data quality; for any system it is only a matter of degree.

THE IMPACT OF POOR DATA QUALITY ON ENERGY DECISIONS

A recent global data management survey by Pricewaterhouse Coopers of 600 companies across the United States, Australia and Britain showed 75% reporting significant problems as a result of data quality issues, with 33% of those saying the problems resulted in delays in getting new business intelligence systems running, or in having to scrap them altogether. Industry analysts at the Gartner Group estimate that over 50% of business intelligence projects will suffer limited acceptance, if not outright failure, due to lack of attention to data quality issues.

Data are the foundation of strategy and success, and sound business decisions depend on their quality. But data can only truly be an asset when they are of high quality; otherwise, they're a liability. As the old saying goes, "garbage in, garbage out."

For example, repeated data cleansing for CRM systems is common to ensure customer and sales prospect name and address information is up-to-date and accurate. Professionals depending on that data are aware of the many pitfalls of poor data quality, as the impacts can be serious if parts of information are missing, invalid or inconsistent. The impact can be worse if the data needed are late, inaccurate, or irrelevant to the decisions being made.

For these systems, the costs of poor quality data can be high, in terms of bad decisions, lost opportunities or lost business, damaged relations with partners, suppliers, or customers due to overcharging or underpayment, or even regulatory noncompliance due to faulty indicators. Managers need to have confidence in the reports they are using. Instead of wasting time wondering if an anomaly is the result of a problem with the supporting data, they should be identifying what caused the anomaly.

In the context of energy management, businesses using EEM systems will also have data quality problems, but many might not realize they do (Figure 33-2). The costs of low quality information can mean an inability to take advantage of better real-time pricing, not identifying energy waste, missing a large discrepancy on a utility bill, incorrect sub-billing, incurring an expensive utility demand or power factor penalty, or being issued a fine by a regulating authority for exceeding an emissions standard.

Overpayment, under-billing, or wasted energy due to inability to validate consumption

Disagreements over inaccurate or unreliable reports

Tenants angry about over-billing due to data inaccuracies

Incurring penalties or fines due to unreliable tracking of real-time conditions

Figure 33-2. Impacts of poor energy data quality

Some specific examples of problems caused by poor data quality in EEM systems, in order of increasing impact:

1. **Misleading energy trend reports**: Collected energy data are used to generate a monthly report showing consumption trends for all utilities – if the data are inaccurate or incomplete, you may fail to identify a serious trend toward over-burdening your system, or miss an opportunity to save energy by rescheduling a process or shutting off a load.

2. **Inaccurate billing reports:** Data quality can affect the accuracy of revenue-related applications. Data from "shadow meters" (installed in parallel with the utility's billing meters) are used to generate a "shadow bill" to help verify if the bill you receive from your utility each month is accurate. Poor data quality from the shadow meters can mean potentially missing a large billing error, or falsely accusing your energy provider of making one. If you use sub-meters to bill your tenants for the energy they consume, poor data quality can cause you to under bill or over bill your tenants, either of which can cause problems.

In the two scenarios above, if your utility billing tariff includes a real-time pricing (RTP) component, you will need to integrate an RTP data feed from the *independent system operator* that is responsible for setting energy pricing for a given region. If there are missing data in that feed, your shadow bills will be inaccurate and so will your tenant's bills. The potential for problems is compounded with each additional input from different energy metering systems (electricity, gas, steam, etc.), building automation systems, and ERP systems.

3. **Reduced confidence in critical business decisions:** Effective long-term energy management can include dynamic procurement strategies and contract negotiation. These require comprehensive modeling and projections of energy requirements based on a depth and breadth of information that includes all of the data described in the preceding examples as well as other data inputs such as weather, occupancy, etc. With this increased complexity and greater potential impact of the resulting decisions on your profit margins, data quality can seriously affect confidence.

All of these examples in turn represent the overall effects on the return on investment you achieve from your EEM system and the energy management program it supports.

DATA QUALITY PROBLEMS AND WHERE THEY COME FROM

To address data quality problems it is first important to understand what is meant by data quality. As each category of business intelligence system has its own data types, and in turn its own data quality criteria, this discussion will be restricted to the quality of data in EEM systems. However, data quality concepts, in general, are applicable across all business systems.

Data quality can be considered in terms of three main categories of criteria:

Validity. Not only does each data location need to contain the information meant to be there, but data also need to be scrutinized in terms of whether they are reasonable compared to established patterns. The data must be within the allowable range expected for that parameter. For example, if a monthly total energy value is being viewed for a facility, there will be a maximum to minimum range that one would expect the usage to fall within, even under the most extreme conditions. If a value is "out of bounds," it probably indicates unreliable data.

Accuracy. The data gathered and stored need to be of high enough accuracy to base effective decisions on. This not only requires that metering equipment be rated adequately for its accuracy, but that every internal and external input to the system is considered in terms of its accuracy, including third-party data feeds. For enterprise-wide systems, it is also important to accurately record the time at which each measurement is taken. When an aggregate load profile is being developed for multiple facilities across geographically dispersed locations, the measurements need to be tightly time-aligned. This is also true for sequence-of-events data being used to trace the propagation of a power disturbance.

Completeness. For any business intelligence system, incomplete data can seriously compromise the precision of trends and projections. There must be a complete data set; each recorded channel of information must contain all the records and fields of data necessary for the business needs of that information. For example, if interval energy data are being read from a tenant submeter, there can be no empty records. Such gaps might be mistakenly interpreted as zero usage, and in turn the tenant could be under-billed for energy that month.

In EEM systems, the above types of data quality issues can come from a number of sources, and for a number of reasons (Figure 33-3). For example, data that are out of range might be the result of an energy meter being improperly configured when it was installed, or a meter that has been improperly wired to the circuit it is measuring. There may also be inconsistencies between how a number of meters on similar circuits are configured, or differences between how the meters and the head-end software are set up.

Another source might be the "rollover" characteristic of registers inside most energy meters. Most energy meters have a specific maximum energy value they can reach, for example 999,999,999 kilowatt-hours. The registers will then rollover and start incrementing again from a count of zero (000,000,000). A system reading the information from the meter may not recognize this behavior and instead interpret values as being in error, or worse, interpret it as a negative value which produces large errors in subsequent calculations.

When there are gaps in data records, the source might be a loss of communications with a remote meter or other device or system due to electrical interference, cable integrity, a power outage, equipment damage or other reasons. Some communication methods are inherently less reliable than others; for example, a dial-up modem connection over a public telephone network will

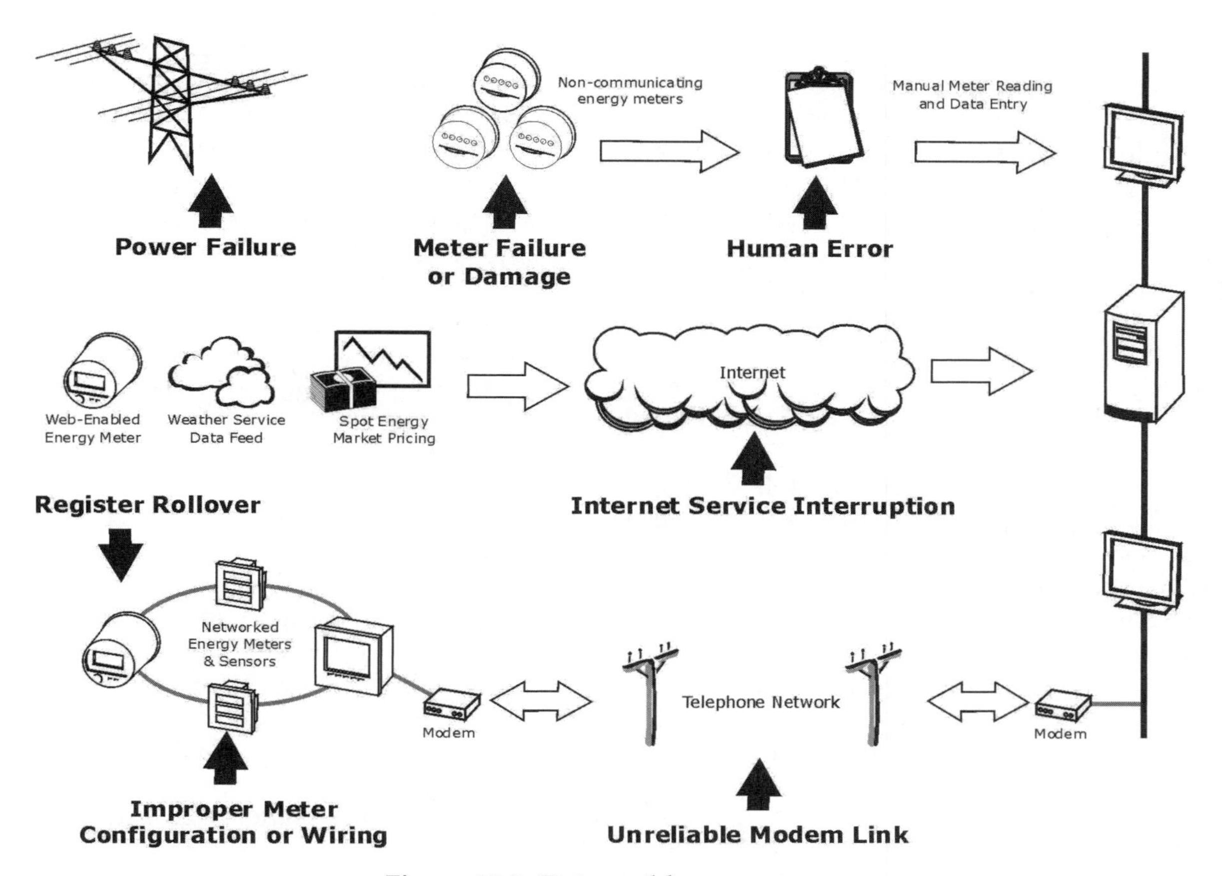

Figure 33-3. Data problem sources

likely be less reliable than a permanently hardwired Ethernet connection. Some meters offer onboard data logging that allows saved data to be uploaded after a connection has been restored, reducing the possibility of gaps. But an extended communication loss can still cause problems.

Other breaks in communication can include the interruption of an Internet connection over which weather or utility rate information is being imported, or the failure of the network feeding information from a third-party building or process automation system. As additional, diverse sources of real-time and historical information are integrated into an EEM system, the possibility of communications problems increases.

A remote meter, sensor, or other instrument may operate incorrectly or fail altogether, the latter condition causing a continuous interruption in data flow until the device is repaired or replaced. Finally, in cases where some remote meters are not permanently connected by a communications link, their data might be collected manually with a dedicated meter reading device or laptop computer, and then manually entered into the head-end system. Anytime this kind of human intervention is required there is room for error.

Ultimately, users judge the reliability of a system by the delivery point of information; they do not know or care where the data originated. If incorrect information is being displayed they simply consider the entire system to be at fault. Thus, the success of the EEM system as a whole is very dependent upon the quality of its data.

NEW DATA QUALITY TOOLS AND HOW THEY HELP

Identifying and correcting data quality problems takes the right tools, as well as the people to use them and a regular, repeatable process to ensure maximum effectiveness.

Electric and gas utilities have traditionally used tools to compensate for data quality problems in their metering and billing systems. The need for these tools was driven by the fact that the revenue billing meter and its data represent a utility's "cash register." Without accurate, reliable and complete information an energy supplier cannot be sure it is being properly compensated for the energy product delivered to the customer.

Utilities use what is commonly referred to as *validation, editing, and estimation tools*, collectively known as *VEE*. Until recently, these tools were not readily available for use by commercial and industrial businesses doing their own internal energy metering and management. New data quality capabilities are now emerging, specifically customized for energy consumer applications and expanding the concept of VEE to encompass *total data quality*. Though they borrow from the capabilities of utility VEE, new data quality tools take into account the wide range of input types that EEM systems leverage to develop a complete understanding of energy usage across an enterprise.

Data quality tools help ensure that data meet specific expectations and are fit for each purpose. The data quality process first identifies anomalies in each data channel, whether that channel is a remote meter, a third party automation system or Internet feed, or data input manually. Specific tools are then provided to correct for errors or gaps so that data are *cleansed* before they are analyzed, reported or otherwise put into action. The entire process is achieved through a sequence of validation, estimation and editing steps.

Validation

To validate that data are of high quality, a set of internal *standards* is created that defines the level of data quality required for each purpose. For example, a property manager may decide that data for sub-billing is acceptable with lower quality than the data used for utility bill verification.

Based on the quality standards, a set of rules is constructed that the data quality tools use to automatically check the quality of energy-related data coming into the EEM system (Figure 33-4). Examples of these rules include the following:

- **Constraints checks**. As mentioned, incoming data representing a particular parameter must meet a *bounds check* to see if they fall within reasonable values, such as between an acceptable minimum to maximum range. Those that fall outside predefined constraints are flagged as "out of bounds." In the case of meter energy register rollovers, a *delta check* can be done to see if the previous and currently read values differ by too much to be reasonable. Similarly, checks can be run to verify data correspond to an established set of allowable values. For example, a record indicating alarm status should either show an active or inactive state, possibly represented as a 1 or a 0. No other value is acceptable.

  Note: Though not an error check, tests will also be run on some data to find which sub-range a measurement falls in within the acceptable boundaries of values. This is needed when the value of a measurement determines how it is used in subsequent calculations. For example, some utilities charge for energy by applying different tariff charges to different levels of energy demand measured over specific demand intervals (usually 15 minutes), or to energy consumed at different times of the day, week or year. If a recorded demand level is greater than or equal to "x demand," a different tariff may be applied than if the value is less than "x demand."

- **Duplicate check**. The system will check for consecutive records having exactly the same data in them. This can include situations where both the value and the timestamp are the same for both records, or where the timestamps are the same but the values are different. Normally this will indicate an error due to a communication problem, improper system settings or metering logging configuration, time jitter, or other issue.

- **Completeness (gap) check**. When verifying interval energy data, where records are expected at specific time intervals (e.g. every 15 minutes), gaps in data are flagged. These can be due to message transfer issues, power outages, communication issues, etc. Missing records can then be compensated for in a number of ways, as described in the following section.

- **Dead source detection.** If a gap in data is long enough the system can flag it as a dead source so that appropriate steps can be taken to investigate the cause.

- **Zero, null and negative detection.** If an energy meter is showing "zero" energy usage for a particular facility, load or other metered location, when that condition is not expected, it will be flagged as a possible error. This can include either a zero reading for an individual interval, or a delta (difference) check between two consecutive readings of a totalizing register. It can then be investigated to see if there may have been a major power outage event. If so, the error indication can be manually *overridden*. Null readings (e.g. no value) can also indicate a problem. Finally, negative

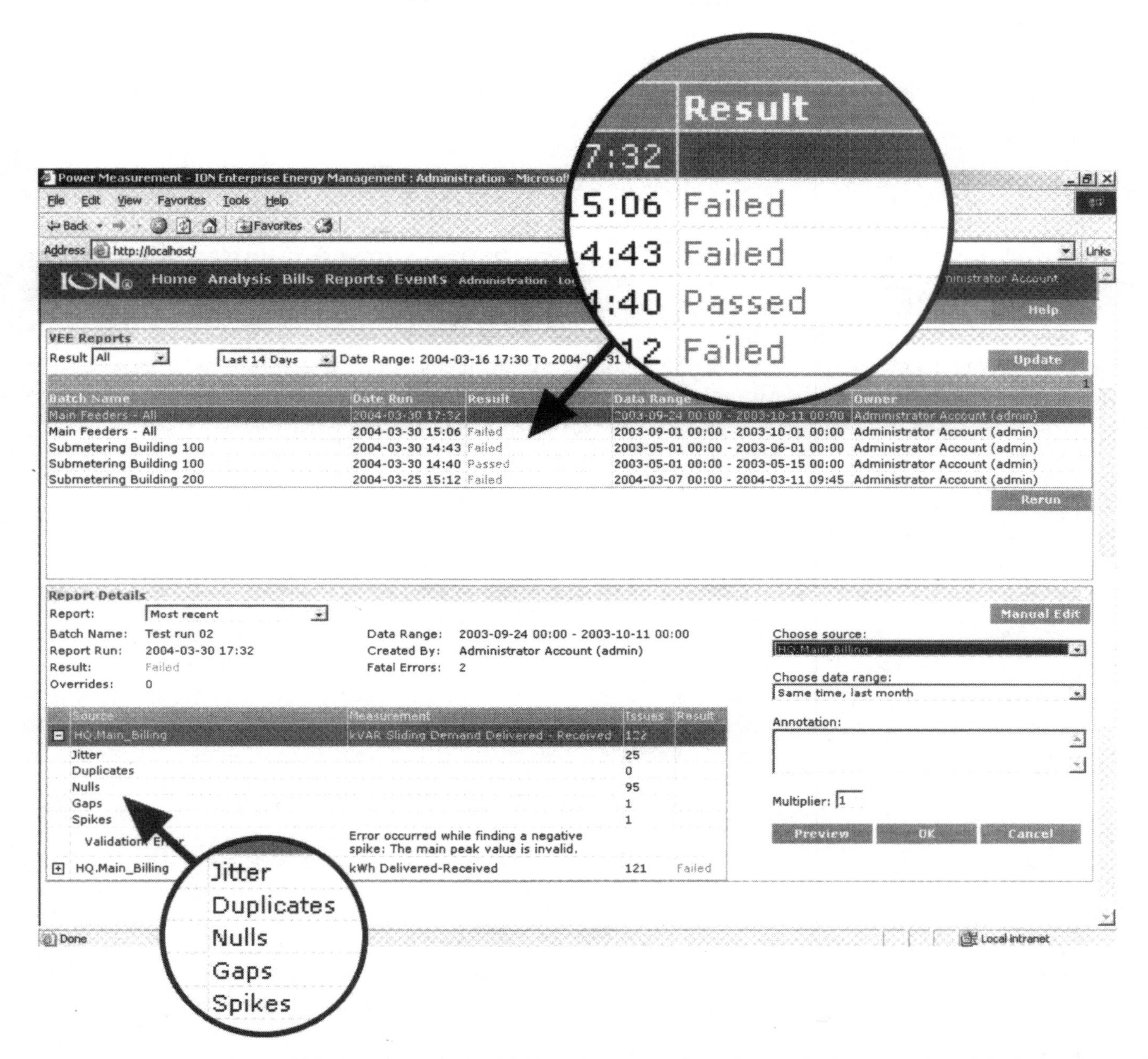

Figure 33-4. Data quality validation screen. Highlights show a series of records for passed and failed tests, and a total count for each error type found when the selected test was run.

checks are done to ensure consecutive readings in a cumulative register are not decrementing instead of incrementing. This can catch conditions such as meter resets, register rollovers, and other issues.

- **Timestamp check.** As mentioned earlier, it is important that measured values being aggregated together are accurately time-aligned. The data quality tools will verify if all timestamps for all values being summed are within an acceptable proximity of each other, often referred to as *time jitter*. Excessive jitter can sometimes be the result of delays caused by a gateway device or software polling a remote meter.

- **Other tests.** Further tests can be applied to help determine if data input is reasonable. For example, a *spike check* (commonly done by utilities) will compare the relative variance between the first and third highest peak energy readings during a specific period. If the readings differ by more than a predefined acceptable amount the variance is flagged as a possible data error, indicating that one or both of the readings are in error.

Estimation & Editing

After the data quality system has been used to verify the validity of incoming data, suspected errors will be flagged. The operator then has a variety of options from which to choose.

Based on the data quality standards the company has defined, the operator may in some cases opt to ignore a particular problem with a data element, if it is not of high enough importance. For critical data, tools are available to correct or compensate for errors.

First, exact duplicates are typically deleted. Rules can be set to deal with *near* duplicates; some may simply be deleted in the same way, some may need to be analyzed further to determine which is the correct record.

Second, automated estimation tools allow erroneous data to be replaced, or missing data to be completed, by "best guess" calculated values that essentially bridge over those records. A variety of preset standard algorithms are provided by the data quality system for this task, with each being optimized for the specific data type and situation. For example, an estimation algorithm for kilowatt-hour measurements will be different than the treatment for humidity or real-time pricing data.

The data quality system will make recommendations as to how the data should be corrected, and may incorporate exogenous factors, such as weather, to make those recommendations more feasible.

One of the most common and simple examples of estimation is straight-line *averaging*. In this case, a bad data point for a particular energy interval reading is replaced with a value representing the straight-line average of the data point values on either side of it. This kind of point-to-point linear interpolation can be applied to multiple contiguous data points that are either missing or otherwise in error. Rules can be set defining the maximum time span allowable for interpolation to be applied.

If a time span of suspect data exceeds the allowable duration, estimation can be performed using data from other similar days. Typically, a number of selected reference days are chosen and their data averaged to produce the replacement data. Reference days need to closely represent the day whose data are being estimated, for example by being the same day of the week, weekend, or holiday as close as possible to the day in question. The data used for estimating would also need to be data that were originally valid; in other words, estimated data cannot be generated from already estimated data. In addition, days that experienced an unusual event, such as a power failure, could not be used for this purpose.

Finally, records representing missing or corrupt data can have new data inserted or their data replaced through manual input or direct editing. This may be appropriate if, for example, a communication failure with a remote meter has caused a gap in data. A technician may be able to retrieve the data by visiting the remote site and downloading the data from the meter's on-board memory into a laptop or meter reader, which in turn can be manually imported into the head-end system to fill missing records, or to replace estimated data. Editing individual records may be appropriate where rules allow, such as when a known event has corrupted a group of records and their correct values can be presumed with a high degree of certainty.

TYPICAL DATA QUALITY RESPONSIBILITIES AND WORKFLOW

Addressing energy-related data quality issues takes a combination of the right tools and the right process. Using a data quality software application can solve specific problems. Commitment from management and availability of proper resources are also needed to ensure that data quality assurance is an ongoing process.

Data quality needs a champion to drive the program and one or more data stewards to execute the necessary steps. Given the importance of energy as a key commodity for an organization, and in turn its impact on profits, the champion can be anyone from an executive through middle management level, including corporate energy managers, operations or facilities managers, or engineering managers.

In terms of day-to-day execution, data cleansing tasks are typically assigned to one or more people within a facility management group, someone with a title such as *data administrator,* or clerical staff specially trained in the data quality tools and rules. Often, it makes more sense for a business to concentrate on their core competencies and outsource the data quality function to an energy management services company.

In general it is always best to fix data problems up front rather than later. That is why the data cleansing process should be positioned at the point where collected data first enter the enterprise energy management system, before they makes it through to where data are

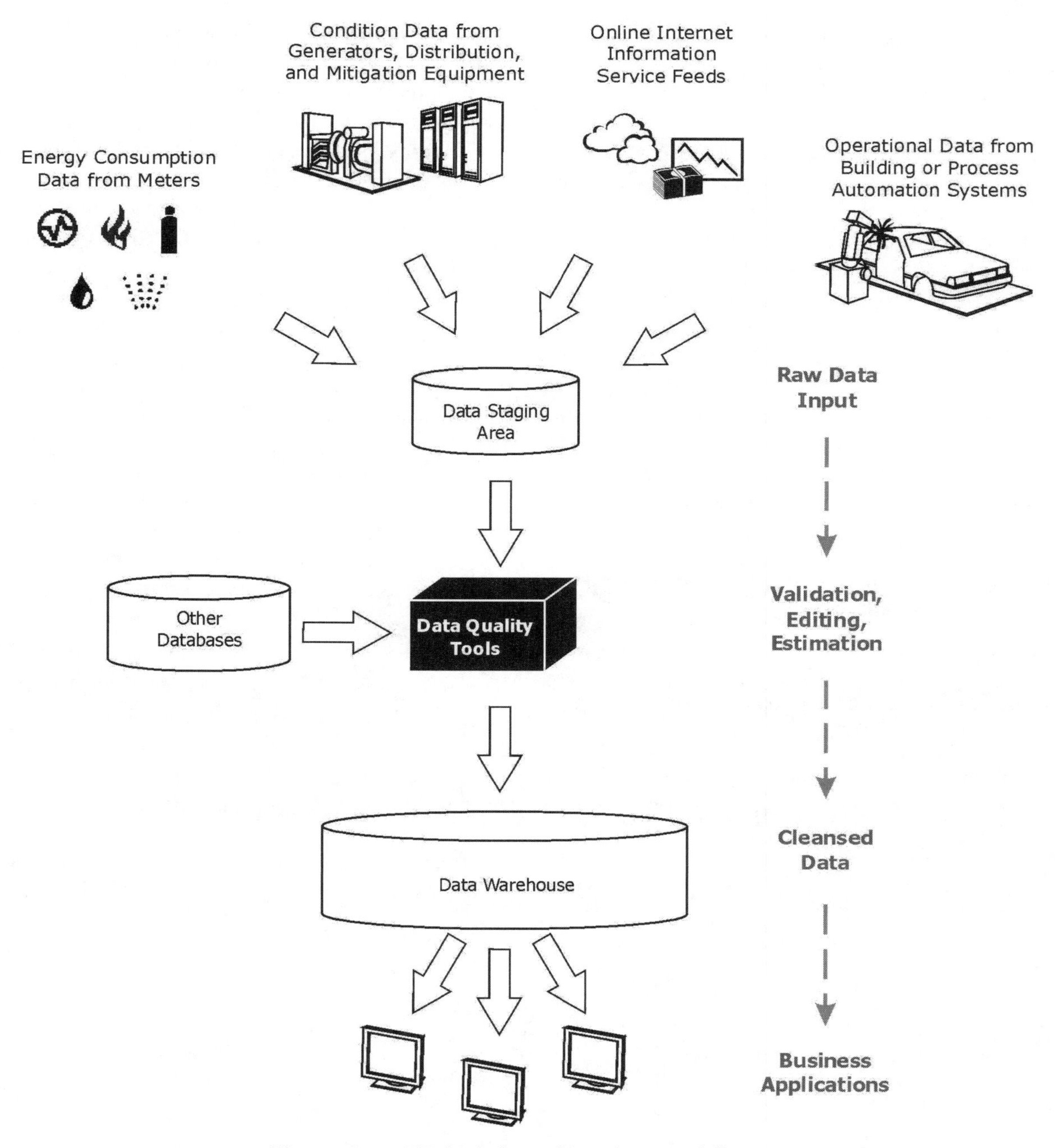

Figure 33-5. Typical data cleansing workflow

ultimately stored in a data warehouse on a data server (Figure 33-5). If this is not done, data problems within the data warehouse start to affect critical calculations and decisions, and can propagate further problems before they are isolated and corrected.

To be most effective, an EEM system is configured to include a front-end data staging area. Data in this area have already been broken out from combined data packets from remote devices or other data steams and translated into the proper units as necessary. The staging area acts as the *raw data* input to the data quality process. In most cases, only after the data are validated or corrected as necessary they are passed on through to the EEM system's data warehouse.

In some special cases, it may be desirable to allow data entering the staging area to be passed on to the data mart without validation, despite the potential for data errors or gaps. This will allow some users that require near "real-time" data to benefit, even if there is an issue with a few readings. In this case, the data quality process can be run on the data at the next scheduled time. If data issues are identified at that point, they will be addressed, and the cleansed data will then propagate to the data mart.

Beyond the real-time inputs to the EEM system, the data quality tools can also be used to validate and cleanse data in previously stored databases before it is integrated within the EEM system database.

During the data quality validation process data problems will be highlighted visually on screen and, if desired, through alarm annunciations. The user can then decide on the best course of action based on the options described above.

The system can help identify persistent data quality issues from a particular incoming data feed, such as a faulty remote meter or other device, an Internet interruption, or a communications network problem. A maintenance protocol can be set up to flag the appropriate technical staff to investigate the source of the problem. If a meter or other data source exhibits an intermittent problem, a decision can be made on whether to repair or replace by comparing that cost to the ongoing man-hours and cost of repeated error correction using data quality tools. The data quality system may also uncover recurring problems with a particular data entry method or other process.

How often the data quality tools need to be used to cleanse data depends on a number of system conditions and the workflow preferences of the user:

- **System size.** The greater the number of data sources (e.g. number of metering points), the higher the probability of data problems, and the harder it is to identify and correct problems. Data problems are often compounded due to sheer size of the data. The user may wish to run

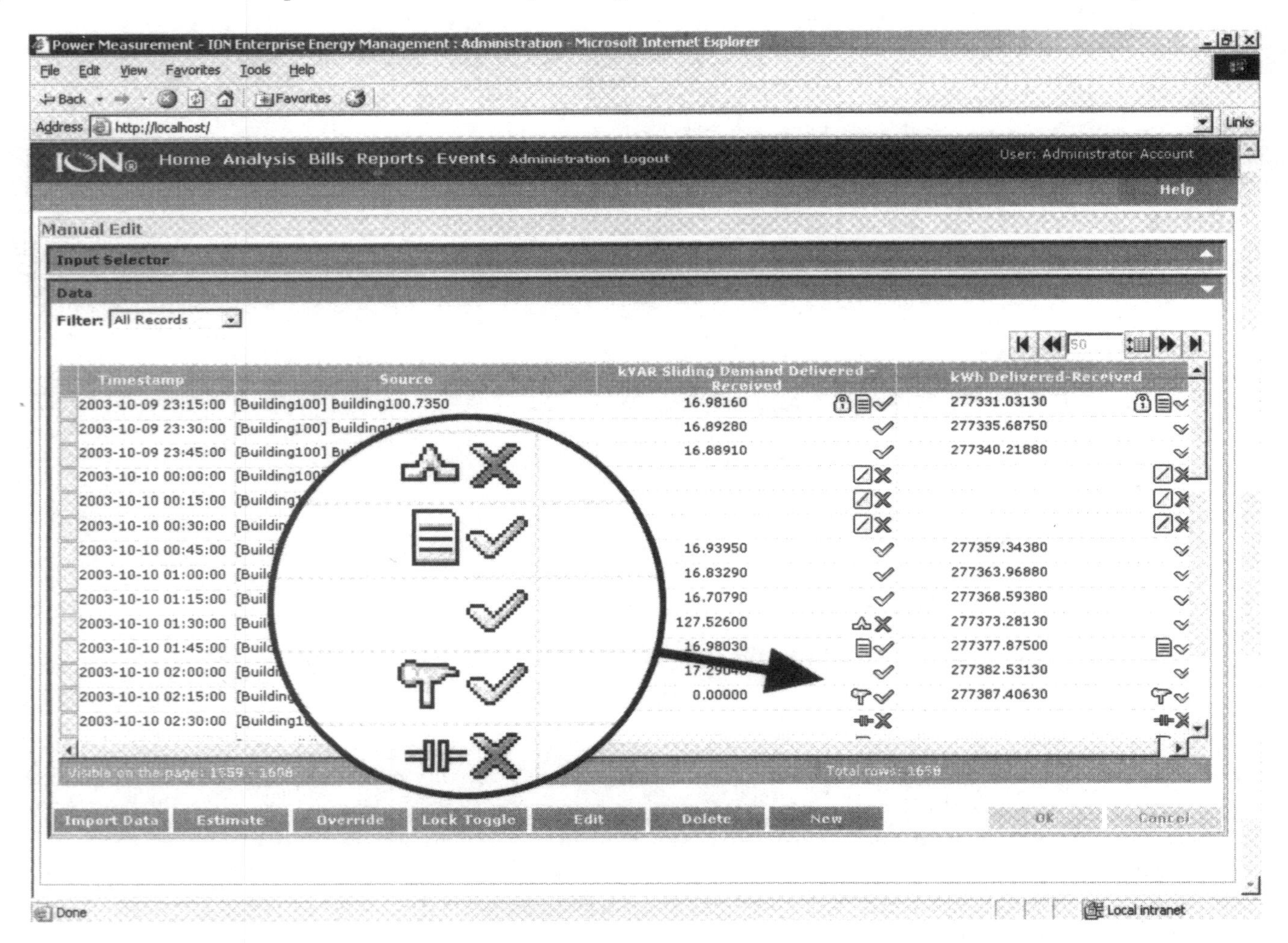

Figure 33-6. Typical data quality estimation and editing screen. Highlighting (from top to bottom) test result indicators for: spike test failure, edited and passed, validated with no errors, failed test override, and gap test failure.

the data quality process more often (daily instead of weekly) to keep on top of the workload.

- **Real-time data requirements**. Monthly data cleansing, well before the billing date, may be sufficient for tenant billing. For real-time applications such as load management, energy procurement, etc., data may be required more frequently and thus the data will need to be cleansed often to deliver up-to-date reliable data.

- **System topology**. Some types of communications (e.g. modem links), may be inherently less reliable than others (e.g. hardwired). The geographical breadth of a system may also affect the reliability of data collected from remote points. Both of these issues may create higher frequencies of data problems that need to be addressed more often.

- **Variety of data types**. The number of distinctly different sources of data (electric meter, gas meters, weather feed, RTP feed, etc.) will add to the complexity of the EEM system and, in turn, influence the expected rate of data errors or gaps.

- **Efficiency.** As mentioned above, the sooner a problem is discovered the easier it is to fix; therefore, the data quality process should be run more often rather than less.

CONSIDERATIONS WHEN CHOOSING A DATA QUALITY SOLUTION

Data quality tools can be effective in addressing data quality problems, but only if they are well designed. If a business intends to have in-house responsibility for the data cleansing process, a number of criteria should be considered when choosing a solution:

- **Flexible and modular.** The data quality system should be flexible enough to align with utility standards to support bill verification and energy procurement. It should also be able to adapt to evolving internal business standards. A modular architecture is an additional advantage, as it allows for sub-components or features to be added or engaged as required for testing of different kinds of problems.

- **Applicable to** ***all*** **EEM data sources.** Data quality tools should be designed for a "whole system approach," available to cleanse not only metered electrical energy data, but also data representing other energy sources like gas, steam, etc. Further, they should be able to validate external data feeds such as weather, real-time pricing (RTP) rate forecasts from the ISO/RTO, etc.

- **Notification system.** The system should allow data administrators to subscribe to desired information and be notified when necessary, for example when a dead source is detected, or when a data quality report has been run and shows data issues. Notification methods should include email, pager, and other convenient options.

- **Report generation.** Reporting tools should provide scheduled or on-demand reporting, listing details on data quality problems and summary "roll-ups" of data quality performance metrics. The data quality reports and, ideally, the bills generated by the EEM system should both reflect valid data statistics.

- **Audit trails and raw data backup.** A complete audit trail should be provided for any data that have been edited. This should indicate the user that executed the change, what was changed and how. For data that have been changed, the complete raw data set should be retained in a backup file and be accessible in case a particular data cleansing step needs to be reversed.

- **Security.** Different password-protected access levels should be provided. This can include "view only" access for some users, while "administrator" access allows viewing and the ability to make changes to the data.

- **Override capability.** The system should allow an administrator to override an error indication for what first appears as an error but may be valid data (e.g. a meter is showing zero energy usage due to a known power outage.)

- **Ease of use.** The data quality process must be cost effective, so the tools must be efficient and easy to use. A number of features can help in this regard. For example, error indicators on data quality screens should provide quick links to view the supporting data. Ideally, this should also be a feature of bills generated by the EEM system. Data should be clearly marked to differentiate between valid, estimated or corrected data.

CONCLUSIONS

Enterprise energy management systems represent the key to energy-related savings and productivity improvements, but their effectiveness is significantly influenced by the quality of data they deliver. As with all business intelligence systems, the right tools and processes must be in place to avoid data quality issues that could otherwise seriously affect business decisions, tenant relations and return on investment. New data quality tools are available for industrial, commercial and institutional energy consumers help ensure the intelligence delivered by EEM systems is accurate, complete and timely.

Whether businesses choose to dedicate in-house staff to the data quality process or outsource it, the design features of the data quality application are critical. Due to typical EEM system breadth, the variety of networking methods, and the number and types of data sources, a comprehensive set of data quality tools is needed to identify and compensate for all potential data quality problems. The data quality solution chosen should provide the flexibility and modularity needed to adapt to evolving business rules and needs. It should also be applicable to all EEM data sources beyond energy metering, including external feeds such as weather and real-time energy pricing. Finally, to be cost-effective, data quality tools must be easy and efficient to use.

Chapter 34

Connecting Energy Management Systems to Enterprise Business Systems Using SOAP and the XML Web Services Architecture

Ron Brown, Gridlogix, Inc.

ABSTRACT

TODAY, COMMERCIAL and governmental energy consuming facilities are discretely managed by sophisticated building automation systems. Energy suppliers rely on sophisticated automation systems to manage the delivery of energy to their customers. The Enterprise Energy Management (EEM) industry emerged in the late nineties with new technology and services designed to help energy consumers and suppliers better understand energy consumption patterns and find curtailment solutions. Internet and web-based technologies have made a modest impact on enterprise energy and information management, but mostly in a reactive way and not in a proactive way.

The automation systems used by energy consumers, suppliers, and EEM providers, which have driven more efficiency into the chain of energy supply and consumption, remain disconnected by design. Some EEM companies have developed costly custom interfaces between their energy management solutions and consumer-owned building management systems.

Now it is time to consider the notion of automation system connectivity between any combination of energy consumers, suppliers, and EEM providers. This level of connectivity should be a standards based open information technology framework that allows secure and trusted communication between any of these separate entities. Such a framework should also allow additional entities such as regulators, ISO's, and equipment manufacturers to connect and access data from consumer, utility, or EEM owned energy management systems.

The IT world has created and defined such a framework for integration and interoperability; it is known as the "XML Web Services Architecture." XML web services is not a human-machine interface via a web browser, but rather a machine-to-machine interface using the transport protocols of the Internet - independent of hardware, operating systems and programming languages. The next generation energy management solutions are now embracing XML web services as the standard for integration with enterprise business systems.

INTRODUCTION

More and more energy and building management systems are utilizing Ethernet technology for controller and device communication. As a consequence, many of these device networks are using existing corporate networks. Corporate business systems connected to these networks are now discovering these device networks and recognizing them as potential data sources for improving operational efficiency and business intelligence so as to improve the bottom line.

The only problem is that while these device networks use Ethernet as a communications backbone, the devices and controllers use protocols encapsulated inside Ethernet that are specific to those devices and controllers. Device protocols such as Bacnet, Lontalk, and Modbus are not inherent to corporate business systems. In other words, a computerized maintenance management system (CMMS) typically cannot directly connect to a critical asset such as a chiller and know its operating state. But if that CMMS system could connect to critical assets in real time, then more timely maintenance is possible - meaning less downtime and less overall maintenance cost.

This chapter explains an IT approach to device network integration. Rather than rely on the vendors of automation systems and the consortiums that specify open protocols to develop interoperability standards, a computing paradigm in use by the IT world known as

XML Web Services and a protocol known as SOAP is introduced. The concept presented here argues that the IT solution of wrapping an XML Web Services interface and the SOAP protocol around device network protocols will extend energy and building management systems far beyond their fundamental use.

HOW SOAP IS USED IN DEVICE COMMUNICATION

SOAP is defined as follows:

> SOAP is a lightweight protocol intended for exchanging structured information in a decentralized, distributed environment. SOAP uses XML to define an extensible messaging framework, which provides a message construct that can be exchanged over a variety of underlying protocols. The framework has been designed to be independent of any particular programming model and other implementation specific semantics.

SOAP provides a way to move XML messages from point A to point B (see Figure 34-1). It does this by using an XML-based messaging framework that is 1) extensible, 2) usable over a variety of underlying networking protocols, and 3) independent of programming models.

Figure 34-2 expands on the prior diagram and shows the process of a web services capable client application reading from, and writing to an energy management system. The client application and its user do not need to have any in-depth knowledge of the device communications protocol,. The process shown is an IT-centered approach, not a device-protocol centered approach for communicating with the energy management system.

This process illustrates the following:

- There is a protocol translation software application connected directly to the energy management system via its device communications protocol such as Bacnet or Modbus.

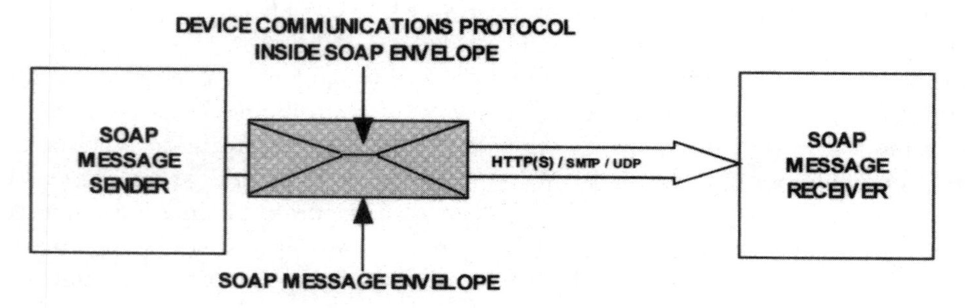

Figure 34-1

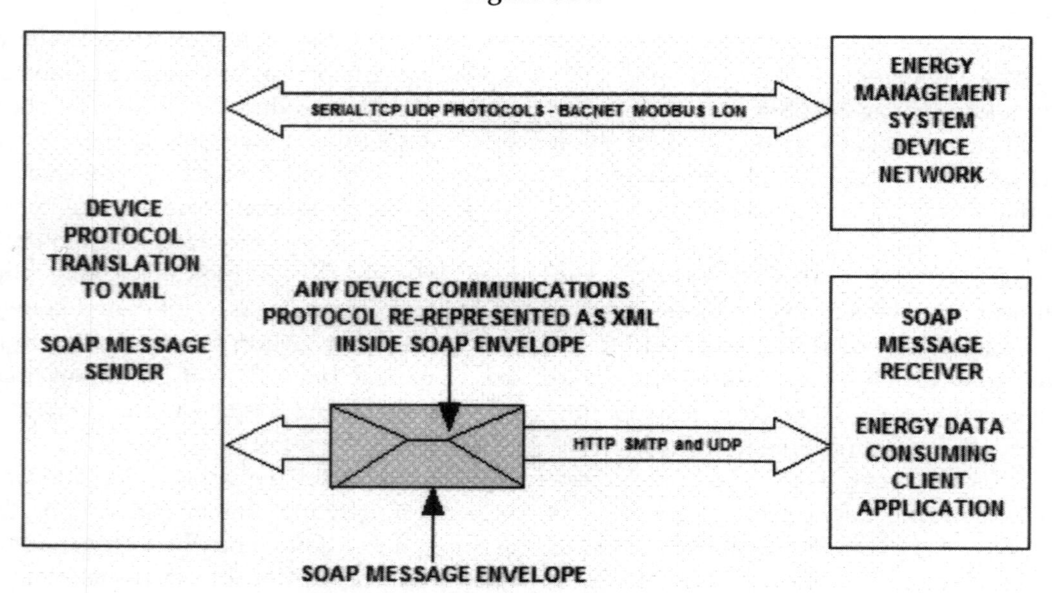

Figure 34-2

- The protocol translation application can fully convert or reformulate the messaging and data transport capabilities of the device protocol as XML messages and data. The XML representation of the device protocol becomes the payload for the SOAP message.

- The client application can connect to a web service that contains or publishes the business rules for sending/receiving the SOAP message and populating/interpreting its payload.

- The business rules for using the web service are typically described by a published XML schema known as the Web Services Description Language or WSDL interface.

Obviously, the application that performs device protocol translation to SOAP needs to exist. However, the translation application can exist as a separate application connected to the energy management system and its devices, it can reside in the energy management system software, or it can reside on the energy management equipment - meters, controllers, sensors, etc.

WEB SERVICES

The application that creates the SOAP envelope containing the XML reformulation of the device protocol is the web service. Microsoft defines web services as follows:

> XML Web services are the fundamental building blocks in the move to distributed computing on the Internet. Open standards and the focus on communication and collaboration among people and applications have created an environment where XML Web services are becoming the platform for application integration. Applications are constructed using multiple XML Web services from various sources that work together regardless of where they reside or how they were implemented.
>
> There are probably as many definitions of XML Web Service as there are companies building them, but almost all definitions have these things in common:
>
> - XML Web Services expose useful functionality to Web users through a standard Web protocol. In most cases, the protocol used is SOAP.
>
> - XML Web services provide a way to describe their interfaces in enough detail to allow a user to build a client application to talk to them. This description is usually provided in an XML document called a Web Services Description Language (WSDL) document.
>
> - XML Web services are registered so that potential users can find them easily. This is done with Universal Discovery Description and Integration (UDDI).

The web services architecture necessary for client applications to connect to energy management systems has either been developed, or is currently being developed by the device vendors of energy management equipment, and by IT companies involved in energy information management, asset management, and enterprise resource planning.

A web service used for communicating with energy management device networks will have an interface that describes the functions that can be called by the client application. The functions, also known as web methods, are used to read or write a value. For example, writing to a temperature set-point or reading the kW on a circuit. Typically the interface that describes the web methods is handled by an XML schema known as the Web Services Description Language, or WSDL.

ADVANTAGES OF WEB SERVICES

1. Language and operating system agnostic
   - Not based on a programming language
   - Java, .Net, C, C++, Python, Perl, ...
   - Not based on a programming data model
   - Objects vs. non-objects environments
   - Convergence of SOA (Service-Oriented Architecture) and Web
   - Web services can run on any OS

2. Relatively simple
   - Based on Web technologies
   - Web technologies proved their scalability.
   - Using an HTTP-based protocol, data can move easily through firewall
   - Protocol may be HTTP or other transport protocols, and now UDP
   - Do not necessarily require a huge framework in memory;
   - A small amount of code could expose a service as a Web Service. (Why do you want to expose a service as a Web Service? What does this mean?

- Can be used easily in today's Web Interface
- Any application/component can be exposed as a Web Service Same comment as above
- Allows clean integration across: departments, agencies, companies, device networks, etc.
- Loose coupling, i.e. doesn't require tightly-coupled, custom-programmed applications
- Client and Service can use different platforms and programming languages
- Only need to agree on interface, use of the WSDL
- Client and Service exchange uses XML
- Client and Service are often front ends for existing applications/systems
- One Web Service can make use of others

3. Basic usages
   - Business to business (B2B)
   - Remote controlled devices
   - Internal/External application communications
   - Machine to Machine interaction: No human assistance or intervention.

WEB SERVICES DESCRIPTION LANGUAGE (WSDL)

The World Wide Web Consortium (W3C) standards organization defines WSDL as follows:

> WSDL is an XML format for describing network services as a set of endpoints operating on messages containing either document-oriented or procedure-oriented information. The operations and messages are described abstractly, and then bound to a concrete network protocol and message format to define an endpoint. Related concrete endpoints are combined into abstract endpoints (services). WSDL is extensible to allow description of endpoints and their messages regardless of what message formats or network protocols are used to communicate.

Figure 34-3 shows an XML segment of simple WSDL describing web methods for connecting to a network of Modbus TCP devices. A software developer uses the description of the web methods for writing a client application that can read and write from/to

```
<?xml version="1.0" encoding="utf-8" ?>
- <definitions xmlns="http://schemas.xmlsoap.org/wsdl/" xmlns:xs="http://www.w3.org/2001/XMLSchema" name="IModbusSlaveEZservice"
    targetNamespace="http://www.gridlogix.com/WebServices/" xmlns:tns="http://www.gridlogix.com/WebServices/" xmlns:soap="http://sch
    xmlns:soapenc="http://schemas.xmlsoap.org/soap/encoding/" xmlns:mime="http://schemas.xmlsoap.org/wsdl/mime/">
  + <message name="PingRequest">
  + <message name="PingResponse">
  + <message name="PingDeviceRequest">
  + <message name="PingDeviceResponse">
  - <message name="ReadHoldingRegistersRequest">
      <part name="UserName" type="xs:string" />
      <part name="Password" type="xs:string" />
      <part name="Host" type="xs:string" />
      <part name="DeviceID" type="xs:int" />
      <part name="PortID" type="xs:int" />
      <part name="Start" type="xs:int" />
      <part name="Count" type="xs:int" />
    </message>
  - <message name="ReadHoldingRegistersResponse">
      <part name="return" type="xs:string" />
    </message>
  + <message name="WriteHoldingRegistersRequest">
  + <message name="WriteHoldingRegistersResponse">
  + <message name="ReadInputRegistersRequest">
  + <message name="ReadInputRegistersResponse">
  - <message name="ReadCoilsRequest">
      <part name="UserName" type="xs:string" />
      <part name="Password" type="xs:string" />
      <part name="Host" type="xs:string" />
      <part name="DeviceID" type="xs:int" />
      <part name="PortID" type="xs:int" />
      <part name="Start" type="xs:int" />
      <part name="Count" type="xs:int" />
    </message>
  + <message name="ReadCoilsResponse">
  + <message name="WriteCoilsRequest">
  + <message name="WriteCoilsResponse">
```

Figure 34-3

Modbus device holding registers, binary coils (on/off), input registers, etc.

The web methods for reading and writing the holding registers use parameters. In order to make the web method call "ReadHoldingRegistersRequest," the WSDL says the client application must supply "string" parameters for the following:

- User Name - Used by the web service to verify the user is authorized
- Password - Used by the web service to verify the user is authorized
- Host - The IP address where the web service resides
- Device ID - The unique Modbus device ID for holding registers we're reading
- Port ID - Describes the Modbus TCP port, typically "502"
- Start - Indicates the starting register number (0-254)
- Count - Indicates how many holding registers to read from Start parameter

Once the web method is invoked or called by the client application, a response message called "ReadHoldingRegistersResponse" is returned to the client application with the result. The result data type is defined as a "string."

THE SOAP ENVELOPE

Figure 34-4 shows the XML SOAP envelope and message payload for the web method called "ReadHoldingRegistersRequest." This web method call is requesting that "10" holding registers be read on Modbus device "1" at the IP address of "192.168.1.19" via network port "502."

Figure 34-5 shows the XML SOAP envelope and message payload for the return value defined by the web method called "ReadHoldingRegistersResponse." The value returned is a single string consisting of 10 values representing the current values of holding registers 100 through 109.

WRITING CLIENT APPLICATIONS

The web services architecture and its communications protocols are independent of the operating system, the programming language, and the hardware. Web services can be written in any language and hosted on any platform that can support the underlying communications protocols such as SOAP, HTTP, SMTP, MIME, etc. Client applications can be written in any language capable of using web services. In other words, a client application written in Visual Basic and running under the Windows operating system can use a web service that is written in JAVA running under LINUX. The user of the client application typically would not know or care that his application is interacting with a JAVA application over a network.

Client applications that use web services are typically written with respect to the published WSDL. Almost all development systems used by programmers today can connect to and utilize the WSDL interface. Once the WSDL is accessed, the development system

```
<?xml version="1.0" ?>
- <SOAP-ENV:Envelope xmlns:SOAP-ENV="http://schemas.xmlsoap.org/soap/envelope/"
   xmlns:xsd="http://www.w3.org/2001/XMLSchema"
   xmlns:xsi="http://www.w3.org/2001/XMLSchema-instance" xmlns:SOAP-
   ENC="http://schemas.xmlsoap.org/soap/encoding/">
 - <SOAP-ENV:Body SOAP-ENV:encodingStyle="http://schemas.xmlsoap.org/soap/encoding/">
   - <NS1:ReadHoldingRegisters xmlns:NS1="Gridlogix.EnNET.Types.IModbusSlaveEZ">
       <UserName xsi:type="xsd:string">ennetdemo@gridlogix.com</UserName>
       <Password xsi:type="xsd:string">Test</Password>
       <Host xsi:type="xsd:string">192.168.1.19</Host>
       <DeviceID xsi:type="xsd:int">1</DeviceID>
       <PortID xsi:type="xsd:int">502</PortID>
       <Start xsi:type="xsd:int">100</Start>
       <Count xsi:type="xsd:int">10</Count>
     </NS1:ReadHoldingRegisters>
   </SOAP-ENV:Body>
 </SOAP-ENV:Envelope>
```

Figure 34-4

Figure 5 shows the XML SOAP envelope and message payload for the return value defined by the web method called "ReadHoldingRegistersResponse". The value returned is a single string consisting of 10 values representing the current values of holding registers 100 through 109.

```
  <?xml version="1.0" ?>
- <SOAP-ENV:Envelope xmlns:SOAP-ENV="http://schemas.xmlsoap.org/soap/envelope/"
    xmlns:xsd="http://www.w3.org/2001/XMLSchema"
    xmlns:xsi="http://www.w3.org/2001/XMLSchema-instance" xmlns:SOAP-
    ENC="http://schemas.xmlsoap.org/soap/encoding/">
  - <SOAP-ENV:Body SOAP-ENC:encodingStyle="http://schemas.xmlsoap.org/soap/envelope/">
    - <NS1:ReadHoldingRegistersResponse xmlns:NS1="Gridlogix.EnNET.Types.IModbusSlaveEZ">
        <return xsi:type="xsd:string">237,241,238,22,34,29,59,92,73,6397</return>
      </NS1:ReadHoldingRegistersResponse>
    </SOAP-ENV:Body>
  </SOAP-ENV:Envelope>
```

Figure 34-5

automatically creates a source code interface to the web service. This process is a huge time saver for the programmer.

For example, Figure 34-6 shows a screen shot from Microsoft Visual Studio.NET a development system for programmers. This windows form is launched from inside the Visual Studio.NET development system when the programmer initiates a connection to a web service running on a network. In this case, the programmer has entered the URL address of the WSDL location and the source code is automatically generated for the web methods published by the web service called "IModbusSlaveEZservice."

With the interface and source code automatically generated inside the Visual Studio.NET programming environment, the programmer can quickly write a client application that communicates with the connected Modbus devices without knowing anything about the technical nuances of the Modbus communications protocol.

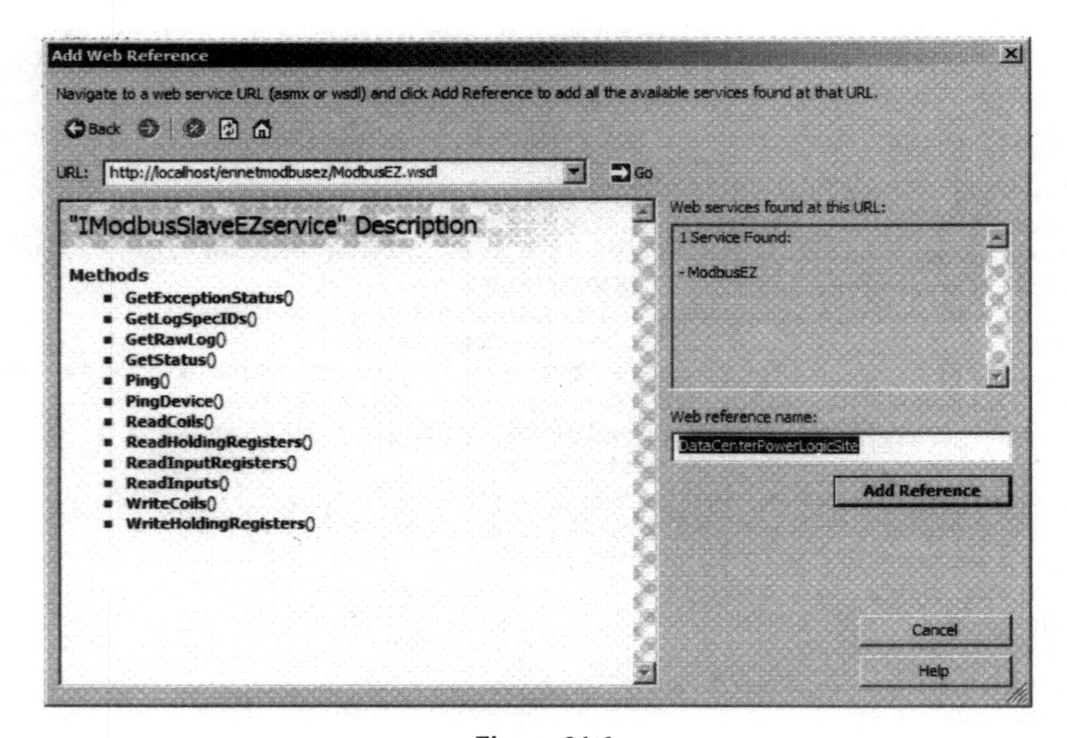

Figure 34-6

USING SOAP AND WEB SERVICES FOR DEVICE PROTOCOL TRANSLATION

Energy and building management systems typically connect to device networks via a single communications protocol that is either proprietary, or open such as Modbus and Bacnet. Larger commercial and industrial facilities often have more than one device network used in the overall operation of the facility. For example, a facility might have a Bacnet building management system, a LON security system, an OPC PLC system, and a Modbus power management system. Each of these systems is operated independently of the other. Device protocol translation makes it possible to consolidate or centralize management and control of these different device networks.

Figure 34-7 expands on the diagram in Figure 34-2 and illustrates the use of SOAP and the web services architecture for translating one device protocol to another device protocol. For example, an energy management head-end program can utilize this framework to connect to manage power management devices on a Modbus device network.

Obviously, such a device protocol translation framework using SOAP and web services must be developed. Figure 34-8 shows a screen shot from such an application. The application is the EnNET Centralized Monitoring Control (CMC) protocol translation system from Gridlogix. The software is built on their SOAP and web services framework called EnNET. The screen shot shows the EnNET Device Explorer application which is used for navigating varying device networks. Data collected by the devices can be read and written in the same protocol, or read and translated to another protocol such as Bacnet.

Industry Acceptance of Soap and Web Services

The automation and controls industry which is at the heart of energy management systems is making great strides in the adoption and commercialization of SOAP and Web Services for integrating their device networks with other business systems. Companies such as Echelon have developed a complete implementation of a web services interface for their i.LON 100 gateway. The oBix organization is an industry-wide initiative to define XML- and Web services-based mechanisms for building

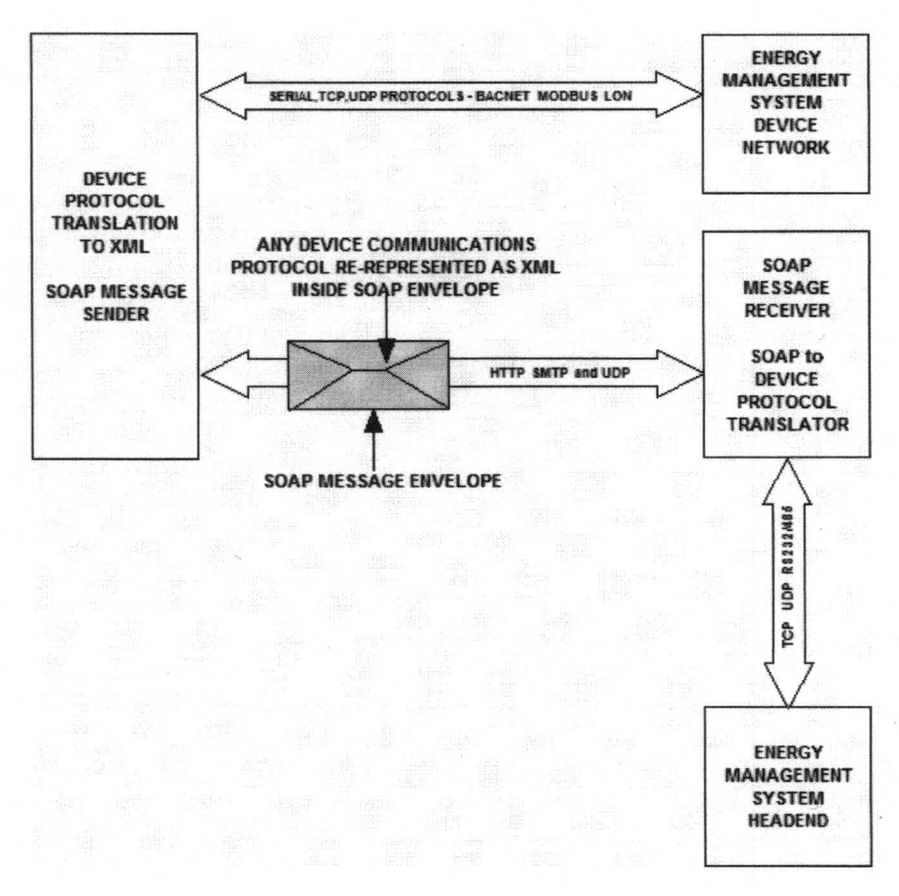

Figure 34-7

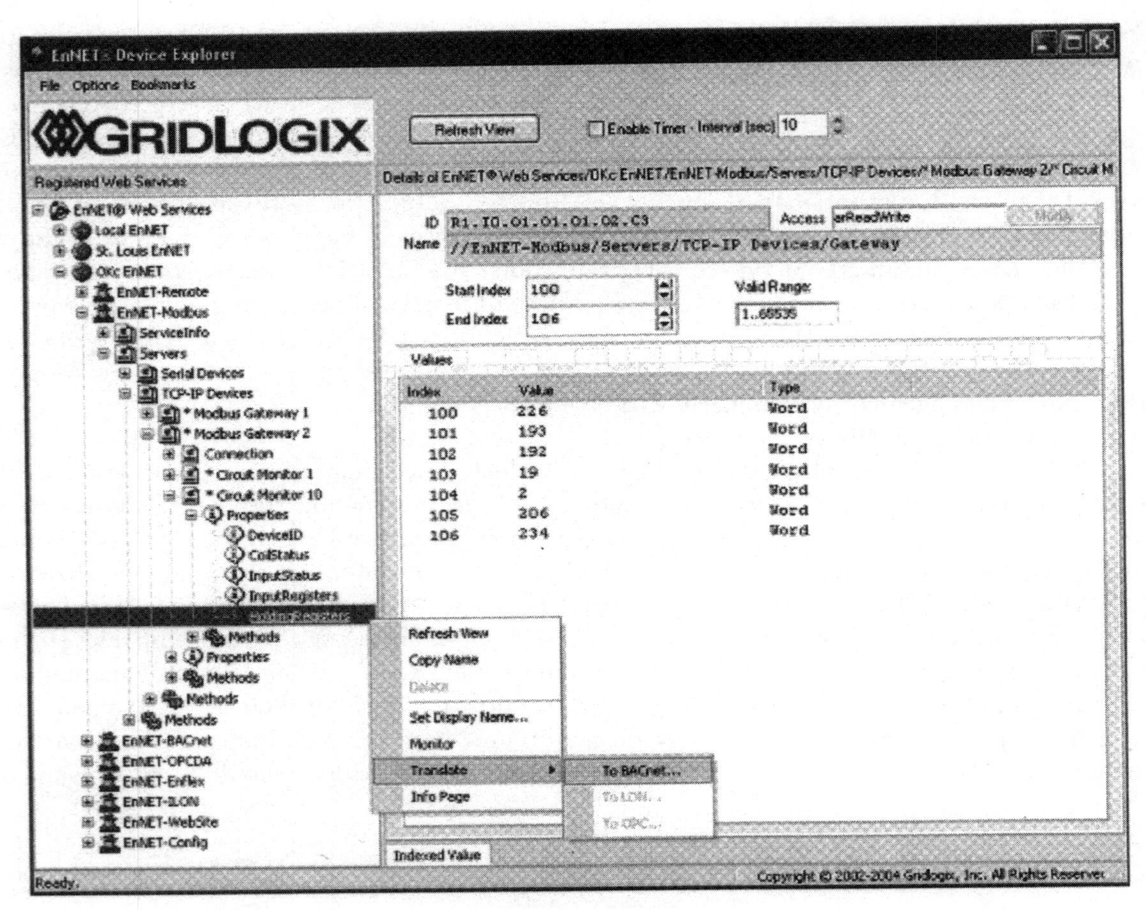

Figure 34-8

control systems. ASHRAE has begun developing a standardized web services interface for the Bacnet protocol. There are many more initiatives underway, and in a few years many open and proprietary device communications protocols will have web services interfaces.

Example Soap and Web Services Framework—EnNET®

The EnNET framework from Gridlogix is a commercial product consisting of a single standard interface to the many different facility and industrial automation systems in use today. This standard interface is independent of any one device communications protocol or system. EnNET is also client application independent. Since EnNET implements the functions normally available in the native versions of a protocol, client applications can access valuable data in a protocol-independent manner. Application developers can use EnNET to transform an application originally only intended to take advantage of the simple features of a protocol into a fully featured application that can take advantage of alarming, calendars, schedules and trends. For example, an enterprise application designed to support a preventive maintenance program can now support any number of different vendor systems. A standard software call to EnNET is translated to the non-standard protocol of the managed equipment. Run time hours or any other operational parameter that might trigger maintenance can be retrieved in a common manner from disparate systems. EnNET is designed to expose real-time data sources, but it can also be used to supplement the weaknesses found in many legacy systems. Developers can now add functionality and value not found in the underlying system.

EnNET was invented to provide a common, standardized interface to real-time data sources and communication protocols of all kinds. EnNET does for automation system connectivity what ODBC did for the way we connect to databases. In much the same way that software developers migrated away from database-specific APIs, automation system integrators can migrate away from creating code that is specific to a single control system or smart device. Solutions can be created that become portable and reusable for any automation system.

Figures 34-9 and 34-10 show block diagrams that illustrate what the EnNET framework is designed to glue together. Business systems with no inherent interface to device networks can now connect to those networks using the IT integration standard of web services that EnNET is built on.

THE WEB SERVICES ARCHITECTURE AND SPECIFICATIONS

The specifications that make up the web services architecture is continually evolving. Energy management systems and the device networks that implement web services architectures are eligible to adopt every aspect of the web services specifications such as security. As energy management systems start to use corporate networks and the internet for communication, use of a web services infrastructure can inherently be more secure and reliable than devising proprietary inter-networked systems without web services.

Since the inception of web services, several companies such as Microsoft and IBM have collaborated through the W3C to refine and enhance the web services architecture. As of this writing the web services specification includes many sub-specifications that address the following:

- Messaging
  - — XML
  - — SOAP
  - — Message Exchange Patterns
  - — Transport Independence
  - — Addressing
- Metadata Security
  - — Message Integrity and Confidentiality
  - — Trust Based on Security Tokens
  - — Secure Sessions
  - — Security Policies
  - — System Federations
  - — Dynamic Discovery of web services - UDDI
  - — Reliable Messaging and Transaction Management
  - — Enumeration, Transfer, and Eventing

CONCLUSION

When web services technology becomes an inherent capability of real-time automation and control systems used in the delivery, management, and consumption of energy, energy producers and consumers will be able to implement many more ways to minimize energy consumption and maximize energy reliability.

For example, if web services were more pervasive in energy management technologies, the following scenarios would be possible:

- The northeast United States blackout in August 2003 might have been prevented. If utilities in-

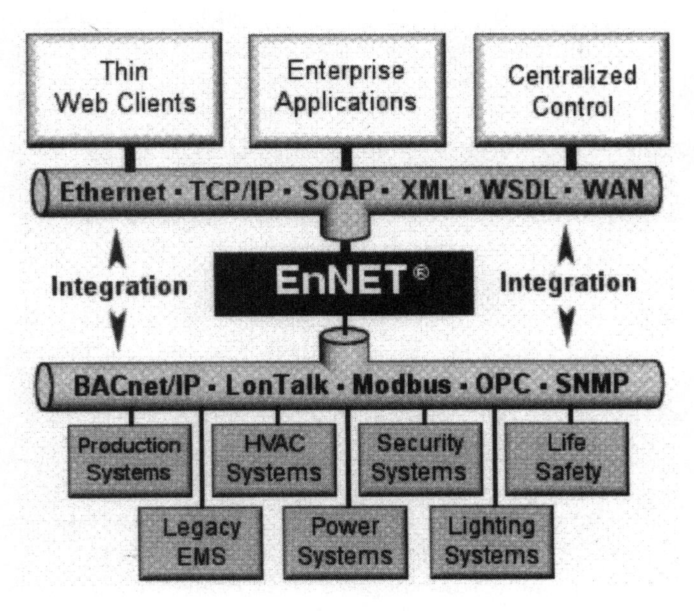

Figure 34-9

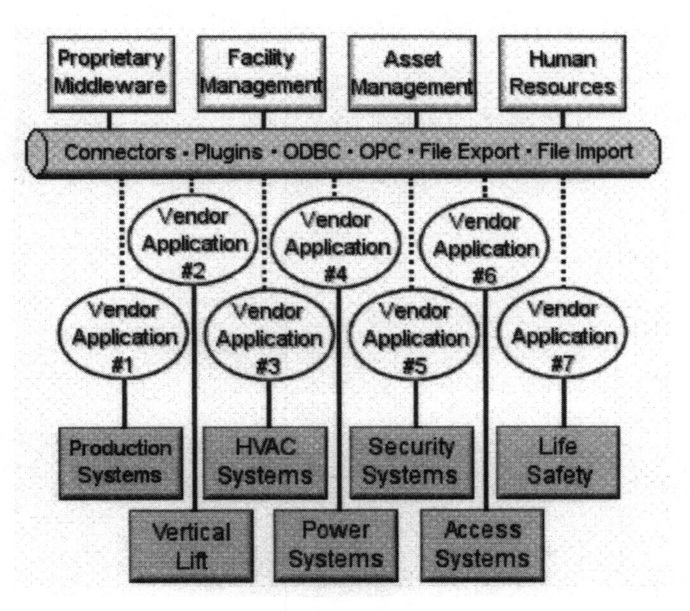

Figure 34-10

volved in the blackouts had publicly accessible read-only access to web services containing real-time energy data for the monitored points on the grid, then external organizations such as state and federal agencies, other utilities, and large power consumers, could also monitor the health of the electrical grid.

- Major equipment suppliers that sell expensive energy consuming equipment such as chillers, generators, and air handling systems could remotely monitor their products as a service to their customers by securely using the customer's network and internet. Such services can reduce energy costs by improving reliability and efficiency, while improving customer satisfaction.

- Commercial property owners and property management companies can use web services technologies to securely centralize command and control of multi-vendor and multi-protocol building management systems. Centralizing management of these systems can increase energy efficiency by providing a more desirable aggregate style of building management.

- Because web services is a machine-to-machine computing paradigm, enterprise energy management systems using sophisticated energy curtailment and demand response algorithms can directly connect to building management systems and automatically control energy-consuming building sub-systems during peak usage.

Bibliography

I. Companies and Organizations

ASHRAE - BACNET Web Services Specification	*www.ashrae.org*
oBix	*www.obix.org*
W3C	*www.w3c.org*
Echelon	*www.echelon.com*
Gridlogix	*www.gridlogix.com*
Microsoft and Web Services	*msdn.microft.com*

II. Core Specifications

ARP: An Ethernet Address Resolution Protocol [RFC826]. David C. Plummer. November 1982. Internet Engineering Task Force.

HTML: HTML 4.01 Specification. Ed. Dave Raggett, et al. 24 December 1999. W3C.org.

HTTP: Hypertext Transfer Protocol – HTTP/1.1 (RFC 2616). Ed. R. Fielding, et al. June 1999. The Internet Society.

KERBEROS: The Kerberos Network Authentication Service (V5). J. Kohl and C. Neuman. September 1993. Internet Engineering Task Force.

MIME: Multipurpose Internet Mail Extensions (MIME) Part One: Format of Internet Message Bodies. Ed. N. Freed, et al. November 1996. Internet Engineering Task Force.

REL: Information technology – Multimedia Framework (MPEG-21) – Part 5: Rights Expression Language. International Organization for Standardization (ISO/IEC 21000-5:2004).

RFC1630: Universal Resource Identifiers in WWW (RFC 1630). Ed. T. Berners-Lee. June 1994. Internet Engineering Task Force.

SAML: Assertions and Protocol for the OASIS Security Assertion Markup Language (SAML) V1.1. Ed. Eve Maler, et al. 2 September 2003. OASIS-Open.org.

SMTP: Simple Mail Transfer Protocol. Ed. J. Klensin. April 2001. Internet Engineering Task Force.

TCP/IP: Transmission Control Protocol. Ed. Jon Postel. September 1981. Defense Advanced Research Projects Agency. Internet Protocol. Ed. Jon Postel. September 1981. Defense Advanced Research Projects Agency.

TLS: The TLS Protocol. T. Dierks, et al. January 1999. The Internet Society.

UDP: User Datagram Protocol. J. Postel. August 1980. Internet Engineering Task Force.

X509: Data Networks and Open System Communications Directory (ITU-T Recommendation X.509). June 1997. International Telecommunication Union.

XSLT-20: XSL Transformations (XSLT) Version 2.0. Ed. Michael Kay. 12 November 2004. W3C.org

XML-Infoset: XML Information Set (Second Edition). Ed. John Cowan, et al. 4 February 2004. W3C.org.

XML-10: Extensible Markup Language (XML) 1.0 (Third Edition). Ed. Tim Bray, et al. 4 February 2004. W3C.org

XMLENC: XML Encryption Syntax and Processing. Ed. Takeshi Imamura, et al. 10 December 2002. W3C.org.

XML-Query: XQuery 1.0: An XML Query Language. Ed. Scott Boag, et al. 23 July 2004. W3C.org

XML-Schema: XML-Schema Part 0: Primer. Ed. David Fallside. 2 May 2001. W3C.org.

XML-Schema Part 1: Structures. Ed. Henry Thomson, et al. 2 May 2001. W3C.org.
XML-Schema Part 2: Datatypes. Ed. Paul Biron, et al. 2 May 2001. W3C.org.

XMLSIG: XML Signature Syntax and Processing. Ed. Donald Eastlake, et al. 12 February 2004. W3C.org.

Web Services Specifications

SOAP: Simple Object Access Protocols (SOAP) 1.1. Ed. Don Box, et al. 8 May 2000. W3C.org.

SOAP-UDP: SOAP-over-UDP. Harold Combs, et al. September 2004. BEA, Lexmark, Microsoft, and Ricoh.

MTOM: SOAP Message Transfer Optimization Mechanism. Ed. Noah Mendelsohn, et al. 8 July 2004. W3C.org.

UDDI: UDDI Version 2.04 API Specification. Ed. Tom Bellwood. 19 July 2004. OASIS-Open.org.

WSDL: Web Service Description Language (WSDL) 1.1. Ed. Erik Christensen, et al. 15 March 2001. W3C.org.

WS-Addressing: Web Services Addressing (WS-Addressing). Don Box, et al. August 2004. BEA, IBM, and Microsoft.

WS-AT: Web Services Atomic Transaction (WS-AtomicTransaction). Luis Felipe Cabrera, et al. September 2003. BEA, IBM, and Microsoft.

WS-BA: Web Services Business Activity Framework (WS-BusinessActivity). Luis Felipe Cabrera, et al. January 2004. BEA, IBM, and Microsoft.

WS-Coord: Web Services Coordination (WS-Coordination). Luis Felipe Cabrera, et al. September 2003. BEA, IBM and Microsoft.

WS-Discovery: Web Services Dynamic Discovery (WS-Discovery). John Beatty, et al. February 2004. Microsoft Corporation.

WS-Enum: Web Service Enumeration (WS-Enumeration). Don Box, et al. September 2004. Microsoft Corporation.

WS-Eventing: Web Services Eventing (WS-Eventing). Luis Felipe Cabrera, et al. September 2004. BEA, Microsoft, and TIBCO.

WS-Federation: Web Services Federation Language (WS-Federation). Siddharth Bajaj, et al. 8 July 2003. IBM, Microsoft, BEA, RSA Security, and VeriSign.

WS-FedActive: WS-Federation: Active Requestor Profile. Siddharth Bajaj, et al. 8 July 2003. IBM, Microsoft, BEA, RSA Security, and VeriSign.

WS-FedPassive: WS-Federation: Passive Requestor Profile. Siddharth Bajaj, et al. 8 July 2003. IBM, Microsoft, BEA, RSA Security, and VeriSign.

WS-MEX: Web Services Metadata Exchange (WS-MetadataExchange). Keith Ballinger, et al. March 2004. BEA, IBM, Microsoft, and SAP.

WS-Policy: Web Services Policy Framework (WS-Policy). Don Box, et al. 3 September 2004. BEA, IBM, Microsoft, and SAP.

WS-PA: Web Services Policy Attachment (WS-PolicyAttachment). Don Box, et al3 September 2004. BEA, IBM, Microsoft, and SAP.

WS-RM: Web Services Reliable Messaging (WS-ReliableMessaging). Ruslan Bilorusets, et al. March 2004. BEA, IBM, Microsoft, and TIBCO.

WS-SecureConv: Web Services Secure Conversation Language (WS-SecureConversation). Steve Anderson, et al. May 2004. BEA Systems, Inc., Computer Associates International, Inc., International Business Machines Corporation, Layer 7 Technologies, Microsoft Corporation, Netegrity, Inc., Oblix Inc., OpenNetwork Technologies Inc., Ping Identity Corporation, Reactivity Inc., RSA Security Inc., VeriSign Inc., and Westbridge Technology.

WS-Security: Web Services Security: SOAP Message Security (WS-Security). Ed. Anthony Nadalin, et al. March 2004. OASIS-Open.org.

WS-SecurityPolicy: Web Services Security Policy Language (WS-SecurityPolicy). Giovanni Della-Libera, et al. 18 December 2002. IBM, Microsoft, and VeriSign.

WS-SecUsername: Web Services Security: Username Token Profile V1.0. Ed. Anthony Nadalin, et al. March 2004. OASIS-Open.org.

WS-SecX509: Web Services Security: X.509 Token Profile V1.0. Ed. Phillip Hallam-Baker, et al. March 2004. OASIS-Open.org.

WS-Transfer: Web Service Transfer (WS-Transfer). Ed. Don Box, et al. September 2004. Microsoft Corporation.

WS-Trust: Web Services Trust Language (WS-Trust). Steve Anderson, et al. May 2004. BEA Systems, Inc., Computer Associates International, Inc., International Business Machines Corporation, Layer 7 Technologies, Microsoft Corporation, Netegrity, Inc., Oblix Inc., OpenNetwork Technologies Inc., Ping Identity Corporation, Reactivity Inc., RSA Security Inc., VeriSign Inc., and Westbridge Technology, Inc.

XOP: XML-binary Optimized Packaging (XOP). Ed. Noah Mendelsohn, et al. 8 June 2004. W3C.org.

III. Interoperability Profiles

WSI-BP10: Basic Profile Version 1.0. Ed. Keith Ballinger, et al. 16 April 2004. The Web Services-Interoperability Organization.

WSI-BSP10: Basic Security Profile Version 1.0 (Working Group Draft). Ed. Abbie Barbir, et al. 12 May 2004. The Web Services-Interoperability Organization.

WS-DP: **Devices Profile for Web Services**. Shannon Chan, et al. May 2004. Microsoft Corporation.

IV. Other Resources

NAICS: North American Industry Classification System (NAICS). NAICS Association.

SIC: Standard Industrial Classification (SIC).

UBR: UDDI Business Registry.

WS-I: The Web Services-Interoperability Organization (WS-I).

Gray & Reuter: Transaction Processing: Concepts and Techniques. Jim Gray and Andreas Reuter. Morgan-Kaufmann, 1993.

Chapter 35

Facility Total Energy Management Program: A Road Map for Web Based Information Technology in the Egyptian Hotels Industry

Khaled A. Elfarra, Msc, CEM, DGCP
General Manager, Engineering/Projects Department
National Energy Corporation – Egypt
khfarra@link.net

ABSTRACT

ENERGY RESOURCES with their utilization and performance in any activity/service that is delivered to the customers are playing the most significant role besides the applied technologies. This role for energy mangers means the importance of energy and how to improve the facility performance from the energy utilization and management point of views as well as the activity/service technologies. Then, all factors affecting the energy management must be considered such as the impact of energy prices, policies, financial mechanisms, standards, programs, and the incentives.

The high wave of new web-based technologies as well as the information technologies is empowered the need for energy information and control solutions. This is encouraging the technology merging in many activities and showing the interlink between the existing energy management programs and the new proposed technologies. The author in this article does his best in order to propose an updated concept for energy management that is entitles as "total energy management program". The components of this program are demonstrated in a simple form that can help facility energy managers to improve their facility performance in a way where cost savings can be achieved. Moreover, the description for the components of the web-based technologies and their interconnections with the total energy management programs are presented showing the merits of such implementation and the barriers which are encountered.

INTRODUCTION

Energy is a costly resource and needs to be managed as well and as wisely as other expensive commodities. Therefore, energy management is considered a cost saving program for any facility since the improvement of facility energy efficiency is a key parameter to financial gain. In general, energy costs play a major role with respect to the other costs that are associated with human activities in the major economic sectors: industrial, residential, commercial, public, transportation, etc. Due to these growing activities, the demand on energy is rapidly increasing even though there is some shortage in the availability of energy which affects its prices. The demand for energy has directed the attention of all interested parties toward:

1. Establishing and setting country, regional, and worldwide energy policies.

2. Setting up energy indices or benchmarks to help in assessing the potential for energy conservation and efficiency measures (ECEMs).

3. The interests in energy management programs (EMP) that monitor, track, and control the energy at any facility.

4. The interlinks between energy utilization and all other enterprise activities, such as ISO standards, regulations, legislation, and information technology (IT).

5. Setting up the required financial firms, institutional and national roles for implementation of energy management programs.

6. Setting up national and international programs for capacity building, promotion, and awareness for considering ECEM technologies, IT, and web-based technologies (WBT) for energy information and control solutions (EICS).

Nowadays, the use of WBT/IT and its interface with many enterprise activities allows the facilitation of worldwide communication which encourages outsourcing, expertise exchange, and open discussions. This approach makes all facility information available for any facility staff, and helps in focusing on the problems encountered with its activity and providing quick response for the required corrective actions that improve the facility performance. Improvement in facility performance, through reducing its energy utilization and increasing efficiency, leads to cost savings as well as market share and competition. This WBT/IT trend, with the advances in methods of communication, brings attention to the modification and improvement in EMP that started 15 to 20 years ago. The author will present a comprehensive principle for a new attitude that can successfully utilize the new wave of WBT/IT. This principle will focus on defining the concept of a Total Energy Management Program (TEMP) in conjunction with its operational procedures, communication channels, the interlinks between different facility departments and applied systems. This new program should motivate the facility manager / energy manager to exert much more effort and thought on energy.

WHY IS THE TEMP CONCEPT ESSENTIAL?

Whether electricity, fuel oil, or gas, the intensive consumption of energy in industrial facilities, commercial buildings, public and governmental buildings/utilities represents a large drain on the available energy resources. Because of this consumption and its associated energy costs, companies are hiring energy managers to help reduce energy consumption and increase energy utilization efficiency. The TEMP is driven by the concept of producing energy cost savings as well as reductions in emissions. The implementation of TEMP can achieve sustainable savings which increase competitiveness, provide better quality/service, increase productivity, improve working conditions, and increase marginal profits.

The TEMP will significantly create an effective positive improvement in the national income. This helps protect against future increases in energy prices and in continual reduction of air pollution, as well as make a significant contribution to the carbon trade market. The TEMP implementation aims to enhance energy efficiency and to improve the specific energy consumption/energy unit indices (EUI) without affecting the facility's production level, product quality, operational safety, or environmental performance. The outcome of implementing the concept of TEMP is its cost effectiveness when it is feasible, if the financial investment, time and manpower exerted are justified by the achieved energy savings.

TEMP CONCEPTUAL DESIGN

The design for such TEMP systems must fulfill the improvement cycle/loop of management that is illustrated in Figure 35-1. As shown in this figure, the design of the TEMP can be divided into four main phases that are:

- Planning (Plan)
- Implementation (Do)
- Monitoring and tracking (Check)
- Management review and corrective actions (Act)

The TEMP core element is that it is composed of teams that are represented by all levels of the company staff, from the top management to the shop floor. This can be primarily initiated through a committee headed by top management representative and other members including the facility staff and energy managers. This committee is first responsible for getting the company/corporate commitment to the TEMP, and secondly to form the different teams of the TEMP system. The formation of the TEMP system teams will be structured along the lines of the four TEMP phases. A "Steering Committee" will handle the whole TEMP system, ensuring its enhanced performance in collaboration with all interested parties.

Planning Phase

The TEMP planning phase is the intrinsic tool for coordinating the interaction of the facility working environment, its staff, and TEMP team in order to achieve the targeted success. This phase sets the company/corporate policy that reflects all tasks within the TEMP system including consideration of all energy aspects, objectives, targets, and any other issues relevant to en-

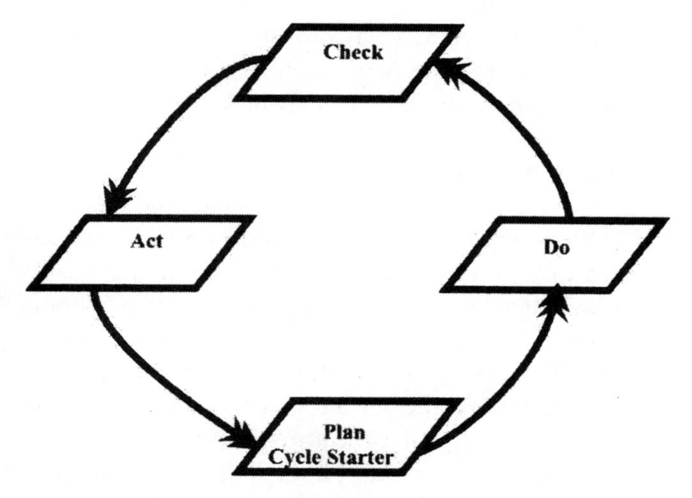

Figure 35-1. TEMP—System Cycle

ergy utilization and efficiency. The company/corporate policies should be:

1. Appropriate to the process/activity and the operations undertaken.
2. Readily understood by all facility levels.
3. Committed to continual improvement and enhancement of energy efficiency.
4. Possessing a clear framework for objectives and targets.
5. Commitment to comply with current EUI (or other benchmarks) of similar facilities.
6. Available for all interested parties whether insiders or outsiders.

In order to set the basic foundation for the TEMP, some essential tasks and outcomes must be identified in the planning phase:

Baseline Register

It is mandatory for the TEMP to assess the potential for ECEMs that might be applicable inside the facility. The creation of the baseline register is the outcome of a standard energy audit that involves the following:

1. Collect energy consumption data for all departmental activities/services as well as the overall consumption. This includes readings for the main meters and the sub-meters.
2. Gather utility bills for electricity, fuel oil, and gas consumption for at least 12 consecutive months. Also, the rate structure for billing must be known for each source.
3. Snapshot and continuous measurements for predetermined periods in order to assess the current energy performance and efficiency.
4. Analyze facility operations, performance mode, volume of production/service, and energy costs to total production/service costs.
5. Set up a comprehensive inventory list for all large energy-using equipment and systems in the facility to assess individual efficiencies and EUIs.
6. Establish an overall energy balance for the facility as a bench mark.

Having completed the standard energy audit, the baseline register will be set to specify the existing situation at the facility and to serve as the baseline for comparison of the effect of future ECEMs implementation. Then, the facility energy manager's role is to coordinate between the steering committee and the TEMP teams to prioritize the necessary actions for implementing the potential ECEMs. These actions must include, but not limited to, framework, human and financial resources, and responsibilities.

Setting Objectives and Targets

The objectives are best set after the completion of the baseline register. These must match the company/corporate policies to meet its goals that lie in the company domain of capability and ability to implement its short and long term plans and strategies. The objectives must be action-oriented and give rise to specific targets that can be measured. An effective target relies on the creation of detailed energy consumption calculations, and coordination between different centers and sections of the entire facility. For instance, a typical objective for a boiler tune up program might include the following hierarchy of considerations:

- First priority is to increase the boiler efficiency.
- Second priority is to recover the condensate from the process whenever possible either by installing a return system, or by installing a heat recovery system to feed some other activities.

- Third priority is to recover heat from the stack losses.

The targets are set considering the potential ECEMs and must be defined by time and benefits. The target settings are:

- Significant: It is high priority, and it has a large positive impact on energy efficiency and utilization.

- Measurable: It reflects the actual cost/benefit ratio.

- Responsible: The TEMP implementation team and the steering committee need to know those results.

- Timely and traceable: There is a predetermined timeframe and full documentation for pre and post ECEMs operations.

Management Program (MP) Identification:

The program must be identified in a way to drive the company/corporation towards the achievement of its objectives and targets. The program is effectively a portfolio of projects supported by the necessary documentation that will systematically drive the company to its objectives and targets through activities of the pre-assigned TEMP working team. This team will coordinate its activities with the other teams that are; a) ECEM implementation team, b) commissioning and startup team, c) monitoring team, and d) departmental facility operations teams.

First, each ECEM is assessed through the baseline register activity to see what impact it will have on the facility energy performance in addition to reducing energy costs. Next, the energy manager, collaborating with the members of the TEMP team, can set a priority list for implementing the ECEMs. Then, each ECEM must be treated as an individual project considering the following:

1. Required financial investment associated with the estimated cost, estimated benefits, and the cost/benefit analysis.

2. Assignment of the human resources for the implementation and the outsourcing requirements, if needed.

3. Implementation schedule and post-implementation measurement and verification protocol that will be executed.

4. Project documentation control including the training needs for the new implementation, preparedness and response procedures for unforeseen problems that might occur, and follow up and proposed corrective actions.

In this regard, the clear identification of the management program will facilitate the role of the facility energy manager in handling the TEMP system without extra burdens associated with his other responsibilities.

The planning phase of the TEMP system just described can follow the structure of Figure 35-2. This figure shows a logical procedure for identifying the proper company/corporate policy, following a step by step procedure starting from the baseline register.

Implementation Phase

Implementing the TEMP system requires someone to devise, organize, and evaluate the energy management plan, and to consider the responsibilities of the different teams. The energy manager is responsible for this phase to make sure the TEMP requirements must match the policy and to ensure proper facility performance from the perspectives of energy and production. Consequently, successful implementation leads to cost savings as well as a reduction in pollution sources.

Once the planning phase is finished, operational procedures must be created to implement the following tasks:

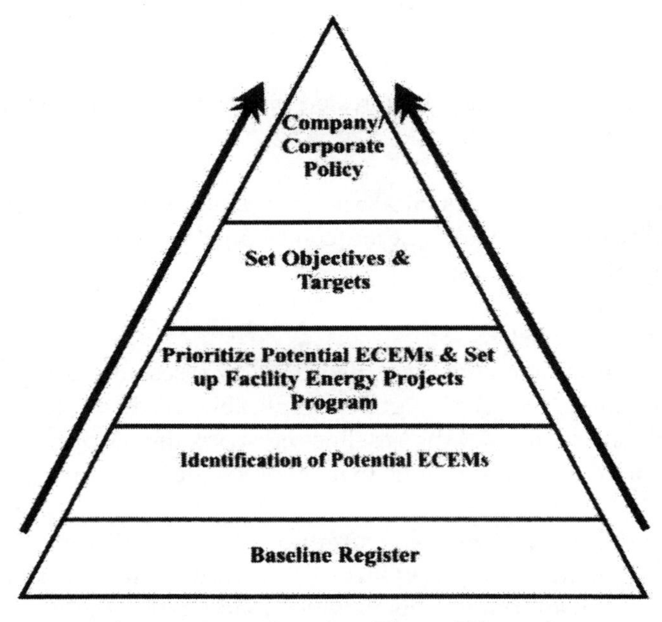

Figure 35-2. Planning Phase Hierarchy

Training and Awareness

The awareness and exposure to the expertise of others are key elements to understand the energy efficiency and performance of any delivered activity/service from some new technology. The successful implementation of any technology needs awareness and training of the facility staff on using it in an efficient way. The startup driver at any facility is to encourage the training department to develop procedures for evaluating its staff in order to assess the training needs considering all facility aspects. This will help the energy manager collaborate with the TEMP teams to build an appropriate training program for using the new technology. The training needs activity will enhance the facility's capacity building and performance, and accelerate the achievement of its objectives and targets. The needs plan has to consider subjects to be covered, area of training, local/outsourced trainers, training timeframe, and trainees' evaluations. Therefore, the contents of the training needs plan must be recorded, documented, and maintained in order to help assess the future needs within a certain framework concerning the implementation of the TEMP system.

To enhance the benefits of the technologies resulting from use of the TEMP system, and to generally enhance the facility performance, the training material shall cover, at least, the following topics:

1. Energy resources, availability, tariffs, and energy utilization.
2. Process equipment and utilities performance, and maintenance.
3. Codes and standards regarding installations, commissioning and startup.
4. Energy information solutions, analytical procedures, and WBT/IT tools.
5. Communication channels and documentation flows.
6. Energy monitoring, tracking, and accounting.
7. Energy efficiency measures, applied technologies, and performance improvement.

In addition to the training needs, awareness of the TEMP system could be handled through short, group lessons for facility employees, leaflets, brochures, wall-mounted posters and instructions, periodical newsletters, and membership in relevant associations.

Internal and External Communication Channels

A successful response to any TEMP system depends mainly on how the involved people can communicate with each other and with outside interested parties. Creating communication channels needs full support from the steering committee and the facility's TEMP system teams to help disseminate the information that generates appropriate corrective actions for system upgrades and modifications. The availability of distinct information from different resources through the system communication channels enables decision makers to make more accurate decisions regarding their facility TEMP system.

Obtaining information through many communication channels will reveal better opportunities for any facility to move significantly ahead in maintaining its system properly performing and ensuring its sustainability. This information can be obtained through media, energy council groups, stakeholders, and organizations who set standards, regulations and protocols. This approach will raise: a) the company's public image and its commitment towards the available resources, b) the company's competitiveness and market share due to the TEMP implementation, c) the progress in achieving company objectives and targets, and d) the company's ability to learn about other facility's experiences that it can use.

Documentation and Control

Any TEMP system requires full documentation in order to facilitate the interlink and the interface between the facility systems and the WBT/IT. Then, the documentation process has to be timely ordered and traceable between all facility staff. Such documents support and facilitate the follow-up for the predetermined management programs and motivate the potential actions that strengthen the system implementation.

The energy manager's responsibility is to set the documentation cycle, its routing, and its controls. As a result, all operational procedures and working instructions regarding the implementation of all management program ECEMs must be communicated to all involved staff, and controlled by the document control section.

Monitoring and Tracking Phase

Once the planning phase is completed, the implemented ECEMs must be verified to cross check the cost/benefit analysis that has been previously performed. The successful energy manager must evaluate the implemented ECEMs, and assess if the company/corporate targets were met. The post-implementation records of energy consumption patterns will guide the attention to other areas in the facility that might need further investigation and will help decision makers by offering reliable and sustainable corrective actions. The records are the outcome of implementing the energy monitoring and tracking scheme. This scheme requires not only the participation of facility manager and his working team but also full coordination between the different departments and sections of the entire facility, and needs hardware, software, human and financial resources to build up a strong facility system.

The monitoring and tracking system relies on continuous data logging for all energy aspects in each consumption center with individual sub-meters and the main meters. Such data logging is best accomplished with the WBT/IT system, which will generate reports that show the energy patterns with their associated costs. Then, they can disseminate information to all interested parties. This in turn will strengthen the comparison and the verification against the pre-implementation of ECEMs and also prove the savings of the proposed technology. This comparison will reveal either improvements in energy efficiency or a fall off in performance levels. In this regard, the recorded information by monitoring, tracking and accounting is the basis for continuous evaluation of the facility performance and how energy control can be conducted. With WBT/IT systems, this monitoring and tracking is usually performed in 15 minute intervals. The reporting of monitoring has to be either daily, weekly, or monthly depending on the company size.

Management Review and Corrective Actions Phase

After specifying the monitoring and tracking system, the implementation phase will handle some of the other tasks needed for the next phase of the TEMP system - the management review phase. In this phase, the company management must review the progress of its TEMP implementation by having periodic meetings and group discussions within the entire company. This action will keep the focus on the objectives and targets and consequently minimize any deviation from the established company policy. If any deviation has occurred, the following corrective actions will be taken:

1. Restructuring the priorities of the TEMP projects with respect to the implementation progress and modifications required.
2. Updating the program objectives and targets.
3. Setting a clear vision regarding future training needs.
4. Making sure of the TEMP system continual improvement.

The energy manager must collaborate with the steering committee to handle all ties and interlinks with the TEMP system to facilitate and coordinate all tasks. The energy manager must also keep a track record of the TEMP system improvements. However, the TEMP system is a dynamic one that must be able to respond quickly, so the energy manager must be prepared to conduct and implement the actions within the TEMP domain and goals. These actions necessitate a robust organizational structure with interlinks between the TEMP system teams associated with tasks and responsibilities. This structure is demonstrated in a simple form in Figure 3, which shows each team role and how each team can communicate with the others. The main condition to be considered in setting up such a structure is to make sure that no interruptions, interference in tasks, or conflicts of interest will take place between the working teams of the TEMP system.

WHY ARE COMPANIES STILL CLINGING TO CONVENTIONAL ENERGY MANAGEMENT PROGRAMS (EMP)?

The conventional EMP system is a merging of the control system, either semi or fully automated, and the manual interface for data analysis and manipulation. Therefore, errors in managing such systems can easily occur and draw negative attention to the company's energy performance. Nevertheless, despite the advances in controls and communication tools, and the problems that might be encountered with the conventional systems, most companies are still holding on to their old systems. This is largely because of the investment required to update or replace their old systems, combined with their lack of knowledge about the new systems.

Regardless of the problems with conventional systems and the advances in new systems, the author has focused the previous sections on an updated concept for EMP that is entitled TEMP. The TEMP system discussed will improve the energy performance at any facility, and will provide an opportunity for integration with other facility systems through fully open systems using the new information and control technologies available. With such open systems, these energy information and control solutions (EICS) will be made available to the facility. Thus, the exchange of expertise will help the facility energy manager to properly assess the different

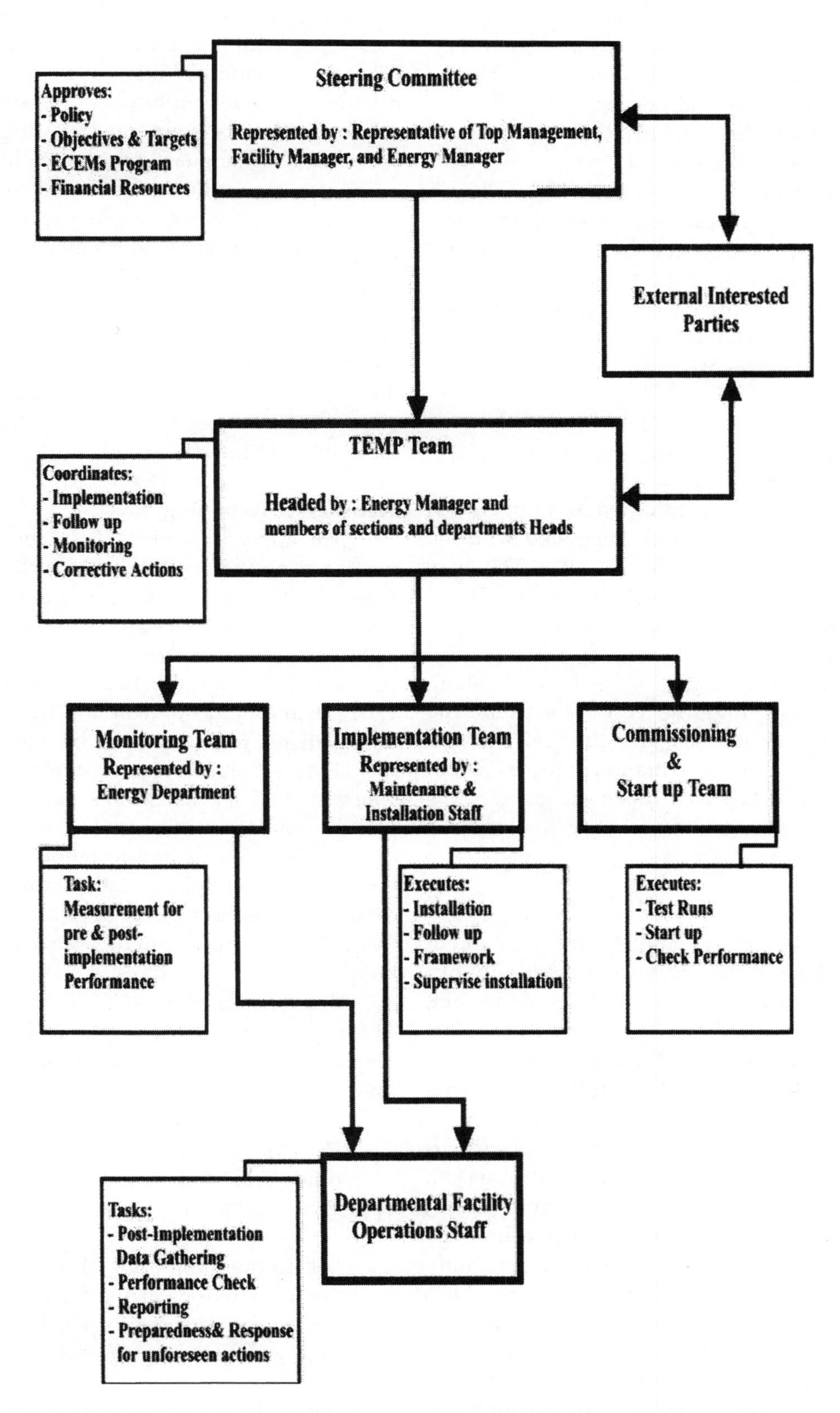

Figure 3. TEMP Organizational Structure, Responsibilities, and Communication

control systems suppliers and set his facility requirements correctly. In addition, the energy manager will be able to: a) improve the future forecasting for his facility from aspect of performance and productivity; b) have a clear picture of the scheduled rehabilitation process for the facility equipment; c) interlink between the TEMP and the other certification procedures like ISO 9000 and 14000 series; d) improve the competitiveness of the company in market share, quality, and facility performance as well as efficient energy utilization; and, e) receive full encouragement for outsourcing consultation activities that help in system review and guidelines for its update and continuous improvement.

INTERLINK BETWEEN TEMP AND WBT/IT

The proposed TEMP is considered as an open system for any facility, and it must be integrated with the facility WBT/IT in order to handle the EICS. This way the open system enables full accessibility to the TEMP system and enhances the facility performance as soon as the system is implemented. The energy manager can simulate company performance of its TEMP system, using automated data logging, control actions, the adapted communication protocol, and the WBT/IT software and hardware. A brief presentation of these components of the TEMP and WBT/IT systems is given below to help any energy manager in assessing the required system in his facility.

Controls-Brief Background

Prior to the 1970's, most controls were either pneumatic or electric. Each system was controlled individually by independent standalone controllers. After that, transducers were introduced that had the ability to transmit data from various sensors of the automation system to a master microprocessor-based controller. This development produced the first computerized energy management system (EMS). The advances in control systems are shown in Figure 35-4. The distributed control system (DCS) shown in this figure came into use within the last decade and the cost of a DCS dropped. This technology had made many advances and the established communication protocols enabled interconnection between a DCS and the new WBT/IT. Moreover, the development in communication protocols started to focus on merging the different protocols into a unified one that could communicate with the existing ones. This approach allows an integrated system to be connected with different sites/locations from any point of a web interface.

Open Communication and Control Systems

Open standard protocols facilitate the integration of systems made by different manufacturers. The key difference for such protocols is the software used. Many software developers are now tailoring their software to use a standard language which can either be understood by all devices in the network or converted into a common language by use of gateways. Therefore, the cost becomes very effective which allows the facility/energy manager to invest in some small pieces which will enhance the existing facility management system. The standard protocols enable the interface between the TEMP and the WBT/IT and allow monitoring and control of the TEMP system from any remote location with internet access. This elevates convenience to new levels, making these internet interfaces a revolutionary trend.

Web-Based Technology

The appearance of WBT/IT and the increasing dependence on sophisticated production processes have dramatically increased the importance of the EICS. The deployment of WBT/IT and its integration of the facility TEMP must be consistent with the defined facility management goals for the EICS. The implementation of WBT/IT in conjunction with the TEMP begins with local data gathered to establish the baseline register on which the TEMP system is based. Moreover, the advances in controls, WBT/IT, hardware, and software require further investigations in order to ensure the compatibility of the WBT systems and to reduce the investment cost thus making its implementation more cost-effective. Successful data gathering, logging, manipulation, and control actions are the key foundations on which the EICS is built.

For the use of a WBT/IT system, the concept of TEMP presented in this chapter is the roadmap for planning a suitable system that considers the following factors:

1. Develop the baseline register to assess the facility objectives and targets, policies, and potential ECEM program.

2. Set up the monitoring and tracking system for all energy activities at the facility. This needs proper setting for data points and time intervals for logging with respect to predetermined accuracy level.

3. Follow up and cross check the monitored data with respect to the operational standards to correct any deviations immediately.

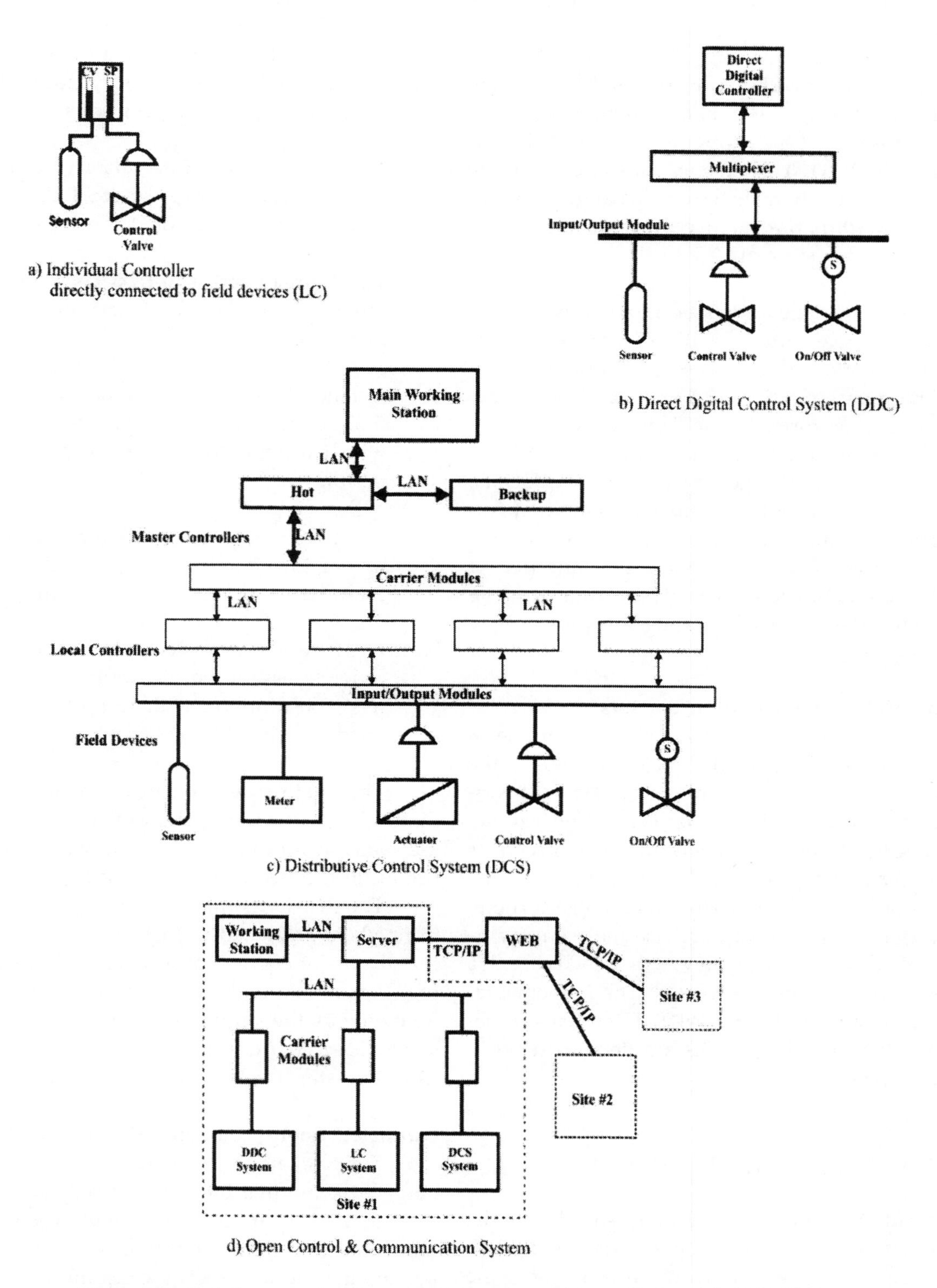

Figure 35-4. Advances in Control Systems and Interlink with WBT/IT

4. Record results in reduction in resources and pollution levels, and increases in quality and productivity.

5. Generate reports within pre-specified time intervals to facilitate periodic management review and verify continual improvement of the TEMP through the WBT. This enables outsourcing, information dissemination, expertise exchange, and discussions within the facility and the interested parties.

These factors can be addressed in a simple way to ease the user interface with the WBT/IT and controls for system documentation. This process is illustrated in Figure 35-5. The data collection and analysis shows how much the facility performance and efficiency have improved. Furthermore, the functionality of the TEMP and all other implemented management systems at a facility, such as ISO 9000 and 14000, can be coherently assembled. This assembly requires a central software system that is capable of managing the interconnection of different users, databases, data analyses, control using network connectivity, and communication protocols to the web.

TEMP and WBT implementation barriers

The lack of knowledge, awareness and training in developing countries is still the major barrier to the implementation of TEMP and advanced WBT/IT systems. In addition, the subsidy for certain energy resources leads the facility managers to other objectives which achieve a higher return on investment than do the implementation of energy efficient technologies. Therefore, many facilities intend to purchase more energy irrespective of its price since they will pass the extra energy costs along to their customers. The implementation barriers to both TEMP and WBT/IT systems in developing and developed countries are demonstrated in the matrix on page 470.

THE HOTEL INDUSTRY IN EGYPT

The rapid growth in the tourist industry during the last decade has encouraged many investors to enter this field. This required a large infrastructure to meet the operational requirements of the hotel industry for such essential resources as water, electricity, and fuels of different types. The statistics for current hotel capacity are available but the consumption regarding energy has not been accurate until now. The author has proposed the following analytical approach for determining the consumption of both electricity and fuel. The data available for calculating the current energy consumption in the Egyptian hotel industry was calculated and cross-checked using the following information:

1. The registered hotel activity in the statistical yearbook of Central Agency for Public Mobilization and Statistics (CAPMAS).

2. The Tourism Sector in Egypt – August 2002, study report prepared by the American Chamber of Commerce in Egypt.

3. Detailed hotel audits conducted by the author for more than 15 hotels distributed through Egypt on which benchmarking for the current situation is performed.

4. Simulation and Analysis for equipment performance in order to assess average demand, maximum demand, and energy consumption.

5. Field survey for the commercial sector in Egypt in order to assess the distribution pattern of the hotels' consumption with respect to the main loads inside the hotels.

The total capacity of rooms in 2003 is presented in Figure 35-6. This figure shows that 227,698 guest rooms are available for tourists in Egypt with the major concentration in the Red Sea and South Sinai areas.

The distribution pattern of electricity consumption in hotels is presented in Figure 35-7. This figure shows that the major energy-consuming equipment is the Heating, Ventilation, and Air Conditioning System (HVAC), following by the lighting system of the hotel. These data were gathered from a field survey in different hotels from two-star hotels to five-star ones.

Estimated Energy Consumption in Hotels

The field audits in the hotels of Egypt and the data gathered show that the annual average occupancy level was about 73% in 2003. Per guest room, the average annual electricity consumption is 30,050 kWh, the average annual thermal energy consumption is 49.2 GJ, the average installed capacity of HVAC is 2.62 tons of refrigeration (TOR), and the average thermal-to-electrical ratio is 0.55. These indices were calculated with respect to the average occupancy level of 73%. The fuel consumption

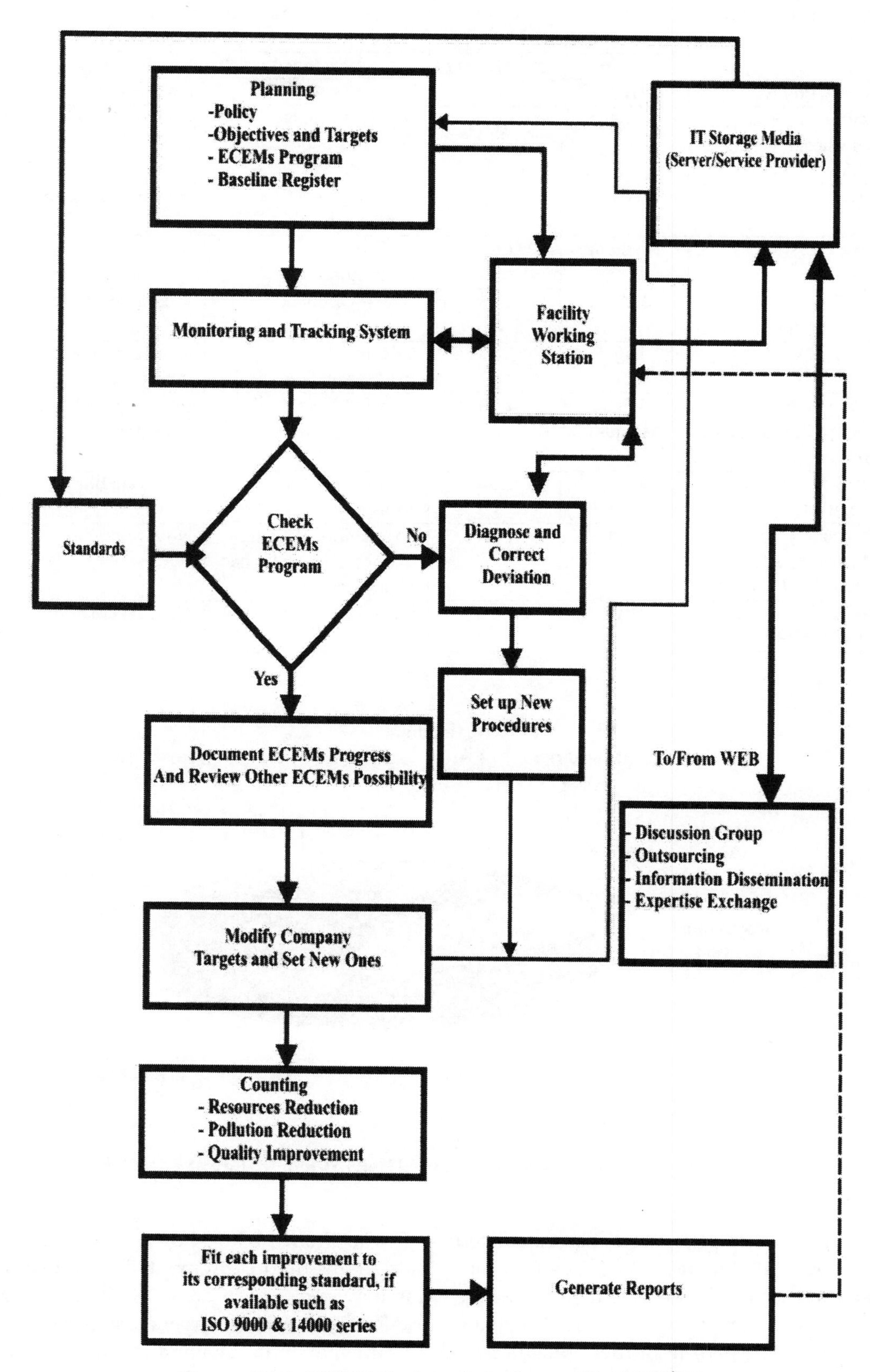

Figure 35-5. TEMP System Interface with WBT/IT

Comparison Item	Item Availability Potential		Barrier Remark	
	Developed	Developing	Developed	Developing
Energy Policy	High	Low	Incentives Availability	No Consideration
Training Programs and Capacity Building	High	Low	Highly Available	Lack of Knowledge
Energy Resources	Medium	High	Encouragement for implementation	No Consideration
Awareness	Medium	Low	Technology Understood	Low Concern
WBT/IT Infrastructure	Medium	Low	Low Investment required	More Investment Needed
Energy Costs	High	Low	Incentive for Implementation	Low Return on Investment
Technology Transfer	High	Low	Easy to Understand	Lack of Knowledge
Energy Technologies Promotion	Medium	Low	Well Handled	Poor Promotion
EICS Local manufacturing	High	Low	Low Investment required	No Availability
Financial Mechanisms for EICS and WBT	High	Low	Availability of Energy Service Companies and Financiers.	No Availability
Energy Programs Sustainability	High	Low	Long Term Programs	No Concern

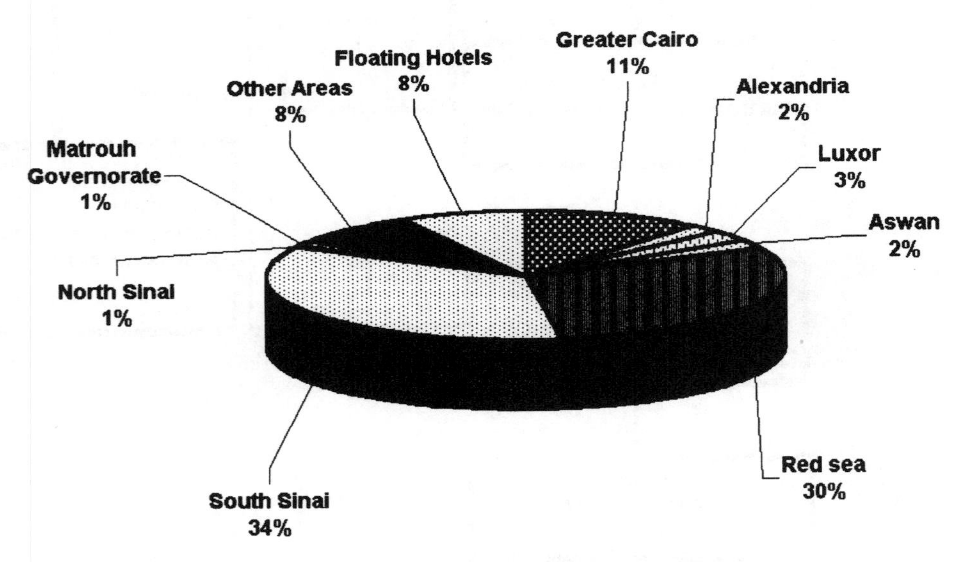

Figure 35-6. Hotel Capacity Distributed in Egypt

and electrical energy consumption were also calculated and the results are tabulated in Table 35-1. The fuel consumption is listed in tons of oil equivalent (TOE) where the conversion factors with respect to each fuel type are shown in Appendix A. The percent of fuel utilization by type for hotels in each area of Egypt is presented in Figure 35-8.

As Figure 35-8 shows, some areas do not have natural-gas-fired equipment. This is because there is no gas network in such areas as Sinai, Luxor, Aswan, and the Red Sea area. The average annual electrical demand and the maximum demand for the different loads inside the hotels is calculated based on the following:

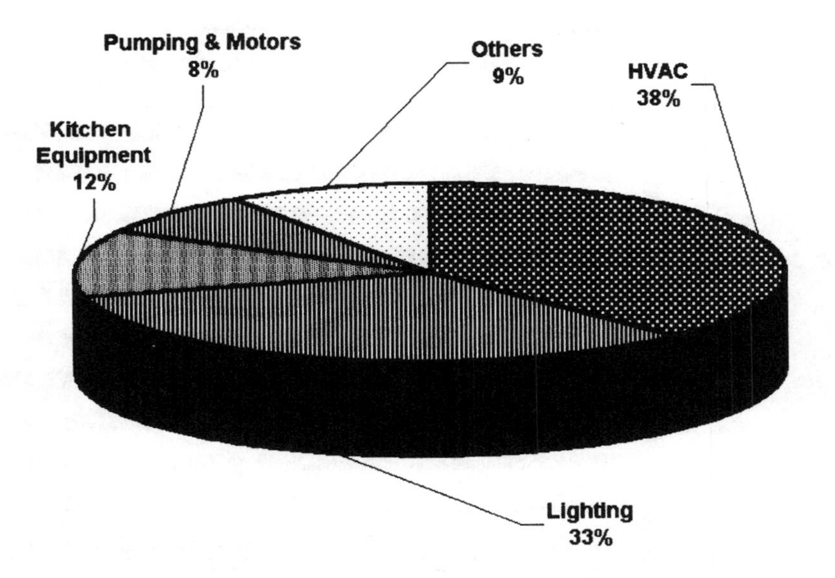

Figure 35-7. Distribution of Electricity Consumption Pattern

1. The annual consumption as shown in Table 35-1.
2. Energy indices that are calculated and presented in this section.
3. The distribution pattern of electrical energy consumption shown in Figure 35-7.
4. The individual load factors of different load types that are calculated based on the following formula:

$$kWh = kW_{\P} \times \sum_{i=1}^{n} LF_i \times h_i$$

Table 35-1. Energy Consumption in Hotels

District	Annual Electricity Consumption (GWh)	Annual Fuel Consumption (1000, TOE)
Greater Cairo	731.8	28.59
Alexandria	153.1	5.98
Luxor	172.5	6.74
Aswan	112.9	4.41
Red Sea	1,555.0	60.75
South Sinai	1,476.5	57.69
North Sinai	26.8	1.05
Matrouh Governorate	87.0	3.40
Other Areas	381.6	14.91
Floating Hotels	0514.8	20.11
Total	5,212.1	203.63

where:

kWh = Electrical Energy Consumption.
kW_f = Equipment rated full load.
LF_i = Equipment load factor at certain duration time i.
h_i = Equipment operating hours at certain duration time i.

Table 35-2 summarizes the calculations for the total average annual and maximum demands for the different major electrical loads in all hotels located in Egypt.

There is a great potential for ECEMs in the Egyptian hotel industry as shown by the magnitude of both the demand and energy consumption. The potential ECEMs can be implemented with a national program

Table 35-2. Annual Average and Maximum Demands for Egyptian Hotel Electrical Loads by Category

Load Identification	Annual Average Demand (MW)	Annual Maximum Demand (MW)
HVAC	226	359
Lighting	196	269
Kitchen Equipment	71	146
Pumping & Motors	46	76
Other Loads	55	100
Total Load	595	850

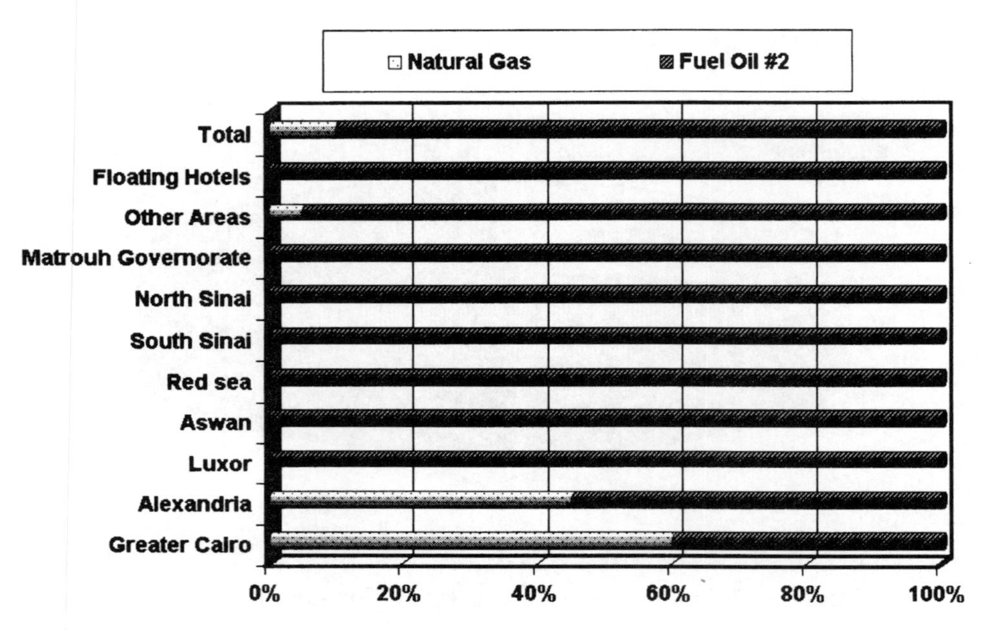

Figure 35-8. Area-wide Hotel Fuel Utilization by Type

that aims to achieve predetermined energy use and demand reduction targets through the energy policy. Such targets could be:

1. Reduction in electrical demand which in turn reduces the need for new generation capacity. This reduction can be achieved through: a) replacement of vapor compression systems (electrically driven) with direct fired/hot water absorption systems; b) replacement of electrically driven equipment with fuel engine driven systems; c) replacement of inefficient lighting systems with the new efficient compact fluorescent lamps (CFL); and d) the use of occupancy sensors, energy savers, and energy management systems.

2. The installation of standby engine-generator sets in hotels. The capacity of the generators should be about 25%-30% of the maximum demand of the hotels. This investment could be encouraged through an incentive program with the utility which either pays the hotel to reduce the facility demand during on peak periods or allows the hotel to connect these generators in parallel with the utility system to use in predetermined periods as payment for demand reduction. Also, the hotels can implement cogeneration or upgrade of their systems with waste heat recovery systems to improve the performance of such equipment and reduce the running costs.

3. Encouragement for switching fuel oil to natural gas whenever possible for hotels that are connected to the gas network.

4. Initiation of an incentive program for utilization of WBT/IT and help with setting up the infrastructure needed for implementation of such technology. This will help in standardizing benchmarking and updating the best practice indices.

APPLICABLE ECEMS IN THE HOTEL INDUSTRY

The advances in energy and demand management technologies must be investigated by the Egyptian market and adopted when financially viable. Technical implementation and performance are already feasible. ECEMS which are likely to be practicable in Egypt include:

- HVAC Absorption Systems
- Engine Driven Vapor Compression Systems
- Distributed Generation and Cogeneration System
- Occupancy Sensors and Energy Savers in Guest Rooms
- High Efficiency Lamps
- Fuel Switching to Natural Gas

HOW WBT/IT HELPS IN IMPLEMENTING ECEMS FOR THE HOTEL INDUSTRY

Rapid growth in energy demand characterizes the energy consumption pattern in the Egyptian hotel industry and shows the importance of adopting energy efficiency technologies. Although implementation of the technologies mentioned above can reduce both demand and energy, several barriers can be encountered. The barriers to implementation in Egypt can be summarized as follows:

1. The subsidy and cross subsidy of energy prices which can affect the economic feasibility of the implemented technology.
2. Lack of awareness and experience regarding the energy conservation and efficiency technologies.
3. Absence of regulations for energy indices.
4. Lack of awareness by local financiers of the value of investing in these technologies.
5. Limited availability of specialized expertise in Web-based technology that helps in information dissemination and case studies.

Among these obstacles to implementation of ECEMs in the hotel industry, it should be recognized that a lack of knowledge of WBT/IT is a critical barrier. WBT/IT is an essential tool for encouraging implementation through a robust setup in which all hotels should be included. This setup has to focus on a clear energy policy for hotels which should include the following factors:

1. Energy indices regarding the consumption per guest night and the base load of the hotel.
2. A recommended hotel load factor and implementation of the available tools for demand side management.
3. Incentives for using green energy sources such as new and renewable energy whenever possible.
4. Participation by all members of the hotel industry in a national program for energy efficiency for hotels.
5. Incentives for meeting energy reduction goals and fines for exceeding selected energy use levels.
6. A clear energy program for each hotel that allows the interested parties to monitor and reduce the demand on energy.

Implementation of ECEMs could be much easier and flexible since WBT/IT can provide an open communication tool that enables all users of energy, i.e. electricity, fuel and gas, to have accurate forecasting and to manage the demand for energy. Because the hotels in Egypt are grouped in districts where the tourism activities are available, they could take advantage of their location by implementing district-wide ECEMs for cogeneration, cooling systems and heating systems. Also, each district could implement a local area network that would provide the base for WBT/IT to be available for the whole hotel industry and facilitate communication worldwide. Developing such a system could achieve the following:

1. Outsourcing of other expertise.
2. Availability of time and concentration of hotel operators for better performance and higher quality services to the guests.
3. Periodical update for new technologies and exposure to others' achievement in similar industries worldwide.
4. Encouragement in the field of competition and market share with respect to service provided.

Today, there are many advances in the web-based technologies dealing with energy efficiency. Implementation of the energy conservation and efficiency technologies presented in this chapter has many potential benefits. These benefits are:

1. Implementation of high efficiency lighting could reduce the energy demand and consumption by 40 – 60 %. Moreover, the return on investment is high and it encourages either self-financing or the financing through Energy Service Companies (ESCO).
2. Energy management systems prove to be most effective in reducing the energy consumption of the HVAC system, with reductions ranging between 24 and 65%. Further, a release in HVAC system capacity could be achieved, about 5 – 10 %, to meet the future extensions.

3. Cogeneration improves the overall system efficiency. The efficiency of the packaged units (reciprocating engines) could reach 85%. Moreover, the utilization of standby engine generators which operate at least during the on peak periods will be very effective in reducing energy demand in the hotel industry.

4. The utilization of engine driven or absorption chillers can reduce the demand by not less than 30% which in turn releases this capacity to be employed for other activities without any extra burden on the country's development plan.

Implementation of energy conservation and efficiency technologies in the hotel industry of Egypt requires the following actions:

1. Quick setup for WBT/IT considering the required infrastructure.

2. Encourage efforts in data gathering, manipulation, verification, and analysis for hotel industry in Egypt.

3. Reform of the Egyptian energy policy and energy tariff structure to drive a sustainable market for energy conservation and efficiency.

4. Increase the awareness of financiers and bankers of the benefits of investing in energy conservation market.

5. Establish a national program for measurement and verification to support the implementation and monitoring of energy conservation technologies.

6. Enhance the role of Energy Service Companies (ESCO) and build a strong relationship among all parties to the national energy conservation program.

7. Establish incentives and penalties for meeting or failing to meet the energy indices.

8. Support the private sector with incentives in its retrofit programs to improve the services while using more efficient equipment.

9. Encourage programs for public awareness regarding energy conservation and efficiency technologies and energy labeling on national scale.

CONCLUSION

The approach of a Total Energy Management Program is still in the developmental stages for many facilities. Implementation of such programs and the interlink with new web-based and information technologies is under investigation by manufacturers and end users. Encouraging implementation of such systems is the responsibility of many interested parties such as energy groups/councils, program initiators, policy developers, technology manufacturers, customers, and governmental energy plans.

The facility management is the key group for successful implementation of web-based technologies. This can be achieved through full commitment by the managers, availability of capacity building to its staff, continual review of the Total Energy Management Program, outsourcing, and expertise exchange. Hence, it is important to overcome the obstacles against the implementation of the web-based technologies together with the energy management programs. Such implementation is essential since it can help facilities improve their productivity and reduce energy costs, as well as help the facility deal with future energy cost increases and their impact on the facility product/service.

Energy conservation and efficiency technologies have a strong market in the hotel industry of Egypt. This market could be driven through setting the proper financing mechanism and clear energy policy in order to meet/reduce the forecast growth rate of the demand for energy. The principal element that should be set first is awareness and knowledge about the WBT.

Section Eight

Future Opportunities for Web Based EIS and ESC; and Conclusion

Chapter 36

History of Enterprise Systems: Where They Are Headed Now

Keith E. Gipson, CEO/CTO
Impact Facility Solutions

ABSTRACT

The purpose of this chapter is to briefly introduce the concept of a Total Enterprise Operation Management System (TEOMS), and discuss some of the main features that form the basic functions of any TEOMS. A TEOMS is the next evolutionary step following the development of Computerized Facility or Energy Management Systems. The author presents his experience in this rapidly evolving technology through his description of the history of Enterprise Systems. Once this history is presented, the focus shifts to an assessment of where the state of the art in Enterprise Systems is headed. This discussion of the history of Enterprise Systems is intended to help users and developers of other IT and Web Based Systems understand where some of the new technological progress came from, and guide them in continuing this development of IT and Web/Internet technology for the improvement of facility operation and energy efficiency.

A TOTAL ENTERPRISE OPERATION MANAGEMENT SYSTEM

A Total Enterprise Operation Management System is the next step in the evolution of Facility or Energy Management systems. This new approach relates generally to a system and method for managing the total enterprise operations of a company or corporation and in particular to a system and method for automatically managing inter- and intra-relationally all the operational functions within a company or corporation having one or more locations or sites around the globe. The system is connected via the World Wide Web (WWW) and/or other TCP/IP Ethernet-based networks. The system also includes a near real-time, data retrieval, archiving and trending system, which permits automated diagnostics data and messages to be communicated within the system. Figures 36-1 and 36-2 illustrate a typical TEOMS setup. Figure 36-3 represents the system architecture of the TEOMS system.

HISTORY OF ENTERPRISE SYSTEMS

As enterprise businesses began to expand significantly after the end of the Second World War, they were faced with constant demands to create more goods and services, to streamline and make their day-to-day operations more efficient, to improve the quality of their customer service, and to reduce expenses in an effort to remain competitive. Their ability to achieve these rather lofty goals was hampered by the state of the available technologies, particularly in digital electronics (including computers) and telecommunications.

The Public Switched Telephone Network (PSTN) has been evolving since Alexander Graham Bell made the first voice transmission over wire in 1876. From a simple one-way voice transmission to a bi-directional one and then to the development of a "centralized switch" and finally digital voice signals using Pulse Code Modulation (PCM), the infrastructure of PSTN dominated the telecommunication scene until the early 1980's when deregulation took place in the telecommunication industry. The Enterprise Telephony (ET) network emerged shortly thereafter and while it has coexisted with the PSTN network ever since, its rapid technological development has gradually made it the dominant favorite for use in the business arena. Today ET is widely accepted as a business telephone system because it provides basic business features, such as hold, three-way calling, call transfer, and call forwarding. The capability to provide advanced features is what differentiates ET from its big brother PSTN. Enterprise Telephony has finally begun to help businesses enterprises achieve their operational objectives.

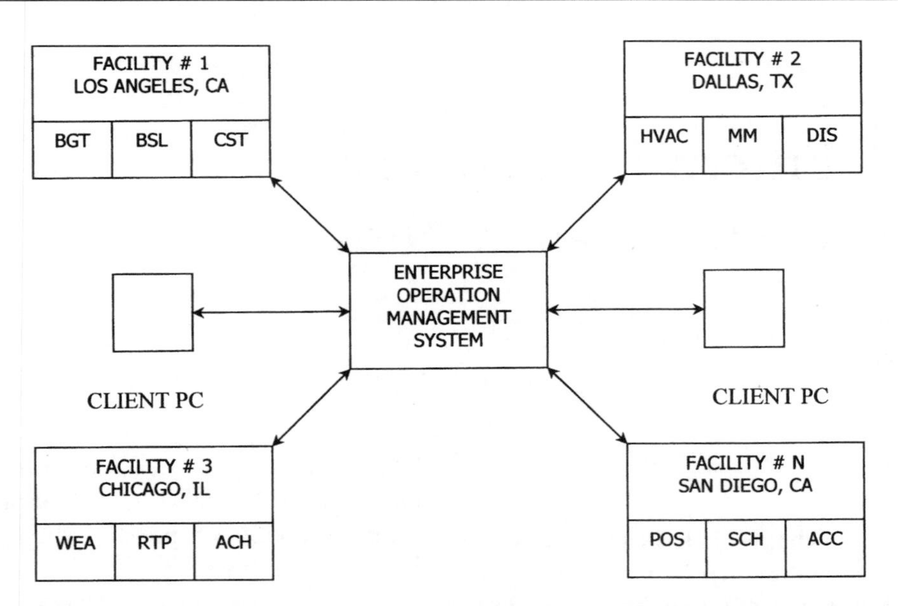

Figure 36-1. A typical Enterprise Operation Management System layer (acronyms are explained in Figure 36-2).

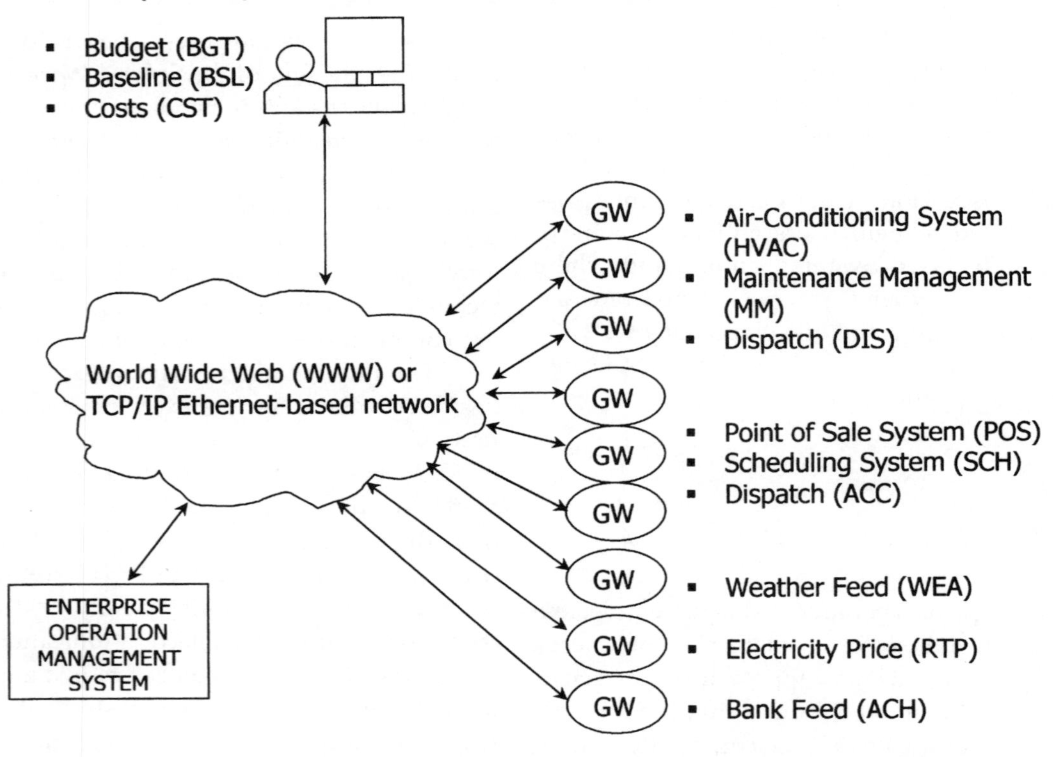

Figure 36-2. This layer connects the Enterprise Operation Management System to a network,

As Enterprise Telephony (ET) continues to provide more and more advanced business networking features, sometimes in collaboration with PSTN, such as the Centrex Line (a key-system or PBX), Virtual Private Network (VPN) etc., businesses react strongly and favorably to these new networks. By far, the most popular option for ET is for businesses to purchase their own key-system or PBX network. Up until now, the most popular network protocol was the Transmission Control Protocol (TCP). But it was the advent of the Internet in the mid 1980's, along with the so-called Internet Protocol or IP and the establishment of the Open Systems Inter-

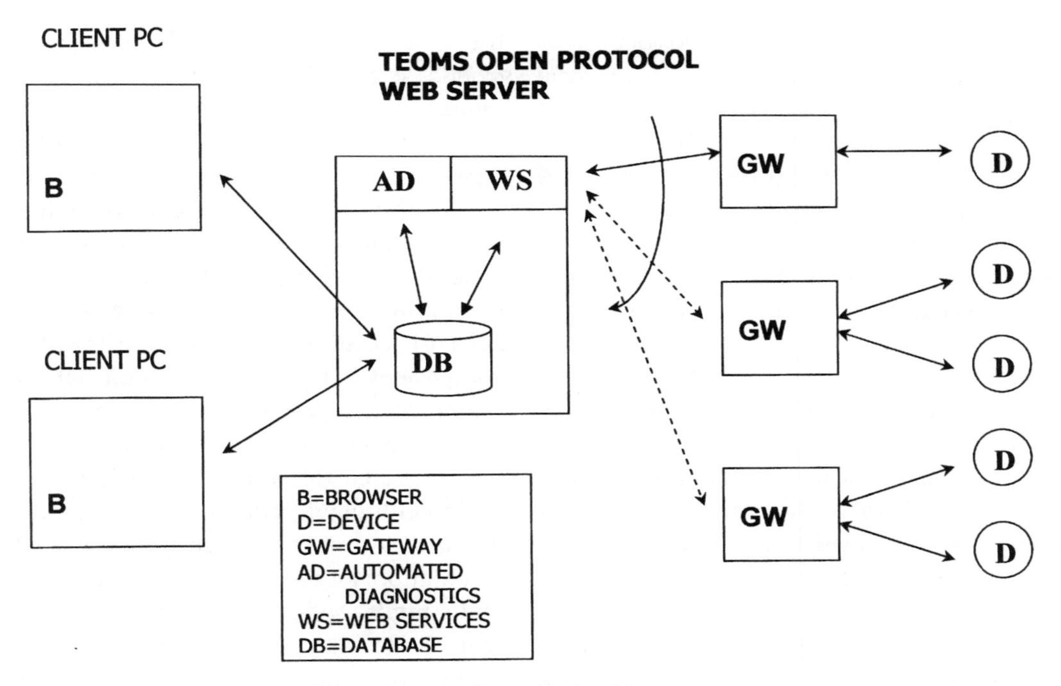

Figure 36-3. System Architecture

connection (OSI) Reference Model by the International Organization for Standardization (ISO) that opened the gate for businesses to totally embrace and exploit these new and advanced networks to their overall business advantage. IP is a connectionless protocol and TCP can "sit" on top of IP and add a host of advanced features such as flow control, sequencing etc. without having to deal with common data link issues like Ethernet, Asynchronous Transfer Mode (ATM), Frame Relay, Token Ring, Synchronous Optical Network (SONET), copper and fiber. This makes the TCP/IP network virtually ubiquitous.

Since the late 1980's, businesses have been building networks based on Transmission Control Protocol/Internet Protocol (TCP/IP) to take advantage of the power of TCP/IP networking and the many services it can provide. These services include Internet access for remote users, easy-to-use Web browsers, internal corporate Intranets and Web servers, Java applications, and Extranets with trading partners and suppliers. All these services make it easier for enterprise businesses to build new business applications, enable Web-browsers access to information databases, and provide new services to both internal and external customers. Examples of well-known businesses taking advantage of the TCP/IP networking technology are the Airline Reservation System for the airline industry, the Rental Car Reservation System for the rental car industry, PeopleSoft for the human resource planning and management applications, companies like SAP and QAD for supply chain management applications, financial institutions such as Freddie Macs, EDS and Quicken respectively for housing mortgages, administration of government contracts and banking applications. Last but not least in the HVAC industry, Building Automation Systems (BAS) and Energy Management Systems (EMS) also take advantage of networking tools such as LonWorks, BACNet and ModBus, to name just a few.

In 1995, monitoring of the consumption of utilities at business premises began to develop. Using a local network, a central computer could receive consumption data from the zones of a group and calculate total utility consumption within the group as well as conduct further analysis on the consumption data. The computer also supplied control data to the zones for controlling utility consumption.

In 1996, an energy management and building automation system including a local area network or home automation data bus was developed. Each load was connected to the network via a control module. Separate current monitoring control modules measured load current, and separate power monitor modules measured power consumed by selected loads. A first (external) microcomputer and a second (internal) microcomputer located within the customer's premise communicated with each other and with the various modules via the network/data bus. The first computer further communicated with the utility company via an appropriate communication link. The second computer served, in part, as an input/output terminal for the system, allowing the customer to set parameters and query the system as to power usage, displaying reports requested by the customer and also messages transmitted by the utility company and by either computer. This energy management system took advantage of the Open System Interconnection (OSI) Reference Model standard for developing protocols that enable computers to communicate to one another.

In 1997, an appliance interface apparatus and automated residence management system was developed. This system included a method for bringing an appliance and/or an electrical or mechanical system of a residence into communication with another, or with a control device within the residence, or with another communication source outside the residence, so as to establish a home automation system, or enlarge upon an existing automation system. An appliance interface module apparatus was also provided for facilitating communication between an appliance, and the automation system as a whole. The appliance interface module could control, upon command, the specific detailed operations of the appliance to which it is attached, and could, upon inquiry, transmit data which had been recorded, stored and/or calculated by the module.

Home Automation started advancing in 1998. An energy management and home automation system included one or more controllers in each facility being managed and one or more energy consuming devices attached to each controller. Each controller responded to digital paging signals from a central command center which had established a schedule of event affecting the operation of each device and the controller scheduled each device to be operated pursuant to the programmed schedule. The user of the system, by appropriate communication with the command center, could cause a paging message to be provided at any time changing the pre-programmed schedule. Set-point temperatures controlled the heating/cooling system, for each of a variety of occupancy modes, whereby the most energy is consumed when the facility is occupied and lesser amounts of energy are consumed when the facility is empty. Intermediate amounts of energy are consumed when the facility is not fully operational because people are asleep or maintenance or cleaning is being done. Prior to a mode change, which demands more energy, the controller calculated the time required to bring the facility to the set point for the new mode and operated the heating/cooling unit at the calculated time prior to the scheduled mode change. The calculation is based upon the time required for the heating/cooling being controlled to change the temperature in the facility being managed. The energy consumption of the facility could also be reduced by pages from the utility company as a part of its emergency load reduction program.

Computer Associates, with their next generation interface called "TNG" (The Next Generation), presented a network management system using virtual reality techniques. A network management system allows a network administrator to more intuitively manage all components of a heterogeneous networked computer system using screen views (graphic visualizations) of any component or any set of components. These views are generated in a multi-dimensional, virtual reality environment. Navigation tools are provided that allow an operator to travel through the network hierarchy's representation in the virtual environment using an automatic flight mode. The system is capable of managing a worldwide network; views of network components may be organized by continent, wide area network, city, building, subnet, segment, and computer. A view may also display internal hardware, firmware, and software of any of any network components. Views of network components may be filtered so only components pertaining to a specific business or other interest are displayed.

Finally, in 2001, John Woolard, Dale Fong, Keith Gipson (the author) and Pat Dell Era developed an innovative energy and facilities management system. This was the beginning of "Enterprise Energy Management" and Silicon Energy Corp (now Itron, Inc.) led the market in providing EEM solutions. Such a system provides energy users who have large physical complexes with a comprehensive understanding of the energy consumption of their facility and with the ability to manage it in a way that makes sense for their business. The system may include three dimensional facilities navigation tools, powerful energy consumption analysis processes, TCP/IP communication capabilities and World Wide Web (WWW)-based interface. The system also includes a real-time data retrieval and dissemination process, which permits real-time energy data to be communicated within the facility complex.

WHERE ARE ENTERPRISE SYSTEMS HEADED NOW?

Prior inventions and implementations of Enterprise systems all have a lot in common, yet each in their own way has advanced the field of enterprise management systems and methods. Each uses various communication technologies and middleware to structure and construct their management systems and methods. They all also seem to be related to the facility management systems and methods for controlling energy consumption and achieving energy savings in both the business and residential sectors.

As the enterprises of tomorrow focus on new modes of operation, systems would automatically analyze and diagnose the historical trended data instead of relying on the human operator to perform a manual analysis. The Automated Diagnostics Engine can command the appropriate enterprise system to take a reasonable course of action based on expert rules embedded in the system. For example, if the Air Conditioning Fan breaks down, the system can automatically analyze the fan system components via the Air Conditioning (HVAC) system interface (Gateway) and generate a work order ticket on the Maintenance Management system. Thus, an air-conditioning mechanic can be automatically dispatched to investigate because the system has diagnosed the "root-cause" rather than providing a simple alarming function which requires manual intervention to determine the cause.

Thus, as enterprise businesses enter the 21st century, they are becoming even more aware that their data network or enterprise management system is an all-important mission-critical piece of their business. Knowing how to effectively utilize their data network or enterprise management system is tantamount to gaining a competitive advantage over their counterparts in their businesses. However, a scrutiny of what is currently available in the overall methodology of enterprise management systems and methods shows that businesses still need a lot more advancement in enterprise management system concepts and methodology.

What we need to develop next is a "total enterprise operation management system and method" aimed at providing not just energy management, but encompassing all necessary operational aspects of a multi-national and multi-divisional corporate entity. Such operations would include at least the following: energy, maintenance, scheduling, external information feeds, banking, budget, dispatch and analytics. Such a total enterprise operation management system need not be a real-time system but could rely upon the use of "trended" data for performing the analytics for the majority of the functions. It should also use only open web-based standards such as XML and SOAP instead of generating its own "proprietary middleware." It should only use "commercially off the shelf" (COTS) communication technology instead of proprietary or patented counterparts.

Last but not least, such a system must be largely automated and able to perform well all the key functions of an enterprise system such as:

1. Normalization (protocol translation and standardization via gateways),

2. Integration (the use of open middleware for tying all the system components together), and

3. Enterprise visualization ("view") and "total access."

The new generation Total Enterprise Operation and Management system is a first step toward meeting and fulfilling the aforementioned performance goals.

CONCLUSION

Technological progress in both Information Technology and use of the Internet and World Wide Web will continue at a rapid rate. Applying these advancements to computerized facility and energy management systems requires the innovative skills of many people in both the IT and the Energy Management fields. If the history in this area is a good indicator of the future, this powerful technological wave should provide continuing innovations for the energy management industry.

Chapter 37

Building Control Systems & the Enterprise

Toby Considine
University of North Carolina, Facilities Services
Chapel Hill, NC

ABSTRACT

MANY CURRENT journal articles and most current building control system product literature make claims about Building Controls interacting with the Enterprise. Far from being ready to interact with the Enterprise, today's control systems have not yet matured into enterprise functions and so are not ready to interact with other enterprise functions. Merely using the current enterprise protocols such as XML and SOAP are not enough. It is only by developing the operation of building automation systems into an enterprise function that control systems can ready themselves for a role as an enterprise asset. This chapter describes a path to this mature function, and introduces the term Building Automation for the Enterprise (BAE) to describe it.

INTRODUCTION

The controls industry today, is abuzz with talk of opening up control systems to the enterprise. Every marketing brochure, every position paper, every salesman talks of it, and has some verbiage about it on their PowerPoint presentations.

These messages are misleading, often to those who speak them as well as those that hear them. They have only the vaguest notion of the enterprise, or what value the enterprise might find in controls. All too often, it is implied that merely using a modern protocol like SOAP or Web Services is sufficient without thinking about what information is needed by the enterprise. The model for securing the information in these systems is rudimentary or is bolted on as an afterthought. Sometimes these new systems put good engineering at risk, threatening the reliability and safety so carefully built into those systems.

What is needed is a clear understanding of enterprise architectures, and the requirements of the protocols underneath them. These architectures require common standards to develop classifications (taxonomies) that deal with the big picture rather than the minute details. They must also have a refined (nuanced) concept of security that interacts well with the enterprise. In other words, these systems must be ready to be participants in an enterprise Service Oriented Architecture (SOA).

Today's discipline of Building Automation Systems (BAS) deals only with the Building Automation Control Systems (BACS). The language, and methods, and design are all best suited to isolated control systems installed by controls engineers and maintained by facilities personnel. The discipline is control-centric and BACS requires practitioners who are trained in the intricacies controls and the physical nature of the systems they control. There is another aspect of BAS, what I call Building Automation for the Enterprise (BAE) which completes BAS, and makes interaction with the BACS safe for the enterprise programmer, one who is trained in business process and enterprise protocols rather than in control systems and building operations.

Today's enterprise systems are assembled from coherent modular systems. These systems have well defined interfaces that have abstracted their internal operations to suit the needs of the enterprise. This abstraction serves three strong purposes. (1) It exposes the internal functions of each system in ways that make sense to the enterprise. (2) It hides the inner complexity of the systems to protect the inner operations of the systems from inappropriate interference at the enterprise level. (3) It enables different systems to be competitively swapped out at as modules, allowing competition to drive innovation and competence within each system.

To support BAE, BAS must develop in ways that other business functions have already. It must develop an abstract lexicon, wherein intricate interactions necessary to BACS are described in terms of higher-level business functionality. It must embrace objects, which provide defined interfaces while hiding their inner complexity, both to simplify higher level programming and to protect complex operations from meddling. BAE must support the protocols of the enterprise, complete with standard taxonomies, or ways of talking about the operations, that persist across the control system life-cycle. When BAE has developed, the full value of control system operations will be made available to the Enterprise. Then and only then will the BAS function be ready for the Enterprise.

THE ENTERPRISE

The Enterprise is, quite simply, the normal business of a corporation or agency. Management guru Peter Drucker once observed that no one should work in a role for a corporation from which one could never rise to be an executive of that corporation. Any employee in such a position would never find his work valued. Drucker said that such activities should be outsourced, and he further noted that this would only improve life for that employee who would spend his working life in an enterprise that valued his work, and whose core values supported him in that work. For example, John Aker, CEO of IBM in the '70s and '80s started his career on the IBM loading dock.) What this means to me is that the Enterprise identifies all the core activities from which a business derives its value.

On the edge of the Enterprise are what I call "hygiene processes." These processes are ones that are needed to keep the Enterprise operating smoothly, such as janitorial services, security, or communications, but they are not part of the core mission of the Enterprise. No one is hired for their personal hygiene, although people have been fired for neglecting it. A cottage in the country can get by with an outhouse; but a town that does not upgrade its hygiene requirements as it grows smells bad. Hygiene processes must simply be dealt with, and their costs minimized. We rarely describe people, systems, or companies in terms of their hygiene unless that hygiene is sub-standard. As an enterprise grows in sophistication, it requires that those hygiene processes be upgraded to meet the expanding needs of the enterprise.

Today, BAS is merely a hygiene process. No one likes to think about it. The only time that the Enterprise thinks hard about BAS is when it fails. There are low expectations for strategic value of BAS. Rarely does the quality of service (QOS) of the BAS get discussed in the Enterprise. Rarely does the Enterprise think of building control systems at all except when the elevator jars the tenant, the too-hot/too-cold call comes in, or the security system fails.

Systems that do not interact with others in the Enterprise are often referred to as process silos. A silo has thick walls to prevent outside influences (insects, weather and livestock) silo from getting in. A process silo goes straight up and down, not interacting with any layers of the organization; a silo stands alone, a monolith on the farm, but somehow not part of the farm. Figure 37-1 shows business process silos in an organization prior to the integration of across silos. Note that there are no interactions between these silos.

At one time, each of the process silos shown in Figure 37-1 might have been considered a hygiene process. When the cost of carrying capital was less significant, the primary purpose of inventory control was to make sure that manufacturing didn't stop. Some companies treated workers as interchangeable parts with no strategic function. Salesmen kept their information to themselves as long as they made their quotas. These functions were what the company had to do, not what it "did."

These hygiene processes took direction only from the top of the corporation. Like the farm silo, they were tall and thin, with well-defined walls that preclude any interaction with other business processes. The narrative of the last 40 years of information technology is a tale of hygiene processes moving to the enterprise, by opening up interactions outside their silo of operations. Figure 37-2 shows how the communication between processes has integrated the Enterprise.

BAS has yet to interoperate with the enterprise. It still remains isolated, with thick walls of process, or terminology, and often of networking infrastructure in place between BAS and the Enterprise.

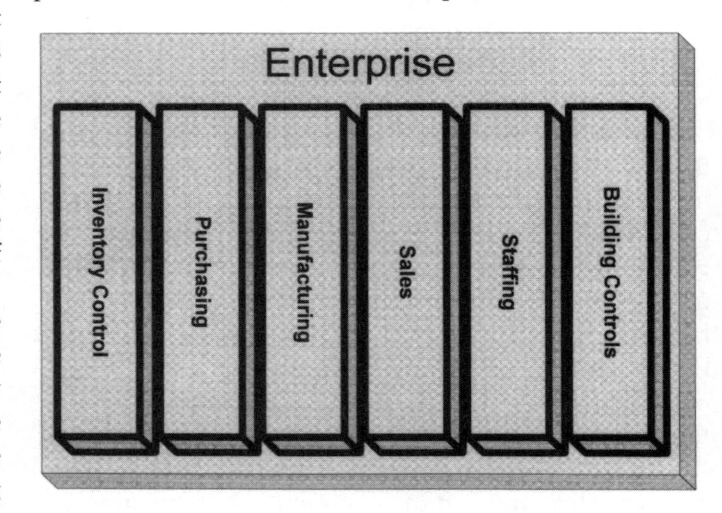

Figure 37-1. Isolated Process Silos

Today, BAS, with a few exceptions such as verified environment systems for pharmaceutical operations, are treated as little more than hygiene processes. As such, their operations are relegated to the dankest room in the back office, with little respect and outsourced if at all possible. BAS are noticed only when they fail (and perhaps by the odor). BAS are certainly not managed as a strategic enterprise function.

But what can control systems provide the enterprise?

The simplest, most elegant answer I know is by example. There is no control whose complete operation is simpler to understand than that of the black mat at the store to open the door for each customer when it is trod upon. One large retailer has enterprise-enabled its door openers to provide live foot-traffic information to its sales staffing operations. This information is also used by its advertising and marketing groups to analyze the direct effects of their activities. A simple convenience control with very little smarts is now a corporate asset to three separate enterprise activities.

A LITTLE HISTORY LESSON

When a business process is first computerized, little more is attempted than some efficiency in doing the same old thing. Paper time sheets and accounting systems were little different from the computerized payroll and accounting systems that followed them. In Boston, where I began my career, the downtown streets are laid out in the same confusion as the original settlement. This process is called "paving the cow paths."

For me, professional computing began with automated purchasing systems in the late 70's. The systems I wrote logged purchase orders, printed invoices, and worked the factory stock room. Some paper handling was eliminated, but even more paper was printed since the printed receiving documents were now filed. Some efficiencies were gained, but no changes were made to the core business processes; and this core process did not interact with any other business process. We stayed in the silo and continued to pave the cow paths.

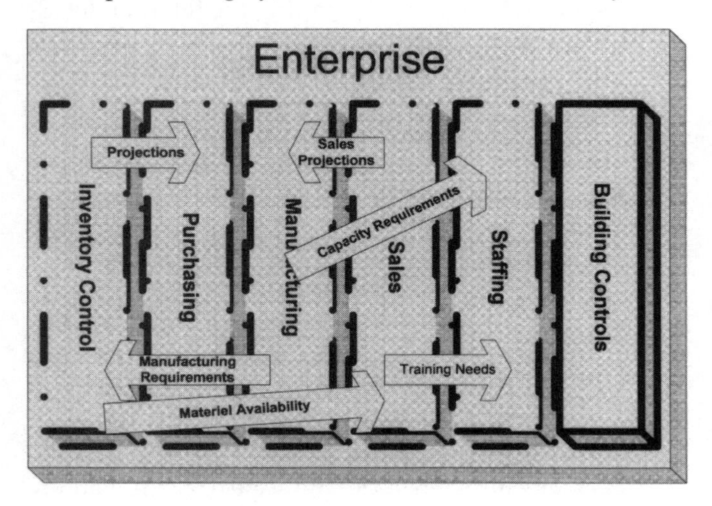

Figure 37-2. The integrated Enterprise

During the eighties, these purchasing systems were expanded, exploding the silo. Detailed work flows and manufacturing schedules dictated exactly what was purchased and when. Off-shore suppliers built materials to meet detailed sales projections and deliver them to the warehouse just as they were needed for shipment to customers. Material Requirements Planning (MRP) grew through several iterations, including more and more processes (MRP II), until we have today's Enterprise Requirements Planning (ERP), encompassing almost all of the core needs and processes of the enterprise. Each activity—sales, manufacturing, distribution, personnel, and fulfillment—interacts with all the others.

To me, BAS remains, for the most part, in the first phase. A janitor used to turn on the furnace in the basement—and tenants on each floor could turn valves on their radiators. We have made that process more efficient, by installing control systems to replace those turncocks, and bringing the set-points and decisions to a man in the control room. BACS is more efficient, perhaps more precise, but essentially unchanged. We have only paved the cow path. We have automated rather than innovated; the process is more efficient. But BAS has not brought any real change to the process; BAS remains BACS.

DOES THE ENTERPRISE NEED BAS?

Health and safety concerns, Green Buildings, and sustainability are beginning to make BACS a strategic asset. Qualities of Services (QOS) requirements are increasingly being written into leases, not only for comfort control but also for power management, for business intelligence, and even for support of new areas of enterprise regulatory requirements.

New regulatory mandates bring new requirements to building automation systems. For example, medical facilities are being required to protect patient information as never before under the Health Insurance Portability and Accountability Act (HIPPA). This means that access to records must be carefully monitored. This can include limiting unsupervised access to records areas, say on weekends or evenings. This requirement naturally drives medical facilities from simple key access to automated access control, able to limit employee access outside of their work shift and to automatically log that access.

Providing access to user-friendly information enables people to make better decisions. At present, much of the information available about our building and utility systems is generated for a specific user group focused on a narrow area of interest. The data are provided at a level of detail, and in a terminology, that is not useful to enterprise audiences. This unnecessarily visible complexity actually reduces the utility of the system.

Putting in place a monitoring and information system that expands opportunities for access and analysis would improve communication and decision making across a range of operational and academic departments and users.

Current attempts at data gathering often create winners and losers. One silo is made more robust and credible while another is weakened. Providing a data gathering network that expands choices and possibilities for all who might be interested in putting the information to productive use is logical and elegant. Just as the internet has made quick access to ever more detailed and credible information available to anybody with interactive connectivity, an improved campus communication network is needed to better share information among diverse parties interested in improving their decision-making capabilities.

A building may have a card-access control system to control who enters each area. A simple relay may trigger the HVAC system from the access control system. There may be a machine-readable electric meter to track energy use (and thereby cost). If these functions were abstracted, and able to fully interact with each other as well as other enterprise functions, such as Human Resources and Accounting, then the card swipe that opens the door would not only turn on the air handler but could also bill the energy use to the appropriate department, facilitate staff scheduling decisions, reduce peak demand, and improve occupant comfort and productivity. This would create interaction between silos (processes), and would build new highways rather than paving the old cow-paths.

ENTERPRISE-READY SYSTEMS ARE ABSTRACT

Interaction with today's control systems requires too much knowledge of the inner workings of the control system. A control protocol like LON or BACnet may tell you that an actuator is open (these data are called tags); the Enterprise only needs to know whether the system is turned on. Another tag may report temperature as milliamps currently going through a probe. Unless that data are converted to a temperature in degrees Celsius with a known accuracy, the information is useless to the enterprise.

While a mechanic or engineer might be interested to know the fan speed at a particular air handling unit, that data will not facilitate metering, billing, occupant scheduling, demand management, utility expansion plans, or fiscal accountability. The details of the internal operations must be hidden.

You would not want an employee to whom you had to give detailed instructions on every little task every time. Still less would you want to go for a walk with a friend if you had to instruct him on the proper clothes to wear, or worse, every muscle movement required to take each and every step. Instead, we "abstract" the activities of those with whom we interact. "Give me the monthly report on the first Monday of the month." "Meet me at the corner at 6:00 AM and we will go for a jog." We assume the person we are talking to will handle the details and we will see them at the appointed time. Many processes, particularly biological ones, are ones we do not wish to know about at all, unless, perhaps, if that friend is stricken ill on that run. We want those complex tasks encapsulated, hidden in a black box, with an abstract higher interface that signifies the activity desired.

Control system interfaces today are not abstract enough; and they do not hide enough. This lack of abstraction has at last five bad effects:

- Systems that are not abstract require too much of the integrator; no one can interoperate with today's systems without understanding the internal details of the system. Programmers at the enterprise level, however, do not routinely have engineering training and expertise in control systems.

- Systems that are not sufficiently abstract are able to pass data, but not information. In today's installed systems, you can find a number representing milliamps passing through an analog sensor named "temperature." Without calibration data, this number provides no information.

- An incompletely abstracted system provides information that is not fully qualified. A fully qualified temperature data point presents a number in degrees, *and* it also includes the scale (Celsius, Fahrenheit, Kelvin, etc.) used, with a known accuracy.

- Without abstraction, and the boxing of functionality that such abstraction enables, it is unsafe for a system to expose its processes to the enterprise. Such systems can only have a simple concept of security and cannot distinguish between managing set-points and managing configuration. This is dangerous not only because it enables the enterprise programmer, not trained to understand the control system, to reconfigure the system, but also because it relieves the engineer of responsibility for the outcomes of the system. Without the ability to have levels of access, we cannot allow the occupant to use the network to adjust the thermostat to change the temperature of the room where she works. Abstraction enables a security systems with different levels of access based upon user role and prevents the diffusion of responsibility.

- Without abstraction, each system integration is unique; true plug and play interoperability is not achievable. If abstraction is somehow managed during the initial system installation, the feat is not likely to be repeated during sub-system swap-out.

System functions must be encapsulated and kept safe from the enterprise. Enterprise programmers and business managers have no understanding of system configuration and calibration. If they could modify a control system's sequence of operations, they could damage the system or even endanger the occupants of a facility. Control systems should display their operations to the enterprise as a car displays its operations on a dashboard to a driver: a few simple abstract measures of speed and engine health, a few simple controls to cause the car to speed up, slow down, or turn. All other functions are made available only to the trained mechanic.

In the Toronto airport today, gate lighting is brightened, and HVAC is increased when a plane is cleared to land by the control tower. This integration occurs because the gate's control system understands enterprise protocols and exposes its complex operations as simplified abstract interfaces. The system integrator does not need to know about the details of the control tower process, about the safety margins between planes, about transponders and post-9/11 communications requirements. The integrator merely wants to know that flight 1374 is coming in. In the same way, the integrator does not want to know about sensor calibration, or loop tuning, or variable speed fans. The system integrator wants to know what the system can do and how to ask it to do it.

ENTERPRISE SYSTEMS HAVE COMMON TAXONOMIES.

With abstractions come taxonomies. When enterprise programmers speak of taxonomies, they mean hierarchically arranged classifications of information, in which meaning is abstracted in standard ways (and with standard nomenclatures) to allow interoperability and discovery. If we are going to abstract systems away from their atomic activities, we have to give things names that are well defined and based upon well-known standards. Such standards are developed using open processes, usually by consortia with representatives from all communities who need the standard.

The closest that most control systems get today to a common taxonomy is in tagging standards. For a given building, the point tags may be those used by the original owner, the contractor, the integrator, or the designer. A vigilant owner may have a detailed tagging standard in place for use in all control points. A good designer may diligently apply that standard in each and every drawing. A top-notch integrator may apply those standard tagging requirements. An especially vigilant owner will verify the application of this standard during a commissioning process for the building. To the extent that each of these activities has been consistently performed, a building may have a tagging schema that is similar to that of a biological taxonomy today.

A tag-based schema is not easily reusable for purposes other than those for which it was initially designed. A chilled water coil return tag on the third floor might look like CWR342, or 3CWR42 or CW3R42. Requesting access to all the flow meters on the third floor, or all chilled water tags in the building, or all return temperatures is difficult. This means that any interaction with those tags is arbitrary and labor intensive.

Control protocols such as LON and BACnet provide solid control-level taxa (names) for many control functions. Because the functions these taxa describe are not abstract enough for the enterprise, they do not provide the taxonomy needed by the enterprise.

ENTERPRISE PROTOCOL SOUP

Today, product sales literature for control systems and the sales forces that use them sprinkle IT acronyms across their product descriptions. While the enterprise practitioner expects a layered implementation wherein each of protocols named by the acronyms fits together, too often both the sales force and the product engineer, as well as the installer, implement only part of one of them, and claim that the entire standard has been delivered.

Real life implementations of standards-based protocols include many protocols, stacked one on top of another, to meet the need. Users may often simplify conversation by referring to only one of them. But if any of the rest fail, the whole stack falls.

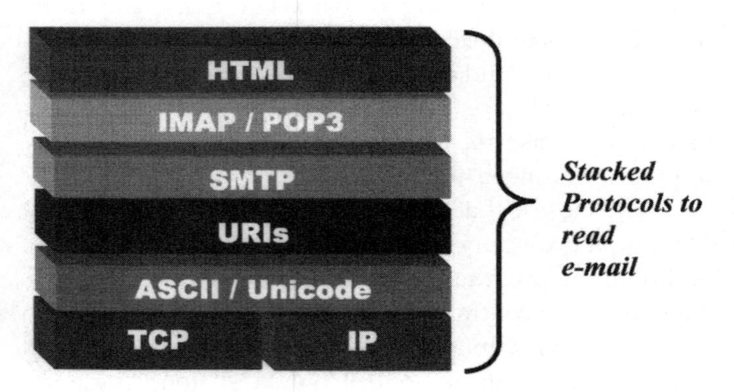

Salesmen and controls practitioners alike often wonder if XML, Web Services, and SOAP are the same thing, or they claim all are in use when only one is. These standards are merely three faces of the same standard. XML (Extended Mark-Up language) is a way to describe data in text. XML extends to data the techniques that are used to describe formatting in HTML (Hypertext Mark-up Language), the language of the World Wide Web. SOAP (Simple Object Access Protocol) is the emerging open standard for requesting and receiving self-describing XML data from point-sources of information on an as-needed basis in the same way one can find and request documents from the World Wide Web on an as-needed basis. SOAP need not be on web servers, nor need it use the well-known web protocols HTTP (Hypertext Transfer Protocol) and HTTPS (the secure form used for transfers such as credit purchases on the web). If it does, then it is known as a web service.

XML and Web Services specify only how the data looks and how to interact with distributed systems across a network. For each vertical industry and business process, consortia are defining the taxonomy of services and objects within that process. Purchase orders, invoices, personnel actions, even inter-agency work flows for governments, each have developing standard taxonomies and XML schema. If the taxonomy is understood, then a networked system can contact a service point, query that service point about its capabilities and functions, and begin interacting. This negotiation is considered characteristic of web services.

This means that by their nature, Web Services are uniquely appropriate for integrating building automation systems. Rarely are two building control systems identical, as rarely are two buildings identical. Many interests in the enterprise would benefit from access to information from building control systems. When a common XML schema is used as the representation of a common taxonomy across building control systems, then these systems can be queries about the functionality they offer as well as their current operating state. SOAP also provides us with a high-level standard for interoperability between building control systems, now isolated in individual control silos.

In the discussion below, I often refer to Scalable Vector Graphics (SVG). Every CAD drawing is a database of linear equations. SVG is an XML schema for linear equations of the type used in CAD. SVG is a graphics standard developed by the World Wide Web consortium, the standards body for core web protocols (*www.w3.org*). Because W3 is vector based, it can be manipulated like a CAD drawing. Because SVG is XML-based, it can be delivered in SOAP, as can data that are more traditional. This offers the possibility of self-describing web services that provide schematics and drawings as well as data and methods.

DISINTEGRATION OF CONTROL SYSTEMS

As this process of abstracting and "black boxing" systems becomes applied to each of the systems in a building, both traditional and non-traditional, it is no longer necessary to tightly couple the control systems. Individual building functions can be isolated behind a clean well-defined interface for interoperating with all of the enterprise processes, whether they are enterprise business functions or other BACS systems.

This will break current large expensive systems into several smaller coordinated systems, each able to be upgraded and tuned based upon its own unique operational requirements. This will lead in time to better performance of each module with its own better defined requirements, a benefit that will more than outweigh any loss of tight coupling. New types of systems can be brought into a building, then, without disturbing existing systems; all can be loosely coupled at a more abstract level.

Instead of tightly integrating control systems, we can disintegrate them. When we do so, we open the door to loose integrations across new classes of control systems.

At the University of North Carolina, I have tenant refrigeration equipment used for research. From a maintenance standpoint, I want my refrigeration mechanics to have the same operational information available from this portable tenant equipment as they do from the embedded built-in systems that are part of the building. From a day-to-day monitoring and operations viewpoint, these systems are the responsibility of the tenant (in this case the researcher who needs to control how the temperature of the samples is maintained).

In a similar way, life-safety systems need to be focused on their mission, and perform it flawlessly, without interference from the energy management systems or the access control systems. Fume hoods, for example, despite moving a lot of air, have a life-safety mission rather than an HVAC mission.

This way of thinking leads to a confederated systems model. Each system is optimized to perform its own mission, and to defend its own business and system needs. Each system is able to provide information to and interact with any number of other systems. All information exchange is abstracted, so similar functions from dissimilar systems are presented to other systems, including enterprise systems, in similar ways. The resulting systems are neither monolithic, with a single architecture, mission, and purpose, nor stand-alone, with little information exchange; they present a united front along with their allies for any external functions, and remain independent in their internal operations.

If we achieve this, it will cause a great flowering of modular control systems. Owners will be more willing to upgrade modules within the enterprise controls than they are to upgrade any part of a monolithic system. We can imagine tactical upgrades, where a package system is added just for the summer, or during a renovation, and that system will be managed by the enterprise system.

Freed from the need to span large networks, controls system engineers will be able to focus better on the simpler, more contained control system modules. This will enable a more performance-oriented market, as engineers compete to produce the best-tuned control application, confident that it will fit as a module into the larger structure.

THE INTELLIGENT BUILDING TALKS TO THE ENTERPRISE

Traditional BACS systems provide silos of interaction, summarized at the top, and locked into homogenous actors. In the past, these silos interacted only through communications at the central level. Now, we begin to see buildings filled with autonomous networked control objects, each supervised by the old silo, but able to interact through abstract communications with the other control systems in the building. BACS systems interact with Security Systems, with Access Control Systems, with Elevators, with Fume Hoods. Each is able to interact with its peers by providing and requesting abstract information.

- Someone has entered the building.
- Someone is authorized, so disable the intrusion detection.
- Someone is allowed to go to the third floor.
- Someone has left the building, leaving the widows open, so summon security.

Sensors of interest to multiple systems can be placed on nodes, and thus be available through the network to each of those systems directly.

Non-traditional sources of data such as SOAP-enabled weather stations and Green Building sensors can provide additional information outside the traditional silo of control. For example, in our research program, special purpose sensors in laboratories, tracked with SOAP, will be provide information to tenant applications that are also able to draw information directly from the BAE interface to the BACS sensors in the BAS.

Maintenance personnel with proper authorization will receive just-in-time operations data delivered directly from the building systems. This information will include XML-based schematic information delivered in SVG directly from the system controllers. These same personnel will receive identical information from tenant-owned equipment if they are responsible for that maintenance.

BAS TO THE ENTERPRISE

Today, BAS is not ready for the Enterprise. The component BACS have only achieved the first step of automating their pre-existing functions. Today's systems are not abstract enough, are not encapsulated enough, are not intelligent enough, and do not even have enough of a common taxonomy to support even the needs of the Operator, mush less the tenant or the owner. Without this, we will not get to the enterprise.

There are six discrete business processes in acquiring and operating a building: programming, design, construction, commissioning, operation, and condition assessment. Only rarely is there any relationship between performing one of these steps and performing any of the others. When we have integrated these steps, we will have created an enterprise function ready to interoperate with other enterprise activities.

Before a building is built, there are design goals, i.e., the intents for performance that should provide the basis for all steps of owning and operating a building. All too often, these goals are ignored after the initial design phase. If done properly, this information about the design goals would be codified in both an abstract machine and human-readable format that would provide the base description of the building and the structure under which all design and equipment decisions should be made.

The actual design of the systems usually pays brief lip service to the design intent, and then launches into a sequence of steps that is only rarely structured properly. Large CAD documents are produced with detailed schematics, circuit designs, sequences of operations, and tag lists. While these all appear on the same sheet of paper, they often have only the vaguest relationship to each other. The schematics and the circuit diagram are two different drawings, each made from scratch. The sequence of operations is prepared separately in a word processor and pasted into the CAD drawing. The tag list often comes from a separate spreadsheet maintained in a separate process. Close inspection of these tags and operations reveals that they were really merely copied, without correction, from the drawing for the floor above or below.

Given the quality of the design project, it is surely no surprise that construction does not match the design. I am reminded of the role of the medieval dissectionist. In medieval medical schools, the professor of anatomy would never touch a cadaver but would read from a classical text. As these texts were prepared by dissecting non-human animals, they often did not describe what was actually being found during the dissection. A dissectionist was successful to the extent that he could anticipate these problems and, at the right time, root around inside the cadaver, and pull out some other bit to match the flawed description. Some of the same skills are required of today's mechanical contractor and his controls subcontractor working from today's flawed design documents.

Following after these processes, commissioning can be a messy, expensive affair. The commissioning process begins without a clear catalog of the components and systems that should be in the building. Rarely are the tag standards of the design documents verified. This is due in equal parts to inaccuracies in the initial drawings and the lack of an automatable way to recapture this information. Even figuring out what to commission can be a daunting task.

When commissioning is performed, it is often limited to making sure that things really work, (e.g., that air is actually moving) rather than ensuring that the initial performance goals of the design intent were met. As described above, there are rarely clear descriptions remaining from the initial design intent of what performance standards are expected from each component system. Pre-construction performance models of the building systems rarely exist except for such government mandates as LEEDS or Green Buildings. Even if we can figure out what to commission, we do not know to what standard we should hold the systems.

The operation of the building from a control center (the process we usually talk about in Building Automation) is no better than it has to be. If we have a single building with a single contractor, it usually works well enough. If there is a marked deficiency in the initial design intent, or in the design, or in the construction, it will be fixed with a service order, as an operating cost. Perhaps that service order will be paid for under warranty. If the facility is part of a larger process with an integrated control system spanning multiple buildings, such as a campus or some type of complex, and even worse, multiple contractors, then a large task looms. These costs will probably be covered by operations.

Condition assessment is external to all the processes above. Underfunded colleges and universities strive to catalog unfunded maintenance deficiencies. Commercial real estate owners try to commoditize and monetize systems to rationalize their markets. But the tools are inadequate. Condition assessment may be based upon the number of "too hot, too cold" calls, or upon obvious dripping pans, or, rarely, complete failures. There are no measures of performance, whether optimal or degraded. There is little basis for analyzing systems that are drifting out of control.

It doesn't have to be this way. Design intent can be codified in machine readable form, and the performance goals for systems can be recorded. Modern CADD systems, if properly used, can produce models to predict performance of the systems under design. These models can generate all of the parts of the mechanical drawing, from schematic to sequence of operations to tag lists as views and reports of the same underlying model, intrinsically unifying the components of the design. With accurate design documents, it will be possible to build what was designed. This is the first part of the problem;

and it requires common data standards and common taxonomies across the processes.

This part, at least, is coming together, thanks to the world's largest landlord, the US General Services Agency (GSA). Beginning in fiscal 2006, the GSA will accept transmissions for the design, construction, and acquisition of buildings only in AECXML, an ISO standard accepted internationally. The taxonomy wars are over; we know what the data looks like.

AECXML is an XML-based taxonomy used to represent information in the Architecture, Engineering, and Construction (AEC) industry. AECXML was developed under the coordination of the National Institute for Building Sciences (NIBS). AECXML is based upon the Industry Foundation Classes (IFCs) data model developed by NIBS to assemble a computer-readable model of a facility that contains all the information on the parts and their relationships to be shared among project participants. The intent is to provide a means of passing a complete, thorough and accurate building data model from the computer application used by one participant to another with no loss of information.

The IFCs (and AECXML) are used to create a Building Information Model (BIM). BIMs create systematized, easily usable data storehouses for 3-D modeling to handle cost, schedule, fabrication, maintenance, energy, and other information across facility lifecycles. Current versions of the major CAD software packages can export AECXML directly from the design. Further information on the IFCs and AECXML can be found at the International Alliance for Interoperability—North America web site (*http://www.iai-na.org/*).

Closely related to AECXML is Green Building XML (GBXML), which is used to model the performance of designed systems. GBXML is a lighter-weight derivative of AECXML developed to enable interoperability between building design models and engineering analysis tools. As of 2004, GBXML generation from CAD is supported by the CADD systems from Autodesk, Graphisoft, and Bentley, eliminating the need to manually transfer information from drawing to model. The exported GBXML can be fed directly into system modeling software. Services that model building performance based upon submitted GBXML are even available over the internet today. Further information on GBXML can be found at *www.gbxml.org*.

With proper procedures, design intent can be stated in GBXML as performance requirements. The results can be described in the AECXML required for US federal contracts. With these inducements, there is no reason not to use those same models, and the tools that create them to prepare construction documents that have higher quality and are automatically internally consistent. We can do this; we have the tools and the taxonomy.

The next part, where most BAS discussions begin, should leverage this work. Commissioning, Building Operation, and Condition Assessment can each leverage this work to address the today's shortcomings. By adopting AECXML taxonomy across the board in the operation of buildings, these three processes become linked with each other. By using performance metrics compatible with GBXML performance modeling, we can verify our processes and systems.

Commissioning should be based upon an inventory that is derived directly from the AECXML inventory taken during the construction process. Because the data are in a standard format, the market place of tools to assist in commissioning will mature. But this is still only the first step, even if it is better than is often done today. Those same tools will be able to read the original design intents, moving commissioning to a higher level.

The first level is seeing if the actuators actually respond when asked, and in the way expected. The next level will be when we automate the answer to the more interesting question; do the systems perform as desired, in conformance with the original design intent? This is the type of abstract question, dealing with performance of function rather than mere movement of parts that is beginning to be more interesting to the enterprise.

The BAE Operations center, then, will run systems based upon the same data structures. The inventory of control points will come from those same data structures, with knowledge derived directly from the commissioning process. The sequence of operations will come from the original design models, as described in AECXML and GBXML. This defines the path to integration of campuses or building portfolios.

Because the modeled performance of the systems is known to and intrinsic to the operations center, tools will be developed to compare the current operations of a building to its original operation as documented during commissioning as well as the original design intent. The current condition of systems then becomes the known equipment (from the original inventory) operating within a known variance of its design specifications. This will lead to standard formulas for assessing the current state of building systems and to predictive systems to guide BACS replacement.

This condition assessment flows from the operations of the building, using known and standard data structures which can be modeled and remodeled. This leads directly to continuous commissioning, wherein the operational status of the building systems is verified and re-verified as a natural outgrowth of normal operations.

Standardizing the assessment of building systems, their performance, and their current state will lead to formulas for monetizing (or quantifying the benefits of) the building control systems. While monetized value is what drives commercial real estate, particularly the REIT (Real Estate Investment Trust) world, it is outside the scope of this discussion. Suffice it to say that once we can monetize the control systems, then control systems will be an enterprise asset.

When we accomplish this integration of the processes of the building systems, we will have created an Enterprise function of Building Automation. This function will have described its internal processes and operations through a common taxonomy, hidden its internal complexity and heterogeneity through abstraction and boxing, and defined its performance metrics through the understanding built by modeling and tuned by performance. This new Building Automation function, now an Enterprise Activity, will be ready to talk to the Enterprise.

EXAMPLES AND POTENTIAL USE CASES

The provost's office sets up class schedules and space assignments and the building automation systems automatically schedule the appropriate spaces for occupancy

1. A research building occupant can access real-time energy use data to take advantage of time-of-day energy rates when running major pieces of equipment

2. A secretary in a conference facility can have a desktop application that enables room scheduling and set-point control for the entire conference facility without any interaction with BACS staff

3. A professor in the School of Business can assign undergraduate research that easily makes use of near real-time data from energy-producing and consuming equipment to develop business models

4. Load profile modeling can provide decision support for configuration and operation of energy-producing utilities such as cogeneration and chilled water distribution systems

5. Hourly electric rate structures analyzed in conjunction with weather models and load profiles enable optimization of combined cycle cogeneration plants and thermal storage systems

6. HVAC analytical programs that rely on historical data for trend analysis and comparison with design models support maintenance, renewal and replacement programs and allow HVAC system optimization using rules-based analysis

7. Construction documents generated using XML and SVG are able to become part of the real-time BACS operator interface after project completion

8. GIS utility maps make use of real-time metering data to identify possible utility delivery problems before customers are impacted

9. Utility billing systems are able to make use of multiple rate structures to allow customers to choose which best suits their operation without additional staff requirements since historical metering data can be automatically integrated into the bill

10. Indoor Air Quality information (both historical and real time) can be accessed directly by Environmental Health and Safety personnel to support healthy buildings and identify unhealthy ones—without the support of BACS staff

12. A maintenance management system that interacts with a building control system can perform predictive maintenance and support run time maintenance models for complex systems. Such a system might track when a process is drifting out of control rather than merely noting a parameter outside its limits. If necessary, the system might generate work requests that are based on real time information. This same system will interact with enterprise functions for materials management to ensure parts and tools required for scheduled maintenance are available.

WHERE DO WE GO FROM HERE?

To become a vital part of the enterprise, to drive market growth and faster replacement cycles, and to provide full value to the owners, operators, and tenants, Building Automation Control Systems need to follow the path of other business support systems and remodel themselves as enterprise functions. To become an enterprise function, BACS must not only accept the technology standards of the enterprise, but also the systems approaches that enable those functions to be full enterprise players. Building operators who fail to heed this call will find themselves increasingly irrelevant to the day-to-day operations of the enterprise.

Today's enterprise systems are assembled from modular coherent systems. BAS must do the same. BAS must have well-defined interfaces that have abstracted their internal operations to a level that suits the needs of the enterprise. This abstraction must expose the internal functions of each in ways that make sense to the enterprise. These systems must hide the inner complexity of the systems to protect the inner operations of the systems from inappropriate interference at the enterprise level. The abstraction must enable different systems to be competitively swapped out at as modules, allowing competition to drive innovation and competence within each system.

The architecture of this integration, today, is called Service Oriented Architecture—it will have other names in the future. The protocol for integrating the enterprise today is the Simple Object Access Protocol (SOAP), which because it is message oriented, standards-based, and self describing, is uniquely positioned for spanning large heterogeneous systems touching every aspect of the enterprise. SOAP messages are encoded in XML. This architecture and protocol will drive the next generation of systems. Protocols and architectures are not enough; BACS must move beyond automation to innovation.

Building operation is just a small part of the overall function of facilities within an enterprise. Building Operation should leverage the taxonomies of design, construction, and contract to provide an open, self-managing asset to the enterprise throughout its life cycle. These taxonomies will provide the framework for abstracting building operations. This abstraction will enable automation of important enterprise functions such as building commissioning and condition assessment.

These common taxonomies and methods overlaid across the underlying technologies will free us up for better and easier integration across building systems. This will reduce the effort spent today on integrating disparate systems that are forced to be too intimate with each other's details. Building systems will become testable provable components working within the larger BACS realm. Such systems can be more easily replaced and upgraded without breaking the whole, thus shortening control system replacement cycles. The new systems will then compete more efficiently in performance.

By becoming abstract systems operating across the building life-cycle as part of an integrated enterprise function, BACS will intrinsically be ready to participate in overall enterprise functions. They will support bi-directional information sharing not only within the BACS realm, but also with non-traditional control and sensor systems as well as with diverse enterprise systems.

Systems providing this level of functionality will be worth upgrading to sooner than today's automated systems. The practitioner who understands this type of system and who can provide this level of functionality will be able to supply higher value to the enterprise than his competitor. This will drive innovation across building control systems to shorten life-cycles and expand market at a time when the underlying controls will increasingly become ever cheaper commodities. For the building system professional, this cannot be ignored.

Chapter 38

Why Can't a Building Think Like a Car? Information and Control Systems Opportunities in New Buildings

Barney L. Capehart, University of Florida
Lynne C. Capehart, Consultant

ABSTRACT

THIS CHAPTER examines the information and control technology used in new vehicles and points out the potential for using similar information and control technology in new buildings. The authors draw on their knowledge of new cars and new buildings to present a list of information and control functions, together with the available sensors, computers, controls and displays used in new cars that can provide significant opportunities for our new buildings. Methods for integrating this new technology into new buildings are also discussed. The use of information and control technology in new cars should serve as a model for new building technology. This potential for new buildings should be recognized, and similar technological improvements should be implemented.

INTRODUCTION

A great deal of new technology is available for buildings. The labels "Smart Buildings" and "Intelligent Buildings" have been around for years. Unfortunately, this wealth of new technology for buildings only exists in pieces and as products from many different companies; virtually no building constructed today utilizes a significant amount of this new technology. Most new buildings operate just like the buildings of the 1970's. Even though new materials, new design and construction methods, and new ASHRAE building codes have greatly improved new buildings, these buildings still look and function much as they did twenty years ago. While most new buildings do have new equipment and better insulation, there is little in the way of new controls and display technology for the building occupants to see and use. Individuals seldom have the ability to control personal comfort and preferences.

In contrast, every new automobile—regardless of its price—is filled with new technology compared to the automobile of the 1970's. A new car typically comes with about fifty separate computers or microprocessors, has around forty to fifty sensors, and provides about twenty electronic display and control functions. It does this for as little as $20,000. This automotive information and control system commonly requires little or no maintenance or repair for a period of three to five years. The technology is often visible, it can be used by the driver and passengers, it is generally standard on all new cars, and it is inexpensive and reliable. There is much fancier technology available if you want to pay for it (Lincoln Navigators, 7-Series BMWs and S-Class Mercedes have around 100 processors on-board), but the majority of new automotive technology is found on every new car.

With all this new technology, today's cars are much more reliable and have significantly reduced maintenance requirements. In the 1970s, an automobile needed a tune up every 10,000 miles. Today, a typical new car does not need a tune up for 100,000 miles. Older cars needed new brakes about every 20,000 miles. Now it's every 50,000 miles. The authors bought a new mini van in 1998, and did not have to take it back to the dealer for any service for 40,000 miles! The vehicle had several oil changes in that period, but it needed no mechanical or electrical work.

In comparison, our buildings need maintenance people from the moment we start using them. We're not talking about janitorial work, but about maintenance of lights, air conditioners, switches, controls, doors and windows. This is like the old days with our cars when we started making a list of things to be fixed as soon as we drove the car off the dealer's lot. We are paying extra for building commissioning just to make sure everything in the building is operating correctly and is fixed if it is not. Why can't a new building operate for six months, a year, or even several years without needing any maintenance? Our cars do.

What is the potential for using reliable, comprehensive, integrated, and inexpensive components in our new buildings to create a transparent and efficient information and control system? And what should we do in terms of buying new buildings? Clearly, progress in adapting and implementing technology for new buildings has a long way to go. Nonetheless, we should demand more technology—a lot more. Technological improvements should be standard features that come with every new building without question rather than options that add significant cost to the building. The only question should be where do we draw the line between standard features and additional new technology that we will pay extra for?

FEATURES OF AUTOMOBILES THAT WE COULD USE IN BUILDINGS

Individual Control Systems

One of the most noticeable features of new automobile technology is how it provides the driver and often the passengers with individual control systems. Compared to a building, a new car has far more sensors, controls and displays for a much smaller space. There are individually controllable air supplies for the driver and the front passenger. Large vehicles often have air controls for the rear seat passengers too. Temperature ranges for heating or air conditioning are individually controllable, often for the front passenger as well as the driver. The air velocity is controllable with a multi-speed fan. The outlet vents are easily reached and can be moved to direct the airflow onto or away from the person. The amount of outside air can be controlled by selecting fresh air or recirculation. Some lights such as headlights and interior dome lights are activated by sensors. Other lights are individually controllable. The driver or the passenger can turn on selected interior lights, can often dim these lights, and can direct the light to the area where it is needed. The moon roof can be opened or closed by the driver or front passenger. Both front seats are individually adjustable for horizontal position, height, tilt, and back support; and many are heated, too. In addition, in some cars, these individual settings or preferences for functions like HVAC and seat positions are provided through a memory setting tied to an electronic key, and settings for more than one person can be stored in memory.

Compare this technology to the control systems currently available in a common new building. A typical room in a new building may have a thermostat with a control setpoint and a temperature display at that location. It also usually has an unseen VAV control function, and in a few instances a humidistat with a setpoint control and a display of the relative humidity at that location. Lighting is controlled with a single light switch or possibly a single occupancy sensor for lighting. Otherwise, the occupants usually have no other sensors, controls or displays in that room.

An example of a new technology that is currently available and that achieves some of the goals of individual control over personal space within a building comes from Johnson Controls. Their Personal Environments system is an easy-to-use, desktop control unit that gives each person the flexibility to adjust temperature, lighting, air flow and acoustic characteristics as often as necessary to maintain personal comfort levels. Individuals can adjust the air temperature and control the amount and direction of air flow at their desktop. They have a heating panel under the desk to adjust the temperature to their legs and feet. The Personal Environments system also allows an individual to control task lighting and to mask background noise. The system has a sensor that turns off all functions when the workstation is unoccupied for more than 10 to 15 minutes. Although this system is being used by a number of companies, it is the exception rather than the rule.

Operational Controls

In addition to personal comfort controls, the new car also has a large number of automatic control systems to optimize and control its own operation. Engine control systems insure fuel efficiency and reduce air pollutants from the combustion process. Sensors for inlet air temperature and relative humidity allow optimum fuel flow control and optimum combustion. System computer modules also control the ABS, transmission, cruise control, and body controller. These microprocessor systems are standard on new vehicles, but new buildings are not built the same way. Operational controls are available for new buildings, but they require special design criteria. No one considers the possibility that they should be standard equipment.

Display Systems

New cars tell the owner about much of the maintenance and repair that needs to be done, and certainly notify the driver whenever one of the major systems is in need of attention. A new car has sensors that report tire pressure, unclosed doors, lights or other controls left on, unfastened seat belts, brake fluid status, and many other operational features related to the safety of the car and the occupants. Even a cursory comparison shows that our new buildings lag very far behind the present use of technology in new cars.

Much of the information on car maintenance and safety is aimed at the driver. What comparable information does a building operator get about the maintenance needs of the building or the various rooms in a building? Things that would be helpful to know include whether the air handling system filters are dirty, whether the refrigerant is at the proper level, whether sensors are working properly, whether lights are burned out, or whether the doors have been left open.

The present system in buildings is essentially a manual system. Filters are checked by maintenance personnel on a time schedule. Maintenance workers often depend on "human" sensors to notify them of burned-out lights, improperly functioning photosensors, or temperature problems in individual rooms.

Options

New cars have options, and new buildings have options—but these mean very different things. An option for a new car is an item or function that is already available and can be installed on the car, but at extra cost. For a building, an option is an item or function that an owner wants to add at extra cost, but expensive additional design, engineering integration and testing work must usually be performed before it can be installed and operated.

Table 38-1 summarizes many of the sensor control and display functions of new cars, and provides a model for desired technology in new buildings.

HOW DID THE AUTOMOTIVE INDUSTRY DO THIS?

We must understand how new automobiles can have so much new technology at such a low cost, and why they are so reliable in order to know how to utilize similar innovations in the building industry.

Engineering Analysis and Design

A significant amount of engineering analysis and design goes into both the structural and operational features of a new car. In addition, significant engineering analysis and design also goes into the manufacturing and production processes for assembling the new cars. A major benefit of this approach is that the car's entire system and subsystems, as well as each of the car's components, are carefully engineered. For example, the electrical power consumption of the components and systems in a new car are carefully analyzed, built and selected to make sure that the total power demand is not greater than the capacity of the electrical power supply system, i.e., the 12-volt battery. Thus, with cars, the need for energy efficient electrical systems is built in from the start.

When a building is designed, the electrical load is specified first, and then a power supply system is specified that is big enough to handle the load of the building. Little or no thought is given to minimizing the electrical load itself because there are generally no constraints on the amount of power a utility will supply to the building.

Overall Quality Control Programs

A new car is reliable because a significant amount of engineering goes into both the car design and its manufacturing process. Quality and quality control start with the engineering design, and are strongly emphasized throughout the manufacturing and assembly of the car. Individual components are designed and made with quality and reliability as major goals. Subsystems and final systems—including the entire car—are similarly produced. Ordinary and accelerated life testing are conducted on the car's components, subsystems and systems. These extensive tests include the effects of temperature, moisture, mechanical and thermal stress, and other factors. As a result, most of the car's components and systems will last at least three years or 36,000 miles. Warranties on some new cars are now available for seven years or 70,000 miles.

Quality control and warranties in building design and construction are very different. Auto manufacturers provide the warranty for the entire vehicle (with the possible exception of the tires); the systems in new buildings are likely to be under several different warranties. HVAC manufacturers cover the HVAC system; flooring manufacturers guarantee the carpet/flooring; plumbing manufacturers guarantee plumbing fixtures; etc. There is usually no centralized quality control or warranty for a new building as there is with cars.

Widespread Use of Microprocessors and Computers

Much of the technology and operational features of our new cars comes from the use of microprocessors and microcomputers. A new car may have as many as 50

Table 38-1. Sensor, Control and Display Comparison for Cars and Buildings

S=Sensor C=Control DI=Display
D=Driver FP=Front Passenger RP=Rear Passengers
CBO=Controllable by Occupant NCBO=Not CBO

Function	Cars			Buildings	
	All	Mid-cost	Luxury	All	Some
I. Comfort and convenience					
A. Climate Control (HVAC)				Yes	
1. Zone of Control					
Single Zone	D	D	D	Few	
Dual Zone		D, FP	D, FP	Very Few	
Multi-Zone			D, FP, RP	Most	Individual Zone, CBO
2. Temperature				Yes	
Lever setting (C, DI)	D	D, FP	D, FP	Some	
Thermostat (S, C, DI)	–	D, FP	D, FP	1 per zone	Individual Zone, CBO
3. Air Supply					
Directional Vents (C, DI)	D, FP	D, FP, RP	D, FP, RP	Partial	
Multi-Speed Fan (C, DI)	D	D, FP	D, FP, RP	No	
Ventilation (S, C, DI)	D	D, FP	D, FP	Yes, NCBO	
Recirculation (C, DI)	D	D	D	Yes, NCBO	
4. Humidity (S, C)	–	–	Yes	Some, NCBO	Yes, CBO
5. Air Quality (S, C) (CO, NO_2)	–	–	Yes	Yes, NCBO	Yes, On/Off
6. Advanced Features					
Reheat Operation			Some	Most, NCBO	Yes, CBO
Window Fog Control			Some	No	
Air Filters			Some	Yes	
Sun Sensors			Some	No	
B. Seating BT=Back Tilt					
1. Basic – Mech or Elec (C)	D,	D, FP	D, FP, RP	Yes	–
2. Horiz Position + Back Tilt (BT)	FP		(BT)		
3. Six-Way (C)	–	D, FP	D, FP		Yes
4. Back Support (C)	–	D, FP	D, FP		Yes
5. Heated (C)	–	D	D, FP		No
6. Memory Function (C)		D	D, FP	No	No
C. Visual (Inside Lighting)					
1. Dome Light (C)	Yes	Yes	Yes	Yes	
With Occupancy Sensor (S,C)	Yes	Yes	Yes	No	Yes
Delayed Dimming (S, C)	No	Yes	Yes	No	No
2. Overhead/Task (C)	No	Yes	Yes	Some	Yes
Directional (C)				Few	Yes
3. General					
Door, Glove Box (S, C)	Yes	Yes	Yes	No	?
Visor, Map (C)	No	Yes	Yes	No	?
D. Windows					
Power Windows	No	Yes	Yes	No	
Power Sunroof	No	Yes	Yes	No	Yes
II. Normal Operation					
A. Speedometer (S, D)	Yes	Yes	Yes	No	*
Cruise Control (S, C, DI)	No	Yes	Yes	No	*
B. Odometer (S, C, DI)	Yes	Yes	Yes	No	*
C. Tachometer (S, DI)	No	Yes	Yes	No	*
D. Fuel (S, DI)	Yes	Yes	Yes	No	*

Table 38-1. (*Continued*)

III.Safety and Maintenance					
A. Engine					
1. Engine Control Module (S, C)	Yes	Yes	Yes		
Temperature, Oil Pressure, Check Engine					
Status Light (S, DI)	Yes	Yes	Yes		
Gauges (S, DI)	No	Yes	Yes		
B. Auxiliary Systems					
1. Electrical					
Generator (Charge)					
Status Light (S, DI)	Yes	Yes	Yes		
Voltage Gauge (S, DI)	No	Yes	Yes		
Lights					
Headlights (C)	Yes	Yes	Yes		
Backup Lights	Yes	Yes	Yes		
2. Brakes					
ABS (S,C)	No	Yes	Yes		
Status Light (S, DI)	Yes	Yes	Yes		
Brake Light Out (S, DI)	No	No	Yes		
3. Others					
Seat Belt (S, DI)	Yes	Yes	Yes		
Turn Signals On (S, DI)	No	Yes	Yes		
Headlights On (S, DI)	No	Yes	Yes		
Low WW Fluid (S, DI)	No	Yes	Yes		
Door Not Closed (S, DI)	No	Yes	Yes		
Exterior Temp (S, DI)	No	Yes	Yes		
Tire Pressure Sensor					
Parking/Backup Sensor					
Automatic Day/Night Mirrors					
IV. Pleasure & Entertainment					
A. Audio System					
Radio	Yes	Yes	Yes		
Satellite Radio			YES		
CD Player	No	Yes	Yes		
B. Video System					
TV	No	No	Yes		
VCR, DVD	No	No	Yes		
C. Computer	No	No	Some		
Internet	No	No	Option		
D. Communications					
Cell phone	No	Yes	Yes		
Internet	No	No	Option		
V. Advanced Systems					
A. Navigation Systems	No	No	Yes		
B. Collision Avoidance Systems	No	No	Option		
C. Rain Sensing Wipers	No	No	Option		
D. DewPoint & Glass Temp Sensors	No	No	Option		
E. Voice Commands	No	No	Option		

separate microprocessors and 11 major computer-based systems. Some new luxury cars have up to 90 microprocessors. It is often said that a new car has more computer power in it than our first manned space capsule. Computer-based systems are found in the System Modules for new cars, and account for much of the engine performance, reduced emissions, sophisticated diagnostics, and many of our comfort and convenience features. The Engine Control Unit (ECU) is the most powerful computer in the car, and it has the demanding job of controlling fuel economy, emissions from the engine and the catalytic converter, and determining optimum ignition timing and fuel injection parameters. These computers, microprocessors and system modules greatly simplify the diagnostic job of finding problems with the car, and providing information on what kind of repair or replacement work is needed.

While a large new building with a sophisticated BAS or Building Automation System may well contain 50 or more microprocessors, this does not match the new car in terms of having equal computing power per room or per group of rooms with 2 to 4 occupants. The rooms and offices in our buildings do not have monitoring and self-diagnostic features. They could, because the technology, equipment and systems exist, but they are not supplied as a standard item, and they are not available in the same way that options are available on new cars.

System Modules

As discussed above, the System Modules are where the computer-based systems reside in new cars. These System Modules are highly complex, and highly important systems in new cars. Many of our highly desirable performance and comfort features are provided by System Modules. Typical System Modules in a new car are: the Engine Control Unit, the Instrument Panel Module, the Climate Control Module, the Transmission Control Module, the Power Distribution Box Module, the Airbag Module, the Driver's Door Module, the ABS Module, the Body Controller Module, and the Cruise Control Module. These are the System Modules on every basic car. Additional System Modules are options for lower priced cars, or standard features of higher priced cars. These include Navigation Control Modules, Entertainment System Modules, Advanced Comfort Control Modules and Communication Control Modules for computers, cell phones and internet access.

Communications Buses

Using standardized communications buses with these System Modules makes both designing and building new cars much easier than it was in the old days. Two major services must be accessible to every area of a new car—electric power and the communications bus. All of a car's System Modules must be able to communicate with each other, receive signals from most of the sensors in the car, and send signals to the control components, systems and actuators. Using a communications bus greatly simplifies the wiring, reduces the number of data sensors, and implements additional features at very little additional cost. Without the communications bus, the job of wiring up a car during the assembly operation would simply be too labor and time consuming to have a reasonable cost product. Also, the speed of communications is so important now that only a digital bus has the speed and capacity to handle the data collection and data transfer load for a new car.

The communications bus and the System Modules work together to make the design and building of the car much easier. Information is sent over the communications bus in a standard communications protocol—usually the SAE J1850 standard, or the Controller-Area Network (CAN) standard, although some manufacturers are using FlexRay, which is a faster and more sophisticated communications bus. Data are sent in packets with a standard structure—a label and some data. For example, an information packet with Speed for the label and 52.5 for the speed data in MPH is picked up by the Instrument Control Module, which refreshes the indication on the speedometer with this new data. The standard communications bus makes the design of the various System Modules much more straightforward. In addition, the sensors in the car only need to send packets of data to the communications bus; therefore, the carmaker does not have to deal with the problem of a particular sensor putting out a strange voltage or current signal that must be converted somewhere into a true physical parameter of the car's operation. In our example, the alternative is to tell the Instrument Panel Module maker that the signal for speed was going to be a 4-20 mA current loop value, and that 10 mA was equivalent to 40 MPH.

The use of the standardized communications bus also makes it easy to use outside suppliers and sources for many of the components and systems in a new car. The carmakers do not have to worry about how a specific sensor or module works internally; they only need to know that the data will be transmitted in a known, standardized manner, and that it will have a known, standardized structure. Much of the success with using modern technology in cars, and much of the reliability of that technology comes from using the simplified approach of a standardized communications bus.

This same type of technology is essentially available for our new buildings. BACnet, LONWorks, and TCP/IP are the most common standard communication protocols. TCP/IP may be the ultimate answer, but another level of standardization is also needed to insure that data that comes across TCP/IP means the same thing to each different piece of equipment in a facility. Most buildings are being wired for a Local Area Network (LAN) with either coaxial cable or fiber optic cable. Thus, the hardware and software are available, but there is no organization responsible for requiring or enforcing the standardized interconnection of all of the building components, subsystems and systems like the automakers have. Without a standardized communications bus running through the entire facility—together with accessible electric power—buildings will never have the kind of technology that cars have, and we will never have the cost benefit or the reliability that this kind of technology can bring to our buildings.

Smart Sensors

Most of the basic automobile sensors that were used in the past to read continuous physical parameters such as temperatures, pressures, flows and levels operated on the principle of producing a voltage or current output proportional to the real value of the parameter. The output of these sensors was almost always nonlinear, and also varied with the temperature or other physical parameters. This resulted in poor measurements, or required using more equipment and processing power to correct the sensor reading for the nonlinearity and to provide temperature compensation curves to get accurate readings. Today, smart sensors are used to provide these functions and to output data to a microprocessor or System Module. The sensor output is input to the microprocessor, and the sensor reading is digitized, corrected, temperature compensated and sent out over the standardized communications bus.

These smart sensors interface directly to the communications bus, and provide fast and accurate measurements. Since the sensor package contains a microprocessor, much of the load is removed from the System Module that the smart sensor is supporting. Designed and built as an integrated package, the smart sensor fulfills its mission reliably with a low initial cost.

The sensors for buildings are expensive, and many of them are not very reliable. They are certainly not reliable in comparison to those in cars. In particular, the relative humidity sensors and CO_2 sensors are notoriously unreliable, and require frequent cleaning, calibration and general maintenance. That level of performance would be unacceptable for these sensors in a car. Why shouldn't the sensors in buildings work reliably for a period of three to five years before they need any significant attention?

Wiring Harnesses and Standard Connectors

The use of pre-assembled wiring harnesses and standard connectors has made the task of wiring up a new car much easier. It is important to use stranded, not solid, wire cable. Each length of stranded wire consists of a twisted bundle of very thin thread-like wires. Solid wire, on the other hand, is a single thick wire segment. The advantage of stranded wire is that it is much more flexible than solid wire, and also less susceptible to breakage. One thread of a stranded wire can break without affecting the performance of the connection, but if a solid wire breaks the connection is lost. Also, if there is one weak link in the reliable performance of any electrical or electronic system, it is the connectors. With this in mind, the importance of carefully and correctly built standardized connectors cannot be overemphasized.

Use of Skilled Assembly Workers

The auto industry has a large supply of skilled workers at its design, engineering and assembly operations. These skilled workers receive training in their specific jobs as well as training in quality control and process improvement techniques. Many of the manufacturing and design improvements in new cars have come from the production workers themselves. In addition, skilled workers have made a great improvement in the overall reliability and quality of the new cars. Auto workers are usually paid more than those working in other industries or services.

Problems with the construction of new buildings often come from the use of workers with minimal or insufficient skills for the job. Finding skilled workers may be difficult, and is certainly expensive. The nature of building construction often impedes the retention of skilled workers. As a result there may not be a large pool of highly qualified building construction workers available when a particular building is being built.

One of the most common problems in building structures is the roof, which is the subject of the greatest number of lawsuits in building construction. Most roofs leak, and leak from the day the building is occupied. Roof leaks are the result of poor installation and construction rather than problems with roofing technology and materials. When a roof leaks, water leaks into the walls and may not be noticed until mildew and rot are visible; by then the building may be significantly damaged. Mold, mildew and IAQ problems in the building will require more time and money to fix. Using sensors in new buildings to identify roof and wall leaks when they occur is a critical application of automotive type technology in our new buildings. New cars use infrared reflectance sensors to identify rainfall on windshields, and automatically start up the windshield wipers. These sensors, or other types of moisture sensors, if installed throughout our new buildings, would quickly identify leaks and moisture buildup and alert building operational people to this serious problem.

Poor workmanship can cause many other problems in buildings. Even the HVAC system can be affected since random testing has shown that many air conditioning systems are installed with an improper charge of refrigerant. In economic terms, the problem of workers with insufficient skills and quality training results in the need to commission buildings to check and see if the building components and systems work as they should. (See discussion on Commissioning below.) This expense is clearly attributable to lack of adequate engineering, lack of quality control measures, and especially lack of highly trained workers.

Why Doesn't New Building Construction Include More New Technology As Standard Equipment And Systems?

Automobiles are built according to a standard plan; building architects on the other hand reinvent the wheel each time they design another building. This lack of standardization in buildings impedes the introduction of new technology in new building construction. Other factors also influence this difference in approach.

Unlike new cars, most new buildings are site built, and are built to "cookie cutter" specifications that emphasize lowest first cost of construction.

Even "custom built" buildings are held hostage to the lowest first cost syndrome. Thousands of different construction companies build residential and commercial buildings. Hundreds of different companies build fairly large commercial buildings. These companies range in size from small businesses to major architectural and engineering firms and major construction firms. It is extremely difficult to implement standards of technology when this many individual companies are involved.

The fact that most buildings are site built impedes the assembly line and systems approach to installing new technology that is used in the auto business. One area of building construction that is immediately amenable to the assembly line approach of the carmakers is the construction of prefabricated or modular buildings. This manufacturing sector could easily incorporate the knowledge from the automotive assembly sector to produce buildings with the same level of technology and reliability as new cars. The engineering and quality control functions are much more cost effective in this sector. This sector could easily use more computers, more microprocessors, more System Modules, more Smart Sensors, and a standardized communications bus.

Cars are constructed in a factory assembly line and moved to their ultimate market and user.

The factory environment makes it easier to train workers in installing the equipment in new cars as well as training them in quality control procedures. Buildings, however, are constructed at the point of use. Construction workers may work for a long time on a single building doing all types of work. Their training is not likely to be technology specific. Auto assembly workers typically specialize in some part of the assembly process, and therefore can be trained on this more limited work task. In addition, they become quite knowledgeable on this part of the assembly operation, and soon become able to add value to the company by suggesting improved methods of designing and constructing components and systems that they assemble. Quality control is more easily stressed in this environment, and many of the workers actually see the final result of their work drive off the assembly line, which serves to positively reinforce the need for a high skill level and the need to perform high quality work. In fact, these workers often own and drive cars produced by the company they work for. They are more likely to reject poor quality parts, components, systems and assembly procedures.

More new cars are sold each year than new buildings, so there is a larger market for the technology, and the price can be reduced due to bulk purchase of the equipment.

This is certainly true at face value, but when the scale of use of technology for buildings is considered, the numerical superiority of the cars goes away. If we consider that the unit of interest in buildings is rooms, and that we are interested in having the same technology level in each room that we have in a car, we now have a very different perspective. There may very well be more rooms than cars built each year. Thus, the comparison of a room to the car, rather than a building to a car, will lead to a much greater economy of scale for new building construction, and should provide a strong economic incentive to move in this direction for buildings.

Cars have a shorter lifetime than buildings, so new technology can be introduced faster, and the customers can develop a faster appreciation for what it does.

Cars do have a shorter lifetime than buildings, but most buildings end up being refurbished, or equipment and systems retrofitted, so there is still a lot of opportunity to use new technology in older buildings. Sensors, controls, System Modules and many of the other features of new car technology can be added to older buildings when they are needed. In general, the most cost effective way to build an energy-efficient and functionally superior building is to do it right the first time, rather than retrofit it later. However, new equipment, and especially new information technology, can be added to rooms and to the entire building. It would have been easier and cheaper to install coaxial or fiber optic cable in a building when it was built, but we still have managed to find a way to get the LAN cable and connections into our rooms and offices so we could network our PCs.

Purchasers of new cars are influenced by features they have seen on other cars. Therefore, consumer demand is important in increasing the marketability of new technology options.

This is one reason we need to start installing some of this new technology in buildings. Once building owners, managers and occupants start seeing what has been done in other buildings, and how much more enjoyable and productive it is to work in buildings with this advanced technology, they will start to demand more technology as a result. It is somewhat amazing that the people who drive cars with all this new technology will go happily to work in buildings that do not come close to providing similar comfort and operational features of automobile technology!

Cars are designed for use by individuals; buildings are designed for use by companies.

The motivation of the designers and the manufacturers of cars is frequently different from that of people who design and build buildings. Car manufacturers build a car to attract a buyer; they add bells and whistles to make their car different. They encourage innovation and thinking outside the box. Architects and construction companies are building a box, so their thinking often stays in the box. They may work on the exterior design; they may make the interior appearance pleasing; but they do not think very hard about what goes on inside the box, and they don't consider the needs of the individuals living and working in the box. A car designer should consider safety when drawing up plans for a new car; beyond putting in emergency exits and sprinkler systems, a building designer may not think about how to make the building safer because that is not part of the job. Among the questions that building designers should be asking are "How can this building be made more comfortable, safer, and more user-friendly?" "How can occupants interact with this building to increase their comfort and safety levels?" "How can we make this a building of the future?" With a little imagination and an increased use of information and controls technology, building designers can make significant changes in the comfort level of the occupants.

What Does The Building Construction Industry Need To Do?

Establish an integrated design-and-build engineering and management structure. The amount of engineering work that goes into a new building must increase significantly. The building structure should be designed with high technology use in mind, and should utilize new technology to deliver the performance and comfort features that we want in our new buildings. In addition, quality control and reliability should be designed and engineered into the building from the start of the project. Then, quality management techniques should be employed so that the building is actually constructed to provide the quality and reliability features that we expect.

Use equipment and system modules in new buildings.

This approach has facilitated the use of most new technology in new cars at a reasonable cost, and with extremely good reliability. However, the standardized communications bus has made the most dramatic difference. By using a standardized communications bus and system modules, car technology could be transferred to buildings relatively easily. Individual HVAC modules for occupants, individual lighting modules, other comfort modules such as for seating, and building operation and maintenance modules could all be used to greatly increase the performance and reliability of new buildings and yet allow us to build them at reasonable costs. Certain sectors such as the residential manufactured housing sector, the hotel/motel sector, and many office buildings could easily adopt this approach.

Even site-built homes could incorporate some of these features. Residences are often pre-wired for intercoms, telephones, security systems, cable, and high-speed internet connections. Designing a central integrated system for monitoring and controlling the performance and comfort of a home and pre-wiring the house for such a system is well within the realm of feasibility. It is possible to envision a home with a central control panel that was accessible from the internet. Homeowners could monitor their homes from work. They could receive security or fire alarms. They could make changes to thermostat settings if they knew they were going to be early or late getting home. They could get alarms if there was a water leak or if the refrigerator stopped running.

Build more modular buildings.

The solutions to providing greater use of technology in new buildings and providing quality and reliable buildings are much easier for modular buildings with significant pre-site construction performed in a factory or controlled environment. High-tech components and equipment can be installed more easily in prefabricated and modular buildings within a controlled environment and with highly skilled and quality control trained workers.

Impose standards on equipment and system suppliers.

Most major construction companies are already in a position to do this. They have the financial leverage to specify components and equipment that meet their exact requirements. The residential manufactured housing sector in particular could do this quite easily. The federal sector, states and large companies also have excellent opportunities to set these standards. One of the most important standards is to require a standardized communications bus in a building with all sensors and controls interfacing directly with that communications bus.

Support codes, standards, or legislation to increase the use of new technology in buildings.

Building codes and standards have been responsible for many of the improvements in standard buildings. With minimum equipment efficiencies, minimum thermal transfer levels, and minimum structural standards in place, companies that construct buildings must meet these minimum standards—regardless of whether it increases the first cost of the building. Without minimum standards such as the ASHRAE 90.1 standard, many buildings would still have inferior equipment and poor insulation, because it was cheaper to put in initially. Other programs like LEED and EnergyStar could incorporate requirements for adding new comfort and control technology in buildings. The standards for utilizing new technology could be set voluntarily by large companies and big purchasers of buildings like the federal sector, states, schools, and the hotel/motel sector. The auto industry has certainly incorporated many of the new technological features without needing government intervention.

Integrate new building technology with the desktop computers and BAS (Building Automation Systems) that are already being installed in new buildings.

The types of smart sensors, system modules and standardized communications buses that the authors have been recommending for use in new buildings should be considered an integral part of the overall Building Automation System. All of these components, systems and equipment must work together seamlessly to provide the expected level of performance and comfort and all the desktop computers should be tied in to these systems through a Local Area Network.

The desktop computer could be the equivalent of the car dashboard or instrument panel, and it should be the personal interface to an expanded BAS. It could tell what the space temperature is and how much ventilation is being provided. It should allow occupants to set their personal preferences for lighting levels, seat positions, window or skylight openings, etc. It should also let them enter new desired values of these space parameters.

Benefits of Standardized Commissioning of Buildings

Commissioning a building is defined in ASHRAE Guideline 1—1996 as: The processes of ensuring that building systems are designed, installed, functionally tested over a full range, and capable of being operated and maintained to perform in conformity with the design intent (meaning the design requirements of the building). Commissioning starts with planning, and includes design, construction, start-up, acceptance and training, and can be applied throughout the life of the building.

Commissioning a building involves inspection, testing, measurement, and verification of all building functions and operations. It is expensive and time consuming, but it is necessary to insure that all building systems and functions operate according to the original design intent of the building. Commissioning studies on new buildings routinely find problems such as: control switches wired backwards; valves installed backwards; control setpoints incorrectly entered; time schedules entered incorrectly; bypass valves permanently open; ventilation fans wired permanently on; simultaneous heating and cooling occurring; building pressurization actually negative; incorrect lighting ballasts installed; pumps running backwards; variable speed drives bypassed; hot and cold water lines connected backwards; and control dampers permanently fully open. And this is only a short list!

The process of commissioning a building constructed like a new car, and using the new car-type technology would be far quicker and simpler, as well as much less expensive. The use of standardized components, subsystems and systems could actually eliminate the need to check and test these items each time they are used in a new building. A factory or laboratory, standardized commissioning test could well determine their acceptability with a one-time procedure. The use of a standardized communications bus would dramatically shorten the time and effort of on-site testing of the building components, subsystems and systems. Data from all sensors and controls would be accessible on the communications bus, and would allow a significant amount of automated testing of basic functions and complex control actions and responses in the building. A Commissioning Module could also be added to the building systems, and would even further automate and speed up the commissioning process. This Commissioning Module would remain as a permanent building system, and would not only aid in the initial commissioning process, but also the recommissioning process, and the continuous commissioning process.

Presently, the cost of commissioning a new building is around 2 to 5 percent of the original cost of construction. The use of Standardized Commissioning tests, and the use of a Commissioning Module, would greatly reduce this cost. Although commissioning is a cost effective process—usually having a payback time of one to two years—many building owners do not want to spend the additional money for the commissioning effort. A prevailing attitude is "I have already paid to have the job done correctly. Why should I have to be the one to pay to check to see that it has actually been done correctly?" This is a difficult attitude to overcome, and it is often a hard sell to convince new building owners that they will actually come out ahead by paying to verify that their building does work as it was designed to work.

One final note on commissioning is that from one of the author's energy audit experience. Many problems found when conducting audits of existing buildings are clearly ones where the problem has been there since the building was constructed. For example, in the audit of a newspaper publishing company it was found that the cost of air conditioning was excessive. Further checking showed that the heating coil and the cooling coil of the major air handling unit were both active during the hottest part of the summer. The control specifications specifically called for that simultaneous heating and then cooling! Once that original problem was corrected, not only did the air conditioning bill go down dramatically, but the building occupants reported that they thought the air conditioning system was working much better since they were much more comfortable.

DO NEW BUILDINGS NEED "DASHBOARDS?"

The dashboard and instrument panel is the heart of the driver—car interface. Status information on the car's operation is presented there in easily understood form. A similar feature in a new building would make sense. Not the complex HMI or GUI from a BAS, but a simplified display for average building occupants, and maybe even one for the building engineer or maintenance supervisor. Each floor of a building could have a "dashboard" type of display. It could be located in a visible place, and occupants could see the status of energy use in terms of peak cost or off-peak cost, daily use of kWh and therms of gas. They could also see temperature and RH conditions at a glance, and could get red light/green light indicators for energy use and maintenance actions. Several of these "dashboards" could be provided to the operation and maintenance staff. These simplified "dashboard" type of displays could also be available on the PCs of the occupants and operating personnel. Cars provide a powerful model to use in many building technology applications.

CONCLUSION

New buildings have not kept up with technological advances, especially when compared to automobiles. All we need to do is to make one trip in a typical new car, and then make one visit to a typical new building to see this for ourselves. Comfort levels, safety levels, reliability levels, quality control levels and automation levels are all much higher in new cars than in buildings. The imagination and creativity that goes into new car technology and manufacture should be harnessed for our new buildings as well. We really do need to start building our new buildings like we build our new cars.

ACKNOWLEDGMENT

An earlier version of the material in this chapter appeared in a paper titled, "If Buildings Were Built Like Cars—The Potential for Information and Control System Technology in New Buildings," by Barney L. Capehart, Harry Indig, and Lynne C. Capehart, *Strategic Planning for Energy and the Environment*, Fall 2004.

Bibliography

Argonne National Laboratory program on sensor development, *www.transportation.anl.gov/ttrdc/sensors/gassensors.html*

Automated Buildings website, *www.automatedbuildings.com*

Court Manager's Information Display System, *www.ncsc.dni.us/bulletin/V09n01.htm*

Delphi Automotive Electronics Website, *www.delphi.com/automotive/electronics*

How Car Computers Work, *www.howstuffworks.com/car-computer.htm*

"Motoring with Microprocessors," by Jim Turley, *http://www.embedded.com/showArticle.jhtml?articleID=13000166*

New car features, *www.autoweb.com.au*

"Sensors," Automotive Engineering International Online, *http://www.sae.org/automag/sensors/*

Smart Energy Distribution and Consumption in Buildings, CITRIS—Center for Information Technology Research in the Interest of Society, *www.citris.berkeley.edu/SmartEnergy/SmartEnergy.html*

"Today's Automobile: A Computer on Wheels," Alliance of Automobile Manufacturers, *http://www.autoalliance.org/archives/000131.html*

Chapter 39

Conclusion

Barney L. Capehart, Ph.D., CEM, University of Florida
Lynne C. Capehart, JD, Consultant

THE NEW WAVE of Information Technology for facility energy management and for enterprise energy management not only continues to roll in, but the wave is growing in size and growing in speed. It is increasingly aiding those early adopters who are seeing that it is giving them a competitive advantage over their non-adopting business and institutional counterparts, and it is also increasingly taking its toll on those who do not embrace it. There is still time for most facilities to get on the wave and ride it to new heights of facility and enterprise performance.

The case studies and applications presented in this book show conclusively that this new wave of technology in the form of web based energy information and control systems for facilities is feasible, cost effective, and can create a significant improvement in the energy related performance of a facility. Improved energy efficiency, reduced energy costs, and better maintenance are only a few of the demonstrated benefits of these new systems.

Most of the case studies and applications presented in this book have described monitoring and control systems for facility energy management. Only a few have gone farther into the area of enterprise energy management. This more comprehensive monitoring and control of overall facility functions that are energy related ñ enterprise energy management ñ will be the next strong advancement in this overall embracing of information technology. Here, energy, maintenance, life safety, security, and operational aspects of facilities are all involved, and will be monitored and controlled by the enterprise energy management system.

Finally, the next logical extension of these IT/web based systems is to tackle the monitoring and control of all operations of a facility or group of facilities. These systems will integrate the energy related operations of a facility to the business side of the facility ñ including financial, purchasing, sales, personnel, manufacturing and service delivery. These will now require the implementation of a complete Enterprise Resource Management System.

Facility energy managers, maintenance managers, and building managers are already having to come to grips with the new web based facility energy management systems. Next, they will have to learn to interact with all of the other energy related activities in order to successfully participate in using enterprise energy management systems. And finally, they will have to become true total team players in order to successfully participate in using enterprise resource management systems.

The editorial team, and all of the chapter authors in this book, hope that this book helps the energy managers, maintenance managers, and facility managers get prepared to both understand what this new technology is, and to aggressively adopt this new technology as quickly as possible to start reaping its benefits.

Glossary

THIS GLOSSARY is a compilation of the glossaries initially included by several authors as appendices to their chapters. A large part of this glossary was provided by Mary Ann Piette, Osman Sezgen, David S. Watson and Naoya Motegi of Lawrence Berkeley National Laboratory, Berkeley, CA, Christine Shockman of Shockman Consulting, and Laurie ten Hope, Program Manager, Energy Systems Integration under sponsorship of the California Energy Commission. Ron Brown of GridLogix was also a major contributor, as was Carla Fair-Wright of Cooper Cameron Corporation. Other contributors to the glossary were: Gaymond Yee of the California Institute for Energy and Environment and Tom Webster, P.E., of UC Berkeley, Center for the Built Environment, Lawrence Berkeley National Laboratory, Berkeley, CA, as well as Paul Allen of Walt Disney World, Rich Remke of Carrier Corporation, and Steve Tom of Automated Logic Corporation.

The first book in this series—*Information Technology for Energy Managers: Understanding Web-based Energy Information and Control Systems*—has an extensive glossary and may include some terms not defined in this volume.

ACRONYMS DEFINITIONS, AND TERMINOLOGY

3G. The third generation International Telecommunications Union standard for cellular phones.

Abstraction. In this document, the term "abstraction" is used with regard to translations between different communication protocols or I/O point mappings. In abstraction, details of one protocol or data format are added or removed in order to function properly in the other. Through translation and abstraction, low level protocols (such as EMCS protocols) can be changed into a common protocol or language (such as TCP/IP and XML). This process is often bi-directional (See *translation, protocol, I/O, point mapping, EMCS, TCP/IP and XML*).

Active Requestor. An active requestor is an application (possibly a Web browser) that is capable of issuing Web services messages such as those described in WS-Security and WS-Trust.

Architectural design. The process of defining a collection of hardware and software components and their interfaces to establish the framework for the development of a computer system. The structure of the components of a program/system, their interrelationships, and principles and guidelines governing their design and evolution over time

ARCNET. Not really an acronym, this is a communications network standard commonly run over twisted pair and coaxial cable.

ASHRAE. American Society of Heating, Refrigerating, and Air Conditioning Engineers

ASP. Applications service provider

Authentication. The process of validating security credentials.

Authorization. The process of granting access to a secure resource based on the security credential provided.

Auto-DR. Automated Demand Response

BACnet. An open data communication protocol for building automation and control networks. A data communication protocol is a set of rules governing the exchange of data over a computer network. The rules take the form of a written specification (in BACnet's case they are also on compact disk) that spells out what is required to conform to the protocol. BACnet was developed under the auspices of ASHRAE. The BACnet standard is also an ANSI standard (*http://www.bacnet.org/FAQ/HPAC-3-97.html, BACnet.org 2004*).

BACnet MSTP. Short for BACnet Master/Slave Twisted Pair. BACnet MSTP describes the characteristics of the BACnet communication protocol when used in a Master/Slave architecture over twisted pair wiring.

Bandwidth. The amount of data that can be transmitted in a fixed amount of time. For digital devices, the bandwidth is usually expressed in bits per second (bps) or bytes per second. For analog devices, the bandwidth is expressed in cycles per second, or Hertz (Hz) (Webopedia, 2004).

Business Logic. In the Auto-DR tests, business logic determines the EMCS actions to be implemented based on price and business rules.

Business Rules. A term used in software architecture to signify a software component, layer (or tier) of software functionality, software library or similar that performs operations on some kind of data passed back and forth through it. It refers to the logic rather than the view of data or storage of data. It is part of a software application. It is usually associated with the three-tier software architecture and has been conceptually popularized by Rational Software and Microsoft.

Canonicalization. The process of converting an XML document to a form that is consistent to all parties. Used when signing documents and interpreting signatures.

CBE. Center for Built Environment, University of California, Berkeley

Claim. A claim is a statement made about a sender, a service or other resource (e.g. name, identity, key, group, privilege, capability, etc.).

Class. Classes describe the rules by which objects behave; those objects, described by a particular class, are known as "instances" of said class. A class specifies the data, which each instance contains; as well as the methods (functions) which the object can perform; such methods are sometimes described as "behavior." A class is the most specific type of an object in relation to a specific layer.

Client (computer). The client part of a client-server architecture. Typically, a client is an application that runs on a personal computer or workstation and relies on a server to perform the operations requested by the client. For example, an e-mail client is an application that enables you to send and receive e-mail. In the Auto-DR tests the clients at each site polled the server to get current pricing information.

Client-Server. A network application architecture which separates the client (usually the graphical user interface) from the server. Each instance of the client software connects to a server or application server.

CMMS. Computerized maintenance management system

Code. The transforming of logic and data from design specifications (design descriptions) into a programming language.

Co-lo. See *Co-Location*.

Co-Location. A server, usually a Web server, that is located at a dedicated facility designed with resources which include a secured cage or cabinet, regulated power, HVAC sufficient to cool all the electronic equipment, dedicated Internet connection, security and support. These co-location facilities offer their customers a secure place to physically house their hardware and equipment as opposed to locating it in their offices or warehouse where the potential for fire, theft or vandalism is much greater. Most co-location facilities offer high-security, including cameras, fire detection and extinguishing devices, multiple connection feeds, filtered power, backup power generators and other features to ensure a level of high-availability which is mandatory for all Web-based, virtual businesses. Co-location sites are being built at various points around the world to provide services to the rapidly expanding Web hosting and e-commerce marketplace. The term co-location is also known as *co-lo* (Webopedia, 2004).

Computerized Maintenance Management System (CMMS). Computerized method of controlling the planning of all the tasks involved in maintaining a facility or plant. This includes planned maintenance scheduling, the recording of breakdown information. It also has many other functions including stock control, inventory and purchasing.

Condition-Based Maintenance (CBM). A maintenance method providing for the installation of remote sensors and wireless technology to monitor the health of selected equipment, including both current condition and failure prediction. As a result, replacement of maintenance significant components can be accomplished in a more cost effective manner just prior to their indicated failure, rather than perpetuating the current costly practice of making the replacement either at a prescribed interval or after failure has occurred.

Control network. A network of controllers, data gathering panels and other devices that measure values from sensors and send commands to actuators. Control networks have been designed and optimized for the requirements of these systems including low installed cost and small communication packet sizes. Historically, many control networks have been based on RS-485 communications using proprietary protocols. Increasingly open protocols are being used including BACnet and LonTalk over RS-485 and Internet Protocols (IP). Control networks are generally separate from enterprise networks.

Coordination Context. The unique identifier for a set of work to be performed by a group of coordinated services.

Data logging. The process by which data from I/O points are logged into a database.

Deserialization. The process of constructing an XML Infoset from an octet stream. It is the method used to create the Infoset representation of a message from the wire format for a message.

Dewpoint. The temperature at which water vapor is saturated.

Digest. A digest is a cryptographic checksum of an octet stream.

Digital outputs (DO). In an I/O controller, digital outputs are used to command equipment on or off. Physically, a digital output consists of an automatically controlled relay contact. Constant volume fans and pumps and lights can be commanded on or off with a digital output (see *I/O controller*).

DMZ. Short for demilitarized zone, a computer or small subnetwork that sits between a trusted internal network, such as a corporate private LAN, and an untrusted external network, such as the public Internet. Typically, the DMZ contains devices accessible to Internet traffic, such as Web servers. The term comes from military use, meaning a buffer area between two enemies (Webopedia, 2004).

Originally used to describe a buffer between hostile countries, the IT world uses it to describe a network segment sandwiched between two firewalls. Web servers are commonly placed in the DMZ, as one firewall protects them from the Internet. If a hacker manages to breach the firewall and take over the server, the second firewall protects the main network from the server.

Domain. A security domain represents a single unit of security administration or trust.

DR. Demand Response.

DRS. Demand Response System

Durable Two Phase Commit. The protocol used for transactions on durable resources, such as files or databases.

EBMS. Enterprise Building Management System is a software based facilities management system including EMCS functionality but with added IT integration features (e.g. SOAP and XML) and multiple data base management capabilities.

ECM. Energy Conservation Measure

EEMS. Enterprise Energy Management System

EIS. Energy Information System.

Embedded devices. Special purpose computers with the following attributes:

1. Targeted functionality with little, if any, flexibility for the user to add different programs or customize the device.
2. User interfaces are usually limited to allow targeted functionality only. May include small LCD screens, LEDs, buttons switches and knobs. QWERTY keyboards and Cathode Ray Tube display screens are generally not included.
3. Memory is usually cost optimized for the targeted functionality. Read only memory (ROM) and flash memory chips are usually used in lieu drives with spinning disks.
4. Form factor is specially designed for the targeted functionality. Examples of embedded devices include Internet routers, automotive engine computers and cell phones.
5. Often designed for reliable continuous operation with little or no human interaction.

Energy Management and Control System (EMCS). A control system (often computerized) designed to regulate the energy consumption of a building by controlling the operation of energy consuming systems, such as the heating, ventilation and air conditioning (HVAC), lighting and water heating systems. (same as EMS).

Enterprise. A business organization. In the computer industry, the term is often used to describe any large organization that utilizes computers. An intranet is an example of an enterprise computing system (Webopedia, 2004).

Ethernet. A local-area network (LAN) architecture developed by Xerox Corporation in cooperation with DEC and Intel in 1976. Ethernet uses a bus or star topology and supports data transfer rates of 10 Mbps. The Ethernet specification served as the basis for the IEEE 802.3 standard, which specifies the physical and lower software layers. Ethernet uses the CSMA/CD access method to handle simultaneous demands. It is one of the most widely implemented LAN standards. A newer version of Ethernet, called 100Base-T (or Fast Ethernet), supports data transfer rates of 100 Mbps. The newest version, Gigabit Ethernet supports data rates of 1 gigabit (1,000 megabits) per second (Webopedia, 2004)

Exchange Pattern. The model used for message exchange between services.

Extensible Markup Language (XML). A standard text format for the transmission of data and its structure. Its primary purpose is to facilitate the sharing of structured text and information across the Internet.

Factory. A factory is a Web service that can create a resource from an XML representation.

FDD. Fault Detection and Diagnosis

Federation. A federation is a collection of trust domains that have established mutual pair-wise trust. The level of trust may vary, but typically includes authentication and may include authorization.

Firewall. A system designed to prevent unauthorized access to or from a private network. Firewalls can be implemented in either or both hardware and software. Firewalls are frequently used to prevent unauthorized Internet users from accessing private networks connected to the Internet, especially intranets. All messages entering or leaving the intranet pass through the firewall, which examines each message and blocks those that do not meet the specified security criteria. There are several types of firewall techniques:

1. Packet filter: Looks at each packet entering or leaving the network and accepts or rejects it based on user-defined rules. Packet filtering is fairly effective and transparent to users, but it is difficult to configure. In addition, it is susceptible to IP spoofing.
2. Application gateway: Applies security mechanisms to specific applications, such as FTP and Telnet servers. This is very effective, but can impose performance degradations.
3. Circuit-level gateway: Applies security mechanisms when a TCP or UDP connection is established. Once the connection has been made, packets can flow between the hosts without further checking.
4. Proxy server: Intercepts all messages entering and leaving the network. The proxy server effectively hides the true network addresses.

In practice, many firewalls use two or more of these techniques in concert. A firewall is considered a first line of defense in protecting private information. For greater security, data can be encrypted (Webopedia, 2004).

Flash Memory. Flash memory stores information on a silicon chip in a way that does not need power to maintain the information in the chip. This means that if you turn off the power to the chip, the information is retained without consuming any power. In addition, flash offers fast read access times and solid-state shock resistance.

Fully Automated Demand Response. Fully Automated Demand Response is initiated at a building or facility through receipt of an external communications signal—facility staff set up a pre-programmed load shedding strategy which is automatically initiated by the system without the need for human intervention.

Gateway. Gateways used in building telemetry systems provide several functions. First, they connect two otherwise incompatible networks (i.e., networks with different protocols) and allow communications between them. Second, they provide *translation* and usually *abstraction* of messages passed between two networks. Third, they often provide other features such as *data logging*, and control and monitoring of I/O points.

Generation. In electronics, computer equipment and software, the term "generation" is used to describe a major upgrade for which previous versions may or may not be compatible.

GIF. Graphics Interchange Format. A commonly used standard for graphics displayed on web pages.

GSA. General Services Administration

Graphical User Interface GUI. A visual interface that enables information to be passed between a human user and hardware or software components of a computer system

High Availability. Used to quantify the "uptime" for computer servers and systems. High availability is a requirement for operation of mission critical systems. High availability systems are often described in terms of the number of "nines" of availability (i.e., four 9s or 99.99% means less than one hour of unscheduled downtime per year).

Human Machine Interface (HMI). Nowadays humans interact more with computer-based technology than with hammers and drills. Unlike tools, the visible shape and controls of a computer do not communicate its purpose. The task of an HMI is to make the function of a technology self-evident. Much like a well-designed hammer fits the user's hand and makes a physical task easy; a well-designed HMI must fit the user's mental map of the task he or she wishes to carry out (International Engineering Consortium (*www.iec.org*), 2004). In control systems, the HMI usually takes the form of a computer display screen with specially designed graphical representations of the mechanical systems or processes including real-time data from sensors.

HVAC. Heating, Ventilation, and Air Conditioning.

HyperText Markup Language (HTML). The most commonly used standard for web pages. A non-proprietary format based upon SGML, a meta-language in which one can define markup languages for documents. HTML can be created and processed by a wide range of tools, from simple plain text editors—you type it in from scratch- to sophisticated WYSIWYG authoring tools. HTML uses tags such as <h1> and </h1> to structure text into headings, paragraphs, lists, hypertext links etc."

Hypertext Transfer Protocol (HTTP). The network transmission protocol used by the Web. A request/response protocol between clients and servers. The standard for packaging and transporting messages (often written in HTML) over the Web. An HTTP client, such as a web browser, typically initiates a request by establishing a TCP/IP connection to a particular port on a remote host. An HTTP server listening on that port waits for the client to send a request string, such as "GET/HTTP/1.1" (which would request the default page of that web server), optionally followed by an email-like MIME message which has a number of informational header strings that describe aspects of the request, followed by an optional body of arbitrary data. Upon receiving the request string (and message, if any), the server sends back a response string, such as "200 OK," and a message of its own, the body of which is perhaps the requested file, an error message, or some other information.

HyperText Transfer Protocol, the underlying protocol used by the World Wide Web. HTTP defines how messages are formatted and transmitted, and what actions Web servers and browsers should take in response to various commands. For example, when you enter a URL in your browser, this actually sends an HTTP command to the Web server directing it to fetch and transmit the requested Web page (Webopedia, 2004).

Identity Mapping. Identity mapping is a method of creating relationships between identity properties. Some Identity Providers may make use of identity mapping.

Identity Provider (IP). Identity Provider is an entity that acts as an authentication service to end requestors. An Identity Provider also acts as a data origin authentication service to service providers (this is typically an extension of a security token service).

Internet Information Services (IIS). This is the name given to the Windows Web server. Originally supplied as part of the Option Pack for Windows NT, it was subsequently integrated with Windows 2000 and Windows Server 2003).

I/O. Abbreviation for Input/Output. Commonly used in the controls industry to describe the hardwired sensor inputs and actuator outputs to a controller, or their corresponding software points. (see abstra*ction, point mapping and translation*).

I/O controller. A device that measures input values from sensors and commands outputs such as temperature control valves, usually to maintain a defined setpoint. An I/O controller has the internal intelligence to perform local control.

I/O module. A device that measures input values from sensors and commands outputs such as temperature control valves based on instructions from a remote controller. An I/O module usually does not have the internal intelligence to perform local control.

Internet. A global network connecting millions of computers. More than 100 countries are linked into exchanges of data, news and opinions. Unlike on-line services, which are centrally controlled, the Internet is decentralized by design. Each Internet computer, called a host, is independent. Its operators can choose which Internet services to use and which local services to make available to the global Internet community. Remarkably, this anarchy by design works exceedingly well. There are a variety of ways to access the Internet, including through on-line services, such as America On-line, and commercial Internet Service Providers (ISP) (Webopedia, 2004). References to the public Internet, as described here, are spelled "Internet" (capital "I"). The lesser used term, "internet" (lower case "i"), refers to a network of intranets that are usually privately operated (see *intranet*).

Intranet. A network based on TCP/IP protocols (an internet) belonging to an organization, usually a corporation, accessible only by the organization's members, employees, or others with authorization. An intranet's Web sites look and act just like any other Web sites, but the *firewall* surrounding an intranet fends off unauthorized access. Like the Internet itself, intranets are used to share information (Webopedia, 2004).

IP. Internet Protocol. A popular messaging and addressing system used on Ethernet networks. An Internet protocol is a standard way of formatting and transporting messages over the Internet. Probably the most well-known Internet protocol is HTTP, or Hyper Text Transfer Protocol. HTTP is used to transfer web pages over a network and to give browsers instructions on how to display them. MIME or Multipurpose Internet Mail Extensions is another common protocol, used to transfer email messages with graphics, audio, and video files. Internet protocols are not confined to the Internet, and are often used within an Intranet such as a small computer network within a building.

IP I/O device. A device that measures inputs (e.g., electric meter data) and controls outputs (e.g., relays) that can be measured and actuated remotely over a LAN, WAN or Internet using Internet Protocols (IP).

IP relay. A device with a relay or relays that can be actuated remotely over a LAN, WAN or Internet using Internet Protocols (IP).

ISO. Independent System Operator.

ISO. International Organization for Standardization. The premier organization for standards, even if they can't get their acronym in the right order.

IT. Commonly used to describe the mysterious world of wizards who work on networks, routers, servers, firewalls, etc. Short for Information Technology, and pronounced as separate letters, the broad subject concerned with all aspects of managing and processing information, especially within a large organization or company. Because computers are central to information management, computer departments within companies and universities are often called IT departments. Some companies refer to this department as IS (Information Services) or MIS (Management Information Services) (Webopedia, 2004).

JavaScript. A scripting language commonly used to incorporate dynamic content on HTML web pages.

LAN. Short for Local Area Network. A computer network that spans a relatively small area, like a home, office or small group of buildings such as a college. The topology of a network dictates its physical structure. Most LANs are confined to a single building or group of buildings. Most LANs connect workstations and personal computers. Each node (individual computer) in a LAN has its own central processing unit (CPU) with which it executes programs, but it also is able to access data and devices anywhere on the LAN. This means that many users can share devices, such as laser printers, as well as data. Users can also use the LAN to communicate with each other such as by sending e-mail. There are many different types of LANs with Ethernets being the most common for PCs (Webopedia, 2004).

LGR. An Automatic Logic Corporation line of BACnet routers. LGR devices can also be used to transfer information to and from other communication protocols.

Load balancing. Distributing processing and communications activity evenly across a computer network so that no single device is overwhelmed. Load balancing is especially important for networks where it's difficult to predict the number of requests that will be issued to a server. Busy Web sites typically employ two or more Web servers in a load balancing scheme. If one server starts to get swamped, requests are forwarded to another server with more capacity. Load balancing can also refer to the communications channels themselves (Webopedia, 2004).

LonTalk™. An open communications protocol used in building control systems and other industries. Publicly published under ANSI/EIA—709.1—(American National Standards Institute/Electronic Industries Alliance). Since 1999, LonTalk protocol may be implemented on any microprocessor.

LonWorks™. Control and communication products available from Echelon Corporation and other companies using Echelon products. LonWorks products use LonTalk open protocol for communications (see *LonTalk*).

Machine-to-Machine (M2M). Machine to Machine (M2M) is a term used to describe the technologies that enable computers, embedded processors, smart sensors, actuators and mobile devices to communicate with one another, take measurements and make decisions—often without human intervention.

Manual Demand Response. Manual demand response involves manually turning off lights or equipment; this can be a labor-intensive approach.

MCC. Motor Control Center.

ME. An Automatic Logic Corporation line of multi-equipment control modules.

Message. A message is a complete unit of data available to be sent or received by services. It is a self-contained unit of information exchange. A message always contains a SOAP envelope, and may include additional MIME parts as specified in MTOM, and/or transport protocol headers.

Message Path. The set of SOAP nodes traversed between the original source and ultimate receiver.

Middleware. A software agent acting as an intermediary, or as a member of a group of intermediaries, between different components in a transactional process. The classic example of this is the separation which is attained between the client user and the database in a client/server situation.

Modbus. An open standard control network protocol originally developed by Modicon, a simple master/slave protocol over RS-485. Current protocol also supports TCP/IP over an Ethernet. Modbus is a common interface with electrical equipment such as meters and generators. It is not typically used for whole building EMCSs.

Modem. A hardware device that allows computers to communicate with one another over any phone-based network.

Monolithic Configuration. A program built or shipped as one large program file or executable image. That is, it is not sectioned into a main program and supporting libraries.

MS/TP. Master Slave/Token Passing. This is a network standard created by BACnet. It's a low cost network that runs over twisted pair.

NMS. Network Management System

NOC. Short for network operations center, the physical space from which a typically large telecommunications network is managed, monitored and supervised. The NOC coordinates network troubles; provides problem management and router configuration services; manages network changes; allocates and manages domain names and IP addresses; monitors routers, switches, hubs and UPS systems that keep the network operating smoothly; and manages the distribution and updating of software and coordinates with affiliated networks. NOCs also provide network accessibility to users connecting to the network from outside of the physical office space or campus (Webopedia, 2004).

OAT. Outside air temperature.

oBIX. Open Building Information eXchange is an industry-wide initiative to define open XML- and Web services-based mechanisms for building control systems. An industry consortium which seeks to promote standards for the evolution of the Web and interoperability between WWW products by producing specifications and reference software.

Object-Oriented Analysis and Design (OOA&D). Object-oriented analysis (OOA) is concerned with developing software engineering requirements and specifications that expressed as a system's object model. Object-oriented analysis builds a model of a system that is composed of objects. The behavior of the system is achieved through collaboration between these objects, and the state of the system is the combined state of all the objects in it. Collaboration between objects involves them sending messages to each other.

Object-oriented design (OOD) is concerned with developing an object-oriented model of a software system to implement the identified requirements. A design method in which a system is modeled as a collection of cooperating objects and individual objects are treated as instances of a class within a class hierarchy.

Object-Oriented Programming (OOP). A computer programming paradigm. OOP defines a computer program as a composite of individual units, or objects which can function like sub-programs. To make the overall computation happen, each object is capable of receiving messages, processing data, and sending messages to other objects. In short, the objects can interact through their own functions (or methods) and their own data.

Object Management Group. A consortium aimed at setting standards in object-oriented programming. In 1989, this consortium, which included IBM Corporation, Apple Computer Inc. and Sun Microsystems Inc., mobilized to create a cross-compatible distributed object standard. The goal was a common binary object with methods and data that work using all types of development environments on all types of platforms. The created the standard for UML.

Occasionally Connected Computing (OCC). The application architecture that provides mobile users a consistent computing experience, independent of the network connection status.

ODBC. Short for Open DataBase Connectivity, a standard database access method developed by Microsoft Corporation. The goal of ODBC is to make it possible to access any data from any application, regardless of which database management system (DBMS) is handling the data. ODBC manages this by inserting a middle layer, called a database driver, between an application and the DBMS. The purpose of this layer is to translate the application's data queries into commands that the DBMS understands. For this to work, both the application and the DBMS must be ODBC-compliant—that is, the application must be capable of issuing ODBC commands and the DBMS must be capable of responding to them. Since version 2.0, the standard supports SAG SQL (Webopedia, 2004).

ODBC is a standard software application programming interface (API) for connecting to database management systems (DBMS). This API is independent of any one programming language, database system or operating system. ODBC is based on the Call Level Interface (CLI) specifications from SQL, X/Open (now part of The Open Group), and the ISO/IEC. ODBC was created by the SQL Access Group and first released in September, 1992.

OLAP. On-line analytical processing

Onboard. Refers to electronic components that are mounted on the main printed circuit board as opposed to components that are mounted remotely and connected via wires.

Open protocol. A communications protocol that is used to communicate between devices of any compliant manufacturer or organization. Open protocols are published in a public forum for use by all interested parties (see *proprietary protocol*).

P. Proportional

Passive Requestors. A passive requestor is an HTTP browser capable of broadly supported HTTP (e.g. HTTP/1.1).

PDA. Personal digital assistant

PI. Proportional plus integral

PID. Proportional plus integral plus derivative

Point mapping. The process by which I/O points are mapped to another system or protocol (see *abstraction, I/O and translation*).

Point of sale (POS). Refers to all components in the POS system including bar code scanners, cash registers, printers and all communications equipment that is used to transfer information about the sale of each item back to a central database. At some sites in the Auto-DR tests, the point of sale communications infrastructure was used to command the shed of electric loads based on electricity pricing information.

Policy. A policy is a collection of policy alternatives.

Policy Alternative. A policy alternative is a collection of policy assertions.

Policy Assertion. A policy assertion represents a domain-specific individual requirement, capability, other property, or a behavior.

Poll. A method by which one computer requests information from another.

Polling Client. In the Auto-DR tests, the polling client is the software used to poll the server to get the price signal.

Price Server. In the Auto-DR tests, the price server is the common source for the current price information.

Principal. Any system entity that can be granted security rights or that makes assertions about security or identity.

Proprietary protocol. A communications protocol that is used to communicate between devices of one manufacturer or organization while effectively disallowing all other devices to exist on the same network. Proprietary protocols are not published in a public forum (see *open protocol*).

Protocol (data communication). A data communication protocol is a set of rules governing the exchange of data over a computer network.

Protocol Composition. Protocol composition is the ability to combine protocols while maintaining technical coherence and absent any unintended functional side effects.

Public Switched Telephone Network (PSTN). International telephone system based on copper wires carrying analog voice data. This is in contrast to newer telephone networks based on digital technologies (Webopedia 2004).

Pull architecture. In a client-server architecture, the client "pulls" information from the server by polling (see *poll*).

Pulse. Contact closures that are measured by an I/O device. Pulses are often produced by electric meters and other devices to indicate a given unit of measurement (for example, 1 pulse = 1 kWh).

Query. A process that allows an information store, such as a database, to be questioned and searched.

RDBMS (relational database management system). Database technology that was designed to support high volume transaction processing (OLTP) and is typically the foundation for a data warehouse. Examples include Oracle, MS SQL Server and IBM DB2.

Real-time. In real-time control and monitoring systems, data is measured, displayed and controlled at a rate fast enough that the system latencies are negligible compared with the process at hand. Acceptable latency can vary substantially based on the type of process (for example, from 1 millisecond to several minutes).

Resource. A resource is defined as any entity addressable by an endpoint reference where the entity can provide an XML representation of itself.

RF. Radio Frequency

RS. An Automated Logic Corporation line of intelligent room sensors.

Sandbox. A sandbox is a security measure used by Java. It is a set of rules used when creating an applet that prevents certain functions when the applet is sent as part of a Web page. The sandbox places strict limitations on what system resources the applet can

request or access. Sandboxes are used when executable code comes from unknown or untrusted sources and allow the user to run untrusted code safely.
(Webopedia, 2004)

SE. An Automated Logic Corporation line of single-equipment control modules.

Security Context. A security context is an abstract concept that refers to an established authentication state and negotiated key(s) that may have additional security-related properties.

Security Context Token. A Security Context Token (SCT) is a wire representation of the security context abstract concept, and which allows a context to be named by a URI and used with [WS-Security].

Security Token. A security token represents a collection of claims.

Security Token Service. A security token service (STS) is a Web service that issues security tokens (see [WS-Security]). That is, it makes assertions based on evidence that it trusts, to whoever trusts it (or to specific recipients). To communicate trust, a service requires proof, such as a signature to prove knowledge of a security token or set of security token. A service itself can generate tokens or it can rely on a separate STS to issue a security token with its own trust statement (note that for some security token formats this can just be a re-issuance or co-signature). This forms the basis of trust brokering.

Semi-Automated Response. Semi-Automated Response involves the use of building energy management control systems for load shedding, where a pre-programmed load shedding strategy is initiated by facilities staff.

Sensor. a device that responds to a physical stimulus (heat, light, sound, pressure, magnetism, or a particular motion) and transmits a resulting impulse that can be used for measurement or operating a controller (Merriam-Webster On-line Dictionary, 2004).

Sequence Diagrams. In UML, Sequence diagrams document the interactions between classes to achieve a result, such as a use case. Because UML is designed for object-oriented programming, these communications between classes are known as messages. The Sequence diagram lists objects horizontally, and time vertically, and models these messages over time.

Serialization. The process of representing an XML Infoset as an octet stream. It is the method used to create the wire format for a message.

Server (computer). Servers are often dedicated, meaning that they perform no other tasks besides their server tasks. On multiprocessing operating systems, however, a single computer can execute several programs at once. A server in this case could refer to the program that is managing resources rather than the entire computer. In the 2003 Auto-DR tests, pricing information was "served" from a Web services server hosted by Infotility, Inc.

Service. A software entity whose interactions with other entities are via messages. Note that that a service need not be connected to a network.

Setpoint. The target value that an I/O controller attempts to maintain. Setpoint values (for example, temperature or pressure) are maintained through adjustments of the final control elements (temperature control valves, dampers, etc.).

SGML. Standard Generalized Markup Language

Signature. A signature is a value computed with a cryptographic algorithm and bound to data in such a way that intended recipients of the data can use the signature to verify that the data has not been altered and has originated from the signer of the message, providing message integrity and authentication. The signature can be computed and verified with either symmetric or asymmetric key algorithms.

Sign-Out. A sign-out is the process by which a principal indicates that they will no longer be using their token and services in the domain can destroy their token caches for the principal.

Silo. A term used to describe a non-integrated way of providing processes or services, within a business or enterprise. A key to the concept of the silo is that there is little or no planning, coordination, or communication between processes or services.

SIMM. Acronym for single in-line memory module, a small circuit board that can hold a group of memory chips. Typically, SIMMs hold up to eight (on Macintoshes) or nine (on PCs) RAM chips. On PCs, the ninth chip is often used for parity error checking. Unlike memory chips, SIMMs are measured in bytes rather than bits. SIMMs are easier to install than individual memory chips. The bus from a SIMM to the actual memory chips is 32 bits wide. A newer technology, called dual in-line memory module (DIMM), provides a 64-bit bus. For modern Pentium microprocessors that have a 64-bit bus, you must use either DIMMs or pairs of SIMMs (Webopedia, 2004)

Single Sign On (SSO). Single Sign On is an optimization of the authentication sequence to remove the burden of repeating actions placed on the requestor. To facilitate SSO, an element called an Identity Provider can act as a proxy on a requestor's behalf to provide evidence of authentication events to 3rd parties requesting information about the requestor. These Identity Providers (IP) are trusted 3rd parties and need to be trusted both by the requestor (to maintain the requestor's identity information as the loss of this information can result in the compromise of the requestors identity) and the Web services which may grant access to valuable resources and information based upon the integrity of the identity information provided by the IP.

SNMP. Simple Network Management Protocol

SOAP. A standard way of packaging XML files for transport across an IP network. Short for Simple Object Access Protocol, a lightweight XML-based messaging protocol used to encode the information in Web service request and response messages before sending them over a network. SOAP messages are independent of any operating system or protocol and may be transported using a variety of Internet protocols, including SMTP, MIME, and HTTP (Webopedia, 2004).

SOAP Intermediary. A SOAP intermediary is a SOAP processing node that is neither the original message sender nor the ultimate receiver.

Software life cycle. The period of time that begins when a software product is conceived and ends when the software is no longer available for use. The life cycle typically includes a concept phase, requirements phase, design phase, implementation phase, test phase, installation and checkout phase, operation and maintenance phase, and sometimes, retirement phase. These phases may overlap or be performed iteratively, depending on the software development approach used.

SQL (Structured Query Language). A standard interactive and programming language for extracting information from and updating a database. Queries take the form of a command language that allows users to select, insert, update and locate data.

SSL. Secure Socket Layers. A 128-bit encryption scheme used to secure Internet transactions.

Symmetric Key Algorithm. An encryption algorithm where the same key is used for both encrypting and decrypting a message.

System. A collection of services implementing a particular functionality. Synonymous with Distributed Application.

Systems Integrator. An individual or company that specializes in building complete computer systems by putting together components from different vendors. Unlike software developers, systems integrators typically do not produce any original code. Instead they enable a company to use off-the-shelf hardware and software packages to meet the company's computing needs (Webopedia, 2004).

TCP/IP (Transmission Control Protocol/Internet Protocol). Internet Protocol that specifies the format of packets and the addressing scheme. Most networks combine IP with a higher-level protocol called Transmission Control Protocol (TCP), which establishes a virtual connection between a destination and a source.

Telemetry. A communications process that enables monitoring and/or control of remote or inaccessible sensors and actuators. Telemetry often uses radio frequency signals or Internet technologies for communications.

Three-tier Architecture. The three tier software architecture emerged in the 1990s to overcome the limitations of the two tier architecture. The third or middle tier is sandwiched by the user interface (client) and the data management (server) components. This middle tier provides process management where business logic and rules are executed and can accommodate hundreds of users by providing functions such as queuing, application execution, and database staging.

TIEMS. Tenant Interface for Energy and Maintenance Systems

Time stamp. A digital message that indicates the time in which a given computer transaction occurred. The time stamp message is usually associated with and stored with the original transaction record. Time stamps are also to note when data is logged or stored.

Top-down Structured Programming. A technique for organizing and coding computer programs in which a hierarchy of modules is used, each having a single entry and a single exit point, and in which control is passed downward through the structure without unconditional branches to higher levels of the structure. Three types of control flow are used: sequential, test, and iteration.

TP. Twisted Pair

Translation. The process by which I/O points are translated to another system or protocol. Translation changes messages in one protocol to the same messages in another (see *abstraction, I/O and point mapping*).

Transmission Control Protocol and Internet Protocol or TCP/IP. The suite of communications protocols used to connect hosts on the Internet. TCP/IP uses several protocols, the two main ones being TCP and IP. TCP/IP is built into the UNIX operating system and is used by the Internet, making it the de facto standard for transmitting data over networks.

Trust. Trust indicates that one entity is willing to rely upon a second entity to execute a set of actions and/or to make set of assertions about a set of subjects and/or scopes.

Trust Domain. A Trust Domain is an administered security space in which the source and target of a request can determine and agree whether particular sets of credentials from a source satisfy the relevant security policies of the target. The target may defer the trust decision to a third party (if this has been established as part of the agreement) thus including the trusted third party in the Trust Domain.

TS. Terminal Service

UDDI. Universal Description, Discovery, and Integration. UDDI is a Web-based directory that allows businesses to list their available Web services. Think of it as the "yellow pages" for the Web services. Not every business chooses to list its services in the UDDI, but those that do will post a WSDL on the UDDI.

Universal Serial Bus (USB). An external bus standard that supports data transfer rates of 12 Mbps. A single USB port can be used to connect up to 127 peripheral devices, such as mice, modems, and keyboards. USB also supports Plug-and-Play installation and hot plugging.

Starting in 1996, a few computer manufacturers started including USB support in their new machines. It wasn't until the release of the best-selling iMac in 1998 that USB became widespread. It is expected to completely replace serial and parallel ports.

Unified Modeling Language (UML). A language for specifying, constructing, visualizing, and documenting the artifacts of a software-intensive system.

Use Case. An object-oriented modeling construct that is used to define the behavior of a system. Interactions between the user and the system are described through a prototypical course of actions along with a possible set of alternative courses of action.

UTC. "Coordinated Universal Time" (abbreviated UTC), is the basis for the worldwide system of civil time. This time scale is kept by time laboratories around the world, including the U.S. Naval Observatory, and is determined using highly precise atomic clocks (U.S. Naval Observatory 2004 *http://aa.usno.navy.mil/faq/docs/UT.html*). UTC differs from Greenwich Mean Time (GMT) in that GMT is based on the rotation of the Earth which is substantially less accurate than atomic clocks. When GMT differs from UTC by more than 0.9 seconds, UTC is re-calibrated by adding a "leap second" so that it is closer to the (less precise and uncontrollable) rotation of the Earth.

Volatile Two Phase Commit. The protocol used for transactions on volatile resources, such as caches or window managers.

VPN. Short for virtual private network, a network that is constructed by using public wires to connect nodes.

For example, there are a number of systems that enable you to create networks using the Internet as the medium for transporting data. These systems use encryption and other security mechanisms to ensure that only authorized users can access the network and that the data cannot be intercepted (Webopedia, 2004).

An Internet standard used to restrict web site access to computers which are pre-loaded with an encryption code. With SSL, the receiving code obtains the key from the web site and uses that to decrypt subsequent traffic. Any computer may access the web site and obtain an SSL key for use in that session, but other computers can't "listen in" on the traffic. With VPN, only computers with the special VPN key can access the web site

WAN (Wide Area Network). A computer network that spans a relatively large geographical area. Typically, a WAN consists of two or more local-area networks (LANs). The largest WAN in existence is the Internet, which is open to the public. Private and corporate WANs use dedicated leased lines or other means of assuring that the network is only available to authorized users of the organization (Webopedia, 2004).

WAP. Wireless Application Protocol. A communication protocol specifically designed for wireless devices like cellular phones.

WBP. Whole building power.

Web Service. A Web service is a reusable piece of software that interacts exchanging messages over the network through XML, SOAP, and other industry recognized standards.

Web Services. The infrastructure of the Auto-DR System is based on a set of technologies known as Web Services. Web Services have emerged as an important new type of application used in creating distributed computing solutions across the Internet. Properly designed Web services are completely independent of computer platform (Microsoft, Linux, Unix, Mac, etc.). Web pages are accessed by people to view information on the Internet. Web services are used by computers to share information on the Internet. Since human intervention is not required, this technology is sometimes referred to as "Machine-to-Machine" or "M2M." XML is often used to enable Web services. M2M is a superset of technologies that includes some XML/Web services based systems (see *XML, Machine to Machine*).

Webopedia™ (*www.webopedia.com*) defines Web services as "a standardized way of integrating Web-based applications using the XML, SOAP,WSDL and UDDI open standards over an Internet protocol backbone."

WML. Wireless Markup Language. An XML application specifically designed for low-bandwidth wireless devices like cellular phones.

WSDL. Web Services Description Language. WSDL is a standard way of describing a Web service's capabilities, including any data that can be read or written through the service. WSDL is designed to make it easy to find out what information is available from a particular application or Website. It is probably not a surprise to learn that WSDL is written in the XML format.

W3C. World Wide Web Consortium

XML (Extensible Markup Language). XML is a "meta-language," a language for describing other languages that allows the design of customized markup languages for different types of documents on the Web (Flynn, 2003). It allows designers to create their own customized tags, enabling the definition, transmission, validation, and interpretation of data between applications and between organizations (Webopedia, 2004). A standard for creating internally documented files used to exchange data between computer applications.

XML files are primarily used to transfer data between applications. It is an especially useful way to package data that may be transferred to multiple applications, including custom applications and applications that are unknown at the time the file is created. There is no specific format for an XML file; instead, the file contains includes a description of its own data structure. This makes it very easy to modify or extend the data file. XML is stored as a text file, making it easy for a human programmer to read the file, find the tags for the data he is interested in, and program another application to read or write data to this file.

ZN. An Automated Logic Corporation line of zone (terminal equipment) control modules.

About the Authors

Paul J. Allen is the chief energy management engineer at Reedy Creek Energy Services (a division of the Walt Disney World Co.) and is responsible for the development and implementation of energy conservation projects throughout the Walt Disney World Resort. Paul is a graduate of the University of Miami (BS degrees in physics and civil engineering) and the University of Florida (MS degrees in civil engineering and industrial engineering). Paul is also a registered Professional Engineer in the State of Florida. Paul was admitted into the Association of Energy Engineers (AEE) Hall of Fame in 2003.

(E-mail: *paul.allen@disney.com*)

Michael Bobker, M.Sc., CEM, has been working in and analyzing New York City buildings for over 25 years, in a career that encompasses stints as community organizer, boiler mechanic, continuing education instructor, energy auditor, engineering manager, and energy services company principal. His expertise includes building mechanical and electrical systems, especially boilers and heating, small-scale cogeneration, and turnkey construction. In his current position with the Association for Energy Affordability, he has been developing advanced technology applications to housing that can be put in the hands of community-based energy service providers. Michael is a Certified Energy Manager and holds graduate degrees in sociology-anthropology (Oberlin College), energy management (NY Institute of Technology) and business (New York University). He is the current president of the NYC chapter of the Association of Energy Engineers and chairs the environmental science section at the New York Academy of Sciences. Contact: *Mbobker@aeanyc.org*, 212-279-3902 ext. 6828.

Michael R. Brambley, Ph.D., manages the building systems program at Pacific Northwest National Laboratory (PNNL), where his work focuses on developing and deploying technology to increase the energy efficiency of buildings and other energy using systems. His primary research thrusts in recent years have been in development and application of automated fault detection and diagnostics and wireless sensing and control. He has been with PNNL for nearly 17 years before which he was an assistant professor in the Engineering School at Washington University in St. Louis.

Michael is the author of more than 60 peer-reviewed technical publications and numerous research project reports. He holds M.S. (1978) and Ph.D. (1981) degrees from the University of California, San Diego, and the B.S. (1976) from the University of Pennsylvania. He is an active member of the American Society of Heating, Refrigerating, and Air-Conditioning Engineers (ASHRAE) for which he has served on technical committees for computer applications and smart building systems. He has been the organizer of numerous seminars and symposia at ASHRAE's semi-annual meetings and is a member of ASHRAE's Program Committee. In addition to several other professional organizations, Michael is also a member of the Instrumentation, Systems, and Automation Society (ISA) and Sigma Xi, The Scientific Research Society.

David L. Brooks, P.E., is employed by Affiliated Engineers in Gainesville, Florida, and currently serves as the systems integration group (SIG) market leader and project manager at this location. In this capacity, Mr. Brooks works directly with owners and contractors on all aspects of the intelligent building system and control system design from master planning, detailed design, and installation management to commissioning. Mr. Brooks works closely with mechanical, electrical and plumbing/fire protection engineers to ensure coordination and technology has been properly integrated into projects. Mr. Brooks has 15 years of experience designing, managing and commissioning all aspects of the intelligent building systems package (i.e. enterprise management systems, laboratory controls, HVAC controls, building automation systems, fire systems, security systems and a multitude of other intelligent building systems). Mr. Brooks is a registered controls engineer, is an active member of ASHRAE and a review board member for the ASHRAE Guideline 13-2000 Specifying Direct Digital Control Systems.

Ron Brown is the CTO and co-founder of Gridlogix, Inc. Mr. Brown is the primary architect behind Gridlogix technology strategies and solutions that include EnNET® an XML web services remote device management and integration framework. Prior to co-founding Gridlogix, Mr. Brown co-founded Automated Energy, Inc. (AEI) where he served as president and chief information officer. During his tenure, he recruited and led the team that designed and engineered AEI's industry leading Enterprise Energy Management (EEM) System. He has several patents pending.

Mr. Brown's 20 plus years of experience have been in the management, consulting, software development, implementation and integration of GIS, energy management, building management, distribution automation, and MIS projects for various corporations and utilities.

Mr. Brown holds a B.S. Degree in engineering physics from the University of Central Oklahoma, and an A.A.S. Degree in general engineering from Oklahoma State University. He is a senior member of the Association of Energy Engineers (AEE), member of the Institute of Electrical and Electronics Engineers (IEEE), and a member of the Geospatial Information & Technology Association (GITA).

Anto Budiardjo, president of Clasma, Inc., is a seasoned marketing and product development professional specializing in the HVAC, security and IT disciplines. Anto has more than two decades experience in these industries and has fashioned his expertise into an energetic, visionary and dynamic approach to business management. He has held executive-level marketing and product development positions with various controls companies where he was responsible for product management and marketing communications. His rare combination of marketing and technology practices has enabled him to fine tune and soften the often daunting task of transitioning the product development process from an engineering-centric focus to a market-centric focus.

Barney L. Capehart, Ph.D., CEM is a Professor Emeritus of Industrial and Systems Engineering at the University of Florida in Gainesville, FL. He has broad experience in the commercial/industrial sector having served as the founding director of the University of Florida Energy Analysis and Diagnostic Center/Industrial Assessment Center from 1990 to 1999. He personally conducted over 100 audits of industrial and manufacturing facilities, and has helped students conduct audits of hundreds of office buildings, small businesses, government facilities, and apartment complexes. He regularly taught a University of Florida course on energy management, and currently teaches energy management seminars around the country for the Association of Energy Engineers (AEE). He is a Fellow of IEEE, IIE and AAAS, and a member of the Hall of Fame of AEE. He is the editor of *Information Technology for Energy Managers: Understanding Web-Based Energy Information and Control Systems*, Fairmont Press, 2004. He is the co-author of *Guide to Energy Management, 4th Edition*, author of the chapter on Energy Management for the *Handbook of Industrial Engineering*, and is co-author of the chapter on Energy Auditing for the *Energy Management Handbook, 5th Edition*. He can be reached at *Capehart@ise.ufl.edu*

Lynne C. Capehart, BS, JD, is a consultant in energy policy and energy efficiency, and resides in Gainesville, FL. She received a B.S. with High Honors in mathematics from the University of Oklahoma, and a JD with Honors from the University of Florida College of Law. She is co-author of *Florida's Electric Future: Building Plentiful Supplies on Conservation*; the co-author of numerous papers on PURPA and cogeneration policies; and the co-author of numerous papers on commercial and industrial energy efficiency. She was project coordinator for the University of Florida Industrial Assessment Center from 1992 to 1999. She is a member of Phi Beta Kappa, Alpha Pi Mu, and Sigma Pi Sigma. She is president of the Quilters of Alachua County Day Guild, and has two beautiful grandchildren. Her email address is *Lynneinfla@aol.com*

Gregory Cmar is cofounder and CTO of Interval Data Systems, Inc. Greg is one of the most knowledgeable people on the planet when it comes to how interval data can be used to manage energy systems. He brings 30 years of experience in facility operations, energy conservation, energy analytics, energy auditing, monitoring and control systems, and utility billing, as well as database and software technologies to IDS. Greg leads the product definition and development effort as well as the energy management services team.

Greg was a cofounder and director of engineering at ForPower, an energy conservation consulting firm; engineering manager at Coneco, an energy services company and subsidiary of Boston Edison; vice president of Enertech Systems, an energy monitoring and control systems contractor; and various roles at Johnson Controls, the Massachusetts Energy Office, and Honeywell.

Greg holds patent #5,566,084 for the process for identifying patterns of electric energy, effects of proposed changes, and implementing such changes in the facility to conserve energy.

Bruce K. Colburn, Ph.D., P.E., CEM, has 34 years of engineering experience. He developed his own engineering consulting firm in College Station, TX, and then combined that work in Houston, TX, with Texas Energy Engineers/ccrd partners where he was CEO for 10 years.

He is executive vice president and COO for an international ESCo company, where he is responsible for the development and successful performance of projects in 14 countries. Prior to entering the consulting field, Dr. Colburn was Associate Professor of Electrical Engineering at Texas A&M University and a Visiting Adjunct Professor of Engineering at Baylor University. He is a widely published author with over 50 publications to his record and is associate editor, reviewer, awards committee member, director and advisor for numerous technical journals and engineering organizations. He was inducted in the AEE Hall of Fame in 2004. E-mail: *bcolburn@epscapital.com* Web: *www.epscapital.com*

Toby Considine has been playing with computers since the New England Time Share in the 60s and first worked professionally with computers when microcomputers required user-written device drivers in the late 70s. He has developed systems in manufacturing, distribution, decision support, and quality assurance for clients who ranged from Digital Equipment Corporation to Reebok. Mr. Considine helped develop and support what grew into Boston Citinet, the largest free public access system of its day, in the mid 1980s.

For the last 17 years, Mr. Considine has worked as an internal consultant to the facilities services division of UNC-Chapel Hill. The difficulty of supporting current control systems in a wide area environment and in bringing information from those systems to the enterprise have been a constant challenge. For the last four years, he has been working to build interfaces to make control systems transparent to the enterprise based on internet standards-based protocols.

Rajesh Divekar, M.Tech, has over 10 years of IT experience in analysis, design, development, testing and implementation of the SCADA and other application software. He completed his Masters in systems and control engineering at Indian Institute of Technology (IIT) in Mumbai in 1992. Since then he has worked as an IT consultant working with electric utilities in the US, India, UK, Czech Republic, Estonia, dealing with implementation of SCADA systems. He has handled various assignments for companies including National Grid Company, Power Grid Corporation, GE Harris and GE Network Solutions. E-mail: *rdivekar@alsysinc.com*

Khaled A. Elfarra, CEM, DGCP, received a B.Sc. of electrical engineering, Aleppo University, Syria, 1984, a M. Sc. of engineering, Cairo University, 1989, and a Diploma of environmental engineering, American University in Cairo, 1995.

Khaled is currently general manager of engineering/projects dept. at the National Energy Corporation - Egypt (NECE). His role is the technical studies and designs of energy efficiency technologies. The focus of his current studies is on distributed generation, cogeneration and the gas fired technologies. He has extensive experience in the field of energy and environmental technologies. As a former technical manager assistant of Energy Conservation and Environment project (ECEP), USAID funded, he was involved in techno-economic feasibility studies, capacity building, technologies implementation, and project management. Mr. Elfarra conducted more than 80 audits, 20 feasibility studies, and supervised projects construction for more than $14 million. He gave training in energy auditing, environmental auditing, energy technologies, energy management systems and pollution control for many of the industrial facilities staff. He also received training on energy and environmental issues in the USA in multi disciplines. Mr. Elfarra has participated in the Egyptian environmental policy and the energy reforming policy. He has conducted many market study surveys on gas sales strategy, the potential for the energy and environmental market in Egypt, and solid waste management programs. He worked as short-term consultant for DANIDA JAICA, USAID and UNDP in many Egyptian Projects. Mr. Elfarra is a Certified Energy Manager by US Association of Energy Engineers as well as a Distributed Generation Certified Professional. He was also certified as second party auditor of environmental management systems (ISO 14000) by British Excel Partnership, Inc. Mr. Elfarra is certified by AEE as a local instructor for the certified energy managers course in Egypt. He is chair of the Egypt CEM Board.

Carla Fair-Wright is a consultant, writer, and educator in the areas of object-oriented systems and methodology.

Throughout most of her 18-year career, Ms. Fair-Wright has been heavily involved in developing better computer applications for mainframe, mini, PC and LAN based systems. She is currently a senior service maintenance planner for the maintenance technology services department at Cooper Cameron Corporation, a leading international manufacturer of oil and gas pressure control equipment, Formerly, she was owner of Optimal Consulting, a small consulting company specializing in visual studio solutions. Her former clients represented many business sectors and include major companies, such as Shell Oil, Pitney Bowes, NEC, EPS software, and Eagle Global Logistics.

Ms. Fair-Wright is a Microsoft Certified Professional (MCP) and holds a BS in computer science, Associate Degrees in electronic technology and technical management. She has also carried out graduate work at Troy State University and served as a technical reviewer for ReviewNet Corporation, an internet-based provider of pre-employment testing for IT personnel. Raised in Dayton, Ohio, Carla now lives with her two children in Houston.

Carla Fair-Wright can be reached at Cooper Compression, 11800 Charles Street, Houston, TX, 77077; email: *fairc@ccc-ces.com* (713) 856-1615

Keith E. Gipson has been a technologist for more than a decade. Starting out as a technician with Honeywell Inc. in 1987, graduating to an engineer at Johnson Controls in the mid-90's and at Pacific Gas and Electric in 1997.

A successful entrepreneur and business professional, Mr. Gipson co-founded in 1997 the world's first internet-based enterprise energy management company, Silicon Energy Corp (*www.siliconenergy.com*). The privately held, multi-million dollar company grew from three to 120 employees. Itron Corp. acquired Silicon Energy in March 2003 for $71M.

Mr. Gipson was awarded U.S. Patent number 6,178,362, Jan 23, 2001 as co-inventor of an energy management system and method utilizing the internet to perform facility and energy management of large corporate enterprises. This was the first EEM or "enterprise energy management" system.

Mr. Gipson is the CEO/CTO and co-founder of Impact Facility Solutions, Inc. (*www.myfacility.com*), formed in connection with the founders of NetZero Corp and other industry veterans, Impact Facility Solutions is dedicated to delivering internet-based, enterprise facility management solutions to large corporate customers.

Mr. Gipson mentors young, business professionals through University of Southern California "100 Black Men" program and minority business students at the California State University "Upward Bound" program. He is the director of the children's ministry and Sunday school teacher at The Roger Williams Baptist Church, Los Angeles, CA. Mr. Gipson enjoys music, computers and spending time with his wife and four children.

Bill Gnerre is the cofounder and CEO of Interval Data Systems, Inc. With an engineering background and 20-plus years of enterprise sales, marketing, and entrepreneur experience, Bill leads the overall company management and growth activities. Bill has an exemplary record of bringing enterprise software applications to market and helping customers the value and accomplishments possible through the use of data and the adoption of technology.

His previous roles include being a partner at Monadnock Associates, a consulting organization specializing in assisting startup software companies; co-founder of ChannelWave Software; director of sales & marketing at Wright Strategies; and product marketing roles at Formtech and Computervision, both vendors of CAD technologies. Earlier in his career Bill worked in various mechanical engineering positions.

David C. Green has combined experience in intranet/internet technology and database queries and has developed programming for energy information systems. David has been the president of his own consulting company, Green Management Services, Inc., since 1994. He has a Bachelor of Science degree in chemistry and a Master of Arts degree in computer science. David is also a lieutenant colonel in the Illinois Army National Guard and has 18 years of military service. David has successfully completed major projects for The ABB Group, Cummins Engine Company, ECI Telematics, M.A.R.C. of the Professionals, Walt Disney World and The Illinois Army National Guard. (*dcgreen@dcgreen.com*)

Daniel Harris, M.Eng., is a graduate mechanical engineer with a Masters degree in control theory and embedded logic from the University of Washington. He has worked in a variety of settings, including production engineering for Boeing Aerospace and in the product development lab for H-Power, a developer of fuel cell technologies and systems. Since joining AEA in 2003, he has been responsible for the start-up and commissioning of the E-Master control system for a large electrically heated complex, for the development, implementation roll-out and testing of the E-Master peak-limiting and demand response algorithm for room unit air-conditioning control and is creating a digital data acquisition system for the boiler training lab at AEA Bronx Energy Management Training Center. In addition to his work at AEA, Daniel is a volunteer development director of GreenHome Inc., a not-for-profit undertaking that promotes access to high performance green building materials and methods for the small to medium residential market. Contact *Dharris@aeanyc.org*, 718-292-6733 ext. 210

Michael Ivanovich has been editor-in-chief, of *HPAC Engineering Magazine* since 1996, when he made a career change from research to publishing. Under his direction, the magazine has been revitalized, a website established, and two regular supplements initiated (*Boiler Systems Engineering* and *Networked Controls*). Prior to joining HPAC Engineering, Mr. Ivanovich was a senior research scientist at Pacific Northwest National Laboratory, working on projects involving internet-technology development, green buildings and residential building codes. His background also includes working on the ozone hole project for NOAA Aeronomy Laboratory, developing a network of solar-powered weather stations for Colorado and development of a protocol for investigating IAQ problems in Minnesota homes. He has a graduate degree in energy engineering and undergraduate degrees in computer science.

Safvat Kalaghchy is the program director for the computing and information technology group at the Florida Solar Energy Center (FSEC). He is responsible for the design, development, and implementation of energy related information technology and scientific computing projects at FSEC. He is the architect for the *www.infomonitors.com* and the backend engine, the experimental management eatabase system (EDBMS) that enables automated field-monitoring project. He co-developed the first version FlaCom, state of Florida's commercial energy code compliance software, now known as EnergyGauge/FlaCom. He has also developed a number of other complex scientific software to analyze the behavior of thermal systems. Safvat has a BS and MS in mechanical engineering from the Florida Institute of Technology.

Srinivas Katipamula, Ph.D., got his M.S. and Ph.D. in mechanical engineering in 1985 and 1989, respectively, from Texas A&M University. He has been working as a senior research scientist at Pacific Northwest National Laboratory, in Richland, WA, since January 2002. He managed the analytics group at the Enron Energy Services for 2 years (2000 through 2001). Before joining EES, he worked at PNNL for 6 years and prior to that he worked for the energy systems lab at the Texas A&M University from 1989 to 1994.

He has authored or co-authored over 60 technical publications, over 25 research reports, and made several presentations at national and international conferences. He has recently written a chapter, "Building Systems Diagnostics and Predictive Maintenance," for *CRC Handbook on HVAC*. He is an active member of both ASHRAE and the American Society of Mechanical Engineers (ASME).

Michael Kintner-Meyer, Ph.D., has been a staff scientist at the Pacific Northwest National Laboratory since 1998. His research focus is on building automation technology for optimal control strategies of HVAC equipment for improving the energy-efficiency of buildings and to enhance the reliability during emergency conditions on the electric power grid. At PNNL, he leads the "Load-As-A-Reliability Resource" research activity that focuses on technology development and analyses of Grid-friendly Appliances™ and load management strategies.

Michael holds a M.S. (1985) from the Technical University of Aachen, Germany and a Ph.D. (1993) from the University of Washington in Seattle, WA. He is an active member of ASHRAE for which he serves on technical committees as well as in the local chapter. He is member of the American Society of Mechanical Engineers (ASME) and the German Engineering Society, Verein Deutscher Ingenieure (VDI). He has authored and co-authored numerous papers and reports in U.S. and international technical journals.

Bill Kivler, director of Global Engineering for Walt Disney World, is a 30-year veteran of facilities construction and operation. Bill has been with the Walt Disney World Company since 1993. He has held positions of increasing responsibility through the present-day role of technical director of Global Engineering. In his current role Bill supports the WDW property in several ways. Global contract administration, I.T. administration of the computerized maintenance management system, technology initiatives, metrics reporting, critical communications and support, hurricane coordination support, communications strategies, productivity initiatives support, energy and utility conservation strategies. Operationally Bill has 7 departments reporting to him supporting the maintenance of technologies such as, Office Machine Systems, I.T. Hardware, Video systems, Radio systems, Support Systems, Key Control Systems, Access Control Systems, Alarm and Monitoring Systems, Energy Management Systems. Bill is also responsible for leading the Resorts Engineering and Downtown Disney Engineering Divisions.

Prior to coming to WDW, Bill spent 13 years in the U.S. Virgin Islands as executive director of engineering and program manager for the largest resort community on the island. His responsibilities included facilities maintenance, power plant design, construction, and maintenance, new construction, and capital renewal. He was also responsible for all governmental regulations local, state, and federal regarding air, water, fuel, and building code permits. The resort operated an autonomous power plant which produced power, chilled water, steam, drinking water from sea water, and sewage treatment.

Preceding the Virgin Islands, Bill managed the facilities for over 30 hotels/resorts along the U.S. East Coast. Bill attended Franklin and Marshall College, The Center for Degree Studies, and RCA Institute.

Mr. Jim Lewis is the CEO and co-founder of Obvius, LLC, in Portland, OR. He was the founder and president of Veris Industries, a supplier of current and power sensing products to BAS manufacturers and building owners. Prior to founding Veris, Mr. Lewis held several positions at Honeywell including Branch Manager. He has extensive experience in knowing the needs of building owners, integrating existing metering and sensing technologies and developing innovative products for dynamic markets.

For more information or a demonstration, contact Obvius Corporation at (503) 601-2099, (866) 204-8134 (toll free), or visit the website at: *http://www.obvius.com*

Fangxing Li is presently a senior consulting R&D engineer at ABB Inc. He received his B.S. and M.S. degrees in electric power engineering from Southeast University, China, in 1994 and 1997 respectively. He received his Ph.D. degree in computer engineering from Virginia Tech in 2001. His areas of interests include Web applications in power systems, power distribution analysis, and energy market simulation. Dr. Li is a member of IEEE and Sigma Xi. He can be reached at *fangxing.li@us.abb.com* or *fangxing.li@ieee.org*.

Joe LoCurcio has a Bachelor of engineering degree in mechanical engineering, specializing in machine systems, from Stevens Institute of Technology, Hoboken, NJ. He has worked for Merck since 1997, and started as the Rahway Site BAS Engineer and Administrator for four years. He is currently working for the Merck Central Engineering division as an HVAC design engineer with world wide responsibilities. He has been working with BAS systems since 1987.

For more information regarding Andover Controls Corporation please use the following information:

United States of America (World Headquarters)
Andover Controls Corporation
300 Brickstone Square
Andover, MA 01810
Telephone: +1-978-470-0555
fax: +1-978-470-0946
http://www.andovercontrols.com

Dirk Mahling Ph.D., is the chief technology and chief information officer at WebGen Systems, Inc. As the co-inventor and patent-holder of WebGen's IUE® System, he has demonstrated the value that advanced information technology can bring to energy management, controls, and facilities.

Dirk has contributed to the fields of artificial intelligence, knowledge management, and IT strategy in industry and academia. He is a professor at the University of Pittsburgh and was a practice leader at A.T. Kearney Management Consultants in Germany and the U.S.

Dirk is an alumnus of the Brunswik Institute of Technology (Germany) and was a Fulbright Scholar at the University of Massachusetts, Amherst. He is past chairman of the Association of Computing Machinery (ACM) chapter on Groupware. He is a licensed psychologist in Germany.

John Marden is marketing manager for Honeywell Energy Services. John is responsible for developing Honeywell's business strategies and new services related to energy. Before joining Honeywell, John ran an independent new product development and marketing strategy practice focusing on growth strategies. He has also served as marketing manager for 3M/Imation Medical Imaging. John is certified in Six Sigma for Growth and DFSS. John received his M.B.A. in marketing and management information systems from the University of Minnesota in 1992 and his B.A. in communications from the University of Minnesota in 1989.

He can be reached at *John.Marden@Honeywell.com* or *John@Marden.com*

Jim McNally P.E., manager of utility information services at Siemens Building Technologies, Inc., in Buffalo Grove, Illinois, is an expert on utility metering and rates applications. He is an applied researcher whose technical innovations include development of an on-line meter reporting system, the pre-packaged wattmeter, and the multi-variant, non-linear [MVNL] load forecasting technology. He consults with various universities, hospitals, and manufacturers in the development of their metering systems. He may be reached at: *jim.mcnally@siemens.com*.

Gerald Mimno has seven years experience developing wireless internet applications for energy measurement, information management, and controls. Mr. Mimno has a BA and MCP from Harvard University followed by experience in economic development, real estate development, and business development. He has

20 years of practical experience in building systems. He is a Licensed Construction Supervisor in the Commonwealth of MA. He is responsible for developing new markets and relationships based on wireless and internet energy technologies and has written extensively on the value of interval data.

Gerald Mimno, General Manager
Advanced AMR Technologies, LLC
285 Newbury Street
Peabody, MA 01960
TEL(978)826-7660
FAX(978)826-7663
gmimno@AdvancedAMR.com

Naoya Motegi is a graduate student research assistant in the Commercial Buildings Systems Group in the Building Technologies Department at LBNL. He is currently a graduate student in the Department of Architecture, University of California, Berkeley. He has a Bachelor of architecture and Master of engineering in architecture and civil engineering from Waseda University in Tokyo, Japan.

David E. Norvell is the energy manager at the University of Central Florida. He manages the energy consumption in over 6 million square feet of building space. He is a registered Professional Engineer with the state of Florida. He received his B.S. in mechanical engineering from the University of Central Florida. He is the current vice-president of the Central Florida chapter of the Association of Energy Engineers (AEE). He is the principle developer of the Open Energy Information System.

Mark A. Noyes is president and COO of WebGen Systems, Inc. He is also a co-inventor and patent-holder of the IUE® System. Mark has worked in the energy and controls industry his entire career, progressing through positions of increased responsibility. Prior to joining WebGen, Mark worked for Consolidated Edison of New York as vice president of asset management and domestic acquisition. During his tenure at Consolidated Edison he also served as chief operating officer and board of director of several subsidiaries. He has been recognized by Crain's business magazine as a "Top 40 under 40" executive. Mark earned his MBA from Rutgers University, and holds a Bachelor's Degree in electrical engineering from Worcester Polytechnic Institute.

Sarah E. O'Connell is an associate consultant at ICF Consulting. Ms. O'Connell has over six years of experience in energy and environmental consulting. She currently supports energy efficiency efforts of the U.S. Environmental Protection Agency Energy Star program in the commercial and industrial building sectors primarily supporting service and product providers.

William O'Connor is a graduate of Northeastern University with a Bachelors degree in mechanical engineering and holds his certificate as a Certified Energy Manager (CEM). Before joining WebGen Systems, he worked for large and small automatic temperature control contractors in the Boston area for over 20 years.

In working for both large and small firms, William has developed a unique combination of experiences in every aspect of the control industry from control technician to operations manager. Whether for clean rooms, laboratories, office buildings or schools, William has been successful for many years in designing and installing control systems.

Patrick J. O'Neill, Ph.D., co-founded NorthWrite, and leads corporate operations. Before joining NorthWrite, Patrick spent 10 years at Honeywell International, where he most recently served as vice president of Technology and Development for e-Business. Patrick defined technology strategy, prioritized developments, allocated resources, and operated the infrastructure for Honeywell's stand-alone e-ventures.

Patrick also co-founded and acted as chief technology officer for Honeywell's *myFacilities.com*, an application service provider targeting the facility management and service contracting industries. Previously, Patrick was director of development for Honeywell's Solutions and Service business, managing global product research and development worldwide, with development teams in the U.S., Australia, India, and Germany.

Before joining Honeywell, Patrick worked at the Department of Energy's Pacific Northwest National Laboratory and the University of Illinois at Urbana-Champaign. He holds Bachelor's, Master's and Doctoral degrees in mechanical and industrial engineering from the University of Illinois, Urbana-Champaign. Patrick is a member of numerous professional organizations including ASHRAE, where he has held leadership positions in the computer applications, controls, and smart building systems technical committees. He has written and published many articles on software, systems and controls, and building operations and management.

Richard Paradis is a certified energy manager, and has been in the energy efficiency industry since 1978.

Rick has worked for utilities, design/consultant firms, non-profit management consultant firms and energy services companies working primarily in the commercial and industrial market sectors. This work included writing technical assistance audit reports, developing design alternatives for HVAC, lighting, thermal storage, and alternative energy projects, providing construction observation and review services as well as monitoring and verification protocols.

Rick has also managed and supervised technical potential studies and various technical assessments of end use equipment for the natural gas utilities in Massachusetts and New Jersey for developing utility Demand Side Management (DSM) programs. For his work in this area and setting the technical evaluation standard for DSM programs in Massachusetts in the early 1990s, Rick received the nomination and induction into the Marquis *Who's Who in Science and Engineering* 1994/1995 edition as well as the millennium edition.

Klaus E. Pawlik is a manager with Accenture working in the natural resources industry. Over the past several years, he has been involved in several enterprise wide net-centric and analytical-engine systems implementations. He is the author of the *Solution Manual for Guide to Energy Management* as well as several periodical publications concerning modeling and analyzing facility energy use. He holds a Master of business administration and a Bachelor of Science in industrial and systems engineering graduating with highest honors from the University of Florida. While at the University of Florida, Klaus worked in the Industrial Assessment Center leading teams of undergraduate and graduate students performing energy and waste minimization, and productivity improvement assessments for manufacturing facilities. Additionally, for two years, he assisted Dr. Barney Capehart with teaching energy management. Klaus has served six years in the United States Navy, where he worked as an electrical operator on nuclear power plants. For two of those years, he served as an instructor training personnel on the electrical operations for nuclear power plants. (*klaus.e.pawlik@accenture.com*)

Mary Ann Piette is the research director of the California Energy Commission's PIER Demand Response Research Center and the deputy group leader of the Commercial Building Systems Group. She has been at Berkeley Laboratory for more than 20 years, with research interests covering commercial building energy analysis, commissioning, diagnostics, controls, and energy information system. Her recent work has shifted toward developing and evaluating techniques and methods to improve demand responsiveness in buildings and industry. She has a Masters in mechanical engineering from UC Berkeley, and a Licentiate in building services engineering at the Chalmers University of Technology in Sweden.

Rich Remke is commercial controls product manager for Carrier Corporation in Syracuse, NY. Rich has been the product manager for Carrier for the past four years and is responsible for controls product marketing and new product development. He holds a B.S. in information system management from the University of Phoenix. Rich has been in the HVAC and controls industry for over 20 years. Rich started his control work as a SCADA technician for Reedy Creek Energy Services at Walt Disney World, FL. He then moved into controls system engineering, project management, sales, and technical support for United Technologies/Carrier Corporation. Rich also spent several years supporting Carrier's Marine Systems group, providing controls technical support and system integration engineering. Rich has created several custom user applications, including a facility time schedule program, a DDE alarm interface, integration of Georgia Power real time pricing data to Carrier CCN, and a custom tenant billing application. (*richard.remke@carrier.utc.com*)

Richard Rogan, P.E., is director, engineering services for Honeywell Integrated Energy Services. Rich is responsible for developing solutions which allow customers to actively manage their entire energy supply chain via integrated energy information systems. In his 18 years with Honeywell International, Rich has been involved in the design, programming, installation, and commissioning of complex building management systems which integrate fire management, security, CCTV, building automation and energy management systems. His experience includes high rise hotel/casino, tunnels, animal/fume hood laboratories, industrial and health care facilities. More recently, Rich has been involved in the development of unique energy savings solutions for the U.S. Federal Government at Fort Richardson, AK, and Fort Bragg, NC. Rich is a licensed professional engineer in the state of Pennsylvania as well as a Certified Energy Manger. He received his B.S. in mechanical engineering technology from Spring Garden College in 1985. He can be reached at *Richard.M.Rogan@Honeywell.com*

Sandra C. Scanlon, P.E., founded Scanlon Consulting Services, Inc., an engineering design and consulting firm, in 1997 after eight years of experience at Amoco Corporation. She has balanced running her business while raising her 8-year-old son; volunteering for the Society of Women Engineers (SWE), Abiding Hope Lutheran Church, and other organizations. She is a registered professional engineer, a senior member of the Institute of Electrical and Electronics Engineers (IEEE), and has a BS in electrical engineering from Valparaiso University. She is a senior life member of SWE and received the SWE Distinguished New Engineer Award in 1998. She was co-chair of the 2001 SWE National Conference, chair of the national Electronic Communications Committee and national Membership Committee, and past president of the SWE Rocky Mountain Section. She is active in the local SWE section chairing the annual Girls Exploring Science, Engineering & Technology Event for middle school girls. In addition, she serves on the board of directors for the Denver School of Science and Technology, a Denver Public Schools charter high school, for which she is also co-chair of the IT committee. She is also a founder and president of the board of directors for a new private K-8 school, Aspen Academy, in Highlands Ranch, Colorado. She lives in Littleton, Colorado, with her son Paul. She loves to play golf, cook, and volunteer in the community.

Blanche Sheinkopf has been the national coordinator of the United States Department of Energy's EnergySmart Schools program since 2001. An educator and curriculum writer for more than 25 years at levels ranging from pre-kindergarten through university, she has been a college of education faculty member at the University of Central Florida, the George Washington University, and American University, and was the coordinator of education and training programs at the Florida Solar Energy Center. She was founder and CEO of Central Florida Research Services, a full-service marketing research company for 11 years. She currently serves on the boards of several organizations including the American Solar Energy Society and the Educational Energy Managers Association of Florida.

Travis R. Short, BSME, is president of Integrated Building Solutions, Inc. (IBSI) a multifaceted systems integrator. He has a Bachelor of Science degree in mechanical engineering from the University of Missouri - Rolla. He has a diverse background in BMCS installations, mechanical/electrical engineering, and integrated systems commissioning in various types of mission critical facilities. He has worked with computer site engineering, a nationally renowned expert in data center design and commissioning. He has implemented strategic concepts in regards to proper BMCS commissioning in all data center projects worked on with computer site engineering. He is truly dedicated to the ideology of "open" protocol control systems and the advancement of web based building monitoring and control systems. He has authored several articles detailing the proper execution of the mission critical BMCS for the *Data Center Journal.com*, Automated Buildings.com, and IBSI's website.

E-mail: *travis@integrated-buildings.com*
Website: *www.integrated-buildings.com*

Ken Sinclair has been in the building automation industry for over 35 years as a service manager, building owner's representative, energy analyst, sub-consultant and consultant. Ken has been directly involved in more than 100 conversions to computerized control. Ken is a founding member and a past president of both the local chapter of AEE and the Vancouver Island chapter of ASHRAE. The last five years his focus has been on *AutomatedBuildings.com*, his online magazine. Ken also writes a monthly building automation column for *Engineered Systems* and has authored three industry automation supplements: *Web-Based Facilities Operations Guide*, *Controlling Convergence* and *Marketing Convergence.*

Greg Thompson is the application architect of the ION® EEM Solution for Power Measurement. Greg began his affiliation with Power Measurement in 2001 and has also been involved as a senior software engineer in the engineering services division. He obtained his electrical engineering degree from the University of Kentucky in 1995 and has been extensively involved in developing and providing enterprise energy management systems for a number of years.

Terrence Tobin is the corporate communications manager for Power Measurement. He began his affiliation with Power Measurement in 1988 and has held a variety of positions including director of marketing communications and brand manager. Terry obtained his diploma of electronics technology in Victoria in 1983 and has worked extensively in the high technology sector in the fields of research, development and communications.

Steve Tom, PE, Ph.D., is the director of technical information at Automated Logic Corporation, Kennesaw, Georgia, and has more than 30 years experience working with HVAC systems. At ALC Steve has coordinated the training, documentation, and technical support programs, and frequently works with the R&D engineers on product requirements and usability. Currently Steve is directing the development of *www.CtrlSpecBuilder.com*, a free web-based tool for preparing HVAC control system specifications. Prior to joining Automated Logic, Steve was an officer in the U.S. Air Force where he worked on the design, construction, and operation of facilities (including HVAC systems) around the world. He also taught graduate level courses in HVAC design and HVAC controls at the Air Force Institute of Technology. (*STom@automatedlogic.com*)

Jason Toy is currently an undergraduate student at the College of Computer and Information Science at Northeastern University, Boston. He is studying for a double major degree in computer science and mathematics. He has several years experience working in the data mining and analysis field. He may be reached at *toy@jtoy.net*.

John Van Gorp is the marketing manager for industrial markets at Power Measurement. He received his B.A.Sc. in electrical engineering from the University of British Columbia. John gained experience building monitoring systems for the Power Smart program at BC Hydro and for utility and industrial customers as an applications engineer at Power Measurement. In January 2004, John received his Certified Energy Manager designation from the Association of Energy Engineers.

Bill Von Neida is a senior engineer at the U.S. Environmental Protection Agency. For the past 12 years he has managed ENERGY STAR technical support programs to assist businesses select the most efficient and profitable commercial building design, technology, application, operations solutions.

Rahul Walawalkar, CEM, CDSM, has over 7 years experience as an IT and energy management consultant, product manager, program manager and coordinator for corporate energy management and ergonomic programs. He is currently pursuing doctoral studies in engineering and pubic policy at Carnegie Mellon Universitiy. He obtained a Master's in energy management and advanced certification in energy technology at New York Institute of Technology. He is associated with Customized Energy Solutions as a research analyst. He has also worked as an energy & IT analyst with various companies including EPS Capital Corp, Alliance to Save Energy, American Public Power Association and Tata Infotech Ltd. He conceptualized and developed an award winning energy efficient lighting design software, Eco Lumen and is the recipient of numerous awards including Computer Society of India's "Young IT Professional Award" and NYIT's "Energy Management Graduate Faculty Award." Rahul has written over 35 technical papers and was editor of *Energy Productivity News*, the newsletter of Council of Energy Efficiency Companies in India in 2000-01. Rahul is member of various professional organizations including AEE, ASHRAE, IEE, IEEE and IAEE. E-mail: *rahul@walawalkar.com Web: www.walawalkar.com*

David Watson has over 15 years experience designing, programming, and managing the installation of control and communications systems for commercial buildings, industrial processes and remote connectivity solutions. At LBNL, he is working with innovative building technologies such as demand response systems, energy information systems and wireless control networks. Prior to joining LBNL, David held engineering, project management and product development positions at Coactive Networks, Echelon, York International and Honeywell. He designed and managed the installation of hundreds of projects including: internet based control and monitoring of thousands of homes and businesses, communication systems for micro turbine based distributed power generation systems and industrial process controls for NASA wind tunnels and biotech manufacturing. Mr. Watson graduated from California Polytechnic University, San Luis Obispo with a degree in mechanical engineering.

John Weber is president and founder of Software Toolbox Inc. Prior to founding Software Toolbox in 1996, John spent 6 years with GE Fanuc Automation and their distribution channel in a variety of technical and commercial field positions. He has been working with communications systems and developing software for over 15 years. He has spoken at numerous ISA and other shows domestically and internationally on subjects including communications, OPC, HMI configuration, and others. John holds a Bachelor of Science in industrial and systems engineering (1989) from the University of Florida and a Masters Degree in business from Clemson University (1995).

Tom Webster, PE is a research specialist at the Center for the Built Environment (CBE) at the University of California-Berkeley. He has been engaged in building energy, controls, and communications R&D for almost 30 years. His experience includes commercial and residential HVAC systems engineering, distributed control systems design, digital controls product development, analysis of energy services in the utility industry, building energy analysis and simulation, and experimental studies of various building technologies. Mr. Webster is currently conducting research on alternative building conditioning technologies as well as continuing EMCIS technology assessment work with FEMPNTDP through LBNL.

Gaymond Yee is a research coordinator with the California Institute for Energy and Environment where he manages the Demand Response Enabling Technology Development Project funded by the California Energy Commission's Public Interest Energy Research Program. Prior, he was a contract researcher with the Lawrence Berkeley National Laboratory and conducted an assessment of current trends in energy management systems for the Department of Energy's Federal Energy Management Program.

Over the past 12 years, Mr. Yee has held project management positions at several energy sector companies. He was the development manager with Silicon Energy Corporation and participated in successful residential electricity curtailment projects at Puget Sound Energy, Northeast Utilities, and Long Island Power Authority. Prior, at Diablo Research, he managed the software development of the Whisper wireless Automatic Meter Reading system and at EnergyLine Systems, he managed the Energy Information Services home automation project funded by Pacific Gas & Electric, Microsoft and TCI. Mr. Yee began his career as a mechanical engineer and spent 13 years working on continuum mechanics numerical simulation analyses.

Mr. Yee received his Bachelor of Science degree in mechanical and nuclear engineering and Master of engineering degree in mechanical engineering from the University of California at Berkeley.

Jeff Yeo, P.Eng., is a senior software architect for Power Measurement. He received a Bachelor of Science from Acadia University and a Bachelor of Electrical engineering from the Technical University of Nova Scotia. Jeff worked in electrical system maintenance, planning and forensics within the mining industry prior to joining Power Measurement in 1993. Since then he has held positions in product testing, field service engineering, and software development.

Index